IET ELECTROMAGNETIC WAVES SERIES 52

Series Editors: Professor P. J. B. Clarricoats
Professor E. V. Jull

Thermal Microwave Radiation: Applications for Remote Sensing

Other volumes in this series:

Volume 10 **Aperture antennas and diffraction theory** E. V. Jull
Volume 11 **Adaptive array principles** J. E. Hudson
Volume 12 **Microstrip antenna theory and design** J. R. James, P. S. Hall and C. Wood
Volume 15 **The handbook of antenna design, Volume 1** A. W. Rudge, K. Milne, A. D. Oliver, P. Knight (Editors)
Volume 16 **The handbook of antenna design, Volume 2** A. W. Rudge, K. Milne, A. D. Oliver, P. Knight (Editors)
Volume 18 **Corrugated horns for microwave antennas** P. J. B. Clarricoats and A. D. Oliver
Volume 19 **Microwave antenna theory and design** S. Silver (Editor)
Volume 21 **Waveguide handbook** N. Marcuvitz
Volume 23 **Ferrites at microwave circuits** A. J. Baden Fuller
Volume 24 **Propagation of short radio waves** D. E. Kerr (Editor)
Volume 25 **Principles of microwave circuits** C. G. Montgomery, R. H. Dicke, E. M. Purcell (Editors)
Volume 26 **Spherical near-field antenna measurements** J. E. Hansen (Editor)
Volume 28 **Handbook of microstrip antennas** J. R. James and P. S. Hall (Editors)
Volume 31 **Ionospheric radio** K. Davies
Volume 32 **Electromagnetic waveguides: theory and application** S. F. Mahmoud
Volume 33 **Radio direction finding and superresolution** P. J. D. Gething
Volume 34 **Electrodynamic theory of superconductors** S.-A. Zhou
Volume 35 **VHF and UHF antennas** R. A. Burberry
Volume 36 **Propagation, scattering and dissipation of electromagnetic waves** A. S. Ilyinski, G. Ya. Slepyan and A. Ya. Slepyan
Volume 37 **Geometrical theory of diffraction** V. A. Borovikov and B. Ye. Kinber
Volume 38 **Analysis of metallic antennas and scatterers** B. D. Popović and B. M. Kolundžija
Volume 39 **Microwave horns and feeds** A. D. Olver, P. J. B. Clarricoats, A. Kishk and L. Shafai
Volume 41 **Approximate boundary conditions in electromagnetics** T. B. A. Senior and J. L. Volakis
Volume 42 **Spectral theory and excitation of open structures** V. P. Shestopalov and Y. V. Shestopalov
Volume 43 **Open electromagnetic waveguides** T. Rozzi and M. Mongiardo
Volume 44 **Theory of nonuniform waveguides** B. Z. Katsenelenbaum, M. K. A. Thumm, L. Mercader Del Río, M. Pereyaslavets and M. Sorolla Ayza
Volume 45 **Parabolic equation methods for electromagnetic wave propagation** M. Levy
Volume 46 **Advanced electromagnetic analysis of passive and active planar structures** T. Rozzi and M. Farina
Volume 47 **Electromagnetic mixing formulas and applications** A. Sihvola
Volume 48 **Theory and design of microwave filters** I. Hunter
Volume 49 **Ridge waveguides and passive microwave components** J. Helszajn
Volume 50 **Channels propagation and antennas for mobile communications** R. Vaughan
Volume 51 **Asymptotic and hybrid methods in electromagnetics** F. Molinet, I. Andronov and D. Bouche

A complete list of publications in this series may be obtained from The Institution of Engineering and Technology.

Thermal Microwave Radiation: Applications for Remote Sensing

Edited by
C. Mätzler

The Institution of Engineering and Technology

Published by: The Institution of Engineering and Technology, London, United Kingdom

The Institution of Engineering and Technology
Michael Faraday House,
Six Hills Way, Stevenage,
Herts. SG1 2AY, United Kingdom

British Library Cataloguing in Publication Data

Mätzler, Christian
Thermal microwave radiation: applications for remote sensing
1. Microwave remote sensing
I. Title II. Institution of Engineering and Technology
621.3′678

ISBN (10 digit) 0 86341 573 3
ISBN (13 digit) 978-086341-573-9

Typeset in India by Newgen Imaging Systems (P) Ltd, Chennai, India
Printed in England by MPG Books Ltd, Bodmin, Cornwall

Contents

Cover images: AMSR-E 36 GHz, horizontally polarized, brightness temperatures, from descending passes, December 31, 2005, gridded to the Northern (Southern) Equal-Area Scalable Earth Grid (EASE-Grid), courtesy of National Snow and Ice Data Center, Boulder, CO, USA.

The brightness temperature ranges from 115 K (violet) to 294 K (red), with intermediate values of 160 K (dark blue) 180 K (bright blue), 225 K (green), 260 K (yellow), 275 K (orange)

Foreword

This book is a result of a wide collaboration between experts in the field of microwave radiometry and its applications. Originally the work began in 1996 by the European Cooperation in Scientific and Technical Research (COST), http:/www.cost.esf.org/. As described in the 'Memorandum of Understanding for the Implementation of a European Concerted Research Action, Designated as COST Action 712 Application of Microwave Radiometry to Atmospheric Research and Monitoring': 'the aims of the action were to improve the application of microwave radiometry to the understanding and monitoring of the hydrological cycle and tropospheric–stratospheric exchange. This includes research for understanding and modelling the atmospheric processes involved in these phenomena. These objectives can be achieved through developments in the following areas:

(1) improved models of the interaction of microwave radiation with the Earth's atmosphere and surface,
(2) improved retrieval, analysis and assimilation techniques, through which atmospheric and surface parameters are estimated from the data,
(3) verification and validation studies, through which the accuracy and characteristics of the data analysis techniques may be assessed,
(4) improved measurement facilities and techniques, including ground-based, aircraft-borne and space-based systems.'

COST Action 712 supported the above activities from 1996 to 2000. Area 1 was addressed as the topic of Project 1 Development of Radiative Transfer Models. Its focus was the development of physical models, fast physical and semi-empirical models and the improvement of critical elements in radiative transfer (e.g. microwave dielectric properties of water, ice, soil, vegetation and of various mixtures). Progress was discussed at several workshops and documented in two reports:

MÄTZLER, C. (ed.): 'Development of radiative transfer models', COST Action 712, Report from Review Workshop 1, held at EUMETSAT, Darmstadt, Germany, April 8 to 10, 1997, revised Oct. (1997).

MÄTZLER, C. (ed.): 'Radiative transfer models for microwave radiometry', COST Action 712 'Application of microwave radiometry to atmospheric research and monitoring', Meteorology, Final Report of Project 1, European Commission, Directorate General for Research, EUR 19543, ISBN 92-828-9842-3 (2000).

The final report, especially, found a wide interest in the scientific community and by students who used it as a supplement to textbooks on the topic of microwave radiometry. Four years later we observed significant advances in radiative transfer and spectroscopy of microwave radiation in natural media. The progress was partly a result of the momentum obtained by the COST 712 Community to perform focused research on identified problem areas. In addition the launch of new satellites, carrying microwave radiometers, also pushed the advancements of microwave radiative transfer. A discussion among members of the former Project 1 of COST 712 led to the proposal in January 2004 to revise, update and widen the report from 2000, and to publish it as a textbook. Many colleagues promised to contribute to this work. We are proud to present the result here.

The introduction, Chapter 1, starts with a historical overview on radiative transfer and microwave radiometry. Kirchhoff's law of radiation in the form generalised by Planck is then used to relate the scene brightness to the spatial distribution of temperature and absorptivity. Simple examples are used for illustration, based on solutions of the radiative-transfer equation. The final part of Chapter 1 is an introduction to the quantities for describing the polarisation of thermal radiation culminating in the fully polarimetric radiative-transfer equation.

Chapter 2 is concerned with the absorption and emission spectra of atmospheric gases. The emitted energy, measured for example by satellites, can be used to determine atmospheric temperature and moisture or other constituents. Interpretation of surface phenomena, on the other hand, requires consideration of attenuation by the atmosphere. The published literature contains many atmospheric absorption models, so in Chapter 2 emphasis is placed on describing relatively recent developments and on atmospheric measurements that have been done to determine the accuracy of models. The latest versions of the GEISA and HITRAN line-parameter databases are described and some relevant laboratory and theoretical work that may influence future modelling efforts is reviewed. Validation experiments in the atmosphere have the advantage of propagation path lengths much longer than in the laboratory, but to be useful in a decision between alternative absorption models, they require very accurate *in situ* measurement of the atmosphere, which is not easily achieved. The concluding section makes some recommendations for directions of future research.

Chapter 3 concentrates on the interaction of solid and liquid hydrometeors, suspended in the atmosphere, with microwave radiation. Especially in the window regions of the microwave electromagnetic spectrum hydrometeors dominate the signals and make them accessible via remote sensing from the ground and from airborne and satellite platforms. Since scattering by hydrometeors becomes an increasingly important process with increasing frequency, a closer look at the related polarisation effects and the influence of three-dimensional structures in vector radiative transfer will be taken. This chapter reviews first the state of the art in deriving the single scattering properties of atmospheric hydrometeors. Then the links of single scattering parameters derived from available codes with the general properties of the vector radiative-transfer equations and its solution in a heterogeneous emitting and scattering atmosphere is elucidated, followed by a general description of the usually applied solution methods. After describing existing and available exact codes to solve the vector radiative-transfer equation some effort is put on the description of efficient approximations useful for operational remote sensing and data assimilation. The chapter closes with new results on the signatures of clouds and precipitation in passive microwave observations.

Chapter 4 presents recent developments which have been made in the radiative-transfer modelling of the microwave surface emission, including studies made over ocean, bare soil (mainly in arid regions), vegetation-, snow- and ice-covered areas. In most cases, the research activities were carried out in the framework of existing or near future space missions.

Over the ocean, the first contribution reviews recent efforts made on improving ocean emissivity models by handling azimuthal variation and the polarimetric phase signal. These studies were carried out with the objective of retrieving instantaneous wind vectors from existing passive microwave observations (SSM/I and polarimetric Windsat instrument launched in 2002 mainly). The second contribution was written in the framework of future space missions which will attempt to globally monitor sea surface salinity (SSS): the ESA mission SMOS (soil moisture and ocean salinity) that will provide dual-polarisation and multi-angular observations and the NASA mission Aquarius. These missions should be launched, respectively, in 2007 and 2009. This contribution reviews models at L-band and the requirements that should be met to retrieve SSS with sufficient accuracy.

Over the land surfaces modelling the emission from bare soil surfaces is analysed in a specific section that includes (1) a review of recent improvements, (2) a new approach: the air-to-soil transition model and (3) an analysis of the signatures from arid regions as seen from space. All these contributions describe not only recent improvements but also problems to be solved in accounting correctly for the effects of both volume and surface scattering. A specific section is dedicated to the modelling of the influence by vegetation. Most of these studies were carried out in the framework of the near future L-band space missions designed to monitor surface soil moisture (the ESA SMOS and NASA HYDROS missions). Correction of vegetation effects was improved by a better accounting for the vegetation structure of crops and forests. Considering the SMOS configuration system, the dependence of these effects on incidence angle and polarisation is a new challenge in the retrieval process.

Even though most of the contributions in Chapter 4 concern the microwave signature of soil, vegetation and ocean, other sections consider more specific aspects. Recent developments in the modelling of the effects of volume scattering, relief and snow over the land surfaces and of ice over the ocean are considered in specific sections. Improvement in the modelling of these effects, which is rarely addressed in the literature, is a major issue in the understanding of the spatial observations made by current instruments (SSM/I, AMSU, etc.) and by near future instruments operating at L-band (SMOS, HYDROS and Aquarius).

Chapter 5 is devoted to the dielectric properties of important materials found at the terrestrial surface. Since magnetic effects can be ignored for the materials to be discussed the relative dielectric constant is equivalent to the square of the complex refractive index. For homogeneous components, such as ice and water, the main emphasis has been laid on the search for accurate data and on the optimisation of empirical models. The work led to new analytical expressions for the complex dielectric constant of fresh water, saline water, fresh-water ice and slightly saline ice as functions of frequency, temperature and salinity. New measurements of dielectric constants of minerals and rocks were also collected. Natural materials often are a heterogeneous mixture of more basic components. Unfortunately the effective dielectric constant of the mixture is not just a volumetric or mass averaging of the permittivities of its components. In fact mixing rules usually are not strict, but they depend on the shape of the mixing particles. A special section is devoted to present the most relevant mixing models. Dielectric properties of heterogeneous media (clouds, snow, vegetation and soil) are presented and interpreted in the light of these models.

Appendices give complementary information, including original data tables.

Acknowledgements

This book would not have been possible without the volunteering work of all authors and chapter coordinators. Their contributions and support in establishing this work are gratefully acknowledged. A list of the authors and their affiliation is given in the next section. Furthermore we would like to thank Barbara Kindler for the secretarial support, Dietrich Feist for setting up a webpage for downloading the draft manuscripts and the IEE for publishing this work.

The work of C. Mätzler was supported by Armasuisse under contract No. 4500312987. The work of P. W. Rosenkranz was supported in part by the US Department of Commerce, National Oceanic and Atmospheric Administration under Contract No. DG133E-02-CN-0011. The work of S. A. Buehler was funded by the German Federal Ministry of Education and Research (BMBF), within the AFO2000 project UTH-MOS, grant 07ATC04 and the DLR project SMILES, grant 50EE9815. It is a contribution to COST Action 723 'Data Exploitation and Modelling for the Upper Troposphere and Lower Stratosphere'. The research of N. Jacquinet-Husson on GEISA/IASI is funded by CNES (Centre National d' Etudes Spatiales -France) and EUMETSAT (EUropean organization for the exploitation of METeorological SATellites) in the frame of the EPS/METOP project. The work of J. Pardo on the ATM code has been supported in recent years by Spanish MCyT grants ESP2002-01627, AYA2002-10113-E and AYA2003-02785-E. Operations of the Fourier Transform Spectrometer at the Caltech Submillimeter Observatory have been supported by NSF grants ATM-9616766, AST-9615025 and AST-9980846. The development of the BEAMCAT database was supported by the Swiss National Science Foundation under Grant 2000-063793.00. The work of J. Boutin was supported by the French CNES/TAOB (50T207) and CNES/TOSCA (504T12) contracts and by the ESA contract 14273/00/NL/DC. The chapter on sea ice emission modelling was sponsored by the 5th framework programme of the European Commission, Integrated

Observing and Modelling of the Arctic Sea Ice and Atmosphere (IOMASA), Project No. EVK3-2001-00116.

April 2005

Christian Mätzler (Editor, Coordinator for Chapters 1 and 5)
Philip W. Rosenkranz (Coeditor and Coordinator for Chapter 2)
Alessandro Battaglia (Coeditor and Coordinator for Chapter 3)
Jean-Pierre Wigneron (Coeditor and Coordinator for Chapter 4)

Curricula

Editor

Christian Mätzler is Professor in applied physics and remote sensing at the Institute of Applied Physics, University of Bern. He has conducted research at the NASA Goddard Space Flight Center in Greenbelt, Maryland and at the ETH, Zurich. Returning to the University of Bern in 1978 he now leads the Project Group *Radiometry for environmental monitoring* on the propagation, emission and scattering of electromagnetic radiation in snow and ice, soil and vegetation, and in the atmosphere for the advancement of remote sensing with emphasis on microwave radiometry. Christian Mätzler is a member of the International Astronomical Union, of the International Glaciological Society, the Geoscience and Remote Sensing Society of IEEE, the Swiss Society of Astronomy and Astrophysics, the Swiss Commission of Remote Sensing, and the Swiss Commission of Space Research. He also is an active member of advisory groups at the European Space Agency (ESA) and EUMETSAT.

Coeditors

Philip W. Rosenkranz is a Principal Research Scientist in the Research Laboratory of Electronics at the Massachusetts Institute of Technology. Born in Buffalo, New York; he received a Ph.D. in Electrical Engineering from the Massachusetts Institute of Technology in 1971, then did postdoctoral research at Caltech's Jet Propulsion Laboratory in Pasadena, California. In 1973 he joined the staff of M.I.T.'s Research Laboratory of Electronics. Examples of his work are theoretical models for absorption of electromagnetic waves by molecular oxygen and water vapour, and studies of hurricane phenomenology, such as the warm core and scattering in rainbands, using microwave radiometers. He participated in the Advanced Microwave Sounding Unit and Geosynchronous Microwave Sounder Working Groups for NOAA, and currently is a member of the Science Team for the Atmospheric Infrared Sounder Facility on NASA's Earth Observing System.

Alessandro Battaglia is Assistant Professor, Meteorological Institute, University of Bonn, Germany, Born in 1972 in Italy. Alessandro Battaglia studied physics at the

University of Padova with a master thesis in particle physics (1996) and at the University of Ferrara where he completed his PhD with a thesis on microwave scattering from hydrometeors and radiative transfer in clouds and precipitation. After postdoctoral research at Colorado State University and at University of Bologna and Ferrara, he joined the Group of Remote Sensing and Meso-Scale Modelling at the University of Bonn headed by Prof. C. Simmer. His main interests are in modelling interactions between electromagnetic radiation and hydrometeors with particular focus on microwave active and passive remote sensing applications. Since 2000, has been a reviewer for OSA.

Jean-Pierre Wigneron is currently a senior research scientist at INRA (Institut National de la Recherche Agronomique) and head of the remote sensing group at EPHYSE (Functional Ecology and Environmental Physics), Bordeaux. Born in 1963 in Aix en Provence, France, Jean-Pierre Wigneron received the engineering degree from ENSAE, Toulouse, and the Ph. D. degree from the University of Toulouse (1993). His research interests are in microwave remote sensing of soil and vegetation. In the framework of the Soil Moisture and Ocean Salinity (ESA-SMOS) Mission, he is responsible for the vegetation modelling within the Expert Support Laboratory developing the Level-2 inversion algorithm.

List of contributors

Søren Andersen
Danish Meteorological Institute
Lyngbyvej 100, 2100 Copenhagen
Denmark
Tel: +45 39 15 75 00
email: dmi@dmi.dk

Alessandro Battaglia
Meteorological Institute
University of Bonn
Auf dem Huegel 20, 53121 Bonn
Germany
Tel: +49 228 73 5181/5779
email: batta@uni-bonn.de

Jacqueline Boutin
LODYC/UPMC, Tour 45-55
Case 100 4, place Jussieu
75252 Paris Cedex05, France
Tel: +33 1 44 27 47 65
email:
jacqueline.boutin@lodyc.jussieu.fr

Stefan A. Buehler
University of Bremen/FB 1
PO Box 330440, 28334 Bremen
Germany
Tel: +49 421 218 4417
email: sbuehler@uni-bremen.de

Jean-Christophe Calvet
METEO-FRANCE/CNRM
42, Av. Coriolis
31057 Toulouse Cedex 1, France
Tel: +33 (0)561 07 93 41
email: calvet@meteo.fr

André Chanzy
INRA, Station Science du Sol
Domaine Saint Paul, B.P. 91
84143 Montfavet Cedex, France
Tel: +33 4 90 31 61 29
email: achanzy@avignon.inra.fr

Kun-Shan Chen
National Central University No. 38
Wuchian Li, Chung-Li 320 TAOYUAN, CHINA
email: dkschen@csrsr.ncu.edu.tu

Susanne Crewell
Meteorologisches Institut
University of Munich
Theresienstr. 37, 80333 Munich
Germany
Tel: +49 (0) 89/2180 4210
email: crewell@meteo.physik.uni-muenchen.de

Harald Czekala
RPG Radiometer Physics GmbH
Birkenmaarstrasse 10
53340 Meckenheim
Germany
Tel: +49 2225 99981 34
email: czekala@radiometer-physics.de

Emmanuel P. Dinnat
NASA/GSFC, Code 614.6
Greenbelt, MD 20771, USA
Tel: +1 301 614 6871
email: emmanueldinnat@yahoo.fr

William Ellison
Laboratoire de Physique des Interactions Ondes-Matière
Ecole Nationale Supérieur de Chimie et de Physique de Bordeaux
33607 Pessac, France
Tel: +33 5 40 00 89 18
email: w.ellison@enscpb.fr

Claudia Emde
DLR, Oberpfaffenhofen
D-82234 Wessling
Germany
Tel: +49 8153 283031
Fax: +49 8153 281841
email: Claudia.Emde@dlr.de

Stephen J. English
Met Office
FitzRoy Road, Exeter
Devon EX1 3PB, UK
Tel: +44 (0)1392 884652
email: stephen.english@metoffice.com

Maria José Escorihuela
CESBIO: Centre dÉtudes Spatiales de la Biosphere
Toulouse (unite mixte de recherche UPS-CNRS-CNES-IRD)
email:
maria-jose.escorihvela@cesbio.cnes.fr

Dietrich G. Feist
Institute of Applied Physics
University of Bern
Sidlerstrasse 5, 3012 Bern
Switzerland
Tel: +41 31 631 86 78
email: dietrich.feist@mw.iap.unibe.ch

Paolo Ferrazzoli
Università di Roma Tor Vergata
Via del Politecnico 1
00133 Roma, Italy
Tel: +39 06 72 59 74 21
email: ferrazzoli@disp.uniroma2.it

Leila Guerriero
Università di Roma Tor Vergata
Via del Politecnico 1
00133 Roma, Italy
Tel: +39 06 72 59 74 21
email: guerriero@disp.uniroma2.it

Martti Hallikainen
Helsinki Univ. of Technology
Laboratory of Space Technology
Otakaari 5A/ room SC 228b
02150 Espoo (Otaniemi)
Finland
Tel: +358 9 451 2371
email: martti.hallikainen@hut.fi

Tim Hewison
Met Office, University of Reading Meteorology Building 1U20
PO Box 243 Earley Gate
Reading RG6 6BB, UK
Tel: +44 (0)118 378 7830
email: tim.hewison@metoffice.gov.uk

Georg Heygster
Universität Bremen
Institut für Umweltphysik
Otto-Hahn-Allee
28359 Bremen
Deutschland
Tel: +49 421 218 3910
email: heygster@uni-bremen.de

Thomas R.H. Holmes
Hydrological Sciences Branch
NASA Goddard Space Flight Center
Greenbelt, MD, 20771, USA
email: holt@hsb.nasa.gsfc.gov

Brian K. Hornbuckle
Dept. of Agronomy
Dept. of Electrical and
Computer Engineering
3007 Agronomy Hall
Iowa State University
Ames, IA 50011-1010
USA
Tel: +1 515 294 9868
email: bkh@iastate.edu

Nicole Jacquinet-Husson
Laboratoire de Meteorologie
Dynamique, Ecole Polytechnique
Route Departementale 36
91128 Palaiseau Cedex, France
Tel: +33 1 69 33 48 02
email: nicole.jacquinet@
lmd.polytechnique.fr

Adrian Jupp
Met Office, Satellite Applications
Exeter, Devon, EX1 3PB, UK
Tel: +44 (0)1392 88 6298
email: adrian.jupp@metoffice.com

Yann H. Kerr
CNES/CESBIO,
18 Avenue Edouard Belin
31401 Toulouse Cedex 9, France
Tel: +33 5 61 55 85 22
email: yann.kerr@cesbio.cnes.fr

Laurent Laguerre
CESBIO
Bpi 2801, 18 Avenue E. Belin
31401 Toulouse Cedex 4, France
Tel: +33 5 61 55 85 11
email: laurent.laguerre@cesbio.cnes.fr

Frank Marzano
University of L'Aquila
Via Vetoio Località Coppito
67100 L'Aquila, Italy
Tel: +39 0644585406
email: marzano@ing.univaq.it
& Univ. 'La Sapienza' of
Rome, Via Eudossiana 18,
00184 Rome
Tel: +39 0862 434412
email: marzano@die.uniroma1.it

Christian Mätzler
Institute of Applied Physics
University of Bern
Sidlerstrasse 5, 3012 Bern,
Switzerland
Tel: +41 31 6314589
email: matzler@iap.unibe.ch

Christian Melsheimer
University of Bremen/FB 1
PO Box 330440
D-28334 Bremen
Tel: +49 421 218 2584
email: melsheimer@uni-bremen.de

Michael I. Mishchenko
NASA Goddard Institute
for Space Studies
2880 Broadway, New York
NY 10025, USA
Tel: +1 212 678 5590
email: crmim@giss.nasa.gov

Juan R. Pardo
Departamento de Astrofisica
Molecular e Infrarroja
Serrano 121, 4th floor
28006 Madrid, Spain
Tel: +34 91 5616800
ext 2416
email: pardo@damir.iem.csic.es

Leif Toudal Pedersen
Danish Meteorological Institute
Lyngbyvej 100
2100 Copenhagen, Denmark
Tel: +45 39 15 75 00
email: dmi@dmi.dk

Thierry Pellarin
CNES/CESBIO
18 Avenue Edouard Belin
31401 Toulouse Cedex 9 France
Tel: +33 5 61 55 85 22
email: thierry.pellarin@cesbio.cnes.fr

Catherine Prigent
LERMA, Observatoire de Paris
61, avenue de l'Observatoire
75014 Paris, France
Tel: +33 (0) 1 40 51 20 18
email: catherine.prigent@obspm.fr

Jouni Pulliainen
Helsinki Univ. of Technology
Dept. of Electrical Eng.
Lab. of Space Technology
Otakaari 5A, 02150 Espoo, Finland
Tel: +358 9 451 2373
email: jpulliai@avasun.hut.fi

K. Suresh Raju
Wichita State University
Dept. of Aerospace Engineering
Wallace Hall, Room 200
1845 North Fairmount, Wichita
Kansas 67260 0044, USA
Tel: +1 316 978 3410
email:
suresh.keshavanarayana@wichita.edu

Philip W. Rosenkranz
Massachusetts Institute of
Technology/26-343
77 Massachusetts Av., Cambridge
MA 02139-4307, USA
Tel: +1 617 253 3073
email: pwr@mit.edu

Patricia de Rosnay
CESBIO 18
av. Edouard Belin BPI 2801
31401 Toulouse Cedex 9, France
Tel: +33 5 61 55 85 24
email: patricia.derosnay@cesbio.cnes.fr

Kauzar Saleh Contell
INRA – EPHYSE, B.P.81
33883 Villenave d'Ornon CEDEX
France
Tel: +33 5 57 12 24 15, sec: 24 08
email: ksalemco@bordeaux.inra.fr

Roger Saunders
Met Office Hq
Fitzroy Rd., Exeter EX1 3PB, UK
Tel: +44 1392 886295
email: roger.saunders@metoffice.com

Mike Schwank
Soil Physics, Institute of
Terrestrial Ecology ETHZ
CHN E29, 8092 Zurich, Switzerland
Tel: +41 1 633 6014
email: mike.schwank@env.ethz.ch

Jiancheng Shi
Institute for Computational
Earth System Sciences
University of California
Santa Barbara, CA 93106, USA
Tel: +1 805 893 2309
email: shi@icess.ucsb.edu

Ari Sihvola
HUT, Electromagnetics Laboratory
P.O. Box 3000, 02015 HUT, Finland
Tel: +358 9 4512261
email: ari.sihvola@hut.fi

Clemens Simmer
Meteorological Institute
University Bonn
Auf dem Huegel 20, 53121 Bonn
Germany
Tel: +49 228 73 5181/5190
email: csimmer@uni-bonn.de

Bertrand Thomas
CCRLC – Rutherford Appleton Laboratory
Space Science & Technology Dept.
Chilton Didcot
Oxfordshire OX11 0QX, UK
Tel: +44 1235 446343
email: b.thomas@rl.ac.uk

Rasmus T. Tonboe
Ice and Remote Sensing Division
Danish Meteorological Institute
Lyngbyvej 100, 2100 Copenhagen
Denmark
Tel: +45 39 15 73 49
email: rtt@dmi.dk

Adriaan van de Griend
Faculty of Earth & Life Sciences
Vrije Universiteit Amsterdam
De Boelelaan 1085
1081 HV Amsterdam, NL
Tel: +31 20 444 7331
email: griend@geo.vu.nl

Philippe Waldteufel
IPSL/Service d'Aéronomie
91371 Verriéres Le Buisson
France
email: philippe.waldteufel@aerov.jussieu.fr

Andreas Wiesmann
Gamma Remote Sensing
Worbstrasse 225
3073 Gümligen
Switzerland
Tel: +41 31 951 70 05
email: gamma@gamma-rs.ch

Jean-Pierre Wigneron
INRA – EPHYSE
B.P. 81
33883 Villenave d'Ornon CEDEX
France
Tel: +33 5 57 12 24 19
email: wigneron@bordeaux.inra.fr

Chapter 1
Radiative transfer and microwave radiometry

Christian Mätzler and Christian Melsheimer

1.1 Historical overview

Radiometry of spectrally resolved thermal radiation in the microwave range has been a valuable tool for the remote sensing of the terrestrial surface and atmosphere with concentration on the hydrosphere. Satellite sensors with continuously improving capabilities have been in operation for more than 25 years. Similar instruments have been used for monitoring atmospheric properties from the surface. The progress in recent developments might give the impression that the method is young. However, when looking at the history we realise that radiometry and radiative transfer are rooted in the oldest human disciplines. Radiative transfer describes the emission, propagation, scattering and absorption of electromagnetic radiation. Since almost everything that we know about the universe has been brought to men by radiation, radiative transfer is as old as vision. The following text is an attempt at a historical overview with relevance to the topic of this book.

The basis of today's knowledge on emission and absorption of thermal radiation was laid by Josef von Fraunhofer (1787–1826), Robert Bunsen (1811–1899), Gustav Robert Kirchhoff (1824–1887) and others at and before the time James Clark Maxwell (1831–1879) stated that magnetism, electricity and light are different manifestations of the same fundamental laws. Application of the first and second laws of thermodynamics let Kirchhoff formulate his law of radiation in 1860, stating in today's language that under local thermodynamic equilibrium the emissivity of a given object is equal to its absorptivity. Even before the Planck function was found in 1900 (Max Planck, 1858–1947), radiative transfer and spectroscopy enabled the identification of atoms in the solar atmosphere. Furthermore, the rules of Bunsen revealed the properties of optically thin and thick radiation. What was missing at that time was the understanding of the spectral line positions, strengths and widths. The driving questions led to the development of quantum theory in the early twentieth century.

The term 'light rays' originated from the observation that light propagates radially from the source along straight lines. This applies to homogeneous media. More complex is the situation where the medium is inhomogeneous; then the radiation suffers refraction, reflection, diffraction and scattering. Some of these phenomena had already been investigated by Willebrord Snell (1591–1626) and Augustin Fresnel (1788–1827). The Fresnel equations, describing the reflection of polarised radiation at a plane, are a basic law to describe surface emissivity in a simple situation. Scattering by reflecting spheres was described by Alfred Clebsch in 1863. Light scattering by refractive spheres was described by Ludvig V. Lorenz in 1890. Better known is the work on the same topic carried out by Gustav Mie (1867–1957) in 1908. Microwave radiometer observations of precipitating clouds can be explained by Lorenz–Mie scattering and absorption. Scattering of radiation by particles, which are small with respect to the wavelength, is named after Baron Rayleigh, John William Strutt (1842–1919). Rayleigh scattering of radar waves by raindrops has become a standard in radar meteorology to relate rain rate with radar intensity. Rayleigh scattering has also been used in microwave radiometry to describe the emission of clouds and snow.

Multiple scattering of radiation in a cloudy atmosphere was first described by Sir Arthur Schuster (1851–1934) in 1905. Technical applications in the paper and paint industry followed (Kubelka and Munk, 1931). However it was mainly in the field of astrophysics that radiative transfer methods made substantial progress. Two important contributors were Subrahmanyan Chandrasekhar (1910–1995) and Hendrik Christoffel van de Hulst (1918–2000), both working for some time at the Yerkes Observatory of the University of Chicago. They developed mathematical methods in the 1940s to describe radiative transfer in stellar and planetary atmospheres under the assumption that the refractive index of the medium is equal to one. Although not necessary, this assumption was used in the majority of works on this topic. Chandrasekhar's collected papers on radiative transfer were published in book form in 1950, and van de Hulst's work led to a book dealing with scattering by small particles in 1957 and one on multiple scattering in 1980. Approximate solutions of the radiative-transfer equation for multiple scattering are named after Sir Arthur Eddington (1884–1944); see Joseph *et al.* (1976), Meador and Weaver (1980) and Thomas and Stamnes (1999).

More general treatments were required when it was realised that strong refractive effects occur for radiation in conductive media, especially at frequencies near the plasma frequency. A formulation of radiative transfer, including the inhomogeneity and anisotropy of the propagating medium, was presented by George Bekefi (1925–1995) in 1960 in his textbook on radiation processes in plasmas, introducing the ray-refractive index. Refractive treatments were also applied to radiative transfer in natural waters (Mobley, 1994). They appeared to be so relevant that they were formulated in the fundamental theorem of radiometry (Wyatt, 1978), as simply expressed by Equation (1.8). The theorem had already been formulated by David Hilbert (1862–1943) in 1912, and even Kirchhoff was aware of it because it follows from Snell's law.

Invisible radiation at infrared wavelengths was known at the time of Kirchhoff. This radiation was called dark rays. Microwaves, at still longer wavelengths, have been known since the validation of Maxwell's electromagnetic wave theory by

Heinrich Hertz (1857–1894) in 1888. Natural radio radiation at a wavelength of 14.5 m was first discovered from the centre of the Milky Way by Karl Jansky (1905–1950) in 1931. The level of radiation was sufficiently high to be detectable because it is of non-thermal origin. Today the Jansky ($=10^{-26}$W/m^2/Hz) is the unit used in radio astronomy to describe the spectral flux density of radio sources.

The pioneer in the development of microwave radiometry was Robert Henry Dicke (1916–1997), a brilliant physicist in theoretical and in experimental work. The radiometer type named after him was a great invention in 1944 to measure the low power levels associated with thermal microwave radiation (Dicke, 1946). A first radiometer operated at the MIT in Massachusetts at a wavelength of 1.25 cm; this is close to the 22 GHz water-vapour line. From theoretical work by John H. Van Vleck (1899–1980) in 1942, Dicke realised that the enhanced radiation was due to emission by water vapour. He and others went to Florida with three radiometers (at wavelengths of 1.0, 1.25 and 1.5 cm) to measure the rather weak atmospheric absorption in this frequency range for the first time (Dicke *et al.*, 1946). In this work they introduced the concepts of antenna temperature T_A and the mean atmospheric temperature T_m relevant for the observed radiation and the tipping-curve method. Dicke also measured the surface temperature of the moon and of the sun. Later he played a key role in the discovery of the cosmic thermal background radiation.

Before microwave radiometry was exploited for the investigation of the earth, emphasis was put on extra-terrestrial objects, leading to the evolution of radio astronomy (Kraus, 1966). But even radio astronomers had to deal with the terrestrial atmosphere to identify the nature of their signals. Therefore they paved the way in spectroscopy and radiative transfer.

Special interest in terrestrial applications of microwaves was based on the capability to penetrate clouds and smoke. Shortly before, during and after the Second World War, the priority in the advancement of microwave methods was given to radar. This active technique is excellent for locating and ranging remote objects, but it is very limited in terms of radiometric accuracy. In contrast to radars, radiometers are passive devices, meaning that they only receive radiation. Their virtue is high radiometric accuracy, but the ranging capability of radiometers is either non-existent or indirect, for instance by using stereo techniques or knowledge on the spectral properties of the medium to be sensed. It is evident that the properties of radars and radiometers are complementary.

The first efforts to measure the microwave thermal radiation of terrestrial surface types were made at the University of Texas by Straiton *et al.* (1958) at a wavelength of 4.3 mm. Only ten years later, the Russian Earth Observing Satellite, COSMOS-243, carried four radiometers at frequencies of 3.5, 8.8, 22.2 and 37 GHz with nadir-viewing horns to investigate the terrestrial surface and atmosphere (Basharinov *et al.*, 1971). Radiometers with different antenna types and view geometry were tested on satellites in the following years. A significant advancement was achieved with the introduction of conically scanning radiometers to image the earth at a constant angle of incidence. This property is important because it enables data comparison and interpretation. A first system, the scanning multichannel microwave radiometer (SMMR) was launched on the NASA Nimbus 7 satellite in 1978 (Gloersen and Barath, 1977) with dual linear polarisation at a set of frequencies between 6 and 37 GHz. Its problem, a rotation of the polarisation plane, was eliminated in the next generation of imaging radiometers, the special sensor microwave imagers (SSM/Is), flown with the Defense Meteorological Satellite Program (DMSP) since 1987 (Hollinger *et al.*, 1987). A conical scan without polarisation rotation has been achieved by rotating the entire antenna and front end around the vertical axis. This type of radiometer has become a standard for imaging sensors.

The combination of the microwave radiometer with a high-resolution spectrometer enables the investigation of atmospheric trace gases, including the determination of the altitudinal distribution. The same method allows the determination of temperature profiles by observing the optically thick radiation of strong spectral lines. The first spaceborne microwave spectrometer to investigate the atmospheric temperature profiles was flown on the Nimbus 5 satellite (Waters *et al.*, 1975), launched in 1972. Today, the advanced microwave sounding units A and B on the US National Oceanic and Atmospheric Administration (NOAA) polar orbiting satellites provide key information for the operational meteorological services worldwide.

The datasets collected from microwave radiometers now cover decadal time scales. They contain information on the global scale of long-term changes and thus are of importance for climatology, and especially for the monitoring of climate change.

Together with the technical and experimental progress, there has been a steady advancement in the theoretical understanding of the observations as documented in the scientific literature. Microwave spectroscopy and the understanding of spectral lines of gases evolved from quantum theory, leading to a famous book by Townes and Schawlow (1955). A treatise on microwave remote sensing, including radiometry, was written by Ulaby *et al.* (1981, 1982, 1986). More specialised books were dedicated to theoretical aspects (Tsang *et al.*, 1985, Fung, 1994) or to special applications, like passive microwave remote sensing of the atmosphere (Janssen, 1993) and of the ocean (Cherny and Raizer, 1998). Very recently, a treatise appeared by Sharkov (2003) on passive microwave remote sensing of the earth, with emphasis on work done in Russia.

The present book concentrates on models to simulate microwave radiation. The state of the art on microwave spectroscopy of gases relevant for the terrestrial atmosphere is reviewed in Chapter 2. The effects of clouds and precipitation are the topic of Chapter 3. Chapter 4 deals with the terrestrial surface. The dielectric behaviour of liquids, solids and mixtures is described in some detail in Chapter 5.

The development of surface- and space-borne microwave radiometry continues at an accelerated pace. The backbone is an adequate technology, on the one hand, and a physical understanding of the observed radiation, on the other. This book deals with the latter, in particular the modelling of thermal microwave radiation in the terrestrial environment.

Accurate physical radiative transfer models are required in order to establish the relationship between the physical state of the earth and the observable radiation. The optimisation of remote sensing methods relies on these models. The same is true for advancing the understanding of processes in the atmosphere, at the ocean and on land surfaces. Physical radiative transfer models are based on the principles of thermal radiation, on the electromagnetic and geometrical properties of matter and its interaction with electromagnetic fields. The properties of matter in nature are very complex. Their investigation is an ongoing research task. Therefore there is a need for empirical data, and many gaps still have to be closed. The present book contributes to this task.

Reliable and fast models are demanded by data assimilation methods in forecasting and nowcasting of the states of the atmosphere and surface, including the water cycle. Fast radiative transfer models can be constructed from the underlying physical models. Some examples will be presented.

Finally, an issue of concern for microwave radiometry is the active use of the electromagnetic spectrum for communication, radar and other technical and commercial applications. The spectral power density of man-made radiation can be much higher than that of natural radiation. Useful radiometer measurements rely on spectral bands which are kept free of man-made noise. The frequency allocation for all uses is controlled by the International Telecommunication Union (ITU) in the Radio Regulations, which are updated every three years. A number of frequency bands are allocated to passive services, the name for radiometry in ITU. New commercial applications exert a high pressure to use the 'free' bands. It is not easy to defend the continued interests required by radiometer operations. Pioneering steps were made by radio astronomers. In Europe the Committee on Radio Astronomy Frequencies (CRAF) of the European Science Foundation (ESF) has been very active since the mid 1990s. The earth observation community began to defend its requirements some years ago. Due to the global nature of earth observation it is of utmost importance that all users follow the internationally agreed rules.

The following sections give an introduction to radiative transfer with focus on microwave radiometry. More detailed discussions are found in the later chapters.

1.2 Kirchhoff's law of thermal radiation

Let us start with Kirchhoff's law of thermal radiation, the basic law relating emission and absorption of radiation in local thermodynamic equilibrium, meaning that the emission is governed by thermal conditions as defined by a unique temperature at any given location. A very useful form of the law for microwave radiometry was provided by Planck (1966, §43–46) in his lectures on heat radiation. Assume that the space where the radiation propagates can be divided into N volumes of temperatures T_i where $i = 1$ to N. The temperature inside each volume is homogeneous. Note that N may be very large so that the radiation from all temperatures and all regions accessible by the radiation can be handled. The situation with $N = 2$ was described by Dicke *et al.* (1946). A radiometer with a given reciprocal antenna and with given spectral properties observes the radiation. The received power must be a linear combination of the blackbody radiances corresponding to the different temperatures. If the power is expressed in terms of the antenna temperature T_A (Dicke *et al.*, 1946; Ulaby *et al.*, 1981), the linear combination follows as, using the Rayleigh–Jeans approximation

$$T_A = \sum_{i=1}^{N} e_i T_i \tag{1.1}$$

To determine the emissivities e_i ($i = 1$ to N), the reciprocal condition is considered in which the radiometer is a stationary transmitter with the same spectral and polarisation properties. Let us call a_i the fraction of power absorbed in volume V_i. Then a_i is the absorptivity of this volume for the given configuration. Energy conservation requires that all radiation be absorbed somewhere, which means that

$$\sum_{i=1}^{N} a_i = 1 \tag{1.2}$$

Kirchhoff's law states that in local thermodynamic equilibrium $e_i = a_i$ for all $i = 1$ to N. We thus have

$$T_A = \sum_{i=1}^{N} a_i T_i \tag{1.3}$$

Since (1.3) is true for any radiometer (with any spectral properties, polarisation and angular pattern), it also holds for an ideal monochromatic pencil-beam radiometer, and thus for the brightness temperature T_b (ν, p) at frequency ν and polarisation p:

$$T_b(\nu, p) = \sum_{i=1}^{N} a_i(\nu, p) T_i \tag{1.4}$$

where $a_i(\nu, p)$ is the monochromatic absorptivity in volume V_i for radiation propagating in the reciprocal (opposite) direction of the polarised radiation considered. Equation (1.2) is also required to hold for $a_i(\nu, p)$. In the more general situation of thermal radiation expressed by the Planck function, the physical temperatures in (1.4) are replaced by the Planck functions for the respective temperatures, and T_b becomes the radiance.

To prove these expressions, Kirchhoff used thought experiments with the aim of showing that any deviation leads to violations of thermodynamic principles. By special insulation an object is allowed to interact with the environment only by radiation. First assume that the emissivity of the object is larger than its absorptivity. Under isothermal conditions, the object has to cool because it loses more energy than it gains. The second law of thermodynamics does not allow the volume to cool. On the other hand if the emissivity is smaller than the absorptivity the object is heated. This violates energy conservation and thus the first law of thermodynamics. The solution is to accept Kirchhoff's law. Kirchhoff's main effort was to show that this law is valid independently at each frequency and polarisation. For this purpose the path between the object and the environment can be restricted to radiation with the selected properties. All other radiation is reflected, e.g. by using perfect filters, polarisers and single-mode waveguides, so that the only way to exchange energy is at the selected frequency and polarisation. Since they are arbitrary, the law must hold for all frequencies and polarisations.

Kirchhoff's law, in the form of Equation (1.3) with condition (1.2), is applicable to radiometers with all kinds of reciprocal antennas and not only to radiation represented by plane waves or by infinitesimally diverging beams. Therefore it is more general than the radiative-transfer equation. Furthermore, boundary conditions, diffraction and other coherent effects, arising from special geometries, are automatically included if correct absorptivities are taken into account. However, their determination requires special tools. One of them is the radiative-transfer equation. Electromagnetic tools have to be used to solve problems with coherent effects.

1.3 The radiative-transfer equation

The radiative-transfer equation describes how radiation is modified when it propagates over a path interval in a given volume under the assumption of incoherent interactions. Integration along the path s together with appropriate boundary conditions at both ends (s_1 and s_2) gives the radiance or brightness temperature at a given location. The path is usually determined by geometrical optics. The derivation and detail of the transfer equation are described in many textbooks (see for instance Chandrasekhar, 1950; Ishimaru, 1978; Mobley, 1994; Thomas and Stamnes, 1999).

We allow the complex refractive index $n(\mathbf{r})$ to depend on the position $\mathbf{r}$ in space. In order to keep the ray path s tractable, a smoothness condition for the gradient of the real part $n'(\mathbf{r})$ of n is imposed:

$$\nabla n'(\mathbf{r}) \ll k = 2\pi\nu/c = 2\pi/\lambda \tag{1.5}$$

Here k, λ and $c = c_0/n'$ are the wave number, wavelength and phase velocity, respectively, in the medium, and $c_0 = 299792.5$ km/s is the speed of light in vacuum. Condition (1.5) means that n' changes only over distances much greater than a wavelength, ensuring that the ray follows a single path without reflection. Reflections are allowed at the ends of the interval, where they can be handled by suitable boundary conditions. For each ray and for each interval separate and coupled transfer equations have to be solved.

The radiation is expressed by the radiance: it is the radiative, directed, spectral power contained in a unit frequency interval, crossing perpendicularly through a unit area with directions contained in a unit solid angle. For thermal radiation the radiance can be expressed by the brightness temperature $T_b = T$; i.e. the temperature T of thermal radiation with the radiance I_ν. This radiance is determined by the Planck function B_ν:

$$I_\nu = B_\nu(T) = \frac{2h\nu^3}{c^2(\exp(h\nu/k_b T) - 1)} \tag{1.6}$$

where $h = 6.62607 \cdot 10^{-34}$ Js is the Planck constant, ν is the frequency and $k_b = 1.38065 \cdot 10^{-23}$ J/K is the Boltzmann constant. The Rayleigh–Jeans approximation, often used in microwave radiometry, is obtained from the Planck function in situations where $h\nu/k_b T \ll 1$ to get

$$B_\nu = \frac{2k_b T \nu^2}{c_2} \tag{1.7}$$

Apart from the linearity between radiance and brightness temperature in the Rayleigh–Jeans approximation, the main motivation for expressing thermal radiation by the brightness temperature is the fact that the equation of radiative transfer for a refractive medium is formally the same as for the vacuum. This will be shown below.

1.3.1 No scattering and no absorption

In the simplest situation the ray propagates freely without absorption and scattering. This applies if the refractive index is real. According to the fundamental theorem of radiometry (Mobley, 1994), the new quantity, defined by

$$I_{1\nu} := \frac{I_\nu}{n'^2} = \text{constant along } s \tag{1.8}$$

is a constant along the propagation path. The specific intensity I_ν changes along the path as a result of refraction. However the quantity $I_{1\nu}$ is conserved. In the more general situation of an anisotropic medium, n' in (1.8) has to be replaced by the ray-refractive index (Bekefi, 1966). Equation (1.8) also means that the Planck function is not conserved, but the following quantity is

$$I_{1\nu} = B_{1\nu} := \frac{B_\nu(\mathbf{r}, T_b)}{n'(\mathbf{r})^2} = \frac{2h\nu^3}{c_0^2(\exp(h\nu/k_b T_b) - 1)} = \text{constant} \tag{1.9}$$

Since the quantities in the right-most expression in (1.9) are either fundamental constants (h, k_b, c_0), an independent but fixed variable (ν) or the brightness temperature T_b, T_b does not change along the path of propagation. Thus $I_{1\nu} = B_{1\nu}$ and T_b are conserved quantities. This is a first important result. If the brightness temperature T_b did change, it would violate principles of thermodynamics. Consider the medium to be isothermal at temperature T and perfectly insulated from interactions with the rest of the world. Radiation originates in a region inside the medium where absorption and emission occur. The brightness temperature must be $T_b = T$ as well. If it changes along the loss-free path, it ends up with a different value. At the end of the path, the radiation is absorbed by a black body, which will be heated or cooled to the new temperature. This is not possible in an insulated, isothermal medium. Therefore T_b must be conserved along the path.

1.3.2 Including absorption and emission

In the next step, absorption and emission are taken into account but scattering is omitted. The absorption coefficient γ_a is related to the imaginary part n'' of the complex refractive index $n(\mathbf{r}) = n' + in''$ by

$$\gamma_a(\mathbf{r}) = 2k \cdot n''(\mathbf{r}); \quad (1/\text{m}) \tag{1.10}$$

The radiative-transfer equation considers the changes that occur to the quantity $I_{1\nu}$ along the propagation path s. Over an infinitesimal step ds it is increased by emission and reduced by absorption. The resulting differential equation is

$$\frac{\mathrm{d}I_{1\nu}(s)}{\mathrm{d}s} = \gamma_\mathrm{a}(s)(B_{1\nu}(s) - I_{1\nu}(s)) \tag{1.11}$$

where Kirchhoff's law was used to express the positive source term by the absorption coefficient multiplied by the normalised Planck function $B_{1\nu}$. The path dependence of $B_{1\nu}$ arises from the dependence on the physical temperature $T(s)$. By the use of the normalised quantities ($I_{1\nu}$ and $B_{1\nu}$), the refractive index does not appear explicitly. Due to the linearity between radiance and temperature in the Rayleigh–Jeans approximation, the radiative-transfer equation can also be expressed as

$$\frac{\mathrm{d}T_\mathrm{b}}{\mathrm{d}s} = \gamma_\mathrm{a}(s)(T(s) - T_\mathrm{b}(s)) \tag{1.12}$$

This is the most common form used in microwave radiometry. It is identical for refractive and non-refractive media.

1.3.3 Including absorption, emission and scattering

In the final step towards the complete radiative-transfer equation, volume scattering in the path interval ds is included. The losses contain contributions from absorption and scattering (with scattering coefficient γ_s), the sum

$$\gamma_\mathrm{e} = \gamma_\mathrm{a} + \gamma_\mathrm{s} \tag{1.13}$$

is being called the extinction coefficient, and the ratio

$$\varpi_0 = \frac{\gamma_\mathrm{s}}{\gamma_\mathrm{e}} \tag{1.14}$$

is called the single-scattering albedo. The radiative-transfer equation now reads

$$\frac{\mathrm{d}I_{1\nu}(s)}{\mathrm{d}s} = -\gamma_\mathrm{e}(s) \cdot I_{1\nu}(s) + \varepsilon_{1\nu}(s) \tag{1.15}$$

The source term

$$\varepsilon_{1\nu}(s) = \gamma_\mathrm{a}(s) \cdot B_{1\nu}(s) + \frac{\gamma_\mathrm{e}(s)}{4\pi} \int_{4\pi} p(\theta,\phi,\theta',\phi') \cdot I_{1\nu}(s,\theta',\phi') \cdot \mathrm{d}\Omega' \tag{1.16}$$

consists of thermal emission (first term) and of radiation scattered from other directions into the considered ray path (second term). The integral in the second term contains the normalised radiance $I_{1\nu}$ at path position s propagating in all directions, as defined by the spherical coordinates (θ', ϕ'). The so-called phase function $p(\theta, \phi, \theta', \phi')$ describes the transfer of radiance from direction (θ', ϕ') to the direction (θ, ϕ) of the considered propagation path. The form of Equation (1.16) requires the normalisation

$$\frac{1}{4\pi} \int_{4\pi} p(\theta,\phi,\theta',\phi') \cdot \mathrm{d}\Omega = \frac{\gamma_\mathrm{s}}{\gamma_\mathrm{e}} = \varpi_0 \tag{1.17}$$

Equivalent expressions apply for the Rayleigh–Jeans approximation to the brightness temperature T_b. To this end, $I_{1\nu}$ is replaced by T_b, and $B_{1\nu}$ by T:

$$\frac{dT_b(s)}{ds} = -\gamma_e(s)\cdot T_b(s) + \gamma_a T(s) + \frac{\gamma_e(s)}{4\pi}\int_{4\pi} p(\theta,\phi,\theta',\phi')\cdot T_b(s,\theta',\phi')\cdot d\Omega' \tag{1.18}$$

This is an integro-differential equation not only for T_b in the propagation direction, but also including the brightness temperatures propagating in all other directions. The coupling of radiation propagating in different directions causes inherent difficulties in finding a practical solution. However, a formal solution is obtained as shown in the next section.

1.3.4 *A formal solution*

Let us recall that we are looking for an expression to describe how T_b or $I_{1\nu}$ changes along the path from a starting point at s_1 to an end point at s_2. First the equation of radiative transfer is simplified by introducing two new quantities (τ and $J_{1\nu}$). The optical depth $\tau(s)$ at a variable path position s as seen from the end position s_2 is defined by

$$\tau(s) := \int_s^{s_2} \gamma_e\,(s')ds' \tag{1.19}$$

Note that $d\tau = -ds\gamma_e$ and ds are in opposite directions. The source function $J_{1\nu}$ is defined by

$$J_{1\nu} := \frac{\varepsilon_{1\nu}}{\gamma_e} = (1-\varpi_0)B_{1\nu}(T) + \frac{1}{4\pi}\int_{4\pi} p(\theta,\phi,\theta',\phi')\cdot I_{1\nu}(s,\theta',\phi')\cdot d\Omega' \tag{1.20}$$

With these quantities the radiative-transfer equation is simplified to

$$\frac{dI_{1\nu}}{d\tau} - I_{1\nu} = -J_{1\nu} \tag{1.21}$$

which can also be written as

$$\frac{d}{d\tau}(e^{-\tau}I_{1\nu}(\tau)) = -e^{-\tau}J_{1\nu} \tag{1.22}$$

Integration from $\tau = 0$ to $\tau = \tau_2 = \int_{s_1}^{s_2}\gamma_e ds$ gives the change of the quantity $\exp(-\tau)\cdot I_{1\nu}$ in the considered medium, leading to a formal solution of the radiative-transfer equation at the output position s_2:

$$I_{1\nu}(s=s_2) = I_{1\nu}(\tau=0) = I_{1\nu}(s_1)e^{-\tau_2} + \int_0^{\tau_2} e^{-\tau}J_{1\nu}(\tau)\,d\tau \tag{1.23}$$

where $I_{1\nu}(s_1)$ is the input, and τ is given by (1.19). In the Rayleigh–Jeans approximation this becomes

$$T_b(s_2) = T_b(s_1)e^{-\tau_2} + \int_0^{\tau_2} \left[(1 - \varpi_0)T(\tau) + \frac{1}{4\pi}\int_{4\pi} p(\theta,\phi,\theta',\phi') \cdot T_b(\tau,\theta',\phi') \cdot d\Omega'\right] e^{-\tau} d\tau \tag{1.24}$$

Again, this form of solution is the same for refractive and non-refractive media. It reduces to the scatter-free situation for $\gamma_s = 0$ and thus $p = \varpi_0 = 0$, in which case $d\tau = -ds\gamma_a$ and

$$T_b(s_2) = T_b(s_1)e^{-\tau_2} + \int_0^{\tau_2} T(\tau)e^{-\tau} d\tau \tag{1.25}$$

where τ is given by (1.19), but with γ_e exchanged by γ_a. The application of Equation (1.24) to microwave radiation in scattering clouds and precipitation will be the topic of Chapter 3, and applications to situations with scattering surface layers will be discussed in Chapter 4.

1.3.5 Special situations

Here we consider simple situations where scattering can be ignored. They typically apply to the atmosphere without precipitating particles, but also to surface layers (sand, snow and vegetation canopies) at wavelengths much larger than the correlation length characterising the structure of the medium.

1.3.5.1 A single layer at constant temperature

If the ray passes through a single layer at a constant temperature T_2, the integration of (1.25) is straightforward, yielding

$$T_b(s_2) = T_b(s_1)e^{-\tau_2} + T_2(1 - e^{-\tau_2}) \tag{1.26}$$

and $\tau_2 = \int_{s_1}^{s_2} \gamma_a ds$. Comparing (1.26) with Kirchhoff's law (1.4), it is obvious that $N = 2$, with Volume 1 being the background, $T_1 = T_b(s_1)$, and $a_1 = \exp(-\tau_2)$, Volume 2, being the homogeneous layer at T_2, with $a_2 = 1 - a_1$, and thus Equation (1.2) is fulfilled. In optically thin media ($\tau_2 \ll 1$), we have $a_1 \cong 1 - \tau_2$, $a_2 \cong \tau_2$, and $T_b(s_2)$ changes linearly with τ_2. On the other hand, for optically thick situations ($\tau_2 \gg 1$), the background radiation is hidden, and $T_b(s_2) = T_2$.

1.3.5.2 A further volume at another temperature

Assume in the above example that the ray path continues beyond s_2 crossing a third volume at temperature T_3 for $s_2 < s < s_3$. The input brightness to this volume is $T_b(s_2)$, and the modification by Volume 3 is equivalent to the second term in (1.26), but with the index increased to 3:

$$T_b(s_3) = T_b(s_2)e^{-\tau_3} + T_3(1 - e^{-\tau_3}) \tag{1.27}$$

and $\tau_3 = \int_{s_2}^{s_3} \gamma_a ds$. Kirchhoff's law is again confirmed, now with $N = 3$, $a_1 = \exp(-\tau_2 - \tau_3)$, $a_2 = \exp(-\tau_3) \cdot [1 - \exp(-\tau_2)]$ and $a_3 = 1 - \exp(-\tau_3)$. This type of solution can be iterated to any number of layers.

1.3.5.3 The concept of an effective mean temperature

The solution of (1.25) in the form of Equation (1.26) is very practical, even in situations where T is not constant. There exists an effective mean temperature $T_m = T_2$ for which (1.26) is valid:

$$T_m = \frac{\int_0^{\tau_2} T(\tau) e^{-\tau} d\tau}{1 - e^{-\tau_2}} \tag{1.28}$$

and τ is still given by (1.19) for $\gamma_e = \gamma_a$. Dicke *et al.* (1946) had already used this concept. A frequent situation is a linear temperature profile

$$T = T_c + T_d \tau \tag{1.29}$$

Inserting (1.29) in (1.28) gives

$$T_m = T_c + T_d \left[1 - \frac{\tau_2 e^{-\tau_2}}{1 - e^{-\tau_2}} \right] \tag{1.30}$$

Two different cases are to be distinguished. First, in an optically thin layer ($\tau_2 \ll 1$), T_m converges to the average physical temperature $T_c + 0.5 T_d \tau_2$. Second, for high optical depth ($\tau_2 \gg 1$), T_m converges to the physical temperature at position $\tau = 1$, giving $T_m = T(\tau = 1) = T_c + T_d$. Geometrically, the position is situated at a distance, $\Delta s = 1/\gamma_a$, from s_2. This property has been used for temperature profiling in spectral bands with high absorption.

1.4 Polarisation and Stokes parameters

In several of the subsequent chapters, the fact that electromagnetic radiation can be polarised is relevant. Here we give some details on the polarisation of radiation, the definition of the Stokes parameters, the definition of antenna polarisation and the radiative-transfer equation for polarised radiation.

1.4.1 Polarisation directions

Electromagnetic waves in homogeneous, isotropic media are transverse waves, i.e. their oscillating electric and magnetic fields are in a plane perpendicular to the propagation direction. The choice of two basis vectors – we shall call them polarisation directions here – that span the transverse plane is arbitrary; often they are called 'horizontal' and 'vertical' and correspond to some horizontal and vertical direction of the particular setting. Nevertheless, what is meant by horizontal/vertical, or parallel/perpendicular, is purely a matter of definition.

In many contexts, there is a preferred direction in space that is fixed. In earth remote sensing, this is typically the vertical axis perpendicular to the earth's surface.

In this case it is useful to define the polarisation directions with respect to that fixed frame as follows (Mishchenko *et al.*, 2000, 2002; Tsang *et al.*, 2000). We use a coordinate system where the z-axis points towards the local zenith and denote the propagation direction of radiation by the unit vector $\boldsymbol{n} = \boldsymbol{k}/k$, where k is the wave number. The vector $\boldsymbol{n}$ is given by two angles, the zenith angle θ, i.e. the angle between $\boldsymbol{n}$ and the z-axis, and the azimuth angle ϕ, i.e. the angle between the projection of $\boldsymbol{n}$ into the xy-plane and the x-axis:

$$\boldsymbol{n} = \begin{bmatrix} \cos\phi \sin\theta \\ \sin\phi \sin\theta \\ \cos\theta \end{bmatrix} \tag{1.31}$$

We now define the polarisation directions by the partial derivatives of $\boldsymbol{n}$ with respect to θ and ϕ and call them θ-direction or vertical and ϕ-direction or horizontal, respectively, see Figure 1.1.

Their unit basis vectors are

$$\boldsymbol{e}_\theta = \boldsymbol{e}_\mathrm{v} = \frac{\partial \boldsymbol{n}}{\partial \theta} \bigg/ \left\| \frac{\partial \boldsymbol{n}}{\partial \theta} \right\| = \begin{bmatrix} \cos\phi \cos\theta \\ \sin\phi \cos\theta \\ -\sin\theta \end{bmatrix} \tag{1.32}$$

$$\boldsymbol{e}_\phi = \boldsymbol{e}_\mathrm{h} = \frac{\partial \boldsymbol{n}}{\partial \phi} \bigg/ \left\| \frac{\partial \boldsymbol{n}}{\partial \phi} \right\| = \begin{bmatrix} -\sin\phi \\ \cos\phi \\ 0 \end{bmatrix} \tag{1.33}$$

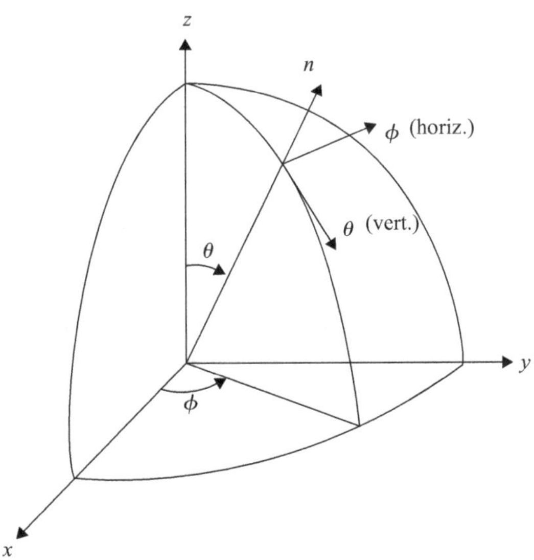

Figure 1.1 The definition of the polarisation directions, adapted from Mishchenko et al. (2000)

The vectors $\boldsymbol{n}, \boldsymbol{e}_\text{v}, \boldsymbol{e}_\text{h}$ are mutually orthogonal and define (in this order) a right-handed system, i.e. $(\boldsymbol{n} \times \boldsymbol{e}_\text{v}) \cdot \boldsymbol{e}_\text{h} = 1$ and the same for all cyclic permutations.[1]

1.4.2 *Stokes parameters*

Plane monochromatic electromagnetic waves are commonly written in the form

$$\boldsymbol{E}(\boldsymbol{x}, t) = \begin{bmatrix} E_\text{v} \\ E_\text{h} \end{bmatrix} e^{\text{i}(\boldsymbol{kx}-\omega t)} = (E_\text{v}\boldsymbol{e}_\text{v} + E_\text{h}\boldsymbol{e}_\text{h})e^{\text{i}(\boldsymbol{kx}-\omega t)} \tag{1.34}$$

where $\boldsymbol{E}$ is the electric field vector, and the subscripts v and h denote the components with vertical and horizontal polarisation, respectively. E_v and E_h, the amplitudes, are complex numbers, $\boldsymbol{k}$ and ω are the wave number vector and the angular frequency of the plane wave and $\boldsymbol{e}_\text{v} = (1, 0)^\text{T}$, $\boldsymbol{e}_\text{h} = (0, 1)^\text{T}$ are the unit vectors; $[E_\text{v}, E_\text{h}]^\text{T}$ is called the Jones vector. It is always implicitly understood that the actual, physical electric field is the real part of the above expression (Equation 1.34). Rewriting the complex amplitudes E_v and E_h using real, non-negative amplitudes a_v and a_h, and real phases δ_v and δ_h, yields

$$E_\text{v} = a_\text{v}e^{\text{i}\delta_\text{v}}, E_\text{h} = a_\text{h}e^{\text{i}\delta_\text{h}} \tag{1.35}$$

The actual electric field vector $\tilde{\boldsymbol{E}}$ is then

$$\tilde{\boldsymbol{E}}(\boldsymbol{x}, t) = \text{Re}[\boldsymbol{E}(\boldsymbol{x}, t)] = \begin{bmatrix} a_\text{v} \cdot \cos(\boldsymbol{kx} - \omega t + \delta_\text{v}) \\ a_\text{h} \cdot \cos(\boldsymbol{kx} - \omega t + \delta_\text{h}) \end{bmatrix} \tag{1.36}$$

In general, instruments do not measure the electric or magnetic field vectors of an electromagnetic wave, but rather the time-averaged intensity, i.e. the energy flux, which is the time average of the Poynting vector. For monochromatic plane waves, this is just proportional to the absolute square of the complex amplitudes,

$$I = \frac{1}{2Z_0}(E_\text{v}E_\text{v}^* + E_\text{h}E_\text{h}^*) \tag{1.37}$$

where $Z_0 = \sqrt{\mu/\epsilon}$ is the impedance (or wave resistance) of the medium. In the more realistic case of quasi-monochromatic radiation, i.e. for complex amplitudes $E_\text{v}(t)$ and $E_\text{h}(t)$ slowly fluctuating on timescales longer than $2\pi/\omega$, a time average or ensemble average has to be introduced.

Since the intensity alone does not carry information on the polarisation, three more quantities of the same dimension have to be defined. These four quantities, called Stokes parameters, fully describe the intensity and polarisation of electromagnetic radiation; they are defined as follows (the averaging is indicated by angular

[1] Instead of using partial derivatives, $\boldsymbol{e}_\text{h}$ and $\boldsymbol{e}_\text{v}$ can also be calculated as $\boldsymbol{e}_\text{h} = \boldsymbol{e}_z \times \boldsymbol{n}/\|\boldsymbol{e}_z \times \boldsymbol{n}\|$ and $\boldsymbol{e}_\text{v} = \boldsymbol{e}_\text{h} \times \boldsymbol{n}$.

brackets $\langle..\rangle$):

$$I = \frac{1}{2Z_0}\left\langle E_\mathrm{v}E_\mathrm{v}^* + E_\mathrm{h}E_\mathrm{h}^*\right\rangle \tag{1.38}$$

$$Q = \frac{1}{2Z_0}\left\langle E_\mathrm{v}E_\mathrm{v}^* - E_\mathrm{h}E_\mathrm{h}^*\right\rangle \tag{1.39}$$

$$U = \frac{1}{Z_0}\left\langle \mathrm{Re}\left(E_\mathrm{v}E_\mathrm{h}^*\right)\right\rangle \tag{1.40}$$

$$V = \frac{1}{Z_0}\left\langle \mathrm{Im}\left(E_\mathrm{v}E_\mathrm{h}^*\right)\right\rangle \tag{1.41}$$

An alternative definition, called modified Stokes parameters, keeps the intensities of the vertically and horizontally polarised components separate and replaces I and Q by them:

$$I_\mathrm{v} = \frac{1}{2Z_0}\left\langle E_\mathrm{v}E_\mathrm{v}^*\right\rangle = \frac{I+Q}{2} \tag{1.42}$$

$$I_\mathrm{h} = \frac{1}{2Z_0}\left\langle E_\mathrm{h}E_\mathrm{h}^*\right\rangle = \frac{I-Q}{2} \tag{1.43}$$

Thus,

$$I = I_\mathrm{v} + I_\mathrm{h} \tag{1.44}$$

$$Q = I_\mathrm{v} - I_\mathrm{h} \tag{1.45}$$

In a similar way, the remaining two Stokes parameters, U and V, can also be expressed by differences of intensities in two different polarisations:

$$U = I_{+45^\circ} - I_{-45^\circ} \tag{1.46}$$

$$V = I_\mathrm{LC} - I_\mathrm{RC} \tag{1.47}$$

Here, I_{+45° and I_{-45° denote the intensity of the component linearly polarised at an angle of 45° clockwise and counterclockwise, respectively, with the vertical polarisation direction, and I_RC and I_LC denote the intensities of the right circularly and left circularly polarised components. Note that here, we use the IRE definition of right and left circular polarisation, i.e.: For a right circularly polarised wave, at a fixed time, the tip of the electric field vector describes a left-handed spiral and vice versa. The basis vectors for right and left circular polarisation are

$$\boldsymbol{e}_\mathrm{RC} = \frac{1}{\sqrt{2}}(\boldsymbol{e}_\mathrm{v} + \mathrm{i}\boldsymbol{e}_\mathrm{h}) \quad \boldsymbol{e}_\mathrm{LC} = \frac{1}{\sqrt{2}}(\boldsymbol{e}_\mathrm{v} - \mathrm{i}\boldsymbol{e}_\mathrm{h}) \tag{1.48}$$

Actually, the opposite definition is also commonly used, particularly in optics, so some care about the definitions is neccessary when dealing with circular polarisation.

Written as a row or column vector, $[I, Q, U, V]$ or $[I_{\mathrm{v}}, I_{\mathrm{h}}, U, V]$ are called Stokes vector or modified Stokes vector, respectively.[2]

The Stokes parameters fully characterise the electromagnetic wave and therefore contain the same information as the electric field vector (except for one absolute phase). Since instruments generally measure intensities (fluxes), describing electromagnetic radiation by the Stokes parameters is more practical than describing it by the electric (or magnetic) field vector. Furthermore, the Stokes parameters are always real numbers. Note that the Stokes parameters are sometimes defined with different signs of Q, U or V (the definitions and signs used here are based on Tsang *et al.* (1985, 2000)). Moreover, their normalisation may vary. In particular, the Stokes parameters can be normalised to represent radiance or irradiance (instead of intensity), which is usually done in radiative-transfer contexts. If they are normalised to represent brightness temperatures, note that the conversion function from radiance to brightness temperature (the inverse of the Planck function or of an approximation thereof) for I is different from the conversion function I_{v} and I_{h}. The Planck function to use for converting I to brightness temperature is

$$B_{\nu}(T) = \frac{(2h/c^2)\nu^3}{\exp(h\nu/k_{\mathrm{b}}T) - 1} \tag{1.49}$$

for frequency ν in Hz. The factor of 2 in the numerator exists because there are two polarisation modes to emit in. The Planck function to use for converting I_{v} or I_{h} to brightness temperature is

$$B_{\nu}(T) = \frac{(h/c^2)\nu^3}{\exp(h\nu/k_{\mathrm{b}}T) - 1} \tag{1.50}$$

because there is only one polarisation mode. This explains why when $I_{\mathrm{v}} = I_{\mathrm{h}}$ the radiance I_{v} is half of I, and yet the brightness temperatures for I_{v} and I are the same. A note of caution when converting Stokes parameters from radiance units to brightness temperatures: Even though Q, U and V can have the dimensions of radiances, they are not physical radiances in the strict sense; they can be written as differences of radiances, $Q = I_{\mathrm{v}} - I_{\mathrm{h}}$ etc. (see Equations 1.45–1.47) and they can be negative. The inverse Planck function, however, is not defined for negative radiances. First converting I_{v} and I_{h} and then taking the difference,

$$T_Q = B_{\nu}^{-1}(I_{\mathrm{v}}) - B_{\nu}^{-1}(I_{\mathrm{v}}) \tag{1.51}$$

is not unique since any pair of $(I_{\mathrm{v}} + C, I_{\mathrm{h}} + C)$ with arbitrary C results in the same Q, but the inverse Planck function is non-linear. Therefore, those Stokes parameters which are not true radiances, i.e. Q, U and V must not be converted to Planck brightness temperature at all. A conversion to brightness temperature only makes sense if a linear conversion function, e.g. according to the Rayleigh–Jeans approximation, is used.

[2] Note that sometimes, S_0, S_1, S_2, S_3 is used instead of I, Q, U, V.

In the case of monochromatic radiation, i.e. if the complex amplitudes E_v and E_h are constant, the averages in Equations 1.38–1.43 can be dropped, and the Stokes parameters fulfil the identity $Q^2 + U^2 + V^2 = I^2$. We then speak of fully or completely polarised radiation. The polarisation of such radiation can also be described by the two parameters that determine the polarisation ellipse, i.e. the ellipse that the field vector of the oscillating electric field describes in time at a fixed location. The eccentricity of this ellipse is described by the so-called ellipticity angle, χ, and its orientation by the angle ζ between the major axis and the v-direction. The tangent of the ellipticity angle, $\tan\chi$, is the ratio of the semi-minor to the semi-major axis. The Stokes parameters are related to these angles via (Chandrasekhar, 1960; Tsang *et al.*, 1985; Kliger, *et al.*, 1990)

$$Q = I\cos(2\chi)\cos(2\zeta) \tag{1.52}$$

$$U = I\cos(2\chi)\sin(2\zeta) \tag{1.53}$$

$$V = I\sin(2\chi) \tag{1.54}$$

As mentioned above, in general, the complex amplitudes E_v and E_h (i.e. the real amplitudes and phases) fluctuate since most radiation originates from several sources that do not emit radiation coherently, and since the emission from one source usually has very short coherence times. This means that we usually have a superposition of radiation from several incoherent sources, and that the polarisation state of the radiation from each source fluctuates as well.[3] Typically, such fluctuations have time scales that are longer than the period ($2\pi/\omega$) of the oscillation, but that are still shorter than the integration time of the instrument that measures the radiation. Thus, the instrument measures an incoherent superposition of time averages over the fluctuating polarisation, and one can show (see, e.g. Chandrasekhar, 1960, ch. I.15) that the Stokes parameters in general fulfil the relation

$$Q^2 + U^2 + V^2 \leq I^2 \tag{1.55}$$

If the fluctuations are random for all the sources and if the different sources emit incoherently and are not in any way oriented, then there is no preferred orientation, ellipticity or handedness of the emitted radiation, which is then called unpolarised. In this case, $Q = U = V = 0$. This is the case for radiation from the sun, but also to a large extent for thermal radiation emitted by the atmosphere: While the emission from individual molecules might be polarised depending on the direction in which it is emitted, the random orientation of the molecules causes the resulting radiation to be unpolarised.

If the fluctuations are not completely random, Q, U and V do not vanish, $0 < Q^2 + U^2 + V^2 < I^2$ and the radiation is called partially polarised.

A measure for the degree of polarisation, p, is

$$p = \frac{\sqrt{Q^2 + U^2 + V^2}}{I} \tag{1.56}$$

[3] This does not, of course, apply to coherent sources like lasers or coherent radars.

For completely polarised radiation, $Q^2+U^2+V^2 = I^2$, so $p = 1$, and for unpolarised radiation, $Q = U = V = 0$, so $p = 0$.

Furthermore, it can be convenient to define the polarised component of radiation by

$$I_\mathrm{p}^2 = Q^2 + U^2 + V^2 \tag{1.57}$$

and the unpolarised component as

$$I_\mathrm{u} = I - I_\mathrm{p} \tag{1.58}$$

Thus, partially polarised radiation, described by a Stokes vector (I, Q, U, V), can be regarded as a superposition of completely polarised radiation described by the Stokes vector (I_p, Q, U, V) and unpolarised radiation described by the Stokes vector $(I_\mathrm{u}, 0, 0, 0)$. We see that the Stokes parameter formalism can conveniently deal with partially polarised and with unpolarised radiation, much in contrast to the formalism using the electric field (amplitude and phase).

In addition to the degree of polarisation, p, we can define measures for the circularity and the linearity of the polarisation. Recalling Equations (1.46) and (1.47), we can define the degree of linear polarisation, p_lin, as

$$p_\mathrm{lin} = \frac{\sqrt{Q^2 + U^2}}{I} \tag{1.59}$$

and the the degree of circular polarisation, p_circ, as

$$p_\mathrm{circ} = \frac{V}{I} \tag{1.60}$$

1.4.3 Antenna polarisation

The antenna polarisation of a receiving antenna is the polarisation of the wave radiated by the antenna if it is transmitting. As we have seen above, polarised radiation is described by either the Jones vector or the Stokes vector or by intensity, I, orientation angle, ψ and ellipticity angle, χ. Since the intensity of the radiation is proportional to the absolute square (the squared 'length') of the complex Jones vector, or, in other words, the first Stokes parameter, I, the polarisation alone is defined by

(1) a normalised Jones vector

$$\boldsymbol{e} = \begin{bmatrix} e_\mathrm{v} \\ e_\mathrm{h} \end{bmatrix} \quad \text{where } \boldsymbol{e} \cdot \boldsymbol{e}^* = 1 \tag{1.61}$$

(note that in the scalar product of two complex vectors, the second one has to be complex-conjugated.);

(2) a normalised Stokes vector

$$\boldsymbol{i} = (1, q, u, v) \quad \text{where } q^2 + u^2 + v^2 = 1; \tag{1.62}$$

(3) the two angles ψ and χ. According to Equations (1.52) to (1.54), we have:

$$q = \cos(2\chi)\cos(2\psi) \tag{1.63}$$

$$u = \cos(2\chi)\sin(2\psi) \tag{1.64}$$

$$v = \sin(2\chi) \tag{1.65}$$

Now we can calculate the intensity I' which the antenna measures if radiation of arbitrary polarisation is incident upon it. In terms of the electrical fields, i.e. Jones vectors, we just have to project the Jones vector $\boldsymbol{E}$ of the incident radiation onto the normalised Jones vector $\boldsymbol{e}$ of the antenna,

$$\boldsymbol{E}' = (\boldsymbol{E} \cdot \boldsymbol{e}^*)\boldsymbol{e} \tag{1.66}$$

(this is in effect like passing through a polariser), then take its absolute square and multiply by $1/2Z_0$ and then by the effective antenna area A_e:

$$I' = \frac{1}{2Z_0} A_\mathrm{e} |\boldsymbol{E}'|^2 = \frac{1}{2Z_0} A_\mathrm{e} |(\boldsymbol{E} \cdot \boldsymbol{e}^*)|^2 \tag{1.67}$$

Rewriting this in terms of the Stokes vector $\boldsymbol{I}$ of the incident radiation and the Stokes vector $\boldsymbol{A}_\mathrm{e} = A_\mathrm{e}\boldsymbol{i}$ of the antenna, we get just a scalar product:

$$I' = \frac{1}{2}\boldsymbol{A}_\mathrm{e} \cdot \boldsymbol{I} \tag{1.68}$$

The components of $\boldsymbol{A}_\mathrm{e}$ are sometimes called antenna Stokes parameters.

1.4.4 The scattering amplitude matrix

The starting point for describing the interaction of polarised radiation with particles (e.g. rain drops) is the scattering amplitude matrix which is introduced in the following way: The electric field, $[E_{\mathrm{v}'}^{\mathrm{sca}}, E_{\mathrm{h}'}^{\mathrm{sca}}]^\mathrm{T}$, originating from a single scattering event of an incident electric field $[E_\mathrm{v}^{\mathrm{inc}}, E_\mathrm{h}^{\mathrm{inc}}]^\mathrm{T}$ may in the far field be written as

$$\begin{bmatrix} E_{\mathrm{v}'}^{\mathrm{sca}} \\ E_{\mathrm{h}'}^{\mathrm{sca}} \end{bmatrix} = \frac{\exp(ikR)}{R} \mathbf{S}(\boldsymbol{n}, \boldsymbol{n}') \begin{bmatrix} E_\mathrm{v}^{\mathrm{inc}} \\ E_\mathrm{h}^{\mathrm{inc}} \end{bmatrix} \tag{1.69}$$

where $\mathbf{S}$, the scattering matrix or scattering amplitude matrix is a complex 2×2 matrix depending on both the direction of the incident wave, $\boldsymbol{n}$, and of the scattered wave, $\boldsymbol{n}'$ (corresponding to four angles)

$$\mathbf{S} = \begin{pmatrix} S_{11} & S_{12} \\ S_{21} & S_{22} \end{pmatrix} \tag{1.70}$$

Note that the polarisation directions for the incident wave, $\boldsymbol{e}_\mathrm{v}$, $\boldsymbol{e}_\mathrm{h}$ as well as for the scattered wave, $\boldsymbol{e}_\mathrm{v}'$, $\boldsymbol{e}_\mathrm{h}'$ are defined, as described above, with respect to the fixed z-axis and the propagation directions for the incident and scattered wave, $\boldsymbol{n}$ and $\boldsymbol{n}'$, respectively. This reference frame is often called the fixed frame system.

For particles with symmetries, however, it is more useful to define the polarisation directions with respect to the plane containing incident and scattered directions, the so-called scattering plane (Tsang *et al.*, 1985). Instead of horizontal polarisation, we have polarisation perpendicular to the scattering plane:

$$\boldsymbol{e}_{\perp} = \boldsymbol{e}'_{\perp} = \frac{\boldsymbol{n} \times \boldsymbol{n}'}{\|\boldsymbol{n} \times \boldsymbol{n}'\|} \tag{1.71}$$

The perpendicular polarisation direction is identical for incident and scattered waves.

Instead of vertical polarisation, we have polarisation parallel to the scattering plane:

$$\boldsymbol{e}_{\|} = \boldsymbol{n} \times \boldsymbol{e}_{\perp} \quad \boldsymbol{e}'_{\|} = \boldsymbol{n}' \times \boldsymbol{e}_{\perp} \tag{1.72}$$

Here, of course, we have two different parallel polarisation directions for incident and scattered waves. In this so-called scattering plane system, the z-axis is usually defined as the direction of the incident wave. The important difference to the scattering amplitude matrix in the fixed frame system is that it only depends on the direction of the scattered wave, since the incident wave is defined to be the z-axis.

While the scattering plane system is appropriate to describe the interaction of radiation with single particles,[4] the fixed frame system is more appropriate for describing the interaction of radiation with a distribution of particles, as, e.g. for radiative transfer in the presence of scatterers like rain or clouds.

In both systems, we can switch from the amplitude scattering matrix that transforms the electrical field vector to a matrix that transforms the Stokes vector:

$$\boldsymbol{I}' = \frac{1}{R^2}\mathbf{F}\boldsymbol{I} \tag{1.73}$$

where $\boldsymbol{I} = [I, Q, U, V]^{\mathrm{T}}$ is the Stokes vector of incident and $\boldsymbol{I}' = [I', Q', U', V']^{\mathrm{T}}$ the Stokes vector of scattered radiation, and the matrix $\mathbf{F}$ is a real 4×4 matrix. In the scattering plane system, for a single particle or target, $\mathbf{F}$ is usually called the Mueller matrix, scattering matrix (not to be confused with the scattering amplitude matrix which is sometimes also called the scattering matrix!) or Stokes matrix. In the fixed frame system, the average of this matrix over a distribution of particles is usually called the phase matrix and written as $\mathbf{Z}$. The definitions of the Mueller matrix and phase matrix can be found in Tsang *et al.* (1985) and Mishchenko *et al.* (2002), respectively, and the transformation from one into the other is described by Mishchenko *et al.* (2002). Note that for the modified Stokes vector, the Mueller matrix and the phase matrix definitions obviously are different.

1.4.5 Vector radiative-transfer equation

The phase matrix describes the scattering by particles of polarised radiation characterised by the Stokes vector $\boldsymbol{I}$. In order to extend the scalar radiative-transfer equation (see above) to a vector equation for the Stokes vector, the phase matrix replaces the phase function in the scattering term. Now we still need to replace the extinction

[4] The scattering plane system is also suitable for describing the interaction with an ensemble of randomly oriented particles.

coefficient by an appropriate extinction matrix and the thermal emission term by an appropriate vector.

The extinction matrix can be derived from the scattering amplitude matrix using the optical theorem (Tsang *et al.*, 1985; Mishchenko *et al.*, 2000, 2002). The emission vector is the product of the Planck function $B(\nu, T)$ and the absorption vector $\boldsymbol{a}$ (Mishchenko *et al.*, 2000, 2002). The absorption vector can be derived from the extinction matrix and the phase matrix (Tsang *et al.*, 1985; Mishchenko *et al.*, 2002).

The radiative-transfer equation for polarised radiation, called the vector radiative-transfer equation (VRTE), then reads

$$\begin{aligned}\frac{\mathrm{d}\boldsymbol{I}(\boldsymbol{n},\nu)}{\mathrm{d}s} &= -\langle\Sigma(\boldsymbol{n},\nu,T)\rangle\,\boldsymbol{I}(\boldsymbol{n},\nu,T) + \langle\boldsymbol{a}(\boldsymbol{n},\nu,T)\rangle\,B(\nu,T)\\ &\quad + \int_{4\pi}\mathrm{d}\Omega'\,\big\langle\mathbf{Z}(\boldsymbol{n},\boldsymbol{n}',\nu,T)\big\rangle\,\boldsymbol{I}(\boldsymbol{n}',\nu,T)\end{aligned} \tag{1.74}$$

where $\boldsymbol{I}$ is the Stokes vector, $\langle\Sigma\rangle$ is the ensemble-averaged extinction matrix, $\langle\boldsymbol{a}\rangle$ is the ensemble-averaged absorption vector, B is the Planck function and $\langle\mathbf{Z}\rangle$ is the ensemble-averaged phase matrix. Furthermore, ν is the frequency of the radiation, T is the temperature, $\mathrm{d}s$ is a path-length element of the propagation path and $\boldsymbol{n}$ the propagation direction. Equation (1.74) is valid for monochromatic or quasi-monochromatic radiative transfer. We can use this equation for simulating microwave radiative transfer through the atmosphere, as the scattering events do not change the frequency of the radiation.

References

BASHARINOV, A. E., GURVICH, A. S., YEGOROV, S. T., KURSKAYA, A. A., MATVYEV, D. T., and SHUTKO, A. M.: 'The results of microwave sounding of the earth's surface according to experimental data from the satellite, COSMOS 243' (Space Research, Vol. 11, Akademie Verlag Berlin, 1971)

BEKEFI, B.: 'Radiation processes in plasmas' (John Wiley, New York, 1966)

CHANDRASEKHAR, S.: 'Radiative transfer' (Oxford University Press 1950, reprinted by Dover Publications, New York, 1960)

CHERNY, V., and RAIZER, V. Y.: 'Passive microwave remote sensing of oceans' (Praxis Publishing, Chichester, UK, 1998)

CLEBSCH, A.: 'Ueber die Reflexion an einer Kugelfläche', *Journal für Mathematik*, 1863, **61** (Heft 3), pp. 195–262

DICKE, R. H.: 'The measurement of thermal radiation at microwave frequencies', *Rev. Sci. Instr.*, 1946, **17**, pp. 268–75

DICKE, R. H., BERINGER, R., KUHL, R. L., and VANE, A. B.: 'Atmospheric absorption measurements with a microwave radiometer', *Phys. Rev.*, 1946, **70**, pp. 340–8

FUNG, A.: 'Microwave scattering and emission models and their applications' (Artech House, Boston, MA, 1994)

GLOERSEN, P., and BARATH, F. T.: 'A scanning multichannel microwave radiometer for Nimbus-G and Seasat-A', *IEEE J. Oceanic Engin.*, 1977, **OE-2**(2), pp. 172–8

HILBERT, D.: 'Begründung der elementaren Strahlungstheorie', *Physik. Zeitschrift*, 1912, **XIII**, pp. 1056–64

HOLLINGER, J. P., LO, R., POE, G., SAVAGE, R., and PEIRCE, J.: 'Special sensor microwave imager user's guide' (Naval Res. Lab., Washington D.C., 1987)

ISHIMARU, A.: 'Wave propagation and scattering in random media' (Vol. 1, Academic Press, New York, 1978)

JANSKY, K. G.: *Proc. IRE*, 1933, **21**, pp. 1387–98

JANSSEN, M. A. (Ed.): 'Atmospheric remote sensing by microwave radiometry' (Wiley Series in Remote Sensing, New York, 1993)

JOSEPH, J. H., WISCOMBE, W. J., and WEINMAN, J. G.: 'The Delta-Eddington approximation for radiative flux transfer', *J. Atmosph. Sci.*, 1976, **33**, pp. 2452–9

KIRCHHOFF, G.: 'Über das Verhältnis zwischen dem Emissionsvermögen und dem Absorptionsvermögen der Körper für Wärme und Licht (On the relationship between emissivity and absorptivity of bodies for heat and light)', *Poggendorfs Annalen der Physik und Chemie*, 1860, **109**, pp. 275–301

KLIGER, D., LEWIS, J., and RANDALL, C.: 'Polarized light in optics and spectroscopy' (Academic Press, 1990)

KRAUS, J. D.: 'Radio astronomy' (McGraw Hill, New York, 1966)

KUBELKA, P., and MUNK, F.: 'Ein Beitrag zur Optik der Farbanstriche', *Zeitschrift Tech. Phys.*, 1931, **12**, p. 593

LORENZ, L. V.: 'Lysbevaegelsen i og uden for en af plane Lysbolger belyst Kugle (Upon the light reflected and refracted by a transparent sphere)', *Kongelige Danske Vidensk. Selskabs Shrifter*, 1890, **6**, pp. 1–62

MEADOR, W. E., and WEAVER, W. R.: 'Two-stream approximations to radiative transfer in planetary atmospheres: A unified description of existing methods and a new improvement', *J. Atmosph. Sci.*, 1980, **37**, pp. 630–43

MIE, G.: 'Beiträge zur Optik trüber Medien, speziell kolloidaler Metallösungen (Considerations on the optics of turbid media, especially colloidal metal solutions)', *Ann. Phys.*, 1908, **25**, pp. 377–442

MISHCHENKO, M. I., HOVENIER, J. W., and TRAVIS, L. D., (Ed.): 'Light scattering by nonspherical particles' (Academic Press, 2000)

MISHCHENKO, M. I., TRAVIS, L. D., and LACIS, A. A.: 'Scattering, absorption and emission of light by small particles' (Cambridge University Press, 2002)

MOBLEY, C. D.: 'Light and water – radiative transfer in natural waters' (Academic Press, New York, 1994)

PLANCK, M.: 'Theorie der Wärmestrahlung (The theory of heat radiation)' (Lectures, 6th Edition, Stuttgart 1966, English transl. by Dover Publications)

PLANCK, M.: 'Zur Theorie der Energieverteilung im Normalspektrum', *Verhandlungen der DPG*, 1990, **2**, pp. 237–245

SCHUSTER, A.: 'Radiation through a foggy atmosphere', *Astrophys. J.*, 1905, **21**, pp. 1–22

SHARKOV, E. A.: 'Passive microwave remote sensing of the earth' (Praxis Publishing, Chichester, UK, 2003)

STRAITON, A. W., TOLBERT, C. W., and BRITT, C. O.: 'Apparent temperature distributions of some terrestrial materials and the sun', *J. Appl. Phys.*, 1958, **29**, pp. 776–82

THOMAS, G. E., and STAMNES, K.: 'Radiative transfer in the atmosphere and ocean' (Cambridge University Press, Cambridge, UK, 1999)

TOWNES, C. H., and SCHAWLOW, A. L.: 'Microwave Spectroscopy' (McGraw Hill, New York, 1955, reprinted by Dover, 1975)

TSANG, L., KONG, J. A., and SHIN, R. T.: 'Theory of microwave remote sensing' (Wiley-Interscience, New York, 1985)

TSANG, L., KONG, J. A., and DING, K.: 'Scattering of electromagnetic waves: Theories and applications' (Wiley-Interscience, New York, 2000)

ULABY, F. T., MOORE, R., and FUNG, A. K.: 'Microwave remote sensing, active and passive, vol. I, microwave remote sensing fundamentals and radiometry (1981), vol. II, radar remote sensing and surface scattering and emission theory (1982), vol. III, from theory to applications' (Artech House Inc., Norwood, MA)

VAN DE HULST, H. C.: 'Light scattering by small particles' (New York, John Wiley 1957, reprinted by Dover, New York, 1981)

VAN DE HULST, H. C.: 'Multiple light scattering: tables, formulae and applications' (Academic Press, New York, 1980)

WATERS, J. W., PETTYJOHN, R. L., POON, R. K. L., KÜNZI, K. F., and STAELIN, D. H.: 'Remote sensing of atmospheric temperature profiles with the NIMBUS 5 microwave spectrometer', *J. Atmosph. Sci.*, 1975, **32**, pp. 1953–69

WYATT, C. L.: 'Radiometric calibration: theory and methods' (Academic Press, New York, 1978)

Chapter 2

Emission and spectroscopy of the clear atmosphere

Philip W. Rosenkranz, Stefan A. Buehler, Dietrich G. Feist, Tim Hewison, Nicole Jacquinet-Husson, Juan R. Pardo, and Roger Saunders

2.1 Introduction

The purpose of this chapter is to review the current state and recent developments in the modelling of microwave absorption by atmospheric gases and in the validation of those models by radiometric measurements. Local thermodynamic equilibrium implies that the absorption coefficient also determines the thermal energy emitted by the clear atmosphere. Absorption models generally consist of (1) a database of line parameters (e.g. transition frequencies, intensities, widths), which may be specific to the model or drawn from a large general-purpose database such as HITRAN or GEISA, (2) shape factors for the lines and (3) ancillary model components such as continuum absorption. In general, the absorption coefficient γ_a at frequency ν due to a particular gas can be written in the form

$$\gamma_{\mathrm{a}}(\nu) = N \sum_{i} S_i F_i(\nu) + \text{continuum terms} \tag{2.1}$$

where S_i is the intensity (dependent on temperature) of line i, $F_i(\nu)$ is the shape factor for line i and N is the abundance of the gas, corresponding to the definition of line intensity. In the HITRAN (see Rothman *et al.*, 1998, eq. A5) and GEISA databases, for example, the definition of line intensity requires N to be the molecule number density of the absorbing gas (i.e. relative isotopic abundance is contained in S_i); but this definition is not universally followed in the literature. The selection of lines to be included in the summation of (2.1) may require the exercise of some educated judgement on the part of the user who wishes to compare calculations with a particular set of measurements. The total absorption by a mixture of gases is the sum of absorption coefficients from the individual species under the conditions of pressure, temperature and abundances existing in the mixture. One may combine absorption models for different gases from different sources; hence, the number of possible combinations is large.

The most abundant gases of the atmosphere, H_2O, O_2 and N_2, have a role in determining atmospheric transmittance over a wide range of frequencies; this makes them necessary components in the modelling of observations of surface phenomena, in addition to their intrinsic interest for measurement of atmospheric water vapour and temperature (the latter being done by observation of oxygen lines).

Hill (1986) devised a test criterion that responds to line shape while being insensitive to width or continuum level. He applied this test to the water-vapour absorption data of Becker and Autler (1946) near 22 GHz and found that the Van Vleck–Weisskopf line shape was an acceptable fit to that line, while the Gross and full Lorentz line shapes were rejected. With higher line centre frequencies, the differences between these shape factors would be reduced, and this experimental test of line shape has apparently not been applied to other water lines. The line shape factor of Van Vleck and Weisskopf is

$$F_i(\nu) = \frac{1}{\pi}\left(\frac{\nu}{\nu_i}\right)^2 \left[\frac{w_i}{(\nu - \nu_i - \delta_i)^2 + w_i^2} + \frac{w_i}{(\nu + \nu_i + \delta_i)^2 + w_i^2}\right] \tag{2.2}$$

In the above, ν_i is the line frequency, δ_i is the line shift and w_i is the line half-width; w_i and δ_i depend on temperature. Line shifts are typically much smaller than widths of microwave lines, but they may produce a detectable effect in some measurements.

At atmospheric densities (except very low densities where Doppler broadening is significant, see below), both widths and shifts are proportional to pressure, but the broadening or shifting coefficients can be different for different perturbing gases. The total width or shift is the sum of contributions from all gases in the atmosphere. For practical calculations, one needs a coefficient for broadening by dry air and another for broadening by water vapour, the latter being the most significant variable gas in the atmosphere. Calculation of water vapour absorption also requires inclusion of a model for the H_2O continuum, which will be discussed later in this chapter.

Oxygen is unusual in that it absorbs microwaves by means of a magnetic dipole moment rather than an electric dipole moment. Consequently, the molecules exhibit both resonant and non-resonant absorption in the gas phase. For O_2, pressure line shifts appear to be negligible, but calculation of the absorption requires the addition to (2.2) of first-order line-mixing (sometimes called line coupling) parameters Y_i, which are also dependent on temperature and proportional to pressure:

$$F_i(\nu) = \frac{1}{\pi}\left(\frac{\nu}{\nu_i}\right)^2 \left[\frac{w_i + (\nu - \nu_i)\,Y_i}{(\nu - \nu_i)^2 + w_i^2} + \frac{w_i - (\nu + \nu_i)\,Y_i}{(\nu + \nu_i)^2 + w_i^2}\right] \tag{2.3}$$

Because the Y_i in (2.3) arise through the mixing of lines due to correlation of the molecular state before and after collisions, they have physical meaning only in a summation over lines, such as in (2.1). Correct calculation of the line-mixing effect requires the summation to include all of the terms considered in derivation of the mixing coefficients, including the non-resonant term. It is also necessary to realise that the mixing parameters and widths are associated; thus they should not be obtained from different sources. The non-resonant term has a Debye shape factor:

$$F_0(\nu) = \frac{\nu^2 w_0}{\nu^2 + w_0^2} \tag{2.4}$$

Measured values of the non-resonant broadening parameter w_0 are about half as large as a typical half-width of the resonant O_2 lines. This difference is closely connected with the phenomenon of line mixing, as shown by Gordon (1967). On the wings of the O_2 band, the net effect of line mixing is to reduce absorption, as though molecular collisions were less effective at broadening the lines.

Nitrogen has no microwave lines, but it does exhibit a weak continuum absorption due to collision-induced dipole moments, as do other molecules, including oxygen. Collision-induced absorption is discussed later in the chapter.

Lines of other atmospheric gases are both broad and weak at tropospheric pressures, which makes them difficult to detect against the background of the major gases. However, in the stratosphere, microwave lines of many molecular species are observable. The lines are very narrow in comparison to their centre frequencies. When that condition applies, and the frequency of observation is also close to the line centre, (2.2) reduces to a Lorentz shape factor:

$$F_i(\nu) = \frac{\pi^{-1} w_i}{(\nu - \nu_i - \delta_i)^2 + w_i^2} \tag{2.5}$$

which is sometimes called the simple Lorentz shape factor when it is necessary to distinguish it from the full Lorentz shape factor, which includes a negative-resonant term. However, if Doppler broadening is significant, which can be easily calculated, then a Voigt lineshape is appropriate. The Voigt shape factor results from the convolution of a Lorentzian function, which accounts for line broadening by collisions, with a Gaussian function having half-width (at half-maximum) of $3.6 \times 10^{-7}\,(T/M)^{1/2}\nu_i$, where T is the temperature in Kelvin and M is the molecular weight in g/mole, which describes the Doppler broadening. Accurate series expansions are available for the Voigt shape factor (see, e.g., Schreier, 1992). Modelling of absorption is then largely concerned with values for line frequency, intensity, width and possibly shift. However, particular effects such as Zeeman splitting also need to be considered for a few gases. A more detailed discussion of these points can be found in Rosenkranz (1993) or Melsheimer *et al.* (2005).

When large quantities of data from a single instrument are processed, a parametrised, or rapid, algorithm may be employed. These models (of which RTTOV is an example) approximate the result of a line-by-line summation, but with significantly less computation. This is achieved by means of coefficients that are pre-computed, usually for a specific instrument with defined channels and viewing geometry. Errors in approximation of transmittance are generally less than 10^{-2}, and sometimes much less, which is smaller than the uncertainties in the line-by-line models. The accuracy of a transmittance calculation is thus limited by the line-by-line model from which the coefficients are derived.

The next eight sections of this chapter describe recent developments in databases, models, laboratory measurements and theoretical work. Two sections concerned with the accuracy and state of validation of models follow. The chapter concludes with some recommendations for future work.

2.2 HITRAN (high resolution transmission)

A new archival compilation was made available in the summer of 2004. The compilation brings together the HITRAN line-transition parameters, infrared cross-sections, UV line-by-line parameters and cross-sections, aerosol refractive indices and documentation. This compilation, called HAWKS (HITRAN atmospheric workstation), is available on an anonymous ftp-site. Instructions for accessing the database can be found in the HITRAN web-site (cfa-www.harvard.edu/HITRAN). A summary of the data that have been added, modified, or enhanced since the previous update of 2001 (Rothman *et al.* (2003)) is given below. For a complete and detailed description of HITRAN-2004 see Rothman *et al.* (2005).

2.2.1 Line-by-line parameters archive

The line-by-line portion of the database contains spectroscopic parameters for 39 molecules including 93 of their isotopologues. Twelve molecular species, i.e. SO_2, NH_3, HF, HBr, HI, ClO, H_2CO, N_2, $ClONO_2$, HOBr, C_2H_4, NO^+, have been kept identical since the former edition of HITRAN. There is a new entry, namely, the methanol molecule (CH_3OH). For two of the molecules, i.e. SF_6 and $ClONO_2$, the parameters have been put in a supplemental folder. Details of the updates follow.

H_2O: Important updates have been accomplished, especially for the main isotopologue $H_2^{16}O$. In the pure-rotation region, 952 lines of $HD^{18}O$ have been added. Line positions and intensities derive from the JPL catalog (Pickett *et al.*, 1998). Updates of line positions have been made for $H_2^{17}O$ and $H_2^{18}O$ from 0 to 500 cm^{-1} based on the work of Toth (1992) and of line intensities for $H_2^{17}O$, $H_2^{18}O$ and $HD^{16}O$ from 0 to 500 cm^{-1} based on the work of Pearson (1999). For the main isotopologue $H_2^{16}O$ between 0 and 800 cm^{-1}, the calculations of Coudert (1999), Lanquetin *et al.* (2001) and Flaud *et al.* (2003a) have been used to update line positions and line intensities of lines involving the first eight vibrational states. In addition, a complete update for positions and line intensities has been done for all water vapour isotopologues between 500 and 8000 cm^{-1} based on the work of Toth (2005). From 9600 to 11400 cm^{-1}, the line positions and intensities of the isotopologue $H_2^{17}O$ have been updated too. From 12400 to 14518 cm^{-1}, 918 lines of $H_2^{18}O$ have been incorporated, from the works of Tanaka *et al.* (2002, 2005). Another significant improvement concerns the 9250–9600 cm^{-1}, 11400–12895 cm^{-1} and 13184–25000 cm^{-1} regions for which the line lists generated from the works of Mérienne *et al.* (2003) and of Coheur *et al.* (2002) have been used to replace the former lines of $H_2^{16}O$. Concerning the line-shape parameters, the air-broadened half-widths and line shifts and the self-broadened half-widths, a complete update has been performed for the $H_2^{16}O$, $H_2^{18}O$ and $H_2^{17}O$ isotopologues.

CO_2: The data of the 'Carbon Dioxide Spectroscopic Database at 296 K' (CDSD-296) by Tashkun *et al.* (2003) have been used to update the positions and intensities of the lines of $^{12}C^{16}O_2$, $^{13}C^{16}O^2$, $^{12}C^{16}O^{18}O$ and $^{12}C^{16}O^{17}O$ when no high-quality laboratory observed data (i.e. the data of Miller and Brown (2004), in a large range of the infrared, the works of Ding *et al.* (2002) around 9950 cm^{-1} and of Benner (2003) around 3500 cm^{-1}) were available. Two line lists, from the JPL catalog (Pickett *et al.*, 1998), in the pure-rotational spectral region have been added in this edition for the isotopologues $^{16}O^{12}C^{18}O$ and $^{16}O^{12}C^{17}O$. Line parameters above 9650 cm^{-1} have been newly archived from works of Campargue *et al.* (1994) and Lucchesini (2004).

O_3: In the pure-rotation region, the positions, the intensities and the lower-state energies of the transitions have been updated for the isotopologues $^{16}O^{16}O^{17}O$ and $^{16}O^{17}O^{16}O$ from the JPL catalog (Pickett *et al.*, 1998). Using the line list of the MIPAS (Michelson Interferometer for Passive Atmospheric Sounding) database (Flaud *et al.*, 2003a), line positions and intensities have been updated for 49 bands of $^{16}O_3$ and the ν_1 and ν_3 bands of the $^{16}O^{16}O^{18}O$ and $^{16}O^{18}O^{16}O$ isotopologues. Seventeen bands of $^{16}O_3$, one band of $^{16}O^{16}O^{18}O$ and $^{16}O^{18}O^{16}O$ and three bands of $^{16}O^{16}O^{17}O$ and $^{16}O^{17}O^{16}O$ have been incorporated. The 14 μm region has been updated with the work of Wagner *et al.* (2002). A total of 12 bands for $^{16}O_3$, from the works of Flaud *et al.* (1987, 1989) and Mikhailenko *et al.* (2002), have been added in three spectral regions, i.e. 1613–1903 cm^{-1} (6 bands), 1872–2288 cm^{-1} (3 bands) and 2590–3006 cm^{-1} (3 bands).

N_2O: The line list has been almost completely revised using the extensive work of Toth (2004) and the results of Daumont *et al.* (2002), Johns *et al.* (1996), Lacome *et al.* (1984) and Nemtchinov *et al.* (2003).

CO: Line intensities have been updated for the first overtone of $^{12}C^{16}O$, located near 2.4 μm, using the measurements of Brault *et al.* (2003) and for the second overtone, located around 1.6 μm, using the calculated values of Sung and Varanasi (2004). Details for the updates of air-broadened half-widths, self-broadened line widths, air-broadening temperature-dependence exponent and air pressure-induced line shifts are given in Rothman *et al.* (2005).

CH_4: Some 48 000 methane parameters were completely replaced by updated line lists of over 240 000 lines. Most of the revisions made in the region from 0 to 6185 cm^{-1} are described by Brown *et al.* (2003). In the pure-rotation region, a new line list was taken from the JPL catalog (Pickett *et al.*, 1998). In the dyad region, a few of the air pressure-induced line shifts of $^{12}CH_4$ were revised. The near-infrared parameters of the lines of CH_4 between 4800 and 9200 cm^{-1} have been improved, using the measurements of Brown (2004) in the 4800–5500 cm^{-1} region and high-resolution laboratory spectra recorded with the McMath–Pierce Fourier transform spectrometer between 6180 and 9200 cm^{-1}.

O_2: The A bands of the isotopologues $^{18}O^{16}O$ and $^{17}O^{16}O$ centred at 0.76 μm have been updated using the work of Camy-Peyret *et al.* (2000).

NO: The calculation by Goldman *et al.* (1998), for line positions and intensities of $^{14}N^{16}O$, has been used to replace the previous parameters for line positions and intensities associated with algorithms deduced from Chackerien *et al.* (1998a) and Spencer *et al.* (1994) to update the air-broadened half-widths. The results of Pine *et al.* (1985) and Spencer *et al.* (1997) have been used to update the self-broadened widths and the temperature dependence of the air-broadened widths, respectively.

NO_2: The coding of the rotational quantum numbers N, K_a, K_c, J and F has been standardised according to Perrin *et al.* (1998). The air-broadened half-widths, the air pressure-induced shifts and the temperature dependence of the air-widths were revised throughout the database using empirical expressions and measurements by Benner *et al.* (2004).

HNO_3: The line list for the pure-rotation band which covered the 0–43 cm^{-1} region has been replaced by the line list of the JPL catalog (Pickett *et al.,* 1998) which extends the spectral coverage to 0–84 cm^{-1}. In the far infrared, two hot bands, $2\nu_9 - \nu_9$ and $\nu_5 - \nu_9$, have been added. Line positions are from a simulation based on analyses of rotational spectra by Petkie *et al.* (2005). The intensities of the ν_5 and $2\nu_9$ bands have been normalised based on the work of Toth *et al.* (2003). The former individual line intensities of the ν_5 and $2\nu_9$ bands have been multiplied by 0.879. The two hot bands $\nu_5 + \nu_9 - \nu_9$ and $3\nu_9 - \nu_9$, located at 885.424 and 830.6 cm^{-1}, respectively, have been updated from the results of Flaud *et al.* (2003a,b) and Perrin *et al.* (1994). Some values of the intensities have been rescaled. The self-broadened half-width has been fixed to the constant value of 0.8 cm^{-1} atm^{-1} for all the lines (Zu *et al.*, 2002; Brockman *et al.*, 1978).

OH: In the pure-rotation region, line intensities and positions of the two isotopologues ^{18}OH and OD have been updated based on the JPL catalog (Pickett *et al.,* 1998). For the air-broadened half-widths and their temperature dependence, the data reported in Rothman *et al.* (1998) for ^{16}OH have been used.

HCl: In the pure-rotation region, the actual line list is a combination of the JPL catalog (Pickett *et al.* (1998)) and of what was in the previous edition of HITRAN Rothman *et al.* (2003).

OCS: In the 3–5 μm region of the main isotopologue the line intensities have been updated using works of Régalia-Jarlot *et al.* (2002) and Vander Auwera *et al.* (2003). The line intensities of the $\nu_1 + \nu_3$ band located around 2900 cm^{-1} have been updated according to the work of Régalia-Jarlot *et al.* (2002) and the line intensities in the

$2\nu_3$ band of $^{16}O^{12}C^{32}S$ were also updated according to the work of Bermejo *et al.* (1997).

HOCl: The positions and intensities of the pure rotation lines were updated, and transitions within the (001) level were added according to the work of Flaud *et al.* (1998). The ν_2 region from 1178 to 1321 cm^{-1} was updated for both the $HO^{35}Cl$ and the $HO^{37}Cl$ isotopologues according to the results reported in Vander Auwera *et al.* (2000). The air-broadened half-width of all the lines in the ν_2 band was set to the average of the measured values reported for two lines by Shorter *et al.* (1997), i.e. 0.1 cm^{-1} atm^{-1}. The default value of 0.5 for the temperature dependence of the air-broadened half-widths was replaced by 0.7.

HCN: Line positions and intensities of the three isotopologues present in the pure-rotation region and in the infrared from 500 to 3425 cm^{-1} have been updated thanks to the works of Maki *et al.* (1996, 2000) and Devi *et al.* (2003). The collision parameters (air and self), the temperature dependence of the air-broadened half-width and the air pressure-induced line shifts have been updated with polynomial expressions based on the works of Devi *et al.* (2003, 2004) and Rinsland *et al.* (2003a).

CH_3Cl: A line list for the ν_2, ν_5 and $2\nu_3$ triad of vibrational bands of methyl chloride covering the 6–8 μm region has been added based on the works of Chackerian *et al.* (1998b) and Bouanich *et al.* (2001). The pure-rotation bands for the two isotopologues have also been added to the database, using the positions, the intensities and the lower state energy of the JPL catalog (Pickett *et al.* 1998).

H_2O_2: A new line list for the pure-rotation region from 0 to 1427 cm^{-1}, prepared by Perrin *et al.* (1996), has been included.

C_2H_2: In the 1.5 μm region (from 6448 to 6686 cm^{-1}), a new line list for the $\nu_1 + \nu_3$ bands of $^{12}C^{13}CH_2$ and three associated hot bands, based on the works of El Hachtouki and Vander Auwera (2002) and of Kou *et al.* (1994) has been incorporated in the database. The air- and self-broadened half-widths, their temperature dependence and the air pressure-induced line shifts were calculated using the polynomial expansion and the constants reported by Jacquemart *et al.* (2003).

C_2H_6: A line list prepared by Pine and Rinsland (1999) in the ν_7 band of ethane has completely replaced the former data.

PH_3: The line list of phosphine has been completely revised and expanded. The line positions and line intensities in the 10 μm region are from Brown *et al.* (2002b), while those in the 5 μm region are from Tarrago *et al.* (1992). The air- and self-broadened half-widths from Brown *et al.* (2002b) replace the former default constant values.

COF_2: A new prediction (the number of lines has more than doubled) has been generated by Cohen (2001). A typographical error for the rotational quantum numbers attribution in the 5.1 μm region has been corrected.

SF_6: An update has been performed for the only band present in the database, the ν_3 band. New positions and intensities by Acef *et al.* (2000) and Boudon *et al.* (2002) have replaced the previous spectroscopic parameters. A constant value for the self-broadened half-width, equal to 0.042 cm^{-1} atm^{-1}, based on Tejwani and Fox (1987) has been adopted.

H_2S: Updates have been made for the air- and self-broadened half-widths and for the air pressure-induced line shift based on the works of Sumpf *et al.* (1996, 1997, 2004), Kissel *et al.* (2002) and Waschull *et al.* (1994).

HCOOH: The spectral parameters of the pure-rotation lines observed between 10 and 100 cm^{-1} have been updated thanks to the works of Vander Auwera (1992) and Winnewisser *et al.* (2002). The new spectral line parameters (positions and intensities) generated by Perrin *et al.* (1999) were used to update the data on the ν_6 band. The constant values of 0.1 cm^{-1} atm^{-1} and 0.4 cm^{-1} atm^{-1}, used, respectively, for the air- and self-broadened half-widths, in the ν_6 band, have been applied to the far-infrared region.

HO_2: The rotational quantum numbers format has been revised.

O: Slight changes have been made for the line positions, intensities and lower-state energies of the two transitions of atomic oxygen thanks to newer values in the JPL catalog (Pickett *et al.* 1998).

CH_3OH: The data on methanol make their debut in HITRAN. In the microwave spectral region, line positions and intensities of the rotational transitions between 0 and 34 cm^{-1} from Müller *et al.* (2001) have been used. A recent work of Xu *et al.* (2004) led to the creation of an empirical line list around 10 μm that has been used. The air- and self-broadened half-widths have been fixed, respectively, to the assumed values of 0.1 cm^{-1} atm^{-1} and 0.4 cm^{-1} atm^{-1}. The assumed value for the temperature dependence of the widths is taken equal to 0.75.

The format of the parameters for each spectral line (equivalent to a record in the database) is given in Table 2.1 (from Rothman *et al.*, 2005) where the description of the parameters is in the bottom half. The total length of the record is 160 characters, an increase from the 100 bytes adopted in the former editions of HITRAN.

Table 2.2 (from Rothman *et al.*, 2005) is a summary of the molecular species and spectral regions covered in the current edition of HITRAN. In column 1 is the identification code for the molecular species listed in column 2 with their associated isotopologue in column 3 (AFGL notation, Rothman *et al.*, 2005); the two last columns, 4 and 5, give the spectral coverage and the number of lines, respectively, for each isotopologue of column 3.

2.2.2 *Infrared cross-sections archive*

Many of the infrared cross-sections have been retained from the previous edition of HITRAN (Rothman *et al.*, 2003) and the format has remained the same. New or

Table 2.1 Format for HITRAN parameters in the line-by-line section, 2004 edition

Parameter	M	I	ν	S	A	γ_{air}	γ_{self}	E''	n_{air}	δ_{air}	V'	V''	Q'	Q''	I_{err}	I_{ref}	* (flag)	g'	g''
Field length	2	1	12	10	10	5	5	10	4	8	15	15	15	15	6	12	1	7	7
FORTRAN descriptor	I2	I1	F12.6	E10.3	E10.3	F5.4	F5.4	F10.4	F4.2	F8.6	A15	A15	A15	A15	6I1	6I2	A1	F7.1	F7.1

Parameter	Meaning	Units or comments
M	Molecule number	HITRAN chronological assignment
I	Isotopologue number	Ordering within a molecule by terrestrial abundance
ν	Vacuum wavenumber	cm^{-1}
S	Intensity	$cm^{-1}/(molecule \cdot cm^{-2})$ at standard 296 K (Rothman *et al.* (1998) discuss T dependence)
A	Einstein A-coefficient	s^{-1}
γ_{air}	Air-broadened half-width (HWHM) at 296 K	$cm^{-1}atm^{-1}$
γ_{self}	Self-broadened half-width (HWHM) at 296 K	$cm^{-1}atm^{-1}$
E$''$	Lower-state energy	cm^{-1}
n_{air}	Temperature-dependence exponent for γ_{air}	unitless $\gamma_{air}(T) = \gamma_{air}(T_0) \times (T_0/T)^{nair}$
δ_{air}	Air pressure-induced line shift	cm^{-1} atm^{-1} at 296 K
V$'$	Upper-state 'global' quanta	
V$''$	Lower-state 'global' quanta	
Q$'$	Upper-state 'local' quanta	
Q$''$	Lower-state "local" quanta	
I_{err}	Uncertainty indices	Accuracy for 6 critical parameters (ν, S, γ_{air}, γ_{self}, n_{air}, δ_{air})
I_{ref}	Reference indices	References for 6 critical parameters (ν, S, γ_{air}, γ_{self}, n_{air}, δ_{air})
*	Flag	Availability of program and data for the case of line mixing
g$'$	Statistical weight of the upper state	
g$''$	Statistical weight of the lower state	

Table 2.2 HITRAN-04 individual lines sub-database detailed content

Mol. ID.	Molecule	Isotopologue (AFGL notation)	Spectral coverage (cm^{-1})	Number of lines
1	H_2O	161	0–25233	36114
		181	0–14519	9548
		171	10–11335	6120
		162	0–7514	9628
		182	0–3825	1611
		172	1234–1599	175
2	CO_2	626	442–9649	27979
		636	497–8105	8836
		628	0–8133	13445
		627	0–6962	7739
		638	567–4947	2312
		637	584–3642	1593
		828	615–3670	721
		728	626–2359	288
3	O_3	666	0–4061	183785
		668	0–2114	21718
		686	1–2075	8937
		667	0–2122	65106
		676	0–2101	31935
4	N_2O	446	0–5135	33066
		456	5–5086	4222
		546	4–4704	4592
		448	542–4672	4250
		447	550–4430	1705
5	**CO**	26	3–8465	917
		36	3–6279	780
		28	3–6267	760
		27	3–6339	728
		38	3–6124	712
		37	1807–6197	580
6	CH_4	211	0–9200	187128
		311	0–6070	28793
		212	7–3307	35519
7	O_2	66	0–15927	1430
		68	1–15852	671
		67	0–14537	4326
8	**NO**	46	0–9274	100902
		56	1609–2061	699
		48	1601–2039	679

Table 2.2 (Continued)

Mol. ID.	Molecule	Isotopologue (AFGL notation)	Spectral coverage (cm^{-1})	Number of lines
9	**SO_2**	626	0–4093	38566
		646	2463–2497	287
10	**NO_2**	646	0–3075	104223
11	**NH_3**	4111	0–5295	27994
		5111	0–5180	1090
12	**HNO_3**	146	0–1770	271166
13	**OH**	61	0–19268	41166
		81	0–329	295
		62	0–332	912
14	**HF**	19	41–11536	107
15	**HCl**	15	20–13458	324
		17	20–10995	289
16	**HBr**	19	16–9759	651
		11	16–9758	642
17	**HI**	17	12–8488	806
18	**ClO**	56	0–1208	3599
		76	0–1200	3631
19	**OCS**	622	0–4119	10553
		624	0–4116	4186
		632	0–4013	2283
		623	509–4116	1802
		822	0–4042	1096
20	**H_2CO**	126	0–2999	1772
		136	0–73	563
		128	0–48	367
21	**HOCl**	165	1–3800	8877
		167	1–3800	7399
22	**N_2**	44	1922–2626	120
23	**HCN**	124	0–3424	2955
		134	2–3405	652
		125	2–3420	646
24	**CH_3Cl**	215	0–3173	16411
		217	0–3162	14708
25	**H_2O_2**	1661	0–1500	100781

Table 2.2 (Continued)

Mol. ID.	Molecule	Isotopologue (AFGL notation)	Spectral coverage (cm^{-1})	Number of lines
26	**C_2H_2**	1221	604–6686	3232
		1231	613–6589	285
27	**C_2H_6**	1221	720–2978	4749
28	**PH_3**	1111	770–2472	11790
29	**COF_2**	269	725–2002	70601
30	**SF_6**	29	929–964	22901
31	**H_2S**	121	2–4257	12330
		141	5–4172	4894
		131	5–4099	3564
32	**HCOOH**	126	10–1235	24808
33	**HO_2**	166	0–3676	38804
34	**O**	6	68–159	2
35	**$ClONO_2$**	5646	763–798	21988
		7646	765–791	10211
36	**NO^+**	46	1634–2531	1206
37	**HOBr**	169	0–316	2177
		161	0–316	2181
38	**C_2H_4**	221	701–3243	12967
		231	2947–3181	281
39	**CH_3OH**	2161	0–1408	19899

updated data are as follows:

- the cross-sections for chlorine nitrate, $ClONO_2$, have been totally updated in the 750–1320 cm^{-1} region from the measurements by Wagner and Birk (2003). In the range 1680–1790 cm^{-1}, data from Ballard *et al.* (1988) have been retained;
- data of Wagner and Birk (2003) also include cross-sections of nitrogen pentoxide, N_2O_5, which have been used to replace entirely the former archived data;
- the cross-sections of HNO_4 (peroxynitric acid, PNA) measured by May and Friedl (1993) at 220 K, using spectral resolution of 0.003 cm^{-1} in the 780–830 cm^{-1} region, have been included and they replace the earlier cross-sections;
- absorption cross-sections of SF_5CF_3 (trifluoromethyl sulphur pentafluoride) have been included for the first time. The cross-sections were measured in the 520–6500 cm^{-1} spectral range, at the Pacific Northwest National Laboratory by Rinsland *et al.* (2003b).

The summary of HITRAN-04 archived cross-sections is given later in Table 2.5, along with those of GEISA.

2.2.3 Ultraviolet datasets

New cross-sections, in the ultraviolet and visible spectral regions, for six molecular species have been implemented in the database:

O_3: Absorption cross-sections in the Hartley–Huggins bands, including their temperature dependence have been included for the spectral range 29 164–40 798 cm^{-1}. The data are adapted from the results of Bass and Paur (1981, 1985a,b), Edlén (1966) and Voigt *et al.* (2001).

NO_3: Cross-sections of nitrogen trioxide at 298 K have been taken from Orphal *et al.* (2003) and www.iup.physik.uni-bremen.de/gruppen/molspec.

BrO: Cross-sections for bromine monoxide at temperatures of 228 and 298 K, at 10 cm^{-1} spectral resolution, are from Wilmouth *et al.* (1999).

OClO: Chlorine dioxide cross-sections at two spectral resolutions (1 cm^{-1} and 20 cm^{-1}) and five temperatures (213 to 293 K) have been taken from Kromminga *et al.* (2003) and www.iup.physik.uni-bremen.de/gruppen/molspec.

H_2CO: Cross-sections of formaldehyde have been adapted from Cantrell *et al.* (1990).

O_2–O_2: Cross-sections for the oxygen collision complex at 296 K have been adapted from Greenblatt *et al.* (1990); they have been corrected from air to vacuum wavelengths using Edlén (1966).

2.2.4 Aerosol refractive indices

Refractive indices of water, ice, aqueous sulphuric and nitric acid, solid hydrates (i.e. nitric acid, mono-, di- and tri- hydrate), organic non-volatile aerosol and crustal material (e.g. quartz, haematite and sand) in the previous version of HITRAN are discussed by Massie and Goldman (2003). Recently, Norman *et al.* (2002) published the real and imaginary refractive indices at 220 K for six different ternary compositions, and these data have been added to the current HITRAN compilation.

2.3 GEISA (Gestion et étude des informations spectroscopiques atmosphériques: Management and study of atmospheric spectroscopic information)

Since its 1997 edition (GEISA-97), the GEISA database, extensively described in Jacquinet-Husson *et al.* (1999), has been updated twice: partially in 2001 (Jacquinet-Husson *et al.* 2003a) and extensively in 2003 (GEISA-03) (Jacquinet-Husson *et al.*, 2003b, 2005).

The GEISA-03 system comprises three independent sub-databases devoted, respectively, to

- individual line transition spectroscopic parameters (line-by-line archive);
- infrared absorption cross-sections;
- micro-physical and optical properties of atmospheric aerosols.

All the archived data can be handled through general and user friendly GEISA associated management software facilities which are interfaced on the ARA/LMD group web site at www.ara.lmd.polytechnique.fr. They are also accessible at the GEISA restricted free access ftp site www.ara.lmd.polytechnique.fr/ftpgeisa. First, the potential user has to get a login and a password at www.ara.lmd.polytechnique.fr/registration.

Currently, GEISA is involved in activities related to the assessment of the capabilities of IASI (infrared atmospheric sounding interferometer), as described in Jacquinet-Husson *et al.* (1998).

2.3.1 Subdatabase on line transition parameters

The GEISA-03 subdatabase of line transition parameters involves 42 molecules (98 isotopic species) and contains 1 668 371 entries (321 905 supplementary entries since GEISA-97) in the spectral range from 10^{-6} to 22 656 cm^{-1}. The included molecules are constituents of the atmospheres of earth (major permanent and trace molecules) and of other planets (such as C_2H_4, GeH_4, C_3H_8, C_2N_2, C_4H_2, HC_3N, H_2S, HCOOH and C_3H_4, for the giant planets). It has to be noted that CH_3D is considered as an individual molecule in GEISA and as an isotope of CH4 in HITRAN. Details of the updates follow.

H_2O: Three spectral regions have been reinvestigated. In the 500–2850 cm^{-1} region, the line parameters were replaced with a compilation of 10 755 water transitions obtained by Toth (1998a, 1999a,b, 2000) and Toth *et al.* (1998b). Alternative line parameters from Stewart (2003), related to the H_2O archive in the 10 μm region have been issued as a support study for IASI and archived in GEISA as an alternative file. In the 9650–11 400 cm^{-1} region, the update comes from empirical line parameters of $H_2^{16}O$ obtained by Brown *et al.* (2002a). In the 13 000–26 000 cm^{-1} region the new line parameters are from Carleer *et al.* (1999), Coheur *et al.* (2002) and Fally *et al.* (2003).

CO_2 : The previous line parameters of the transitions belonging to the four most abundant isotopomers $^{12}C^{16}O_2$, $^{13}C^{16}O_2$, $^{16}O^{12}C^{18}O$ and $^{16}O^{12}C^{17}O$ in the 442–2797 cm^{-1} spectral region have been replaced with 48 627 new ones covering the 436–2826 cm^{-1} spectral range, based on the results described in Tashkun *et al.* (1998, 2001) and Teffo *et al.* (2002, 2003).

O_3: Updates occur in four spectral regions: in the 600–1232 cm^{-1} spectral region, there are new and more accurate line parameters for the ν_1 and ν_3 bands of $^{16}O_3$ derived by Wagner *et al.* (2002) and Flaud *et al.* (2003c). A complete new study of the 1300–1500 cm^{-1} spectral range, related to the $2\nu_2$, $3\nu_2 - \nu_2$, $\nu_1 + \nu_3 + \nu_3 - \nu_2$

and $2\nu_3 - \nu_2$ bands of $^{16}O_3$, has been made by Barbe *et al.* (1998). The 1820–2260 cm^{-1} ($2\nu_3$, $\nu_1 + \nu_3$ and $2\nu_1 + \nu_3$ interactive bands of $^{16}O^{18}O^{16}O$) and the 2600–2900 cm^{-1} (triad: $\nu_2 + 2\nu_3$, $\nu_1 + \nu_2 + \nu_3$ with $2\nu_1 + \nu_2$) regions have been reinvestigated by De Backer-Barilly *et al.* (2003) and Mikhaïlenko *et al.* (1999, 2002), respectively.

N_2O: 279 lines in the spectral region 870–1240 cm^{-1}, recently revisited by Daumont *et al.* (2002), have been substituted in two bands, $\nu_3 - \nu_1$ and ν_3–$2\nu_2$, the intensities of which were doubtful. In addition, a technical update has removed 118 duplicated lines in the spectral interval 564–629 cm^{-1}.

CH_4 and **CH_3D:** The CH_4 and CH_3D contents of GEISA have been extensively updated. The spectral range has been extended from 6184.492 cm^{-1} to 9199.285 cm^{-1} and the number of entries increased from 66 883 to 216 196 (weaker transitions of $^{12}CH_4$ and new bands of $^{13}CH_4$ and CH_3D included). Full details about the revision in the spectral interval from 0.01 to 6184.492 cm^{-1} can be found in a review paper by Brown *et al.* (2003). The new CH_4 near infrared data from 4800 to 5500 cm^{-1} and 6180–10 000 cm^{-1} are from an empirical list obtained by Brown (2001, 2002, 2003), from a few selected FTS laboratory spectra.

O_2: The two spectral regions, 7664.726–8064.311 cm^{-1} and 11 483.727–15 927.806 cm^{-1}, have been updated with new results by Goldman *et al.* (2003a) and by Brown and Plymate (2000), respectively.

NO: A new line list has been produced by Goldman (2003) in the first overtone region of the main isotopic species $^{14}N^{16}O$, i.e. between 3547.318 and 3799.155 cm^{-1}. This calculation has been issued from the experimental data from Mandin *et al.* (1997b, 1998) and theoretical results from Gillis and Goldman (1982).

NO_2: A new line list was set up in the spectral region of the $\nu_1 + \nu_3$, $\nu_1 + 2\nu_2$ and $\nu_1 + \nu_2 + \nu_3 - \nu_2$ bands of the $^{14}N^{16}O_2$ main isotopic species. New line parameters come from the works of Mandin *et al.* (1997a), Dana *et al.* (1997), Devi *et al.* (1982) and May and Webster (1990).

NH_3: The line parameters of the interval 0.058–5294.502 cm^{-1} have been totally replaced by those issued by Kleiner and Brown (2003) and are described in Kleiner *et al.* (2003).

PH_3: New data from Kleiner and Brown (2003), described in Kleiner *et al.* (2003), have replaced the whole content of the region from 770.877 to 2478.765 cm^{-1}.

OH: Line parameters in the ultraviolet region from 29 808.500 to 35 877.030 cm^{-1}, from Gillis *et al.* (2001) have been added to the GEISA archive.

HBr: New line parameters of $H^{79}Br$ and $H^{81}Br$, for the spectral regions of the pure rotation band $X^1\Sigma^+$ (0-0) (16.692–396.474 cm^{-1}) and of the fundamental band (1-0) (2123.812–2790.533 cm^{-1}), have been derived for spectroscopic databases update, as described in Goldman *et al.* (2003b).

HI: The description of the updates in the spectral regions of the $X^1\Sigma^+$ (0-0) (12.842–319.820 cm^{-1}) and of the (0-1) (1950.689–2403.162 cm^{-1}) bands is given in Goldman *et al.* (2003b).

C_2H_6: In the spectral region 2975.331–2977.926 cm^{-1}, a modified list by Rinsland *et al.* (1998), for the $^PQ_3$ subbranch in the ν_7 band, has replaced the previously archived line parameters.

C_2H_2: The updates of the database concern three spectral regions: the 13.6 and 5 μm regions, from work performed by Mandin *et al.* (2000) and Jacquemart *et al.* (2001, 2002) and the 7.5 μm region, from results by Vander Auwera (2000a).

HOCl: A new line list has been created in the spectral interval 1178.673–1320.332 cm^{-1}, corresponding to the ν_2 region, on the basis of the works of Flaud *et al.* (1998), Flaud (2002) and Vander Auwera *et al.* (2000).

CH_3Cl: In the spectral region from 1261.562 to 1645.899 cm^{-1}, a list of 8989 transitions of the [ν_2, ν_5, $2\nu_3$] -vibrational-band triad, prepared by Brown (2001, 2002, 2003) and based upon the work of Chackerian *et al.* (1998b), has been newly archived for $CH_3^{35}Cl$ and $CH_3^{37}Cl$.

COF_2: An update of the 1856.730–2001.348 cm^{-1} region has been made thanks to a list generated by Brown (2001, 2002, 2003) based on an unpublished analysis of the line positions.

HO_2: Spectroscopic parameters for the ground state have been replaced by those generated on the basis of the work of Chance (1997, 2003).

The line parameters for each transition are stored following the new GEISA format standard with an extended number (from 16 to 30) of the selected line parameters, including associated error estimations. The file structure is presented in Table 2.3; symbolic field names are in the first line and field lengths and format descriptors in lines 2 and 3, respectively. The A–Q fields are similar to those described in Jacquinet-Husson *et al.* (1999) but with some extensions in the field lengths and differences in the format description; these differences are identified by a reminder of the former format in italics. The description of each field is given at the bottom of the table.

A summary of the GEISA-03 line transition parameters subdatabase content is given in Table 2.4. The items listed for each molecular species, given in column 1, are as follows: the identification code (ID codes defined for the GEISA management software), the number of lines, the intensity average (in cm molecule^{-1}), the mean half-width at half maximum (cm^{-1} atm^{-1} at 296 K), the identification codes of its various isotopes and for each isotope, the number of lines, the transitions minimum and maximum wavenumbers (in cm^{-1}), and the line intensities minimum and maximum values (in cm molecule^{-1}), in columns 2 to 11, respectively.

Table 2.3 GEISA-03 line parameters record format

Field name	A	B	C	D	E	F	G	I	J	K	L	M	N	O	P	Q
Field length	12	11	6	10	36	4	3	3	1	2	1	10	5	8	3	6
Fortran descriptor	F12.6 *(F10.6)*	D11.4 *(D10.3)*	F6.4 *(F5.3)*	F10.4 *(F10.3)*	A36	F4.2	I3	I3	A1 *(A3)*	I2	I1	E10.3	F5.4	F8.6	I3	I6

R	*A'*	*B'*	*C'*	*F'*	*O'*	*R'*	*N'*	*S*	*S'*	*T*	*T'*	*U*	*U'*
6	10	11	6	4	8	6	5	4	4	8	8	4	4
F6.4	F10.6	D11.4	F6.4	F4.2	F8.6	F6.4	F5.4	F4.2	F4.2	F8.6	F8.6	F4.2	F4.2

A: Wavenumber (cm^{-1}) of the line
B: Intensity of the line in (cm^{-1}/(molecule.cm^{-2}) at 296 K
C: Air broadening pressure half-width (HWHM) (cm^{-1}atm^{-1}) at 296 K
D: Energy of the lower transition level (cm^{-1})
E: Transition quantum identifications for the lower and upper state of the transition
F: Temperature dependence coefficient n of the air broadening half-width
G: Identification code for isotope as in GEISA
I: Identification code for molecule as in GEISA
J: Internal GEISA code for the data identification

K: Molecule number as in HITRAN
L: Isotope number (1 = most abundant. 2 = second…etc) as in HITRAN
M: Transition probability (in debye^2)
N: Self broadening pressure half-width (HWHM) (cm^{-1}atm^{-1}) at 296 K (for water)
O: Air pressure shift of the line transition (cm^{-1}atm^{-1}) at 296 K
P: Accuracy indices for frequency, intensity and half-width
Q: Indices for lookup of references for frequency, intensity and half-width

Fields **A** to **Q** common to the former GEISA-97 edition format (Jacquinet-Husson *et al.* 1999)

R: Temperature dependence coefficient n of the air pressure shift
A': Estimated accuracy (cm^{-1}) on the line position
B': Estimated accuracy on the intensity of the line in (cm^{-1}/(molecule.cm^{-2})
C': Estimated accuracy on the air collision half-width (HWHM) (cm^{-1}atm^{-1})
F': Estimated accuracy on the temperature dependence coefficient n of the air broadening half-width
O': Estimated accuracy on the air pressure shift of the line transition (cm^{-1}atm^{-1}) at 296 K
R': Estimated accuracy on the temperature dependence coefficient n of the air pressure shift
N': Estimated accuracy on the self broadened (HWHM) (cm^{-1}atm^{-1}) at 296 K (for water)
S: Temperature dependence coefficient n of the self broadening half-width (for water)
S': Estimated accuracy on the temperature dependence coefficient n of the self broadening half-width (for water)
T: Self pressure shift of the line transition (cm^{-1}atm^{-1}) at 296 K (for water)
T': Estimated accuracy on the self pressure shift of the line transition (cm^{-1}atm^{-1}) at 296 K (for water)
U: Temperature dependence coefficient n of the self pressure shift (for water)
U': Estimated accuracy on the temperature dependence coefficient n of the self pressure shift (for water)

Fields ***R*** to ***U'*** new for GEISA-03

Fields ***R*** to ***U'*** added to GEISA/IASI-03

The GEISA-97 format is recalled, in italic brackets, if different

Table 2.4 GEISA-03 individual lines sub-database detailed content

Molecule	Mol. ID.	# lines	Intensity average (cm mol.$^{-1}$)	Alpha average (cm^{-1}atm.$^{-1}$)	Isot. ID	# lines	Minimum frequency (cm^{-1})	Maximum frequency (cm^{-1})	Minimum intensity (cm mol.$^{-1}$)	Maximum intensity (cm mol.$^{-1}$)
H_2O	1	58726	1.242E − 21	0.0712	161	36701	0.401	25232.004	1.010E − 32	2.670E − 18
					181	7406	6.785	11246.170	1.440E − 28	5.390E − 21
					171	3874	6.471	11150.790	1.480E − 27	9.830E − 22
					162	10132	0.007	5507.548	1.240E − 32	2.700E − 22
					182	438	1173.772	1684.226	2.030E − 27	5.080E − 26
					172	175	1234.235	1598.765	2.030E − 27	9.320E − 27
CO_2	2	76826	1.470E − 21	0.0713	626	32692	436.123	9648.007	3.440E − 39	3.530E − 18
					636	11122	494.472	8104.666	1.820E − 39	3.750E − 20
					628	17417	498.933	8132.007	1.390E − 36	6.840E − 21
					627	10009	514.430	6961.226	1.000E − 27	1.250E − 21
					638	2312	567.596	4946.384	3.700E − 27	7.230E − 23
					637	1584	584.754	3641.072	3.710E − 27	1.360E − 23
					828	1107	615.974	3669.609	1.760E − 40	1.310E − 23
					728	288	626.438	2358.226	3.870E − 27	2.500E − 24
					838	295	2115.685	2276.481	4.870E − 42	1.760E − 25
O_3	3	319248	5.583E − 23	0.0693	666	190832	0.026	4060.783	3.904E − 29	4.060E − 20
					668	19147	0.921	1177.493	4.692E − 28	7.760E − 23
					686	22211	1.177	2259.896	9.970E − 29	7.560E − 23
					667	58171	0.289	820.380	5.135E − 31	5.356E − 25
					676	28887	0.213	822.795	1.433E − 31	5.827E − 25

N_2O	4	26681	2.696E – 21	0.0748	446	19333	0.838	5131.249	1.280E – 28	1.000E – 18
					456	2128	5.028	3462.689	5.220E – 26	3.670E – 21
					546	2182	4.858	3473.528	4.720E – 26	3.600E – 21
					448	2034	545.179	3463.967	1.230E – 25	2.050E – 21
					447	1004	542.242	3482.917	3.430E – 26	4.150E – 22
CO	5	13515	7.543E – 22	0.0467	26	5908	3.530	8464.882	7.880E – 78	4.460E – 19
					36	4768	3.414	8180.219	3.610E – 73	4.690E – 21
					27	748	3.714	6338.061	8.190E – 40	1.600E – 22
					28	770	3.629	6266.577	7.610E – 39	8.320E – 22
					37	580	1807.871	6196.551	1.030E – 36	1.680E – 24
					38	741	3.462	6123.294	2.580E – 40	8.700E – 24
CH_4	6	216196	8.214E – 23	0.0521	211	187403	0.010	9199.285	4.060E – 34	2.100E – 19
					311	28793	0.032	6069.084	4.100E – 34	2.320E – 21
O_2	7	6290	4.004E – 26	0.0431	66	1433	0.000	15927.806	9.810E – 51	8.830E – 24
					67	4186	0.000	14536.515	8.533E – 51	1.160E – 27
					68	671	1.572	15851.213	1.190E – 35	1.710E – 26
NO	8	99123	4.773E – 23	0.0525	46	97745	0.000	9273.214	1.510E – 85	2.320E – 20
					48	679	1601.909	2038.846	4.190E – 28	1.390E – 22
					56	699	1609.585	2060.462	4.430E – 28	2.550E – 22
SO_2	9	38853	1.065E – 21	0.1145	626	38566	0.017	4092.948	1.020E – 28	6.090E – 20
					646	287	2463.470	2496.088	9.740E – 24	3.430E – 23
NO_2	10	104224	5.979E – 22	0.0671	646	104224	0.498	3074.366	4.240E – 28	1.300E – 19
NH_3	11	29082	1.639E – 21	0.0827	411	27992	0.058	5294.502	8.090E – 39	5.670E – 19
					511	1090	0.375	5179.786	5.460E – 29	1.990E – 21

Table 2.4 (Continued)

Molecule	Mol. ID.	# lines	Intensity average (cm mol.$^{-1}$)	Alpha average (cm^{-1}atm.$^{-1}$)	Isot. ID	# lines	Minimum frequency (cm^{-1})	Maximum frequency (cm^{-1})	Minimum intensity (cm mol.$^{-1}$)	Maximum intensity (cm mol.$^{-1}$)
PH_3	12	11740	2.315E − 21	0.0864	131	11740	17.805	2478.765	1.850E − 28	2.520E − 19
HNO_3	13	171504	6.879E − 22	0.11	146	171504	0.035	1769.982	3.490E − 27	3.020E − 20
OH	14	42866	2.806E − 20	0.044	61	42711	0.005	35877.031	1.500E − 85	6.450E − 17
					62	90	0.010	1.824	2.090E − 31	5.780E − 29
					81	65	0.053	6.325	1.200E − 30	1.200E − 26
HF	15	107	6.773E − 19	0.0407	19	107	41.111	11535.570	1.110E − 26	1.440E − 17
HCl	16	533	3.189E − 20	0.0403	15	284	20.270	13457.841	1.090E − 26	5.030E − 19
					17	249	20.240	10994.721	1.010E − 26	1.610E − 19
HBr	17	1294	4.770E − 21	0.0429	11	642	16.232	9757.189	1.530E − 32	1.180E − 19
					19	652	16.237	9758.565	9.450E − 33	1.210E − 19
HI	18	806	1.361E − 21	0.05	17	806	12.509	8487.305	1.640E − 30	3.420E − 20
ClO	19	7230	1.605E − 22	0.0873	56	3599	0.028	1207.639	1.520E − 29	3.240E − 21
					76	3631	0.015	1199.840	5.090E − 30	1.030E − 21
OCS	20	24922	4.659E − 21	0.0896	622	14500	0.406	4118.004	1.560E − 25	1.220E − 18
					624	4764	0.396	4115.931	6.400E − 27	4.720E − 20
					632	2403	0.404	4012.467	1.720E − 27	1.200E − 20
					623	1802	509.007	4115.588	1.010E − 23	8.430E − 21
					822	1096	0.381	4041.565	2.620E − 28	2.090E − 21
					634	357	1972.188	2032.039	1.010E − 23	5.240E − 22

H_2CO	21	2701	8.604E − 21	0.1074	126	1771	0.000	2998.527	1.020E − 38	7.500E − 20
					128	367	0.035	47.486	1.160E − 30	1.110E − 22
					136	563	0.037	72.745	2.020E − 30	6.290E − 22
C_2H_6	22	14981	7.144E − 23	0.1002	226	8944	765.027	2977.926	5.800E − 27	6.640E − 21
					236	6037	725.603	918.717	1.320E − 28	1.770E − 23
CH_3D	23	35518	2.382E − 25	0.048	212	35518	7.760	3306.810	5.570E − 30	5.710E − 23
C_2H_2	24	3115	1.343E − 20	0.0701	221	2879	604.774	3358.285	9.490E − 27	1.190E − 18
					231	236	613.536	3374.223	3.820E − 26	1.580E − 20
C_2H_4	25	12978	1.411E − 21	0.087	211	12697	701.203	3242.172	6.940E − 26	8.410E − 20
					311	281	2947.832	3180.238	5.060E − 24	1.620E − 21
GeH_4	26	824	4.978E − 20	0.1	411	824	1937.371	2224.570	1.960E − 22	3.680E − 19
HCN	27	2550	1.216E − 20	0.1272	124	2250	2.956	18407.973	1.780E − 28	7.100E − 19
					125	115	2.870	9671.953	5.110E − 26	2.730E − 21
					134	185	2.880	9627.961	4.150E − 26	8.290E − 21
C_3H_8	28	8983	4.139E − 23	0.08	221	8983	700.015	799.930	3.770E − 24	4.310E − 22
C_2N_2	29	2577	2.668E − 21	0.08	224	2577	203.955	2181.690	6.590E − 24	2.580E − 20
C_4H_2	30	1405	3.445E − 21	0.1	211	1405	190.588	654.425	2.650E − 24	6.930E − 20
HC_3N	31	2027	2.693E − 21	0.1	124	2027	474.293	690.860	6.360E − 24	4.420E − 20
HOCl	32	17862	1.867E − 21	0.0689	165	9293	0.024	3799.249	1.650E − 27	3.590E − 20
					167	8569	0.349	3799.682	7.220E − 28	1.140E − 20
N_2	33	120	5.601E − 29	0.0419	44	120	1992.628	2625.497	2.190E − 34	3.420E − 28

Table 2.4 (Continued)

Molecule	Mol. ID.	# lines	Intensity average (cm mol.$^{-1}$)	Alpha average (cm^{-1}atm.$^{-1}$)	Isot. ID	# lines	Minimum frequency (cm^{-1})	Maximum frequency (cm^{-1})	Minimum intensity (cm mol.$^{-1}$)	Maximum intensity (cm mol.$^{-1}$)
CH_3Cl	34	18344	4.370E − 22	0.0825	215	10039	679.050	3172.927	9.050E − 32	1.130E − 20
					217	8305	674.143	3161.830	4.190E − 26	3.540E − 21
H_2O_2	35	100781	4.848E − 22	0.1076	166	100781	0.043	1499.487	5.090E − 29	5.610E − 20
H_2S	36	20788	2.992E − 22	0.08	121	12330	2.985	4256.547	1.450E − 26	1.360E − 19
					131	3564	5.601	4098.234	2.020E − 26	5.990E − 21
					141	4894	5.615	4171.176	2.020E − 26	1.080E − 21
HCOOH	37	3388	5.186E − 21	0.4	261	3388	1060.962	1161.251	2.140E − 22	2.840E − 20
COF_2	38	83750	2.105E − 21	0.0845	269	83750	725.006	2001.348	4.740E − 24	3.940E − 20
SF_6	39	11520	4.551E − 21	0.05	29	11520	940.425	952.238	2.160E − 22	1.500E − 20
C_3H_4	40	3390	4.277E − 22	(*)	341	3390	290.274	359.995	2.020E − 23	3.180E − 21
HO_2	41	38804	6.846E − 22	0.1072	166	38804	0.173	3675.819	1.000E − 26	2.740E − 20
$ClONO_2$	42	32199	8.768E − 23	0.14	564	21988	763.641	797.741	1.240E − 24	3.850E − 22
					764	10211	765.212	790.805	6.340E − 25	1.260E − 22
	TOTAL	1668371	LINES							

2.3.2 Subdatabase on absorption cross-sections

Infrared Cross-Sections: The GEISA-03 archive has been significantly enriched since the GEISA-97 issue. The spectral range has been extended from 200 cm^{-1} to 2000 cm^{-1} (from 556 to 1763 cm^{-1}, previously) and the number of molecules has increased, as well, from 23 to 35. The updated already-archived species are: CFC-11, CFC-12, CFC-14, HCFC-22, HCFC-123, HCFC-124, HFC-125, HFC-134a, HCFC-141b, HCFC-142b, HFC-152a, HCFC-225ca, HCFC-225cb, HFC-32, HFC-143a, HFC-134, N_2O_5, SF_6, $ClONO_2$. No update has occurred for CFC-13, CFC-113, CFC-114, CFC-115. Twelve molecular species are new for the GEISA-03 archive; these are HFC-143, HCFC-21, CCl4, C_2F_6, C_2H_2, C_2H_4, C_2H_6, C_3H_8, C_4F_8, HNO_4, SF_5CF_3, HCH-365mfc.

A summary of the GEISA-03 subdatabase on absorption cross-sections is given in Table 2.5. The molecular species names are listed in columns 1. In the three following columns (2 to 4) are the experimental conditions associated with the data files, i.e. the spectral range (cm^{-1}), the overall temperature range (K) and the total pressure range (Pa). If the pressure is set to zero, it means that the cross-sections have been measured for a pure vapour (non-air broadened) at different pressures for the same point or that the pure vapour cross-sections were extrapolated to zero pressure from a set of laboratory data. For each file, the number of associated temperature/pressure sets is in column 5 and the related references, in column 6. Reference 'GEISA-97' corresponds to molecules not updated since the archive described in Jacquinet-Husson *et al.* (1999).

Table 2.5 summarises the HITRAN-04 infrared cross-sections content too. GEISA specific molecular species or references are identified by an asterisk. Slant characters are used when references are different in HITRAN and in GEISA, for a common molecular species. Note that the original data for cross-sections contain some negative numbers, which was due to the noise in the recorded spectra. HITRAN made the choice to set all of the negative cross-sections to zero; GEISA made the choice to keep the integrity of the original data provided by their authors.

Ultraviolet and Visible Cross-Sections: In the spectral range 230–1333 nm, 11 molecular species, i.e. O_3, NO_2, BrO, OClO, OBrO, SO_2, H_2CO, O_2, $O_2 + N_2$, O_4, CS_2, have been archived, as summarised in Table 2.6 (organised as Table 2.5, with an extra seventh column giving the number of entries). These represent, as a whole, 164 temperature/pressure sets for a total of 12 958 953 entries. 'Undef' in column 4 corresponds to a missing pressure value.

2.3.3 Subdatabase on microphysical and optical properties of atmospheric aerosols

A GEISA aerosol subdatabase has been recently issued (see Jacquinet-Husson *et al.* 2003a for details). It includes microphysical and optical properties from published aerosol data catalogs from Massie (1994, 2003), Rublev (1994), Hess *et al.* (1998), Köpke *et al.* (1997) and Van de Hulst (1981). The overall content of this archive deals with complex refractive indices data and possibly computed optical related

Table 2.5 GEISA-03 and HITRAN-04 infrared cross-sections subdatabase

Molecule	Spectral range (cm^{-1})	Temperature range (K)	Pressure range (Pa)	Number of T,P sets	References
CFC-11	210–2000	296	93325	1	Hurley (2003); Christidis (1997) *
	500–1600	297	0	1	Smith (2003) *
	810–1120	190–296	1000–101325	55	Li and Varanasi(1994); Varanasi (2000)
CFC-12	850–1190	253–287	0	3	Clerbaux (1993) *
	210–2000	296	93325	1	Hurley (2003) *
	800–1200	190–296	1000–101392	57	Varanasi and Nemtchinov (1994); Varanasi (2000)
CFC-13	765–1235	203–293	0	3	GEISA97
CFC-14	220–2000	296	93325	1	Hurley (2003) *
	1250–1290	180–296	1005–101458	55	Nemtchinov and Varanasi (2003a)
CFC-113	780–1232	203–293	0	6	GEISA97
CFC-114	815–1285	203–293	0	6	GEISA97
CFC-115	955–1260	203–293	0	6	GEISA97
HFC-32	204–2000	296	93325	1	Hurley (2003) *; Pinnock (1995) *
	995–1475	203–297	0–100000	17	Smith (2003)
HFC-125	700–1465	287	0	1	Clerbaux (1993)
	495–1504	203–293	0–80000	16	Di Lonardo (2000) *
	208–2000	296	93325	1	Pinnock (1995) *; Hurley (2003) *
HFC-134	210–2000	296	93325	1	Hurley (2003) *; Christidis (1997) *
	600–1700	203–297	0–100000	9	Smith (2003)
HFC-134a	815–1485	253–287	0	3	Clerbaux (1993)
	203–2000	296	93325	1	Pinnock (1995) *; Hurley (2003) *
	600–1600	203–296	0–100000	15	Smith (2003) *
	1035–1340	190–296	2666–101376	33	Nemtchinov and Varanasi (2004)

HFC-143	204–2000	296	93325	1	Pinnock (1995) *; Hurley (2003) *
HFC-143a	694–1504	203–293	0–800000	19	Di Lonardo (2000) *
	200–2000	296	93325	1	Pinnock (1995) *; Hurley (2003) *
	580–1500	203–297	0–100000	9	Smith (2003)
HFC-152a	700–1600	203–293	0–80000	16	Vander Auwera (2000b) *
	840–1490	253–287	0	3	Clerbaux (1993)
	200–2000	296	93325	1	Pinnock (1995) *; Hurley (2003) *
HCFC-21	785–840	296	133	1	Massie *et al.* (1985)
HCFC-22	700–1500	203–293	0–80000	8	Vander Auwera (2003) * Ballard *et al.* (2000) *
	765–1380	253–287	0	3	Clerbaux (1993)
	208–2000	296	93325	1	Pinnock (1995) *; Hurley (2003) *
	760–1195	181–297	2666–101936	51	Varanasi (1994, 2001))
HCFC-123	740–1450	253–287	0	3	Clerbaux (1993)
	204–2000	296	93325	1	Pinnock (1995) *; Hurley (2003) *
HCFC-124	675–1425	287	0	1	Clerbaux (1993)
	208–2000	296	93325	1	Pinnock (1995) *; Hurley (2003) *
HCFC-141b	209–2000	296	93325	1	Pinnock (1995) *; Hurley (2003) *
	710–1470	253–287	0	3	Clerbaux (1993)
HCFC-142b	650–1475	253–287	0	3	Clerbaux (1993)
	200–2000	296	93325	1	Pinnock (1995) *; Hurley (2003) *

Table 2.5 (Continued)

Molecule	Spectral range (cm^{-1})	Temperature range (K)	Pressure range (Pa)	Number of T,P sets	References
HCFC-225ca	695–1420	253–287	0	3	Clerbaux (1993)
	600–2000	296	93325	1	Pinnock (1995); Hurley (2003)
HCFC-225cb	715–1375	253–287	0	3	Clerbaux (1993)
	600–2000	296	93325	1	Pinnock (1995) ∗; Hurley (2003) ∗
N_2O_5	540–1380	205–293	0	5	Wagner and Birk (2003)
SF_6	650–2000	296	93325	1	Hurley (2003) ∗
	925–955	180–295	2693–101350	29	Varanasi (1994, 2001))
$ClONO_2$	500–1330	190–297	0–15580	25	Wagner and Birk (2003)
	1265–1325	201–222	0	3	GEISA97
CCl_4	750–812	208–297	1070–101272	32	Nemtchinov and Varanasi (2003b)
C_2F_6	1061–1285	180–296	3320–101363	43	Zou *et al.* (2004)
	210–2000	296	93325	1	Highwood (1999) ∗, Hurley (2003) ∗
	600–2750	203–293	0–80000	15	Smith (2003) ∗
C_2H_2∗	450–2000	296	93325	1	Highwood (1999) ∗; Hurley (2003) ∗
C_2H_4∗	220–2000	296	93325	1	Highwood (1999) ∗; Hurley (2003)
C_2H_6∗	220–2000	296	93325	1	Highwood (1999) ∗; Hurley (2003)
C_3H_8∗	220–2000	296	93325	1	Highwood (1999) ∗; Hurley (2003)
C_4F_8∗	500–1600	203–297	0–65000	19	Smith (2003) ∗
HNO_4	770–830	268	93	1	Massie *et al.* (1985)
SF_5CF_3	600–2000	296	93325	1	Hurley (2003)
HCH-365mfc∗	665–1480	287	0	1	Clerbaux (1993) ∗

Table 2.6 GEISA-03 ultraviolet/visible cross-sections subdatabase

Molecule	Spectral range (nm)	Temperature range (K)	Pressure range (hPa)	Number of T,P sets	References	Number of entries
O_3	231–794	202–293	undef	5	BU	16500
	230–830	203–293	100–1000	10		200000
	230–1070	203–293	50–900	5		21000
NO_2	231–794	221–293	undef	4		13600
	230–830	223–293	100–1000	10		1358000
	230–900	203–293	100–150	5		16800
	238–1000	294	0	1	BIRA/IASB	13500
	238–1000	220–294	0–1.33	13		225000
	385–725	220–294	0–1013.23	27		9855000
BrO	300–385	203–298	undef	5	BU	25000
OClO	325–425	213–293	undef	10		132500
	290–460	293	900	1		1200
ObrO	385–616	298	undef	1		40500
SO_2	240–395	203–293	100	5		7000
	250–330	294	undef	2	BIRA/IASB	9600
H_2CO	250–400	293	4.10	1	BU	1400
O_2	235–800	203–293	900–945	6		5990
	627–1290	223–290	100–1000	24	MSF/RAL	390360
	246–266	296	200	5		517495
	240–330	287–289	undef	2	BIRA/IASB	15560
$O_2 + N_2$	1000–1333	132–295	668–1009	12	MSF/RAL	50800
O_4	333–666	294	undef	1	BIRA/IASB	12200
	454–667	223–294	1000	8	MSF/RAL	29040
CS_2	290–350	294	undef	1	BIRA/IASB	908

BU: Bremen University *Molecular Spectroscopy and Chemical Kinetics Group* (Orphal, 2003; Vandaele *et al.* 2002).
BIRA/IASB: Belgium Institute for Space Aeronomy (Vandaele *et al.* 2002).
MSF/RAL: Molecular Spectroscopy Facility/Rutherford Appleton Laboratory (Smith, 2003).

properties, for selected basic aerosol components. Software for data management and user-selected aerosol mixtures elaboration are available from the database too.

2.4 BEAMCAT

The well-known spectroscopic databases like the JPL catalog (Pickett *et al.*, 1998), HITRAN (Rothman *et al.*, 1998) or GEISA (Jacquinet-Husson et al., 1999) provide only part of the information that is necessary for the calculation of atmospheric spectra in the microwave region. The JPL catalog, for example, contains a large collection of molecular species and an exhaustive list of their spectral lines in the millimetre and submillimetre domain. However, while it lists centre frequency and line strength with good accuracy, it does not contain any line broadening parameters that are also necessary for these calculations. The reason is probably that the target audience for this catalog are radio astronomers and not necessarily atmospheric scientists. The catalogs for atmospheric scientists, like HITRAN or GEISA, list these parameters but miss a lot of the molecular species that are included in the JPL catalog. Even for the common species they list only a fraction of the lines in the millimetre and submillimetre domain that are included in the JPL catalog. Besides that, the accuracy of the listed centre frequencies in the microwave region is limited by the data format in both these catalogs. The reason for that may be that the catalogs were designed with the needs of the infrared remote sensing community in mind.

The choice for a microwave remote sensing scientist used to be either a catalog like JPL that lacked line broadening parameters or a catalog like HITRAN or GEISA that missed many molecules and spectral lines and listed centre frequencies only with degraded accuracy. The usual workaround for that problem was to combine data from two or more catalogs and create a merged catalog (Chance *et al.*, 1994; Eriksson and Merino, 1997; Bauer *et al.*, 1998). The ideal merged catalog would list all necessary parameters and provide them from the source with the best quality. However, the process of creating such a merged catalog was anything but trivial and included several pitfalls:

Line identification. Since the spectral line databases like the JPL catalog, HITRAN or GEISA are huge, the merging process should be completely automatic. However, the different data formats and different philosophies of the catalogs make it very difficult to build an algorithm that can accurately identify lines in several catalogs. In theory, lines should be identified by their quantum numbers. However, while that approach works well for some species it completely fails for others since the catalogs may list different sets of quantum numbers for the same species.

Quality selection. Even if one is able to correctly identify spectral lines within two different catalogs, it is still another problem to automatically decide which catalog lists a certain parameter with better quality. This might be different for each listed line and it might even change over time as new data from various sources are incorporated into the source catalogs.

Errors and mismatches. The merging process would most likely introduce some errors and mismatches. These mismatches could be corrected manually but the corrections would be lost when the merging process has to be repeated.

Handling updates. Updates in the source catalogs would make it necessary to produce a new merged catalog version. However, since this would be a complex task every time, frequent updates would not be possible.

Lost information. All the existing merged catalogs throw away information during the merging process by preferring parameters from one catalog over the corresponding parameters from the other catalog. This lost information cannot be recovered from the merged catalog any more.

An alternative that avoids most of the problems of a merged catalog is a meta catalog. Unlike a merged catalog, a meta catalog does not contain spectral line parameters

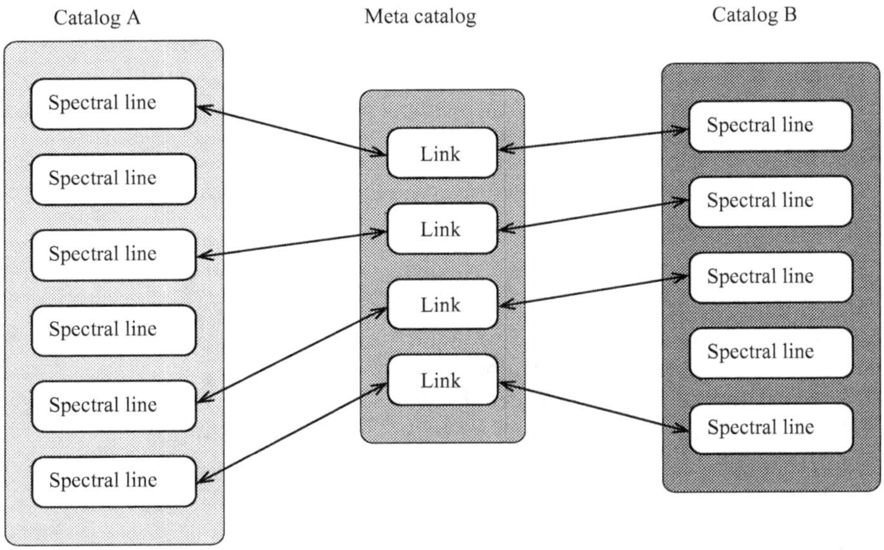

Figure 2.1 *Concept of the meta catalog. Unlike a merged spectral line catalog, the meta catalog provides links between identical quantum-mechanical transitions listed in different catalogs (A and B in this case). The meta catalog itself does not contain any spectral parameters. It rather provides an index to where the parameters for a certain spectral line can be found in each of the catalogs A and B. The concept works with any number of catalogs (reprinted from Dietrich G. Feist: The Bernese Atmospheric Multiple Catalog Access Tool (BEAMCAT): a tool for users of popular spectral line catalogs, Journal of Quantitative Spectroscopy and Radiative Transfer, 85(1), pp. 57–97, Copyright (2004), with permission from Elsevier)*

itself. Instead, it contains information that describes which spectral line in one catalog corresponds to a spectral line in one or more other spectral line catalogs. In other words, the meta catalog provides links between matching spectral lines in different catalogs. Figure 2.1 illustrates the concept. By following these links, merged spectral line catalogs can be generated from the meta catalog on-the-fly. The pitfalls of merged catalogs do not apply to a meta catalog:

Line identification. For the meta catalog, it is not important how a link between two matching lines was established. Therefore, it is not necessary to build an algorithm that can automatically identify lines in several catalogs. Links could even be created or corrected manually.

Quality selection. There is no need to select the quality of parameters when the links are created.

Errors and mismatches. Errors and mismatches can easily be corrected by removing or changing the link between two mismatched lines.

Handling updates. Updates in the source catalogs only affect the spectral line parameters and not the links between the lines in different catalogs. Only additional lines in the source catalogs would have to be checked after an update.

Lost information. No information would be lost since all the information is available from the source catalogs.

Besides that, a meta catalog has several additional advantages over a merged catalog:

- the concept is not limited to two catalogs. A meta catalog could contain link information for any number of spectral line catalogs. This solves the problem of how to best combine a small set of good-quality spectral lines with background spectral lines from the large catalogs;
- most data format issues with spectral line databases become irrelevant as merged databases can be created from the meta catalog in arbitrary formats;
- a meta catalog allows different versions of spectral line databases to be used together.

BEAMCAT is a spectral line meta catalog that was developed at the Institute of Applied Physics at the University of Bern (Feist, 2004). It resides on a MySQL database server and can be accessed either through a web interface or directly through a MySQL client on the user's computer. Since MySQL is free software, clients are available for almost any platform and operating system. Interfaces are available for most programming and scripting languages so users can access the BEAMCAT database server directly from their applications.

BEAMCAT is provided as a free service to the scientific community. The website at www.beamcat.org contains information on how to use the catalog. Users can download merged spectral line data directly from the web site or sign up to receive a free user account for the BEAMCAT MySQL server. With a user account, they can access the MySQL server directly through a local MySQL client or through programs in their favourite programming language. Registered users will also be informed about new developments around the BEAMCAT database.

One of the first applications of BEAMCAT was an extensive intercomparison of spectral line data from the JPL catalog and HITRAN (Feist, 2004). This intercomparison between roughly 148 000 spectral lines that exist both in HITRAN and the JPL catalog points out several problems with the spectral line data in those two catalogs. The article also provides detailed information on how BEAMCAT works internally and how the line matching between the JPL catalog and HITRAN was accomplished.

2.5 Atmospheric radiative-transfer simulator

The atmospheric radiative-transfer simulator, ARTS (Buehler *et al.*, 2005b), is a public domain project which was initiated jointly by the University of Bremen, Germany and Chalmers University, Gothenburg, Sweden. The aim was to create a radiative-transfer code that is as general and flexible as possible, while at the same time being portable and user-friendly. The program itself is written in C^{++}, but is accompanied by interface functions and helper functions in IDL and Matlab. The complete source code, along with extensive documentation, is freely available on the Internet, under www.sat.uni-bremen.de/arts/.

There are two versions of ARTS, a clear-sky version (current version at the time of writing 1-0-173) and a version with scattering (current version at the time of writing 1-1-912). The 1-0-x version simulates scalar radiative transfer (total intensity only) in a one-dimensional spherical atmosphere. The 1-1-y version simulates vector radiative transfer (1 to 4 Stokes components) in a one-dimensional or three-dimensional spherical atmosphere. It is planned that at some time in the future the 1-0-x version will disappear, when the 1-1-y version is considered sufficiently stable for operational work and when all features of absorption calculation that are now available in the 1-0-x version have been ported to the 1-1-y version. The text here describes the 1-0-x version; for information on the 1-1-y version of ARTS see Chapter 3.

The primary vertical coordinate in ARTS is pressure. All other quantities, such as temperature, geometric altitude and trace gas concentrations, are given on pressure grids. The model can be used to simulate measurements for any observation geometry: up looking, down looking or limb looking; and for any sensor position: on the ground, inside the atmosphere or on a satellite. It has been developed having passive emission measurements in mind, but pure transmission measurements are also handled.

ARTS works with arbitrary frequency grids; hence, it can be used both for the simulation of high-resolution sensors and for the simulation of broad frequency ranges. The applicable spectral range is from the microwave up to the thermal infrared, but so far it has been mostly used for the frequency region roughly below 1 THz. In that frequency range, particular care has been taken to make the absorption calculation consistent with state-of-the-art continuum models for water vapour and nitrogen, and with continuum and line mixing models for oxygen.

Absorption coefficients can be calculated in ARTS by a combination of line-by-line calculation (using external or built-in spectroscopic data) and various continua. Supported external spectral line catalog formats are JPL (Pickett *et al.*, 1998), HITRAN (Rothman *et al.*, 2003) and MYTRAN (Bauer *et al.*,1998), as well as a native catalog format, as described by Buehler and Eriksson (2000). For the temperature conversion of the line strength, partition functions are calculated from an internal polynomial representation, which is taken from the TIPSdst97 FORTRAN program (Gamache *et al.*, 2000) for those species included in TIPS. For the other species, a polynomial fit to the tabulated partition function values of JPL is used. Where not even that is available for some isotopic species, the partition function from the main isotope is used. For a discussion on the impact of different partition function sources on the forward model and retrieval accuracy see Verdes *et al.* (2005). For a detailed list of the partition function source that is used by ARTS for each species see Buehler *et al.* (2005b).

All standard line shape functions can be used (Lorentz, Doppler, Voigt, Van–Vleck–Weisskopf). The recommended line shape function is a hybrid between Voigt and Van–Vleck–Weisskopf, behaving like the former in the low pressure limit and like the latter in the high pressure limit. It is discussed in more detail in Buehler *et al.* (2005b). In addition to the line spectrum, ARTS can use any of the following continuum absorption models: MPM (Liebe and Layton, 1987; Liebe, 1989; Liebe *et al.*, 1993), PWR (Rosenkranz, 1993, 1998), CKD (Clough *et al.*, 1989), the model by Ma and Tipping (2002) and the model by Cruz Pol *et al.* (1998). For some of these models it is up to the user to use only the continuum part along with an external line catalog or to also use a model-specific built-in line list. The latter is safer for inexperienced users, since line-by-line calculation and continuum model must fit together in order to give the right total absorption and cannot be combined arbitrarily. For details on the available options see Buehler *et al.* (2005b).

The radiative transfer in ARTS-1-0-x is calculated in equal distance steps along the line of sight, assuming a spherically symmetric atmosphere. The absorption coefficients at line-of-sight grid points are interpolated (linearly in log pressure) from absorption coefficients pre-calculated on a sufficiently fine pressure grid. Besides providing sets of spectra, ARTS can calculate Jacobians for a number of variables. Analytical expressions are used to calculate Jacobians for trace gas concentrations, continuum absorption and ground emissivity. Perturbations are used to calculate Jacobians for pointing and frequency offsets and spectroscopic parameters. For temperature Jacobians, the user can choose between an analytical method, which does not assume hydrostatic equilibrium, and a perturbation method, which does assume hydrostatic equilibrium.

ARTS itself calculates only monochromatic pencil beam radiances; no sensor characteristics are included. The sensor part is covered by a set of Matlab functions that make use of a matrix vector formalism (Eriksson *et al.*, 2005a,b).

Melsheimer *et al.* (2005) describe an intercomparison of eight clear-sky radiative-transfer models, including ARTS. The aim of the exercise was to compare a number of radiative-transfer models for atmospheric sounding in the millimetre and submillimetre wavelength range and to check their consistency and mutual deviations. The intercomparison dealt with three different aspects of the radiative-transfer models: (1) The inherent physics of gaseous absorption lines and how they are modelled, (2) the calculation of absorption coefficients and (3) the full calculation of radiative transfer for different geometries. The correctness and consistency of the implementations was tested by comparing calculations with pre-defined input such as spectroscopic data, line shape, continuum absorption model and frequency grid. It was found that the absorption coefficients and brightness temperatures calculated by the different models are generally within about 1 per cent of each other. In a second step, the variability or uncertainty of the model results is estimated if – except for the atmospheric scenario – the input such as spectroscopic data, line shape and continuum absorption model was varied. Here, the models deviate from each other by what amounts to about 10 per cent around the centre of major absorption lines. The main sources of such discrepancies are the variability of the reported spectroscopic data for the line absorption and in the continuum absorption model. Further possible causes of discrepancies are different frequency and pressure grids and differences in the corresponding interpolation routines as well as differences in the line shape functions used.

2.6 Atmospheric transmission at microwaves

The atmospheric transmission at microwaves (ATM) model has been in continuous development for 20 years (see the first report, Cernicharo 1985). It was first developed for ground-based astronomy applications and later extended for remote sensing applications. Besides model development, the ATM team has devoted a great effort to validation and measurement of continuum terms (Pardo *et al.*, 2001a, 2005a) in the submillimetre domain. Attention has been paid to modelling the real part of the atmospheric refractivity because of its impact on ground-based interferometry. The ATM model was part of an intercomparison including nine microwave models (Garand *et al.*, 2001). Scattering of atmospheric radiation by hydrometeors has been included in the model and several studies have been conducted (Prigent *et al.*, 2001, 2005; Wiedner *et al.*, 2004).

The gaseous absorption algorithm and its coupling to specific 'cloud' and 'surface' routines in ATM have been programmed in FORTRAN 90. Information on the code can be found at www.damir.iem.csic.es/ATM/atmmain.htm. The executable is available upon request. The gaseous absorption and phase delay in ATM was described in detail in Pardo *et al.* (2001b), whereas the scattering part was presented and used for the first time to study real data in Prigent *et al.* (2001). The potential contribution of the surface, ocean or land, can also be calculated. For the ocean, a geometric optics model is adopted (Guillou *et al.*, 1996) coupled to recent measurements of the sea water dielectric properties (Guillou *et al.*, 1998). The land-surface emissivities between 15 and 200 GHz can be deduced from pre-calculated emissivity atlases derived from satellite observations (Prigent *et al.*, 1997, 2000).

Frequency and pressure grids, geometry and surface type can be fully defined by the user. If due to one or more of these parameters (ocean surface or proximity to Zeeman-affected O_2 lines for example) different polarisations need to be treated separately, the code will use the appropriate transfer equation. Some additional information may then be necessary. For example, geographical location and orientation of the antenna are necessary to use the correct geomagnetic field and its angle with respect to the path for Zeeman calculations near O_2 lines. The description of this part of the code can be found in Pardo *et al.* (1995, 1998, 2002). The line-by-line opacity is calculated as described in this chapter, with Lorentz, Gaussian, Voigt, Van Vleck–Weisskopf and Zhevakin–Naumov line shapes available, and using a line data base created by J. Cernicharo using the most recent Hamiltonians found in the literature (see Table 1 in Pardo *et al.*, 2001a), although for the dominant (in the troposphere) H_2O and O_2 contributions, specific results obtained by the ATM team's ground-based Fourier transform spectroscopy (FTS) experiment (see Pardo *et al.*, 2001a, and Section 2.11.2) have been incorporated.

Due to its primary development for use in opacity estimates and calibration procedures at ground-based millimetre and submillimetre observatories, the ATM team has defined a set of basic atmospheric parameters that can easily be measured or estimated for real-time use at the observatory: ground temperature, ground pressure, ground relative humidity, water vapour scale height, tropospheric temperature lapse rate and water vapour zenith column. Due to the impossibility of measuring vertical profiles of minor gases in real time and their much smaller impact on the calibration of millimetre and submillimetre astronomical observations, standard profiles for the given latitude are assumed. Based on these parameters, a series of calibration equations have been described (Serabyn *et al.*, 1998; Pardo *et al.*, 2005b) that allow reaching absolute flux calibrations better than 5 per cent in the submillimetre, necessary for future instruments such as the Atacama Large Millimetre Array (see www.alma.nrao.edu). Specific routines have been developed for retrievals of water vapour zenith column from radiometric measurements (Pardo *et al.*, 2004) and its use to predict atmospheric opacities and phase delays over a wide range of frequencies. The phase delay is derived from the real part of the refractivity, which, in turn, is simply obtained from the imaginary part (related to absorption) by applying the Kramers–Kronig theory (see Section 5.1).

The most specific contribution of the ATM team to the current status of atmospheric absorption models has been the study of excess absorption (compared to the resonant or line-by-line absorption), especially at high frequencies (~300–1600 GHz). Figure 2.2 illustrates real data of the atmospheric spectrum seen from the ground in the 150–1080 GHz frequency range and the different opacity terms that account for the measurements, according to an analysis made with the ATM model.

2.7 RTTOV-8

RTTOV is a radiative-transfer model for very rapid calculations of top-of-atmosphere radiances for a range of space-borne infrared and microwave radiometers as listed in Table 2.7. The original basis for the RTTOV fast computation of transmittances is described by Eyre and Woolf (1988). This was successively modified for later versions of RTTOV by Eyre (1991), Rayer (1995), Saunders and Matricardi (1999) and most recently Matricardi (2003). The recent developments of RTTOV (versions 6–8) have been carried out as part of a collaboration between the Met Office (UK), MétéoFrance and ECMWF in the framework of the EUMETSAT funded NWP Satellite Application Facility. The latest version of RTTOV is RTTOV-8 and was released in mid 2004. In summary RTTOV-8 has the following features:

- It comprises forward, tangent linear, adjoint and K (full Jacobian matrices) versions of the model, the latter three modules for variational assimilation or retrieval applications.
- Top-of-atmosphere radiances, brightness temperatures and layer-to-space plus surface-to-space transmittance for each channel are output for a given input atmospheric profile. There are also other layer-to-space and layer-to-surface radiances output for computing cloudy radiances. In the microwave range, RTTOV-8 approximates the line-by-line models of Liebe (1989) for H_2O absorption and Liebe *et al.* (1992) for O_2 absorption.
- It takes about 0.5 ms to compute radiances for 20 HIRS channels for one profile on a HP workstation.

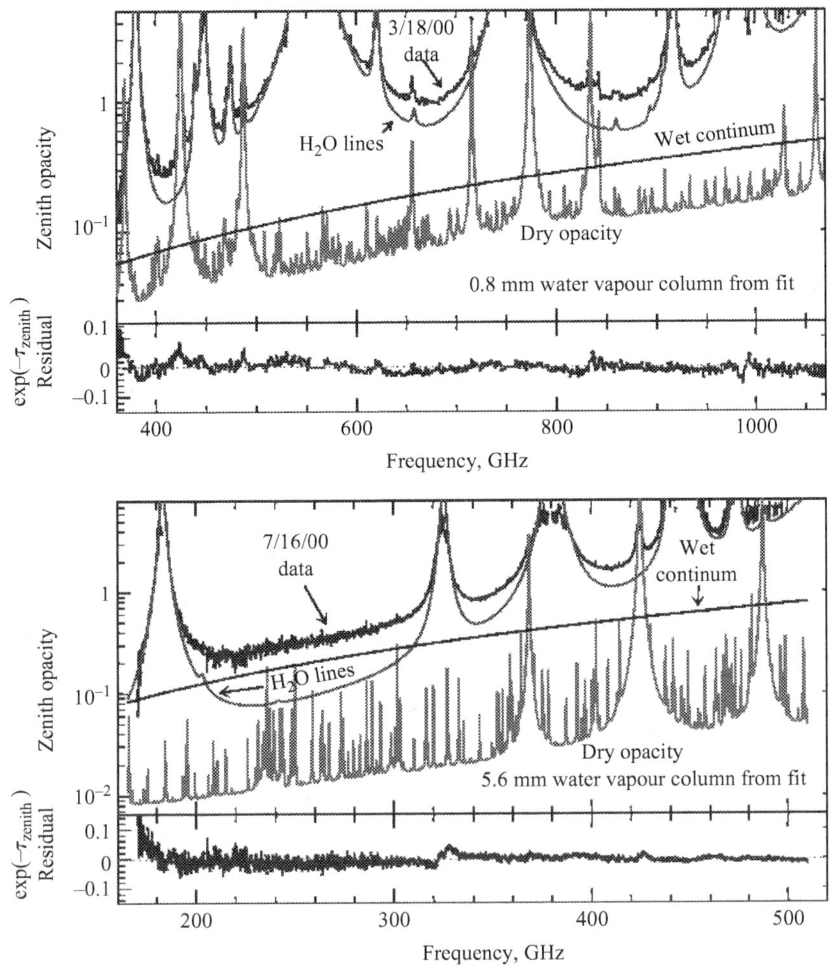

Figure 2.2 *FTS zenith atmospheric opacity spectra obtained in 2000 from the top of Mauna Kea, HI at 4200 m and opacity contributions (binned to the resolution of the spectra, 0.9 Ghz in 3/18/00 and 0.115 Ghz in 7/16/00) that fit the measurements. Meteorogical data from a weather station at the site and radiosonde data from nearby Hilo airport were used as a priori information for a fit with only one free parameter, the zenith water vapour column. The resulting values from each fit are stated in the figure*

- The input profile must have temperature and water vapour concentration. Optionally ozone and carbon dioxide can also be variable gases.
- It can compute sea surface emissivity for each channel internally or use a value provided by the user. For the infrared the ISEM-6 model is used. For the microwave the FASTEM model (see Chapter 4) is used.

- Cloud top pressure and effective cloud amount can also be specified for simple single-layer cloudy radiance calculations.
- There is a 'wrapper code' which allows RTTOV to be used to compute more complicated multi-layer cloudy radiances (RTTOV_CLD) with various overlap assumptions.
- There is a new 'wrapper code' which allows RTTOV to be used to compute rain affected microwave radiances (RTTOV_SCATT).
- It can reproduce the results for RTTOV-7 (except for RTTOV_CLD).
- It supports all the sensors given in Table 2.7 for all the platforms the sensor has flown on.
- It is written in Fortran-90 and run under unix or linux.
- It has been tested on a range of platforms including linux PCs.

All the details on the model and how to request a free copy of the software are at www.metoffice.com/research/interproj/nwpsaf/rtm.

2.8 MPM and MonoRTM

A widely cited model for the microwave frequency range is MPM (millimetre-wave propagation model) developed by H. J. Liebe and co-workers (1977–93). The model underwent numerous revisions during the years that Liebe was active and Table 2.8 summarises the major changes with dates of publication. The code includes absorption by N_2 and cloud liquid water/ice, in addition to that of O_2 and H_2O. Zeeman splitting is included in a related but separate code, ZPM (Zeeman propagation model). Rosenkranz (1998) examined the laboratory measurements of H_2O continuum absorption that were available at the time, and recommended a further revision, referred to below as R98, which combines the foreign-broadened component of the H_2O continuum from MPM87 and the self-broadened component of the H_2O continuum from MPM93 with the O_2 line parameters from MPM92.

MonoRTM (Clough *et al.*, 2005) is a code with H_2O continuum based on the modelling work of Clough *et al.* (1989). It thereby achieves self-consistency across the microwave, submillimetre and infrared spectral regions. The code also includes absorption by O_2, O_3, N_2, N_2O, CO, SO_2, NO_2, OH and cloud liquid water.

2.9 Laboratory and theoretical work

2.9.1 *Line parameters*

As a general rule, the large spectroscopic databases contain calculated values of line intensities. Different calculations typically agree to within ~1 per cent. Line frequencies are derived from fitting a theoretical model to measured frequencies, which generally yields agreement much better than required for atmospheric applications, often as close as tens of kHz, depending on the molecule. Calculation of

Table 2.7 Instruments supported by RTTOV-8 as of 1 June 2004. Sensors in italics are not yet supported by RTTOV-8 but will be in the future

Sensor	RTTOV id	Sensor Channel #	RTTOV-8 Channel #
HIRS	0	1 to 19	1 to 19
MSU	1	1 to 4	1 to 4
SSU	2	1 to 3	1 to 3
AMSU-A	3	1 to 15	1 to 15
AMSU-B	4	1 to 5	1 to 5
AVHRR	5	3b to 5	1 to 3
SSMI	6	1 to 7	1 to 4
VTPR1	7	1 to 8	1 to 8
VTPR2	8	1 to 8	1 to 8
TMI	9	1 to 9	1 to 5
SSMIS	10	1 to 24*	1 to 21
AIRS	11	1 to 2378	1 to 2378
HSB	12	1 to 4	1 to 4
MODIS	13	1 to 17	1 to 17
ATSR	14	1 to 3	1 to 3
MHS	15	1 to 5	1 to 5
IASI	16	1 to 8461	1 to 8461
AMSR	17	1 to 14	1 to 7
MVIRI	20	1 to 2	1 to 2
SEVIRI	21	4 to 11	1 to 8
GOES-Imager	22	1 to 4	1 to 4
GOES-Sounder	23	1 to 18	1 to 18
GMS imager	24	1 to 3	1 to 3
FY2-VISSR	25	1 to 2	1 to 2
FY1-MVISR	26	1 to 3	1 to 3
CriS	27	TBD	TBD
CMISS	28	TBD	TBD
VIIRS	29	TBD	TBD
WINDSAT	30	1 to 10	1 to 5

*Channels 19–21 are not simulated accurately.

line widths and shifts involves the interaction between an absorbing molecule and a perturbing molecule and is therefore more complex. Modern calculations often yield agreement with measured widths to within a few per cent, but this depends on the quality of the model for the intermolecular potential. It is probably debatable whether calculations or measurements are more reliable as a source for line width parameters in models. Traditionally modellers have tended to favour measured values when they are available, although calculations have the advantage of being able to yield

Table 2.8 Evolution of O_2 and H_2O line and continuum parameters in the millimeter-wave propagation model of Liebe (1977–1993). The O_2 parameters include line-mixing coefficients

Year of publ.	O_2 parameters	H_2O lines	H_2O continuum
1977	M	O	O
1978	M	–	–
1981	R	O	–
1984	–	O	M
1987	–	O	M
1989	R	O	–
1992	M	–	–
1993	R	O	O

M: new measurements by Liebe and co-workers.
R: revised analysis of earlier measurements by Liebe and co-workers.
O: revision based on measurements by others.
–: no change from previous version.

mutually consistent results for molecules that have numerous lines. The width values actually contained in the large databases come from a variety of sources; in some cases they are merely extrapolated from other lines of the same molecule. Hence, for applications that depend on the precision of line parameters, the modeller needs to trace the source of the values being used. HITRAN and GEISA now provide indices to assist in the lookup of references.

A line-parameter database for selected windows in the 294–626 GHz range was created by Perrin *et al.* (2005), who made an effort to analyse and collect the most reliable values from general-purpose catalogs and the published literature for twenty molecules of interest in a limb-sounding experiment. New measurements of some H_2O, CO, O_3, BrO, ClO, N_2O and HNO_3 lines were also made for this database (Bauer *et al.*, 1998; Demaison *et al.*, 2004).

Table 2.9 compares calculations by Gamache and Fischer (2003) of pressure broadening and shift coefficients for the 183 GHz H_2O line with three recent sets of measurements. The values of the broadening coefficient lie within roughly a 10 per cent range, and the calculated and measured line shifts also show reasonable agreement, considering the difficulty of measuring such small shifts. For air, one would use a (0.79, 0.21) weighted average of the N_2 and O_2 coefficients. Colmont *et al.* (1999) measured N_2- and O_2-pressure-broadening coefficients for the 325 GHz H_2O line, obtaining agreement within 3 per cent with calculated values. However, the relatively small uncertainties associated with their measurement of line widths led to proportionally greater uncertainty in the temperature dependence (represented as T^{-n}) of the widths, e.g. $n = 0.63 \pm 0.22$ for H_2O broadened by N_2. There is a consequent extrapolation uncertainty in modelled absorption when, as in this

Table 2.9 Recent measurements and calculations of the 183 GHz H_2O line half-width and shift, with N_2 and O_2 as perturbing molecules. Values in parentheses are the estimated standard errors in the final digit. HWHM = half-width at half-maximum

	T (K)	N_2-HWHM (MHz/Torr)	N_2-shift (MHz/Torr)	O_2-HWHM (MHz/Torr)	O_2-shift (MHz/Torr)
Bauer *et al.* (1989)	300	4.07(7)		2.54(2)	
Goyette and De Lucia (1990)	300	4.19(15)		2.77(10)	
Golubiatnikov (2005)	302	4.230(7)	−0.092(10)	2.580(5)	−0.097(5)
Gamache and Fischer calculated (2003)	296	4.41	−0.140	2.46	−0.099

Note: 1 Torr = 133.322 Pa.

case, the temperature range of the laboratory measurements does not overlap that of atmospheric temperatures. Pressure-broadening and shift of the 556 GHz H_2O line by N_2 and O_2 was measured by Markov and Krupnov (1995); also, self-broadening and shift parameters were measured by Markov (1994) for the 439 GHz H_2O line and by Podobedov *et al.* (2004) for six strong H_2O lines in the range of 380–1542 GHz.

Tretyakov *et al.* (2005) measured the pressure-broadened widths of oxygen lines near 60 GHz and also obtained revised values for the line-mixing coefficients. Their line widths are 3–4 per cent larger for air broadening than the earlier measurements of corresponding lines by Liebe and Gimmestad (1978) and thus are intermediate between the line-width values used in MPM92 and those in MPM93. Line intensities were also updated to the newer HITRAN values. Neither group of spectroscopists found any measurable pressure-shifting of the O_2 lines, but Tretyakov *et al.* (2004) succeeded in measuring a line-mixing coefficient at 118.75 GHz, where the effect of line mixing is much smaller than in the 60 GHz band. Pressure-broadened widths of the submillimetre lines of O_2 were measured by Golubiatnikov and Krupnov (2003).

2.9.2 Continuum absorption

In the far wings of lines and atmospheric windows, radiative transfer models require a description of what is now known as water-vapour continuum absorption. The continuum is defined as the difference between observed absorption and what can be described by conventional line profiles, such as Van Vleck–Weisskopf. This empirical continuum is dependent on the number of lines considered and their shape, which may or may not incorporate a cutoff; what is not accounted for as line absorption becomes continuum. There is general agreement that the continuum contains two components, one proportional to the square of water-vapour partial pressure and the other proportional to the product of water-vapour and dry-air partial pressures.

However, there are large differences among empirical models for the frequency and temperature dependence of these two components.

Several possible causes of the H_2O continuum have been proposed, among them (1) the inadequacy of analytic line shapes at frequency displacements of hundreds of GHz from the centres of the extremely strong far-infrared lines, (2) a possible spectral contribution from water dimers, clusters of molecules or weakly bound complexes, (3) collision-induced absorption and (4) co-operative absorption by pairs of molecules.

Theoretical work on the continuum has made some progress (e.g. Ma and Tipping, 1990, 2002; Hudis *et al.*, 1991, 1992) but at the present time, theoretical calculations are in the status of being validated by measurements, rather than vice versa, even though the measurements do not present an entirely consistent picture. Working with the far-wing line shape hypothesis for H_2O–H_2O collisions, Ma and Tipping (1990) were able to reproduce the approximate magnitude of the (H_2O vapour)^2 component in MPM89, but the calculated dependence on frequency and temperature was not separable into a product form, as in MPM. At 239 GHz, their calculation was ~15 per cent less than the absorption measured by Bauer *et al.*(1995). For H_2O–N_2 collisions, Ma and Tipping (2002) calculated absorption intermediate between MPM89 and MPM93, with temperature dependence close to that of MPM93.

Bauer and Godon (2001) measured absorption at 239 GHz by H_2O mixed with N_2, CO_2, Ar, CH_4, C_2H_4 and C_2H_6 and found that the H_2O-dependent absorption is proportional to the square root of the collision-induced absorption by the perturber gas alone. They interpret this result as evidence in favour of the collision-induced hypothesis for the H_2O continuum. However, Brown and Tipping's (2003) theoretical calculation of H_2O–N_2 collision-induced absorption in the range 400–2000 cm^{-1} is approximately two orders of magnitude less than the observed continuum, implying that the contribution of induced dipole moments is negligible. These computations have not yet been extended to the microwave range.

Kuhn *et al.* (2002) measured absorption at 350 GHz in pure H_2O vapour and in H_2O mixed with N_2. From these measurements and earlier ones at 153, 213 and 239 GHz, they derived an H_2O continuum model for use with each of several line-parameter databases, with and without cutoff of the line shape function.

Boissoles *et al.* (2003) calculated collision-induced absorption for N_2–N_2, N_2–O_2 and O_2–O_2 molecular pairs for frequencies up to 12 THz (400 cm^{-1}). They obtained an adjustment factor for air versus pure N_2 that was nearly constant at 1.34 up to ~3.6 THz, close to Pardo *et al.*'s (2001a) experimental value of 1.29, which was derived from observations on Mauna Kea, HI.

Calculation of collision-induced absorption is complicated, but Figure 2.3 shows that its frequency dependence can be approximated, up to ~2 THz, by an analytic function. The data points in the figure plot absorption divided by (frequency)^2, calculated from the model of Borosov and Frommhold (1986) for N_2. Including the factor of 1.34 to account for O_2, collision-induced absorption by dry air can be calculated as

$$\gamma_a(\nu) = 8.7 \times 10^{-14} p_{\text{dry}}^2 \nu^2 \left[0.5 + \frac{0.5}{1 + (\nu/450)^2}\right] (300/T)^{3.6} \tag{2.6}$$

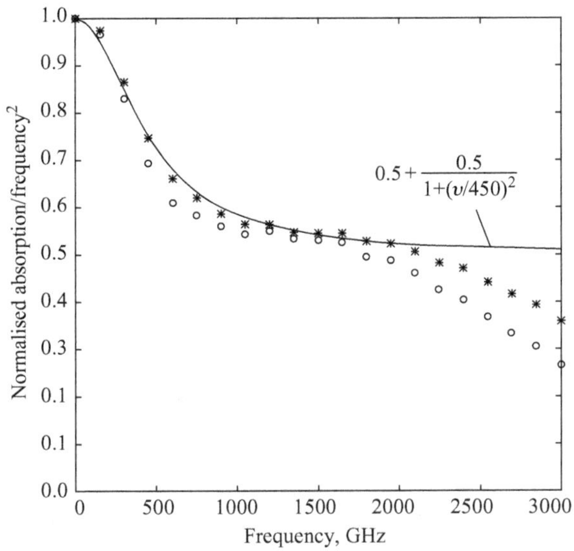

Figure 2.3 Absorption in pure N_2, divided by (frequency)^2 and normalised. Asterisks – 296 K; circles – 220 K

where $\gamma_a(\nu)$ is the absorption coefficient in km^{-1}, ν is frequency in GHz (< 2000), p_{dry} is the pressure of the dry-air components in hPa and T is temperature in kelvin.

2.10 Modelling and validation issues

The 22 GHz line of H_2O in the atmosphere is less than opaque at its centre. During the 1990s, several experiments that compared ground-based radiometer measurements in the vicinity of this line with calculations from radiosondes showed that the line appeared to be stronger than calculated. Some modellers (Liebe *et al.*, 1993; Cruz Pol *et al.*, 1998) adjusted the line intensity to reproduce these measured brightness temperatures. Later, it was determined that on a type of radiosonde widely used during this period, the humidity sensor absorbed contamination from the sonde package material during storage and as a result yielded low-biased relative humidities (Wang *et al.*, 2002). Turner *et al.* (2003) found that correction of radiosonde humidities for this bias resulted in close agreement in the mean with radiometric determination of total precipitable water vapour (PWV), although there remained significant variations of radiosondes within calibration batches. Recent experiments (Cimini *et al.*, 2003) have shown that radiative-transfer calculations using radiosondes of more modern design agree more closely with theory (R98 model). Soden *et al.* (2004), using satellite-based infrared and ground-based lidar measurements, found that both types of radiosonde underestimated humidity at pressures less than 500 hPa, even when known corrections were applied to the older type; however, even if this is true, the results of Cimini *et al.* and Turner *et al.* can be explained by the fact that the upper part of the atmosphere contributes very little to brightness temperatures observed at the ground.

Gamache *et al.* (1994) surveyed and compared ∼4000 published infrared and microwave measurements of H_2O line widths broadened by various perturbing gases. They concluded that experimentalists frequently underestimate uncertainties in measurements, sometimes ignoring the possibility of systematic errors. They estimate that as a general rule, most H_2O line widths are uncertain by 10–15 per cent (2 standard deviations) for broadening by dry air components and 20–25 per cent for self-broadening. Although this rather pessimistic conclusion may be justified as a general rule, it does not consider that measurement techniques tend to improve over time. Gamache *et al.* also identified lines, among them those at 22 and 380 GHz, for which the uncertainties were evaluated as less than 5 per cent.

Although H_2O line and continuum absorption are generally treated separately in models, the derivation of the continuum is linked to a line absorption model. The continuum is really an empirical correction added to the line model to obtain agreement with some particular set of measurements. As a consequence of this re-fitting to data involved in derivation of the continuum, errors in the combination of lines and continuum at window frequencies, where the continuum is dominant, result from the data used and the fitting procedure rather than from the line parameters. The question of cause of the H_2O continuum does not affect the accuracy of these empirical models, except that they usually represent the frequency and temperature dependence of the continuum by separate functions, which may be an oversimplification.

Uncertainties in line and continuum parameters affect calculated brightness temperatures in various ways, depending on frequency, atmospheric state and conditions of observations. For example, an increase in a water line width by x per cent will reduce its opacity at line centre by x per cent but increase the line's opacity in its far wings by x per cent. The latter opacity may be smaller than the continuum contribution, however. Although the greatest sensitivity of upwelling brightness temperature to H_2O opacity occurs in the semi-transparent parts of the spectrum and over an ocean surface, it is also true that for sounding applications, an error in opacity per unit of water translates into an equivalent and opposite error in inferred water vapour at the corresponding sounding altitude.

Oxygen line-mixing coefficients in MPM92 were determined by adjusting them to fit the difference between measurements and absorption calculated using the Van Vleck–Weisskopf line shape with previously-measured line widths. MPM92 was based on a much larger set of laboratory measurements than previous versions of MPM. The measurements were estimated to have an accuracy of ±2 per cent rms for the higher rates of absorption, and in general MPM92 fits the laboratory data to within the experimental scatter; systematic departures of the model from the measurements (by as much as 20 per cent at 50 GHz, though there is considerable scatter in the data at this frequency) are visible in Figures 5 and 6 of Liebe *et al.* (1992), but only at the highest pressure (1013 hPa) and lowest temperature (6°C), consequently the highest density, measured. Some of the atmospheric measurements described in the next section address the question of how well the MPM92 model extrapolates to temperatures lower than 6°C and frequencies outside of 50–70 GHz. In MPM93, the line widths between 50 and 70 GHz were adjusted upward by 5 per cent, with a compensating adjustment of the mixing coefficients; this change slightly improved the fit of the model at low pressures.

There are some anomalies in submillimetre line parameters in MPM. The intensities of H_2O lines at 548, 552 and 923 GHz in MPM93 are stronger than HITRAN values by factors of 10, 82 and 12, respectively, for unknown reasons. In MPM85-93, the intensity of the 774 GHz O_2 line is 17 per cent stronger than in HITRAN, and the temperature dependence of the submillimetre O_2 line widths is $T^{-0.2}$ at constant pressure, compared to $T^{-0.8}$ for the millimetre lines. This temperature dependence was however based on a measurement of the 425 GHz line width by Pickett *et al.* (1981).

2.11 Comparisons of model predictions with atmospheric measurements

Atmospheric absorption models are based on theoretical calculations, approximations and assumptions and laboratory measurements. Laboratory measurements allow the state of a gas to be varied in a controlled way. That makes it easier to isolate the influence of each model input, which is often difficult in nature. As the sampling can be controlled, it is also much easier to ensure that observations of model inputs and outputs are representative in a laboratory environment, though it may not be possible to explore the full range of parameter space that occurs in nature.

Validation experiments, on the other hand, check the predictions of these models against real-world observations. An ideal validation experiment would investigate the model's systematic biases, estimate its accuracy and also check whether it is capable of representing the full range of observed conditions. Validation can be part of an iterative cycle of model development, allowing refinement of certain model parameters, if the validation experiment is designed so as to be able to isolate each mechanism represented in the model.

A well-designed experiment includes *in situ* measurements of all model inputs, together with estimates of their uncertainty. Spatial scales should be sampled at sufficient resolution and coverage to ensure that the observations of model inputs accurately represent them over their full range. The uncertainty introduced by spatial representativeness sometimes dominates the error budget of a validation experiment. The uncertainty in each component of the validation should be analysed – both in terms of random uncertainties, which may be reduced by averaging, and systematic biases, which may not.

2.11.1 Ground-based radiometers

Ground-based radiometer or spectrometer measurements can be used for validation of models, given the *in situ* observations of the model's input variables. When radiosondes provide the *in situ* data, clear sky conditions must be selected as the measurements

*Table 2.10 Calculated minus measured brightness-temperature differences (K) at Wallops Island, VA from England et al. (1993)**

Frequency (GHz)	PWV	Average T_B (K)	MPM89	W76
20.7	Low	14.1	+0.38	+0.80
20.7	High	30.0	+0.15	+1.00
22.2	Low	18.8	+0.56	−0.49
22.2	High	45.0	−0.76	−4.30
31.4	Low	13.5	−0.33	+0.31
31.4	High	20.0	−1.10	+0.22

Low PWV: <1 cm; High PWV: 2.1 cm.
*The original publications present differences with the opposite sign, i.e. measured minus calculated.

are also sensitive to emission from cloud liquid water, which is not measured by the radiosonde sensors. If long-term datasets are available it is possible to validate a model over a wide range of inputs at a single site. Given that both radiometric and *in situ* observations sample small spatial scales, it is usually not difficult to ensure that the *in situ* data are representative of the scene observed by the radiometer.

England *et al.* (1993) compared ground-based radiometer measurements from Wallops Island, VA at 20.7, 22.2 and 31.4 GHz with brightness temperatures computed from radiosonde profiles using either MPM89 or the model of Waters (1976), designated W76, for H_2O absorption. In both cases O_2 absorption was calculated from Rosenkranz (1988). The differences between calculated and measured brightness temperatures are listed in Table 2.10. However, the combined uncertainties of radiometers and radiosondes were larger than the differences between absorption models. The radiosondes used for the experiment carried instrument packages made by three different suppliers, and the authors noted significant differences in the relative humidity profiles from different sensors.

Westwater *et al.* (2003) measured atmospheric emission at 23.8 and 31.4 GHz with radiometers on the island of Nauru (tropical western Pacific) and compared the data with brightness temperatures calculated from radiosondes using MPM87, MPM93 and R98. They applied corrections to the radiosonde humidity measurements which, depending on age of the radiosonde, were as large as 10 per cent. The authors concluded, however, that factors other than age might influence the radiosonde accuracy. After those corrections, they obtained the mean differences listed in Table 2.11.

Marchand *et al.* (2003) used a different approach to correct radiosonde humidity; they multiplied the moisture profile by a factor so as to bring the calculated brightness temperature at 23.8 GHz into agreement with measured brightness temperature. Brightness temperature at 31.4 GHz was then calculated from the adjusted profile for comparison to measurements at that frequency. In effect, this is a test of model

Table 2.11 Calculated minus measured brightness-temperature differences (K) at Nauru Island (tropical western Pacific) from Westwater et al. (2003). Average PWV = 4.7 cm. The values in parentheses are 99 per cent confidence intervals for the final digit

Frequency (GHz)	Average $T_B(K)$	MPM87	MPM93	R98
23.8	65	+0.80 (67)	+3.90 (70)	+0.69 (67)
31.4	32	−0.16 (32)	+3.37 (36)	+0.86 (33)

Table 2.12 Calculated minus measured 31.4 GHz brightness-temperature differences (K) at three sites, from Marchand et al. (2003)

Location	Average PWV (cm)	MPM87	R98	MonoRTM
Nauru	4.0	−0.35	+0.65	−0.51
Oklahoma	1.0	−0.04	+0.42	+0.05
Alaska	0.7	−0.79	−0.34	−0.84

consistency between the two frequencies, relative to the vertical distribution of moisture determined by the radiosondes. Table 2.12 lists the brightness temperature differences obtained from MPM87, R98 and MonoRTM at three field sites of the Atmospheric Radiation Measurement (ARM) program. Some sign errors in the Nauru differences have been corrected (private communication, R. Marchand, 2004). Radiometer calibration accuracy was estimated to be ±0.3 K. The authors also noted that differences between individual components of the models (i.e. O_2 absorption, H_2O line absorption, H_2O continuum) were often larger than the differences between total absorption predicted by the three models.

Mattioli *et al.* (2005) compared 23.8 and 31.4 GHz radiometer measurements made at the ARM site in Oklahoma with calculations using the MPM87, MPM93 and R98 models, and also a model by Liljegren *et al.* (2005), denoted L04, which uses the MT-CKD water continuum (Clough *et al.*, 2005) and in which the 22 GHz line width was reduced by 5 per cent compared with R98. A recent update of R98 was also tested, which yielded results very close to R98 at these two frequencies. The *in situ* atmospheric measurements were obtained from radiosondes of a new design that do not suffer from the dry bias described by Wang *et al.* (2002). Three different radiometers and two radiometric calibration methods were employed. Brightness temperatures covered the range of 12 to 42 K at 23.8 GHz and 11 to 22 K at 31.4 GHz. Table 2.13 lists the mean differences of calculated minus measured brightness temperatures for the C1 radiometer with ARM calibration, followed in brackets by the minimum and maximum values of the difference over the six combinations of radiometer and calibration method. Generally, the C1/ARM value is close to the centre of the ensemble of measurements. Comparison of the 31 GHz results in Table 2.13 with the ones from the same site in Table 2.12 shows that the predictions of R98 are ~0.5 K higher than MPM87 in both cases, but the measurements of Marchand *et al.* are closer to the MPM87 model, while the measurements of Mattioli *et al.* are closer to the R98 model at this frequency. These two papers used different methods to determine water vapour, but the disagreement may also reflect the practically achievable calibration accuracy of radiometric measurements. Table 2.14 lists the slopes obtained by Mattioli *et al.* in a regression of the calculated brightness temperatures against the C1/ARM measurements. Most of the variability is due to water vapour, so the deviation from unity in slope of the regression line tests the H_2O component of the models.

If we estimate atmospheric emission at a window frequency as the measured downwelling brightness temperature minus 2.7 K from the cosmic background, then the comparisons in Tables 2.10–2.13 indicate that the discrepancies with models are, with some exceptions (e.g. MPM93 tends to over-predict H_2O absorption), often within a few per cent in the 20 to 30 GHz range. There are cases where agreement is within a fraction of a per cent, though not consistently.

Keihm *et al.* (2002) utilised GPS measurements of the atmospheric time delay at the Goldstone Tracking Station in California and the ARM site in Oklahoma for model validation instead of *in situ* atmospheric measurements. The time delay was separated into a hydrostatic part that can be determined directly from surface pressure and a second part termed 'wet delay' which is related to the atmospheric water-vapour profile. Downwelling brightness temperatures measured at 20.7, 22.2, 23.8 and 31.4 GHz were converted to atmospheric opacities using an effective radiating temperature. The measured validation data was then obtained as the slope of the regression line of opacity versus wet delay. This measured quantity was compared with a regression line

Table 2.13 Calculated minus measured brightness-temperature differences (K) at the Oklahoma ARM site, from Mattioli et al. (2005). Values within brackets indicate the range of differences*

Frequency (GHz)	R98	L04	MPM87	MPM93
23.8	−0.44 [−0.75,+0.03]	+0.07 [−0.25,+0.55]	−0.60 [−0.91,−0.12]	+0.65 [+0.33,+1.13]
31.4	−0.09 [−0.27,+0.005]	+0.11 [−0.06,+0.21]	−0.64 [−0.81,−0.54]	+0.77 [+0.59,+0.86]

*The original publications presented differences with the opposite sign, i.e. measured minus calculated.

obtained from calculations based on a database of radiosonde profiles for each of the two sites, using a refraction model for time delay and the MPM87 and R98 models for absorption, as well as two other models with line and continuum parameters that had previously been adjusted to fit atmospheric measurements: one from Cruz-Pol *et al.* (1998), denoted C98 here, and the JPL model (Keihm *et al.*, 1995). The comparison, shown in Table 2.15 as percentage differences in the slopes of the two regressions, is thus between statistics of two ensembles rather than between elements of a single ensemble, as in the validation experiments previously discussed. Hence, there are uncertainties associated both with the measurement of correlation between opacity and wet delay and with the statistical representativeness of the profiles used for the model calculations. The estimates by Keihm *et al.* of those two uncertainties are given in the last two columns of Table 2.15. These results indicate that both MPM87 and R98 under-predict H_2O absorption at all four frequencies, whereas Mattioli *et al.* (2005) found slopes greater than unity (see Table 2.14), implying over-prediction of H_2O absorption by amounts ranging from 0.5 to 3 per cent for these two models.

Table 2.14 Slope of modelled versus measured brightness temperatures at Oklahoma ARM site, from Mattioli et al. (2005)

Frequency (GHz)	R98	L04	MPM87	MPM93
23.8	1.0059	1.0247	1.0174	1.0616
31.4	1.0364	1.0079	1.0239	1.1519

Table 2.15 Modelled minus measured slopes of opacity versus wet delay, as percentages of measured slope at two sites, from Keihm et al. (2002). Uncertainties (%) in measurement and model calculations are also given

Frequency (GHz)	Location	JPL	C98	MPM87	R98	Meas. error	Modelling error
20.7	Oklahoma	+3.2	+1.7	−2.0	−3.3	2.1	0.4
20.7	California	0.0	−1.0	−5.0	−6.5	1.2	0.4
22.2	Oklahoma	+3.3	−1.8	−1.9	−4.1	1.6	3.5
22.2	California	+3.2	−1.9	−1.9	−4.2	2.2	1.1
23.8	Oklahoma	+3.4	+2.0	−1.9	−3.2	0.7	0.4
31.4	Oklahoma	+1.5	+1.5	−4.9	−4.9	3.2	4.3
31.4	California	−2.0	−1.6	−8.5	−9.0	3.3	4.6

These different conclusions probably result from the differences in technique of the two papers as well as the known statistical uncertainties. Mattioli *et al.* also intercompared PWV derived from radiosondes, GPS and the microwave radiometers (using inversion techniques based on the absorption models). They found biases of a few hundredths of a cm between the three techniques of measurement, which would correspond to a few per cent in the Oklahoma climate.

Hewison *et al.* (2003) used radiosondes and radiative transfer models to validate coincident brightness temperatures measured at twelve frequencies from 22.2 to 58.8 GHz by a ground-based microwave radiometer at Camborne, England. Although the objective was validation of radiometer measurements, the comparison is also sensitive to bias in the models and radiosondes. A bias in channels near the 22 GHz water vapour line was found to be partly explained by a dry bias in the radiosonde sensors. Other biases in the channels close to the centre of the 60 GHz oxygen complex were identified as a calibration problem in the radiometer, allowing this to be rectified. Work is now in progress to analyse the channels between 51 and 53 GHz, which showed biases that depended on brightness temperature and varied from model to model.

Other validation work in progress includes ground-based experiments at 94 and 151 GHz in Switzerland. The 94 GHz measurements were made with a transmitter–receiver link (Martin *et al.*, 2000). At 151 GHz, downwelling atmospheric emission was measured (L. Martin, private communication).

2.11.2 Ground-based FTS

Extensive studies of the atmospheric transmission spectrum as seen from the ground at millimetre and submillimetre wavelengths have been performed with Fourier transform spectrometers (FTS) mounted at high mountain-top sites, potentially interesting for submillimetre astronomy. The earliest attempts were made at 2400 metres above sea level in Tenerife by Hills *et al.* (1978). In recent years, FTS experiments have been mounted at Chajnantor, Chile between 4800 and 5500 m above sea level (see Matsushita *et al.,* 1999; Paine and Blundell, 1999), Mauna Kea, HI at 4200 m above sea level (Serabyn *et al.,* 1998) and the geographic South Pole at 2900 m elevation (Chamberlin *et al.,* 2003). These FTS experiments differ in frequency coverage, resolution and calibration procedures but together have provided valuable data sets for validating and refining atmospheric absorption models from ~0.1 to 2 THz, at least for the conditions of the high mountain-top sites. Figure 2.4 shows some of the best spectra obtained with the FTS installed at Mauna Kea. More information can be found at www.damir.iem.csic.es/PARDO/latest_fts.html.

Pardo *et al.* (2001a) and Pardo *et al.* (2005a) analysed the line shapes and continuum terms. The results published in Pardo *et al.* (2001a) advise the use of ν^2 laws for both the dry and H_2O continuum terms, with coefficients determined in that work from FTS observations under very dry tropospheric conditions extending from 300 to 1100 GHz. Recently, Pardo *et al.* (2005a) have noticed that the H_2O continuum above 1 THz may be following a ν^x law with $x < 2$, which would agree with Ma and Tipping (2002). In the same work the agreement of the measured dry continuum with the result of Boissoles *et al.* (2003) has been well established for

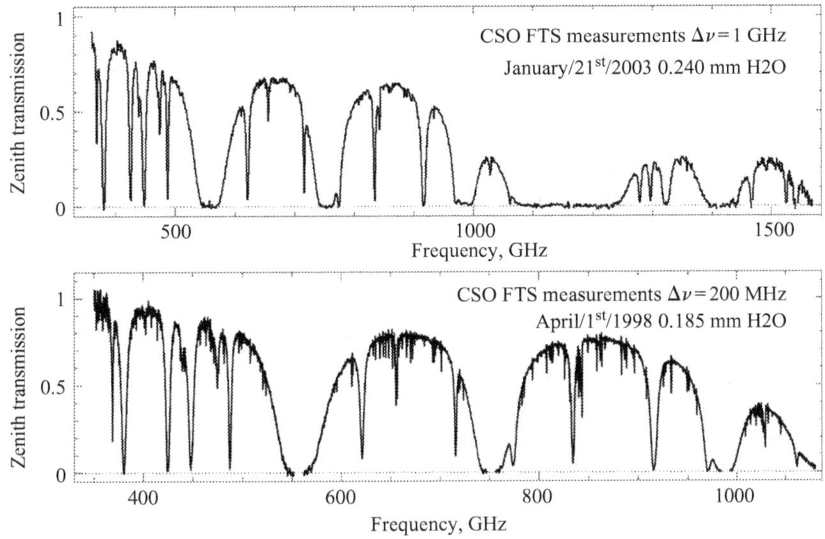

Figure 2.4 *Atmospheric transmission data obtained at Mauna Kea under the driest conditions. Windows centred at ~ 1.04, 1.34 and 1.48 THz are visible. The resolution of the 4/1/98 spectrum (200 MHz) allows separation of the weak and narrow O_3 lines, which get diluted in the 1/21/03 spectrum (1 GHz resolution). These data have allowed separation of the wet- and dry-continuum absorption from the line-by-line absorption at submillimetre wavelengths and therefore their comparison with theory (see Section 2.6 and Pardo et al., 2001a, 2005a)*

the full 300–1600 GHz range: The N_2–N_2 collision-induced model of Borisow and Frommhold (1986) has to be multiplied by a factor 1.34 to reproduce all dry collision-induced absorption mechanisms operating in our atmosphere (N_2–N_2, O_2–O_2 and O_2–N_2).

2.11.3 Airborne radiometers

A radiometer on a research aircraft offers the potential to simultaneously sample the radiative and thermodynamic properties of the atmosphere. With careful experiment design, this can ensure that errors of representativeness are minimised. For example, the aircraft can provide a profile of pressure, temperature and humidity for input to a radiative-transfer model. These calculations can be compared with brightness temperatures measured in clear skies during level runs to validate the absorption models. This is akin to comparing a ground-based radiometer with model calculations from radiosondes, although the analysis can be repeated at different levels in the atmosphere. Errors of representativeness can be further reduced by comparing brightness temperature measurements made *during* an aircraft profile with *in situ* measurements. However, this requires careful control of the radiometer's calibration, which can change with temperature during aircraft ascents/descents.

Measurements of downward-propagating brightness temperatures during ascents and descents of a high-altitude aircraft were made by Schwartz (1998) with two radiometers, one tunable over the range 52.5 to 55.8 GHz and the other having eight fixed double-sideband channels at 0.35 to 2 GHz from the 118.75 GHz O_2 line. The measurements were compared with calculations based on radiosonde profiles using either MPM92 or MPM93. In the 52.5–55.8 GHz band, the measurements agreed within the observational uncertainty of 2 to 5 K with both models, which were indistinguishable at that level of precision. The observational uncertainty corresponds to ~2 per cent of zenith atmospheric opacity for the more transparent channels below 54 GHz. Near 118 GHz, discrepancies from the models were as large as 4 K, with uncertainty ±1.5 K, leading to the conclusion that the width of the 1- line is larger by 5 to 6 per cent in the upper troposphere and stratosphere than given by MPM92 or MPM93. (Both models have the same width coefficients for this line.) That result could be obtained either by increasing the line width by a constant factor at all temperatures or by increasing its temperature exponent from $n = 0.8$ to 0.97. It should be noted that the later laboratory measurements of Tretyakov *et al.* (2004) yielded a room-temperature width 2 per cent higher than the value in MPM92/93, which would leave 3 to 4 per cent of Schwartz's recommended increase to be obtained from temperature dependence.

English *et al.* (1994) used an airborne radiometer to test MPM89 and three older H_2O continuum models in the windows at 89 and 157 GHz. They found that all four of the models underestimated downwelling brightness temperatures in the tropics. In subarctic conditions atmospheric emission was overestimated. They proposed a modification to the temperature dependence of the water vapour continuum in MPM89 to match their observations. An improved version of this radiometer was flown by Hewison (unpublished data) to measure downwelling brightness temperatures at 89, 157, 183 ± 1, 183 ± 3 and 183 ± 7 GHz. Flights took place at various locations with climates ranging from arctic to tropical. Several absorption models are being compared to the measurements; results indicate negative calculation–measurement discrepancies in the window channels for very cold, dry conditions.

2.11.4 *Satellite-based radiometers*

Satellite observations of model predictands are able to represent their global range. However, in practice, careful data selection is needed to isolate the variable of interest from other factors which may also influence the scene radiance. Usually it is necessary to make some correction for antenna sidelobes, which may receive energy from a different part of the earth or from space. The need for *in situ* measurements of model input variables also imposes stringent selection criteria. This may reduce a huge global dataset to a small number of satellite observations co-located with *in situ* observations, e.g. radiosondes and buoys. Co-location is needed both in space and time. With polar-orbiting satellites, this can drastically reduce the data available for comparison. The principal problem in using satellite data to validate radiative-transfer models with *in situ* data is that these observations sample different spatial scales. The co-location criteria must be chosen by consideration of the spatial and temporal variability of the parameters in question. To minimise errors due to representativeness, it is often necessary to impose criteria on the observations – both spatially for satellite data and temporally for *in situ* data.

Another way to exploit satellite data for model validation is to compare it with numerical weather prediction (NWP) analysis fields in lieu of *in situ* data. Such fields are, by definition, the near-optimal analysis of all current and past observations. They have the advantage of global availability and often lower errors of representativeness, as they are sampled on scales similar to the resolution of space-borne microwave radiometers (tens of km). The NWP field is constructed to be consistent with a diverse set of current and past observations. It is inevitable that this technique includes the observations' and the NWP model's biases as well as those of the radiative-transfer model. Thus, if the observation and analysis biases are unknown or not corrected as part of the assimilation process, they will affect the comparison in a way that may vary with location and time.

The 'observation-background' residuals can be monitored as part of the variational assimilation (Rodgers, 2000) of observations into NWP. This process constructs a best estimate of the model state that is consistent with its own short-range forecast and the observations, taking into account the assumed error characteristics of both. These residuals often show systematic biases, which can be further analysed in multivariate space, e.g. by temperature, humidity, pressure, wind speed, scan angle, etc. This technique is generally believed to be the most effective validation of the consistency of radiometric observations and radiative transfer models. It implies the need to involve NWP centres as partners in the calibration/validation for new instruments. The availability of satellite data often comes late in the development of the model needed to retrieve geophysical parameters from the data, so this technique is most applicable to iterative model development. It is also possible to use NWP analysis fields with ground-based or aircraft-based measurements, but this would re-introduce the problem of representativeness due to different sampling scales.

Meissner and Wentz (2003) tested the R98 and MonoRTM models with satellite-based measurements made by the Special Sensor Microwave Imager. PWV, wind speed and cloud liquid water were first retrieved from the 19, 22 and 37 GHz channels using the algorithm of Wentz (1997). The measurements were screened to find low values of liquid water and wind speed. Then sea surface temperature and water-vapour profiles were obtained from analyses of the National Centre for Environmental Prediction, and the water-vapour density was scaled to match the satellite measurement. Brightness temperatures were calculated for comparison with the 85 GHz measurements, which had not been used for the retrieval of atmospheric parameters, using either R98 or MonoRTM for H_2O and MPM92 for O_2 absorption. The calculation is also dependent on a surface emissivity model but with the best choice for that component, calculated-measured differences at 85.5 GHz were within ± 1.2 K for both models, over the range of 0 to 6 cm PWV. At a given value of PWV, the two models could yield results differing by as much as 1 K. However, both H_2O models were considered to be validated within the estimated uncertainty of 3.3 K in each PWV interval. At this frequency and the 53 degree viewing angle of the instrument, the sensitivity of vertically-polarised brightness temperature to PWV is approximately 1 K/mm.

Pumphrey and Buehler (2000) examined measurements made by two satellite limb-sounding instruments, the microwave limb sounder (MLS) and the millimetre-wave atmospheric sounder (MAS). They found that allowing their retrieval algorithm to solve for a pressure-proportional frequency shift of the 183 GHz H_2O line produced improved agreement with the measurements. Their retrieved values for the shift parameter were equal to -0.19 ± 0.04 MHz/Torr from MLS and -0.27 ± 0.05 MHz/Torr from MAS. These are larger than the laboratory and theoretical values in Table 2.9; however, considering other uncertainties associated with the satellite measurements, this result should be interpreted as consistent with the existence of and the need to model the frequency shift.

Buehler *et al.* (2005a) used data from the up-looking submillimetre wave radiometer ASUR to estimate the pressure shift parameter of the HCl line cluster near 626 GHz. Depending on the assumed temperature dependence of the shift, the HCl pressure shift value consistent with the ASUR data is 0.120 to 0.156 MHz/Torr, instead of the 0.040 MHz/Torr reported in HITRAN. This result is in good agreement with very recent independent laboratory work by Drouin (2003), which suggests a value of 0.147 MHz/Torr for the shift.

Buehler *et al.* (2004) compared brightness temperatures from the 183.3 ± 1, ± 3 and ± 7-GHz channels of the Advanced Microwave Sounding Unit (AMSU-B) with calculations based on radiosonde profiles using ARTS with the R98 absorption model. Comparisons were only made for satellite data within 30 min of the nominal time for radiosondes launched from Lindenberg, Germany and where the sonde drifted by less than 50 km before reaching the tropopause. They also filtered the data to reject dense ice cloud. Brightness temperatures in the three AMSU channels around the water vapour line depend to a good approximation only on atmospheric temperature and humidity; therefore the radiosonde gives all necessary input data. The investigated channels are so close to the line that continuum does not play an important role, resulting in a very simple absorption calculation. The comparison yielded agreement for the two outer channels, but for the channel closest to the line, which is sensitive to the atmospheric state in the upper troposphere, a scatter plot of modelled versus measured brightness temperature showed a slope deviating slightly from unity. The error analysis included contributions from the radiometers, *in situ* observations, radiative-transfer model and uncertainty in the representativeness of the two observations. In the case of the radiative transfer model, the authors tried to reproduce the slope by changing various spectroscopic parameters. They concluded that the most likely explanation for the deviation is a dry bias of the radiosonde humidity sensors under very dry conditions (radiosonde giving 0 per cent relative humidity when the true relative humidity is still 4 per cent), even after correction for the previously known biases. Subsequently, this has been suggested as a method of evaluating the accuracy of different radiosonde observations.

Rosenkranz and Barnet (2006) compared measurements made by the Humidity Sounder for Brazil (HSB), on the Aqua satellite, with calculations based on special radiosondes launched from the ARM-TWP and SGP sites just before and after the satellite overpass. Profiles from the two sondes were interpolated for the calculation of brightness temperatures with the R98 absorption model. These radiosondes were of the newer design which is not believed to have a dry bias. Also employed, at a variety of sites, were research radiosondes with a chilled-mirror humidity sensor, which is considered to be the most accurate instrumentation for measurement of humidity. The HSB has channels at 150, 183.3 ± 1, 183.3 ± 3 and 183.3 ± 7 GHz; mean differences between calculations and measurements ranged from 0.1 to 0.8 K. The calculations based on the chilled-mirror sondes were between 0.1 and 0.4 K. For this instrument, no corrections were made for antenna sidelobes, which could account for the measurements being slightly colder than the calculated brightness temperatures. An error of 5 per cent in absorption would correspond to brightness-temperature errors of ~0.5 K for these channels.

In the oxygen band, Rosenkranz (2003) compared measurements from the Advanced Microwave Sounding Unit on the NOAA-15 satellite with brightness temperatures computed from radiosondes using a rapid algorithm based on MPM92, for six channels centred at 52.8, 53.6, 54.4, 54.94, 57.29 and 57.29 ± 0.217 GHz. The measurements were corrected for contributions from sidelobes that miss the earth, although some residual asymmetries indicated that the correction was not perfect. Mean calculated minus measured differences ranged from −0.23 to +0.42 K; all of them were smaller than the instrument's specified 1.5 K accuracy of calibration. Of these channels, 53.6, 54.4 and 54.94 GHz are the most sensitive to atmospheric opacity; an error of 5 per cent in the absorption calculation would have caused brightness temperature errors of 0.7 to 0.9 K. The other channels are less sensitive because of the temperature structure of the atmosphere.

Further validation experiments using aircraft and satellites are described by Wang (2002), Meissner and Wentz (2004) and Ellison *et al.* (2003). Although they were primarily designed to compare water-surface emissivity models, the calculations include modelling of atmospheric absorption and hence test the complete radiative-transfer models, in many cases obtaining agreement of a few degrees Kelvin or less.

2.12 Conclusions and recommendations for future development of models and databases

A careful inspection of the comparisons cited in the previous section does not lead one to identify one particular model as best for all frequencies and atmospheric conditions. More comprehensive testing of models at multiple frequencies and in varied atmospheric conditions is desirable. However, for all measurements in the atmosphere, it is essential to place emphasis on obtaining accurate *in-situ* data for comparison. Especially for water vapour, this is a problem of non-negligible difficulty. Establishment of additional sites with high-accuracy instrumentation in different climates would be useful, particularly in connection with satellite experiments. For the latter, radiosonde launches should be timed to coincide with satellite overpasses, or slightly earlier to allow for rise time of the sonde. Efforts should be made to distinguish between errors in water-vapour absorption and in dry-air absorption when model calculations are compared with measured brightness temperatures. For example, calculated-measured differences at window frequencies could be plotted versus PWV. The regression-line intercept would be related to the dry-air component of absorption, although it would be necessary to use care in interpreting such a regression, because PWV is correlated with temperature. Involvement of forecasting centres for variational assimilation of observations into NWP models can also be helpful for this purpose.

The most successful theoretical work on the H_2O continuum has calculated absorption due to the permanent dipole moment of H_2O molecules perturbed by collisions, as opposed to permanently-bound dimers or induced dipoles. The calculated temperature and frequency dependence of the continuum, both for pure vapour and for mixtures with foreign gases, should be incorporated into models and tested by comparison with laboratory and atmospheric measurements.

The new generation of atmospheric remote-sensing satellite instruments requires unprecedentedly high accuracies of the molecular spectroscopic parameters. HITRAN's progress in the future will also require the determination of line parameters for systems of bands in the near-infrared, extending the applicability of the parameters at all wavelengths to high temperatures (often tantamount to archiving atmospheric weak lines), addition of collision-induced absorption bands, characterisation of line mixing for more molecules and addition of new molecules that are important for terrestrial, planetary and astrophysical applications.

The pressure-broadened line parameters of greatest utility for atmospheric applications are (1) half-widths and shifts for dry air as a perturber (i.e. weighted sum of N_2 and O_2), (2) half-widths and shifts for H_2O as a perturber and (3) line-mixing coefficients for the 50–70 GHz O_2 band, for both dry air and H_2O as perturbers. Self-broadened half-widths are unnecessary for gases other than H_2O.

Many researchers throughout the world are currently making efforts to ensure the quality of the spectroscopic parameters which are needed as input to atmospheric remote sensing investigation. Several studies, partially overlapping, partially complementary, are currently being undertaken. The regularly updated and evolving spectroscopic databases, such as GEISA, still have their limitations and faults, which have to be corrected or improved upon, in order to meet the requirements of a diverse group of users.

As an example of actual improvement needs, the following list, established in the framework of the ISSWG (IASI Science Sounding Working Group) activities, provides the IASI soundings spectroscopy enhancement requirements.

Quality control of spectroscopic parameters for the most abundant species

- Need to systematically validate the existing databases (HITRAN, GEISA) for the stronger absorbers: H_2O, CO_2, CH_4, O_3, N_2O.

- Intercomparison of existing databases is not sufficient. Comparison with real atmospheric spectra recorded in well-documented conditions is necessary to locate deficiencies.
- Patches and add-ons can degrade the database consistency (duplicates or missing lines).
- Existing and new laboratory data should be combined in a consistent manner using suitable theoretical models for the calculation of line positions, intensities and widths.
- Coding of uncertainties by an index is insufficient for proper use in the forward model variance–covariance matrix. Better and more realistic errors on spectroscopic parameters are required.

Urgent needs

- Water vapour continuum studies (foreign and self) and temperature dependence.
- Perform long path laboratory measurements.
- Perform real atmospheric measurements along horizontal paths in well-documented conditions.
- Improve existing parametrisations.
- Elaborate new theoretical models.
- Correct known inadequacies observed when modelling high resolution atmospheric spectra (H_2O, CO_2, CH_4) recorded from the ground, balloon or space.

Specific recommendations

H_2O

- Weak lines in the regions 650–1200 cm^{-1} and 3000–6000 cm^{-1} especially, still have incorrect positions and intensities. There is a need for both measurements and improved theoretical calculations.
- Widths (foreign and self) have not been validated by enough systematic laboratory work. Very strong variations from line to line exist.
- Temperature dependences of widths are poorly measured or calculated.
- Pressure shifts can be important in the lower troposphere.
- Some lines recorded at high resolution in ground-based solar spectra do show profiles inconsistent with current line shape formalisms.
- More detailed line shape measurements and modelling needed for H_2O (in connection with continuum studies).

CO_2

- Validation of line intensities for the isotopomers in ^{13}C, ^{18}O and ^{17}O needed.
- New line shape studies needed in the 15 μm band (laboratory and theoretical).

O_3

- For lower stratospheric and tropospheric ozone retrievals a better knowledge of the temperature dependence of the line widths is needed.

CH_4

- Intensive laboratory and theoretical work still needed for air-broadened widths.
- Line shape studies needed including line interference effects in multiplets and Q branches.

HNO_3

Work on absolute line intensities is still needed.

Heavy molecules described in terms of cross-section instead of line by line parameters

- Establish a priority list with the known scenarios for man-made species (CFCs, HCFCs, perfluorinated compounds) based on existing cross-section data.
- Perform new measurements and establish smooth parametrisations (as a function of temperature and total pressure) to avoid the use of empirical extrapolations of accurately measured cross-sections as substitutes for unavailable data.

References

ACEF, O., BORDÉ, C. J., CLAIRON, A., PIERRE, G., and SARTAKOV, B.: 'New accurate fit of an extended set of saturation data for the ν_3 band of SF_6: Comparison of hamiltonians in the spherical and cubic tensor formalisms', *J. Molec. Spectrosc.*, 2000, **199**, pp. 188–204

BALLARD, J., JOHNSTON, W. B., GUNSON, M. R., and WASSELL, P. T.: 'Absolute absorption coefficients of $ClONO_2$ infrared bands at stratospheric temperatures', *J. Geophys. Res.*, 1988, **93**, pp. 1659–65

BALLARD, J., KNIGHT, R. J., NEWNHAM, D. A., *et al.*: 'An intercomparison of laboratory measurements of absorption cross-sections and integrated absorption intensities for HCFC-22', *J. Quant. Spectrosc. Radiat. Transfer*, 2000, **66**, pp. 109–28

BARBE, A., CHICHERY, Vl. A., TYUTEREV, G., TASHKUN, S., and MIKHAILENKO, S.: 'The 2 ν_2 and $3\nu_2 - \nu_2$ bands of ozone', *Spectrochim. Acta A.*, 1998, **54**, pp. 1935–45

BASS, A. M., and PAUR, R. J.: 'UV absorption cross-sections for ozone: The temperature dependence', *J. Photochem.*, 1981, **17**, p. 41

BASS, A. M., and PAUR, R. J.: 'The ultraviolet cross-sections of ozone: I the measurements,' in ZEREFOS, C. S., and GHAZI, A., (Eds.) 'Atmospheric ozone' (Dordrecht: Reidel D., 1985a), pp. 606–10

BASS, A. M., and PAUR, R. J.: 'The ultraviolet cross-sections of ozone: II results and temperature dependence' in ZEREFOS, C. S., and GHAZI, A. (Eds.) 'Atmospheric ozone' (Dordrecht: Reidel D., 1985b), pp. 611–16

BAUER, A., and GODON, M.: Continuum for H_2O-X mixtures in the H_2O spectral window at 239 GHz; X = C2H4, C2H6', *J. Quant. Spectrosc. Radiat. Transfer*, 2001, **69**, pp. 277–90

BAUER, A., GODON, M., KHEDDAR, M., and HARTMANN, J. M.: 'Temperature and perturber dependences of water vapor line-broadening. Experiments at 183 GHz; calculations below 1000 GHz', *J. Quant. Spectrosc. Radiat. Transfer*, 1989, **41**, pp. 49–54

BAUER, A., GODON, M., CARLIER, J., and MA, Q.: 'Water vapor absorption in the atmospheric window at 239 GHz', *J. Quant. Spectrosc. Radiat. Transfer*, 1995, **53**, pp. 411–23

BAUER, A., BIRK, M., WAGNER, G. *et al.*: 'Study on a spectroscopic database for millimeter and submillimeter wavelengths'. Final Report, ESA contract No. 11581/95/NL/CN, 1998

BECKER, G. E., and AUTLER, S. H.: 'Water vapor absorption of electromagnetic radiation in the centimeter wave-length range', *Phys. Rev.*, 1946, **70**(5/6), pp. 300–7

BENNER, D. C.: College of William and Mary (Williamsburg VA, USA, private communication 2003)

BENNER, D. C., BLAKE, T. A., BROWN, L. R., DEVI, V. M., SMITH, M. A. H., and TOTH, R. A.: 'Air-broadening parameters in the ν_3 band of $^{14}N^{16}O_2$ using a multispectrum fitting technique', *J. Molec. Spectrosc.*, 2004, **228**, pp. 593–619

BERMEJO, D., DOMENECH, J. L., SANTOS, J., BOUANICH, J. P., and BLANQUET, G.: 'Absolute line intensities in the $2\nu_3$ band of $^{16}O^{12}C^{32}S$', *J. Molec. Spectrosc.*, 1997, **185**, pp. 26–30

BOISSOLES, J., BOULET, C., TIPPING, R. H., BROWN, A., and MA, Q.: 'Theoretical calculation of the translation-rotation collision-induced absorption in N_2–N_2, O_2–O_2, and N_2–O_2 pairs', *J. Quant. Spectrosc. Radiat. Transfer*, 2003, **82**, pp. 505–16

BORYSOW, A., and FROMMHOLD, L.: 'Collision-induced rototranslational absorption spectra of N_2–N_2 pairs for temperatures from 50 to 300 K', *Astrophys. J.*, 1986, **311**, pp. 1043–57

BOUANICH, J. P., BLANQUET, G., POPULAIRE, J. C., and WALRAND, J.: 'N_2-broadening for methyl chloride at low temperature by diode-laser spectroscopy', *J. Molec. Spectrosc.*, 2001, **208**, pp. 72–8

BOUDON, V., and PIERRE, G.: 'Rovibrational spectroscopy of sulphur hexafluoride, a review in recent research developments in molecular spectroscopy', in PANDALAI, S. G. (Ed.): 'Transworld research network' (Trivandrum, India 2002), **1**, pp. 25–55

BRAULT, J. W., BROWN, L. R., CHACKERIAN, C., FREEDMAN, R., PREDOI-CROSS, A., and PINE, S.: 'Self-broadened $^{12}C^{16}O$ line shapes in the $v = 2 \leftarrow 0$ band', *J. Molec. Spectrosc.*, 2003, **222**, pp. 220–39

BROCKMAN, P., BAIR, C. H., and ALLARIO, F.: 'High resolution spectral measurement of the HNO_3 11.3-μm band using tunable diode lasers', *Appl. Opt.*, 1978, **17**, pp. 91–9

BROWN, L. R.: (private communications 2001, 2002, 2003)

BROWN, L. R.: 'Near-infrared line parameters of methane' (in preparation 2004)

BROWN, L. R., and PLYMATE, C.: 'Experimental line parameters of the oxygen A band at 760 nm', *J. Molec. Spectrosc.*, 2000, **199**, pp. 166–79

BROWN, A., and TIPPING, R. H.: 'Collision-induced absorption in dipolar molecule-homonuclear diatomic pairs', in CAMY-PEYRET, C., and VIGASIN, A. A. (Eds): Weakly interacting molecular pairs: unconventional absorbers of radiation in the atmosphere (Kluwer Academic, pp. 93–99, 2003)

BROWN, L. R., TOTH, R. A., and DULICK, M.: 'Empirical line parameters of $H_2^{16}O$ near 0.94 μm: positions, intensities and air-broadening coefficients', *J. Molec. Spectrosc.*, 2002a, **212**, pp. 57–82

BROWN, L.R., SAMS, R. L., KLEINER, I., COTTAZ, C., and SAGUI, L.: 'Line intensities of the phosphine dyad at 10 μm', *J. Molec. Spectrosc.*, **215**, 2002b, pp. 178–203

BROWN, L.R., BENNER, D. C., CHAMPION, J. P. *et al.*: 'Methane Line Parameters in HITRAN', *J. Quant. Spectrosc. Radiat. Transfer*, 2003, **82**, pp. 219–38

BUEHLER, S., and ERIKSSON, P. (Eds.): 'Atmospheric millimeter and submillimeter wave radiative transfer modeling' (Berichte aus der Physik, Shaker Verlag GmbH, Aachen, 2000)

BUEHLER, S. A., KUVATOV, M., JOHN, V. O., LEITERER, U., and DIER, H.: 'Comparison of microwave satellite humidity data and radiosonde profiles: a case study', *J. Geophys. Res.*, 2004, **109**, D13103, doi:10.1029/2004JD004605

BUEHLER, S. A., VERDES, C. L., TSUJIMARU, S., *et al.*: 'The expected performance of the SMILES submillimeter-wave limb sounder compared to aircraft data', *Radio Sci.*, 2005a, **40**(3) RS3016, doi:10.1029/2004RS003089

BUEHLER, S. A., ERIKSSON, P., KUHN, T., VON ENGELN, A., and VERDES, C.: 'ARTS, the atmospheric radiative transfer simulator', *J. Quant. Spectrosc. Radiat. Transfer*, 2005b, **91**(1), pp. 65–93, doi:10.1016/j.jqsrt.2004.05.051

CAMPARGUE, A., CHARVAT, A., and PERMOGOROV, D.: 'Absolute intensity measurement of CO_2 overtone transitions in the near-infrared', *Chem. Phys. Lett.*, 1994, **223**, pp. 567–72

CAMY-PEYRET, C.: LPMA, Paris, France, private communication based on CAMY-PEYRET, C., PAYAN, S., JESECK, P., TÉ, Y., and HAWAT, T.: 'High resolution balloon-borne spectroscopy within the O_2 A band: observations and radiative transfer modeling'. Proceedings of the international *Radiation* symposium, Saint Petersburg, Russia, 2000

CANTRELL, C. A., DAVIDSON, J. A., McDANIEL, A. H., SHETTER, R. E., and CALVERT, J. G.: 'Temperature-dependent formaldehyde cross sections in the near-ultraviolet spectral region', *J. Phys. Chem.*, 1990, **94**, pp. 3902–8

CARLEER, M., JENOUVRIER, A., VANDAELE, A. C. *et al.*: 'The near infrared, visible, and near ultraviolet overtone spectrum of water', *J. Chem. Phys.*, 1999, **111**(6), pp. 2444–50

CERNICHARO, J.: 'ATM: A program to compute atmospheric transmission between 0–1000 GHz'. Internal Rep. Institut de Radioastronomie Millimétrique – IRAM, Grenoble, France ,1985

CHACKERIAN, C., FREEDMAN, R. S., GIVER, L. P., and BROWN, L. R.: 'The NO vibrational fundamental band: O_2-broadening coefficients', *J. Molec. Spectrosc.*, 1998a, **192**, pp. 215–9

CHACKERIAN, C., BROWN, L. R., LACOME, N., and TARRAGO, G.: 'Methyl chloride ν_5 region line shape parameters and rotational constants for the ν_2, ν_5 and $2\nu_3$ vibrational bands', *J. Molec. Spectrosc.*, 1998b, **191**, pp. 148–57

CHANCE, K.V.: (private communication 2003)

CHANCE, K., JUCKS, K. W., JOHNSON, D. G., and TRAUB, W. A.: 'The Smithsonian Astrophysical Observatory database SAO92', *J. Quant. Spectrosc. Radiat. Transfer*, 1994, **52**(3/4), pp. 447–57

CHANCE, K. V., PARK, K., EVENSON, K. M. *et al.*: 'Improved molecular constants for the ground state of HO_2', *J. Molec. Spectrosc.*, 1997, **183**, p. 418

CHAMBERLIN, R. A., MARTIN, R. N., MARTIN, C. L., and STARK, A. A.: 'Submillimeter atmospheric FTS at the geographic South Pole', Millimeter and submillimeter detectors for astronomy, in PHILLIPS, T. G. and ZMUIDZINAS, J. (Eds): Proceedings of the SPIE, **4855**, pp. 609–20, 2003

CHRISTIDIS, N., HURLEY, M. D., PINNOCK, S., SHINE, K. P., and WALLINGTON, T. J.: 'Radiative forcing of climate change by CFC-11 and possible CFC replacements', *J. Geophys. Res.*, 1997, **102**, pp. 19597–609

CIMINI, D., WESTWATER, E. R., HAN, Y., and KEIHM, S. J.: 'Accuracy of ground-based microwave radiometer and balloon-borne measurements during the WVIOP2000 field experiment', *IEEE Trans. Geosci. Remote Sen.*, 2003, **41**, pp. 2605–615

CLERBAUX, C., COLIN, R., SIMON, P. C., and GRANIER, C.: 'Infrared cross sections and global warming potentials of 10 alternative hydrohalocarbons', *J. Geophys. Res.*, 1993, **98**, pp. 10491–497

CLOUGH, S. A., KNEIZYS, F. X., and DAVIES, R. W.: 'Line shape and the water vapor continuum', *Atmos. Res.*, 1989, **23**, pp. 229–41

CLOUGH, S.A., SHEPHARD, M. W., MLAWER, E. J. *et al.*: 'Atmospheric radiative transfer modeling: a summary of the AER codes', *J. Quant. Spectrosc. Radiat. Transfer*, 2005, **91**, pp. 233–44

COHEN, E. A.: Jet Propulsion Laboratory, (private communication 2001)

COHEUR, P. F., FALLY, S., CARLEER, M. *et al.*:'New water vapor line parameters in the 26000–13000 cm^{-1} region', *J. Quant. Spectrosc. Radiat. Transfer*, 2002, **74**, pp. 493–510

COLMONT, J.-M., PRIEM, D., WLODARCZAK, G., and GAMACHE, R. R.: 'Measurements and calculations of the halfwidth of two rotational transitions of water vapor perturbed by N_2, O_2, and air', *J. Molec. Spectrosc.*, 1999, **193**, pp. 233–43

COUDERT, L. H.: 'Line frequency and line intensity analyses of water vapour', *Mol. Phys.*, 1999, **96**, pp. 941–54

CRUZ-POL, S. L., RUF, C. S., and KEIHM, S. J.: 'Improved 20- to 32-GHz atmospheric absorption model', *Radio Sci.*, 1998, **33**, pp. 1319–33

DANA, V., MANDIN, J.-Y., ALLOUT, M. Y. *et al.*: 'Broadening parameters of NO_2 lines in the 3.4 micron spectral region', *J. Quant. Spectrosc. Radiat. Transfer*, 1997, **57**, pp. 445–57

DAUMONT, L., CLAVEAU, C., DEBACKER-BARILLY, M. R. *et al.*: 'Line intensities of $^{14}N_2^{16}O$: the 10 micrometer region revisited', *J. Quant. Spectrosc. Radiat. Transfer*, 2002, **72**, pp. 37–55

DE BACKER-BARILLY, M.-R., BARBE, Vl. A., and TYUTEREV, G.: 'Infrared spectrum of $^{16}O^{18}O^{16}O$ in the 5 micron range: Positions, intensities, and atmospheric applications', *Atmos. Oceanic Opt.*, 2003, **16–3**, pp. 183–8

DEMAISON, J., BUEHLER, S., KOULEV, N. *et al.*: 'Characterisation of millimeter wave spectroscopic signatures'. ESTEC contract no. 16377/02/NL/FF, 2004

DEVI, V. M., FRIDOVICH, B., JONES, G. D. *et al.*: 'Tunable diode laser spectroscopy of NO_2 at 6.2 μm', *J. Molec. Spectrosc.*, **93**, 179–95

DEVI, V. M., FRIDOVICH, B., JONES, G. D., SNYDER, D. G. S., and NEUENDORFFER, A.: 'Temperature dependence of the widths of N_2-broadened lines of the ν_3 band of $^{14}N^{16}O_2$', *Appl. Opt.*, 1982, **21**, pp. 1537–38

DEVI, V. M., BENNER, D. C., SMITH, M. A. H., RINSLAND, C. P., SHARPE, S. W., and SAMS, R. L.: 'A multispectrum analysis of the ν_1 band of $H^{12}C^{14}N$: I intensities, self-broadening and self-shift coefficients', *J. Quant. Spectrosc. Radiat. Transfer*, 2003, **82**, pp. 319–42

DEVI, V. M., BENNER, D. C., SMITH, M. A. H., RINSLAND, C. P., SHARPE, S. W., and SAMS, R. L.: 'A multispectrum analysis of the $2\nu_2$ spectral region of $H^{12}C^{14}N$: intensities, broadening and pressure-shift coefficients', *J. Quant. Spectrosc. Radiat. Transfer*, 2004, **87**, pp. 339–66

DI LONARDO, G., and MASCIARELLI, G.: 'Infrared absorption cross-sections and integrated absorption intensities of HFC-125 and HFC-143a', *J. Quant. Spectrosc. Radiat Transfer*, 2000, **66**, pp. 129–42

DING, Y., BERTSEVA, E., and CAMPARGUE, A.: 'The $2\nu_1 + 3\nu_3$ Triad of $^{12}CO_2$', *J. Molec. Spectrosc.*, 2002, **212**, pp. 219–22

DROUIN, B. J.: 'Temperature dependent pressure-induced lineshape of the HClJ = 10 rotational transition in nitrogen and oxygen', *J. Quant. Spectrosc. Radiat. Transfer*, 2004, **83**, pp. 321–31

EDLÉN, B.: 'The refractive index of air', *Metrologia*, 1966, **2**, pp. 71–80

El HACHTOUKI, R., and VANDER AUWERA, J.: 'Absolute line intensities in acetylene: the 1.5-μm region', *J. Molec. Spectrosc.*, 2002, **216**, pp. 255–362

ELLISON, W. J., ENGLISH, S. J. LAMKAOUCHI, K. *et al.*: 'A comparison of ocean emissivity models using the Advanced Microwave Sounding Unit, the Special Sensor Microwave Imager, the TRMM Microwave Imager, and airborne radiometer observations', *J. Geophys. Res.*, 2003, **108** (D21), 4663, doi:10.1029/2002JD003213

ENGLAND, M. N., SCHMIDLIN, F. J., and JOHANSSON, J. M., 'Atmospheric moisture measurements: a microwave radiometer – radiosonde comparison', *IEEE Trans. Geosci. Rem. Sens.*, 1993, **31**(2), pp. 389–98

ENGLISH, S. J., GUILLOU, C., PRIGENT, C., and JONES, D. C.: 'Aircraft measurements of water vapour continuum absorption at millimeter wavelengths', *Quart. J. Roy. Meteorol. Soc.*, 1994, **120**, pp. 603–25

ERIKSSON, J. P., and MERINO, F.: 'On simulating passive observations of the middle atmosphere in the range of 1–1000 GHz. Technical Report No. 179'. Department of Radio and Space Science, Chalmers University of Technology, Goteborg, Sweden, 1997

ERIKSSON, P., JIMENEZ, C., and BUEHLER, S. A.: 'Qpack, a general tool for instrument simulation and retrieval work', *J. Quant. Spectrosc. Radiat. Transfer*, 2005b, **91**(1), pp. 47–64, doi:10.1016/j.jqsrt.2004.05.050

ERIKSSON, P., EKSTROEM, M., MELSHEIMER, C., and BUEHLER, S. A.: 'Efficient forward modelling by matrix representation of sensor responses', *Int. J. Remote Sensing*, in press (2005a)

EYRE J. R.: 'A fast radiative transfer model for satellite sounding systems', *ECMWF Research Dept. Tech. Memo.*, 1991, **176**

EYRE, J. R., and WOOLF, H. M.: 'Transmittance of atmospheric gases in the microwave region: a fast model', *Appl. Opt.*, 1998, **27**, pp. 3244–49

FALLY, S., COHEUR, P.-F., CARLEER, M. *et al.*: 'Water vapor line broadening and shifting by air in the 26000–13000 cm^{-1} region', *J. Quant. Spectrosc. Radiat. Transfer*, 2003, **82**, pp. 119–32

FEIST, D. G.: 'The Bernese Atmospheric Multiple Catalog Access Tool (BEAMCAT): a tool for users of popular spectral line catalogs', *J. Quant. Spectrosc. Radiat. Transfer*, 2004, **85**, pp. 57–97

FLAUD, J.-M.: (private communication, 2002)

FLAUD, J.-M., CAMY-PEYRET, C., DEVI, V. M., RINSLAND, C. P., and SMITH, M. A. H.: 'The ν_1 and ν_3 bands of $^{16}O_3$ line positions and intensities', *J. Molec. Spectrosc.*, 1987, **124**, pp. 209–17

FLAUD, J.-M., CAMY-PEYRET, C., RINSLAND, C. P., SMITH, M. A. H., and DEVI, V. M.: 'Line parameters of $^{16}O_3$ in the 7-μm region', *J. Molec. Spectrosc.*, 1989, **134**, pp. 106–12

FLAUD, J.-M., BIRK, M., WAGNER, G., ORPHAL, J., KLEE, S., and LAFFERTY, W. J.: 'The far-infrared spectrum of HOCl: line positions and intensities', *J. Molec. Spectrosc.*, 1998, **191**, pp. 362–67

FLAUD, J.-M., PICCOLO, C., CARLI, B. *et al.*: 'Molecular line parameters for the MIPAS (Michelson Interferometer for Passive Atmospheric Sounding) experiment', *Atmos. Oceanic Opt.*, 2003a, **16**, pp. 172–81

FLAUD, J.-M., PERRIN, A., ORPHAL, J. *et al.*: 'New analysis of the $\nu_5 + \nu_9 - \nu_9$ hot band of HNO_3', *J. Quant. Spectrosc. Radiat. Transfer*, 2003b, **77**, pp. 355–64

FLAUD, J.-M., WAGNER, G., BIRK, M. C. *et al.*: 'Ozone absorption around 10 μm', *J. Geophys. Res.*, 2003c, **108-D9**, p. 4269

GAMACHE, R. R., and FISCHER, J.: 'Half-widths of $H_2^{16}O$, $H_2^{18}O$, $H_2^{17}O$, $HD^{16}O$, and $D_2^{16}O$: I. Comparison between isotopomers', *J. Quant. Spectrosc. Radiat. Transfer*, 2003, **78**, 289–304. (Calculated line shifts available at http://faculty.uml.edu/Robert_Gamache)

GAMACHE, R. R., HARTMANN, J. M., and ROSENMANN, L.: 'Collisional broadening of water vapor lines. I- A survey of experimental results', *J. Quant. Spectrosc. Radiat. Transfer*, 1994, **52**, pp. 481–99

GAMACHE, R. R., KENNEDY, S., HAWKINS, R., and ROTHMAN, L. S.: 'Total internal partition sums for molecules in the terrestrial atmosphere', *J. Mol. Struct.*, 2000, **517–518**, pp. 407–25

GARAND, L., TURNER, D. S., LAROCQUE, M. *et al.*: 'Radiance and Jacobian intercomparison of radiative transfer models applied to HIRS and AMSU channels', *J. Geophys. Res.*, 2001, **106**, pp. 24017–31

GILLIS, J. R., and GOLDMAN, A., 'Nitric oxide IR line parameters for the upper atmosphere', *Appl. Opt.*, 1982, **21**, pp. 1616–27

GILLIS, J.R., GOLDMAN, A., STARK, G., and RINSLAND, C. P.: 'Line parameters for the $A^2\Sigma^+$_$X^2\Pi$ bands of OH', *J. Quant. Spectrosc. Radiat. Transfer*, 2001, **68**, pp. 225–30

GOLDMAN, A.: (private communication, 2003)

GOLDMAN, A., BROWN, L. R., SCHOENFELD, W. G. *et al.*: 'Nitric oxide line parameters: review of 1996 HITRAN update and new results', *J. Quant. Spectrosc. Radiat. Transfer*, 1998, **60**, pp. 825–38

GOLDMAN, A., STEPHEN, T. M., ROTHMAN, L. S. *et al.*: 'The 1-μm CO_2 bands and the O_2 (0-1) $X^3\Sigma_g^-$–$a^1\Delta g$ and (1-0) $X^3\Sigma_g^-$– $b^1\Sigma_g^+$ bands in the earth atmosphere', *J. Quant. Spectrosc. Radiat. Transfer*, 2003a, **82**, pp. 197–205

GOLDMAN, A., COFFEY, M. T., HANNIGAN, J. W., MANKIN, W. G., CHANCE, K. V., and RINSLAND, C. P.: 'HBr and HI line parameters update for atmospheric spectroscopy databases', *J. Quant. Spectrosc. Radiat. Transfer*, 2003b, **82**, pp. 313–17

GOLUBIATNIKOV, G. Yu.: 'Shifting and broadening parameters of the water vapor 183-GHz line (3 1 3-2 2 0) by H_2O, O_2, N_2, CO_2, H_2, He, Ne, Ar, and Kr at room temperature', *J. Molec. Spectrosc.*, 2005, **230**, pp. 196–8

GOLUBIATNIKOV, G. Yu., and KRUPNOV, A. F.: 'Microwave study of the rotational spectrum of oxygen molecule in the range up to 1.12 THz', *J. Molec. Spectrosc.*, 2003, **217**, pp. 282–7

GORDON, R. G.: 'On the pressure broadening of molecular multiplet spectra', *J. Chem. Phys.*, 1967, **46**, pp. 448–55

GOYETTE, T. M., and DE LUCIA, F. C.: 'The pressure broadening of the 31,3-22,0 transition of water between 80 and 600 K', *J. Molec. Spectrosc.*, 1990, **143**, pp. 346–58

GREENBLATT, G. D., ORLANDO, J. J., BURKHOLDER, J. B., and RAVISHANKARA, A. R.: 'Absorption measurements of oxygen between 330 and 1140 nm', *J. Geophys. Res.*, 1990, **95**, pp. 18577–82

GUILLOU, C., ENGLISH, S. J., and PRIGENT, C.: 'Passive microwave airborne measurements of the sea surface response at 89 and 157 GHz', *J. Geophys. Res.*, 1996, **101**, pp. 3775–88

GUILLOU, C., ELLISON, W. J., EYMARD, L. *et al.*: 'Impact of new permittivity measurements on sea surface emissivity modelling', *Radio Sci.*, 1998, **33**, pp. 649–67

HESS, M., KÖPKE, P., and SCHULT, I., 'Optical properties of aerosols and clouds: the software package OPAC', *BAMS*, 1998, **79-5**, pp. 831–44

HEWISON, T. J., GAFFARD, C., and NASH, J.: 'Validation of microwave radiometer measurements in clear air', Proceedings of ISTP2003 conference, Leipzig, Germany, September 2003

HIGHWOOD, E. J., SHINE, K. P., HURLEY, M. D., and WALLINGTON, T. J.: 'Estimation of direct radiative transfer forcing due to non-methane hydrocarbons', *Atmos. Environ.*, 1999, **33**, pp. 159–67

HILL, R. J.: 'Water vapor absorption line shape comparison using the 22-GHz line: The Van Vleck–Weisskopf shape affirmed', *Radio Sci.*, 1986, **21**(3), pp. 447–51

HILLS, R. E., WEBSTER, A. S., ALSTON, D. A. *et al.*: 'Absolute measurements of atmospheric emission and absorption in the range 100–1000 GHz', *Infrared Phys.*, 1978, **18**, pp. 819–25

HUDIS, E., BEN-ARYEH, Y., and OPPENHEIM, U. P.: 'Third-order linear absorption by pairs of molecules', *Phys. Rev. A*, 1991, **43**, pp. 3631–39

HUDIS, E., BEN-ARYEH, Y., and OPPENHEIM, U. P.: 'The contribution of third order linear absorption to the water vapor continuum', *J. Quant. Spectrosc. Radiat. Transfer*, 1992, **47**, pp. 319–23

HURLEY, M. D.: (private communication, 2003)

JACQUEMART, D., CLAVEAU, C., MANDIN, J.-Y., and DANA, V.: 'Line intensities of hot bands in the 13.6 μm spectral region of acetylene $^{12}C_2H_2$', *J. Quant. Spectrosc. Radiat. Transfer*, 2001, **69**, pp. 81–101

JACQUEMART, D., MANDIN, J.-Y., DANA, V., RÉGALIA-JARLOT, L., THOMAS, X., and VON DER HEYDEN, P.: 'Multispectrum fitting measurements of line parameters for 5 μm cold bands of acetylene', *J. Quant. Spectrosc. Radiat. Transfer*, 2002, **75**, pp. 397–422

JACQUEMART, D., MANDIN, J. Y., DANA, V. *et al.*: 'The IR acetylene spectrum in HITRAN: update and new results', *J. Quant. Spectrosc. Radiat. Transfer*, 2003, **82**, pp. 363–82

JACQUINET-HUSSON, N., SCOTT, N. A., CHÉDIN, A. *et al.*: 'The GEISA system in 1996: Toward an operational tool for the second generation vertical sounders radiance simulation', *J. Quant .Spectrosc. Radiat. Transfer*, 1998, **59**, pp. 511–27

JACQUINET-HUSSON, N., ARIÉ, E., BALLARD, J. *et al.*: 'The 1997 spectroscopic GEISA databank', *J. Quant. Spectrosc. Radiat. Transfer*, 1999, **62**, pp. 205–54

JACQUINET-HUSSON, N., SCOTT, N. A., CHÉDIN, A., and CHURSIN, A. A.: 'The GEISA spectroscopic database system updated for IASI direct radiative transfer modelling', *Atmos. Oceanic Opt.*, 2003a, **16-3**, pp. 256–61

JACQUINET-HUSSON, N., SCOTT, N. A., GARCERAN, K., ARMANTE, R., and CHÉDIN, A.: Proceedings of the 13th International TOVS study conference (ITSC-13), 28 October – 4 November 2003b, Sainte-Adèle, Canada

JACQUINET-HUSSON, N., SCOTT, N. A., CHÉDIN, A. *et al.*: 'The 2003 edition of the GEISA/IASI spectroscopic database', *J. Quant. Spectrosc. Radiat. Transfer*, 2005, **95**, pp. 429–67

JOHN, V. O., and BUEHLER, S. A.: 'The impact of ozone lines on AMSU-B radiances',*Geophys. Res. Lett.*, 2004, **31**, L21108, doi:10.1029/2004GL021214

JOHNS, J. W. C., LU, Z., WEBER, M., SIROTA, J. M., and REUTER, D. C.: 'Absolute intensities in the ν_2 fundamental of N_2O at 17 μm', *J. Molec. Spectrosc.*, 1996, **177**, pp. 203–10

KEIHM, S. J., JANSSEN, M. A., and RUF, C. S.: 'TOPEX/Poseidon microwave radiometer (TMR): III. Wet troposphere range correction algorithm and pre-launch error budget', *IEEE Trans. Geosci. Remote Sens.*, 1995, **33**, pp. 147–61

KEIHM, S. J., BAR-SEVER, Y., and LILJEGREN, J. C.: 'WVR-GPS comparison measurements and calibration of the 20–32 GHz tropospheric water vapor absorption model', *IEEE Trans. Geosci. Remote Sens.*, 2002, **40**, pp. 1199–210

KISSEL, A., SUMPF, B., KRONFELDT, H. D., TIKHOMIROV, B. A., and PONOMAREV, Y. N.: 'Molecular-gas-pressure-induced line-shift and line-broadening in the ν_2 band of H_2S', *J. Molec. Spectrosc.*, 2002, **216**, pp. 345–54

KLEINER, I., and BROWN, L. R.: (private communication, 2003)

KLEINER, I., TARRAGO, G., COTTAZ, C. *et al.*: 'NH_3 and PH_3 line parameters: 2000 HITRAN update and new results', *J. Quant. Spectrosc. Radiat. Transfer*, 2003, **82**, pp. 293–312

KÖPKE, P., HESS, M., SCHULT, I., and SHETTLE, E. P.: 'Global aerosol data-set', *Max-Planck-Institut für Meteorologie*. Report No. 243, Hamburg, Germany, 1997

KOU, Q., GUELACHVILI, G., ABBOUTI TEMSAMANI, M., and HERMAN, M.: 'The absorption spectrum of C_2H_2 around $\nu_1 + \nu_3$: energy standards in the 1.5 μm region and vibrational clustering', *Can. J. Phys.*, 1994, **72**, pp. 1241–50

KROMMINGA, H., ORPHAL, J., SPIETZ, P., VOIGT, S., and BURROWS, J. P.: 'The temperature dependence (213–293 K) of the absorption cross-sections of OClO in the 340–450 nm region measured by Fourier-transform spectroscopy', *J. Photochem. Photobiol. A: Chem.*, 2003, **157**, pp. 149–60

KUHN, T., BAUER, A., GODON, M. , BÜHLER, S., and KÜNZI, K.: 'Water vapor continuum: absorption measurements at 350 GHz and model calculations', *J. Quant. Spectrosc. Radiat. Transfer*, 2002, **74**, pp. 545–62

LACOME, N., LEVY, A., and GUELACHVILI, G.: 'Fourier transform measurement of self-, N_2-, and O_2 broadening of N_2O lines: temperature dependence of linewidths', *Appl. Opt.*, 1984, **23**, pp. 425–34

LANQUETIN, R., COUDERT, L. H., and CAMY-PEYRET, C.: 'High-lying rotational levels of water: an analysis of the energy levels of the five first vibrational states', *J. Molec. Spectrosc.*, 2001, **206**, pp. 83–103

LI, Z., and VARANASI, P.: 'Measurement of the absorption cross-sections of CFC-11 at conditions representing various model atmospheres', *J. Quant. Spectrosc. Radiat Transfer*, 1994, **52**, pp. 137–44

LIEBE, H. J.: 'The atmospheric water vapor continuum below 300 GHz', *Int. J. Infrared Milli. Waves*, 1984, **5**(2), pp. 207–27

LIEBE, H. J.: 'An updated model for millimeter wave propagation in moist air', *Radio Sci.*, 1985, **20**, pp. 1069–89

LIEBE, H. J.: 'A contribution to modeling atmospheric millimeter-wave properties', *Frequenz*, 1987, **41**, pp. 31–6

LIEBE, H. J., 'MPM – an atmospheric millimeter-wave propagation model', *Int. J. Infrared Milli. Waves*, 1989, **10**(6), pp. 631–50

LIEBE, H. J., and GIMMESTAD, G. G.: 'Calculation of clear air EHF refractivity', *Radio Sci.*, 1978, **13**(2), pp. 245–51

LIEBE, H. J., and LAYTON, D. H.: 'Millimeter-wave properties of the atmosphere: laboratory studies and propagation modeling'. NTIA Report 87-224, U. S. Dept. of Commerce; NTIS No. PB 88-164215/AF, 1987

LIEBE, H. J., GIMMESTAD, G. G., and HOPPONEN, J. D.: 'Atmospheric oxygen microwave spectrum – experiment versus theory', *IEEE Trans. Ant. Prop.*, 1977, AP-25(3), pp. 327–35

LIEBE, H. J., ROSENKRANZ, P. W., and HUFFORD, G. A.: 'Atmospheric 60-GHz oxygen spectrum: New laboratory measurements and line parameters', *J. Quant. Spectrosc. Radiat. Transfer*, 1992, **48**(5/6), pp. 629–43

LIEBE, H. J., HUFFORD, G. A., and COTTON, M. G.: Propagation modeling of moist air and suspended water/ice particles at frequencies below 1000 GHz, AGARD conference Proc. 542, 'Atmospheric Propagation Effects through Natural and Man-Made Obscurants for Visible to MM-Wave Radiation', pp. 3.1–3.10, 1993

LILJEGREN, J. C., BOUKABARA, S.-A., CADY-PEREIRA, K., and CLOUGH, S. A.: 'The effect of the half-width of the 22-GHz water vapor line on retrievals of temperature and water vapor profiles with a 12-channel microwave radiometer', *IEEE Trans. Geosci. Remote Sens.*, 2005, **43**, pp. 1102–8

LUCCHESINI, A.: *IPCF*, Pisa, Italy (private communication, 2004)

MA, Q., and TIPPING, R. H.: 'Water vapor continuum in the millimeter spectral region', *J. Chem. Phys.*, 1990, **93**, pp. 6127–39

MA, Q., and TIPPING, R. H.: 'Water vapor millimeter wave foreign continuum: A Lanczos calculation in the coordinate representation', *J. Chem. Phys.*, 2002, **117**, pp. 10581–96

MAKI, A. G., QUAPP, W., KLEE, S., MELLAU, G. C., and ALBERT, S.: 'Infrared transitions of $H^{12}C^{14}N$ and $H^{12}C^{15}N$ between 500 and 10000 cm^{-1}', *J. Molec. Spectrosc.*, 1996, **180**, pp. 323–36

MAKI, A. G., MELLAU, G. C., KLEE, S., WINNEWISSER, M., and QUAPP, W., 'High-temperature infrared measurements in the region of the bending fundamental of $H^{12}C^{14}N$, $H^{12}C^{15}N$, and $H^{13}C^{14}N$', *J. Molec. Spectrosc.*, 2000, **202**, pp. 67–82

MANDIN, J.-Y., DANA, V., PERRIN, A. *et al.*: 'The $\{\nu_1 + 2\nu_2, \nu_1 + \nu_3\}$ bands of $^{14}N^{16}O_2$: line positions and intensities; line intensities in the $\nu_1 + \nu_2 + \nu_3$-ν_2 hot band', *J. Molec. Spectrosc.*, 1997a, **181**, pp. 379–88

MANDIN, J.-Y., DANA, V., RÉGALIA, L., BARBE, A., and THOMAS, X.: 'Λ-splittings and line intensities in the first overtone of nitric oxide', *J. Molec. Spectrosc.*, 1997b, **185**, pp. 347–55

MANDIN, J.-Y., DANA, V., RÉGALIA, L., BARBE, A., and VON DER HEYDEN, P.: 'Lambda-splittings and line intensities in the 3 ← 1 hot band of $^{14}N^{16}O$: The spectrum of nitric oxide in the first overtone region', *J. Molec. Spectrosc.*, 1998, **187**, pp. 200–5

MANDIN, J.-Y, DANA, V., and CLAVEAU, C.: 'Line intensities in the ν_5 band of acetylene $^{12}C_2H_2$', *J. Quant. Spectrosc. Radiat. Transfer*, 2000, **67**, pp. 429–46

MARCHAND, R., ACKERMAN, T., WESTWATER, E. R., CLOUGH, S. A., CADY-PEREIRA, K., and LILJEGREN, J. C.: 'An assessment of microwave absorption models and retrievals of cloud liquid water using clear-sky data', *J. Geophys. Res.*, 2003a, **108**(D24), 4773, doi:10.1029/2003JD003843

MARKOV, V. N.: 'Temperature dependence of self-induced pressure broadening and shift of the 6_{43}–5_{50} line of the water molecule', *J. Molec. Spectrosc.*, 1994, **164**, pp. 233–8

MARKOV, V. N., and KRUPNOV, A. F.: 'Measurements of the pressure shift of the 1_{10}–1_{01} water line at 556 GHz produced by mixtures of gases', *J. Molec. Spectrosc.*, 1995, **172**, pp. 211–14

MARTIN, L., MÄTZLER, C., and KÄMPFER, N.: 'Tropospheric transmission measurements at 94 GHz to validate the Liebe model'. AP 2000 conference, Davos, Switzerland.

MASSIE, S. T.: 'Indices of refraction for the HITRAN compilation', *J. Quant. Spectrosc. Radiat. Transfer*, 1994, **52**, pp. 501–13

MASSIE, S. T.: (private communication, 2003)

MASSIE, S. T., and GOLDMAN, A.: 'The infrared absorption cross-section and refractive-index data in HITRAN', *J. Quant. Spectrosc. Radiat. Transfer*, 2003, **82**, pp. 413–28

MASSIE, S. T., GOLDMAN, A., MURCRAY, D. G., and GILLE, J. C.: 'Approximate absorption cross sections of F12, F11, $ClONO_2$, N_2O_5, HNO_3, CCl_4, CF_4, F21, F113, F114, and HNO_4', *Appl. Opt.*, 1985, **24**, pp. 3426–27

MATRICARDI, M.: 'RTIASI-4, a new version of the ECMWF fast radiative transfer model for the infrared atmospheric sounding interferometer'. ECMWF Research Department Technical Memo, **425** (2003). www.ecmwf.int/publications

MATSUSHITA, S., MATSUO, H., PARDO, J. R., and RADFORD, S.: 'FTS measurements of submillimeter-wave atmospheric opacity at Pampa la Bola II: Supra-Terahertz windows and model fitting', *Publ. Astron. Soc. Japan*, 1999, **51/5**, pp. 603–10

MATTIOLI, V., WESTWATER, E. R., GUTMAN, S. I., and MORRIS, V. R.: 'Forward model studies of water vapor using scanning microwave radiometers, global positioning system, and radiosondes during the cloudiness intercomparison experiment', *IEEE Trans. Geosci. Remote Sens.*, 2005, **43**, pp. 1012–21

MAY, R. D., and FRIEDL, R. R.: 'Integrated band intensities of HO_2NO_2 at 220 K', *J. Quant. Spectrosc. Radiat. Transfer*, 1993, **50**, pp. 257–66

MAY, R. D., and WEBSTER, C. R.: 'Laboratory measurements of NO_2 line parameters near 1600 cm^{-1} for the interpretation of stratospheric spectra', *Geophys. Res. Lett.*, 1990, **17**, pp. 2157–60

MEISSNER, T., and WENTZ, F. J.: 'A radiative transfer model function for 85.5 GHz special sensor microwave imager ocean brightness temperatures', *Radio Sci.*, 2003, **38**(4), 8066, doi:10.1029/2002RS002655

MEISSNER, T., and WENTZ, F. J.: 'The complex dielectric constant of pure and sea water from microwave satellite observations', *IEEE Trans. Geosci. Remote Sens.*, 2004, **42**(9), pp. 1836–49

MELSHEIMER, C., VERDES, C., BUEHLER, S. A. *et al.*: 'Intercomparison of general purpose clear sky atmospheric radiative transfer models for the millimeter/submillimeter spectral range', *Radio Sci.*, 2005, **40**, RS1007, doi:10.1029/2004RS003110

MÉRIENNE, M. F., JENOUVRIER, A., HERMANS, C. *et al.*: 'Water vapor line parameters in the 13000–9250 cm^{-1} region', *J. Quant. Spectrosc. Radiat. Transfer*, 2003, **82**, pp. 99–117

MIKHAILENKO, S., BARBE, A., PLATEAUX, J. J., and TYUTEREV, V. G.: 'New analysis of $2\nu_1+\nu_2$, $\nu_1+\nu_2+\nu_3$, and $\nu_2+2\nu_3$ bands of ozone in the 2600–2900 cm^{-1} region', *J. Molec. Spectrosc.*, 1999, **196**, pp. 93–101

MIKHAILENKO, S., BARBE, A., and TYUTEREV, Vl. G.: 'Extended analysis of line positions and intensities of ozone bands in the 2900–3200 cm^{-1} region', *J. Molec. Spectrosc.*, 2002, **215**, pp. 29–41

MILLER, C. E., and BROWN, L. R.: 'Near infrared spectroscopy of carbon dioxide I. $^{16}O^{12}C^{16}O$ line positions', *J. Molec. Spectrosc.*, 2004, **228**, pp. 329–54

MÜLLER, H. S. P., THORWIRTH, S., ROTH, D. A., and WINNEWISSER, G.: 'The Cologne database for molecular spectroscopy', *CDMS. A&A*, 2001, **370**, pp. L49–L52

NEMTCHINOV, V., and VARANASI, P.: 'Thermal infrared absorption cross-sections of CF_4 for atmospheric applications', *J. Quant. Spectrosc. Radiat Transfer*, 2003a, **82**, pp. 461–71

NEMTCHINOV, V., and VARANASI, P.: 'Thermal infrared absorption cross-sections of CCl_4 needed for atmospheric remote-sensing', *J. Quant. Spectrosc. Radiat. Transfer*, 2003b, **82**, pp. 473–81

NEMTCHINOV, V., and VARANASI, P.: 'Absorption cross-sections of HFC-134a in the spectral region between 7 and 12 μm', *J. Quant. Spectrosc. Radiat*, 2004, **84**, pp. 285–94

NEMTCHINOV, V., SUN, C., and VARANASI, P.: 'Measurements of line intensities and line widths in the ν_3-fundamental band of nitrous oxide at atmospheric temperatures', *J. Quant. Spectrosc. Radiat. Transfer*, 2003, **83**, pp. 267–84

NORMAN, M. L., MILLER, R. E., and WORSNOP, D. R.: 'Ternary $H_2O/HNO_3/H_2O$ optical constants: New measurements from aerosol spectroscopy under stratospheric conditions', *J. Phys. Chem.*, 2002, **106**, pp. 6075–83

ORPHAL, J.: (private communication, 2003)

ORPHAL, J., FELLOWS, C. E., and FLAUD, P. M.: 'The visible absorption spectrum of NO_3 measured by high-resolution Fourier-transform spectroscopy', *J. Geophys. Res.*, 2003, **108(D3)**, p.4077

PAINE, S., BLUNDELL, R., PAPA, D. C., BARRETT, J. W., and RADFORD, S. J. E.: 'A fourier transform spectrometer for measurement of atmospheric transmission at submillimeter wavelengths', *Publ. Astron. Soc. Pacific*, 2000, **112**, pp. 108–18

PARDO, J. R., PAGANI, L., GERIN, M., and PRIGENT, C.: 'Evidence of the Zeeman splitting in the 2(1)-0(1) rotational transition of the atmospheric $^{16}O^{18}O$ molecule from ground-based measurements', *J. Quant. Spectrosc. Radiat. Transfer*, 1995, **54/6**, pp. 931–43

PARDO, J. R., GERIN, M., PRIGENT, C., CERNICHARO, J., ROCHARD, G., and BRUNEL, P.: 'Remote sensing of the mesospheric temperature profile from close-to-nadir observations: discussion about the capabilities of the 57.5–62.5 GHz frequency band and the 118.75 GHz single O_2 line', *J. Quant. Spectrosc. Radiat. Transfer*, 1998, **60/4**, pp. 559–71

PARDO, J. R., SERABYN, E., and CERNICHARO, J.: 'Submillimeter atmospheric transmission measurements on Mauna Kea during extremely dry El Nino conditions: implications for broadband opacity contributions', *J. Quant. Spectrosc. Radiat. Transfer*, 2001a, **68**, pp. 419–33

PARDO, J. R., CERNICHARO, J., and SERABYN, E., 'Atmospheric transmission at microwaves (ATM): An improved model for millimeter/submillimeter applications', *IEEE Trans. Antenn. Propag.*, 2001b, **49**, pp. 1683–94

PARDO, J. R., RIDAL, M., MURTAGH, D. P., and CERNICHARO, J.: 'Microwave temperature and pressure measurements with the Odin satellite: I. Observational method', *Can. J. Phys.*, 2002, **80/4**, pp. 443–54

PARDO, J. R., WIEDNER, M. C., and SERABYN, E.: 'Side-by-side comparison of fourier transform spectroscopy and water vapor radiometry as tools for the calibration of mm/submm ground-based observatories', *Astrophys. J. Suppl.*, 2004, **153**, pp. 363–67

PARDO, J. R., SERABYN, E., WIEDNER, M. C., and CERNICHARO, J.: 'Measured telluric continuum-like opacity beyond 1 THz', *J. Quant. Spectrosc. Radiat. Transfer*, 2005a, **96**, pp. 537–45

PARDO, J. R., SERABYN, E., and WIEDNER, M. C.: 'Broadband submillimeter measurements of the full moon center brightness temperature and application to a lunar eclipse', 2005b, Icarus, doi:10.1016/icarus2005.04.005, 2005b, **178**(1), pp. 19–26

PEARSON, J. C.: Jet Propulsion Laboratory, Pasadena CA, USA (private communication, 1999)

PERRIN, A., FLAUD, J.-M., CAMY-PEYRET, C. *et al.*: 'First analysis of the $3\nu_9 - \nu_9$, $3\nu_9 - \nu_5$ and $3\nu_9 - 2\nu_9$ bands of HNO_3: torsional splitting in the ν_9 vibrational mode', *J. Molec. Spectrosc.*, 1994, **166**, pp. 224–3

PERRIN, A., FLAUD, J.-M., CAMY-PEYRET, C. *et al.*: 'Line intensities in the far-infrared spectrum of H_2O_2', *J. Molec. Spectrosc.*, 1996, **176**, pp. 287–96

PERRIN, A., FLAUD, J.-M., GOLDMAN, A. *et al.*: '$NO_2$and SO_2 line parameters: 1996 HITRAN update and new results', *J. Quant. Spectrosc. Radiat. Transfer*, 1998, **60**, pp. 839–50

PERRIN, A., RINSLAND, C. P., and GOLDMAN, A.: 'Spectral parameters for the ν_6 region of HCOOH and its measurement in the infrared tropospheric spectrum', *J. Geophys. Res.*, 1999, **104**, pp. 18661–6

PERRIN, A., PUZZARINI, C., COLMONT, J.-M. *et al.*: 'Molecular line parameters for the MASTER (millimeter wave acquisitions for stratosphere/troposphere exchange research) database', *J. Atmos. Chem.*, 2005, **50**, pp. 161–205

PETKIE, D. T., HELMINGER, P., WINNEWISSER, B. P. *et al.*: 'The simulation of infrared bands from the analyses of rotational spectra: the $2\nu_9 - \nu_9$ and $\nu_5 - \nu_9$ hot bands of HNO_3', *J. Quant. Spectrosc. Radiat. Transfer*, 2005, **92**, pp. 129–41

PICKETT, H. M., COHEN, E. A., and BRINZA, D. E.: 'Pressure broadening of oxygen and its implications for cosmic background measurements', *Astrophys. J.*, 1981, **248**, pp. L49–L51

PICKETT, H. M., POYNTER, R. L., COHEN, E. A., DELITSKY, M. L., PEARSON, J. C., and MÜLLER, H. S. P.: 'Submillimeter, millimeter, and microwave spectral line catalog', *J. Quant. Spectrosc. Radiat. Transfer*, 1998, **60**, pp. 883–90

PINE, A. S., and RINSLAND, C. P.: 'The role of torsional hot bands in modeling atmospheric ethane', *J. Quant. Spectrosc. Radiat. Transfer*, 1999, **62**, pp. 445–58

PINE, A. S., MAKI, A. G., and CHOU, N. Y.: 'Pressure broadening, lineshapes, and intensity measurements in the $2 \leftarrow 0$ band of NO', *J. Molec. Spectrosc.*, 1985, **114**, pp. 132–47

PINNOCK, S., HURLEY, M. D., SHINE, K. P., WALLINGTON, T. J., and SMYTH, T. J.:'Radiative forcing of climate by hydrochlorofluorocarbons and hydrofluorocarbons', *J. Geophys. Res.*, 1995, **100**, pp. 23227–38

PODOBEDOV, V. B., PLUSQUELLIC, D. F., and FRASER, G. T.: 'THz laser study of self-pressure and temperature broadening and shifts of water vapor lines for pressures up to 1.4 kPa', *J. Quant. Spectrosc. Radiat. Transfer*, 2004, **87**, pp. 377–85

PRIGENT, C., ROSSOW, W. B., and MATTHEWS, E.: 'Microwave land surface emissivities estimated from SSM/I observations', *J. Geophys. Res.*, 1997, **102**, pp. 21867–90

PRIGENT, C., WIGNERON, J. P., ROSSOW, W. B., AND PARDO-CARRION, J. R.: 'Frequency and angular variations of land surface microwave emissivities: Can we estimate SSM/T and AMSU emissivities from SSM/I emissivities?', *IEEE Trans. Geosci. Remote Sens.*, 2000, **38**, pp. 2373–86

PRIGENT, C., PARDO, J. R., MISHCHENKO, M. I., and ROSSOW, W. B.: 'Microwave polarized signatures generated within cloud systems: SSM/I observations interpreted with radiative transfer simulations', *J. Geophys. Res.*, 2001, **106**, pp. 28243–58

PRIGENT, C., DEFER, E., PARDO, J. R., PEARL, C., ROSSOW, W., and PINTY, J.-P.: 'Relations of polarized scattering signatures observed by the TRMM microwave instrument with electrical processes in cloud systems', *Geophys. Res. Lett.,*, 2005, **32**, L04810, doi:10.1029/2004GL022225

PUMPHREY, H. C., and BUEHLER, S., 'Instrumental and spectral parameters: their effect on and measurement by microwave limb sounding of the atmosphere', *J. Quant. Spectrosc. Radiat. Transfer*, 2000, **64**, pp. 421–37

RAYER, P. J.: 'Fast transmittance model for satellite sounding', *Appl. Opt.*, 1995, **34**, pp. 7387–94

RÉGALIA-JARLOT, L., HAMDOUNI, A., THOMAS, X., VON DER HEYDEN, P. and BARBE, A.: 'Line intensities of the ν_3, $4\nu_2$, $\nu_1 + \nu_3$, $3\nu_1$, and $2\nu_1 + \nu_2$ bands of the $^{16}O^{12}C^{32}S$ molecule', *J. Quant. Spectrosc. Radiat. Transfer*, 2002, **74**, pp. 455–70

RINSLAND, C. P., JONES, N. B., CONNOR, B. J. *et al.*: 'Northern and southern hemisphere ground-based infrared spectroscopic measurements of tropospheric carbon monoxide and ethane', *J. Geophys. Res.*, 1998, **103:28**, pp. 197–217

RINSLAND, C. P., DEVI, V. M., SMITH, M. A. H., BENNER, D. C., SHARPE, S. W., and SAMS, R. L.: 'A multispectrum analysis of the ν_1 band of $H^{12}C^{14}N$: II. Air- and N_2-broadening, shifts and their temperature dependences', *J. Quant. Spectrosc. Radiat. Transfer*, 2003a, **82**, pp. 343–62

RINSLAND, C. P., SHARPE, S. W., and SAMS, R. L.: 'Temperature-dependent cross-sections in the thermal infrared bands of SF_5CF_3', *J. Quant. Spectrosc. Radiat. Transfer*, 2003b, **82**, pp. 483–90

RODGERS, C. D.: 'Inverse methods for atmospheric sounding. Theory and practice', Series on Atmospheric, Oceanic and Planetary Physics-Vol.2 (World Scientific, Singapore, 2000)

ROSENKRANZ, P. W.: 'Absorption of microwaves by atmospheric gases', (Chap. 2 in Atmospheric Remote Sensing by Microwave Radiometry, JANSSEN, M. A. ed.), John Wiley & Sons, New York, 1993

ROSENKRANZ, P. W.: 'Interference coefficients for overlapping oxygen lines in air', *J. Quant. Spectrosc. Radiat. Transfer*, 1988, **39**(4), pp. 287–97

ROSENKRANZ, P. W.: 'Water vapor microwave continuum absorption: A comparison of measurements and models', *Radio Sci.*, 1998, **33**, pp. 919–28 (also see correction, **34**, p.1025, 1999)

ROSENKRANZ, P. W.: 'Rapid radiative transfer model for AMSU/HSB channels', *IEEE Trans. Geosci. Remote Sens.*, 2003, **41**, pp. 362–8

ROSENKRANZ, P. W., and BARNET, C. D.: 'Microwave radiative transfer model validation', *J. Geophys. Res.,* 2006, **111**, D09S07, doi:10.1029/2005JD006008

ROTHMAN, L. S., BARBE, A., BENNER, D. C. *et al.*: 'The HITRAN molecular spectroscopic database: Edition of 2000 including updates through 2001', *J. Quant. Spectrosc. Radiat. Transfer*, 2003, **82**, pp. 5–44

ROTHMAN, L. S., GAMACHE, R. R., GOLDMAN, A. *et al.*: 'The HITRAN database:1986 edition', *Appl. Opt.*, 1987, **26**, pp. 4058–97

ROTHMAN, L. S., RINSLAND, C. P., and GOLDMAN, A.: 'The HITRAN molecular spectroscopic database and HAWKS (HITRAN atmospheric workstation):1996 edition', *J. Quant. Spectrosc. Radiat. Transfer*, 1998, **60**, pp. 665–710

ROTHMAN, L. S., JACQUEMART, D., BARBE, A. *et al.*: 'The HITRAN 2004 molecular spectroscopic database', *J. Quant. Spectrosc. Radiat. Transfer*, 2005, **96**, pp. 139–204, doi:10.1016/j.jqsrt.2004.10.008

RUBLEV, A. A.: 'Algorithm and computation of aerosol phase functions', *Internal Note IAE-5715/16 of Russian Research Center Kurchatov Institute*, Moscow, Russia, 1994

SAUNDERS, R. W., MATRICARDI, M., and BRUNEL, P.: 'An improved fast radiative transfer model for assimilation of satellite radiance observations', *QJRMS*, 1999, **125**, pp. 1407–25

SCHREIER, F.: 'The Voigt and complex error functions: A comparison of computational methods', *J. Quant. Spectrosc. Radiat. Transfer*, 1992, **48**, pp. 743–62

SCHWARTZ, M. J.: 'Observations and modeling of atmospheric oxygen millimeter-wave transmittance'. Ph.D. thesis, Mass. Inst. of Tech. Dept. of Physics, Cambridge, MA (1998)

SERABYN, E., WEISSTEIN, E., LIS, D. C., and PARDO, J. R.: 'Submillimeter FTS measurements of atmospheric opacity above Mauna Kea', *Appl. Opt.*, 1998, **37/12**, pp. 2185–98

SHORTER, J. H., NELSON, D. D., and ZAHNISER, M. S.: *J. Chem. Soc. Faraday Trans.*, 1997, **93**, pp. 2933–5

SMITH, K. M.: (MSF/RAL: Molecular Spectroscopy Facility/RutherfordAppleton Laboratory, (private communication, 2003)

SODEN, B. J., TURNER, D. D., LESHT, B. M., and MILOSHEVICH, L. M.: 'An analysis of satellite, radiosonde, and lidar observations of upper tropospheric water vapor from the atmospheric radiation measurement program', *J. Geophys. Res.*, 2004, **109**, D04105, doi:10.1029/2003JD003828

SPENCER, M. N., CHACKERIAN, C., GIVER, L. P., and BROWN, L. R.: 'The nitric oxide fundamental band: Frequency and shape parameters for ro-vibrational lines', *J. Molec. Spectrosc.*, 1994, **165**, pp. 506–24

SPENCER, M. N., CHACKERIAN, C., GIVER, L. P., and BROWN, L. R.: 'Temperature dependence of nitrogen broadening of the NO fundamental vibrational band', *J. Molec. Spectrosc.*, 1997, **181**, pp. 307–15

STEWART, B. C.: (Ed.), 'Support study on Water Vapor Spectroscopy for IASI'. Final Report, EUMETSAT Contract EUM/CO/01/939/DK., 2003, pp. 1–159

SUMPF, B.: 'Experimental investigation of the self-broadening coefficients in the $\nu_1 + \nu_3$ band of SO_2 and the $2\nu_2$ band of H_2S', *J. Molec. Spectrosc.*, 1997, **181**, pp. 160–7

SUMPF, B., MEUSEL, I., and KRONFELDT, H. D.: 'Self- and air-broadening in the ν_1 and ν_3 bands of H_2S', *J. Molec. Spectrosc.*, 1996, **177**, pp. 143–5

SUMPF, B., KISSEL, A., and KRONFELDT, H. D.: 'Line-broadening and line-shift in the ν_1, ν_3, and $2\nu_2$ bands of H_2S', (in preparation 2004)

SUNG, K., and VARANASI, P.: 'Intensities, collision-broadened half-width, and collision-induced line shifts in the second overtone band of $^{12}C^{16}O$', *J. Quant. Spectrosc. Radiat. Transfer*, 2004, **82**, pp. 445–8

TANAKA, M., BRAULT, J. W., and TENNYSON, J.: 'Absorption spectrum of $H_2^{18}O$ in the 12,400 – 14,520 cm^{-1} range', *J. Molec. Spectrosc.*, 2002, **216**, pp. 77–80

TANAKA, M., NAUMENKO, O., BRAULT, J. W., AND TENNYSON, J.: ' Fourier transform absorption spectra of $H_2^{18}O$ and $H_2^{17}O$ in the $3\nu + \delta$ and 4ν polyrad region', *J. Molec. Spectrosc.*, 2005, **234**, pp. 1–9, doi:10.1016/jms2005.07.007

TARRAGO, G., LACOME, N., LEVY, A., GUELACHVILI, G., BEZARD, B., and DROSSART, P.: 'Phosphine spectrum at 4-5 μm: analysis and line-by-line simulation of $2\nu_2$, $\nu_2 + \nu_4$, $2\nu_4$, ν_1, and ν_3 bands', *J. Molec. Spectrosc.*, 1992, **154**, pp. 30–42

TASHKUN, S. A., PEREVALOV, V. I., TEFFO, J.-L., ROTHMAN, L. S., and TYUTEREV, VI. G.: 'Global fitting of $^{12}C^{16}O_2$ vibrational-rotational line positions using the effective Hamiltonian approach', *J. Quant. Spectrosc. Radiat. Transfer.*, 1998, **60**, pp. 785–801

TASHKUN, S. A., PEREVALOV, V. I., and TEFFO, J.-L.: 'Global fittings of the vibrational-rotational line positions of the $^{16}O^{12}C^{17}O$ and $^{16}O^{12}C^{18}O$ isotopic species of carbon dioxide', *J. Molec. Spectrosc.*, 2001, **210**, pp. 137–45

TASHKUN, S. A., PEREVALOV, V. I., TEFFO, J. L., BYKOV, A. D., and LAVRENTIEVA, N. N.: 'CDSD-296, the carbon dioxide spectroscopic databank: version for atmospheric applications'. XIV symposium on high resolution molecular spectroscopy, Krasnoyarsk, Russia, July 6–11, 2003. See also ftp://ftp.iao.ru/pub/CDSD-296/

TEFFO, J.-L., DAUMONT, L., CLAVEAU, C., VALENTIN, A., TASHKUN, S. A., and PEREVALOV, V. I.: 'Infrared spectra of the $^{16}O^{12}C^{17}O$ and $^{16}O^{12}C^{18}O$ species of carbon dioxide: The region 500–1500 cm^{-1}', *J. Molec. Spectrosc.*, 2002, **213**, pp. 145–152

TEFFO, J.-L., DAUMONT, L., CLAVEAU, C., VALENTIN, A., TASHKUN S. A., and PEREVALOV, V. I.: 'Infrared spectra of the $^{16}O^{12}C^{17}O$ and $^{16}O^{12}C^{18}O$ species of carbon dioxide: II The region 1500–3000 cm^{-1}', *J. Molec. Spectrosc.*, 2003, **219**, pp. 271–81

TEJWANI, G. D. T., and FOX, K.: 'Calculated self- and foreign-gas-broadened linewidths for SF_6', *J. Quant. Spectrosc. Radiat. Transfer*, 1987, **37**, pp. 541–6

TOTH, R. A.: 'Transition frequencies and absolute strengths of $H_2^{17}O$ and $H_2^{18}O$ in the 6.2-μm region', *J. Opt. Soc. Am. B*, 1992, **9**, pp. 462–82

TOTH, R. A.: 'Water vapor measurements between 590 and 2582 cm^{-1}: Line positions and strengths', *J. Molec. Spectrosc.*, 1998a, **190**, pp. 379–96

TOTH, R. A.: 'Analysis of line positions and strengths of $H_2^{16}O$ ground and hot bands connecting to interacting upper states: (020), (100), and (001)', *J. Molec. Spectrosc.*, 1999a, **194**, pp. 28–42

TOTH, R. A.: 'HDO and D_2O low pressure, long path spectra in the 600–3100 cm^1 region I. HDO line positions and strengths', *J. Molec. Spectrosc.*, 1999b, **195**, pp. 73–97

TOTH, R. A.: 'Air- and N_2-broadening parameters of water vapor: 604 to 2271 cm^1', *J. Molec. Spectrosc.*, 2000, **201**, 218–2200043

TOTH, R. A.: 'Measurements of positions, strengths and self-broadened widths of H_2O from 2900 to 8000 cm^{-1}: line strength analysis of 2nd triad bands', *J. Quant. Spectrosc. Radiat. Transfer* **94**, pp. 51–107 (2005) and see http://mark4sun.jpl.nasa.gov.data/spec/H2O (line lists of water vapour from 500 to 8000 cm^{-1})

TOTH, R. A.: 'Linelist of N_2O parameters from 500 to 7500 cm^{-1}', see http://mark4sun.jpl.nasa.gov.data/spec/N2O (2004)

TOTH, R. A., BROWN, L. R., and PLYMATE, C.: 'Self-broadened widths and frequency shifts of water vapor lines between 590 and 2400 cm^{1}', *J. Quant. Spectrosc. Radiat. Transfer*, 1998b, **59**, pp. 529–62

TOTH, R. A., BROWN, L. R., and COHEN, E. A.: 'Line intensities of HNO_3', *J. Molec. Spectrosc.*, 2003, **218**, pp. 151–68

TRETYAKOV, M. Yu., GOLUBIATNIKOV, G. Yu., PARSHIN, V. V. *et al.*: 'Experimental study of the line mixing coefficient for 118.75 GHz oxygen line', *J. Molec. Spectrosc.*, 2004, **223**, pp. 31–8

TRETYAKOV, M. Yu., KOSHELEV, M. A., DOROVSKIKH, V. V., MAKAROV, D. S., and ROSENKRANZ, W.: '60-GHz oxygen band: precise broadening and central frequencies of fine structure lines, absolute absorption profile at atmospheric pressure, and revision of mixing coefficients', *J. Molec. Spectrosc.*, 2005, **231**, p. 1–14

TURNER, D. D., LESHT, B. M., CLOUGH, S. A., LILJEGREN, J. C., REVERCOMB, H. E., and TOBIN, D. C.: 'Dry bias and variability in Vaisala RS80-H radiosondes: The ARM experience', *J. Atmos. Oceanic Tech.*, 2003, **20**, pp. 117–32

VAN DE HULST, H. C.: 'Light scattering by small particles' (Dove Publications, New-York, 1981)

VANDAELE, A. C., HERMANS, C., FALLY, S. *et al.*: 'High-resolution Fourier transform measurements of the NO_2 visible and near-infrared absorption cross-section: temperature and pressure effects', *J. Geophys. Res.*, 2002, **107** (D18), 4348, doi:10.1029/2001JD000971 and see http://www.oma.be/BIRA-IASB/Scientific/Topics/Lower/LaboBase

VANDER AUWERA, J.: ULB, Brussels, Belgium, private communication (2004), based on VANDER AUWERA, J.: 'High-resolution investigation of the far-infrared spectrum of formic acid', *J. Molec. Spectrosc.*, 1992, **155**, pp. 136–42

VANDER AUWERA, J.: 'Absolute intensities measurements in the $\nu_4 + \nu_5$ band of $^{12}C_2H_2$: Analysis of Herman-Wallis effects and forbidden transitions', *J. Molec. Spectrosc.*, 2000a, **201**, pp. 143–50

VANDER AUWERA, J.: 'Infrared absorption cross-sections for two substituted ethanes: 1,1-difluoroethane (HFC-152a) and 1,2-dichloroethane', *J. Quant. Spectrosc. Radiat Transfer*, 2000b, **66**, pp. 143–51

VANDER AUWERA, J.: (private communication, 2003)

VANDER AUWERA, J., KLEFFMANN, J., FLAUD, J.-M. *et al.*: 'Absolute ν_2 line intensities of HOCl by simultaneous measurements in the infrared with a tunable diode laser and far-infrared region using a fourier transform spectrometer', *J. Molec. Spectrosc.*, 2000, **204**, pp. 36–47

VANDER AUWERA, J., El HACHTOUKI, R., AMARA, K. , FAYT, A.: 'Absolute line intensities for carbonyl sulfide near 4.85-μm'. Poster D23, 18th colloquium on High resolution molecular spectroscopy, Dijon, France, 2003

VARANASI, P.: (private communication, 2000, 2001).

VARANASI, P., and NEMTCHINOV, V.: 'Thermal infrared absorption coefficients of CFC-12 at atmospheric conditions', *J. Quant. Spectrosc. Radiat. Transfer*, 1994, **51**, pp. 679–87

VARANASI, P., NEMTCHINOV, V., LI, Z., and CHERUKURI, A.: 'Spectral absorption-coefficient data on HCFC-22 and SF_6 for remote sensing applications', *J. Quant. Spectrosc. Radiat. transfer*, 1994, **52**, pp. 323–32

VERDES, C., VON ENGELN, A., and BUEHLER, S. A.: 'Partition function data and impact on retrieval quality for a Mm/Sub-mm limb sounder', *J. Quant. Spectrosc. Radiat. Transfer*, 2005, **90**(2), 217–38, doi:10.1016/j.jqsrt.2004.03.012

VOIGT, S., ORPHAL, J., BOGUMIL, K., and BURROWS, J. P.: 'The temperature dependence (203-293 K) of the absorption cross sections of O_3 in the 230–850 nm region measured by Fourier-transform spectroscopy', *J. Photochem. Photobiol. A: Chem.*, 2001, **143**, pp. 1–9

WAGNER, G., and BIRK, M.: 'New infrared spectroscopic database for chlorine nitrate', *J. Quant. Spectrosc. Radiat. Transfer*, 2003, **82**, pp. 443–60

WAGNER, G., BIRK, M., SCHREIER, F., and FLAUD, J.-M.: 'Spectroscopic database for ozone in the fundamental spectral regions', *J. Geophys. Res.*, 2002, **107**:D22

WANG, J. R.: 'A comparison of the MIR-estimated and model-calculated fresh water surface emissivities at 89, 150, and 220 GHz', *IEEE Trans. Geosci. Remote Sens.*, 2002, **40**(6), pp. 1356–65

WANG, J., COLE, H. L., CARLSON, D. J. *et al.*: 'Corrections of humidity measurement errors from the Vaisala RS80 radiosonde – application to TOGA COARE data', *J. Atmos. Oceanic Tech.*, 2002, **19**, pp. 981–1002

WASCHULL, J., KUHNEMANN, F., and SUMPF, B.: 'Self-, air- and helium broadening of the ν_2 band of H_2S', *J. Molec. Spectrosc.*, 1994, **165**, pp. 150–8

WATERS, J. W.: 'Absorption and emission by atmospheric gases', in MEEKS, M. L. (ed.): 'Methods of experimental physics' (Vol. 12, Astrophysics, Part B, Academic Press, 1976)

WENTZ, F. J.: 'A well-calibrated ocean algorithm for special sensor microwave/imager', *J. Geophys. Res.,* 1997, **102,** pp. 8703–18

WESTWATER, E. R., STANKOV, B. B., CIMINI, D. *et al.*: 'Radiosonde humidity soundings and microwave radiometers during Nauru99', *J. Atmos. Oceanic Tech.*, 2003, **20**, pp. 953–71

WIEDNER, M. C., PRIGENT, C., PARDO, J. R. *et al.*: 'Modeling of passive microwave responses in convective situations using output from mesoscale models: Comparison with TRMM/TMI satellite observations', *J. Geophys. Res.*, 2004, **109**, D6, D06214

WILMOUTH, D. M., HANISCO, T. F., DONAHUE, N. M., and ANDERSON, J. G.: 'Fourier transform ultraviolet spectroscopy of the A 2.3/2 . X 2.3/2 transition of BrO', *J. Phys. Chem.*, 1999, **103**, pp. 8935–45

WINNEWISSER, M., WINNEWISSER, B. P., STEIN, M. *et al.*: 'Rotational spectra of cis-HCOOH, trans-HCOOH, and trans-^{13}COOH', *J. Molec. Spectrosc.*, 2002, **216**, pp. 259–65

XU, L. H., LEES, R. M., WANG, P., BROWN, L. R., KLEINER, I., and JOHNS, J. W. C.: 'New assignments, line intensities, and HITRAN database for CH_3OH at 10 μm', *J. Molec. Spectrosc.*, 2004, **228**, pp. 453–70

ZOU, Q., SUN, C., NEMTCHINOV, V., and VARANASI, P.: 'Thermal infrared cross-sections of C_2F_6 at atmospheric temperatures', *J. Quant. Spectrosc.*, 2004, **83**, pp. 215–21

ZU, L., HAMILTON, P. A., and DAVIES, P. B.: 'Pressure broadening and frequency measurements of nitric acid lines in the 683 GHz region', *J. Quant. Spectrosc. Radiat. Transfer*, 2002, **73**, pp. 545–56

Chapter 3

Emission and scattering by clouds and precipitation

Alessandro Battaglia, Clemens Simmer, Susanne Crewell, Harald Czekala, Claudia Emde, Frank Marzano, Michael I. Mishchenko, Juan R. Pardo, and Catherine Prigent

3.1 Introduction, purpose and scope

The strength of the interaction of clouds and precipitation with microwave radiation strongly increases with frequency. While at frequencies around 10 GHz and below radiances are influenced only marginally by large precipitating particles in rain, higher frequency signals emanating from the earth–surface–atmosphere system are increasingly dominated by both clouds and precipitation, especially outside the strong absorption lines of atmospheric gases. While in the infrared and solar spectral ranges clouds completely block the transmission of radiation in most of the microwave region both clouds and precipitation stay transparent at least to some degree. While only the outer surfaces of clouds can be seen in the solar and infrared spectral regions, the whole cloud and its internal structure is in principle accessible to remote sensing in the microwaves. Of importance for the spectral signal in the microwave region is the behaviour of the complex index of refraction of liquid and solid water. At low frequencies liquid precipitation dominates radiative transfer by absorption and emission processes due to the relatively large imaginary part of the index of refraction of liquid water. Atmospheric ice has almost no noticeable absorption due to its very small imaginary index of refraction. With increasing frequency, scattering at both liquid and solid hydrometeors increases dramatically in the transition region from the Rayleigh to the Lorenz–Mie regime, and at frequencies above the molecular absorption band around 60 GHz ice particles dominate the interaction in a way that, e.g. liquid precipitation in the lower parts of a cloud is almost hidden from the satellite view by the scattering effects of ice particles aloft.

Since large atmospheric hydrometeors tend to have non-spherical shapes (e.g. falling raindrops, snow and other ice crystals) polarisation becomes increasingly important. Over water surfaces all four components of the Stokes vector may become activated by radiation–surface interactions. Thus in general to take care of all important interaction processes and also to exploit significant signals for remote sensing purposes fully polarised atmospheric radiative-transfer models have a much higher importance for microwaves as compared with other spectral regions. Finally, the combination of scattering effects, the large spatial variation of the dominating hydrometeors in the atmosphere, the partly reflecting properties of the surface and the typically large footprints of microwave radiometers put weight on the consideration of three-dimensional effects in microwave radiative transfer.

This chapter reviews first the state of the art in deriving the single scattering properties of atmospheric hydrometeors, and then models are described for the computation of radiance distributions in cloudy atmospheres. The chapter finishes with the characterisation of the known signatures of clouds and precipitation in passive microwave observations.

3.2 Basic quantities in RT

We will now introduce some basic quantities of scattering theory aiming at solving the scattering problem for a single particle. The most appropriate methodologies to compute these quantities in the MW regime for typical hydrometeors are then discussed, and examples are provided.

3.2.1 Reference frames and particle orientation

In order to specify the directions of the incident and scattered waves, we will use a laboratory reference frame (indicated by $\mathcal{L}$), identified by the right-handed triad $(\hat{\theta}, \hat{\phi}, \hat{\mathbf{n}})$ with the z-axis directed in the local vertical direction as shown in Figure 1.1, Chapter 1 (see also Mishchenko *et al.*, 2000). The θ and ϕ components of the electric field are denoted as E_θ and E_ϕ so that $\vec{\mathbf{E}}_\theta = E_\theta\hat{\theta}$ (or equivalently $\vec{\mathbf{E}}_v = E_v\hat{\mathbf{v}}$) lies in the meridional or vertical plane (i.e. the plane spanned by the beam and the z-axis) whereas the $\vec{\mathbf{E}}_\phi = E_\phi\hat{\phi}$ (or equivalently $\vec{\mathbf{E}}_h = E_h\hat{\mathbf{h}}$) is perpendicular to this plane and is called the horizontal component.[1] Very often, in order to describe the scattering properties of a particle or of an ensemble of particles, it is useful to define another right-handed reference frame, called the particle reference frame and indicated by $\mathcal{P}$, having the same origin as $\mathcal{L}$. The choice of this frame is generally

[1] The transformation from the θ and ϕ components of a vector (the electric field, for instance) into its x, y and z components is performed by the 3×2 matrix:

$$\begin{bmatrix} E_x \\ E_y \\ E_z \end{bmatrix} = \tilde{\alpha}(\theta,\phi) \begin{bmatrix} E_\theta \\ E_\phi \end{bmatrix} = \begin{pmatrix} \cos\theta\cos\phi & -\sin\phi \\ \cos\theta\sin\phi & \cos\phi \\ -\sin\theta & 0 \end{pmatrix} \begin{bmatrix} E_\theta \\ E_\phi \end{bmatrix}$$

driven by symmetric properties of the scattering body (e.g. with the z-axis coincident with the axis of particle symmetry for axially symmetric particles). As described in Mishchenko (2000) the orientation of the $\mathcal{P}$ frame with respect to the $\mathcal{L}$ frame is typically specified by three Eulerian angles of rotation[2] that transform the coordinates of a vector in the $\mathcal{L}$ frame into the coordinates of a vector in the $\mathcal{P}$ frame.

3.2.2 Amplitude matrix

The solution of the scattering problem of a single particle reduces to the computation of the amplitude scattering matrix that relates the scattered field to the incident field. Throughout most of the chapter, we assume and suppress the time factor $e^{-i\omega t}$ (thus we adopt the + convention for the imaginary part of the refractive index).

In the far-field approximation (conditions $kr \gg 1$, $r \gg a$ and $r \gg ka^2/2$ as shown in Mishchenko *et al.* (2004) where a is the radius of the smallest circumscribing sphere of the scattering object and k is the wave number), the spherical wave-like scattered field $\vec{\mathbf{E}}_s$ going out in the direction (θ_s, ϕ_s) is linearly[3] related to the incident electric field by the complex amplitude matrix $\mathbf{S}$ defined in Equation (1.69), Chapter 1. Here, by following Tsang *et al.* (1985), we will adopt the following notation

$$\begin{pmatrix} S_{11} & S_{12} \\ S_{21} & S_{22} \end{pmatrix} = \begin{pmatrix} f_{vv}(\theta_s, \phi_s, \theta_i, \phi_i,) & f_{vh}(\theta_s, \phi_s, \theta_i, \phi_i,) \\ f_{hv}(\theta_s, \phi_s, \theta_i, \phi_i,) & f_{hh}(\theta_s, \phi_s, \theta_i, \phi_i,) \end{pmatrix} \tag{3.1}$$

The $\mathbf{S}$ matrix, also called the amplitude matrix, has dimension $[L]$ and linearly transforms the electric field vector components of the incident wave into the electric field vector components of the scattered wave; besides the incident and scattering directions it generally depends on the size, morphology and composition of the scattering particle and on its orientation in the lab frame. Once the $\mathbf{S}$ matrix is known, every other scattering characteristic of the particle can be computed. An important property of the amplitude matrix is that it satisfies the reciprocity relation:

$$\mathbf{S}(\theta_s, \phi_s, \theta_i, \phi_i,) = Q\mathbf{S}^{\mathrm{T}}(\pi - \theta_i, \pi + \phi_i, \pi - \theta_s, \pi + \phi_s)Q \quad Q \equiv \mathrm{diag}(1, -1)$$

3.2.2.1 Changing coordinate systems

Generally the amplitude matrix of non-spherical particles is more conveniently computed in the particle coordinate system $\mathcal{P}$. It is therefore necessary to transform the scattering amplitude matrix from the particle frame $\mathcal{P}$ to the lab frame $\mathcal{L}$. Here we briefly summarise the result (details for the derivation are found,

[2] For instance for the electric field

$$\begin{pmatrix} E_x \\ E_y \\ E_z \end{pmatrix}_{\mathcal{P}} = \begin{pmatrix} \cos\alpha\cos\beta\cos\gamma - \sin\alpha\sin\gamma & \sin\alpha\cos\beta\cos\gamma + \cos\alpha\sin\gamma & -\sin\beta\cos\gamma \\ -\cos\alpha\cos\beta\sin\gamma - \sin\alpha\cos\gamma & -\sin\alpha\cos\beta\sin\gamma + \cos\alpha\cos\gamma & \sin\beta\sin\gamma \\ \cos\alpha\sin\beta & \sin\alpha\sin\beta & \cos\beta \end{pmatrix} \begin{pmatrix} E_x \\ E_y \\ E_z \end{pmatrix}_{\mathcal{L}}$$

We will denote with $Eu(\alpha, \beta, \gamma)$ the rotation matrix in the right-hand side.

[3] This is because of the linearity of Maxwell's equations and boundary condition.

e.g. in Mishchenko, 2000):

$$\mathbf{S}^{\mathcal{L}}(\theta_s^{\mathcal{L}}, \phi_s^{\mathcal{L}}, \theta_i^{\mathcal{L}}, \phi_i^{\mathcal{L}}; \alpha, \beta, \gamma) = \hat{\rho}^{-1}(\hat{\mathbf{n}}_s; \alpha, \beta, \gamma)\mathbf{S}^{\mathcal{P}}(\theta_s^{\mathcal{P}}, \phi_s^{\mathcal{P}}, \theta_i^{\mathcal{P}}, \phi_i^{\mathcal{P}}) \times \hat{\rho}(\hat{\mathbf{n}}_i; \alpha, \beta, \gamma) \tag{3.2}$$

where

$$\hat{\rho}(\hat{\mathbf{n}}, \alpha, \beta, \gamma) = \tilde{\alpha}^{\mathrm{T}}(\theta^{\mathcal{P}}, \phi^{\mathcal{P}}) Eu(\alpha, \beta, \gamma)_{\mathcal{L}\to\mathcal{P}} \tilde{\alpha}(\theta^{\mathcal{L}}, \phi^{\mathcal{L}})$$

is the matrix transforming the $\theta - \phi$ components of a vector from $\mathcal{L}$ to $\mathcal{P}$. For instance in the very simple case of a rotation by an angle γ around the z-axis it comes out that $\hat{\rho}$ is the identity matrix and

$$\begin{aligned}\mathbf{S}^{\mathcal{L}}(\theta_s^{\mathcal{L}}, \phi_s^{\mathcal{L}}, \theta_i^{\mathcal{L}}, \phi_i^{\mathcal{L}}; 0, 0, \gamma) &= \mathbf{S}^{\mathcal{P}}(\theta_s^{\mathcal{P}}, \phi_s^{\mathcal{P}}, \theta_i^{\mathcal{P}}, \phi_i^{\mathcal{P}}) \\ &= \mathbf{S}^{\mathcal{P}}(\theta_s^{\mathcal{L}}, \phi_s^{\mathcal{L}} - \gamma, \theta_i^{\mathcal{L}}, \phi_i^{\mathcal{L}} - \gamma)\end{aligned}$$

as expected. The transformation rule (3.2) is very useful to compute the amplitude matrix of non-spherical particles in an arbitrary orientation and for arbitrary illumination and scattering directions with respect to $\mathcal{L}$, provided that the scattering problem has been solved in the $\mathcal{P}$ frame. This technique is particularly useful within the T-matrix context (see Section 3.4.2).

3.2.3 *Scattering amplitude matrix*

In the scattering community instead of dealing with the amplitude matrix, the scattering amplitude matrix is computed. The name for the two matrices are very similar but they should not be confused. The main difference consists in the fact that the electric field components are defined in the scattering amplitude matrix with respect to the scattering plane (the plane spanned by the directions of incidence and scattering) and not to the meridional planes. Moreover the amplitude scattering matrix is computed by assuming that the incident light propagates along the z-axis of the laboratory coordinate system (see Figure 3.3 in Bohren and Huffman (1983)) (i.e. $\theta_i = 0$). Following the notation of Bohren and Huffman (1983) (van de Hulst (1981), has a different notation but the same approach) the scattered field can be written as

$$\begin{pmatrix} E_{\parallel} \\ E_{\perp} \end{pmatrix}_s = \frac{e^{ikr}}{-ikr} \begin{pmatrix} S_2 & S_3 \\ S_4 & S_1 \end{pmatrix} (\theta_s, \phi_s) \begin{pmatrix} E_{\parallel} \\ E_{\perp} \end{pmatrix}_i$$

Also note the presence of the $(-ik)$ term in the denominator (absent in Equation (1.69)) that makes the amplitude scattering matrix non-dimensional. The electric fields are decomposed now into slightly different coordinates $\hat{\mathbf{e}}_{\parallel s} = \hat{\boldsymbol{\theta}}_s$ and $\hat{\mathbf{e}}_{\perp s} = -\hat{\boldsymbol{\phi}}_s$ for the scattered field (which apart from a minus coincide with the coordinates used when dealing with meridional planes) while $\hat{\mathbf{e}}_{\parallel i} = \cos\phi_s \hat{\mathbf{e}}_x + \sin\phi_s \hat{\mathbf{e}}_y$ and $\hat{\mathbf{e}}_{\perp i} = \sin\phi_s \hat{\mathbf{e}}_x - \cos\phi_s \hat{\mathbf{e}}_y$ (which are quite different from the meridional basis $\hat{\boldsymbol{\theta}}_i$ and $\hat{\boldsymbol{\phi}}_i$) for the incident field. By changing the basis (see also Draine and Flatau, 2000a) it can be shown that

$$\begin{pmatrix} S_2 & S_3 \\ S_4 & S_1 \end{pmatrix}(\theta_s, \phi_s) = \frac{i}{k}\begin{pmatrix} -f_{vv} & -f_{vh} \\ f_{hv} & f_{hh} \end{pmatrix}(\theta_s, \phi_s, \theta_i = 0, \phi_i)$$
$$\times \begin{pmatrix} \cos(\phi_s - \phi_i) & \sin(\phi_s - \phi_i) \\ \sin(\phi_s - \phi_i) & -\cos(\phi_s - \phi_i) \end{pmatrix} \quad (3.3)$$

so that the amplitude scattering matrix can be considered as a particular case of the **S** matrix defined in Equation (1.69). In fact by choosing $\phi = \phi_i = \phi_s$ Equation (3.3) becomes

$$\begin{pmatrix} S_2 & S_3 \\ S_4 & S_1 \end{pmatrix}(\theta, \phi) = \frac{i}{k}\begin{pmatrix} -f_{vv} & f_{vh} \\ f_{hv} & -f_{hh} \end{pmatrix}(\theta, \phi, 0, \phi)$$

with the minus sign difference in the off-diagonal elements of the amplitude matrix coming essentially from the use of different RH and LH bases.

The Stokes parameters are usually defined relative to their respective meridional planes. Note that, in the scattering community the Stokes parameters typically refer to the scattering plane. Therefore attention has to be paid when using externally computed single scattering properties as input in RT models (see Mishchenko *et al.*, 2000 for details).

3.2.4 Stokes and scattering matrix

From the definition of the amplitude matrix (see Equation (1.69)) and the Stokes parameters, it is possible to compute the Stokes parameters of the scattered spherical wave from those of the incident plane wave by

$$\boldsymbol{I}_s = \frac{1}{r^2}\mathbf{Z}(\theta_s, \phi_s, \theta_i, \phi_i,)\boldsymbol{I}_i$$

with $\mathbf{Z}$ the Stokes matrix that has a dimension of $[L^2]$; the choice of the Stokes basis is not unique and depends on the kind of application. By using the modified Stokes basis $\boldsymbol{I} = [I_v, I_h, U, V]^T$ (see Section 1.4.2, Chapter 1) the Stokes matrix expressed in terms of amplitude matrix elements is (see Mishchenko *et al.*, 2000, p. 11; Tsang *et al.*, 1985, p. 133)

$$Z_{vh} = \begin{pmatrix} |f_{vv}|^2 & |f_{vh}|^2 & \mathrm{Re}(f_{vh}^* f_{vv}) & -\mathrm{Im}(f_{vh}^* f_{vv}) \\ |f_{hv}|^2 & |f_{hh}|^2 & \mathrm{Re}(f_{hh}^* f_{hv}) & -\mathrm{Im}(f_{hh}^* f_{hv}) \\ 2\,\mathrm{Re}(f_{vv} f_{hv}^*) & 2\,\mathrm{Re}(f_{vh} f_{hh}^*) & \mathrm{Re}(f_{vv} f_{hh}^* + f_{vh} f_{hv}^*) & -\mathrm{Im}(f_{vv} f_{hh}^* - f_{vh} f_{hv}^*) \\ 2\,\mathrm{Im}(f_{vv} f_{hv}^*) & 2\,\mathrm{Im}(f_{vh} f_{hh}^*) & \mathrm{Im}(f_{vv} f_{hh}^* + f_{vh} f_{hv}^*) & \mathrm{Re}(f_{vv} f_{hh}^* - f_{vh} f_{hv}^*) \end{pmatrix} \quad (3.4)$$

In circular (C) or linear (L) basis (as defined in Chapter 1 and in Bohren and Huffman, 1983) the Stokes matrix $\mathbf{Z}_{C,L}$ defined by

$$(\boldsymbol{I}_{C,L})_s = \frac{1}{r^2}\mathbf{Z}_{C,L}(\boldsymbol{I}_{C,L})_i$$

can be easily related to the form (3.4) (see Mischchenko, 1991b, for details). Note that we have a minus sign in the elements 13, 23, 34, 31, 32, 34 with respect to Mishchenko *et al.* (2000)[4] because of the opposite convention for the U element. In general, all 16 elements of the phase matrix are non-zero, but only 7 of the phase matrix elements[5] are independent and there are several relations among the 16 quantities. The reciprocity relation for the phase matrix in the linear basis has the general form

$$\mathbf{Z}_L(-\hat{\mathbf{n}}', -\hat{\mathbf{n}}) = P\mathbf{Z}_L^T(\hat{\mathbf{n}}, \hat{\mathbf{n}}')P \quad P = \mathrm{diag}(1, 1, -1, 1) \tag{3.5}$$

This allows some memory saving when storing the elements of the phase matrix.

Similarly to what we have already done with the amplitude matrix and the scattering amplitude matrix it is possible to define (see Bohren and Huffman, 1983; van de Hulst, 1981) a non-dimensional scattering or Mueller matrix $\mathcal{S}$

$$\boldsymbol{I}_{\mathrm{s}} = \frac{1}{k^2 r^2} \mathcal{S}(\theta, \phi) \boldsymbol{I}_{\mathrm{i}}$$

that relates the incident $\boldsymbol{I}_{\mathrm{i}}$ and the scattered $\boldsymbol{I}_{\mathrm{s}}$ Stokes vector defined with respect to the scattering plane, which is the plane spanned by $\hat{\mathbf{n}}_{\mathrm{i}}$ and $\hat{\mathbf{n}}_{\mathrm{s}}$. Note that apart from the $1/k^2$ factor and a minus sign in the two 2×2 anti-diagonal block the Mueller $\mathcal{S}$ matrix coincides with $\mathbf{Z}(\theta, \phi, 0, \phi)$ (compare for instance Equations 14–29, p. 11 in Mishchenko *et al.* (2000) and Equation (3.16) in Bohren and Huffman (1983)).

3.2.5 Phase matrix

Because of the incoherent addition of Stokes parameters, the phase matrix is equal to the Stokes matrix averaged over the distribution of particles in terms of size, shape and orientation, that is

$$\langle \mathbf{Z} \rangle (\theta_{\mathrm{s}}, \phi_{\mathrm{s}}, \theta_{\mathrm{i}}, \phi_{\mathrm{i}}) = \sum_{\text{size,shape,orientation}} \mathbf{Z}(\theta_{\mathrm{s}}, \phi_{\mathrm{s}}, \theta_{\mathrm{i}}, \phi_{\mathrm{i}}) \tag{3.6}$$

[4] In terms of Stokes parameter definition Mishchenko *et al.* (2000) use the same convention as Bohren and Huffman (1983); the only difference is in terminology, that is when $I = V = 1$, $Q = U = 0$ Mishchenko *et al.* (2000) (like Tsang *et al.*, 1985) refer to a LH state and Bohren and Huffman (1983) to a RH state. This is because RH states are defined as states where the electric field is clockwise rotating by an observer looking in the direction of propagation and by an observer looking towards the source for Mishchenko *et al.* (2000) and Bohren and Huffman (1983), respectively.

[5] In the phase matrix there are only seven independent real constants resulting from the four complex elements of the matrix minus one irrelevant phase.

For example, for n_0 ellipsoids per unit volume with axes a, b, c and orientation Euler angles α, β, γ with respect to the laboratory reference frame, the phase matrix is

$$\langle \mathbf{Z} \rangle(\theta, \phi, \theta', \phi') = n_0 \int \mathrm{d}a \int \mathrm{d}b \int \mathrm{d}c \int \mathrm{d}\alpha \int \sin\beta \mathrm{d}\beta \times \int \mathrm{d}\gamma\, p(a, b, c, \alpha, \beta, \gamma) \mathbf{Z}(\theta, \phi, \theta', \phi')$$

where $p(a, b, c, \alpha, \beta, \gamma)$ is the probability density function for the quantities $a, b, c, \alpha, \beta, \gamma$; if the n_0 particles in the unit volume are all identical in shape and orientation

$$\langle \mathbf{Z} \rangle(\theta, \phi, \theta', \phi') = n_0 \mathbf{Z}(\theta, \phi, \theta', \phi') \tag{3.7}$$

Note that while the dimension of $\mathbf{Z}$ is $[L^2]$ the dimension of $\langle \mathbf{Z} \rangle$ is $[L^{-1}]$. It is the phase matrix which enters the VRTE.

3.2.6 Cross sections

The phase matrix describes the angular distribution of electromagnetic waves scattered by a particle. Two mechanisms change the characteristics of the incident beam when it interacts with the particle: scattering and absorption. The sum of the two is called extinction. Cross sections (that have the dimension of an area) are a measure of the probability that an incident beam (with a fixed direction relative to the orientation of the particle and with a fixed polarisation state) undergoes for one of these processes. The physical interpretation of these quantities is that the product of the extinction, scattering or absorption cross sections and the incident monochromatic energy flux is equal to the total monochromatic power that is removed, scattered or absorbed by the particle from the incident beam.

3.2.7 Extinction matrix

Neglecting scattering effects, non-spherical particles with preferential orientation can change not only the energy characteristic of the incident beam (by reducing the intensity of the beam) but also its state of polarisation. In fact there are different extinction coefficients for the different Stokes components of incident light. This phenomenon is known as dichroism of the scattering medium and has well-known manifestations in interstellar polarisation of star-light by magnetic-field-oriented cosmic dust particles and in cross polarisation of radio waves propagating through partially aligned hydrometeors. So, while for spherical particles the extinction matrix is diagonal, for non-spherical particles this is not true any more. According to the optical theorem, the elements of the extinction matrix for an ensemble of particles in the modified Stokes basis are expressed in the elements of the scattering amplitude matrix evaluated in the forward direction as follows (see Tsang *et al.*,1985, p. 140):

$$\langle \mathbf{\Sigma}_{vh}\rangle(\theta,\phi) = \frac{2\pi n_0}{k}\begin{pmatrix} 2\mathrm{Im}(\langle f_{vv}\rangle) & 0 & \mathrm{Im}(\langle f_{vh}\rangle) & -\mathrm{Re}(\langle f_{vh}\rangle) \\ 0 & 2\mathrm{Im}(\langle f_{hh}\rangle) & \mathrm{Im}(\langle f_{hv}\rangle) & \mathrm{Re}(\langle f_{hv}\rangle) \\ 2\mathrm{Im}(\langle f_{hv}\rangle) & 2\mathrm{Im}(\langle f_{vh}\rangle) & \mathrm{Im}(\langle f_{vv}+f_{hh}\rangle) & -\mathrm{Re}(\langle f_{hh}-f_{vv}\rangle) \\ 2\mathrm{Re}(\langle f_{hv}\rangle) & -2\mathrm{Re}(\langle f_{vh}\rangle) & -\mathrm{Re}(\langle f_{vv}-f_{hh}\rangle) & \mathrm{Im}(\langle f_{vv}+f_{hh}\rangle) \end{pmatrix} \tag{3.8}$$

where the angular brackets denote averages over the orientation and the size distribution of the particles. All amplitude matrix elements are evaluated in the forward direction (so $\theta_\mathrm{i} = \theta_\mathrm{s} = \theta$ and $\phi_\mathrm{i} = \phi_\mathrm{s} = \phi$). All extinction matrix elements have the dimension $[L^{-1}]$. Eigenvectors and eigenvalues of matrix (3.8) are described in Tsang *et al.* (1985, p. 141) and are very useful when solving the VRTE in the limit of no scattering. The reciprocity relation for the extinction matrix in the linear basis has the same form as that for the Stokes matrix given in (3.5). From (3.8) we define $k_\mathrm{e}^v(\theta,\phi) = n_0(4\pi/k)\mathrm{Im}\langle f_{vv}\rangle$ and $k_\mathrm{e}^h(\theta,\phi) = n_0(4\pi/k)\mathrm{Im}\langle f_{hh}\rangle$ as the extinction coefficients related to the v and h components.

3.2.8 *Emission vector*

Kirchoff's law states that, for a body in local thermodynamic equilibrium, the spectral directional absorptance is equal to spectral directional emittance. In the scalar RTE absorption of energy is described by the absorption coefficient, and therefore, due to Kirchoff's law, the proportionality constant for emission of energy is described by the same coefficient. Using conservation of energy arguments, Tsang *et al.* (2000) derived an expression for the emission vector for passive remote sensing of non-spherical particles in general conditions; a much simpler derivation of the emission vector is given in Section 2.10 of the book by Mishchenko *et al.* (2002). The derivation is valid for particles in equilibrium and in non-equilibrium. For a single particle in the modified Stokes basis the emission cross-section vector (with dimension $[L^2]$) in direction $-\hat{r}$ is equal to the absorption cross-section vector $\bar{\sigma_a}$ in direction $\hat{r}$:

$$\begin{aligned}(\sigma_{vh})_{a1}(\hat{r}) &= \frac{4\pi}{k}\mathrm{Im}(f_{vv}(\hat{r},\hat{r})) - \int [|f_{vv}(\hat{r}',\hat{r})|^2 + |f_{hv}(\hat{r}',\hat{r})|^2]\mathrm{d}\Omega' \\ &= \sigma_\mathrm{e}^v - \sigma_\mathrm{s}^v \equiv \sigma_\mathrm{a}^v \end{aligned} \tag{3.9}$$

$$\begin{aligned}(\sigma_{vh})_{a2}(\hat{r}) &= \frac{4\pi}{k}\mathrm{Im}(f_{hh}(\hat{r},\hat{r})) - \int [|f_{hh}(\hat{r}',\hat{r})|^2 + |f_{vh}(\hat{r}',\hat{r})|^2]\mathrm{d}\Omega' \\ &= \sigma_\mathrm{e}^h - \sigma_\mathrm{s}^h \equiv \sigma_\mathrm{a}^h \end{aligned} \tag{3.10}$$

$$(\sigma_{vh})_{a3}(\hat{r}) = -2\mathrm{Re}\left\{\frac{2\pi i}{k}[f_{vh}^* - f_{hv}](\hat{r},\hat{r}) - \int [f_{vv}f_{vh}^* + f_{hv}f_{hh}^*](\hat{r}',\hat{r})\mathrm{d}\Omega'\right\} \tag{3.11}$$

$$(\sigma_{vh})_{a4}(\hat{r}) = -2\mathrm{Im}\left\{\frac{2\pi i}{k}[f_{vh}^* - f_{hv}](\hat{r},\hat{r}) - \int [f_{vv}f_{vh}^* + f_{hv}f_{hh}^*](\hat{r}',\hat{r})\mathrm{d}\Omega'\right\} \tag{3.12}$$

When considering an ensemble of particles, by using the form of the extinction matrix (3.8) and that of the phase matrix (3.7), Equations (3.9)–(3.12) reduce to

$$
\begin{aligned}
(a_{vh})_1(\hat{r}) &= k_e^v(\hat{r}) - \int_{4\pi} [\langle \mathbf{Z}_{vh}\rangle_{11} + \langle \mathbf{Z}_{vh}\rangle_{21}](\hat{r};\hat{r}')\mathrm{d}\Omega' \\
(a_{vh})_2(\hat{r}) &= k_e^h(\hat{r}) - \int_{4\pi} [\langle \mathbf{Z}_{vh}\rangle_{12} + \langle \mathbf{Z}_{vh}\rangle_{22}](\hat{r};\hat{r}')\mathrm{d}\Omega' \\
(a_{vh})_3(\hat{r}) &= -2\langle \mathbf{\Sigma}_{vh}\rangle_{13}(\hat{r}) - 2\langle \mathbf{\Sigma}_{vh}\rangle_{23}(\hat{r}) \\
&\quad + 2\int_{4\pi} [\langle \mathbf{Z}_{vh}\rangle_{13} + \langle \mathbf{Z}_{vh}\rangle_{23}](\hat{r};\hat{r}')\mathrm{d}\Omega' \\
(a_{vh})_4(\hat{r}) &= 2\langle \mathbf{\Sigma}_{vh}\rangle_{14}(\hat{r}) + 2\langle \mathbf{\Sigma}_{vh}\rangle_{24}(\hat{r}) \\
&\quad - 2\int_{4\pi} [\langle \mathbf{Z}_{vh}\rangle_{14} + \langle \mathbf{Z}_{vh}\rangle_{24}](\hat{r};\hat{r}')\mathrm{d}\Omega'
\end{aligned}
\tag{3.13}
$$

that define an absorption vector coefficient $\langle \boldsymbol{a} \rangle$ with dimension $[L^{-1}]$.

3.3 Simplified forms of extinction and phase matrix and of absorption vector

When working with the RTE it is crucial to exploit all possible symmetries in the phase matrix (3.6), in the extinction matrix (3.8) and in the absorption vector coefficients (3.13) appropriate for the system under study. Here we analyse two particular conditions that are usually met when dealing with atmospheric scatterers.

3.3.1 *Macroscopically isotropic and symmetric media*

A very important case when dealing with atmospheric particles is macroscopically isotropic and symmetric (hereafter MIS) media. This term refers to media where particles are randomly oriented and, moreover, when each particle has a plane of symmetry and/or particles and their mirror particles are present in equal numbers.

In this case the scattering matrix for the population of particle, $\mathcal{S}$, is invariant with respect to the choice of the scattering plane and depends only on the angle of scattering Θ, i.e. the angle between the incident and the scattering direction is given by

$$
\mu_\Theta \equiv \cos\Theta = \cos\theta_i \cos\theta_s + \sin\theta_i \sin\theta_s \cos(\phi_i - \phi_s) \tag{3.14}
$$

Therefore, following the considerations of Section 3.2.4 we can actually write

$$
\mathcal{S}(\Theta) = k^2 \langle \mathbf{Z} \rangle (\Theta = \theta_s, 0, 0, 0,) \tag{3.15}
$$

Moreover, for MIS media the scattering and extinction coefficients are independent of the direction and polarisation of the incident beam (let us call them k_{sca}, k_e). In fact in this case $\langle f_{hv}(\mathbf{n},\mathbf{n})\rangle = \langle f_{vh}(\hat{\mathbf{n}},\hat{\mathbf{n}})\rangle = 0$ and $\langle f_{hh}(\hat{\mathbf{n}},\hat{\mathbf{n}})\rangle = \langle f_{vv}(\hat{\mathbf{n}},\hat{\mathbf{n}})\rangle$ for all directions $\hat{\mathbf{n}}$. Then the extinction matrix (3.8) is diagonal and direction independent with $k_e = k_e^v = k_e^h$. It is possible then to introduce the so-called single scattering albedo $\varpi = (k_{sca}/k_e)$ which also is both direction and polarisation independent. The absorption vector (3.13) has only the first two elements different from zero and equal to $k_e - k_{sca}$.

It is very useful to introduce a renormalised scattering matrix by

$$\mathbf{F}(\Theta) = \frac{4\pi}{k_{sca}}\langle\mathbf{Z}\rangle(\Theta = \theta_s, 0, 0, 0) = \frac{\lambda^2}{\pi k_{sca}}\mathcal{S}(\Theta) \tag{3.16}$$

the general structure of which is given by[6]

$$\mathbf{F}(\Theta) = \begin{pmatrix} F_{11}(\Theta) & F_{12}(\Theta) & 0 & 0 \\ F_{12}(\Theta) & F_{22}(\Theta) & 0 & 0 \\ 0 & 0 & F_{33}(\Theta) & F_{34}(\Theta) \\ 0 & 0 & -F_{34}(\Theta) & F_{44}(\Theta) \end{pmatrix} \tag{3.17}$$

The term $P(\Theta) \equiv (F_{11}^{vh} + 2F_{12}^{vh} + F_{22}^{vh})/2 = F_{11}^L$ represents the phase function for unpolarised radiation while $P_v(\Theta) \equiv F_{11}^{vh} + F_{21}^{vh} = F_{11}^L + F_{12}^L$ and $P_h(\Theta) \equiv F_{22}^{vh} + F_{12}^{vh} = F_{11}^L - F_{12}^L$ are the phase functions for vertical and horizontal polarisation.

The quantity

$$g \equiv \langle\cos\Theta\rangle = \frac{1}{2}\int_{-1}^{1} \cos\Theta P(\Theta)\mathrm{d}\mu_\Theta \tag{3.18}$$

is called the asymmetry parameter of the phase function. Its interpretation is that $g > 0$ for particles that scatter predominantly in the forward direction and $g < 0$ otherwise.

The proportionality constant in (3.16) is chosen in such a way that the phase function satisfies the normalisation condition:

$$\int_0^{2\pi}\int_{-1}^{1} \frac{P(\Theta)}{4\pi}\mathrm{d}\mu_\Theta \mathrm{d}\phi = 1 \tag{3.19}$$

[6] This matrix is typically given in the linear basis

$$\begin{pmatrix} F_{11}^L(\Theta) & F_{12}^L(\Theta) & 0 & 0 \\ F_{12}^L(\Theta) & F_{22}^L(\Theta) & 0 & 0 \\ 0 & 0 & F_{33}^L(\Theta) & F_{34}^L(\Theta) \\ 0 & 0 & -F_{34}^L(\Theta) & F_{44}^L(\Theta) \end{pmatrix} = \begin{pmatrix} \frac{F_{11}^{vh} + 2F_{12}^{vh} + F_{22}^{vh}}{2} & \frac{F_{11}^{vh} - F_{22}^{vh}}{2} & 0 & 0 \\ \frac{F_{11}^{vh} - F_{22}^{vh}}{2} & \frac{F_{11}^{vh} - 2F_{12}^{vh} + F_{22}^{vh}}{2} & 0 & 0 \\ 0 & 0 & F_{33}^{vh} & F_{34}^{vh} \\ 0 & 0 & -F_{34}^{vh} & F_{44}^{vh} \end{pmatrix}$$

For spheres $F_{11}^L = F_{22}^L$ and $F_{33}^L = F_{44}^L$ so that $F_{12}^{vh} = 0$, $F_{33}^{vh} = F_{44}^{vh}$, $F_{11}^{vh} = F_{11}^L + F_{12}^L$ and $F_{22}^{vh} = F_{11}^L - F_{12}^L$.

For example the Rayleigh scattering matrix (3.17) becomes

$$\mathbf{F}^{vh}_{\text{Rayleigh}}(\Theta) = \frac{3}{2}\begin{pmatrix} \cos^2\Theta & 0 & 0 & 0 \\ 0 & 1 & 0 & 0 \\ 0 & 0 & \cos\Theta & 0 \\ 0 & 0 & 0 & \cos\Theta \end{pmatrix}$$

3.3.1.1 Rotation of reference frame: connection between phase and renormalised scattering matrix

For MIS media the renormalised scattering matrix carries all information about the scattering properties of the medium. In this section, a relationship between the renormalised scattering matrix and the general form of the phase matrix (needed by the VRTE) is developed. Looking at Figure 3.1, the meridian plane P_1OZ is the plane of the incoming wave $\boldsymbol{I}_{\text{i}}$; the scattering matrix is defined relative to the scattering plane P_1OP_2 and the outgoing intensity $\boldsymbol{I}_{\text{s}}$ is defined in terms of the meridian plane P_2OZ. Therefore, for the VRTE and the Stokes matrix, the polarisation must be rotated into the scattering plane P_1OP_2 by angle i_1 (the angle between the meridian plane through the incident direction $(\theta_{\text{i}}, \phi_{\text{i}})$ and the scattering plane), the scattering matrix is then applied and finally the polarisation is rotated out of the plane by angle i_2 (the angle between the meridian plane through the scattering direction $(\theta_{\text{s}}, \phi_{\text{s}})$ and the scattering plane) to obtain the rotated scattering phase matrix. In mathematical terms, this transformation is expressed by

$$\langle\mathbf{Z}\rangle(\theta_{\text{s}}, \phi_{\text{s}}, \theta_{\text{i}}, \phi_{\text{i}}) \equiv \frac{k_{\text{sca}}}{4\pi}L(\pi - i_2)\mathbf{F}(\Theta)L(-i_1) \tag{3.20}$$

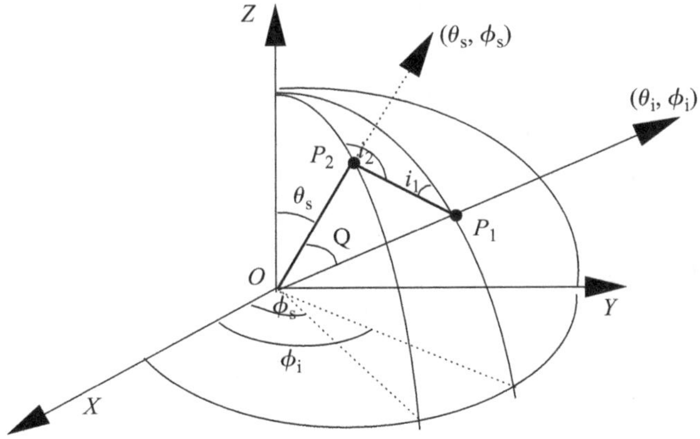

Figure 3.1 Illustration of the relationship between the phase matrix and the renormalised scattering matrix

where the matrix[7] L accounts for the transformation of the Stokes parameter when the reference plane is rotated by an angle φ in the clockwise direction[8] when looking towards the source or equivalently in the anticlockwise direction when looking in the direction of propagation. (Note that the rotation around OP_1 by i_1 in Figure 3.1 is anticlockwise when looking at the source; this is the reason why $L(-i_1)$ is applied.) Note that when the rotation of Stokes parameters is described by Equation (3.20) for all scattering configurations (like in Liou, 2000 or Chandrasekhar, 1960, while Mishchenko *et al.*, 2002, distinguish two cases, see pp. 88–90) i_1 and i_2 are assumed positive in patterns like in Figure 3.1 (on the other hand they will be negative like in Figure 4.3(a) in Mishchenko *et al.*, 2002).

The rotation of Stokes parameters in Equation (3.20) which is necessary to combine scattering theory and radiative transfer theory is often the main source of complication in practical radiative-transfer problems.

3.3.1.2 Geometry calculation

We briefly compile some useful relationships between the angles involved in scattering by providing direct and inverse formulae to compute the different angles of the problem. Taking (u_i, v_i, w_i) and (u_s, v_s, w_s) the vectors of the incident and of the scattered directions, $\hat{\mathbf{n}}_i$ and $\hat{\mathbf{n}}_s$, it is possible to write

$$\begin{aligned}\hat{\mathbf{n}}_s &= (\hat{\mathbf{n}}_s \cdot \hat{\mathbf{n}}_i)\hat{\mathbf{n}}_i + (\hat{\mathbf{n}}_s \cdot \hat{\mathbf{h}}_i)\hat{\mathbf{h}}_i + (\hat{\mathbf{n}}_s \cdot \hat{\mathbf{v}}_i)\hat{\mathbf{v}}_i \\ &= \cos\Theta\hat{\mathbf{n}}_i - \sin\Theta[\cos(i_1)\hat{\mathbf{v}}_i + \sin(i_1)\hat{\mathbf{h}}_i]\end{aligned}$$

so that the scattering directions are computed from the incident directions and the angles Θ and i_1 by

$$\begin{aligned}u_s &= \cos\Theta u_i + \sin\Theta[-\cos i_1 \cos\theta_i \cos\phi_i + \sin i_1 \sin\phi_i] \\ v_s &= \cos\Theta v_i + \sin\Theta[-\cos i_1 \cos\theta_i \sin\phi_i - \sin i_1 \cos\phi_i] \\ w_s &= \cos\Theta w_i + \sin\Theta \cos i_1\sqrt{1 - w_i^2}\end{aligned} \tag{3.22}$$

The angle i_1 is here defined as the angle between the two planes OP_1P_2 and OP_1Z [measured in anticlockwise direction when looking towards the source, i.e. starting from the $-\hat{\mathbf{v}}_i$ direction; same convention as for Equation (3.20)] and can be

[7] When using the modified Stokes basis L assumes the form

$$\mathbf{L}_{vh}(\varphi) = \begin{pmatrix} \cos^2(\varphi) & \sin^2(\varphi) & 0.5\sin(2\varphi) & 0 \\ \sin^2(\varphi) & \cos^2(\varphi) & -0.5\sin(2\varphi) & 0 \\ -\sin(2\varphi) & \sin(2\varphi) & \cos(2\varphi) & 0 \\ 0 & 0 & 0 & 1 \end{pmatrix} \tag{3.21}$$

[8] Here we adopt the same convention as Tsang *et al.* (1985) and Liou (2002) which is opposite to Mishchenko *et al.* (2002). For the two conventions the same L matrices are obviously obtained with opposite angles.

computed as

$$\cos i_1 = -\hat{\mathbf{v}}_\mathrm{i} \cdot \frac{\hat{\mathbf{n}}_\mathrm{s} - (\hat{\mathbf{n}}_\mathrm{i} \cdot \hat{\mathbf{n}}_\mathrm{s})\hat{\mathbf{n}}_\mathrm{i}}{\sin \Theta} = -\frac{\hat{\mathbf{v}}_\mathrm{i} \cdot \hat{\mathbf{n}}_\mathrm{s}}{\sin \Theta}$$
$$= \frac{\cos\theta_\mathrm{s}\sin\theta_\mathrm{i} - \cos\theta_\mathrm{i}\sin\theta_\mathrm{s}\cos(\phi_\mathrm{i} - \phi_\mathrm{s})}{\sin\Theta} \tag{3.23}$$

Note that a similar expression is valid for i_2 with the suffices i and s interchanged. Now if we want to select i_1 as the rotation angle around $\hat{\mathbf{n}}_\mathrm{i}$ measured in an anticlockwise direction from $\hat{\mathbf{v}}_\mathrm{i}$ we do have $\pi < i_1 < 2\pi$ when $\pi < \phi_\mathrm{i} - \phi_\mathrm{s} < 2\pi$ and $0 < i_1 < \pi$ when $0 < \phi_\mathrm{i} - \phi_\mathrm{s} < \pi$ (and, with the convention adopted in Equation (3.20), the same is true for i_2). This is enough to discern the ambiguity when giving the cosine of the angle only, i.e. in formulae

$$i_{1,2} = \begin{cases} 2\pi - \mathrm{acos}[\cos(i_{1,2})] & \pi < \phi_\mathrm{i} - \phi_\mathrm{s} < 2\pi \\ \mathrm{acos}[\cos(i_{1,2})] & 0 < \phi_\mathrm{i} - \phi_\mathrm{s} < \pi \end{cases} \tag{3.24}$$

Note that if you are only interested in the rotation of the Stokes parameter given by Equation (3.20) the angles i_1 and i_2 can be rescaled into the interval $[0, \pi]$ (this is because the L matrix defined in (3.21) is periodic with period π, see Figure 4.3 in Mishchenko *et al.* (2002)). For this kind of applications, $0 < i_1, i_2 < \pi$, and by taking into account the sign notation, they can be obtained (for both of them) from the acos of the cosine of formula (3.23) with the sagacity of changing its sign when $\pi < \phi_\mathrm{i} - \phi_\mathrm{s} < 2\pi$.

Note that in formula (3.23) the denominator becomes zero when $\sin(\Theta) = 0$, that is for forward and backscattering: in these cases, however, there is no need to perform the rotation (3.20).

3.3.2 Axially symmetric media

Another interesting case is populations of axially symmetric particles axially oriented symmetrically around the local vertical direction (ASPASO hereafter). For instance this is the case for falling raindrops with their symmetry axis aligned in the vertical direction with very little deviation; the tilt of the symmetry axis from the vertical direction is called canting. The mean canting angle for raindrops has been estimated to be near 0° with a standard deviation less than 3°. Columnar and planar snow crystals fall with their larger dimension horizontally oriented when larger than 100 μm; deviations from the horizontal exhibit a Gaussian distribution. In the presence of electrification crystals can be even vertically oriented. This kind of situation is particularly relevant when assessing polarisation effects. The first study referring to this configuration was represented by Mishchenko (1991a) (with focus on non-spherical dust grain effects onto polarisation of light), where the T-matrix approach is used to develop an analytical method to average the extinction matrix over orientations (see also in Tsang *et al.*, 1985, Chapter 3, Section 5.8).

For ASPASO particles, $\langle f_{hv}(\hat{\mathbf{n}}, \hat{\mathbf{n}})\rangle = \langle f_{vh}(\hat{\mathbf{n}}, \hat{\mathbf{n}})\rangle = 0$, $\langle f_{hh}(\hat{\mathbf{n}}, \hat{\mathbf{n}})\rangle = \langle f_{hh}(\theta)\rangle$, $\langle f_{vv}(\hat{\mathbf{n}}, \hat{\mathbf{n}}) = f_{vv}(\theta)\rangle\rangle$ so that

$$\langle \mathbf{\Sigma}_{vh}\rangle(\theta) = \begin{pmatrix} k_{\mathrm{e}}^{v}(\theta) & 0 & 0 & 0 \\ 0 & k_{\mathrm{e}}^{h}(\theta) & 0 & 0 \\ 0 & 0 & \dfrac{k_{\mathrm{e}}^{v} + k_{\mathrm{e}}^{h}}{2}(\theta) & -k_{\mathrm{e}}^{c}(\theta) \\ 0 & 0 & k_{\mathrm{e}}^{c}(\theta) & \dfrac{k_{\mathrm{e}}^{v} + k_{\mathrm{e}}^{h}}{2}(\theta) \end{pmatrix} \tag{3.25}$$

where k_{e}^{v}, k_{e}^{h} (defined in Section 3.2.7) and $k_{\mathrm{e}}^{c}(\theta) = -n_0(2\pi/k)\mathrm{Re}(\langle f_{vv}\rangle - \langle f_{hh}\rangle)$ are here the extinction coefficient for vertical, horizontal and circular polarisation, which depend only on the θ direction of incidence. Note that k_{e}^{c} coincides with the specific differential propagation phase shift K_{DP} (usually measured in °/km), a parameter very common in the polarimetric radar community. In this case the linearly and vertically polarised, and the left and right-handed circularly polarised, waves are the eigenvectors of the coherent wave propagation with eigenvalues equal to k_{e}^{v}, k_{e}^{h} and $0.5(k_{\mathrm{e}}^{v} + k_{\mathrm{e}}^{h}) \pm ik_{\mathrm{e}}^{c}$. This is the reason why we have used the $v-h$ basis. Note that now the single scattering albedo (like the extinction) depends both on the direction and on the state of polarisation of the wave. Another important characteristic for ASPASO is that the phase matrix depends only on the difference $\phi_{\mathrm{i}} - \phi_{\mathrm{s}}$ rather than on ϕ_{i} and ϕ_{s} separately.

The general case when the axis of symmetry for the orientation distribution does not coincide with the vertical z-axis (for instance, applicable to hydrometeors in presence of strong wind) has been discussed in Mishchenko (1991a) as well.

3.4 Single scattering parameter computations

The computation of the single scattering properties (the absorption vector, the extinction and the phase matrix) for the population of hydrometeors present at each point of the medium is always the first step in RT calculations. The single scattering parameters will generally depend on the size, shape, orientation distribution, wavelength and refractive index of the particles.

As described in Chapter 5 and references therein distinctive features characterise the refractive index of pure ice (like in ice crystals), pure water (like in cloud and rain droplets), low density ice particles (like in snow) and mixtures of ice-water and air (like in melting snowflakes). Especially for liquid water, the dielectric constant, described by the dielectric relaxation spectra of Debye, shows significant temperature sensitivity. Uncertainties are still large especially in the region of super-cooled water, below 0°C.

General descriptions about the size, shape and orientation distribution of hydrometeors can be found in Pruppacher and Klett (1997) and in Haferman (2000). The observation and parametrisation of these distribution functions still present a major challenge in the field of cloud research. When dealing with radiative transfer inside realistically structured cloud scenarios, the typical approach is to use the output of simulations from cloud resolving models (e.g. Tao and Simpson, 1993; Panegrossi *et al.*, 1998 and Section 3.8.1). These models include, however, also microphysical assumptions; typically they assume spherical hydrometeors and exponential-like size distribution. To identify the microwave signatures of cloud and precipitation the single scattering properties are first computed by using the assumptions coming from the CRM; then sensitivity studies on the impact of realistic size, shape and orientation distributions are performed as well.

For spherical particles, the classical method to calculate scattering and absorption coefficients and the phase functions are the Lorenz–Mie equations (see Westwater *et al.*, 2004 and Section 3.4.1). For a given wavelength and a single particle, the particle contribution is calculated; the bulk coefficients are then obtained by integration over the size distribution of particles. Atmospheric particle dimensions seldom exceed 1 cm, so that even for wavelengths of the order of 1 mm the x Mie parameter never exceeds 60; therefore, in the microwaves, the typical scattering regime is represented by the so-called resonance regime. For sufficiently small particles, the Rayleigh approximation can still be used (see Bohren and Huffman, 1983, for details).

The main difficulties in computing single scattering properties come from the non-spherical shapes of some hydrometeors (raindrops, ice crystals, snowflakes, graupel) and from the inhomogeneity of particles. For instance a melting hydrometeor present in the so-called radar 'bright band' can result from the combination of air, ice and water; because of the strong difference between ice and water refractive indices (particularly at low MW frequency[9]) a very thin skin of liquid water can be sufficient to cause significant absorption and thus emission.

3.4.1 Lorenz–Mie theory

When dealing with homogeneous spheres (or coated spherical objects) the scattered fields can be derived analytically by an infinite expansion in terms of vector spherical harmonics. The expansion coefficients called Mie-coefficients, usually named a_n and b_n, depend only on the relative refractive index of the sphere m (or both on the relative refractive index of the core and of the coating layer) and on the dimensionless Mie parameter

$$x \equiv ka = \frac{2\pi Na}{\lambda} \tag{3.26}$$

with k being the wavenumber in the medium surrounding the sphere (with refractive index N) and a the radius of the sphere. For coated spheres the Mie coefficients depend on size parameters of the inner and outer sphere. For a detailed derivation the interested reader is referred to Bohren and Huffman (1983) (Chapters 4

[9] The dielectric losses of ice have a minimum near 1 GHz, and ice is an almost perfectly loss-free medium over a large frequency range. Therefore microwave emission of pure ice particles can be neglected in most cloud situations.

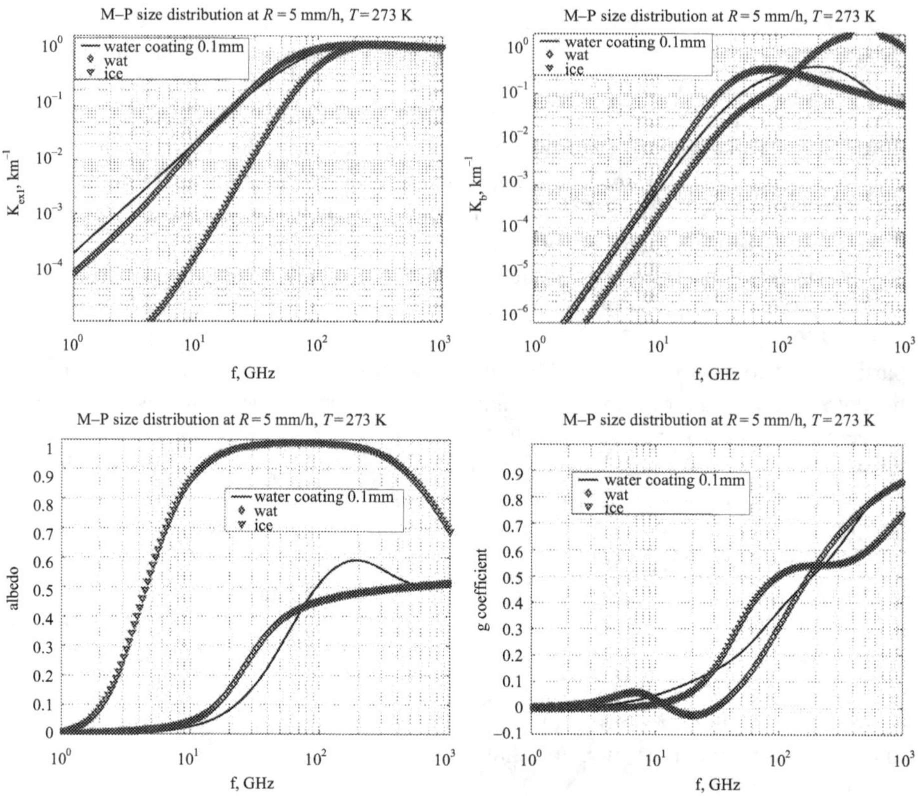

Figure 3.2 Extinction and backscattering coefficients (top panel), asymmetry parameter and single scattering albedo (bottom panel) for a population of water spheres, ice spheres and ice spheres with a 100 μm water coating

and 8 in particular). An updated list of websites having Mie codes with different coding language, available for download, can be found at http://www.iwt-bremen.de/~wriedt/Mie_Type_Codes/body_mie_type_codes.html. Many examples of mono-dispersed and poly-dispersed hydrometeor scattering properties can be found in the literature (e.g. Van De Hulst, 1981; Bohren and Huffman, 1983; Ulaby *et al.*, 1986; Janssen, 1996).

As a simple example we show in Figure 3.2 the scattering properties of a poly-disperse population of spheres corresponding to a 5 mm/h Marshall and Palmer precipitation size distribution. Note that at low frequencies ($f < 50$ GHz) water has, in general, a much greater extinction coefficient than ice and a much lower single scattering albedo. This is the reason why the radiometers working in this frequency range are essentially sensitive to water emission and pertinent algorithms usually referred to as emission methods (Stephens, 1994). At higher frequencies ($f > 50$ GHz) ice extinction becomes comparable to water so that ice particles start to play a major role in influencing the brightness temperatures. In this region the remote sensing of hydrometeors will essentially be driven by their scattering properties due to the high single scattering albedo values. Note that at $f > 200$ GHz there is no more increase of extinction with frequency, and extinction differences between ice and water become lower and lower: both tendencies reflect the fact that the extinction efficiency of the particles is approaching the geometric optic limit of 2 (despite composition and frequency). The backscattering coefficient reveals the f^4 dependence (the same is true for the scattering coefficient) in the Rayleigh regime (indicated by straight curves in log–log plots). Note the scaling factor between ice and water in this regime (driven by the $|K|^2$ factor, see Battan, 1973). The asymmetry parameter is, almost always, greater than 0 (i.e. more scattering in the forward than in the backward direction) and increases with frequency. At very high frequencies ($f > 200$ GHz) water particles scatter radiation even more in the forward direction than ice particles, with g values up to 0.9. Note that in this regime, the RT technique to solve the RTE has to be carefully chosen. Ice scattering is dominant over ice absorption at all microwave frequencies, the reverse being true for rain (see bottom left panel in Figure 3.2).

In Figure 3.2 we have also considered a population of ice spheres with a very thin water coating, thus providing a very rough description about the behaviour of first-stage melting particles. A remarkable result is that a very small water coating makes the particles very similar to pure water spheres (note that at lower frequencies they achieve even higher extinction than pure water spheres). This causes the well known bright band phenomenon in active remote sensing and increases the total optical thickness when a melting layer is present. Note that by simply looking at the two top panels of Figure 3.2, one would expect larger effects of melting particles at lower frequencies. In fact, the radar bright band is becoming less and less accentuate at K_a band with respect to X, S or C band and TMI-like T_B have lower sensitivity to the melting layer with increasing frequencies (see Battaglia *et al.*, 2003).

3.4.2 *T*-matrix method

The T-matrix method (TMM) is based on expanding the incident field in vector spherical wave functions (VSWFs) regular at the origin and expanding the scattered field outside a circumscribing sphere of the scatterer in VSWFs regular at infinity. The T-matrix transforms the expansion coefficients of the incident field into those of the scattered field and, if known, can be used to compute any scattering characteristic of the particle. The TMM was initially developed by Waterman (1971) for single homogeneous objects and was generalised to multi-layered scatterers and arbitrary clusters of non-spherical particles. For spheres, all TMM formulae reduce to those of the Lorenz–Mie theory. The T-matrix for single homogeneous and multi-layered scatterers is usually computed using the extended boundary condition method (EBCM) (see Waterman, 1971). The EBCM can be applied to any particle shape, although computations become much simpler and more efficient for bodies of revolution such as spheroids, circular cylinders and so-called Chebyshev particles. Special procedures have been developed to improve the numerical stability of EBCM computations for large size parameters and/or extreme axial ratios. Recent work has extended the practical applicability of the EBCM to particles without axial symmetry, e.g. ellipsoids, cubes and finite polyhedral cylinders (e.g. Kahnert, 2003). The loss of efficiency for particles with large axial ratios or with shapes lacking axial symmetry is the main drawback of the TMM. The main advantages of TMM are high accuracy and speed, an automatic convergence test and applicability to particles with equivalent-sphere size parameters exceeding 180. Mishchenko (1991b,1993), Khlebtsov (1992) and Mackowski and Mishchenko (1996) have developed analytical orientation averaging procedures which make TMM computations for randomly oriented particles as fast as those for particles with fixed orientation. Further information on the TMM can be found in Mischchenko *et al.* (1996), in Chapter 5 of Mishchenko *et al.* (2002) and in Kahnert (2003). A representative collection of public-domain T-matrix codes is posted on the World Wide Web at http://www.giss.nasa.gov/~crmim. A user guide to these codes can be found in Mishchenko and Travis (1998) and in Chapter 5 of Mishchenko *et al.* (2002).

As an example of possible applications of the T-matrix method in the microwave domain, Figure 3.3 shows the extinction coefficient and the single scattering albedo for raindrops at 280 K perfectly oriented in the horizontal plane as a function of the rain rate. The raindrops are modelled as oblate spheroids with axis ratio depending on the equivalent spherical diameter according to the formula by Chuang and Beard (1990) (see also Haferman, 2000, for a discussion of raindrop shape distributions). The scattering properties are shown for the H and V polarisation at nadir and at side view (i.e. at a zenith angle equal 90°) where differences are greater. As a reference value the equivolume sphere approximation is shown as well. For the given orientation distribution obviously the extinction coefficient at nadir is bigger than at side view, while the vertical extinction coefficient is always lower (equal at nadir) than the horizontal one; spheres provide intermediate values. As demonstrated in Section 3.9 the differences in the extinction coefficient and single scattering albedo between different polarisation states and incidence angles cause the PMW polarisation signatures observed from rain showers.

In Figure 3.4 the upper panel shows the k_{e}^{c} coefficient for the same population of Figure 3.3. In addition an azimuthally averaged phase function versus outgoing polar direction for different incoming directions for a population of $RR = 30$ mm/h at 85.6 GHz is shown as well (note the pronounced scattering in the forward direction).

As another example single scattering properties for mono-disperse spheroidal ice particles (mimicking cirrus crystals) with an equal volume sphere radius of 75 μm are calculated and compared with equivalent Lorenz–Mie calculations (Mätzler, 2002). The results of the calculations for 318 GHz are shown in Figure 3.5. All the six phase matrix elements of Equation (3.17) are presented. The grey lines correspond to the Mie calculation and the black lines correspond to the T-matrix calculations. The solid lines correspond to the normalised scattering matrix for spherical particles.

The matrix element F_{11}^{L}, which corresponds to the phase function, shows that randomly oriented non-spherical particles scatter slightly more radiation into the forward and slightly less radiation into the backward directions compared with spherical

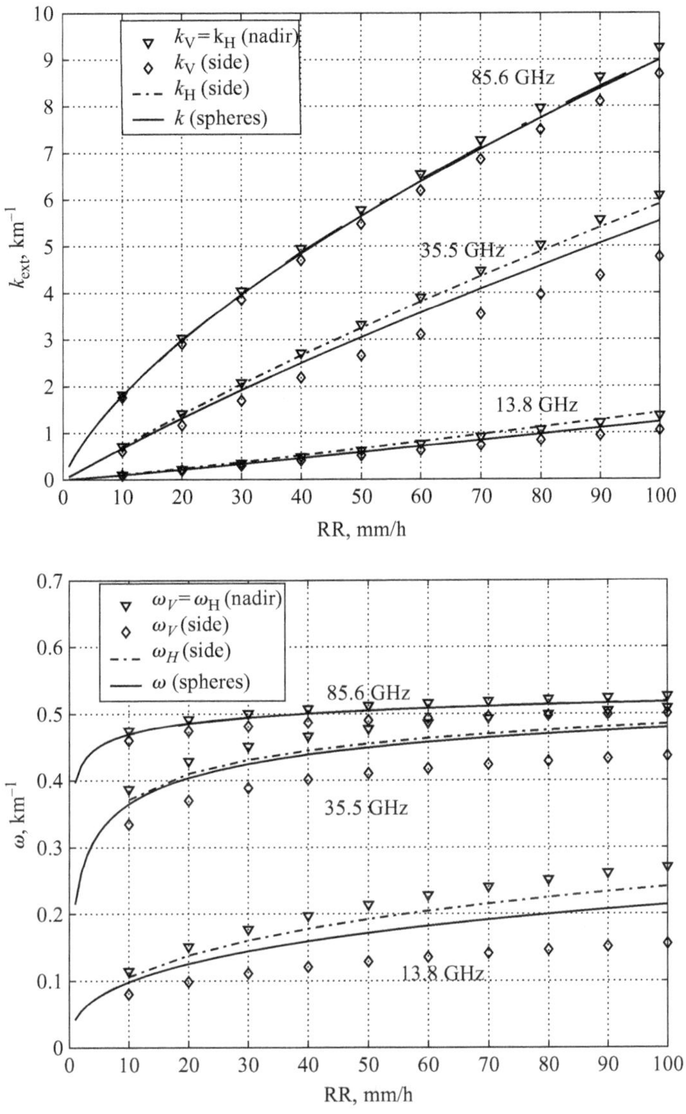

Figure 3.3 Extinction coefficient and single scattering albedo for population of perfectly oriented population of raindrops at 280 K temperature and size distribution according to Marshall and Palmer

particles. The difference increases with increasing deformation. The matrix element $F^L_{21} = F^L_{12}$ mainly determines the polarisation state of the scattered radiation, since, for an un-polarised incident beam, the second Stokes component Q of the scattered beam corresponds to the product of F^L_{21} and the incoming intensity I. The plot shows that maximal polarisation occurs at a scattering angle of about 90° and that Q is

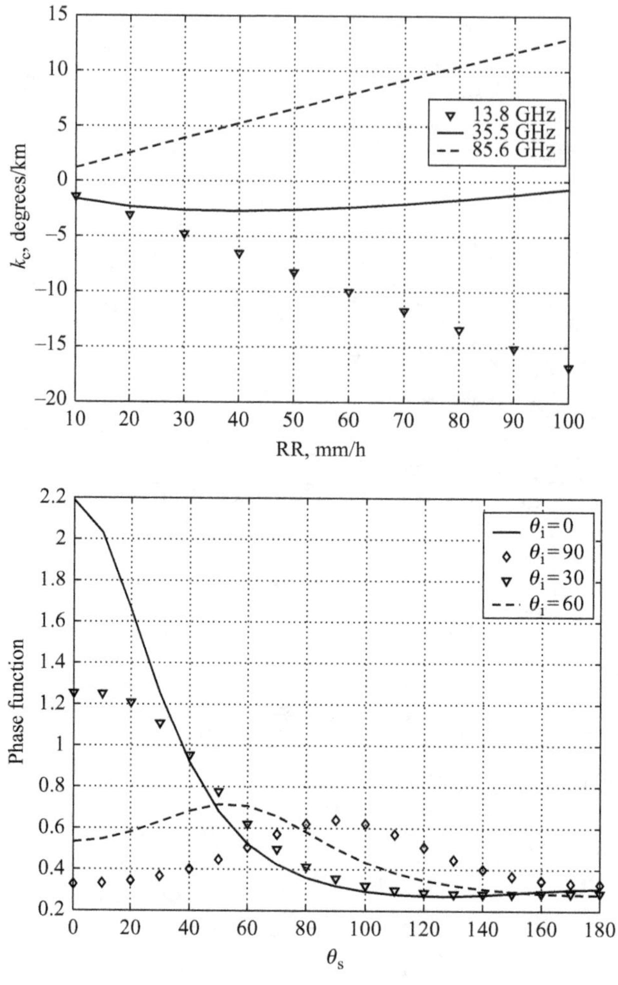

Figure 3.4 Extinction coefficient for circular polarisation and phase function averaged over azimuth angles versus outgoing polar direction for different incoming directions for a population with $RR = 30$ mm/h at 85.6 GHz

negative. The matrix element F_{22}^L, which equals F_{11}^L for spherical particles, is for non-spherical particles smaller than F_{11}^L, especially in the backward direction. The absolute value of the element $F_{34}^L = F_{43}^L$ is very small. F_{33}^L and F_{44}^L deviate only slightly from the result for spherical particles. Overall, the calculations for randomly oriented non-spherical particles show that at 318 GHz for a particle size of 75 μm the phase matrix is very similar to that for spherical particles. Therefore the dependence of simulated radiances on the particle shape at this frequency is expected to be much less than the dependence on the particle size.

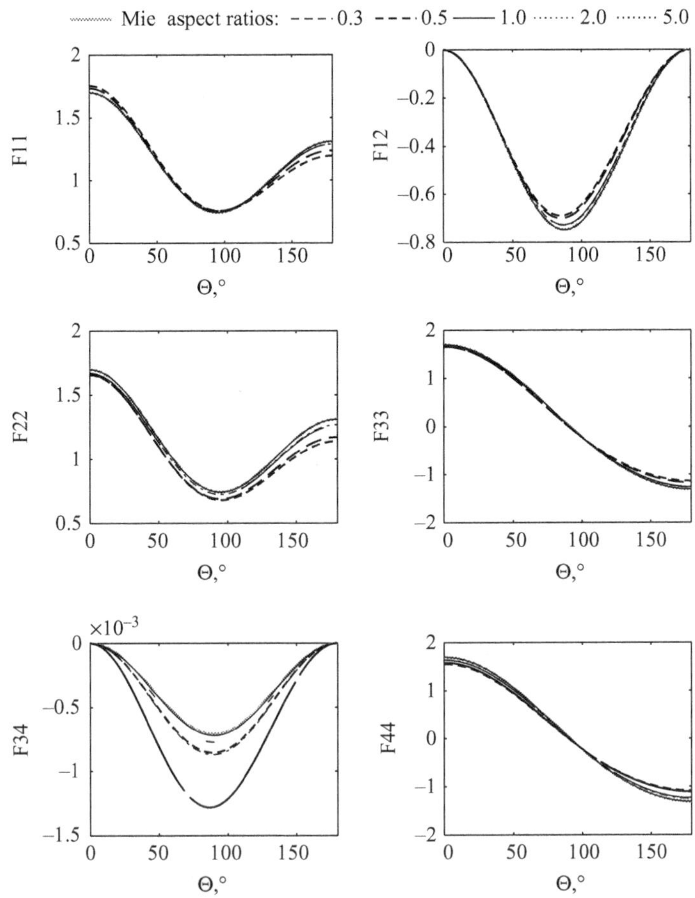

Figure 3.5 Orientation averaged phase matrix elements for spheroidal particles with different axial ratios. The dashed lines correspond to prolate particles with axial ratios of 0.3 (thick) and 0.5 (thin). The dotted lines correspond to oblate particles with axial ratios of 2.0 (thin) and 5.0 (thick). The solid line corresponds to spherical particles (in grey the computation has been repeated with Lorentz-Mie theory). The equal volume sphere radius is 75 μm and the frequency is 318 GHz

The previous results were obtained for randomly oriented particles (Mishchenko, 1991b). In reality the ice crystals in cirrus clouds often tend to be horizontally aligned. Another T-matrix program (Mishchenko, 2000) is applicable for non-spherical rotationally symmetric particles with arbitrary orientation. This program has been used to calculate the phase matrix for cylindrical particles with different axial ratios, which are horizontally oriented with the symmetry axis in the y direction in the laboratory frame. To be able to compare the results with Mie calculations, we have calculated

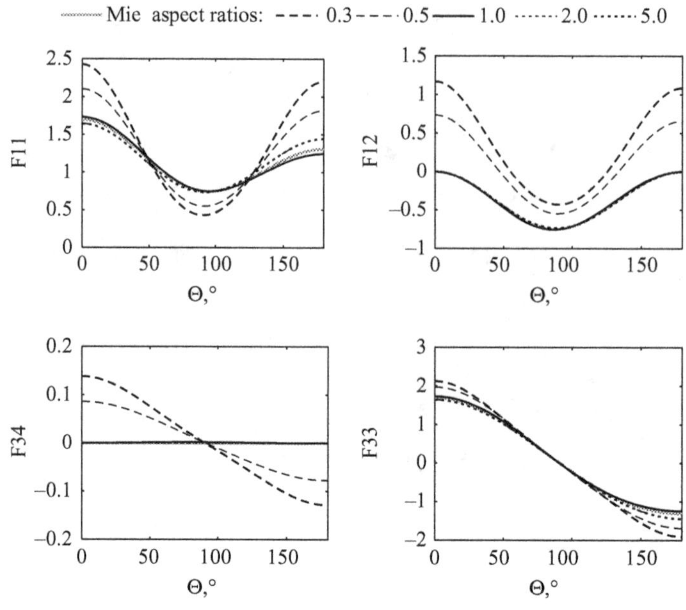

Figure 3.6 *Phase matrix elements evaluated at $\theta_i = 0$, $\phi_i = \phi_s = 0$ for spheres and perfectly oriented cylinders. All particles correspond to equivolume sphere of radius 75 μm; the frequency is 318 GHz*

the phase matrix for $\theta_i = 0$, $\phi_i = 0$ and $\phi_s = 0$, so that it corresponds to the scattering matrix **F** (see Equation 3.16). The results of the calculations are presented in Figure 3.6. A cylinder with an axial ratio of 1.0 gives almost identical results to the Mie calculation, when the symmetry axis corresponds to the z-axis. The results for horizontally oriented plates (dotted lines) are also similar. But the results for oriented cylinders (dashed lines) are very different, which is reasonable considering the symmetry: like a sphere the plate has a circular geometrical cross section when it is seen from the top; the horizontally oriented cylinder has a rectangular geometrical cross section when seen from the top. As the horizontal and the vertical components of the electric field vector are perpendicular to the direction of propagation, the plate must have the same influence on both components for forward and backward scattering, because of the symmetry. Therefore F_{21}^L, which is related to linear polarisation, must be zero for $\Theta = 0°$ and $\Theta = 180°$. Since the cylindrical particle is not symmetric relative to the propagation direction, the scattered radiation is polarised, even for forward and backward scattering directions. For oriented particles the phase matrix does not only depend on the scattering angle but also on the incident and scattered directions with respect to the particle orientation. For different directions, the phase matrix for the plates also deviates strongly from the Mie phase matrix. This very short analysis of these special phase matrices shows that particle shape has a significant impact on the intensity and the polarisation signal of microwave radiation in the atmosphere, especially if the cirrus cloud particles are oriented.

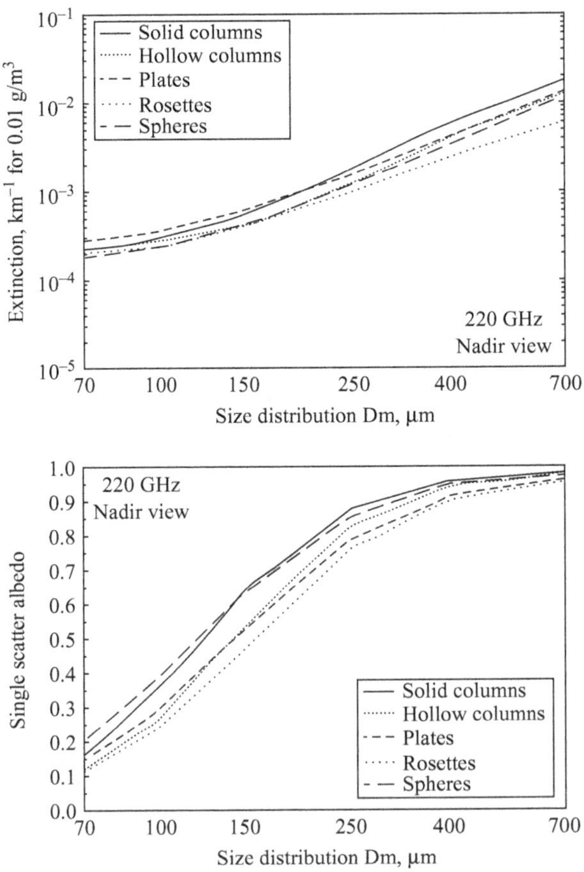

Figure 3.7 *Extinction and single scattering albedo versus median volume diameter at 340 GHz. A Γ size distribution is assumed. Reprinted from Evans and Stephens (1995a)*

3.4.3 DDA method

The discrete dipole approximation (DDA) is an approximation of a continuous target by a finite array of polarisable elements. The general method for computing the scattering parameters has been developed by Purcell and Pennyacker (1973) and reviewed by Draine and Flatau (2000b); a very well-documented code (see Draine and Flatau, 2004) developed by Draine and Flatau (2000a) is freely available at http://arxiv.org/abs/astro-ph/0008151v4. Because of its flexibility regarding the geometry of the target, this code has been extensively used in the MW community to compute single scattering properties of ice crystals (e.g. Evans and Stephens, 1995a; Liu, 2004). Evans and Stephens computed scattering quantities for different ice habits ranging in size from 30 to 2000 μm at 85.5, 157, 220 and 340 GHz. Figure 3.7 shows the extinction and single scattering albedo for different habits: solid and hollow

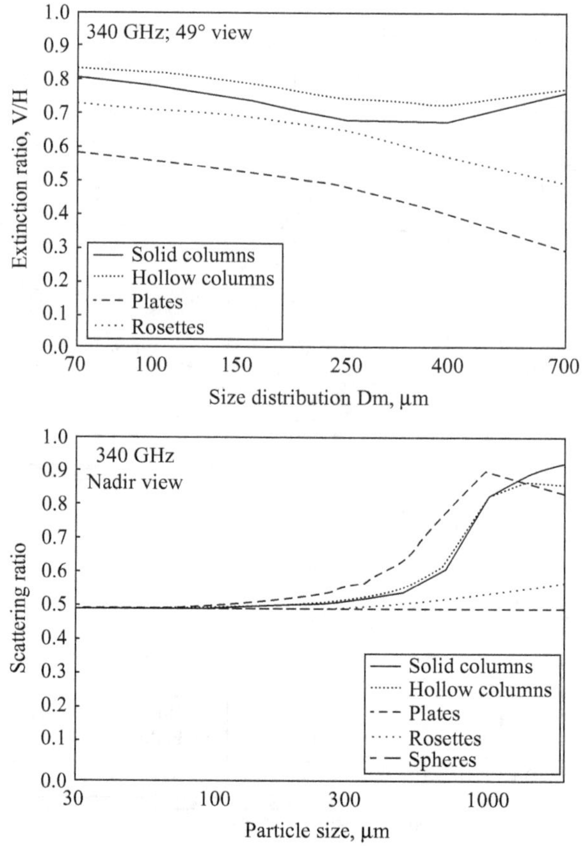

Figure 3.8 Extinction ratio (V / H) at 49° and scattering ratio versus median volume diameter at 340 GHz. Reprinted from Evans and Stephens (1995a)

columns, hexagonal plates, planar bullet rosettes with axial ratio derived by empirical formulae developed by Heymsfield (1978) and equivalent-volume spheres. All particles have their major axis randomly oriented but aligned in the horizontal plane. The extinction and the single scattering albedo increase dramatically with size and/or with frequency. Note that the extinction coefficient of a population with the same ice water content but with different shapes can vary by a factor of three. Rosettes and solid columns have the smallest and the highest extinction, respectively. This means that cirrus clouds will be generally thicker and less absorbing at higher frequencies and/or when containing larger particles.

The extinction ratio (Figure 3.8) is always lower than one because of particle orientation. This parameter characterises the polarisation property regarding radiation upwelling through the ice layer (see Section 3.9.1). Plates polarise most, followed by rosettes and columns. The lower panel of Figure 3.8 shows the scattering ratio

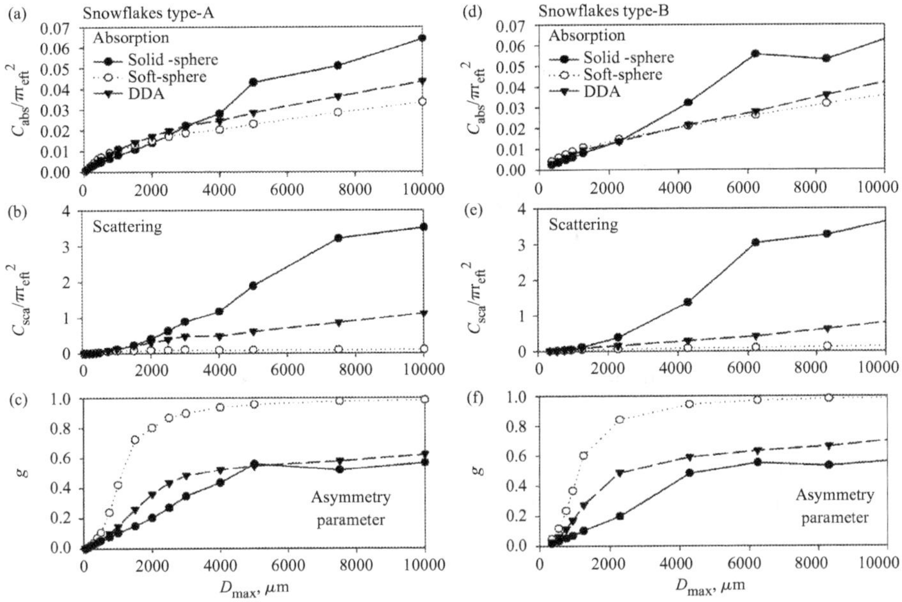

Figure 3.9 Single scattering properties of different snowflake types simulated by DDA (dashed lines) and approximated by solid spheres (solid lines) and soft spheres (dotted lines). Reprinted from Liu (2004)

(see Evans and Stephens (1995a) for the details) that measures the amount of forward scattering (equivalent to the *g* parameter for MIS media). Note that beyond the Rayleigh regime (scattering ratio larger than 0.5) different shapes are characterised by different scattering patterns with the thinner shapes leading to lower scattering ratio. In fact Liu (2004) has shown that solid or soft equivalent spheres cannot reproduce simultaneously extinction, single scattering albedo and asymmetry parameters of randomly oriented large ice aggregates and snow flakes (see Figure 3.9). As a consequence T_Bs computed with the DDA single scattering properties for scenarios including ice clouds and snow are largely different from those based on the solid sphere (that produces too high T_B depression) and the soft sphere approximation (that overestimates T_Bs). This issue is further discussed in Section 3.7.1.

3.4.4 Summary

We have briefly reviewed some of the mostly commonly used methodologies for single scattering property computations of hydrometeors in the MW region. Other techniques applicable to general non-spherical particles are well described by Mishchenko *et al.* (2000). Among the others the finite difference time domain method (FTDM) has demonstrated the capability and flexibility in dealing with ice crystal-like particles (see Yang and Liou, 1998, Tang and Aydin, 1995) (but, to our knowledge, no codes are freely available on the web). With the focus on the next generation of satellite radiometers, a renewed interest has been put in single scattering properties of ice cloud particles in the sub-mm region. As shown by Liu (2004) a correct treatment of the shape and density of these particles is a prerequisite to a full understanding of observed brightness temperatures. Because of the complications introduced by the richness of crystal shapes the FTDM and the DDA techniques, although very time consuming, seem to be more appropriate to tackle these issues. Similar conclusions can be drawn when considering the even more challenging problem of computing scattering properties of melting particles where the heterogeneity of the dielectric constant adds another complication to the problem.

3.5 Simplified forms of the radiative-transfer equation

We will focus in Section 3.6 on different numerical methods useful for the solution of the VRTE (Equation (1.74)) discussed in Chapter 1. Here we provide an overview of some simplifications applied to the VRTE, which will be the basis for the solution methods discussed later.

3.5.1 Cartesian geometry

Using cartesian geometry means that the streaming term on the left-hand side of Equation (1.74), Chapter 1 can be written in cartesian coordinates. The treatment of the spherical geometry becomes important especially for limb sounding applications and will be tackled in Section 3.6.3 when describing the ARTS code. When considering the scanning configuration of typical spaceborne PMW radiometers (e.g. like those deployed on TRMM, SSM/I or AMSU) the effect due to the curvature of the earth can usually be neglected. Some attention to earth curvature has to be paid, however, in distinguishing between satellite scan angle and zenith angle (see, e.g. Key, 2002).

3.5.1.1 Simplification of the VRTE

3.5.1.1.1 1D vertical inhomogeneous atmosphere

With the exception of convective cloud structures the vertical structure in the atmosphere has smaller spatial scales than the horizontal one (the validity of this hypothesis will be further investigated in more detail in Section 3.8.7). Therefore horizontal inhomogeneities are often ignored and all quantities are assumed to depend only on the vertical coordinate z. With $\boldsymbol{I} = \boldsymbol{I}(z, \theta, \phi)$ the VRTE reduces to:

$$\cos\theta \frac{\mathrm{d}\boldsymbol{I}}{\mathrm{d}z} = -\langle \boldsymbol{\Sigma}(z, \theta, \phi)\rangle \boldsymbol{I} + \langle \boldsymbol{a}(z, \pi - \theta, \pi + \phi)\rangle B[T(z)] + \int_{4\pi} \mathrm{d}\Omega' \langle \mathbf{Z}\rangle (z, \theta, \theta', \phi, \phi')\boldsymbol{I} \tag{3.27}$$

In this approximation the medium is treated as a stratification of L layers with the bottom level at $z = 0$ (surface) and the top height at $z = H$ (Figure 3.10).

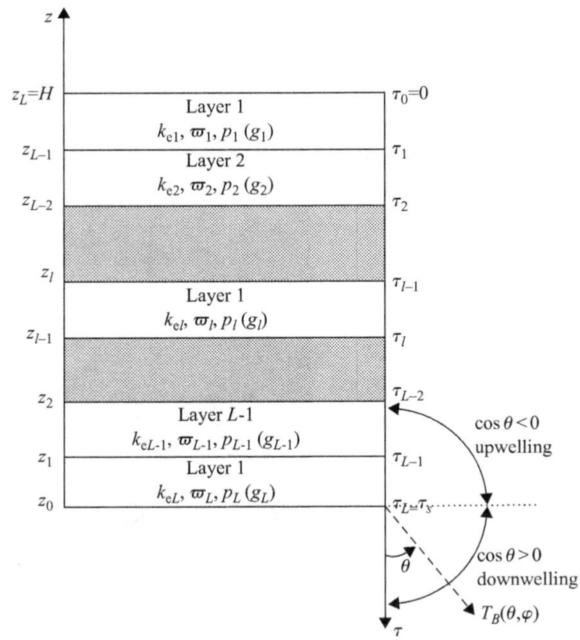

Figure 3.10 *Illustration of a stratified medium along the vertical coordinate z, having a thickness H. The medium is composed of L adjacent layers characterised by three optical parameters: extinction coefficient k_e, single scattering albedo ϖ and phase function p (or asymmetry factor g). The layers are numbered from the top in an increasing order ($l = 1, 2, \ldots, L$) downwards. The optical thickness coordinate τ is opposite z with its origin $\tau = \tau_0 = 0$ corresponding to $z = z_L = H$. The lower boundary is at $\tau_L = \tau_s = 0$ corresponding to $z = 0$*

3.5.1.1.2 MIS media

Another simplification to Equation (1.74) can be achieved when MIS media is considered. In this case the extinction matrix reduces to the scalar extinction coefficient and the absorption vector in the $I - Q$ basis has only the first component different from zero. After introducing the differential optical slant path $d\tau_{sl} = k_e ds$, Equation (1.74) can be rewritten in the $I - Q$ basis as

$$\frac{d\boldsymbol{I}(\vec{r},\hat{s})}{d\tau_{sl}} = -\boldsymbol{I}(\vec{r},\hat{s}) + (1-\varpi)B[T(\vec{r})]\begin{bmatrix}1\\0\\0\\0\end{bmatrix} + \frac{\varpi}{4\pi}\int d\Omega'\langle\mathbf{Z_n}\rangle(\vec{r},\hat{s},\hat{s'})\cdot\boldsymbol{I}(\vec{r},\hat{s'}) \tag{3.28}$$

with $\langle \mathbf{Z_n} \rangle = (4\pi/k_{sca})\langle \mathbf{Z} \rangle$ the dimensionless normalised[10] phase matrix of the medium.

3.5.1.2 Combining plane parallel and MIS media

When the two former assumptions are combined it is possible to introduce as a vertical coordinate the zenith optical thickness τ defined as

$$\tau(z,\nu) \equiv \int_z^H k_e(z,\nu)\mathrm{d}z \tag{3.29}$$

It is also convenient to introduce the variable μ defined as the cosine of the nadir or zenith-angle ϑ (measured from the $-\hat{\mathbf{z}}$ direction) so that $0 \le \mu \le 1$ and $-1 \le \mu \le 0$ hold for downward and upward directions, respectively. With this convention $\mathrm{d}\tau_{sl} = -\cos\theta\mathrm{d}\tau = \cos\vartheta\mathrm{d}\tau = \mu\mathrm{d}\tau$ so that Equations (3.27–3.28) can be properly merged. Hereafter we adopt the notation ${}_{\downarrow}\mu$ and ${}_{\uparrow}\mu$ to indicate downwelling and upwelling directions, respectively. When dealing with MW scenarios it is often possible to further simplify the VRTE by decoupling the first two Stokes parameters from the others. In fact we demonstrate the following:

Theorem: *In a plane-parallel homogeneous medium with no external illumination and no sources for the third and fourth Stokes parameter, for a population of scatterers composed by spheres and/or randomly oriented particles, U and V components have zero solution.*

Since the atmosphere is plane parallel and azimuthally isotropic, the solution has the form $I = I(\mu)$ so that in the scattering term of the VRTE the integration of the phase matrix over $\mathrm{d}\phi'$ can be performed. Using (3.17) for randomly oriented particles it is possible to show that the resulting matrix has a block structure with the first two Stokes parameters decoupled from the other two. In fact the integrals:

$$\int_0^{2\pi} \langle \mathbf{Z} \rangle_{ij}(z,\theta,\theta',\phi-\phi')\mathrm{d}\phi' = 0 \quad \text{for } ij = 13, 14, 23, 24, 31, 32, 41, 42 \tag{3.30}$$

because these terms contain a product between either a $\sin i_1$ or a $\sin i_2$ term [that are proportional to $\sin(\phi - \phi')$] and a term expanded[11] in terms of $\cos\Theta$ [therefore containing $\cos(\phi - \phi')$]. Without a source there is no way to generate U and V and both will remain 0. We cannot neglect U and V in cases with particles with preferential orientation[12] and/or with finite cloud, and/or with U and V sources or if the specific intensity, for some reason, maintains a dependence on azimuthal angle.

[10] $\int \mathrm{d}\Omega' \langle \mathbf{Z_n} \rangle (\vec{r}, \hat{s}, \hat{s}') = 4\pi$ for every incident direction.

[11] Each scattering matrix element can be expanded in a series of Legendre polynomials, see Thomas and Stamnes (1999).

[12] But it is still applicable with axial-symmetric non-spherical particles aligned parallel to the horizontal plane of the laboratory frame and randomly aligned in azimuthal directions.

In this approximation, the VRTE obviously simplifies to

$$\mu \begin{pmatrix} \dfrac{\mathrm{d}I}{\mathrm{d}\tau} \\ \dfrac{\mathrm{d}Q}{\mathrm{d}\tau} \end{pmatrix} = -\begin{pmatrix} I \\ Q \end{pmatrix} + (1-\varpi)\begin{pmatrix} B(T) \\ 0 \end{pmatrix}$$

$$+ \frac{\varpi}{4\pi}\int_{4\pi} \begin{bmatrix} F_{11}^{L} & F_{12}^{L}\cos(2i_1) \\ F_{12}^{L}\cos(2i_2) & \begin{matrix} F_{22}^{L}\cos(2i_1)\cos(2i_2) \\ -F_{33}^{L}\sin(2i_1)\sin(2i_2) \end{matrix} \end{bmatrix} \begin{pmatrix} I \\ Q \end{pmatrix} \mathrm{d}\Omega'$$

In the particular case of Rayleigh scattering (see Chandrasekhar, 1960) the scattering integral in the right side of the former equation becomes

$$\frac{3\varpi}{8}\int_{-1}^{1} \begin{bmatrix} 3-\mu^2-\mu'^2+3\mu^2\mu'^2 & (1-\mu'^2)(1-3\mu^2) \\ (1-\mu^2)(1-3\mu'^2) & 3(1-\mu'^2)(1-\mu^2) \end{bmatrix} \begin{bmatrix} I \\ Q \end{bmatrix} (\mu',\tau)\mathrm{d}\mu'$$

Formulae for spherical particles (in the h–v basis) can be found in Chapter 4 of Tsang *et al.* (1985). As noted by Liu and Simmer (1996) the polarisation Q is decoupled from the intensity I at nadir because in this case the term coupling of I with Q is proportional to

$$\int_0^{2\pi} F_{12}^{L}(\cos\Theta)\cos(2i_2)\mathrm{d}\phi' = F_{12}^{L}(\cos\theta')\int_0^{2\pi}\cos[2(\phi'-\phi)]\mathrm{d}\phi' = 0$$

where we have evaluated Equations (3.14–3.23) at $\theta = 0$. Because of the scattering term in Equation (3.31) and provided that F_{12}^{L} is different from zero (like for isotropic emission), even in the presence of randomly oriented particles, the I and Q channel do not decouple. The intensity I serves as a source for Q resulting in a non-zero polarisation. The non-zero Q parameter, on the other hand, changes I.

3.5.1.3 Reduction to scalar RTE

When considering only the intensity component and neglecting the interaction with the Q term, Equation (3.31) reduces to the scalar RTE. Since the scalar RTE is often used for MIS media it is important to assess the error introduced in the intensity computation by this approximation. For visible reflectance and transmittance computations different authors have reported errors up to 12 per cent for reflectance (Mishchenko *et al.*, 1994). Errors up to 23 per cent for transmittance can be found in Rayleigh atmospheres (see Mishchenko *et al.*, 1994). These errors decrease when phase functions for large spherical particles are considered. The coupling term between I and Q is proportional to the single scattering albedo (therefore in the pure absorption regime there is no difference), which is generally much lower in the MW region than, e.g. in the VIS. The upper panel of Figure 3.11 represents the error introduced by the scalar approximation in the T_{B}s evaluated at a zenith angle of 52.3° for a Rayleigh 4 km thick layer with temperature linearly decreasing from 300 to 273 K with different single scattering albedos and optical thicknesses above a Fresnel surface with

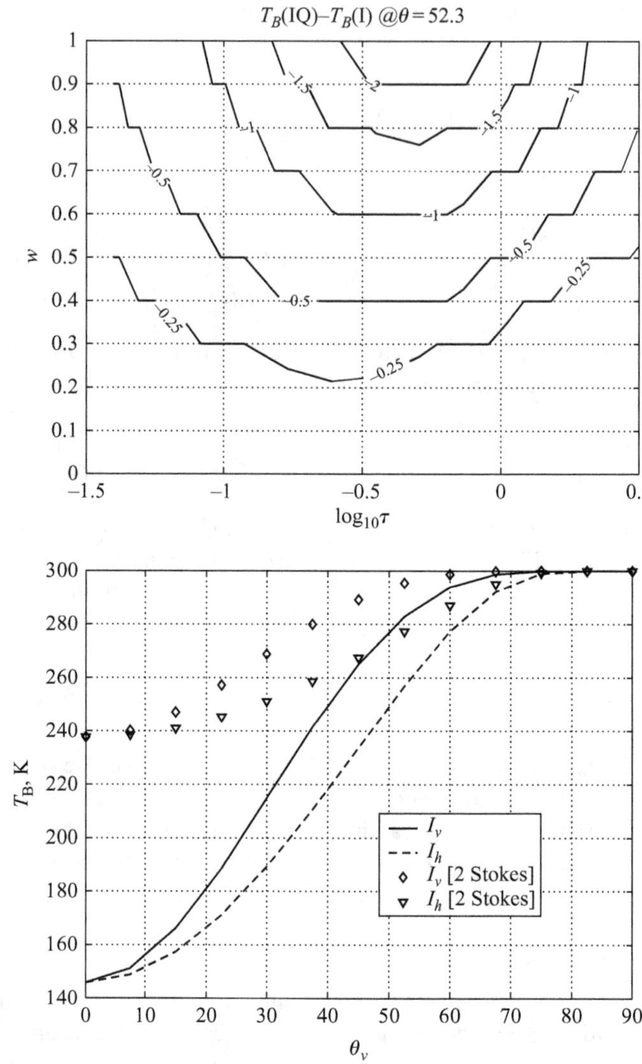

Figure 3.11 *Evaluation of the importance of solving the full VRTE. See text for details*

refractive index 3.724 + i2.212 at 300 K. The error is maximum with high single scattering albedos and at intermediate optical thickness but it never exceeds 2 K. Liu and Simmer (1996) have found a similar order of magnitude when using phase functions appropriate for ice and water spheres in the MW region.

The full VRTE solution (i.e. the coupling with the U and V term) is however needed when particles with preferred orientation are present. Even pure emitting particles can lead to a substantial difference between the solution with only the first two Stokes parameter and the full Stokes solution. As an example the right panel of Figure 3.11 shows the thermal radiation emitted by a layer of small oblate spheroids perfectly aligned along $\hat{\mathbf{y}}$ as described in prob. 4.25 in Tsang *et al.* (1985) and observed in the ι-z plane (ι being the bisetrix of the x–y plane). The solution provided by using the full Stokes vector is completely different than the one found with using only the first two.

3.5.1.4 Scalar RTE

In the scalar approximation and thanks to the Rayleigh–Jeans approximation (see Equation (1.7), Chapter 1), the unpolarised brightness temperature $T_{\mathrm{B}}(\tau, \mu, \phi)$ observed at τ for a frequency ν along a direction determined by the cosine-zenith angle μ and the azimuth angle ϕ satisfies the following integro–differential RTE for microwave passive sensors (Ishimaru, 1978; King and Harshvardhan, 1986):

$$\mu \frac{\mathrm{d}T_{\mathrm{B}}(\tau, \mu, \phi)}{\mathrm{d}\tau} = -T_{\mathrm{B}}(\tau, \mu, \phi) + J(\tau, \mu, \phi) \tag{3.31}$$

where $\mathrm{d}\tau/\mu$ is the slant differential optical thickness and $J(\tau, \mu, \phi)$ is the source function given by

$$\begin{aligned} J(\tau, \mu, \phi) = & \frac{\varpi(\tau)}{4\pi} \int_0^{2\pi} \int_{-1}^{+1} p(\tau, \mu, \mu', \phi, \phi') T_{\mathrm{B}}(\tau, \mu', \phi') \mathrm{d}\mu' \mathrm{d}\phi' \\ & + [1 - \varpi(\tau)] T(\tau) \end{aligned}$$

with T the physical temperature, and the optical parameters are assumed to depend on τ. The first term of J is sometimes referred to as the multiple scattering source, while the second term represents thermal emission. The azimuthal dependence in (3.31) can be due to

(1) anisotropic sources of emitted radiation and multiple scattering;
(2) azimuthal asymmetry of surface reflectivity.

If we can exclude these causes the dependence of T_{B} on ϕ may be omitted. If we integrate (3.31) with respect to ϕ over 2π and define an azimuthally averaged brightness temperature

$$T_{\mathrm{B}}(\tau, \mu) \equiv \frac{1}{2\pi} \int_0^{2\pi} T_{\mathrm{B}}(\tau, \mu, \phi) \mathrm{d}\phi$$

and an azimuthally averaged scattering phase function p_{a} as

$$p_{\mathrm{a}}(\tau, \mu, \mu') \equiv \frac{1}{2\pi} \int_0^{2\pi} \frac{1}{2\pi} \int_0^{2\pi} p(\tau, \mu, \mu', \phi, \phi') \mathrm{d}\phi' \mathrm{d}\phi \tag{3.32}$$

(3.31) can be rewritten in the following form:

$$\mu \frac{\mathrm{d}T_{\mathrm{B}}(\tau,\mu)}{\mathrm{d}\tau} = -T_{\mathrm{B}}(\tau,\mu) + J(\tau,\mu) = -T_{\mathrm{B}}(\tau,\mu) + \frac{\varpi(\tau)}{2}\int_{-1}^{+1} p_{\mathrm{a}}(\tau,\mu,\mu')T_{\mathrm{B}}(\tau,\mu')\mathrm{d}\mu' + [1-\varpi(\tau)]T(\tau) \tag{3.33}$$

The same result can be obtained by expanding the brightness temperature in a Fourier cosine series with respect to ϕ and only retaining the zero-order term (Stamnes and Swanson, 1981).

Referring to Figure 3.10, the downwelling brightness temperature $T_{\mathrm{B}}(\tau_s, {}_{\downarrow}\mu)$, observed at τ_s, can be formally expressed by means of the RTE integral form

$$T_{\mathrm{B}}(\tau_{\mathrm{s}},{}_{\downarrow}\mu) = T_{\mathrm{B}}(0,{}_{\downarrow}\mu)e^{-\tau_{\mathrm{s}}/\mu} + \frac{1}{\mu}\int_{0}^{\tau_{\mathrm{s}}} J(\tau,\mu)e^{-(\tau_{\mathrm{s}}-\tau)/\mu}\mathrm{d}\tau \tag{3.34}$$

while the upwelling $T_{\mathrm{B}}(0, {}_{\uparrow}\mu)$, observed at $\tau = 0$, is given by

$$T_{\mathrm{B}}(0,{}_{\uparrow}\mu) = T_{\mathrm{B}}(\tau_{\mathrm{s}},{}_{\uparrow}\mu)e^{-\tau_{\mathrm{s}}/\mu} + \frac{1}{\mu}\int_{\tau_{\mathrm{s}}}^{0} J(\tau,\mu)e^{-\tau/\mu}\mathrm{d}\tau \tag{3.35}$$

The above expressions are only formal solutions since the source function J is unknown because it depends on T_{B}. These equations are the basis of the iteration and successive order of scattering methods discussed in Section 3.6.2. The solution of (3.33) can be derived once boundary conditions are imposed. The boundary conditions to couple with (3.33) are (e.g. Stamnes and Swanson, 1981)

$$\begin{aligned} T_{\mathrm{B}}(0,{}_{\downarrow}\mu) &= T_{\mathrm{c}}, \\ T_{\mathrm{B}}(\tau_{\mathrm{s}},{}_{\uparrow}\mu) &= e_{\mathrm{s}}(\mu)T_{\mathrm{s}} + 2\pi\int_{0}^{1} \rho_{\mathrm{ds}}({}_{\uparrow}\mu,{}_{\downarrow}\mu')T_{\mathrm{B}}(\tau_{\mathrm{s}},{}_{\downarrow}\mu')\mu'\mathrm{d}\mu' \end{aligned} \tag{3.36}$$

where $T_{\mathrm{c}} = 2.73$ K is the cosmic background brightness temperature at microwave and T_{s} is the surface physical temperature. The surface emissivity e_{s} is related to the surface differential (bi-directional) reflectivity ρ_{ds} by means of the Kirchhoff law

$$e_{\mathrm{s}}(\mu) + 2\pi\int_{0}^{1} \rho_{\mathrm{ds}}({}_{\uparrow}\mu,{}_{\downarrow}\mu')\mu'\mathrm{d}\mu' = e_{\mathrm{s}}(\mu) + r_{\mathrm{s}}(\mu) = 1$$

where r_{s} is the surface reflectivity.

For a rough surface, such as an idealised land surface, a Lambertian surface emissivity model can be adopted. In this case, the differential reflectivity ρ_{ds} can be assumed independent of angle so that the azimuthally isotropic surface boundary condition becomes

$$T_{\mathrm{B}}(\tau_{\mathrm{s}},{}_{\uparrow}\mu) = e_{\mathrm{sL}}T_{\mathrm{s}} + (1-e_{\mathrm{sL}})2\pi\int_{0}^{1} T_{\mathrm{B}}(\tau_{\mathrm{s}},{}_{\downarrow}\mu')\mu'\mathrm{d}\mu' \tag{3.37}$$

with $r_{\mathrm{sL}} = 1 - e_{\mathrm{sL}}$ with e_{sL} the Lambertian surface emissivity.

For a specular surface, such as an idealised ocean, the differential reflectivity ρ_{ds} can be expressed by the polarisation-dependent Fresnel reflection coefficients. In this case, the azimuthally isotropic surface boundary condition can be formulated as

$$T_B(\tau_{s,\uparrow}\ \mu) = e_{sF}(\mu)T_s + [1 - e_{sF}(\mu)]T_B(\tau_{s,\downarrow}\ \mu)$$

with $r_{sF}(\mu) = 1 - e_{sF}(\mu)$ with e_{sF} the Fresnelian surface emissivity, dependent on μ and on the polarisation state. Indeed, the surface emissivity of a partially rough surface (such as a land surface including relief effects or an ocean surface with foam) can be treated by decomposing e_s into a properly weighted specular Fresnel component and a diffused Lambertian component.

From (3.33) and (3.36), it is apparent that, in the case of a scattering atmosphere, solutions for upwelling and downwelling T_B in a closed form cannot be given in general.

3.6 Numerical methods for the solution of the VRTE

Several methods for numerically solving the VRTE for a scattering and emitting atmosphere as required for the MW region are extensively described in the literature in books like Liou (2002), Chandrasekhar (1960), Thomas and Stamnes (1999), Tsang *et al.* (1985, 2000), Lenoble (1985) and Goody and Young (1989) or in reviews like Haferman (2000) and Gasiewski (1993). This section is not intended to be exhaustive on this topic but is provided to give the reader an idea about the radiative-transfer codes most commonly used in the MW community. Hereafter we will briefly describe the basic ideas underlying most of them

- the discrete ordinate method;
- the doubling-adding method;
- the iteration method;
- the Monte Carlo method.

We will try to emphasise their potential caveats and limits by describing to what degree they are able to solve the VRTE extensions to non-plane parallel geometries, including spherical geometries, and non-isotropic media.

3.6.1 The discrete ordinate method

The discrete ordinate method (DOM) is one of the most used approaches deployed in the solution of the RTE. It is very well documented in the literature (Stamnes and Swanson, 1981; Stamnes *et al.*, 1988; Thomas and Stamnes, 1999; Liou, 2002). The general idea is to replace the scattering integral in the RTE by a Gaussian quadrature sum. The angular variation in μ is discretised into $2N_\mu$ points (and a stream remains associated to each point) so that the multiple scattering is transformed into a sum by means of a Gaussian quadrature (which implicitly ensures the normalisation of the phase function). By doing so the RT integro-differential equation is reduced to a system of coupled linear differential equations (as many as the number of streams).

The DOM is generally described first for a simple isotropic scattering layer and then for a general homogeneous anisotropic layer. Finally the solution is extended to 1D inhomogeneous atmospheres consisting of L adjacent homogeneous layers in which volumetric single scattering albedo ϖ, extinction coefficient k_e and phase function p are taken to be constant within each layer. Physical temperature is supposed to be linearly dependent on the vertical coordinate within each layer. A multiple choice of source function and boundary condition types allows for a general and flexible description of several physical environment and observation techniques. Highly accurate numerical procedures are used to ensure convergence and stability of each matrix operation. The DISORT (DIScrete Ordinate Radiative Transfer, Stamnes *et al.*, 1988) is publicly available at ftp://climate.gsfc.nasa.gov/wiscombe/Multiple_Scatt/. Note that both SBDART and STREAMER (visit http://atol.ucsd.edu/ pflatau/index.html), codes designed for the analysis of a wide variety of radiative transfer problems encountered in satellite remote sensing in the VIS and IR region and atmospheric energy budget studies, are based on this code.

Extensions to the solution of the VRTE (VDOM) and to 3D cartesian geometry have been described by Haferman (1995) and references therein. Therein useful benchmark results and intercomparison studies can be found as well.

3.6.2 *Iterative and successive order of scattering method*

Both the iterative method (IM) and the successive order of scattering method (SOS) solve the radiative-transfer equation iteratively as a series of single scattering events. A general description can be found, e.g. in Goody and Young (1989, p. 354f). Both methods are based on the formal solution of the radiative transfer equation as expressed by (3.34, 3.35) with the unknown radiance distribution both on the left-hand side of the equation and within a layer integral of the source function on the right-hand side. The iterative method usually starts with a first guess of the radiance field (usually the solution of the VRTE assuming a scatter-free or a cloud-free atmosphere or any other useful guess), which is then used to evaluate the right-hand side of the equation leading to an improved guess of the radiance field. The latter is then used in the next iteration as a first guess on the right-hand side. This procedure is iterated until no significant change of the guessed radiance field occurs anywhere in the model domain.

The successive order of scattering method starts on the right-hand side with the solution of the RTE without the scattering source function. The solution includes now exactly the contribution of all the photons, which are not scattered. The resulting field is used as the radiance field on the right-hand side for the next iteration, which takes account of the scattering source function but omits the emission source function. The resulting radiance field on the left-hand side then contains the radiance contribution of all photons, which are scattered exactly once. The same procedure in the nth iteration gives the radiance contributions by photons scattered exactly n times. The contributions, which accordingly become smaller with each iteration due to extinction processes, are summed up for the total field until the contributions of higher order scattering become negligible.

It can be shown formally that the changes in the guessed radiance field obtained by the nth iteration of the iterative method are identical to the inclusion of photons which have been scattered exactly n times. Thus both methods are equivalent. The main shortcoming of the IT-SOS schemes is related to the use of a formal solution of the RTE, which requires the integration of the source function over a layer of finite depth, within which only single scattering events are assumed. For optically thick scattering atmospheres this layer obviously must become increasingly thin[13] leading to drastically increasing computing times. In the microwave range the method has probably been applied first by Weinman and Guetter (1977) for spherical particles. An extension to non-spherical oriented particles has been implemented and applied by Czekala and Simmer (1998), Czekala (1998), Czekala (1999), Czekala *et al.* (1999) to study the effects of deformed liquid drops described by Chebychev particles and solid precipitation approximated by spheroids. Emde *et al.* (2004b) extended the method to three dimensions and to a spherical coordinate system in order to study the effects of cirrus clouds on limb sounding measurements in the upper troposphere. Because of its relevance in spherical geometry computations we will now deepen the description of the ARTS models.

3.6.3 The polarised discrete ordinate iterative 3D model ARTS-DOIT

The atmospheric radiative transfer simulator (ARTS) was already introduced in Section 2.5. In this section we focus on the application to scattering atmospheres. ARTS includes two scattering modules: a backward Monte Carlo (MC hereafter) algorithm (ARTS-MC) (Davis *et al.*, 2005) and an iterative method (ARTS-DOIT) (Emde *et al.*, 2004b). Both algorithms work in 3D spherical atmospheres and both include polarisation. ARTS-DOIT also includes a 1D version. The ARTS software package including an extensive documentation is freely available at http://www.sat.uni-bremen.de/arts.

A feature of ARTS-DOIT is that it yields the whole radiation field, which is useful to obtain a basic physical understanding of the polarisation signal. The drawback is that ARTS-DOIT in 3D limb geometry is computationally very expensive. For limb sounding applications, when one is only interested in a few selected viewing directions, DOIT-MC is the preferred algorithm, especially if fine 3D cloud structures need to be resolved.

3.6.3.1 Radiation field and notation

For ARTS the cloudy part of the atmosphere must be contained in the so-called 'cloud box', because only in the cloud box multiple scattering is assumed. If the optical depth between the surface and the cloud is small multiple scattering between cloud and surface needs to be considered. In this case the cloud box is extended to the surface and the surface is treated as a scattering object inside the cloud box. The Stokes vector depends on the position in the cloud box and on the propagation direction specified

[13] Usually an automatic subdivision into sublayers is implemented in such codes.

by the zenith angle (θ) and the azimuth angle (ϕ). All these dimensions are discretised inside the model; five numerical grids are required to represent the radiation field $\mathcal{I}$: the pressure grid $\vec{p} = \{p_1, p_2, \ldots, p_{N_p}\}$, the latitude grid $\vec{\alpha} = \{\alpha_1, \alpha_2, \ldots, \alpha_{N_\alpha}\}$, the longitude grid $\vec{\beta} = \{\beta_1, \beta_2, \ldots, \beta_{N_\beta}\}$, the polar grid $\vec{\theta} = \{\theta_1, \theta_2, \ldots, \theta_{N_\theta}\}$ and the azimuthal grid $\vec{\phi} = \{\phi_1, \phi_2, \ldots, \phi_{N_\phi}\}$. The radiation field is a set of Stokes vectors ($N_p \times N_\alpha \times N_\beta \times N_\theta \times N_\phi$ elements) for all combinations of positions and directions:

$$\mathcal{I} = \{\boldsymbol{I}_1(p_1, \alpha_1, \beta_1, \theta_1, \phi_1), \boldsymbol{I}_2(p_2, \alpha_1, \beta_1, \theta_1, \phi_1), \ldots, \\ \boldsymbol{I}_{N_p \times N_\alpha \times N_\beta \times N_\theta \times N_\phi}(p_{N_p}, \alpha_{N_\alpha}, \beta_{N_\beta}, \theta_{N_\theta}, \phi_{N_\phi})\}$$

In the following we will use the notation

$$\mathcal{I} = \{\boldsymbol{I}_{ijklm}\}$$

with $i = 1 \ldots N_p$, $j = 1 \ldots N_\alpha$, $k = 1 \ldots N_\beta$, $l = 1 \ldots N_\theta$ and $m = 1 \ldots N_\phi$.

3.6.3.2 Vector radiative-transfer equation solution

The first guess field for the iterative solution

$$\mathcal{I}^{(0)} = \{\boldsymbol{I}^{(0)}_{ijklm}\}$$

is partly determined by the boundary condition given by the radiation coming from the clear sky part of the atmosphere travelling into the cloud box. Inside the cloud box an arbitrary field can be chosen as a first guess. Then the scattering integrals are computed:

$$\langle \mathbf{S}^{(0)}_{ijklm} \rangle = \int_{4\pi} \mathrm{d}\boldsymbol{n}' \langle \mathbf{Z}_{ijklm} \rangle \boldsymbol{I}^{(0)}_{ijklm}$$

based on the first guess field. For the integration equidistant angular grids are used. The radiation field, which is generally defined on finer angular grids $(\vec{\phi}, \vec{\theta})$, is interpolated on the equidistant angular grids. The integration is performed over all incident directions $\boldsymbol{n}'$ for each propagation direction $\boldsymbol{n}$. The evaluation of the scattering integral is done for all grid points inside the cloud box. The obtained integrals are interpolated on $\vec{\phi}$ and $\vec{\theta}$. The result is the first guess scattering integral field $\mathcal{S}^0$:

$$\mathcal{S}^{(0)} = \{\langle \boldsymbol{S}^{(0)}_{ijklm} \rangle\}$$

Figure 3.12 shows a propagation path step from a grid point $\boldsymbol{P} = (p_i, \alpha_j, \beta_k)$ into direction $\boldsymbol{n} = (\theta_l, \phi_m)$. The radiation arriving at $\boldsymbol{P}$ from the direction $\boldsymbol{n}'$ is obtained by solving the linear differential equation:

$$\frac{\mathrm{d}\boldsymbol{I}^{(1)}}{\mathrm{d}s} = -\overline{\langle \boldsymbol{\Sigma} \rangle} \boldsymbol{I}^{(1)} + \overline{\langle \boldsymbol{a} \rangle}\, \bar{B} + \overline{\langle \boldsymbol{S}^{(0)} \rangle} \tag{3.38}$$

where $\overline{\langle \boldsymbol{\Sigma} \rangle}$, $\overline{\langle \boldsymbol{a} \rangle}$, $\bar{B}$ and $\overline{\langle \boldsymbol{S}^{(0)} \rangle}$ are averaged quantities. This equation can be solved analytically for constant coefficients. Multi-linear interpolation gives the quantities $\boldsymbol{\Sigma}', \boldsymbol{a}', \boldsymbol{S}'$ and T' at the intersection point $\boldsymbol{P}'$. To calculate the radiative transfer from $\boldsymbol{P}'$ towards $\boldsymbol{P}$ all quantities are approximated by taking the averages between the values

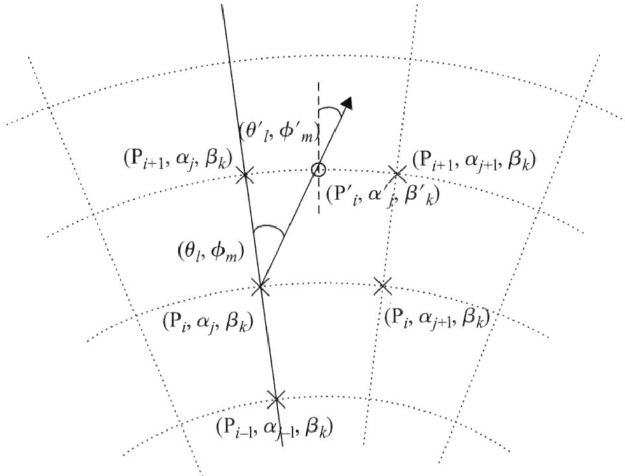

Figure 3.12 *Path from a grid point* $((p_i, \alpha_j, \beta_k) - (\times))$ *to the intersection point* $((p'_i, \alpha'_j, \beta'_k) - (\circ))$ *with the next grid cell boundary. Viewing direction is specified by* (θ_l, ϕ_m) *at* $(\times)$ *or* (θ'_l, ϕ'_m) *at* $(\circ)$

at $\boldsymbol{P}'$ and $\boldsymbol{P}$. The average value of the temperature is used to get the averaged Planck function $\bar{B}$.

The solution of Equation (3.38) is found analytically using a matrix exponential approach:

$$\boldsymbol{I}^{(1)} = e^{-\overline{\langle\boldsymbol{\Sigma}\rangle}s}\boldsymbol{I}^{(0)}(\mathbb{I} - e^{-\overline{\langle\boldsymbol{\Sigma}\rangle}s})\overline{\langle\boldsymbol{\Sigma}\rangle}^{-1}(\overline{\langle\boldsymbol{a}\rangle}\,\bar{B} + \overline{\langle\boldsymbol{S}^{(0)}\rangle})$$

equivalent to (3.34, 3.35), where $\mathbb{I}$ denotes the identity matrix and $\boldsymbol{I}^{(0)}$ the initial Stokes vector. The radiative transfer step from $\boldsymbol{P}'$ to $\boldsymbol{P}$ is calculated; therefore $\boldsymbol{I}^{(0)}$ is the incoming radiation at $\boldsymbol{P}'$ into direction (θ'_l, ϕ'_m), which is the first guess field interpolated on $\boldsymbol{P}'$. This radiative-transfer step calculation is done for all points inside the cloud box in all directions. The resulting set of Stokes vectors ($\boldsymbol{I}^{(1)}$ for all points in all directions) is the first order iteration field $\mathcal{I}^{(1)}$:

$$\mathcal{I}^{(1)} = \{\boldsymbol{I}^{(1)}_{ijklm}\}$$

We can formulate a differential equation for the nth order iteration field. The scattering integrals, before the iteration is terminated, are given by

$$\langle\boldsymbol{S}^{(n-1)}_{ijklm}\rangle = \int_{4\pi} \mathrm{d}\boldsymbol{n}'\langle\mathbf{Z}\rangle\boldsymbol{I}^{(n-1)}_{ijklm}$$

and the differential equation for a specified grid point into a specified direction is

$$\frac{\mathrm{d}\boldsymbol{I}^{(n)}}{\mathrm{d}s} = -\overline{\langle\boldsymbol{\Sigma}\rangle}\boldsymbol{I}^{(n)} + \overline{\langle\boldsymbol{a}\rangle}\,\bar{B} + \overline{\langle\boldsymbol{S}^{(n-1)}\rangle}$$

Thus the nth order iteration field

$$\mathcal{I}^{(n)} = \{\boldsymbol{I}^{(n)}_{ijklm}\}$$

is given by

$$\boldsymbol{I}^{(n)} = e^{-\overline{\langle \boldsymbol{\Sigma} \rangle} s} \cdot \boldsymbol{I}^{(n-1)} (\mathbb{I} - e^{-\overline{\langle \boldsymbol{\Sigma} \rangle s}}) \overline{\langle \boldsymbol{\Sigma} \rangle}^{-1} (\overline{\langle \boldsymbol{a} \rangle}\, \bar{B} + \overline{\langle \boldsymbol{S}^{(n-1)} \rangle})$$

for all cloud box points and all directions defined in the numerical grids.

After each iteration a convergence test is performed. If the absolute difference for all successive Stokes vectors is smaller than the limit ϵ, where ϵ is a vector containing the user specified convergence limits for all four Stokes components, a solution to the VRTE has been found as

$$\mathcal{I}^{(N)} = \{\boldsymbol{I}^{(N)}_{ijklm}\}$$

3.6.3.3 Sequential update

Suppose that there are N pressure levels inside the cloud box. If the radiation field is updated taking into account for each grid point only the adjacent grid cells, at least $N - 1$ iterations are required until the scattering effect from the lowermost pressure level has propagated throughout the cloud box up to the uppermost pressure level. Thus the number of iterations depends on the number of grid points inside the cloud box which would make the solution very ineffective when a fine resolution inside the cloud box is required. A solution to this problem is the 'sequential update of the radiation field', which is shown schematically in Figure 3.13. For simplicity it will be explained for a 1D cloud box. We divide the update of the radiation field, i.e. the radiative-transfer step calculations for all positions and directions inside the cloud box, into three parts: update for 'up-looking' zenith angles ($0° \leq \theta_{\text{up}} \leq 90°$), for 'down-looking' angles ($\theta_{\text{limit}} \leq \theta_{\text{down}} \leq 180°$) and for 'limb-looking' angles ($90° < \theta_{\text{limb}} < \theta_{\text{limit}}$). The 'limb-looking' case is needed, because for angles between 90° and θ_{limit} the intersection point is at the same pressure level as the observation point. The limiting angle θ_{limit} is calculated geometrically. In the 1D case the radiation field is a set of Stokes vectors each of which depends upon the position and direction $\mathcal{I} = \{\boldsymbol{I}(p_{\text{i}}, \theta_l)\}$.

The boundary condition for the calculation is the incoming radiation field on the cloud box boundary $\mathcal{I}^{bd}$:

$$\begin{aligned} \mathcal{I}^{bd} = \{\boldsymbol{I}(p_{\text{i}}, \theta_l)\} \text{ where } & p_{\text{i}} = p_N \;\forall\; \theta_l \in [0, \theta_{\text{limit}}] \\ & p_{\text{i}} = p_0 \;\forall\; \theta_l \in (\theta_{\text{limit}}, 180°] \end{aligned}$$

where p_0 and p_N are the pressure coordinates of the lower and upper cloud box boundaries, respectively. For down-looking directions, the intensity field at the lowermost cloud box boundary and for up- and limb-looking directions the intensity field at the uppermost cloud box boundary are the required boundary conditions, respectively. For each of the three directional sets radiation is propagated from boundary to boundary within one iteration (see also Weinman and Guetter, 1977; Czekala and Simmer, 2002).

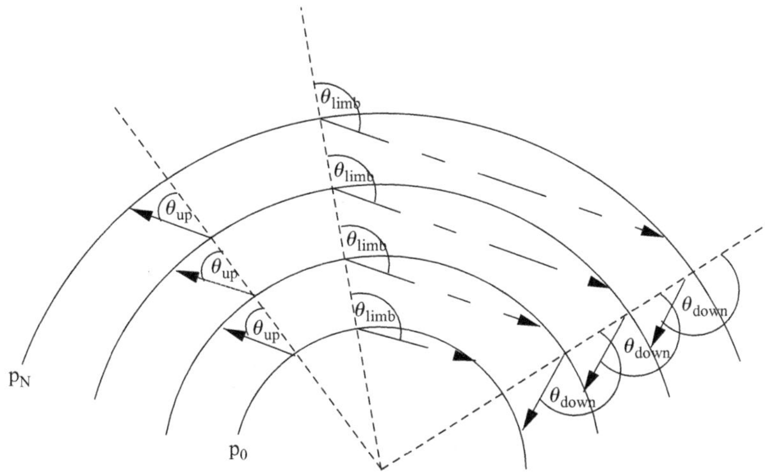

Figure 3.13 *Schematic of the sequential update (1D) showing the three different parts: 'up-looking' corresponds to zenith angles θ_{up}, 'limb-looking' corresponds to θ_{limb} 'down-looking' corresponds to θ_{down}. The arrows indicate the lines of sight, which are opposite to the direction of propagation*

3.6.3.4 Grid optimisation and interpolation methods

The accuracy of the DOIT method depends very much on the discretisation of the zenith angle. The reason is that the intensity field strongly increases at about $\theta = 90°$. For angles below 90° ('up-looking' directions) the intensity is very small compared with angles above 90° ('down-looking' directions), because the thermal emission from the lower atmosphere and from the ground is much larger than thermal emission from trace gases in the upper atmosphere. Figure 3.14 shows an example of an intensity field as a function of the zenith angle for different pressure levels inside a cloud box, which is placed from 7.3 to 12.7 km altitude, corresponding to pressure limits of 411 and 188 hPa, respectively. A mid-latitude-summer scenario including water vapour, ozone, nitrogen and oxygen was used. As the intensity (or the Stokes vector) at the intersection point of a propagation path is obtained by interpolation, large interpolation errors can occur for zenith angles of about 90° if the zenith angle grid discretisation is too coarse. Taking a very fine equidistant zenith angle grid leads to very long computation times. For this reason a zenith angle grid optimisation method has been implemented into ARTS.

For the computation of the scattering integral it is possible to take a much coarser zenith angle resolution without losing accuracy. It does not make sense to use the zenith angle grid, which is optimised to represent the radiation field with a certain accuracy. The integrand is the product of the phase matrix and the radiation field. The peaks of the phase matrices can be at any zenith angle, depending on the incoming and the scattered directions. The multiplication smoothes out both the radiation field increase at 90° and the peaks of the phase matrices. Test calculations have shown that

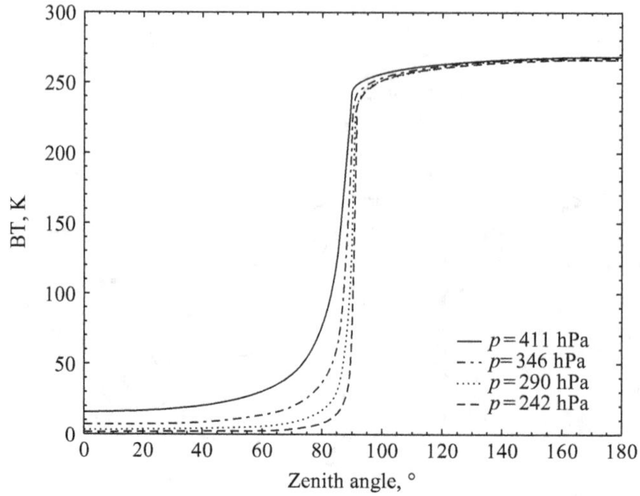

Figure 3.14 Intensity field for different pressure levels at 318 GHz (from Emde et al., 2004b)

an increment of 10° is sufficient. Taking the equidistant grid saves the computation time of the scattering integral to a very large extent, because much fewer grid points are required.

3.6.4 The doubling-adding method

A conceptually simple method for solving the RTE is provided by the doubling-adding method (Thomas and Stamnes, 1999; Hovenier *et al.*, 2004). This method is intrinsically 1D; it practically computes the radiative properties of the medium (that is the reflection and the transmission matrix and the emission vector) rather than the radiance field itself; therefore, radiances exiting the atmosphere for many different conditions may be easily found once the solution is computed. Each input layer is divided into a number of homogeneous sublayers with each sublayer being thin enough for the finite difference initialisation to be accurate. Infinitesimal generator initialisation is used to relate the scattering matrix to the reflection and transmission matrices. The sublayers are integrated with the doubling algorithm. For each desired output level the transmission, reflection and source of the layers above and below the level are combined with the adding algorithm. The internal radiances are computed from the properties of the layers above and below and the incident radiance from the boundaries. Doubling-adding is numerically stable for large optical depths.

For MW applications two codes in FORTRAN are available at the website, http://nit.colorado.edu/evans/polrad.html based on the two papers (Evans and Stephens, 1991, 1995b). Documentation and useful comments can be found inside the source files. They treat the fully polarised RT in 1D vertically inhomogeneous atmosphere and they can accommodate thermal emission and reflection from a Lambertian or Fresnel ground surface. The first one is adapted for randomly oriented particles with a plane of symmetry and it includes beam and/or thermal sources of radiation (hence a radiation field with full angular dependence); the second one is designed for particles having random azimuthal orientation and pure thermal sources. This symmetry implies that the radiation field is azimuthally symmetric (therefore there is only the lowest azimuthal Fourier mode to consider) and that U and V Stokes parameters are zero. This code is particularly fast and accurate so that it can be used both for validation tests and for massive computations.

In the first case the complication of passing from the scattering to the phase matrix in the full-Stokes vector analyses [14] is faced by calculating the phase matrix elements for each pair of quadrature angles and for a number of azimuth angle differences by accomplishing the usual polarisation transformation from the scattering matrix (computed in terms of its Legendre coefficient expansion) to the phase matrix. Then a Fourier transform is done to transform the phase matrices from ϕ space to azimuth mode space. The number of quadrature directions and the order of azimuth expansion are among the input parameters.

In the second case the azimuthally averaged phase matrix (computed at the incident and outgoing quadrature angles), the extinction matrix and emission vector (listed for the quadrature incident angles) are given as an input of the RT procedure. The phase matrix, the extinction matrix and the absorption vector are used to initialise the local reflection and transmission matrices and the source vector for an infinitesimal layer; from these local properties the properties of the whole atmosphere are computed by using the classical doubling-adding method.

3.6.5 *The Monte Carlo method*

Because of their capability of treating geometries other than plane parallel and/or media with irregular boundaries MC techniques have been widely used in VIS and IR RT computations within intrinsic 3D structured clouds and precipitation (e.g. McKee and Cox, 1974; Marchuk *et al.*, 1980; Platnick, 2000). The underlying idea of an MC solution to the RTE consists in interpreting the radiance field in terms of a photon distribution function as described in Chapter 2, Thomas and Stamnes (1999). This allows us to derive the radiance fields by averaging over a large number of trajectories of individual photons simulated by using probabilistic methods (House and Avery, 1969). The MC technique can be extended to the MW regime by introducing 'microwave photons' and by properly tracing them through the medium. In this context, the first solutions for horizontally finite clouds have been provided by Weinman and Davies (1978) and Kummerow and Weiman (1988) . With the goal of describing the radiation field at particular viewing angles of interest for PMW radiometry, a very efficient backward MC code has

[14] In the scalar RT case, this conversion is accomplished by using the addition theorem of the associated Legendre function.

been implemented by Roberti *et al.* (1994). In this approach the only polarisation mechanism resides in the interaction with the surface. To overcome this problem, Liu *et al.* (1996), developed the so called 'backward-forward' scheme that allows for the treatment of polarised radiation, but requires that the extinction matrix be diagonal. By doing so scattering by hydrometeors is taken into account with good approximation.

As a useful example of the MC technique, we show here how it is possible to sample the new propagation direction after a scattering event in the case of MIS media (e.g. see Muinonen, 2004). Note that both Robert *et al.* (1994) and Liu *et al.* (1996) use an approximate method to sample the scattering directions because in the backward approach the polarisation state of the photon is unknown. The angular intensity distribution comes directly from Equation (3.20), i.e.

$$I_s = I_i F_{11}^L(\Theta) + Q_i F_{12}^L(\Theta)\cos(2i_1) + U_i F_{12}^L(\Theta)\sin(2i_1) \tag{3.39}$$

so that it generally depends on the polarisation state of the incoming radiation. The problem consists in generating Θ and i_1 for any incident direction. If we normalise Equation (3.39) we get the *pdf*

$$p(\Theta, i_1) = \frac{1}{4\pi} F_{11}^L(\Theta) + \frac{F_{12}^L(\Theta)}{4\pi}\left[\frac{Q_i}{I_i}\cos(2i_1) + \frac{U_i}{I_i}\sin(2i_1)\right]$$

that gives the density distribution for the two independent variables i_1 and Θ. The two unconditional probabilities associated with i_1 and Θ are

$$\begin{aligned} p(\Theta) &= \int_0^{2\pi} p(\Theta, i_1)\mathrm{d}i_1 = \frac{1}{2} F_{11}^L(\Theta) \\ p(i_1) &= \int_{-1}^{1} p(\Theta, i_1)\mathrm{d}\mu_\Theta = \frac{1}{2\pi} + \frac{1}{4\pi}\int_{-1}^{1} F_{12}^L(\Theta) \\ &\quad \left[\frac{Q_i}{I_i}\cos(2i_1) + \frac{U_i}{I_i}\sin(2i_1)\right]\mathrm{d}\mu_\Theta \end{aligned}$$

so that Θ can be sampled[15] by the phase function and then, according to the Bayes theorem, i_1 can be sampled from the conditional probability

$$\begin{aligned} p(i_1|\tilde{\Theta}) &= \frac{p(\tilde{\Theta}, i_1)}{p(\tilde{\Theta})} = \frac{1}{2\pi}\left[1 + \frac{F_{12}^L(\tilde{\Theta})}{F_{11}^L(\tilde{\Theta})}\left(\frac{Q_i}{I_i}\cos(2i_1) + \frac{U_i}{I_i}\sin(2i_1)\right)\right] \\ &\equiv \frac{1}{2\pi}[1 - A\cos(2i_1 + \gamma)] \end{aligned} \tag{3.40}$$

[15] The best in this case is to use the look up table comparison method, that is to interpolate Θ in a pre-computed table of Θ values as a function of random number r within $]0, 1[$. See House and Avery (1969) for details.

with A and γ defined by the equations

$$A\cos\gamma \equiv -\frac{F^L_{12}(\tilde{\Theta})}{F^L_{11}(\tilde{\Theta})}\frac{Q_{\rm i}}{I_{\rm i}}$$

$$A\sin\gamma \equiv +\frac{F^L_{12}(\tilde{\Theta})}{F^L_{11}(\tilde{\Theta})}\frac{U_{\rm i}}{I_{\rm i}}$$

To generate this kind of distribution, by using the method of inverse function, it is necessary to solve the equation $r = Q(i_1|\tilde{\Theta})$ where r is a uniform random number in $]0,1[$ and where $Q(i_1|\tilde{\Theta})$ is the cumulative distribution function associated to (3.40)

$$Q(i_1|\tilde{\Theta}) = \frac{1}{2\pi}\left[i_1 - \frac{A}{2}\sin(2i_1+\gamma) + \frac{A}{2}\sin(\gamma)\right] \tag{3.41}$$

The equation so found can be easily solved numerically (it is a generalised Kepler's equation). Once Θ and i_1 have been sampled the scattered direction is directly found by applying (3.22).

Backward and backward–forward MC methods are very efficient and are now in use in operational algorithms (e.g. GPROF, see Kummerow *et al.*, 2001). As implemented, these methods cannot deal properly with oriented non-spherical particles which seem to play an important role in polarisation signatures over stratiform precipitation, see Evans and Stephens (1995b). To address this issue two different approaches have been taken: the development of efficient forward MC codes and of importance sampling.

1. Pure forward MC schemes are able to completely describe the polarisation effects of preferentially oriented particles. Roberti and Kummerow (1999), developed a forward MC code that treats oriented non-spherical hydrometeors, but only for plane parallel situations. As an incise, in this code the anisotropic effects are taken into account only in the propagation and not in the scattering action. This limitation forces the introduction of biasing techniques that lead to unphysical photons (i.e. Stokes vector outside the Poincaré sphere, see Tsang *et al.*, 1985). To avoid these drawbacks, Battaglia and Kummerow (2000) and Battaglia and Mantovani (2005) developed a plane parallel polarised code for oriented symmetric particles and extended it to a 3D fully polarised version; anisotropic effects are taken in account in the propagation and in the scattering events as well. A considerable reduction in computational time is obtained thanks to biasing techniques[16] that compute the brightness temperatures at some fixed positions and directions only. By so doing the problem becomes manageable for MW problems where total optical thickness does not exceed several units. Forward MC codes are available at web-sites http://i3rc.gsfc.nasa.gov/Public_codes_3DRT.htm and http://www.meteo.uni-bonn.de/mitarbeiter/battaglia/.

[16] For MW applications biasing techniques have been extensively reviewed by Roberti (1997).

2. Recently, Davis *et al.* (2005) clarified some general misunderstandings about forward-backward schemes by introducing important sampling techniques to correctly treat dichroic media. The code was implemented as a module of ARTS (http://www.sat.uni-bremen.de/arts/).

Although MC-codes have already been developed to treat 3D polarised radiation in non-isotropic media this area remains an active area of research, especially in assessing the polarisation signatures typical of these kinds of clouds.

3.6.6 Test studies and benchmark results

A bunch of benchmark results is available in the literature (Evans and Stephens, 1991; Kuik *et al.*, 1992; Wauben and Hovenier, 1992; Haferman, 1995). They involve either 1D or 3D structures with spherical or randomly oriented particles. One dimensional benchmark results for non-isotropic media are available at the website http://nit.colorado.edu/ evans/polrad.html and at http://www.meteo.uni-bonn.de/mitarbeiter/battaglia. Just as an example, in Figure 3.15 we show the polarisation signatures PD $\equiv I_v - I_h$ in the upwelling and downwelling radiation for a 2 km height 25 mm/h Marshall and Palmer distributed 1D spheroidal rain layer (details in Czekala *et al.* (1999) for the description of the raindrops' non-spherical model). The results are shown in Figure 3.15 at five different frequencies by solving the VRTE with a forward MC (see Battaglia and Mantovani, 2000) and by the well-documented RT4 FORTRAN code by Evans (1991). Polarisation differences are in the worst case within 0.2 K between these two codes and compare well with their analogous situation shown in Figure 3.10, panel (a) in Czekala *et al.* (1999) (computed with the iteration method described in Section 3.6.2).

3.6.6.1 RTE intercomparison for clouds

An intercomparison of microwave multiple scattering radiative-transfer codes used in generating databases for satellite rainfall retrieval algorithms has been carried out to ensure that differences obtained from retrieval techniques do not originate from the underlying radiative-transfer code employed for the forward modelling (Smith *et al.*, 2002). A set of profiles containing liquid water and ice contents of cloud and rain water as well as snow, graupel and pristine ice were distributed to the participants together with a black box routine providing Mie single scattering, atmospheric background absorption and surface emissivity. Simulations were to be carried out for nadir and off-nadir (53.1°) observation angles at frequencies between 10 and 85 GHz. Among the radiative-transfer models were two stream, multiple stream and MC models. The results showed that there were two major sources of differences between the codes.

1. If surface reflection/emission was considered isotropic, simulated brightness temperatures were significantly higher than for specular reflection and this effect was most pronounced at nadir observation and over ocean-type surfaces.

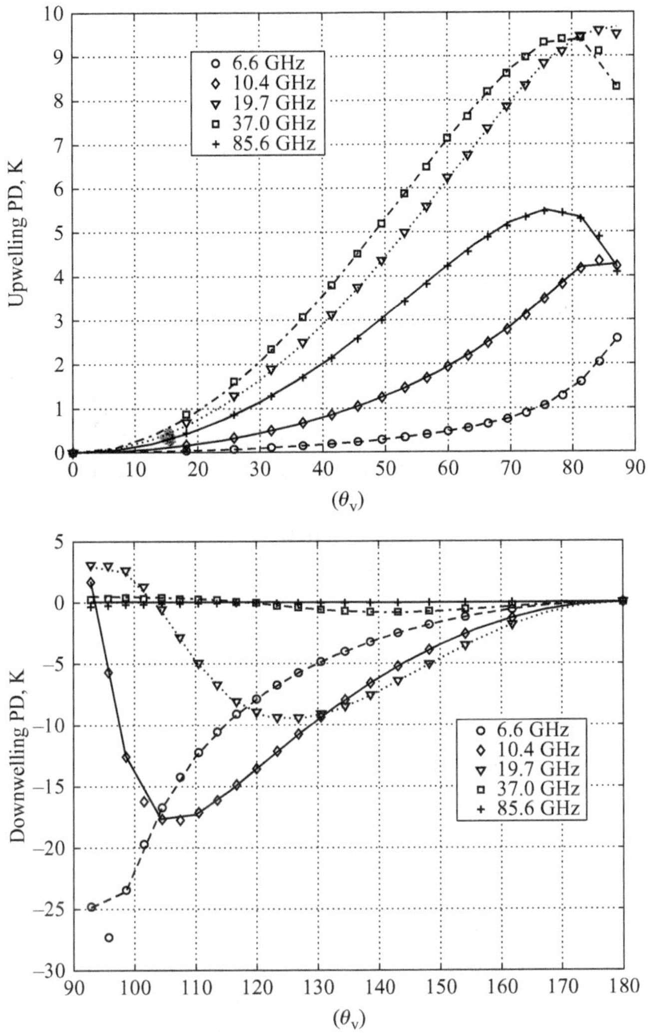

Figure 3.15 Polarisation difference $T_v - T_h$ versus zenith angle of propagation for 5 frequencies at 25 mm/h rain rate with a black-body emission surface. Marks and lines correspond to forward MC and to RT4 results, respectively

2. Flux-type models including delta-scaling could partially compensate for the errors introduced by the two-stream approximation.

The largest discrepancies occurred at high frequencies where atmospheric scattering is most pronounced and at nadir observation. If the same surface boundary conditions, the same multiple-stream resolution and the same scaling procedures are used, the models are very close to each other with discrepancies below 1 K.

3.6.7 *Future developments*

The most challenging issues still unsolved in the MW RT community are connected to the 3D solution of the VRTE. Although many codes are now available to tackle this problem not enough sensitivity and inter-comparison studies have been done yet for full 3D scenarios involving non-isotropic media. In this context an MC-frame seems to be particularly suitable and attractive because of its capability to solve both the 3D geometry problem and the radiative equation for the full Stokes vector. The disadvantages of this numerical tool usually are a lower accuracy and a longer computational time. Recent advances in computer technology coupled with the increasing complexity of the problem under examination (finite cloud models, size and shape and orientation distributions of hydrometeors) make the MC technique appealing and a viable option in the MW region. Nevertheless, the computational aspect is always to be carefully considered. In fact the MC methods seem useful when few radiative quantities are desired, e.g. the radiances at a satellite at a particular viewing angle. On the other hand, methods that explicitly represent the radiance field in the computational domain (like doubling/adding, VDOM) are obviously preferable for plane parallel problems and when the whole radiation field (or related quantities) has to be computed. In fact with explicit methods any desired radiative property can be computed at little extra cost.

3.7 Approximate solution methods

Approximate solution methods play a major role in MW RT. Some of them allow analytical solutions to the RTE, thus providing more understanding and clarity in the physical aspects of the problem. They are typically much faster than accurate solutions (while their accuracy being usually better than the inputs to the model's accuracy), so that they can be used in very time consuming retrieval procedures (e.g. rainfall retrieval). Moreover approximate methods can provide a first guess solution to the RTE (e.g. in the iterative methods). Generally approximate solutions are based on the techniques described in Section 3.6. For instance two-stream approximations use the discrete ordinate method formalism in its simplest version. A computationally efficient scalar two-stream model is provided by Weng and Grody (2000) and references therein while a polarimetric version has been developed by Liu and Weng (2002). Although the two-stream approximation is typically used when computing angular average properties (i.e. fluxes or heating rates) it gives particularly good results also close to the Gaussian point angle $\mu_1 = 0.577$; thus it is adapted for conical scanning radiometers like SSM/I or TRMM. However, there is no a priori method to estimate its accuracy so that ad hoc sensitivities studies need to be performed for each application (e.g. Weng and Grody, 2000).

When dealing with pure absorbing atmospheres the source function simplifies to the pure emitting part, so that the solution to the VRTE is reduced to a simple integration. This is known as the Schwarzschild approximation (see Chapter 1 for examples or Ulaby *et al.*, 1986). Similarly, when dealing with very absorbing and/or optically thin atmospheres with $\varpi\tau^* \ll 1$, the single scattering approximation is applied (e.g. see appendix in Evans *et al.*, 1998). These are just the first two steps of the method of successive orders of scattering described in Section 3.6.2.

Other approximate hybrid techniques combine different methods. Deeter and Evans (1998) developed an Eddington and a single scattering radiative transfer model for computing radiances from thermally emitting atmospheres; Heidinger *et al.* (2004, personal communications) used a doubling method to obtain atmospheric layer reflectance and transmittance properties, combined with SOS to obtain final upwelling radiances for azimuthally symmetric problems. The model (available at http://naftali.aos.wisc.edu/soi/) makes use of the fact that angular and layer accuracy are not independent, hence the number of streams used is tied to the layer accuracy required. It has nominally been found to be accurate to better than 1 K for most cases, even highly scattering ones. The primary motivation for the development of this model was for direct microwave radiance assimilation under cloudy and precipitating conditions for NWP.

Because of their importance in the MW community we will now describe in detail the Eddington and Delta–Eddington approximations.

3.7.1 Eddington approximation for plane-parallel clouds

Consider, for simplicity, a uniform atmospheric layer between $\tau = 0$ and $\tau = \tau_s$ of thickness H. In this circumstance, RTE exhibits constant coefficients, that is the optical parameters become independent of distance, so that $p_a(\tau, \mu, \mu') = p_a(\mu, \mu')$, $\varpi(\tau) = \varpi$ and $k_e(z) = k_e$.

A way to approach the solution of RTE is to expand the angular spectrum of the phase function p in terms of Legendre polynomials[17] P_n of order n, that is under the hypothesis of symmetric scattering (Sobolev, 1975; Ishimaru, 1978):

$$p(\mu, \mu', \phi, \phi') = p(\mu_\Theta) = \sum_{n=0}^{N} (2n+1) h_n P_n(\mu_\Theta) \tag{3.43}$$

μ_Θ being defined by (3.14). The expansion coefficients h_n are given by

$$h_n = \frac{1}{2} \int_{-1}^{+1} p(\mu_\Theta) P_n(\mu_\Theta) \mathrm{d}\mu_\Theta$$

[17] The Legendre polynomials P_n exhibit the orthogonality property in the $(-1,1)$ interval:

$$\int_{-1}^{+1} P_n(\mu_\Theta) P_n(\mu_\Theta) \mathrm{d}\mu_\Theta = 0 \quad \text{with } n \neq m \tag{3.42}$$

with $P_0(\mu_\Theta)$=1, $P_1(\mu_\Theta) = \mu_\Theta$ and recursively for higher-order polynomials. Using the following expansion of $P_n(\mu_\Theta)$

$$P_n(\mu_\Theta) = P_n(\mu)P_n(\mu') + 2\sum_{m=1}^{n} \frac{(n-m)!}{(n+m)!} P_n^m(\mu)P_n^m(\mu')\cos m(\phi - \phi')$$

where P_n^m are the associate Legendre polynomials, we get from (3.43) and (3.32)

$$p_a(\mu, \mu') = \sum_{n=0}^{N} (2n+1)h_n P_n(\mu)P_n(\mu') \tag{3.44}$$

Analogously to (3.43), the angular spectrum of T_B can be also expanded in terms of Legendre polynomials P_n of order n, that is by making use of the Rayleigh–Jeans approximation[18]:

$$T_B(\tau, \mu) = \sum_{n=0}^{N} T_{Bn}(\tau)P_n(\mu) \tag{3.45}$$

where the expansion coefficients and constants have been grouped in the terms $T_{Bn}(\tau)$.

The truncation of Legendre series (3.44) and (3.45) to the first order (i.e. $N = 1$) is called Eddington's approximation, first introduced in radio-astronomy (e.g. King and Harshvardhan, 1986; Liou, 2002). Under this assumption, from (3.43) the scattering phase function becomes

$$p_S(\mu_\Theta) = h_0 + 3h_1\mu_\Theta = 1 + 3g\mu_\Theta \tag{3.46}$$

Thus, from (3.46) and (3.32), the azimuthally averaged scattering phase function is given by

$$p_{aS}(\mu, \mu') = h_0 + 3h_1\mu\mu' = 1 + 3g\mu\mu' \tag{3.47}$$

It is worth mentioning that (3.46) coincides with the Sobolev approximation of p, which is the reason for the subscript S for both (3.46) and (3.47) (Sobolev, 1975).

By substituting (3.45) with $N = 1$ and (3.47) into (3.33) and easily solving the integral of J, it holds

$$\mu P_0(\mu)\frac{\mathrm{d}T_{B0}(\tau)}{\mathrm{d}\tau} + \mu P_1(\mu)\frac{\mathrm{d}T_{B1}(\tau)}{\mathrm{d}\tau} = -P_0(\mu)T_{B0}(\tau) - P_1(\mu)T_{B1}(\tau) + J_E(\tau, \mu) \tag{3.48}$$

where J_E represents the Eddington approximation of the source function J, given by

$$J_E(\tau, \mu) = \varpi[P_0(\mu)T_{B0}(\tau) + 3gP_1(\mu)T_{B1}(\tau)] + (1 - \varpi)(t_0 + t_1\tau) \tag{3.49}$$

[18] The following results for the brightness temperatures are strictly valid for the radiances.

Notice that the layer temperature $T(\tau)$ in (3.49) is supposed to be linearly dependent on τ with coefficients t_0 and t_1:

$$T(\tau) = t_0 + t_1\tau = T(0) + \left[\frac{T(\tau_s) - T(0)}{\tau_s}\right]\tau = T(0) + \left[\frac{T(\tau_s) - T(0)}{k_e H}\right]$$

$$\tau = T(0) + \frac{\Gamma_T}{k_e}\tau$$

where Γ_T is the vertical temperature gradient (or thermal lapse rate) and H is the layer thickness. Scalarly multiplying the differential Equation (3.48) first by $P_0(\mu)$,

$$\int_{-1}^{1} P_0(\mu)\left[\mu P_0(\mu)\frac{dT_{B0}(\tau)}{d\tau} + \mu P_1(\mu)\frac{dT_{B1}(\tau)}{d\tau}\right] d\mu$$

$$= \int_{-1}^{1} P_0(\mu)\left[-P_0(\mu)T_{B0}(\tau) - P_1(\mu)T_{B1}(\tau) + J_E(\tau,\mu)\right] d\mu$$

and then by $P_1(\mu)$,

$$\int_{-1}^{1} P_1(\mu)\left[\mu P_0(\mu)\frac{dT_{B0}(\tau)}{d\tau} + \mu P_1(\mu)\frac{dT_{B1}(\tau)}{d\tau}\right] d\mu$$

$$= \int_{-1}^{1} P_1(\mu)\left[-P_0(\mu)T_{B0}(\tau) - P_1(\mu)T_{B1}(\tau) + J_E(\tau,\mu)\right] d\mu$$

and using the orthogonal property of Legendre polynomials, the differential Equation (3.48) reduces to the following system of ordinary differential equations with constant coefficients:

$$\begin{cases} \dfrac{dT_{B0}(\tau)}{d\tau} = -(1-\varpi g)T_{B1}(\tau), \\[2mm] \dfrac{dT_{B1}(\tau)}{d\tau} = -3(1-\varpi g)[T_{B0}(\tau) - (t_o + t_1\tau)] \end{cases}$$

This system can be conveniently put in a matrix form:

$$\frac{d}{d\tau}\begin{bmatrix} T_{B0}(\tau) \\ T_{B1}(\tau) \end{bmatrix} = \mathbf{A}\begin{bmatrix} T_{B0}(\tau) \\ T_{B1}(\tau) \end{bmatrix} + \mathbf{b} \tag{3.50}$$

where $\mathbf{A}$ is a $[2 \times 2]$ matrix and $\mathbf{b}$ a $[2 \times 1]$ vector. The latter arrays have the following expressions:

$$\mathbf{A} = \begin{bmatrix} 0 & -(1-\varpi g) \\ -3(1-\varpi g) & 0 \end{bmatrix}; \qquad \mathbf{b} = \begin{bmatrix} 0 \\ 3(1-\varpi g)(t_0 + t_1\tau) \end{bmatrix}$$

3.7.1.1 General analytical solution

The solution of the associated homogeneous differential equation of (3.50) can be found by using the eigen-analysis method, which furnishes as a solution:

$$\begin{bmatrix} T_{B0}(\tau) \\ T_{B1}(\tau) \end{bmatrix}_H = C_1 \begin{bmatrix} 1 \\ c \end{bmatrix} e^{-\lambda_1 \tau} + C_2 \begin{bmatrix} 1 \\ -c \end{bmatrix} e^{-\lambda_2 \tau} \tag{3.51}$$

while C_1 and C_2 are the integration constants. The two eigenvalues $\lambda_1 = +\lambda$ and $\lambda_2 = -\lambda$ and the constant c are given by

$$\begin{aligned} \lambda &= \sqrt{3(1-\varpi g)(1-\varpi)} \\ c &= \sqrt{\frac{3(1-\varpi)}{1-\varpi g}} \end{aligned} \tag{3.52}$$

The structure of the known vector **b** suggests looking for a particular solution which is a two-parameter linear function of τ. Substituting this unknown function into (3.50), the particular solution is found to be

$$\begin{bmatrix} T_{B0}(\tau) \\ T_{B1}(\tau) \end{bmatrix}_P = \begin{bmatrix} t_0 \\ -t_1/(1-\varpi g) \end{bmatrix} + \begin{bmatrix} t_1 \\ 0 \end{bmatrix} \tau \tag{3.53}$$

The general solution of (3.50) is the sum of the homogeneous solution (3.51) and the particular solution (3.53), that is:

$$\begin{aligned} T_{B0}(\tau) &= C_1 e^{-\lambda\tau} + C_2 e^{\lambda\tau} + t_0 + t_1 \tau \\ T_{B1}(\tau) &= c C_1 e^{-\lambda\tau} - c C_2 e^{\lambda\tau} - \frac{t_1}{1-\varpi g} \end{aligned} \tag{3.54}$$

The integration constants C_1 and C_2 can be derived by imposing the boundary conditions for T_{B0} and T_{B1}. These constants can be physically obtained by imposing the general boundary in terms of upwelling and downwelling flux densities $F_\downarrow$ and $F_\uparrow$ defined, respectively, by

$$F_\downarrow(\tau) = 2\pi \int_0^1 T_B(\tau, _\downarrow \mu)\mu \, d\mu = \pi \left[T_{B0}(\tau) - \frac{2}{3} T_{B1}(\tau) \right]$$

$$F_\uparrow(\tau) = 2\pi \int_0^1 T_B(\tau, _\uparrow \mu)\mu \, d\mu = \pi \left[T_{B0}(\tau) + \frac{2}{3} T_{B1}(\tau) \right]$$

Substituting T_B's with their Eddington approximation, for a Lambertian boundary condition, it holds

$$\begin{aligned} &T_{B0}(0) + \frac{2}{3} T_{B1}(0) = T_c \\ &T_{B0}(\tau_s) - \frac{2}{3} T_{B1}(\tau_s) = e_{sL} T_S + (1 - e_{sL}) \left[T_{B0}(\tau_s) + \frac{2}{3} T_{B1}(\tau_s) \right] \end{aligned} \tag{3.55}$$

where the right-hand side is derived by using (3.45) with $N = 1$.

Note that, for a specular surface, (3.55) can be also written properly taking into account the dependence on μ and polarisation. Inserting the expressions of T_{B0} and T_{B1} into previous equations, after some algebraic manipulation the following linear equation system is deduced for C_1 and C_2:

$$\begin{aligned} &\begin{bmatrix} 1 + \dfrac{2c}{3} & 1 - \frac{2}{3}c \\ e^{-\lambda \tau_s} \left[e_{sL} - \dfrac{2c}{3}(2 - e_{sL}) \right] & e^{\lambda \tau_s} \left[e_{sL} + \frac{2c}{3}(2 - e_{sL}) \right] \end{bmatrix} \begin{pmatrix} C_1 \\ C_2 \end{pmatrix} \\ &\quad = \begin{pmatrix} T_c - t_0 + \dfrac{2}{3} \dfrac{t_1}{(1 - \varpi g)} \\ e_{sL} T_S - e_{sL}(t_0 + t_1 \tau_s) - \dfrac{2 - e_{sL}}{3} \dfrac{2t_1}{(1 - \varpi g)} \end{pmatrix} \end{aligned} \tag{3.56}$$

The previous linear system can be solved either analytically in the simple case of a single layer or numerically. It is worth noting that the matrix to be inverted on the left-hand side of (3.56) contains exponentials with a positive argument whose values, for a large optical thickness, can give rise to unexpected ill-conditioning in the inversion procedure (Stamnes and Swanson, 1981).

In order to tackle this problem, a scaling transformation of the unknowns C_1 and C_2 can be performed by exploiting the structure of the matrix itself. We can simply rewrite (3.56) as

$$\begin{bmatrix} c_{11} & c_{12} \\ c_{21} e^{-\lambda \tau_s} & c_{22} e^{\lambda \tau_s} \end{bmatrix} \begin{bmatrix} C_1 \\ C_2 \end{bmatrix} = \begin{bmatrix} d_1 \\ d_2 \end{bmatrix} \tag{3.57}$$

where the constants c_{ij} and d_i ($i = 1, 2$) are clearly identified by comparing (3.57) with (3.56). If we let $\tau_0 = 0$ and define

$$\begin{bmatrix} C_1 \\ C_2 \end{bmatrix} \equiv \begin{bmatrix} \tilde{C}_1 e^{\lambda \tau_0} \\ \tilde{C}_2 e^{-\lambda \tau_s} \end{bmatrix} = \begin{bmatrix} \tilde{C}_1 \\ \tilde{C}_2 e^{-\lambda \tau_s} \end{bmatrix} \tag{3.58}$$

then the linear system (3.57) becomes

$$\begin{bmatrix} c_{11} & c_{12}e^{-\lambda\tau_s} \\ c_{21}e^{-\lambda\tau_s} & c_{22} \end{bmatrix} \begin{bmatrix} \tilde{C}_1 \\ \tilde{C}_2 \end{bmatrix} = \begin{bmatrix} d_1 \\ d_2 \end{bmatrix} \tag{3.59}$$

which is inconditionally stable for any optical thickness value. The solution of the linear algebraic system (3.59) is

$$\begin{bmatrix} \tilde{C}_1 \\ \tilde{C}_2 \end{bmatrix} = \frac{1}{c_{11}c_{22} - c_{12}^2 e^{-(\lambda\tau_s)^2}} \begin{bmatrix} c_{22}d_1 - c_{21}e^{-\lambda\tau_s}d_1 \\ -c_{12}e^{-\lambda\tau_s}d_1 + c_{11}d_2 \end{bmatrix}$$

By coupling (3.58) and (3.59) and once the physical and optical parameters of the scattering layer are known, the RTE is analytically solved under the Eddington approximation in a plane-parallel geometry. In fact, the azimuthally average brightness temperature $T_B(\tau,\mu)$ can be reconstructed from the knowledge of $T_{B0}(\tau_\nu)$ and $T_{B1}(\tau_\nu)$ from (3.45) with $N = 1$:

$$T_B(\tau,\mu) = T_{B0}(\tau) + \mu T_{B1}(\tau) \tag{3.60}$$

3.7.1.2 The *g* effect

A known limitation of (3.60) is due to the first-order expansion of the azimuthally averaged scattering phase function in (3.47) which may be inadequate for strongly scattering conditions when $g > 0.5$, e.g. in the presence of precipitating ice within a rainfall cloud (note that in Kummerow, (1993) accuracy of the Eddington approximation is studied for g values lower than 0.5).

This is easily proved in Figure 3.16 where results from a very simple computation with a single layer and with linearly decreasing temperature between surface temperature at 293.0 K and a top temperature of 273.0 K have been assumed. The surface has been supposed Lambertian with an emissivity of 0.9. In the case of the exact computations (with RT3/MC), a Heyney–Greenstein phase function (see (3.63)) is assumed with the number of quadrature directions $N_\mu = 32$ in the RT3 code. Three optical thicknesses ($\tau = 0.1, 1, 10$) are considered. Note that, as expected, the T_Bs become warmer when increasing the g parameter: a highly scattering cloud ($\varpi = 0.75$) with optical thickness equal to 10 can cause a quite consistent scattering signature (≈ 50 K) or practically no scattering signature at all if its g parameter varies between 0 and 0.9. Moreover the Eddington solution becomes less accurate with increasing g (when compared with exact solutions like the RT3 or the MC solution). On the other hand the Eddington works fine when viewing angles close to the first Gaussian quadrature point are considered. This reveals how important is the correct determination of single scattering properties; in particular, the phase function details (here described in a rough way by the asymmetry parameter) can lead to substantial change in the T_B pattern (see also the discussion in Gasiewki, 1993).

3.7.1.3 The Delta–Eddington approximation

In order to extend the validity of (3.60), a Delta–Eddington approach can be adopted (Joseph *et al.*, 1976).

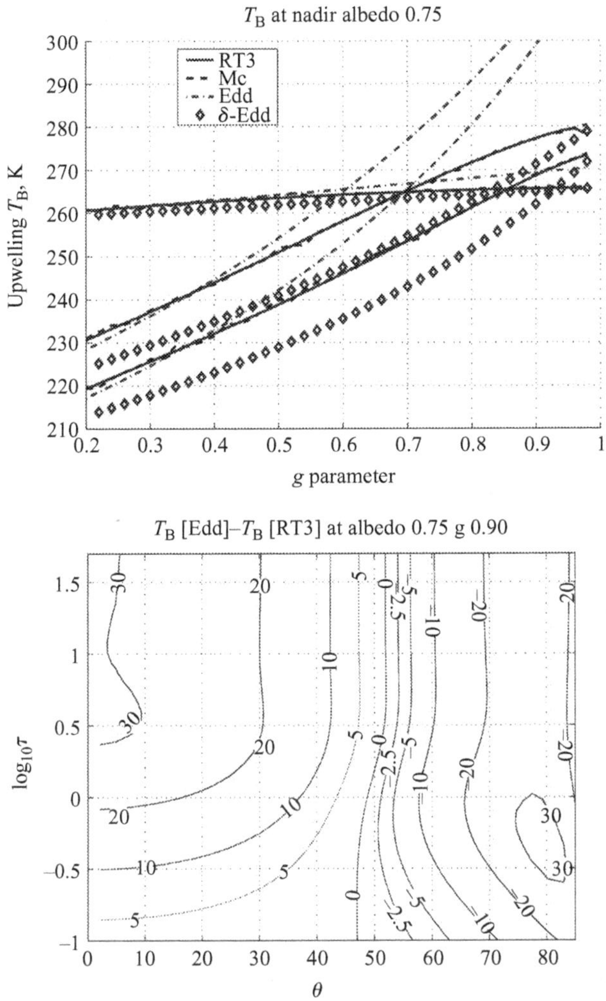

Figure 3.16 Upper panel: upwelling T_Bs as computed from different RT models for a simple one-layer atmosphere. Lower panel: contour levels of the computational error in the Eddington approximation as a function of the viewing angle and the optical thickness for an 'extreme' scattering scenario

Within this approach, the truncated Delta–Eddington phase function p_δ is defined as an approximation of the phase function p including a forward scatter peak through a Dirac delta function δ and a two-term Legendre expansion of p itself, that is (Joseph *et al.*, 1976):

$$p_\delta(\mu_\Theta) \equiv 2f\delta(1-\mu_\Theta) + (1+3g'\mu_\Theta) - f(1+3g'\mu_\Theta) \tag{3.61}$$

where f is, by definition, the fraction of forward scattering peak and $g\prime$ is the asymmetry factor of p_δ. We note that p_δ is correctly normalised (i.e. it satisfies (3.19)). The new factors f and g' can be determined by imposing the values of the first and second moments of the Delta–Eddington phase function p_δ. If p_δ is required to have the same asymmetry factor g as the original phase function p, from (3.18), it yields

$$\frac{1}{2}\int_{-1}^{1} \mu_\Theta p_\delta(\mu_\Theta)\,\mathrm{d}\mu_\Theta = f + (1-f)g' \equiv g \tag{3.62}$$

from which $g\prime$ is determined once f is known. The value of f is obtained by requiring that the second moment of p_δ is equal to the second moment of p, approximated by the Heyney–Greenstein phase function p_{HG} (Heyney and Greenstein, 1941):

$$p_{\mathrm{HG}}(\mu_\Theta) \equiv \frac{1-g^2}{(1-g^2-2g\mu_\Theta)^{-3/2}} = \sum_{n=0}^{\infty}(2n+1)\; g^n \; P_n(\mu_\Theta) \tag{3.63}$$

A property of the Heney–Greenstein phase function worth mentioning is that its nth moment is equal to nth power of g. It has been shown in the literature that p_{HG} is a fairly good approximation of the Mie numerical phase function. Thus, f can be determined from (Joseph *et al.*, 1976; Wiscombe and Joseph, 1977)

$$\frac{1}{2}\int_{-1}^{1} F_{12}^{L}(\mu_\Theta)p_\delta(\mu_\Theta)\mathrm{d}\mu_\Theta = f \equiv g^2 \tag{3.64}$$

which, coupled with (3.62), provides the explicit expression of $g\prime$:

$$g' = \frac{g}{1+g} \tag{3.65}$$

Notice that g' is always less than 0.5, as requested.

As a final step, it is possible to evaluate the consequences of using the Delta–Eddington phase function p_δ by inserting the latter into the azimuthally averaged RTE. From (3.61), it can be easily shown that the azimuthally average Delta–Eddington phase function $p_{a\delta}$ is given by

$$p_{a\delta}(\mu,\mu') \equiv 2f\delta(\mu-\mu') + (1-f)(1+3g'\mu\mu') \tag{3.66}$$

where f and g' are given by (3.64) and (3.65). By inserting (3.66) in (3.33), it is readily deduced that the Delta–Eddington RTE is formally identical to the original RTE if, besides transforming g into g', the optical parameters τ and ϖ are modified by

$$\begin{aligned} \tau' &= (1-\varpi g^2)\tau \\ \varpi' &= \frac{(1-g^2)\varpi}{1-\varpi g^2} \end{aligned} \tag{3.67}$$

which can be interpreted as a similarity transformation in RTE. In the following section we will show the significant impact of using the Delta–Eddington approach in terms of accuracy of the RTE solution.

3.7.2 Antenna brightness temperature in the Eddington approximation

The general analytical solution, described in the previous section, has been derived, for simplicity, considering a homogeneous (uniform) single atmospheric layer. Realistic structures of clouds and precipitation generally show a strong vertical inhomogeneity.

The finite horizontal extent of clouds cannot be considered in a rigorous manner within this theoretical framework. However, on one hand, highly directive antennas used in microwave and millimetre-wave applications partially minimise the problem of non-uniform beam filling, and, on the other hand, parametric approaches (such as weighting raining cloud and clear-air areas by means of semi-empirical factors or equivalent 1D slant geometries) can still be applied to obtain an analytical solution to the radiative transfer problem. If the goal of a closed-form solution is not required, more sophisticated models can be used to numerically solve RTE in the case of horizontally finite structures.

Polarisation is the other limitation of the proposed model. However, the proposed analytical model could take into account polarisation-dependent boundary conditions (i.e. different horizontal and vertical surface reflectivity) coupled with a polarisation diversity of atmospheric extinction and phase function (i.e. different horizontal and vertical optical parameters).

The vertically inhomogeneous scattering atmosphere modelling clouds and precipitation is divided into L layers. The associated L+1 vertical levels are indicated by optical depths $\tau_0, \tau_1, \tau_2, \ldots, \tau_{L-1}, \tau_L$, where the layer number and optical depth increase from the top to bottom of the atmosphere, that is, $\tau = \tau_0 = 0$ for $z = H$ and $\tau = \tau_L = \tau_s$ for $z = 0$. The basic assumption is that each layer is homogeneous so that optical parameters of lth layer k_{el}, ϖ_l and g_l (and their Delta–Eddington transform) are constant. In accordance with the previous parametrisation, the temperature T_l within each layer is assumed linearly dependent on τ:

$$T_l(\tau) = t_{0l} + t_{1l}(\tau - \tau_{l-1}) = T(\tau_{l-1}) + \frac{\Gamma_{Tl}}{k_{el}}\tau \quad \text{with } \tau_{l-1} < \tau < \tau_l$$

where the thermal lapse rate Γ_{Tl} may be different from layer to layer. The thermal structure of the stratified atmosphere is thus determined by $2L$ constants. Exploiting the results already obtained for a single layer, under the Eddington approximation, the analytical solution of RTE is for the lth layer ($l = 0, 1, 2, \ldots, L$):

$$\begin{cases} T_{\mathrm{B0}}^{(l)}(\tau) = C_{1l}e^{-\lambda_l\tau} + C_{2l}e^{\lambda_l\tau} + t_{0l} + t_{1l}\tau \\ T_{\mathrm{B1}}^{(l)}(\tau) = c_l C_{1l}e^{-\lambda_l\tau} - c_l C_{2l}e^{\lambda_l\tau} - \dfrac{t_{1l}}{1 - \varpi_l g_l}\tau \end{cases} \quad \text{with } \tau_{l-1} < \tau < \tau_l$$

where c_l and λ_l are the terms c and λ of Equation (3.52) specific for the lth layer. The $2L$ integration constants C_{1l} and C_{2l} (l=1,2, …, L) are relative to lth layer and can be determined by imposing not only the boundary conditions at the top ($\tau = \tau_0 = 0$) and bottom ($\tau = \tau_L = \tau_s$) of the atmosphere but also the continuity equations at the $L-1$ intermediate boundaries ($\tau = \tau_l$ with $l \neq 0,\ L$). Generalising (3.55), in terms of flux densities and previous boundary and continuity conditions, yields

$$\begin{aligned}
&T_{B0}^{(1)}(\tau_0) + \frac{2}{3}T_{B1}^{(1)}(\tau_0) = T_c \\
&T_{B0}^{(l)}(\tau_l) = T_{B0}^{(l+1)}(\tau_l) \\
&T_{B1}^{(l)}(\tau_l) = T_{B1}^{(l+1)}(\tau_l) \\
&T_{B0}^{(L)}(\tau_L) - \frac{2}{3}T_{B1}^{(L)}(\tau_L) = e_s T_S + (1-e_s)\left[T_{B0}^{(L)}(\tau_L) + \frac{2}{3}T_{B1}^{(L)}(\tau_L)\right]
\end{aligned} \tag{3.68}$$

with $l = 1, 2, \ldots, L-1$. After substituting (3.54) into (3.68), the previous linear system represents a set of $2L$ equations in $2L$ unknowns which are the integration constants C_{1l} and C_{2l} ($l = 1, 2, \ldots, L$). A numerical solution of the previous linear system provides the complete expression of $T_{B0}(\tau)$ and $T_{B1}(\tau)$ for each layer of the stratified atmosphere.

Notice that also in this case a scaling transformation is appropriate in order to avoid unwanted singularities during the inversion procedure (Stamnes and Swanson, 1981; Stamnes *et al.*, 1988). By generalising (3.58) with $\tau_0 = \tau_{l-1}$ and $\tau_l = \tau_s$, the unknown constants C_{1l} and C_{2l} can be transformed as follows:

$$\begin{aligned}
&C_{1l} = \tilde{C}_{1l}e^{\lambda\tau_{l-1}} \\
&C_{2l} = \tilde{C}_{2l}e^{-\lambda\tau_l}
\end{aligned} \tag{3.69}$$

where $l = 1, 2, \ldots, L$. The transformed linear system via (3.69) can be verified to be inconditionally stable.

3.7.2.1 Ground-based antennas

As mentioned, the knowledge of $T_{B0}(\tau)$ and $T_{B1}(\tau)$ completely solves the RTE under Eddington's approximation through (3.60). Another form of the illustrated RTE analytical solution, which is physically meaningful and very useful for practical applications, is obtained by evaluating the source function $J_E(\tau,\ \mu)$, given in (3.49), using the obtained expressions of $T_{B0}(\tau)$ and $T_{B1}(\tau)$. In fact, $J_E(\tau,\ \mu)$ can be substituted in (3.34) and (3.35) to explicitly express both downwelling $T_B(\tau_s,\ {}_\downarrow\mu)$ at ground and upwelling $T_B(0,\ {}_\uparrow\mu)$ at the top of the atmosphere, respectively.

For ground-based antennas, the integral form of the Eddington solution for a single homogeneous layer is

$$\begin{aligned}
T_B(\tau_s,\ {}_\downarrow\mu) = T_B(0,\ {}_\downarrow\mu)e^{-\tau_s/\mu} + \frac{1}{\mu}\int_0^{\tau_s} &[\varpi T_{B0}(\tau) + 3\varpi g\mu T_{B1}(\tau) \\
&+ (1-\varpi)(t_0 + t_1\tau)]e^{-(\tau_s-\tau)/\mu}\,d\tau
\end{aligned}$$

After the evaluation of the integral using (3.60), the closed-form Eddington approximation of downwelling T_B for a single homogeneous layer is

$$\begin{aligned} T_B(\tau_s, {}_\downarrow\mu) &= T_c e^{-\tau_s/\mu} + \left[t_0 + t_1\mu - \frac{3\varpi g\mu}{3(1-\varpi g)} t_1 \right](1 - e^{-\tau_s/\mu}) \\ &+ t_1\tau_s - \varpi C_1 \frac{1 + c\varpi g}{1 - \lambda\mu}(e^{-\tau_s/\mu} - e^{-\lambda\tau_s}) \\ &- \varpi C_2 \frac{1 - c\varpi g}{1 + \lambda\mu}(e^{-\tau_s/\mu} - e^{-\lambda\tau_s}) \end{aligned} \tag{3.70}$$

where the optical parameters τ_s, ϖ and g may be transformed by Delta–Eddington similarity equations, given in (3.65) and (3.67).

As a simple test, for a homogeneous slab with no scattering having $\varpi = 0$ and uniform temperature $T_0 = b_0$ and $b_1 = 0$, the previous equation reduces to

$$T_B(\tau_s, {}_\downarrow\mu) = T_c e^{-\tau_s/\mu} + T_0(1 - e^{-\tau_s/\mu})$$

which is the well-known RTE solution for ground-based radiometric observations of absorbing media.

Previous expressions can be extended to a stratified atmosphere in a straightforward manner using a recursive approach. With respect to the single slab, we can simply transform $\tau_0 = 0$ into τ_{l-1} and $\tau_L = \tau_s$ into τ_l. The set of recursive equations to determine the downwelling $T_B(\tau_l, \mu)$ at the lth atmospheric level τ_l is then the following:

$$\begin{aligned} T_B(\tau_0 = 0, {}_\downarrow\mu) &= T_c \\ T_B(\tau_l, {}_\downarrow\mu) &= T_B(\tau_{l-1}, {}_\downarrow\mu)e^{-(\tau_l - \tau_{l-1})/\mu} + \left[t_{0l} + t_{1l}\mu - \frac{3\varpi_l g_l\mu}{3(1-\varpi_l g_l)} t_{1l} \right] \\ &\quad \times (1 - e^{-(\tau_l - \tau_{l-1})/\mu}) \\ &\quad + t_{1l}(\tau_l - \tau_{l-1}) - \varpi_l C_{1l} \frac{1 + c_l\varpi_l g_l}{1 - \lambda_l\mu}[e^{-(\tau_l - \tau_{l-1})/\mu + \lambda_l\tau_{l-1}} - e^{-\lambda_l\tau_l}] \\ &\quad - \varpi_l C_{2l} \frac{1 - c_l\varpi_l g_l}{1 + \lambda_l\mu}[e^{-(\tau_l - \tau_{l-1})/\mu + \lambda_l\tau_{l-1}} - e^{-\lambda_l\tau_l}] \end{aligned} \tag{3.71}$$

with $l = 1, 2, \ldots, L$ where all terms have been already defined.

3.7.2.2 Satellite-based antennas

For satellite-based antennas, the argument is analogous to that illustrated for ground-based ones. In this case the Eddington solution for a single homogeneous layer can be written as

$$\begin{aligned} T_B(0, {}_\uparrow\mu) = T_B(\tau_s, {}_\uparrow\mu)e^{-\tau_s/\mu} + \frac{1}{\mu}\int_{\tau_s}^{0} [\varpi T_{B0}(\tau) + 3\varpi g\mu T_{B1}(\tau) \\ + (1-\varpi)(t_0 + t_1\tau)]e^{-\tau/\mu}\,d\tau \end{aligned}$$

After the evaluation of the integral using (3.60), the closed-form Eddington approximation of upwelling T_B for a single homogeneous layer is

$$T_B(0, {}_\uparrow\mu) = T_B(\tau_s, {}_\uparrow\mu)e^{-\tau_s/\mu} + \left[t_0 + t_1\mu + \frac{3\varpi g\mu}{3(1-\varpi g)}t_1\right](1 - e^{-\tau_s/\mu})$$
$$- t_1\tau_s + \varpi C_1\frac{1 - c\varpi g}{1+\lambda\mu}(1 - e^{-\tau_s/\mu-\lambda\tau_s})$$
$$+ \varpi C_2\frac{1 + c\varpi g}{1-\lambda\mu}(1 - e^{-\tau_s/\mu+\lambda\tau_s}) \tag{3.72}$$

where $T_B(\tau_s, {}_\uparrow\mu)$ is given by inserting (3.70) into the surface boundary condition (3.37) and, again, the optical parameters τ_s, ϖ and g may be transformed by Delta–Eddington similarity equations.

Previous expressions can be generalised to a stratified atmosphere using the same recursive approach as before. It is worth mentioning that the upwelling $T_B(\tau_s = \tau_L, {}_\uparrow\mu)$ at the surface is obtained from the downwelling $T_B(\tau_s = \tau_L, {}_\downarrow\mu)$ by applying the surface boundary condition. The set of recursive equations to determine the upwelling $T_B(\tau_{l-1}, {}_\uparrow\mu)$ at the $(l-1)$th atmospheric level τ_{l-1} is then the following:

$$T_B(\tau_L, {}_\uparrow\mu) = e_s(\mu)T_s + 2\pi\int_0^1 \rho_{ds}({}_\uparrow\mu, {}_\downarrow\mu')T_B(\tau_L, {}_\downarrow\mu')\mu'\,d\mu'$$

$$T_B(\tau_{l-1}, {}_\uparrow\mu) = T_B(\tau_l, {}_\uparrow\mu)e^{-(\tau_l-\tau_{l-1})/\mu} + \left[t_{0l} + t_{1l}\mu - \frac{3\varpi_l g_l\mu}{3(1-\varpi_l g_l)}t_{1l}\right]$$
$$\times (1 - e^{-(\tau_l-\tau_{l-1})/\mu}) - t_{1l}(\tau_l - \tau_{l-1})$$
$$+ \varpi_l C_{1l}\frac{1 - c_l\varpi_l g_l}{1+\lambda_l\mu}[e^{-\lambda_l\tau_{l-1}} - e^{-(\tau_l-\tau_{l-1})/\mu-\lambda_l\tau_l}]$$
$$+ \varpi_l C_{2l}\frac{1 + c_l\varpi_l g_l}{1-\lambda_l\mu}[e^{-\lambda_l\tau_{l-1}} - e^{-(\tau_l-\tau_{l-1})/\mu+\lambda_l\tau_l}]$$

with $l = L, L-1, \ldots 1$ and where $T_B(\tau_L, {}_\downarrow\mu)$ is determined by inserting (3.71) into the surface boundary condition (3.37) at $\tau_l = \tau_L$.

3.7.2.3 General accuracy of Eddington-like solution

The validation of the analytical radiative tranfer model, described above, can be performed by resorting to RTE numerical solutions, well tested and documented in the literature. Among the various RTE codes available, here we have selected the doubling-adding RT3 (see Section 3.6.4) and the discrete ordinate radiative transfer (DisORT, see Ishimaru, 1978) as one of the most accurate and widely used for inter-comparison purposes.

Other works have been dealing with the validation of the Eddington solution (Wiscombe and Joseph, 1977; Smith *et al.*, 2002). Hereafter we will follow two distinct approaches in order to perform these numerical tests on the accuracy of the Eddington solution. On one hand, we'll let the optical parameters vary independently within the whole of their definition range: hereafter the optical thickness is varied between 0.1 and 50, the single scattering albedo between 0.5 and 0.99, and the g parameter between 0 and 0.96. This test is an extreme one, but it stresses the mathematical conditions where the solution denotes its limitations even though these limits may be not of physical importance. A worthy fallout of this approach is that the proposed solution could be used in a general context of passive observations of an absorbing and scattering medium, not necessarily related to atmospheric applications (e.g. radio-propagation through vegetation). On the other hand, we will select a simple rain slab model, compute the optical parameters and verify the related errors. The rain slab model is widely used in the radio-propagation literature and it represents a key reference model for many applications. The adopted one-layer atmosphere is the same as in Figure 3.16; in the case of DisORT/RT3 computations, a Heyney–Greenstein phase function (see (3.63)) is assumed with the number of streams $N_\mu = 16$. Errors as a function of the viewing angle are plotted in Figure 3.17 for the Eddington and the δ–Eddington approximation. The grey tones of the different lines are modulated by the single scattering albedos, extinction and asymmetry parameter. The white line corresponds to $\varpi = 0.99$, $g = 0.96$ and $\tau = 30$. As already observed Eddington is doing quite well close to 50° (this is the reason of its extensive use for TMI or SSM/I-like radiometers); on the other hand it is generally overestimating T_Bs at nadir and underestimating at very oblique angles (except for strong scattering media). This overestimation at nadir can reach up to 50 K in the worst case ($\varpi = 0.91$, $g = 0.96$ and $\tau = 6.7$) and is generally large when the blue component is strong (i.e. for high g values). On the other hand the δ–Eddington is doing quite well also at nadir (generally underestimating T_Bs) while it is overestimating at very oblique angles.

Figure 3.18 shows the percentage T_B error as a function of the vertical optical thickness τ, single scattering albedo ϖ and asymmetry factor g and zenith angle θ for ground-based antennas. For each plot, the error results for six values of the optical thickness $\tau = 0.1, 1, 5, 10, 25, 50$, four values of the single scattering albedo $\varpi =$ 0.1, 0.3, 0.6, 0.9, four values of the scattering asymmetry factor $g = 0$, 0.3, 0.6, 0.9, four values of the zenith angle, i.e. 0°, 25°, 50° and 75° are shown in order to give the overall impact. These results refer to the case of avoiding the Delta–Eddington transform for T_B calculations. We note that the errors are larger for smaller optical thickness (i.e. smaller extinction), higher single scattering albedo and asymmetry factor and smaller zenith angles. An average error of about 25 per cent is obtained with peaks more than 50 per cent in some critical circumstances. A major source of discrepancy may be attributed to the behaviour of the scattering phase function which, in the case of a Sobolev approximation (see Equation 3.46), is significantly different from that of Heyney–Greenstein.

Figure 3.19 shows the same as in Figure 3.18, but using the Delta–Eddington transform. It is apparent that in all cases the overall percentage error is drastically reduced to less than 5 per cent, still with the same trend with respect to the optical parameters. Indeed, very low optical thickness values are rarely associated with large volumetric single scattering albedo and scattering asymmetry for natural media.

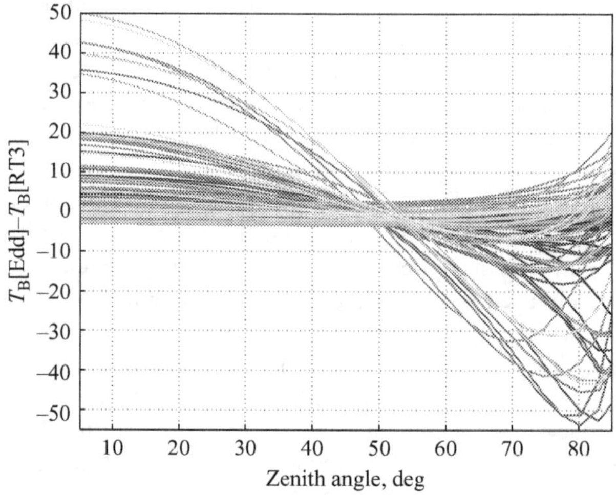

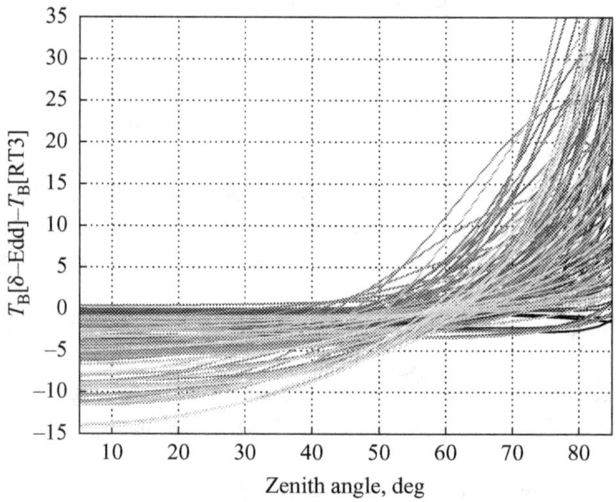

Figure 3.17 Brightness temperature error as a function of the zenith angle θ for satellite-based antennas for the Eddington and the δ−Eddington code

As a more realistic model of a scattering medium, we have referred to a uniform rain slab: a Marshall–Palmer drop size distribution has been used to compute the optical parameters under the assumption of spherical liquid particles through a standard Mie code. The Delta–Eddington transform has been adopted for T_B calculations using (3.70) or (3.72). The frequency range extends from 10 to 90 GHz including well-known beacon frequency bands (i.e. 12.5, 18.7, 29.6, 39.6, 49.5, 90.0 GHz) while rain rates are between 0.01 and 100 mm/h (i.e. 0.01, 1, 5, 10, 20, 30, 50, 75 and 100 mm/h). Figure 3.20 shows the percentage T_B error as a function of the vertical optical thickness τ and brightness temperature T_B for both ground-based

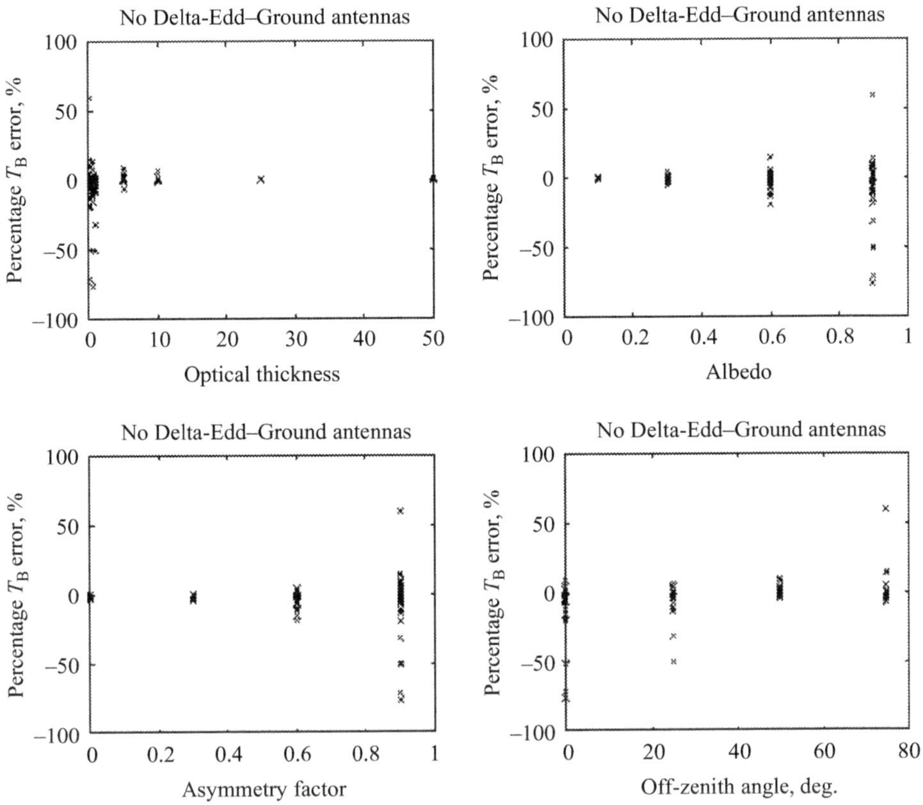

Figure 3.18 *Percentage T_B error as a function of the vertical optical thickness τ, single scattering albedo ϖ and asymmetry factor g and zenith angle θ for ground-based antennas*

and satellite-based antennas, while the two bottom panels show the single scattering albedo and asymmetry factor for a Marshall–Palmer polydispersion of raindrops. It is worth mentioning that in this case the natural variability of the optical parameters is much less than the simulated one. The nadir optical thickness τ can go up to 30 (being the layer of 3 km) while the single scattering albedo is confined between 0 and 0.5 and the asymmetry factor between −0.1 and 0.4. Of course, larger values are attributed to increasing frequency and rain rate. Using the rain slab model, the overall percentage errors are less than 2 per cent with a tendency to underestimate for ground-based antennas and less than 1.5 per cent for satellite-based antennas. The larger errors for small values of optical thickness where there is a maximum underestimation may be explained by looking at the behaviour of the volumetric single scattering albedo which tends to rapidly increase for increasing values of the extinction. Since the albedo weights the multiple scattering effects (see (3.33)) mostly governing the T_B angular spectrum, in this region the errors due to the first-order approximation of T_B

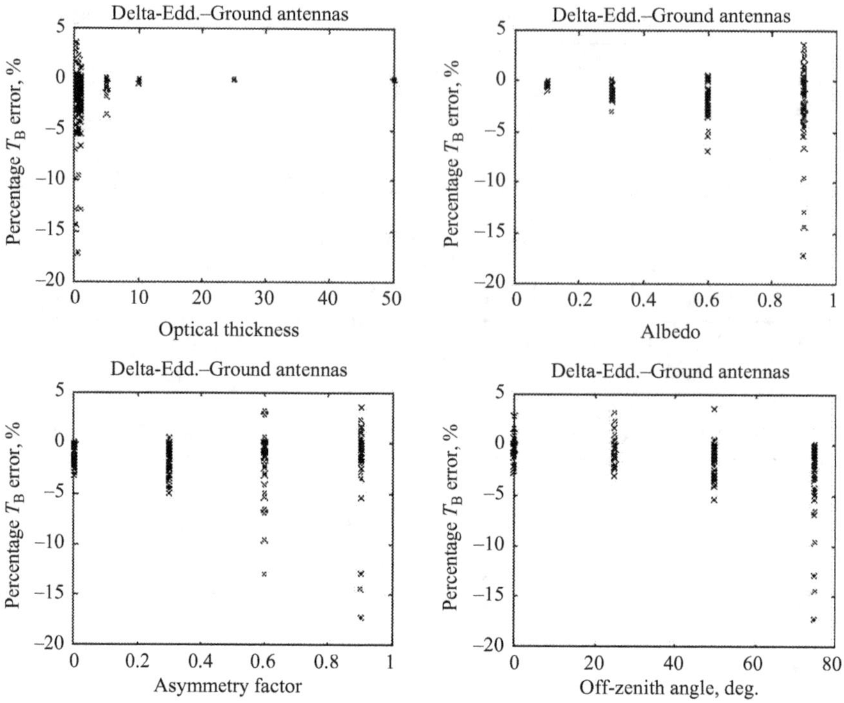

Figure 3.19 Same as in Figure 3.18, but using the Delta–Eddington transform

zenith variations are not compensated for by the extinction through the medium and its emission.

3.8 Microwave signatures of clouds and precipitation

The possibility of remotely sensing the properties of clouds and precipitation by passive sensors, at present, relies on using visible, infrared or microwave channels. Visible (VIS) and infrared (IR) sensors cannot see inside the clouds, and information on their properties can only be gained indirectly by using the albedo and the temperature of the cloud top, which these sensors are able to give. On the other hand the multi-channel microwave observation of the T_Bs of clouds has been recognised to be promising because these sensors respond to many more details of the cloud properties (Weinman and Davies, 1978; Smith and Mugnai, 1989; Smith *et al.*, 1992; Mugnai *et al.*, 1993; Kummerow and Giglio, 1994). The radiative contribution to the observed PWM T_Bs originated from the different components of the scene under observation can be visualised by the emission source and generalised weighting functions WF (see Mugnai, *et al.*, 1993 for their definition) or by the Jacobians (e.g. Bauer and Mugnai, 2003).

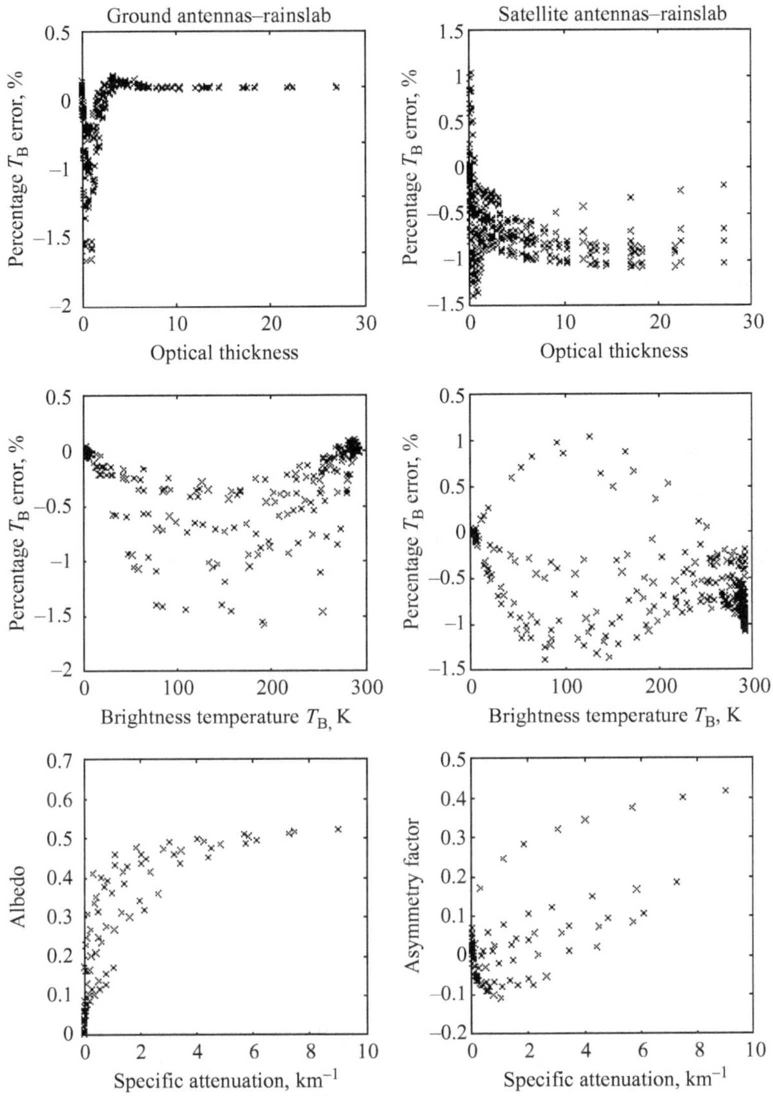

Figure 3.20 *Percentage T_B error as a function of the vertical optical thickness τ and brightness temperature T_B for both ground-based and satellite-based antennas. Bottom panels show the single scattering albedo and asymmetry factor for a Marshall–Palmer polydispersion of raindrops*

As a simple example Figure 3.21 shows the weighting function for TMI-like windows channels when looking at two different points at the surface of a raining scene over sea. Note that, similarly to Bauer *et al.* (1988), the weighting functions extend to negative altitudes, which refer to signals reflected from the surface. Lower frequencies

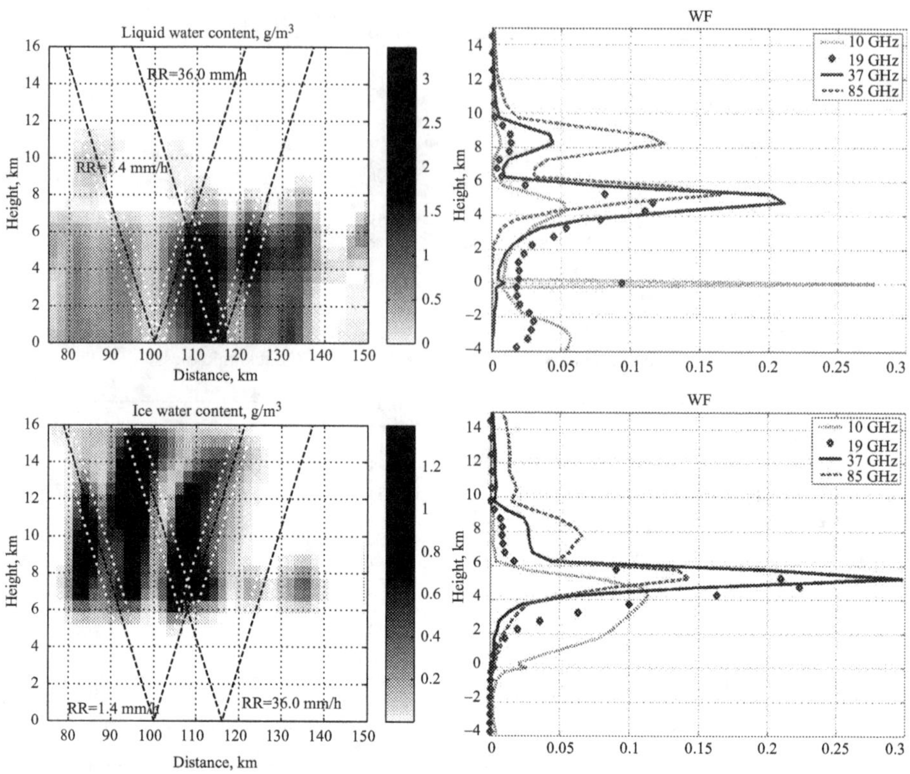

Figure 3.21 *Hydrometeor vertical cross section of water and ice content along a squall line over the ocean simulated by GCE-CRM and TMI channel weighting function (in km^{-1}) for two different observation cases. The weighting functions in the top-right and bottom-right panels refer to a rain rate at the ground equal to 1.4 mm/h and 36 mm/h, respectively*

sound deeper in the troposphere; for instance, in the presence of light precipitation the 10 and 19 GHz channels have a significant contribution from the surface's emission. On the other hand for high rain rates the contribution from the surface becomes negligible even at low frequencies and the centre of mass of the weighting function moves upwards. Finally, note how the 85 GHz WF has a relevant contribution from the upper ice layer of the cloud while the two lower frequencies are practically insensible to it. Similar basic examples can be found in Gasiewski (1993) in Section 3.4.3 for sounding channels and in Section 3.4.5 for window channels.

Microwave T_Bs are highly non-linear functions of surface emissivity, surface temperature, vertical distribution of hydrometeor, temperature, humidity and gas profiles, radiometer footprint and viewing angle. As such it is always a complicated problem to identify general signatures of clouds and precipitation. For instance, a strong emission signal connected to precipitation at low MW frequencies over the cold background ocean surface is completely obscured in the presence of an underlying warm land surface. Similarly 85 GHz scattering signatures can be related to thick ice clouds but also to snow covered or cold land (Spencer *et al.*, 1989). To avoid this problem many statistically-based algorithms or empirical methods have been developed: measurements of both brightness temperatures and parameters to be retrieved have been collected to train retrieval algorithms. On the contrary, here we will focus on physically-based methods based on PMW observations, i.e. on methodologies that try to infer some cloud property (be it the rain or the snow-rate at the ground or the integrated liquid/ice water or the cirrus mean size) by constructing a forward model, able to reproduce the cloud physics and from this to perform the RT. Among all, Bayesian techniques have been proved to have a large potential and flexibility for precipitation profiling (e.g. Evans *et al.*, 1995; Kummerow *et al.*, 1996, 2001; Marzano *et al.*, 1999a; Tassa *et al.*, 2003; Di Michele *et al.*, 2005). Their unique feature is that they constitute a rigorous statistical framework in which to develop cloud-model-based inversion methods and combined multi-sensor approaches. The model-based approaches resort to refined physical models[19] to simulate the measurements (e.g. Smith *et al.*, 1992, 1994; Mugnai *et al.*, 1993; Kummerow *et al.*, 1996). In fact it is well known that few experimental data exist in the literature for setting up a reliable database (Pruppacher and Klett, 1997). We will now briefly describe the main characteristics of the cloud resolving models mostly used in the MW community.

3.8.1 Cloud resolving models

One of the main problems to simulate cloud and precipitation is to model their three-dimensional aspect and their intrinsic variability (Rogers, 1979). Simple *ad hoc* models, such as uniform slabs, can be useful to appraise the main radiative features of the radio-propagation issues or of some particular topic (like cirrus studies), but can be misleading when trying to interpret and quantify real beacon and sky noise measurements (Ishimaru and Cheung, 1980; Marzano *et al.*, 1999b). A way to approach this problem is to resort to the use of mesoscale cloud-resolving three-dimensional (3D) dynamical models which can provide output grids at a given time along a simulation of a selected meteorological event. The inherent advantage of this technique is to deal with physically-consistent cloud distribution in space due to the explicit microphysics describing the cloud formation and evolution. Moreover, in this way, we physically constrain the vertical correlation of hydrometeor contents.[20] On the other hand, the choice of a certain microphysics and rain event might represent a limitation when trying to generalise the extraction of cloud structure from the output grids.

[19] Although sometimes it is better to go back to basic understanding by using simple models (e.g. see Weinman and Davies, (1978); Wilheit *et al.* (1991)).

[20] For instance, this allows us to retrieve rainfall at the ground from space-borne measured radiances that are indeed affected either by the whole hydrometeor profile or even only by the upper layers.

Among the most used mesoscale dynamical models for PMW T_B simulations, we cite here the University of Wisconsin Non-hydrostatic Modelling System (UW-NMS) described in Tripoli (1992), the Meso-NH model (see Lafore, 1998), the Goddard Cumuls Ensemble (GCE) model developed by Tao and Simpson (1993) and the MM5 model (Dudhia, 1993). These CRMs typically include five/six[21] categories of hydrometeors (cloud droplets, raindrops (precipitating liquid), ice graupel (precipitating ice), ice crystals (non-precipitating small ice), snow aggregates (precipitating mixed air-ice) and hail). Conversions of water substance from one form/hydrometeor to another can occur through many different processes (like melting, riming, collection, condensation, deposition, etc.). These microphysical processes are described in terms of bulk quantities (see Chapter 3 in Houze, 1993). Therefore there is no explicit microphysics,[22] and the particle size distribution (PSD) is usually characterised by monodispersion (cloud droplets and ice crystals) and by fixed inverse-exponential in the form

$$N_h(r) = N_0(L_h)e^{-\Lambda(R_h)r} \tag{3.73}$$

where the subscript 'h' stands for the hth hydrometeor, r is the radius, N_0 (m^{-3} mm^{-1}) is the intercept in the logarithmic plane, Λ is the slope, L_h (g/m^3) is the equivalent liquid water content and R_h (mm/h^1) the precipitation rate. Equation (3.73) resembles the well-known Marshall–Palmer PSD, but only one parameter (the slope Λ) is parameterised being the intercept N_0 fixed by the hydrometeor content L_h (Marshall and Palmer, 1948; Gasiewski, 1993) or vice versa. Therefore, once the vertical profile $L_h(z)$ with z the altitude is assigned, only one of the two PSD parameters is free. If Λ is known, then it results in:

$$L_h = \int_{r_{m_h}}^{r_{M_h}} \frac{4}{3}\pi r^3 N_h(r)\rho_h \, dr = \frac{4}{3}\pi N_{0h}\rho_h \left[-\frac{\gamma_2(0, r_{Mh})}{\Lambda(R_h)} + \frac{\gamma_2(0, r_{m_h})}{\Lambda(R_h)} \right] \tag{3.74}$$

where r_{m_h} and r_{M_h} are the minimum and maximum radius of each hydrometeor,[23] ρ_h their density[24] and γ_2 is the incomplete Γ function of order 2. The slope parameter Λ of the size distributions is typically parametrised to the rain rate using a relation $\Lambda = aR_h^b$ with a [mm^{-1} mm^{-b} h^b] and b constant empirical coefficients (e.g. $\Lambda = 4.1R^{-0.21}$ for Marshall–Palmer raindrops PSD, $\Lambda = 2.29R^{-0.45}$ for Sekhon–Srivastava ice crystals/graupel PSD, $\Lambda = 25.5R^{-0.48}$) for Gunn–Marshall snow aggregates PSD. More specific details about the microphysical assumptions can be found in Kummerow *et al.* (1996), Panegrossi *et al.* (1998).

[21] At least four hydrometeors are generally accepted as being sufficient for a description of the vertical structure of hydrometeor relevant for PMW: rain drops, cloud water droplets, graupel particles and ice crystals (Smith *et al.*, 1992).

[22] Explicit microphysical models are now also being developed (e.g. see Lynn *et al.*, 2005).

[23] For instance radius size ranges of cloud droplets, raindrops, ice graupel, ice crystals and snow can be fixed to 0.001–0.01, 0.1–3.0, 0.1–5, 0.1–1.5 and 0.1–5.0 mm, respectively.

[24] Density of ice graupel, ice crystals and snow is set to 0.5, 0.2 and 0.1 g/cm^3 in some models and to 0.4, 0.92 and 0.1 g/cm^3 in others.

All these models, initiated with the appropriate boundary conditions, can describe the temporal evolution of different mesoscale systems (e.g. squall lines, hurricanes, cold and warm fronts, snow storms, etc.). Therefore they can help in providing useful statistics of the background conditions (temperature and humidity profiles) and of the vertical distributions of hydrometeors that have occurred under a certain case study (e.g. a 2001 blizzard over New England, see Skofronick-Jackson *et al.*, 2004; a COHMEX hail storm on 11 July, 1986, see Smith *et al.*, 1992; a snow storm in the East Coast of the US, see Mugnai *et al.*, 2005a; a Baltic Sea area frontal system, Mugnai *et al.*, 2005b) or over a general meteorological condition where the attention is focused (e.g. cold fronts in the Mid-Atlantic, hurricanes or squall lines in the tropical region). As an example Figure 3.34 shows a cross section of the total hydrometeor content for a tropical squall line simulated by the GCE; the same model has been used to produce the integrated water and ice contents for a Mid-Atlantic cold front as represented in Figure 3.35. It is a challenging task of the cloud physicists to produce a three-dimensional structure of hydrometeor as realistically as possible.

3.8.2 Hydrometeor scattering computation and simulated T_B*s*

As seen in Section 3.3, from a radiative transfer point of view, an MIS medium is completely characterised by the knowledge of its optical parameters k_e, ϖ and p. In the CRMs the hydrometeor shapes are assumed spherical or, better, sphere-equivalent. This implies that the Lorenz–Mie theory (Section 3.4.1) can be applied in an efficient way for both homogeneous and spherically-coated inhomogeneous particles (Ishimaru, 1978). Large raindrops and ice particles cannot verify this assumption, but the non-sphericity can be considered in the first stage as a second-order effect (see discussion in Section 3.9). When considering ice particles with low density (like snow or graupel) their dielectric constant is typically derived by Maxwell–Garnett formula for inclusions of air in an ice matrix. Note that generally CRMs do not include melting hydrometeors. Because of their importance in enhancing the backscattering and the extinction properties this has been recognised as a major lack of the CRMs. To remedy this drawback simplified melting layer models have been produced (Bauer *et al.*, 2000, Olson *et al.*, 2001a) and bright band signatures evaluated (Olson *et al.*, 2001b; Battaglia *et al.*, 2003) with a particular emphasis laid on the MW electromagnetic behaviour of the melting particles (Fabry and Szymer, 1999 and references therein). In general, the single scattering properties of heterogeneous particles have to be carefully validated by using cross checks with experimental measurements. As examples, Olson *et al.* (2001a), Battaglia *et al.* (2003) have used precipitation radar mirror images and profiler observations, respectively, to further constrain the electromagnetic modelling of melting particles, while Skofronick-Jackson *et al.* (2004) have used attenuation measurements to characterise the dimension of the equivolume spheres having the same scattering properties as snowflakes.

The vertical profile of k_e, ϖ and g is a sufficient input to the Eddington analytical model (Section 3.7.1) devised to compute the atmospheric contribution to the antenna noise temperature for ground- and satellite-based antennas. Figure 3.22 shows these optical parameters as a function of the precipitation rate for raindrops,

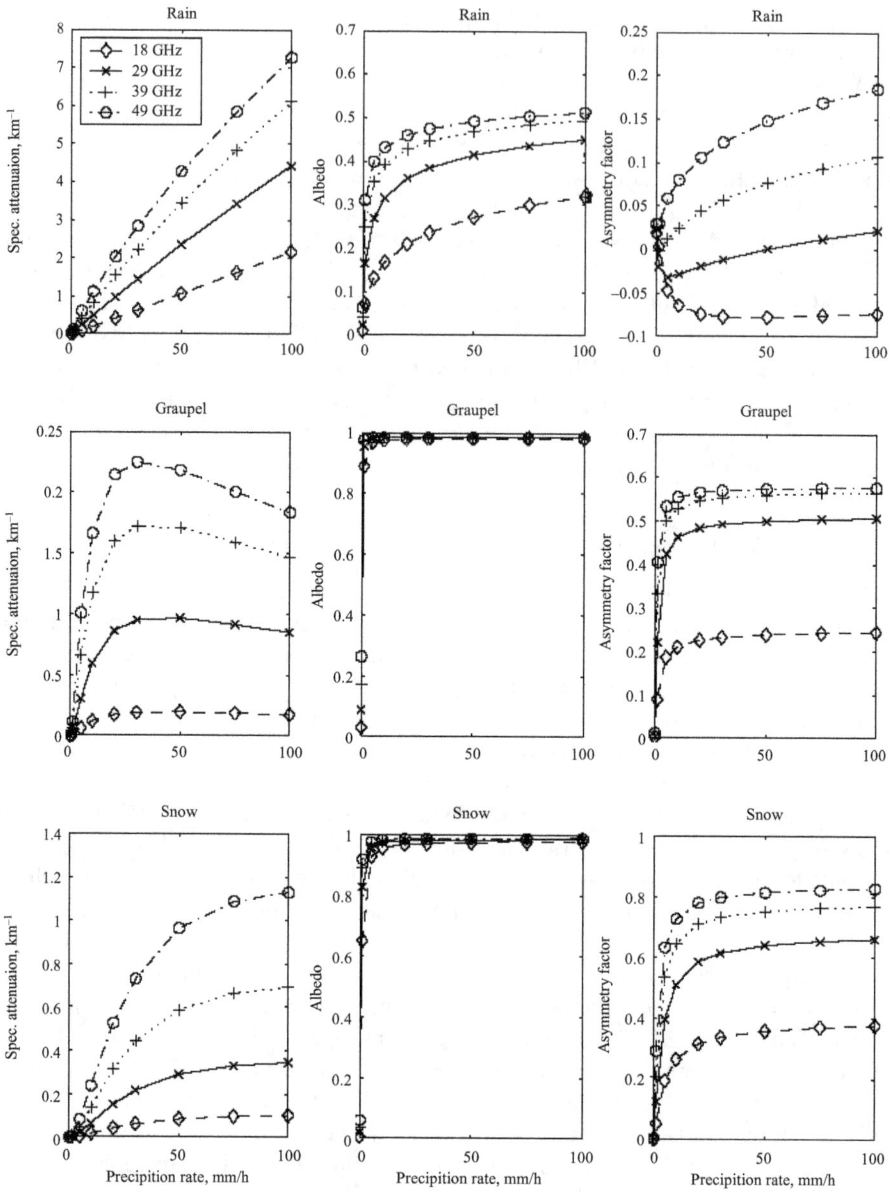

Figure 3.22 *Optical parameters in terms of specific attenuation (or extinction coefficient), volumetric single scattering albedo and scattering asymmetry factor as a function of the precipitation rate for raindrops (top panels), graupel particles (middle panels) and snow (bottom panels) at 18, 29, 39 and 49 GHz. Ambient temperature has been set to 10°C for cloud droplets and raindrops, 0°C for ice graupel and −10°C for ice crystals and snow aggregates*

graupel particles and snow at 18, 29, 39 and 49 GHz. Single scattering albedo of ice hydrometeors, i.e. graupel and snow, is close to 1 even for low rain rates, as opposed to that of raindrops which is at most 0.5. However, the latter exhibits a higher specific attenuation which is, for a given precipitation rate, twice the graupel one (which is, in its turn, double the snow one). The asymmetry factor for graupel and snow is also much higher than that of raindrops which can be at lower frequencies even negative (Gasiewski, 1993). The implication of these considerations is that raindrop layers will mostly contribute to determine signal fade, while graupel, and to a minor extent snow, will give rise to incoherent effects weighted by volumetric single scattering albedo (Marzano and Roberti, 2003).

Once the single scattering properties have been computed[25] the T_Bs emerging at the ground or at the top of the atmosphere corresponding to each CRM simulation can be evaluated via one of the methodologies illustrated in Sections 3.6 and 3.7. Since typically CRMs have spatial resolution of a few kilometres (thus much higher than satellite footprints), T_Bs are generally computed at model resolution and then spatially filtered in order to reach the effective resolution by taking into account the radiometer antenna pattern at each different channel. Many authors have followed this procedure when trying to explain observations during airborne field campaigns (e.g. Katsumata *et al.* (2000) snow clouds over ocean at 36 and 89 GHz; Skofronick-Jackson and Wang, 2000 wide band observations between 10 and 220 GHz over an oceanic squall line during TOGA-COARE experiment; Skofronick-Jackson *et al.*, 2003 radiometer observations in the interval between 10 and 340 GHz over Hurricane Bonnie) or when focusing on case studies oriented to TMI and SSM/I applications like tropical squall lines (Hong *et al.*, 2000; Olson *et al.*, 2001b), rainfall systems (Alder *et al.*, 1991), hail storm over land (Smith *et al.*, 1992) and hurricanes (e.g. Tassa *et al.*, 2003). More recently great emphasis has been laid on snow storm PMW signatures over land: Mugnai *et al.* (2005b) have analysed the role played by different surface emissivities in both absorption (channels in the 60 and 118 GHz oxygen bands) and window channels (up to 150 GHz) while Skofronick-Jackson *et al.* (2003) have focused on AMSU-B channels at 89 and around the 183 GHz water vapour absorption line. All these works can actually provide hints and suggestions about the most significant PMW signatures characteristic of the different storm features (main core, rear anvil, flanks, updraft region) of each of the particular mesoscale system under study.

3.8.3 Consistency between predicted and observed T_Bs

According to the particular application this kind of forward modelling can produce datasets of T_Bs that can be used as inputs for retrieval procedures. As noted by many authors (e.g. Panegrossi *et al.*, 1988; Bauer, 2001; Wiedner *et al.*, 2004), before performing the retrieval, model-based approaches have to tackle the critical issue of tuning simulations to the measurement manifolds in order to be statistically

[25] Microwave gaseous absorption on the other hand can be computed by means of the different models described in Chapter 2.

significant[26] and representative of real radiative observations of the meteorological event under investigation. On a rigorous basis, a retrieval based on a pre-generated database may be performed only if the manifold of the measured T_Bs is completely overlapped by that of the simulations. On the contrary, when the two manifolds are largely different, the precipitation event is not adequately represented by the database (Panegrossi *et al.*, 1998) and therefore no retrieval should be attempted. At this point it is always important to refer to the actual satellite/ground-based microwave multi-channel radiances and their joint variability. For instance, principal component analysis of observed T_Bs (like in Petty, 2001b for the special sensor microwave/imagers (SSM/Is) flown aboard spacecrafts of the US Defense Meteorological Satellite Program (DMSP) and in Bauer *et al.*, 2002) for the TMI sensor on board TRMM can be considered.

It is obviously by looking at inconsistencies between predicted and observed T_Bs that deficiencies in CRM can be highlighted. For instance Skofronick-Jackson *et al.* (2004) have underlined the need of a more accurate specification of local optical properties in order to match the AMSU-B observation with simulated T_Bs from cloud modelling. By using SSM/I data Chevallier and Bauer (2003) individuated problems in the advection scheme of the ECMWF operational short range forecast system.

3.8.4 *Sensitivity studies*

As noticed by Tassa *et al.* (2003) extreme care must be paid to the significance and the limitations of the database itself, the construction of which requires a number of additional information: error distributions, profile accuracy and range of applicability of the database within different meteorological conditions.[27] In particular the error covariance matrix takes into account all modelling as well as instrumental errors. The modelling errors are computed by evaluating the simulated T_B sensitivity to main parameters and/or approximations used within the cloud-radiation database generation (e.g. drop size distribution, sea surface wind speed, presence of partially melted hydrometeors; Tassa *et al.*, 1999). Obviously according to the kind of application some approximations/parametrisations will be more important than others. For instance when focusing on cirrus clouds ice crystal shape plays a very important role: *ad hoc* sensitivity studies have been performed by Evans and Stephens (1995a,b), Evans *et al.* (1998,1999,2002), Liu and Curry (2000); the variability induced by raindrops size distribution variability has to be assessed in rain retrievals (e.g. Viltard *et al.*, 2000) while snow density assumptions will drive the sub-millimetre T_B depressions when looking at snowfalls (e.g. Bennartz and Petty, 2001; Petty, 2001a; Skofronick-Jackson *et al.*, 2002; Bennartz and Bauer, 2003; Wiedner *et al.*, 2004). In this last case, when over land, a major source of variability will be introduced by the emissivity variability of the surface as well (Mugnai *et al.*, 2005b).

[26] This can be evaluated by considering a statistical parameter like the database matching index (DMI) introduced by Tassa *et al.* (2003).

[27] For instance a mid-latitude rainfall algorithm should not be based on hurricane simulations.

3.8.5 Cloud genera

To illustrate the general procedure described in Sections 3.8.1–3.8.2, the numerical outputs of a three-dimensional time-dependent cloud mesoscale model have been exploited for generating a cloud-genera dataset, explicitly describing the gross vertical distribution of five species of hydrometeors: cloud droplets, raindrops, graupel particles, ice crystals and snow particles. The hydrometeor scattering parameters have been calculated for a large set of beacon channel frequencies and the slant-path attenuation and antenna noise temperatures both for ground- and satellite-based antennas at vertical and horizontal polarisation have been simulated. The effect of the radiometer antenna pattern has been evaluated as well.

A mesoscale cloud model simulation of a mid-latitude summer rain event has been used for training the entire scheme. In order to reduce the number of hydrometeor parameters consisting of equivalent water content within each layer (i.e. 42 levels by 5 hydrometeors), the cloud structures have been vertically resolved in seven homogeneous layers with a variable vertical resolution of about 1.5 km. A classification of cloud structures into meteorological cloud genera has been applied in order to derive both mean vertical profiles and covariance matrices of hydrometeor contents of each cloud class (see Auria *et al.*, 1998). By using seasonal statistical values of meteorological profiles, mean structures of cloud genera have been tuned to some key parameters: surface temperature, humidity and pressure and their vertical gradients. The latter can be characterised by a linear decrease for temperature and by an inverse-exponential law for both pressure and humidity. The freezing and the glaciation levels have been chosen as the driving parameters to determine the melting layer and the presence of solid ice. Other free parameters of cloud genera are the cloud droplet bottom level and the ice crystal top level in order to specify the cloud vertical extension. In this way, even though derived from a specific rain event simulated by the cloud-resolving model, we are able to generalise the synthetic structures to a wide variety of meteorological conditions with the only basic limitation of assuming a given correlation among hydrometeor contents.

The classified cloud dataset has been then extended by means of a Monte Carlo statistical procedure, based on the use of a truncated Gaussian multi-variate generator defined by the mean and covariance matrix of the hydrometeor contents of each class. Meteorological variables have been supposed to be uniformly variable around their mean values within a given percentage. In this section we have considered both stratiform and cumuliform non-precipitating clouds (i.e. stratus, stratocumulus, cumulus, alto-stratus and altocumulus), stratiform rainfall (i.e. nimbostratus) and convective rainfall (i.e. cumulonimbus and cumulus congestus). As a result, a dataset of 10 000 cloud structures has been statistically generated, retaining the physical and statistical features of the input microphysical cloud model.

Figure 3.23 shows vertical profiles of liquid (left panels) and ice (right panels) hydrometeors for clouds classified as stratocumulus (Sc), nimbostratus (Ns) and cumulonimbus (Cb) in the top, middle and bottom panels, respectively. Profiles are expressed in terms of the integrated equivalent water content within each layer (so that the vertical sum provides the total columnar content). The symbols refer to the middle

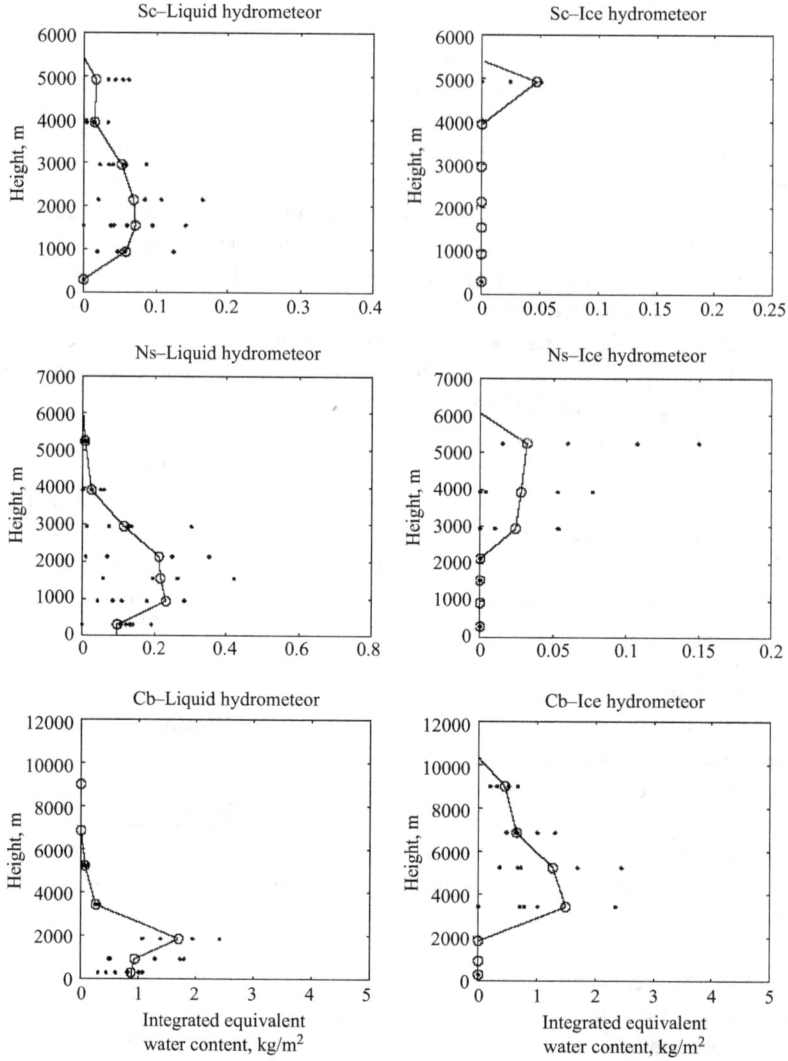

Figure 3.23 Vertical profiles of liquid (left panels) and ice (right panels) hydrometors for clouds classified as strato-cumulus (Sc), nimbostratus (Ns) and cumulonimbus (Cb) in the top, middle and bottom panels, respectively. See text for details

of each layer, while the point dispersion around them gives an example of all possible realisations within the generated ensemble.

The profiles of Figure 3.23 refer to a summer mid-latitude meteorological condition characterised by an average clear-air state. The latter was specified by a surface temperature, relative humidity and pressure equal to 22°C, 90 per cent and 1010 hPa, and a temperature gradient, humidity scale and pressure scale equal to $6.5\,\mathrm{km}^{-1}$, $1.3\,\mathrm{km}^{-1}$ and $7\,\mathrm{km}^{-1}$, respectively. A uniform random variation has been supposed equal to about 10 per cent of the mean meteorological value for each level. The vertical extension of Sc clouds is between 0.5 and 5.5 km, while Ns and Cb genera can extend up 6 km and 10 km, respectively. We stress again the fact that Sc clouds represent not only stratus but also cumulus clouds and are not strictly related to stratocumuli, but more generally to non-precipitating clouds.

3.8.5.1 Results for ground-based and satellite-based antennas

We have basically transformed the input cloud physical profile into microwave optical parameters, which describe the absorption and scattering properties of the hydrometeor polydispersion. In order to take into account the transport of radiation through such a medium, we can apply the Eddington approximation illustrated in Section 3.7.1. The main feature of this model is the ability to describe in a fairly simple but accurate way both emission and multiple scattering phenomena affecting the sky noise temperature received by an antenna. However the model can treat only planar stratified structures. This means that we are forced to lose the 3D features of the mesoscale model, but this limitation may be not dramatic when considering directive antennas with a field of view comparable to or less than the horizontal scale of the observed raining cloud (Marzano *et al.*, 1999b).

Apart from the optical parameters, under the assumption of azimuthal symmetry the input RTE parameters are as follows:

(1) the beacon frequency ν;
(2) the observation angle θ;
(3) the surface emissivity model e_S.

With respect to item (1), this section being oriented to microwave and millimetre-wave applications, we have selected typical communication frequencies. The analysis has included the frequency bands of OLYMPUS and ITALSAT beacons, i.e. 12.5, 18.7, 29.7, 39.6 and 49.5 GHz and the most common channel frequencies of ground-based radiometers, that is at 13.0, 20.6, 22.3, 23.8, 31.7, 36.5, 50.2, 53.8 and 90 GHz in order to characterise the entire spectrum of interest. For brevity, only the results at 18.7 GHz (later, also called 18 GHz) and at 39.6 GHz (later, also called 39 GHz) will be shown.

With respect to item (2), we have considered all zenith (nadir) angles for ground-based (satellite-based) antennas, but for brevity we will show only the intermediate case of 45° observation angle.

With respect to item (3), we have at our disposal a radiative model characterised by a Lambertian emissivity e_{sL} which can be effectively applied to land surfaces. However, other surfaces (such as oceans and lakes) are better described by a specular Fresnelian emissivity e_{sF}, depending on the polarisation state (Gasiewski, 1993). The analytical Eddington model can be generalised to polarised radiation by treating them separately (the validity of this assumption is reasonable in MIS media (see Section 3.5)). Sea surfaces can play an important role in the satellite-based antenna measurements. Thus, the land-surface emission has been characterised by a Lambertian emissivity model, depending on the surface humidity. The humidity value has been supposed randomly variable in order to cover a large variety of surface conditions. On the other hand, the sea-surface emissivity has been parametrised to wind velocity with a dielectric constant depending on the salinity. A Lambertian foam contribution can be also included by selecting a fraction f_v (between 0 and 1), depending on the wind velocity itself and the polarisation state (Marzano *et al.*, 2001; Smith *et al.*, 2002):

$$e_{\mathrm{s}}(\mu) = e_{\mathrm{sF}}(\mu)(1 - f_v) + e_{\mathrm{sL}} f_v$$

where μ is the cosine of the incident zenith angle θ (i.e. $\mu = \cos\theta$). Similarly to land, the sea-surface wind velocity has been supposed randomly variable in order to cover several sea-surface states.

Finally, using the coupled rainfall cloud and radiative transfer model illustrated before, a large dataset, consisting of 10 000 cloud structures together with related columnar hydrometeor contents, surface rain-rate, slant-path attenuations and received brightness temperatures at given frequencies and observation angles, has been simulated.

As a first result, Figure 3.24 shows the slant-path attenuation at 45° elevation angle and at 18 and 39 GHz, simulated from the cloud structure ensemble shown in Figure 3.23, for Sc, Ns and Cb genera as function of the surface rain rate and columnar equivalent water content. Left panels refer to the case where all hydrometeors are considered, while right panels show the relative errors due to the case where only liquid hydrometeors (i.e. cloud droplets and raindrops) are taken into account. Note that the slant-path attenuation is defined here by

$$A(\theta, \nu, H) \equiv \frac{4.343}{\mu} \int_0^H k_e(z, \nu)\, \mathrm{d}z \tag{3.75}$$

H being the cloud vertical extension. For completeness, all values of path attenuation are shown thus reaching values much higher than 50 dB at 39 GHz for a 45° elevation angle.

At all regimes a fairly high correlation is noted between either columnar water or rain rate and slant path attenuation. As expected, the higher the frequency the larger the total path attenuation for a given surface rainfall. At 39 GHz, slant path attenuation larger than 50 dB is obtained for rainfall less than 40 mm/h. The role of

---

Figure 3.24 Slant-path attenuation at 45° elevation angle and at 18 and 39 GHz, simulated from the cloud structure ensemble shown in Figure 3.23, for Sc (top panels), Ns (middle panels) and Cb genera (bottom panels) as function of the surface rain rate and columnar equivalent water content. Left panels refer to the case where all hydrometeors are considered, while right panels show the relative errors due to the case where only liquid hydrometeors (i.e. cloud droplets and raindrops) are taken into account

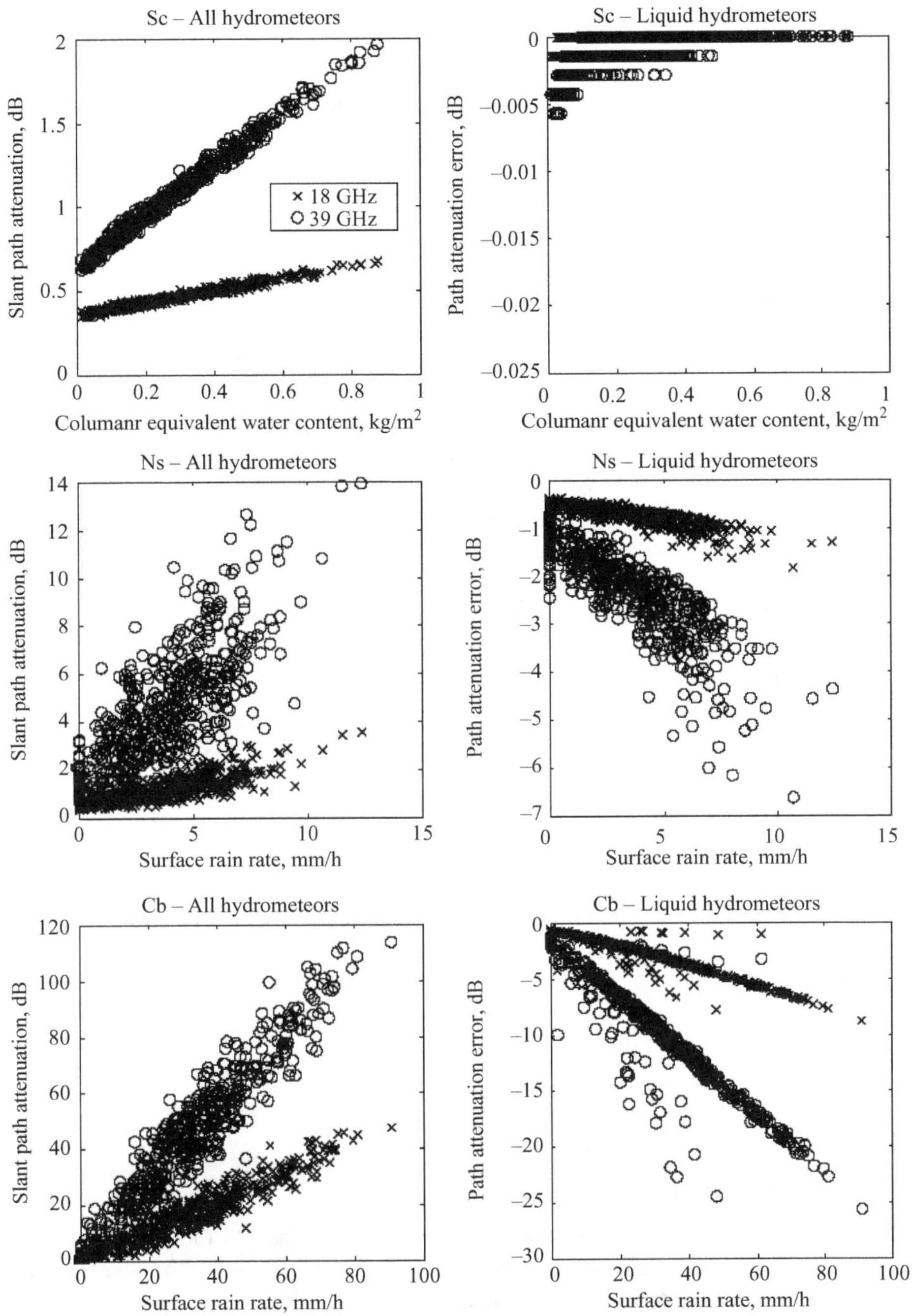
Sc – All hydrometeors
Slant path attenuation, dB
Columanr equivalent water content, kg/m2
× 18 GHz
○ 39 GHz
Sc – Liquid hydrometeors
Path attenuation error, dB
Columanr equivalent water content, kg/m2
Ns – All hydrometeors
Slant path attenuation, dB
Surface rain rate, mm/h
Ns – Liquid hydrometeors
Path attenuation error, dB
Surface rain rate, mm/h
Cb – All hydrometeors
Slant path attenuation, dB
Surface rain rate, mm/h
Cb – Liquid hydrometeors
Path attenuation error, dB
Surface rain rate, mm/h

ice hydrometeors in determining total attenuation is appreciated looking at the right panels. For convective rain and at 39 GHz, neglecting precipitating ice can give rise to an underestimation of several tens of decibels. The underestimation is lower for stratiform and non-precipitating rain due to the smaller contents of precipitating ice associated to the lower rain rates.

3.8.5.1.1 Ground-based antennas

For simulating ground-based antenna measurements, we have assumed a land surface with an average soil moisture saturation of 0.9. The atmospheric contribution has been assumed unpolarised thus setting $k_{eh} = k_{ev} = k_e$ and $p_{ahv} = p_{avh} = p_{aS}/2$ where p_{aS} is the unpolarised Sobolev phase function.

As an example, Figure 3.25 shows the ground-based brightness temperature at 45° zenith angle and at 18 and 39 GHz over a Lambertian land surface, simulated from the cloud structure ensemble shown in Figure 3.22, for Sc, Ns and Cb genera as a function of the slant path attenuation and columnar liquid water content. The stronger the rain regime the faster the saturation of T_{B}s as a function of path attenuation. For non-precipitating clouds the relation between A and T_{B} is almost linear independently of frequency, while for Ns a saturation effect is noted for larger path attenuation values. For convective rain, the frequency dependence is almost lost for a path attenuation larger than 15 dB or columnar water larger than 2 kg/m^2 being the asymptotic values of T_{B}s almost identical.

We have plotted the results in terms of observed brightness temperature T_{B}. It would be interesting to evaluate the effect of the antenna pattern when observing a rain medium in order to quantify the differences with respect to the received antenna noise temperature T_{A}. The latter is given by:

$$T_{\mathrm{A}}(\Omega_0) \equiv \frac{1}{\Omega_A} \int_{4\pi} |f_n(\Omega_0, \Omega)|^2 \, T_{\mathrm{B}}(\Omega) \, \mathrm{d}\Omega \tag{3.76}$$

where Ω_0 is the antenna pointing angle, $\mathrm{d}\Omega$ is the elementary solid angle and Ω_{A} is the radiation solid angle (see Ulaby *et al.*, 1986 for details). For a Gaussian antenna, (3.76) becomes

$$T_{\mathrm{A}}(\theta_0) = \frac{2 \ln 2}{\theta_{\mathrm{HP}}^2 \Omega_n} \int_0^{\pi} T_{\mathrm{B}}(\theta) e^{-(\ln 2)(\theta - \theta_0/\theta_{\mathrm{HP}})^2} \sin\theta \, \mathrm{d}\theta \tag{3.77}$$

In order to test the difference $T_{\mathrm{B}} - T_{\mathrm{A}}$, using (3.77), we have selected a simple rain slab model of 3 km and a variability rain rate. We have considered an antenna with a diameter of 30 and 120 cm to set up its field of view. Figure 3.26 shows the effects of the antenna pattern for ground-based antennas on the observation of a 3 km rain slab with a 10 and 50 mm/h precipitation rate. Panels show the angular spectrum (off-zenith angle) of the incident brightness temperature T_{B} and antenna noise temperature (T_{A}) and brightness temperature differences. Note that values of the error plot for the smallest and largest value of θ are affected by border integration and should be disregarded. As θ tends to 90° (horizon), T_{B} tends to saturate as the

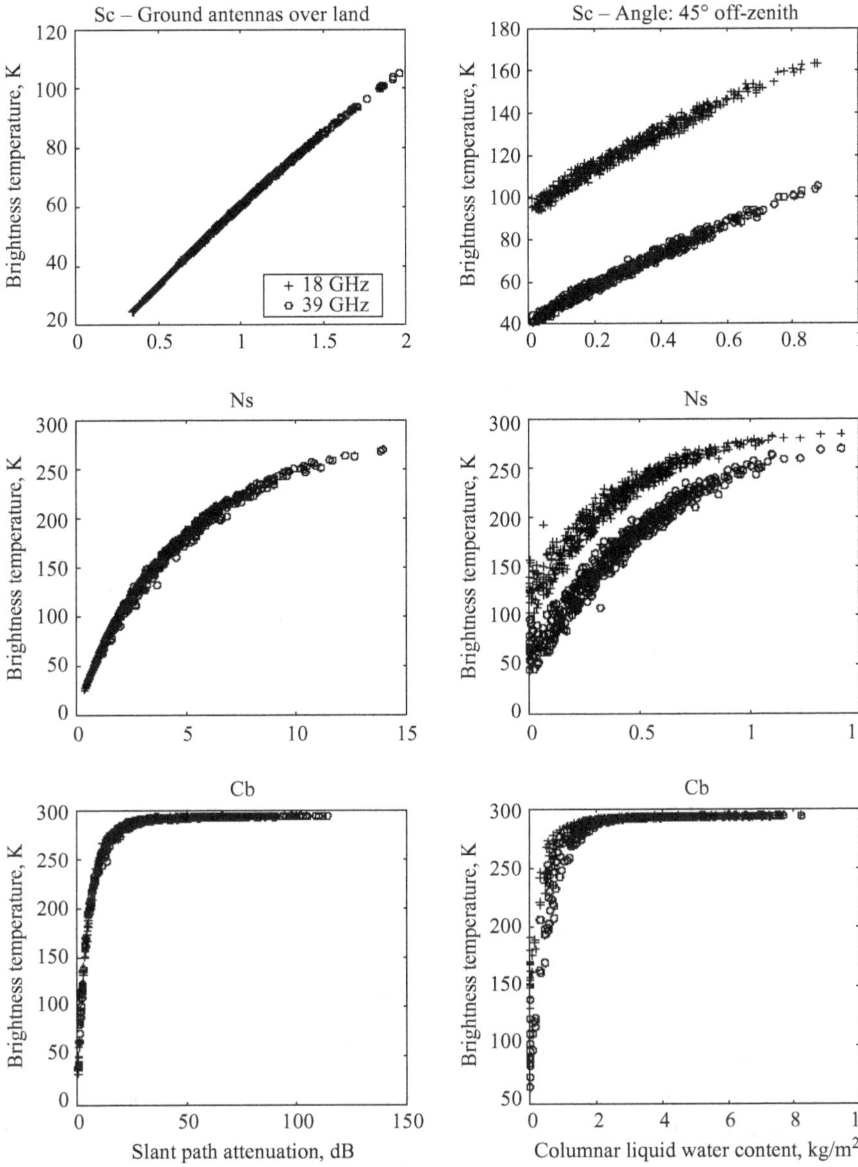

Figure 3.25 Ground-based T_Bs at 45° zenith angle and at 18 and 39 GHz over a Lambertian land surface, simulated from the cloud structure ensemble shown in Figure 3.23, for Sc (top panels), Ns (middle panels) and Cb genera (bottom panels) as function of the slant path attenuation and columnar liquid water content

rain slab is horizontally stratified and its path attenuation becomes larger and larger. The behaviour of $T_B(\theta)$ is quite smooth and this explains the very small effects of antenna pattern integration. As expected, the latter is larger for smaller antennas and higher frequencies (the corresponding antenna beam-width being larger). These plots support the choice to confuse T_B with T_A for ground-based antennas. In fact their difference is less than 0.2 K which is much less than the typical instrumental noise (around 1 K).

3.8.5.1.2 Satellite-based antennas

For simulating satellite-based antenna observations, we have assumed both a land surface with an average soil moisture saturation of 0.9 and a sea surface with an average wind speed of 10 m/s, governing a foam contribution (Gasiewski, 1993; Marzano *et al.*, 2001). Again, for simplicity, the atmospheric contribution has been assumed unpolarised thus setting $k_{eh} = k_{ev} = k_e$ and $p_{ahv} = p_{avh} = p_{aS}/2$.

As an example, Figure 3.27 shows the satellite-based antenna pointing at 45° nadir and at 18 and 39 GHz over a Lambertian land surface, simulated from the cloud structure ensemble shown in Figure 3.22, for Sc, Ns and Cb genera as function of the slant-path attenuation and columnar liquid water content, but for the satellite-based antenna pointing at 45° nadir angle over a Lambertian land surface.

A very small dynamics of the satellite radiometric signatures is noticed for non-precipitating clouds. This is due to the fact that cloud emission is over a land background with fairly high emissivity (about 0.9) thus producing a small signature enhancement. When rain clouds are considered, the satellite radiometric signal is dominated by hydrometeor scattering which tends to diminish T_Bs with respect to its clear-air background. This decrease is larger for higher frequencies and more intense regimes as either path attenuation or columnar water content increase.

The effects of a polarised and radiometrically cold surface, such as sea (with emissivity between 0.3 and 0.6), can be appreciated in the next two figures. Figure 3.28 shows the same effects as in Figure 3.27, but for a satellite-based antenna pointing at 45° nadir angle and at vertical polarisation over a Fresnel sea surface. Figure 3.29 shows the same as in Figure 3.28, but with a sea emmissivity equal to 0.6.

The striking difference with respect to simulations over a land background is the opposite behaviour of T_Bs with respect to either path attenuation or columnar water content, except for developed convective clouds. This difference can be explained by considering that, for clear-air conditions, T_Bs are radiometrically cold, being the measured values due to surface cold emission and gaseous absorption in a frequency dependent fashion. As cloud and rainfall are present as radiometric sources then emission mechanisms tend to be more and more efficient and tend to provide warmer T_Bs as noted for Sc and Ns genera.

When path attenuation is beyond a value typical of convection rainfall, then scattering phenomena become dominant and T_Bs tend to decrease. Beyond this threshold, the surface emission is completely extinguished by cloud attenuation and the radiometric signature at the satellite height is almost independent of the surface

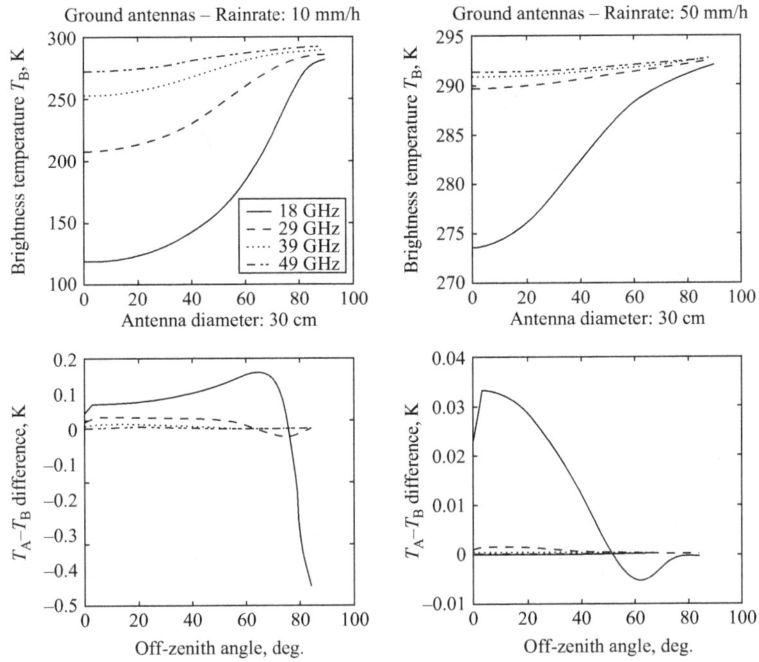

Figure 3.26 *Effects of the antenna pattern for ground-based antennas on the observation of a 3 km rain slab with a 10 mm/h (left panels) and 50 mm/h (right panels) precipitation rate. Top panels show the angular spectrum (off-zenith angle) of the incident brightness temperature* T_B*. Antenna noise temperature* (T_A) *and brightness temperature differences are shown for a parabolic antenna with a diameter of 30 cm (bottom panels). When a 120 cm antenna is considered differences (not shown) are an order of magnitude smaller*

characteristics. This explains the similar behaviour of T_B scatterplots for land and sea background after path attenuation larger than about 10 dB or columnar water larger than about 1 kg/m^2. The difference between H and V polarisation is mostly due to the Fresnel coefficients which provide, for a given incidence angle, a horizontal emissivity lower than that at vertical polarisation (recall that emissivity is complementary to reflectivity with respect to unity). We note that at 45° we are closer to the Brewester angle and that this H–V emissivity difference is more apparent for clear-air scenarios (negligible atmospheric scattering). Finally, we note that the effect of the antenna pattern can also be evaluated for a satellite antenna once the scanning geometry and platform altitude is defined. The antenna radiation pattern can be still described as in the previous section. However, since ground resolution is of the order of several tens of kilometres, 3D features of cloud field can play a significant role (Marzano and Roberti, 2003).

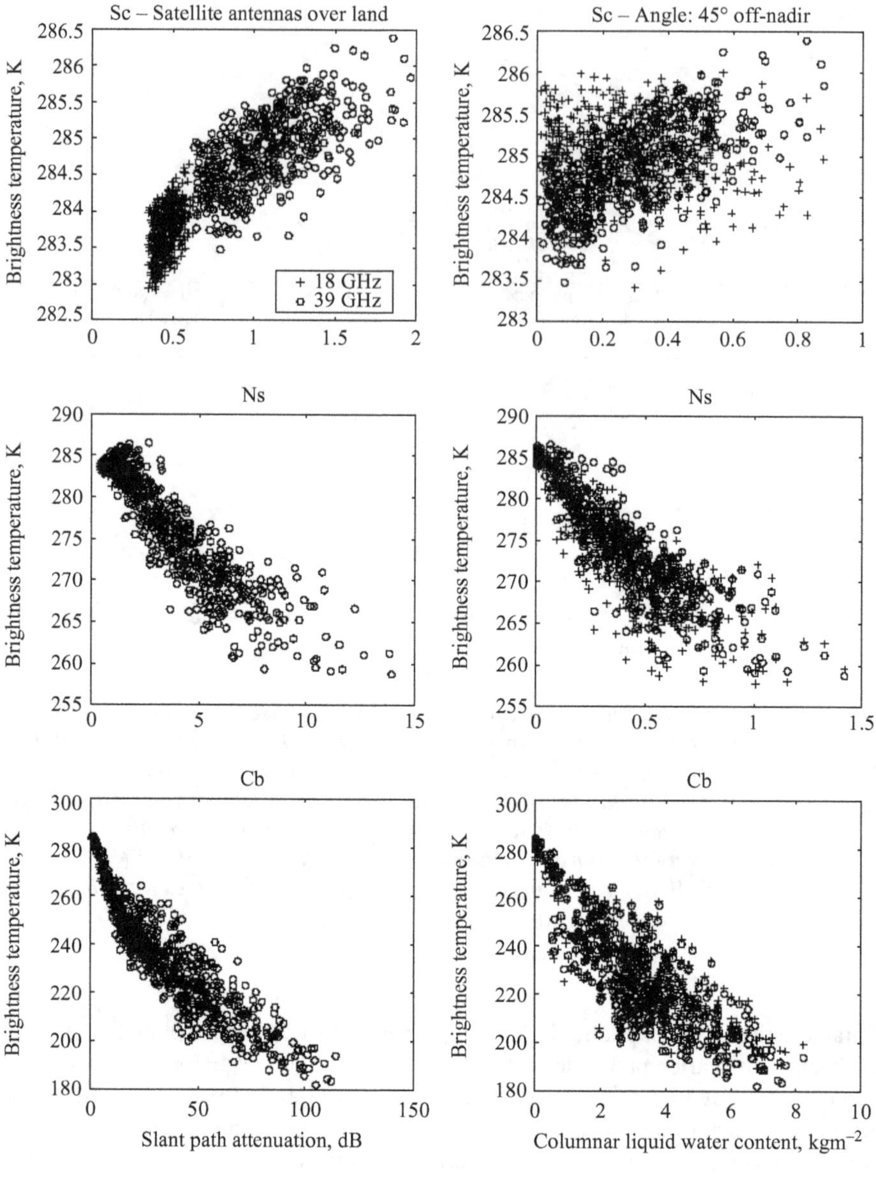

Figure 3.27 *Same as in Figure 3.25, but for satellite-based antenna pointing at 45° nadir angle over a Lambertian land surface*

3.8.5.2 Estimation of liquid water path

Ground-based, airborne or spaceborne microwave radiometers measure the thermal emission which is typically given as brightness temperature in kelvin. Water vapour has distinct spectral features at 22.235 and 183 GHz (see Figure 3.30). Since the

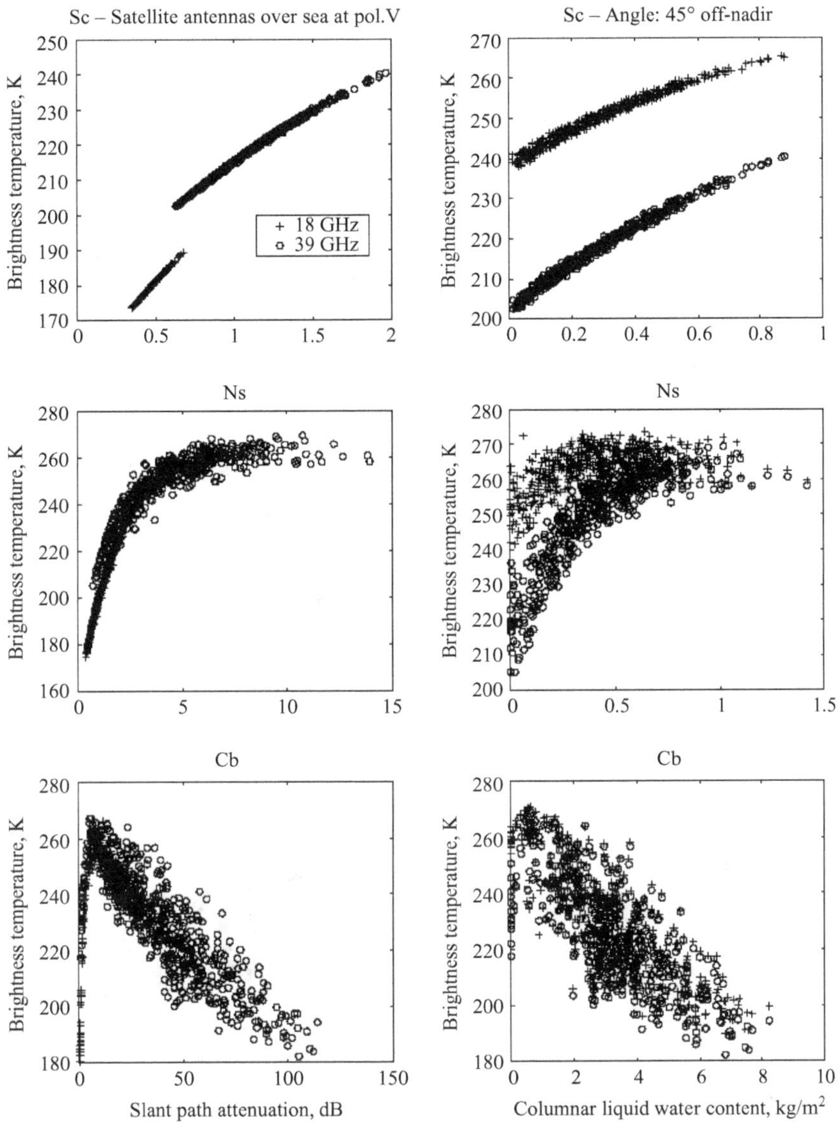

Figure 3.28 *Same as in Figure 3.25, but for satellite-based antenna pointing at 45° nadir angle and at horizontal polarisation over a Fresnel sea surface. Sea emissivity is equal to 0.3*

introduction of satellite remote sensing, dual channel methods have made use of the rotational line at 22.235 GHz to derive the vertically integrated water vapour over oceans (Grody, 1993). Within this method one channel measures at a frequency close to the line centre where water vapour absorption is almost independent of height. The

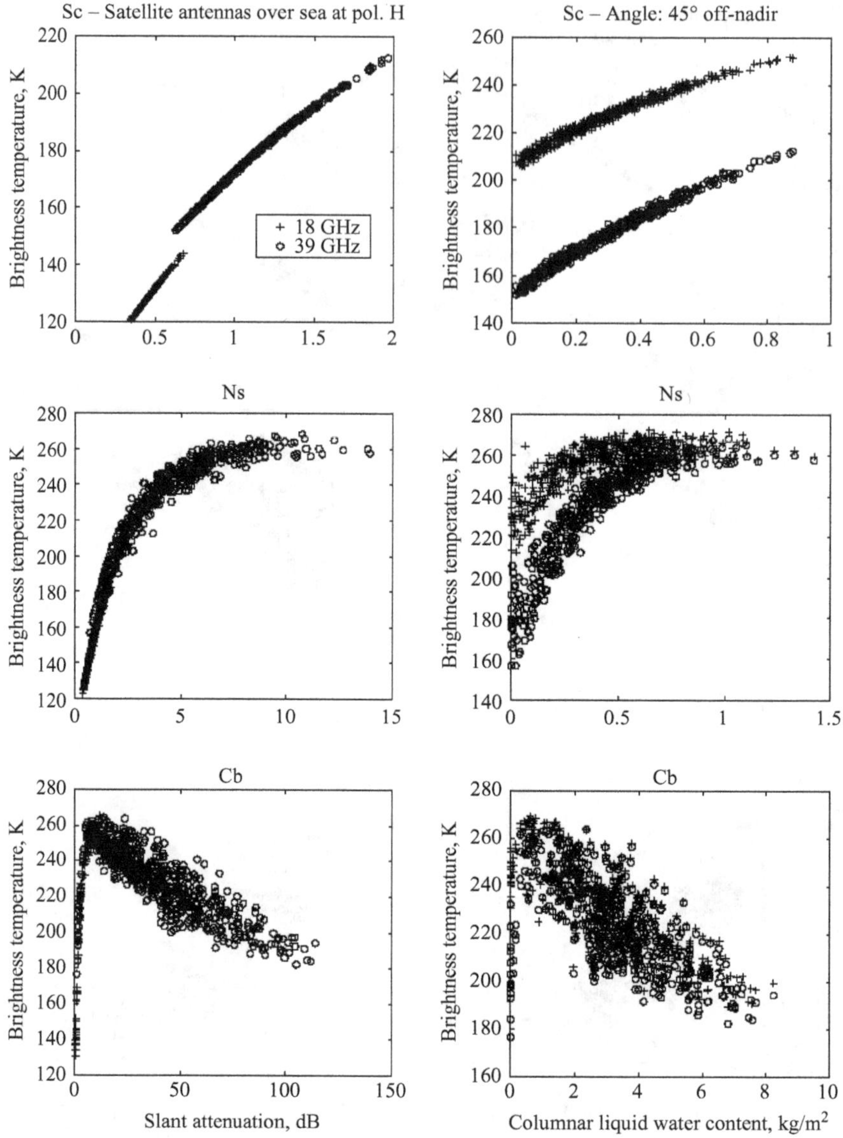

Figure 3.29 Same as in Figure 3.23, but for satellite-based antenna pointing at 45° nadir angle and at horizontal polarisation over a Fresnel sea surface. Sea emissivity is equal to 0.6

observed brightness temperature is approximately proportional to IWV. The second channel is located in a window region where water vapour and oxygen absorption are low (e.g. 31.4 GHz). Since microwave emission of clouds increases approximately with the square of the frequency (Figure 3.30) the second channel can be

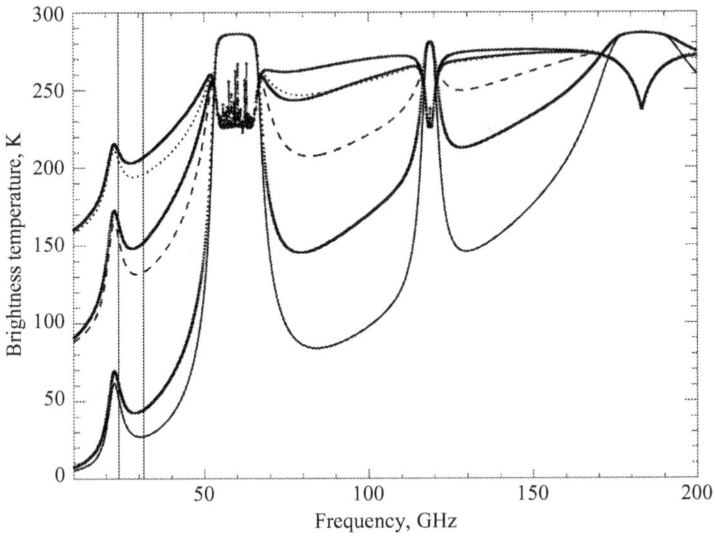

Figure 3.30 *Effect of a liquid water cloud with LWP = 250 g/m² (thick solid lines) added to the clear sky standard atmosphere for different viewing geometries: ground-based (thin solid)and satellite (horizontal polarisation is dashed and vertical polarisation is dotted) observation over ocean with 52 deg*

used to characterise the cloud's contribution to the intensity observed at the first frequency. This combination is extremely powerful as the IWV and the liquid water path (LWP), which is the vertically integrated liquid water density, can be retrieved simultaneously. For the past few decades this principle has been used for ground-based instruments (Westwater, 1978) as well as on the first satellites employing microwave radiometers at similar frequencies. For the retrieval of LWP the channel close to the water vapour absorption line corrects for the changing water vapour concentration of the atmosphere. Such observations are, with the exception of expensive and rather limited aircraft measurements, the most accurate method to observe LWP with an estimated accuracy of better than 30 g/m^2. A rough estimate shows that about 10 g/m^2 are caused by the measurement error while the rest can be attributed to the under-determined retrieval problem. The additional use of the 90 GHz channel can further constrain the problem and improve accuracy to less than 15 g/m^2 (Crewell and Löhnert, 2003).

3.8.5.3 Influence of cloud model assumption on LWP retrieval

Most LWP retrieval algorithms employ statistical information about the vertical distribution of cloud liquid water content (LWC). Due to the difficulty in observing LWC (rare aircraft observations), however, no systematic observations exist. Therefore approaches of different complexity are used to tackle that problem. One can distinguish between three approaches: idealised box models, methods which diagnose

LWC from radio soundings and cloud model output. The latter certainly have the advantage in that they provide profiles which are physically consistent with those of temperature and humidity: furthermore, 3D fields are available. But as clouds are strongly parametrised in atmospheric models the quality of the hydrometeor forecast is rather uncertain and many efforts are undertaken to carefully evaluate those predictions.

The simplest model to estimate the LWC profile from radiosonde measurements places the cloud in an altitude range while the relative humidity exceeds a certain threshold (mostly 95 per cent is assumed). Because the commonly used RS80 sensor shows a dry bias (Leiterer *et al.*, 2005) this threshold often produces very few clouds so that the threshold is lowered in order to get a more realistic cloud cover. After determining the cloud boundaries' LWC needs to be calculated as a function of altitude within the cloud boundaries' adiabatic assumption. Generally, the liquid water content as calculated for an adiabatic ascent (LWC_{ad}) (see, e.g. Rogers and Yau, 1989) is assumed to be the maximum possible LWC. It can be corrected for effects of dry air entrainment, freezing drops or precipitation using a modified adiabatic approach (Karstens *et al.,* 1994) which uses an empirical correction function derived from aircraft observations in different types of clouds (Warner, 1955):

$$\mathrm{LWC} = \mathrm{LWC_{ad}}[1.239 - 0.145\ln(h)] \tag{3.78}$$

with h in m indicating the height above cloud base and h within the range between 1 m and 5140 m.

An alternative approach for deriving cloud boundaries from radiosonde ascents is the gradient method proposed by Chernykh and Eskridge (1996) (in the following referred to as CE). A cloud is modelled into layers which satisfy the following conditions:

$$\frac{\mathrm{d}^2 T}{\mathrm{d}z^2} \geq 0 \qquad \frac{\mathrm{d}^2 RH}{\mathrm{d}z^2} \leq 0 \tag{3.79}$$

T denoting the temperature. These conditions can be interpreted as a region of an RH maximum due to saturation and a region of weaker temperature decrease within the cloud due to a pseudo-adiabatic lapse rate. After cloud boundary detection, Chernykh and Eskridge (1996) used a diagram according to Arabey (1975) to determine the dependency of cloud fraction on cloud temperature and dew point depression which can be modified to other climatological conditions (Löhnert and Crewell, 2003). For generating vertical LWC profiles cloud fraction was interpreted as the probability of a cloud to exist above the measurement site. The LWC profile is then calculated with the modified adiabatic method as described above.

As an example for a physically-based cloud model a 1.5 dimensional, microphysical dynamic cloud model (DCM) by Issig (1997) is considered. The DCM calculates LWC in 40 logarithmic radius classes every 250 m from ground to 10 km height (model top). This convective model is initialised with radiosonde profiles. Convection is initialised diabatically via a radiation module. This type of cloud modelling accounts for the temporal evolution of a typical cumulus cloud. The model always

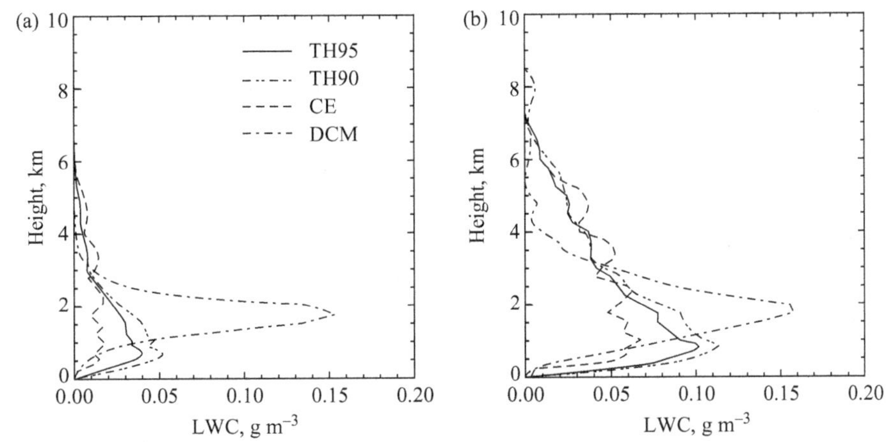

Figure 3.31 *Vertical distribution of liquid water content (LWC) derived from a radio sounding dataset using different methods to diagnose LWC: THx – 0 threshold in relative humidity of x per cent; CE – gradient method after Chernykh and Eskridge (1996); DCM – dynamic cloud model after Issig (1997). Figure from Löhnert and Crewell (2003)*

generates clouds, and hence, clear-sky cases are not well represented. A more detailed description of the model can be found in Löhnert *et al.* (2001).

To investigate the influence of the different cloud models Löhnert and Crewell (2003) prepared databases for each of the cloud models described above and used them to develop statistical LWP algorithms for different frequency combinations (23, 28, 50 and 90 GHz). The mean vertical LWC distribution (Figure 3.31) of the different cloud models shows a rather different behaviour ranging from a rather homogeneous height distribution (gradient approach) to a well-peaked one (DCM). In general, the DCM profiles have a peak at higher altitudes (and thus colder temperatures) where also the standard deviation is highest. The influence of the cloud model on retrieval performance was analysed by applying algorithms developed from one model to a data set where another cloud model was assumed. The effects observed strongly depend on the frequency channels used. For the common dual channel algorithms (20/30 GHz) a maximum of 5 g/m^2 bias occurred. More pronounced effects occur when more (and higher frequency) channels are incorporated. Specifically, when the 50.8 GHz channel is combined with the lower two channels, the algorithm accuracy depends very much on the cloud model used. In terms of LWP retrieval accuracy, the 90 GHz channel and the lower two channels complement one another better than the combination of the 50 GHz channel with the lower two channels. BIAS errors due to the different cloud model data sets are reduced and additionally the correlation between modelled and retrieved LWP is higher: the relative unexplained variance between 'true' and retrieved LWP is reduced from 12 per cent for a common two-channel algorithm to 4 per cent.

3.8.5.4 Influence of drizzle

In general, the radiative transfer is drop size distribution (DSD) dependent on the microwave region, although DSD is mostly significant at higher frequencies (>90 GHz) or in rainy conditions. The occurrence of drizzle particles (larger drops with significant terminal velocities) can lead to large uncertainties when using cloud radar reflectivities alone to determine microphysical cloud properties. However, it is known from concurrent cloud radar and lidar observations that drizzle occurs quite frequently. Therefore, it is important to know how sensitive microwave radiometer retrieval algorithms are to clouds containing moderate amounts of drizzle (e.g. drizzle which does not reach the surface). To study this effect the drop size distribution calculated by a spectrally resolved microphysical cloud model can be used.

From a large data set Löhnert and Crewell (2003) identified the strongest drizzle case where the drizzle water path (DWP) defined as the liquid water path due to drops between 50 and 400 μm was 32 g/m^2. With a total LWP of about 110 g/m^2 about one-third of the water was in the form of drizzle drops. Brightness temperatures were simulated using the full spectrum given by the cloud model. Then a standard retrieval algorithm which was developed assuming a standard cumulus DSD containing no drizzle drops was applied to these brightness temperatures. The differences in LWP (<2 g/m^2; <2 per cent) is not significant. Thus, moderate amounts of drizzle are not expected to have a significant impact on the accuracy of LWP retrievals and do not need to be considered when deriving algorithms.

3.8.5.5 Ground-based observations

Much effort has been made in the last few years to develop ground-based microwave radiometers (Solheim *et al.*, 1998; Rose *et al.*, 2005) which are able to continuously observe the downwelling thermal emission. They are now deployed on an operational basis (Liljegren, 2000; Crewell *et al.*, 2002) as well as in specific campaigns (e.g. Crewell *et al.*, 2004) to observe the liquid water path as well as the profiles of water vapour and temperature profiles. A good overview of different radiometer capabilities is given by Westwater *et al.* (2004). Satellite measurements (SSM/I, AMSU, TRMM, AMSR) have been available for more than two decades. Their main disadvantage is that reasonable estimates can only be made above the oceans and only with poor (several 10 km) resolution. The high variability of clouds can cause errors in LWP due to inhomogeneous beam filling of the satellite field of view, especially in the case of precipitation. Over land the ground-based microwave radiometers can observe LWP with high temporal resolution at one point (Figure 3.31). These capabilities were used by setting up a network of ground-based microwave radiometers within the BALTEX Cloud Liquid Water Network (Crewell *et al.*, 2002) in North Central Europe to evaluate and improve the forecast of clouds in atmospheric models.

The difficulty in observing clouds lies in their high temporal and spatial variability. Figure 3.32 gives an example of the temporal variability of LWP observed by three different microwave radiometers which were operated in zenith direction close to each other. The radiometers have very different spatial and temporal offsets. Obviously the radiometer with highest resolution can resolve much higher variability which is

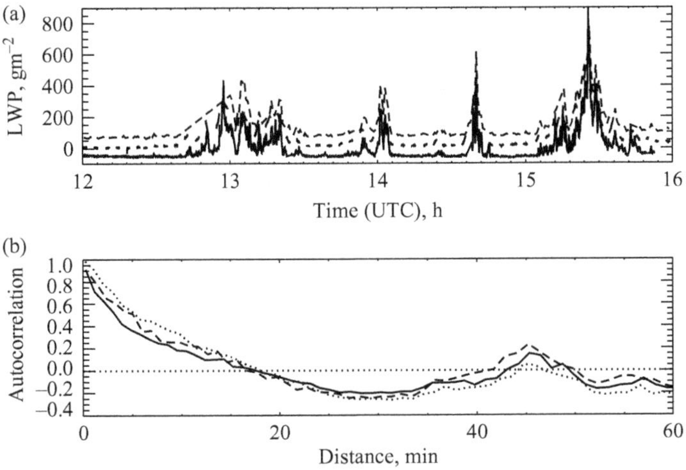

Figure 3.32 Top panel: Time series of LWP simultaneously observed with three different microwave radiometers: WVRA (top, dashed) with an integration time of 30 s and a beam width of about 5°; DRAKKAR (middle, dotted) with 1 s integration time and a beam width of about 12° and MICCY (bottom, solid) with an integration time of 1 s and a beam width of <1°. The theoretically derived accuracies are 29, 23 and 15 g/m^2, respectively. For better comparison the curves have an artificial LWP offset. Bottom panel: Autocorrelation of the LWP time series in the top panel

of interest for some applications like solar radiative transfer. This can be quantified using the autocorrelation function and by calculating the respective correlation length. In this case the correlation length determined from the highest resolution radiometer is 5 min while it is 9 min when the radiometer with the 12° beam is used. This example is rather typical and illustrates the need for high temporal and spatial resolution when cloud structures are to be studied. It is also of importance when LWP probability distributions are calculated: long integration times will reduce the number of high LWP events. Furthermore, the non-linearity of the retrieval problem will lead to errors if T_B averaged over long integration times are considered.

3.8.5.6 Angular scanning observations of cloud liquid

Most of the time microwave radiometers are operated in zenith direction which is most suitable to combine the measurements with other remote sensing observations in synergetic algorithms (e.g. Löhnert *et al.*, 2004). However, the representativity of the observed column for a larger area, for example, a model grid box or a satellite pixel, is unclear. For example, in model evaluation the observations are often aggregated over a specific time interval to imitate the model grid width (see van Meijgaard and Crewell, 2005). One approach to investigate the spatial cloud inhomogeneity is to perform hemispheric imaging by employing a radiometer with relative broad

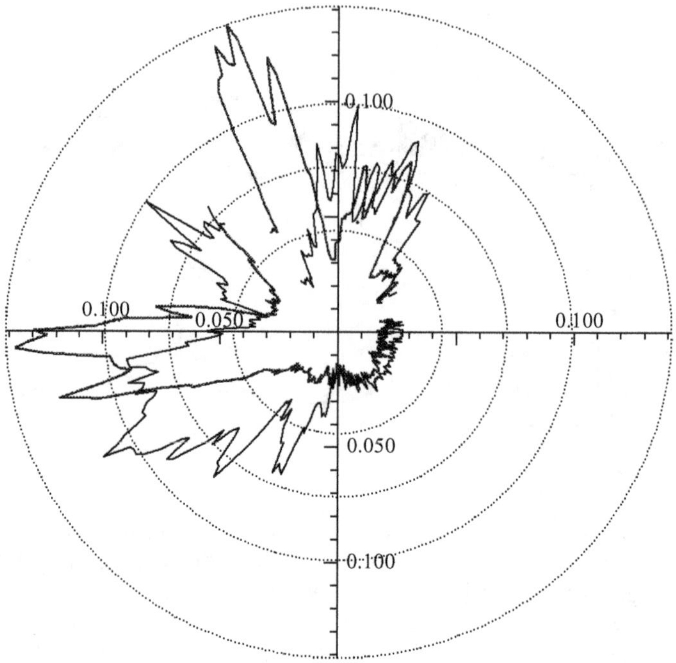

Figure 3.33 LWP observed during two azimuth scans performed at 30° elevation with the MICCY radiometer

antenna beams allowing a fast overview of the whole sky. Information about cloud boundary temperature (height) can be gained by simultaneous observations with an infrared radiometer which is realised by the All-sky multi-wavelength radiometer ASMUWARA (Martin *et al.*, 2003). Another approach uses scanning measurements with high temporal (0.1 s) and angular (less than 1°) resolution to obtain small-scale information on the cloud structure. Examples of azimuth scans taken by the microwave radiometer for cloud cartography (MICCY) (Crewell *et al.*, 2001) shown in Figure 3.33 reveal the strong anisotropy of the cloud fields. Such information is important for studies concerning three-dimensional radiative transfer of solar radiation through clouds. Time series of cloud liquid water are strongly correlated on all spatial and temporal scales. Thus, for measurements in a fixed direction, it normally does not make sense to measure faster than 1/s; this would only mean that you are measuring the same volume multiple times. By scanning, a new cloud volume is observed every time. Furthermore, by scanning, the measurements are much better decorrelated; because the average distance between two points will be larger. Thus, a more representative, broader cloud water distribution will be found. The example shows that zenith measurements of the clouds that drift by on the wind would show a correlation length that is not representative of the field. Systematic observations need to be performed to investigate the importance of this effect.

3.8.6 3D radiative-transfer effects

Clouds and precipitation, especially in a convective region, are actually characterised by strong horizontal and vertical inhomogeneities, which will reflect into 3D patterns of microwave scattering properties. When MW radiometer observations are considered the particular inhomogeneity of the scene is observed from a particular point of view (e.g. from a satellite or from the ground), at a certain viewing angle and with a particular field of view.

The interplay between the observing system and the observed scene results in what are usually called (in a broad sense) 3D effects. To have an insight into such aspects the RTE have to be solved in full 3D geometry. However, as we will show hereafter, to account for these 3D effects it is not always necessary to solve the 3D RT since it is possible to reduce the problem to surrogate 1D profiles.[28]

Although other techniques allow 3D RT computations (e.g. the VDOM Haferman *et al.*, 1997, 1993) or the SHDOM (Evans, 1998), the MC techniques (see Section 3.6.5), in their different implementations; backward (Roberti *et al.*, 1994; Davis *et al.*, 2005); forward-backward (Liu *et al.*, 1996); forward (Roberti and Kummerow, 1999; Battaglia and Mantovani, 2000), are more suitable (because of their simplicity) to investigate MW RT through the complicated 3D cloud structure. In fact, when dealing with MW problems, differently from what often happens in the VIS and NIR (e.g. with the need of computing fluxes and heating rates at different altitudes), only a few quantities need to be computed (typically radiances at some particular viewing angles at the sensor location). Therefore, there is no need of explicit methods which represent the radiance field in the computational domain and which require much more memory and computational time. In any case, fairly substantial computational time is necessary, but nowadays certainly manageable for off-line applications.

On the other hand, many times, the 3D structure of clouds does not preclude the use of 1D RT approaches, which allow much faster solution techniques for the RTE. For instance, as far as pure absorbing atmosphere and Fresnel-like surfaces are involved the 3D structure is not effective because the radiation paths are straight. Therefore, the only complication is introduced by extracting a 1D profile from a 3D geometry. While the treatment of nadir/zenith viewing angle is straightforward, for oblique viewing angle the 1D profile (input for the RT) has to be reconstructed from the 3D structure by following the radiation slant path with the sagacity of producing an independent 1D profile for the reflected rays as well (therefore, different reconstructions are needed for different viewing angles). RT computations are usually performed to simulate the T_{B}s as measured by an oblique viewing PMW radiometer over the FOV of the instrument; in this case the spatial average should be derived by dividing the FOV into many subgrids and by attributing to each subgrid the slant profile obtained in correspondence to the centre of the subgrid. By decreasing the

[28] The same is generally not true in the visible regime. The main reason for that has to be ascribed to the much lower scattering capability of microwave radiation (compared to visible) from atmospheric constituents.

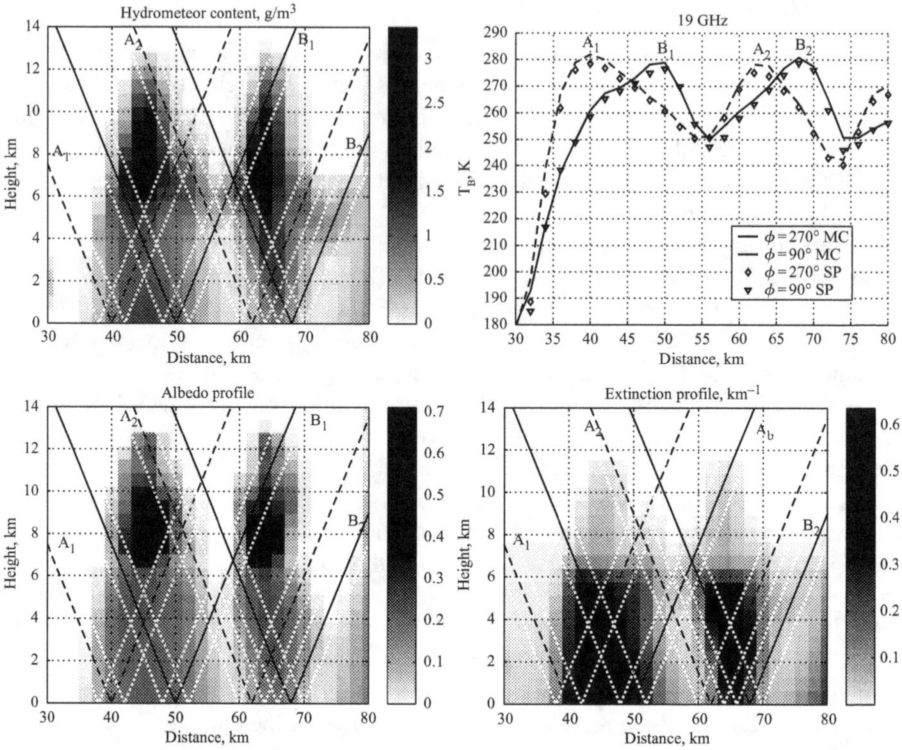

Figure 3.34 Total hydrometeor content (top left), 19_H GHz T_B profile (top right), single scattering albedo and extinction coefficient at 19 GHz (bottom left and right). The dotted white lines indicate the footprint at which T_Bs are computed. See text for details

size of the subgrids the values of the spatially averaged T_Bs converge to the sought value. This procedure[29] perfectly accommodates cloud edges so that a 3D structure over water surfaces at pure emission channels can be definitively treated as a 1D problem. In the review chapter Battaglia *et al.* (2005) indicate all these 3D effects as geometrical ones. They include inhomogeneity inside the radiometer footprint (the so-called beam filling effect see Kummerow (1998) and reference therein), warm side emission and reflections from surfaces.

Figure 3.34 represents the hydrometeor content, the scattering properties and the 19 GHz T_Bs as measured by two radiometers downward looking at a viewing

[29] Schematic picture and examples can be found in Liu *et al.*(1996); Roberti *et al.*(1994). These latter authors discuss also the difference between the 1D slant path and the 1D plane parallel approximation.

angle of 50° in the forward and backward direction[30] (i.e. at azimuthal viewing angle equal to 90° and 270°) overpassing two close rain shafts (located at $y = 46$ km with RR = 29 mm/h and at $y = 66$ km with RR = 16.5 mm/h) of a squall line system over the ocean simulated by the GCE-CRM (Tao and Simpson, 1993) at a resolution of 2×2 km. The combination of the single scattering albedo and extinction panels shows that the scene is dominated by emission features. Accordingly the MC and slant path solutions are always close by. The rain shafts are revealed by two peaks ($A_1 - B_1$ and $A_2 - B_2$) associated with the different viewing angles. However, the positions of the peaks do not correspond with the maximum of the rain shaft but with the observation of the side of the cloud (see Figure 3.34). This occurs because of the radiation leaked from the side of the rain shaft and because of the high reflectance properties of the sea surface. Note that the combination of the two viewing angles T_Bs can provide hints with regard to the 2D structure of the system under observation (like tilted convective systems, Hong *et al.*, 2000). For instance in Figure 3.34 the distance between the peaks corresponding to the two looking directions gives an idea of the maximum dimension of the cell, while we can reasonably expect that the position of the highest rainfall rate is located in between the two maxima (i.e. A_1 and B_1).

On the other hand when Lambertian-like surfaces are involved and/or scattering plays an important role, the 3D structure has to be faced in all its complexity. In fact, in the presence of diffuse radiation, the received photons at a particular viewing angle can originate from outside the geometrical slant beam. Let us consider a warm T_B sector of an ocean scene like the emission-dominated area of heavy rainfall at low frequencies (e.g. 10.7 and 19.3 GHz) as observed by a nadir-looking spaceborne radiometer. If a 1D approximation is used, the warm radiation emitted inside each subcloud is trapped inside the cloud and cannot exit the sides of the subcloud itself. This means that the 1D results are warmer in the heaviest raining portion of the cloud. Thus warm becomes warmer when neglecting 3D effects. On the other hand, when considering a cold T_B sector of a scattering ice deck at higher frequencies (e.g. 85, 150 GHz), the warm radiation emitted by water vapour and cloud liquid water surrounding the cold sector is prevented from entering the cloud (and perhaps subsequently from exiting the top of the cloud) in a 1D approach. This means that the 1D results tend to be colder in the highly scattering regions of the cloud. Thus cold becomes colder when neglecting 3D effects. But notice this is true only for the very cold region and the very warm region; so we could say extreme temperatures are enhanced in 1D approximations. On the other hand, when 3D is considered, cold and warm centres are less intense but tend to spread more out.

The top panels in Figure 3.35 show the integrated liquid and ice content of a cold front system over the mid-Atlantic simulated by the GCE-CRM (Tao and Simpson, 1993) at a resolution of 4×4 km. The 85.5_H GHz T_Bs plotted in the left bottom panel clearly show the scattering signatures in those regions where high amounts of ice

[30] The recently launched US Navy Coriolis and future planned spacecraft (EGPM drone, see Mugnai, 2003) host or are going to host conically equipped spaceborne MW radiometers able to provide fore/after views of the swath.

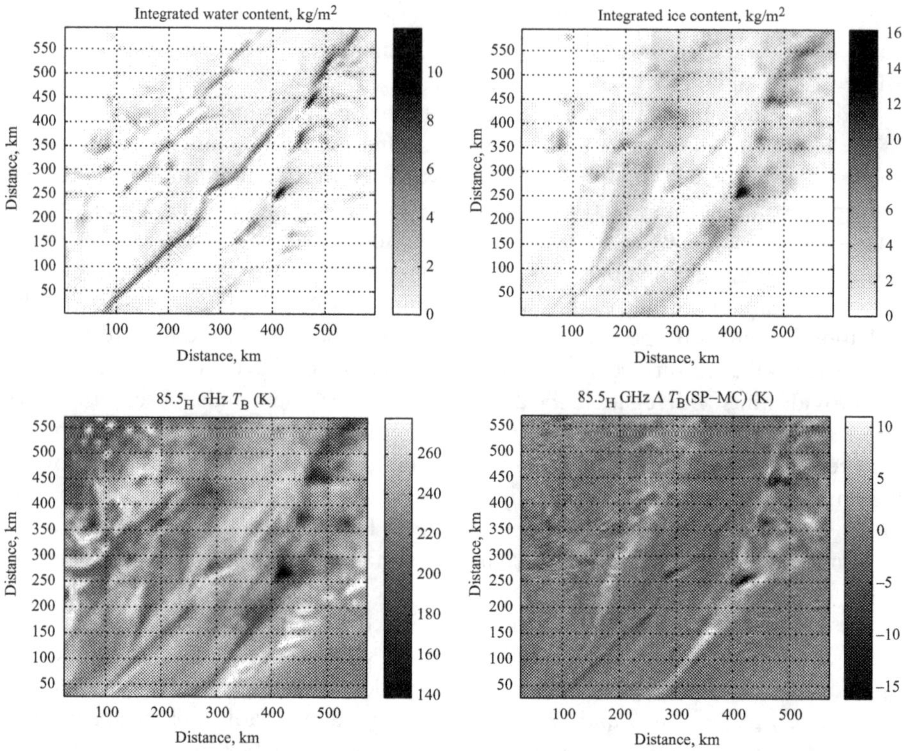

Figure 3.35 *Top panels: integrated liquid (left) and ice (right) path. Bottom panels:85.5*$_H$ *GHz* T_B *and* T_B *difference between the SP and MC solution to the RTE at 85.5*$_H$*. See text for details*

(highly scattering) particles are present. In the bottom right panel the differences between a slant path corrected and the 3D MC $T - B$ at 85.5 GHz range between −15 and +10 K. So this gives an idea of the maximum error introduced by the 3D scattering effect. Figure 3.36 shows the regions where the radiation field has strong horizontal variation (left panel) and where the scattering is relevant.[31] Otherwise if these two conditions are not simultaneously satisfied (for instance in a homogeneous thick ice deck or in the presence of a strong absorbing inhomogeneous water vapour field like region 1 and 2 in Figure 3.36) no scattering 3D effects are really detectable.

[31] Note that, at this frequency, the average number of photons undergone by an MW photon before being absorbed (in a backward MC vision) can be as high as 12. Actually a better variable is represented by the fraction of radiation coming from outside the instrument FOV. In fact this quantity takes into account the fact that for a highly forward scattering phase function photons remain inside the field of view despite the scattering number.

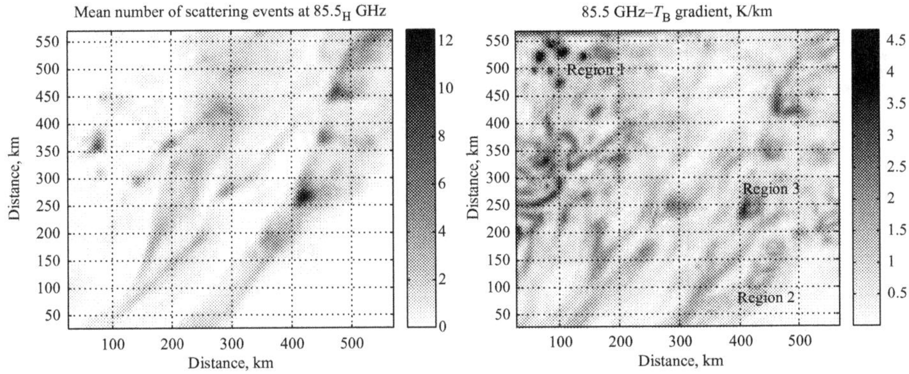

Figure 3.36 *Average number of scattering and 85.5*$_H$ T_B *gradient for the CRM simulation of Figure 3.35*

As a summary, in 1D model approximations, radiation remains trapped by construction in the slant tube: no contribution from outside the tube is allowed, and abrupt not-physical variations may happen for contiguous pixels. 3D MC radiation field can be considered a sort of redistribution of the 1D radiation pattern. Therefore, as already noticed by Kummerow (1998) and consistent with earlier findings in Roberti and Kummerow (1999), the 1D modelling, compared with the 3D correct one, introduces an rms more than a bias error. The amplitude of this difference is a function of

(1) the number of scattering events (the higher the frequency the higher the number of scattering events);
(2) the shape of the phase function (higher values of the asymmetry parameter g maintain the radiation inside the slant beam, therefore cancelling the redistribution of radiation);
(3) the T_B gradients (higher close to cloud boundaries and in the presence of cold backgrounds) that are strictly correlated to the resolution to which T_Bs are averaged (with higher resolution having higher 3D effects).

As a prospective for better instantaneous RR retrievals, the 1D SP (not 1D PP) model partially accommodates for the geometrical problems due to surface scattering and oblique viewing angle but still cannot take into account diffusion effects due to cloud scattering. Therefore it reproduces T_B quite accurately for emission dominated frequencies (10 and 19 GHz), even at high CRM resolution but not at higher frequencies. If convolved to TMI resolution the result remains acceptable for 37.0 GHz but seems to deteriorate at 85.5 GHz. Therefore this last frequency and higher MW frequencies have to be treated more carefully by performing dedicated 3D RT simulations especially with high equivalent water path and in coincidence with highly vertically developed and strongly horizontally inhomogeneous raining systems. This is still an active field of research.

3.8.7 *Microwave signatures of clouds in limb geometry*

3.8.7.1 Unpolarised 1D simulations for the MASTER instrument

Using the 1D unpolarised version of ARTS-DOIT limb spectra were calculated for the MASTER (Millimeter Wave Acquisitions for Stratosphere/Troposphere Exchange Reasearch) instrument. The impact of various cloud parameters is presented in Emde *et al.* (2004a). It is shown that the simulated brightness temperatures most strongly depend on particle size, ice mass content and cloud altitude. The impact of particle shape is much smaller, but still significant.[32] Increasing the ice mass content has a similar effect as increasing the particle size; this complicates the prediction of the impact of clouds on microwave radiances without the exact knowledge of these parameters.

Figure 3.37 shows simulations for five different cloud scenarios, where we assumed that the cloud particle size is correlated with the cloud ice mass content in limb geometry, for a tangent altitude of 11.5 km (inside the cloud layer) and for nadir geometry. Compared to nadir radiances, limb radiances are more complex. In nadir geometry, cirrus clouds always lead to a brightness temperature depression compared with the clear sky radiances. Nadir instruments look at the tropopause and the ground, where the major source of microwave radiation is located. Clouds absorb and scatter part of the radiation out of the LOS. When we consider the lower atmosphere and the emitting ground as major sources of radiation, the propagation direction of the radiation is upwards. In nadir geometry we measure the upwelling radiation; there cannot be an enhancement in this direction due to scattering according to the law of energy conservation. In limb geometry, clouds usually lead to a brightness temperature depression if the tangent altitude lies below the cloud. But this can also lead to a T_B enhancement if the tangent altitude is inside the cloud, because part of the upwelling radiation from the earth's surface and the lower atmosphere is scattered into the LOS. In the clear sky case the sensor does not see thermal emission from the lower atmosphere at all. The impact of cloud in nadir geometry is smaller, as the path-length through the cloud is much shorter. Of course the path-length through the cloud for a limb measurement depends on the horizontal extent of the cloud.

3.8.7.2 Polarised 3D simulations for 3D box-type cloud

The 3D version of the model was applied for simulating limb radiances for a cloud of finite extent embedded in a horizontally homogeneous tropical atmosphere. The height of the cloud box was 7.3 to 12.5 km, and the vertical extent of the cloud was from 9.4 to 11.5 km. The horizontal extent was approximately 64×32 km. Simulations were performed for an IMC of 0.02 g/m^3 corresponding to a limb optical depth of approximately 0.5. It was assumed that the cloud consists of spheroidal ice particles

[32] Previous studies on this subject by Czekala (1998) suggested that without knowledge of particle shape and size and knowledge of whether the cloud is in front, behind or directly at the tangent point, a usage of the spectra for trace gas retrieval is ambiguous. Again polarisation information could help to identify such problems.

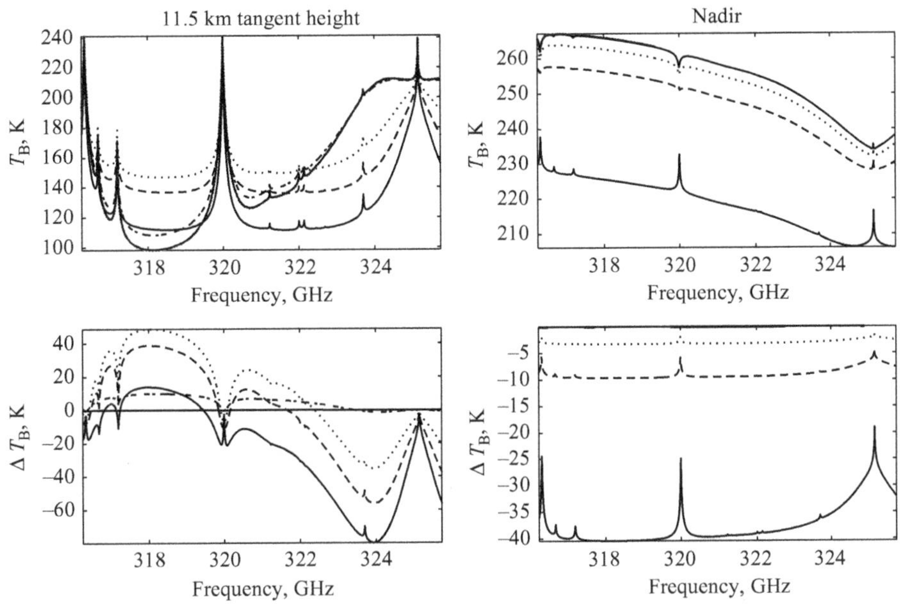

Figure 3.37 Left panel – limb spectrum at 11.5 km tangent altitude. Right panel – equivalent nadir spectrum. Clear sky spectrum (grey) and cloudy spectra for IMC = $4 \cdot 10^{-5}$ g/m³ (—), $1.6 \cdot 10^{-3}$ g/m³ (– · – · –), $8 \cdot 10^{-3}$ g/m³ (· · · · · ·), 0.016 g/m³ (- - -) and 0.04 g/m³ (—) and the corresponding particle sizes. The top plots show absolute BTs, the bottom plots show the differences between cloudy and clear sky spectra

with a particle size of 75 μm and an aspect ratio of 0.5. Calculations were performed for completely randomly oriented particles and for horizontally aligned particles with azimuthally random orientation. The sensor was assumed to be placed on board a satellite following a polar orbit at 820 km altitude. At each sensor position tangent altitudes from 0 to 13 km were simulated. Figure 3.38 shows corresponding lines of sight (LOS). The figure shows that the cloud is seen from different sides from the top, from the bottom or from the left side. When the satellite is at a latitude of 25° the cloud is only seen for low tangent altitudes (from 0 to 6 km). The cloud is seen at higher tangent altitudes at 27.5°. For even greater latitudes the sensor sees the cloud from the bottom. Note that the tangent point in the first plot is behind the cloud, in the second plot in the middle of the cloud and in the last plot in front of the cloud. In order to compare the 1D and the 3D model versions, simulations for a 1D cloud layer with an equivalent limb optical depth at 10 km tangent height was performed. The IMC for the equivalent 1D cloud was only 0.005 g/m³, since the path length through the cloud is very large. The results shown here and additional simulations for a cloud case with larger optical thickness are presented in Emde *et al.* (2004b).

Figures 3.39 and 3.40 show the simulated radiances plotted as a function of tangent altitude and the sensor position for randomly oriented particles. The top panels

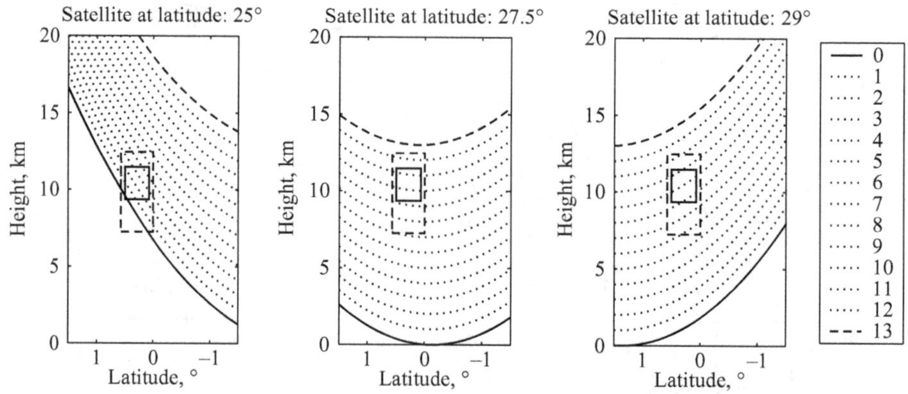

Figure 3.38 *Lines of sight (LOS) for different sensor positions and tangent altitudes (km). The solid line corresponds to an LOS for a tangent altitude of 0 km and the dashed line to an LOS for a tangent altitude of 13 km. Dotted lines correspond to LOS for tangent heights between 0 and 13 km. Inside the solid rectangle the single scattering properties are defined, and the dashed rectangle labels the cloud box. Courtesy of Claas Teichmann*

show the intensity differences between the clear sky calculation and the cloudy sky calculation ΔI. The bottom panels show the polarisation difference Q. The contour plots on the left hand side are the 3D results. Bright colours indicate a brightness temperature enhancement due to the cloud, and dark colours indicate a T_B depression. A cloud effect can only be seen at tangent heights for which the corresponding lines of sight intersect with the cloud. The intensity plot shows that up to a latitude of 27° there is a T_B depression due to the cloud. The reason is that in those cases the tangent point, from where the major source of thermal radiation emerges, is behind the cloud. The cloud scatters part of the radiation away from the line of sight. For latitudes above 28° a T_B enhancement is observed. In these cases the tangent point is in front of the cloud. The sensor measures all radiation emerging from the tangent point and additionally the back-scattered radiation from the cloud behind the tangent point. If the tangent point is inside the cloud, between 26.5° and 28°, a T_B enhancement can be observed for high tangent altitudes because part of the upwelling radiation from the lower atmosphere is scattered into the direction of the LOS. For lower tangent points the scattering away from the LOS dominates, hence, a BT depression is observed in this latitude range. The polarisation plots show that in the 3D case as well as in the 1D case there is only a very small polarisation for randomly oriented particles. The intensity plot in Figure 3.40 for horizontally aligned particles looks similar to that for completely randomly oriented particles. The polarisation is much larger for horizontally aligned particles.

The comparison between 1D and 3D shows similar results at tangent heights inside the cloud, where the optical depth is approximately equivalent. For other tangent heights, the optical depths are different, and therefore the results deviate

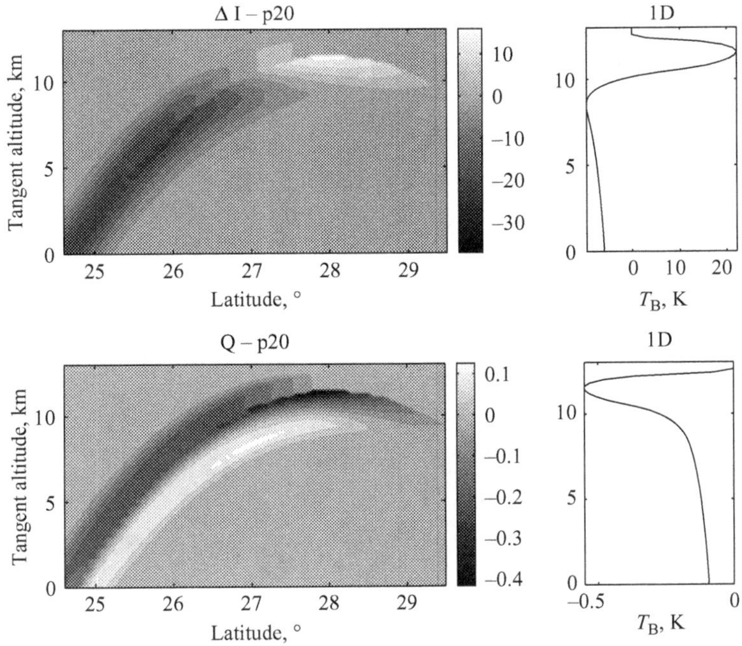

Figure 3.39 *Left panels: scattering signal of a 3D box-type cloud embedded in a 1D atmosphere as a function of sensor position and tangent altitude for 318 GHz. The cloud consists of completely randomly oriented spheroidal particles of 75 μm size and with an aspect ratio of 0.5. The IMC is 0.02 g/m³. Right panels: 1D result for a cloud with an equivalent optical depth in limb (IMC = 0.005 g/m³). The upper plots show the intensity I and the lower plots, the polarisation Q*

strongly. The scattering signal in 3D depends very much on the sensor position with respect to the cloud. Hence it is very important to use the 3D model where the cloud extent is not very large, as in this example, or where the clouds are horizontally inhomogeneous.

3.9 Polarisation effects of particle orientation

Non-spherical and orientation effects[33] are very often neglected in MW RT computations. Although this can be a reasonable approximation when dealing with intensities, it becomes completely misleading when polarisation signatures are considered. We will briefly review the main causes for polarisation in the microwave region and then

[33] As seen in Section 3.3.1 randomly oriented non-spherical particles often behave like MIS media. The peculiarity of non-sphericity is really enhanced in the presence of preferential alignments.

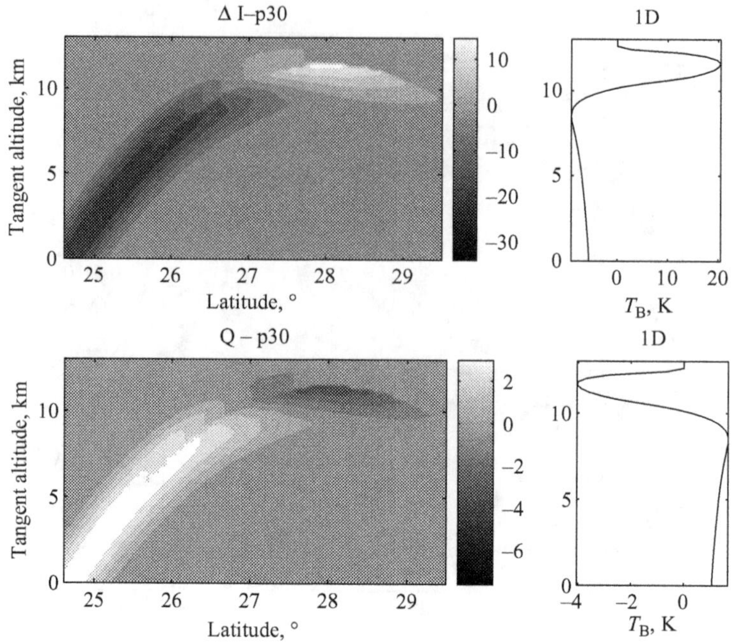

Figure 3.40 *Left panels: scattering signal of a 3D box-type cloud embedded in a 1D atmosphere as a function of sensor position and tangent altitude for 318 GHz. The cloud consists of horizontally aligned spheroidal particles with a size of 75 μm and with an aspect ratio of 0.5. The IMC is 0.02 g/m^3. Right panels: 1D result for a cloud with an equivalent optical depth in limb (IMC = 0.005 g/m^3). The upper plots show the intensity I and the lower plots the polarisation difference Q*

give some examples of observations that cannot be explained with models involving spherical particles.

3.9.1 *Theoretical studies on polarisation signatures*

Many theoretical studies have been carried out on this topic. A detailed review and references can be found in Haferman (2000). Here we will just give some general explanations and examples. First of all, we briefly summarise the possible sources for polarisation when dealing with microwave signals:

- presence of polarised emission sources, such as from dichroic material or Fresnel surfaces (strongly vertically polarised with no polarisation at 0 and 90° and with a maximum around 70°). For upwelling radiances this last contribution can be very relevant so that it is crucial to understand how much of the total signal is coming from the Fresnel-like surface;

- scattering from surface, if Fresnel-type. In this case, since more vertically polarised photons are absorbed, scattered photons are preferentially horizontally polarised;
- propagation in dichroic material (i.e. in media with non-diagonal extinction matrix). For the medium composed by perfectly/partially aligned non-spherical particles extinction coefficients are different for vertical and horizontal radiation so that the more radiation propagates in the medium the more it enhances one component with respect to the other. In the radar community this is known as differential absorption. In typical atmospheric scenarios this phenomenon is more pronounced at far-nadir propagation angles, where differences between h and v extinction/scattering coefficients are extreme;
- the scattering from hydrometeor particles into the line of sight because of the $I-Q$ coupling term introduced in the RTE by the scattering matrix. This effect has been discussed in depth by Liu *et al.* (1996) when dealing with MIS media. They have shown that, in the absence of any polarisation source, even spherical particles in a 1D homogenous cloud can cause polarisation (although never greater than a few kelvin). This can be seen already at the first order of scattering, as a result of the angular dependence of the impinging radiation field and of the anisotropicity of the phase matrix. The coupling is driven by the sign of $F_{11} - F_{22}$ in Equation (3.17): when this element is negative (positive) for all scattering angles positive (negative) Q are produced. For instance, Rayleigh scatterers always produce positive Q. Since this is not always the case the polarisation will generally depend on the interplay between the structure of the radiation field and of the phase matrix. Liu *et al.* (1996) have shown that rain clouds modelled as MP sphere populations give a polarisation signature which is always positive for the frequency in the microwave region (see their Figures 5a,6a). Besides, considering single scattering albedo lower[34] than 0.5 (i.e. assuming rain to be more absorptive than scattering) the behaviour of the polarisation difference $T_v - T_h = 2Q$ can reach values as high as 8 (at very far from nadir view $\sim 80°$) at intermediate rain rates, while it never exceeds 4 K at 54°. Generally it increases by increasing the single scattering albedo while it has a different behaviour, increasing the optical thickness for different single scattering albedo. On the other hand, for spherical ice clouds (see their Figures 5b, 6b) the polarisation difference is always positive at low frequencies, but negative for high ice water content and high frequencies (i.e. for clouds involving particles with high size parameter).

When dichroic materials are considered, things are even more complicated; in fact, besides the $I - Q$ coupling in the scattering event generally the radiation that is scattered has already a preferential polarisation due to different cross sections for vertically and horizontally polarised photons. Therefore, the properties of radiation scattered into the line of sight will generally depend in a complicated way on the angular distribution and polarisation of the incoming radiance and on the phase matrix.

34 This is a good assumption for frequencies lower than 85 GHz.

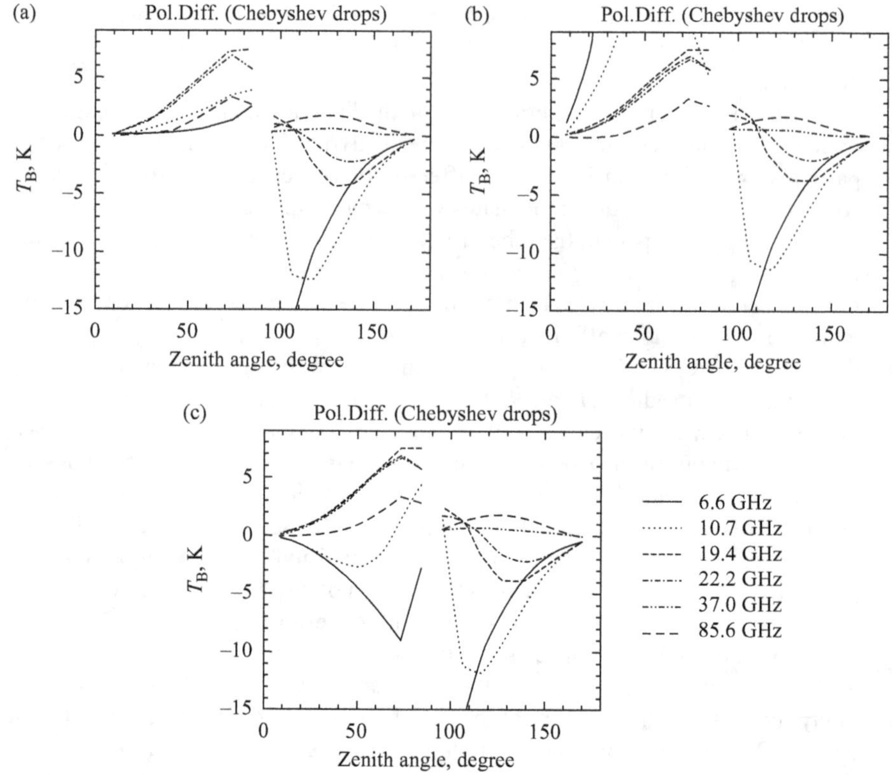

Figure 3.41 *Polarisation difference $PD \equiv T_V - T_H$ versus zenith angle for six frequencies at 25 mm/h rain rate and three different surface conditions: (a) black body emission, (b) sea surface (Fresnel-like), (c) lowered emissivity ($e = 0.5$)*

3.9.1.1 An example: polarisation signatures above and below a raining cloud

We will now put together all the pieces for a simple case, that is, the polarisation signatures observed at the top and at the bottom of a raining layer, as modelled in Czekala *et al.*(1999). Figure 3.41 is extracted from that paper.

We can first evaluate the combined effect of absorption and emission in a dichroic material. If we use a simple model such as in Section 1.3.3, Chapter 1, the radiation emitted from a single isotherm layer at temperature T_{eff} will contribute at the boundaries with a polarisation

$$(T_V - T_H)(\mu) = T_{\text{eff}}[e^{-\tau_{\text{sl}}^H}(1 - \varpi_H(\mu) - e^{-\tau_{\text{sl}}^V}(1 - \varpi_V(\mu)) + \varpi_H(\mu) - \varpi_V(\mu)] \tag{3.80}$$

Note that in the limit of high optical thickness (for instance when moving towards the side view) or of small optical thickness (3.80) gives a polarisation proportional

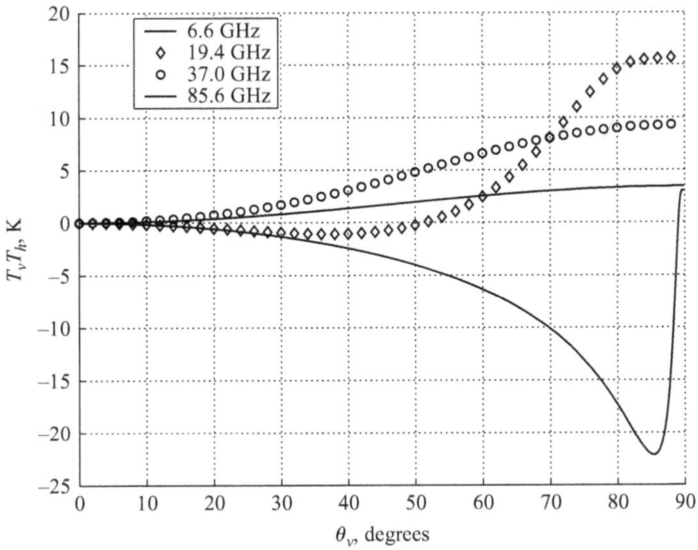

Figure 3.42 Downwelling polarisation signatures from an isothermal 25mm/h rain layer at 278 K

to $\varpi_H(\mu) - \varpi_V(\mu)$ and to $\tau_{\rm sl}^V(1 - \varpi_V(\mu)) - \tau_{\rm sl}^H(1 - \varpi_H(\mu))$, respectively. The results corresponding to (3.80) have been plotted in Figure 3.42 with the scattering parameters computed for horizontally perfectly aligned spheroids[35] (axial ratio from Chuang and Beard, 1990).

However, if a source of radiation at temperature $\tilde{T}_V$, $\tilde{T}_H$ is present beyond the layer (like emission from a surface) the contribution

$$\tilde{T}_V e^{-\tau_{\rm sl}^V} - \tilde{T}_H e^{-\tau_{\rm sl}^H} \tag{3.81}$$

has to be added to that in Equation (3.80). When $\tau_{\rm sl}^V < \tau_{\rm sl}^H$ (such as in our example) this term is always positive.

Since at 6.6 GHz the scattering is practically negligible ($\varpi \approx 0.05$ for 25 mm/h rain) we can explain the 6.6 GHz lines in Figure 3.41 by using Equations (3.80–3.81) only. The downward polarisation signal is always negative and practically the same as in Figure 3.42 despite the surface type. Equation (3.80) explains the polarisation signatures; on the other hand the upwelling polarisation signature strongly depends on the surface type. In particular note how by changing the emissivity from 1 to 0.5 (left and right panel, respectively) the polarisation passes from positive to negative values; this is because the term (3.81) gives a substantial contribution to the total $T_{\rm B}$.

[35] Czekala *et al.* (1999) noticed that the results for raindrops modelled as oblate spheroids and oblate Chebyshev both show large deviations from results obtained with spheres. On the other hand the Chebyshev shape (compared with the oblate spheroid) adds only little changes.

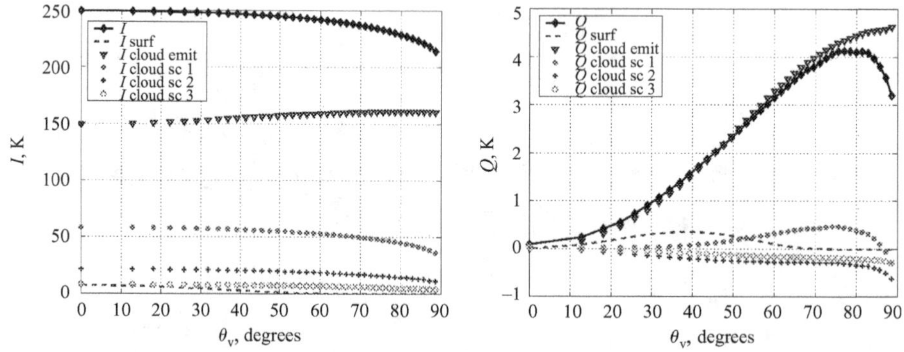

Figure 3.43 Contributions to intensity and polarisation in the upwelling radiances for the test case at 37 GHz

In panel (b) note that large PD are produced by Fresnel surfaces that are not obscured by the thin rainy layer ($\tau_{\text{nadir}} \approx 0.07$), so that the results have been clipped.

By increasing the frequency the general trend shows an increase of the single scattering albedo (0.27, 0.44 and 0.5 at nadir for 19.4, 37 and 85.6 GHz, respectively) and of the optical thickness of the layer (1.16, 3.65 and 7.11 at nadir for 19.4, 37 and 85.6 GHz, respectively). Therefore, in the downwelling radiation at the ground the contribution from pure emission will be driven by the positive $\varpi_H(\mu) - \varpi_V(\mu)$, ranging between 0 at nadir and a maximum positive value at the side view (see Figure 3.42). Moreover (see Figures 3.43–3.44) the first and successive order scattering contributions start becoming important. For instance at 37 GHz the third order of scattering contributes to the total T_B up to 10 K. The different higher scattering orders tend to be more and more unpolarised, while even the polarisation of the first scattering order can have both signs (on the other hand spheres produce always a positive polarisation). For instance Figures 3.43–3.44 demonstrate how the polarisation contribution of radiation scattered once is always negative in downwelling radiation while it is positive at almost all downward looking viewing angles. The total polarisation signal has to take into account all these effects so that it can switch from positive to negative according to the variability of the radiation field structure and of the scattering parameters (i.e. the angular variability of h and v extinction and single scattering albedo and the phase function pattern). (A similar example is shown in Figures 1–2 in Troitsky *et al.*, 2003.) As a rule of thumb, remember that polarisation has to go to zero at nadir and at side view in the pure absorption regime (due to the blackening of the atmosphere, i.e. black body radiation is unpolarised); generally the polarisation differences tend to be much lower when radiation is similar in all directions of propagation (Czekala, 1998).

Finally, we observe that in the high-frequency upwelling signal the effect of the surface becomes lower and lower (practically no difference between the different surface types is seen for frequencies greater than 19 GHz in Figure 3.41). In Figure 3.43 the small positive contribution from the surface is still visible. Although not relevant

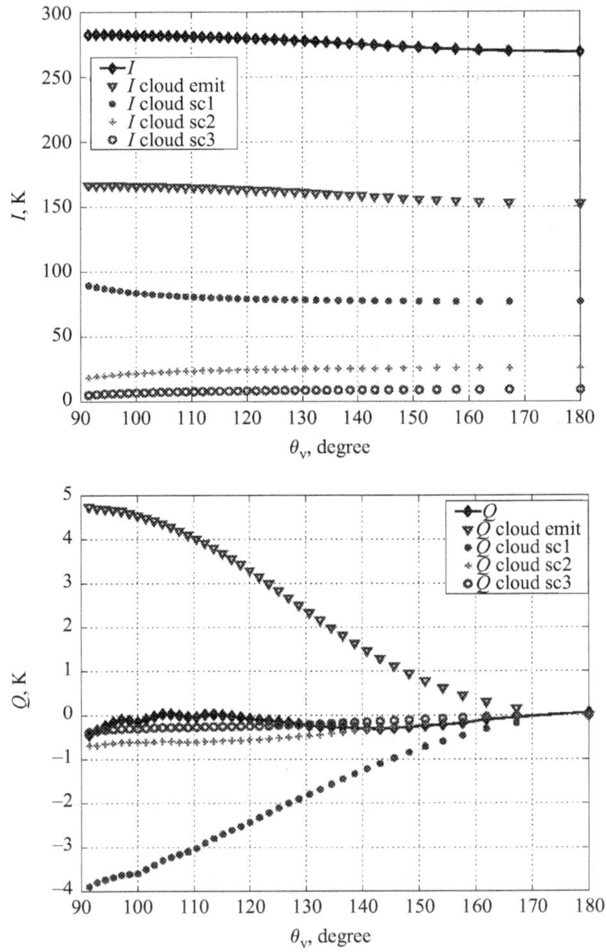

Figure 3.44 *Contributions to intensity and polarisation in the downwelling radiances for the test case at 37 GHz*

here, this contribution can be extremely important in the presence of a highly scattering ice layer composed by horizontally oriented crystals. If the radiation emitted below the ice layer is unpolarised, at the top of the layer the radiation will be partially vertically polarised (positive Q) just because the v-component interacts less than the h-component with the ice layer. On the other hand, radiation entering the scattering process will have a predominant h-component. The scattering event itself is generally less important in terms of the polarisation effect. Therefore, in the case of the ice layer, part of the scattered radiation will overpass the layer (so decreasing the whole polarisation effect); on the other hand, the downward scattered radiation (below the ice layer) is characterised by a negative polarisation signal: in this case a greater contribution than from the upward radiation comes from scattered photons (because of the upward/downward asymmetry of the radiation field). Scattered photons are preferentially h-polarised: this is not due to the scattering itself but mostly to the fact that interacting photons are preferentially h-polarised. Examples relative to ice crystal clouds can be found in Evans and Vivekanandan (1990), Evans and Stephens (1995b) and Czekala (1998). Polarisation signatures in the downwelling radiation have been simulated by Troitsky *et al.* (2003) for mixed phase clouds and for rain clouds by Hornbostel and coauthors (see Haferman, 2000, pp. 496–7, and references therein) and by Czekala and Simmer (1998) and Czekala *et al.*(1999).

3.9.2 Experimental observations of polarisation signatures

In the last decade, some authors (Spencer *et al.*, 1989; Heymsfield and Fulton, 1994) have speculated that the shape of the hydrometeors influences MW polarisation signatures, attributing features observed in systems like mesoscale convective systems and in tornadic storms over land or heavy precipitation in tropical cyclones over the ocean to non-spherical hydrometeors with a preferred orientation. More recent observations of polarised scattering signatures related to cloud systems have finally assessed that these results can be interpreted only in association with partially oriented non-spherical particles. We will give here some examples of such observations both for ground-based and for spaceborne radiometers.

3.9.2.1 Downwelling polarisation signal from raining clouds

Ground-based observations with a 19.2 GHz dual polarisation radiometer at a fixed elevation angle (30°) have been performed in southern Germany during 5 months in 1996 and continuously from November 1998 to December 1999 (see Czekala *et al.*, 2001b for details) for a total of 513 observation days.

The mean PD observations (with corresponding standard deviation) sorted according to the T_B values (see Figure 3.45, middle line) show a typical negative signal (first decreasing PD with increasing T_B, then saturation around 200 K, then increasing PD towards zero for T_B >220 K). Note that the large amount of data allows for a complete coverage of all possible T_Bs (up to 280 K). Model calculations (similar to those discussed previously) with a variety of atmospheric and cloud profiles to reproduce those found in observations have been performed as well with two different shapes. The PD versus T_B is clearly divided into two branches: spherical particles (diamond line) result in small positive PD; oblate perfectly oriented droplets (squares line) produce results much more similar to the observed ones (except for a slight underestimation of PD above 150 K).

Comparisons involving only small portions of the data were carried out as well (with a subset of the model output matching the range of atmospheric conditions). Examples drawn in Figure 3.46 demonstrate an example of very bad and good agreement (left and right panel, respectively) between model and observations within the same raining event. In the left panel in particular the strong negative signal cannot be simulated at all. The observed high rain rates (36 mm/h) suggest a convective precipitation type, which is typically of a small horizontal scale. Therefore 3D effects (not

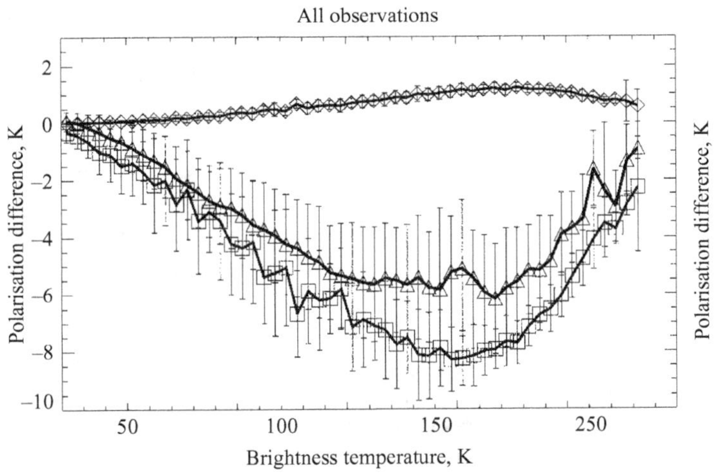

Figure 3.45 *Measurements (triangles) and model calculated (diamonds for spherical and squares for oblate particles) PD as a function of* T_B*s in histogram mode (with within 5 K classes); from Czekala et al. (2001b)*

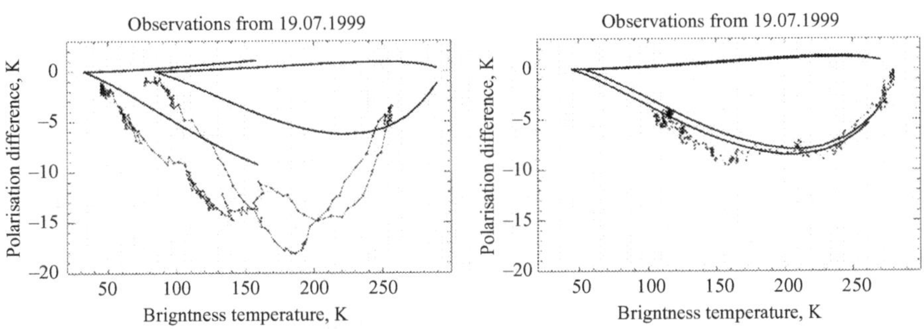

Figure 3.46 *Comparison of model results and observations for the warm (left) and cold (right panel) part of the same event, occurred on 19/7/1999; from Czekala et al.(2001b)*

accounted for by the 1D model) are very likely to be present in this situation. Although additional model refinements (like inclusion of melting particles, introduction of drop oscillations and canting angles and 3D effect evaluation) can be introduced, this study supports the use of the polarisation signal in ground-based radiometer retrieval.

Initial studies by Czekala *et al.*(2001a) indicate the potential to discriminate cloud and rain liquid water paths by ground-based polarised microwave radiometry. Although ground-based microwave radiometry is a fairly established technique to retrieve vertically-integrated liquid water path (LWP), water vapour profiles and temperature profiles (e.g. Janssen, 1996), current LWP retrievals (not using polarised observations) are limited in accuracy by the presence of drizzle and rain, which introduces a substantial change in the proportionality between water mass and T_B. Cloud droplets have a different T_B per water mass ratio than larger raindrops (Rayleigh scattering versus Mie scattering) as shown in Figure 3.47. Since larger drops exhibit a polarisation signature in the downwelling microwave radiation, such ambiguities can be significantly reduced with polarised observations of clouds and rain.

Czekala *et al.* (2001b) support the idea of a retrieval methodology for rain and cloud LWP by using simultaneous measurements of downwelling PD and T_B. For instance, when looking at the simulated signals at 19 GHz in the upper panel of Figure 3.47, a hypothetical measurement of $T_B = 110$ K and PD $= -5$ K will result in a total LWP $= 1.6$ kg/m^2 with 1.1 kg/m^2 of cloud water and 0.5 kg/m^2 of rain water. On the other hand, a simple T_B measurement can be assigned to a wide range of LWP values (from 0.8 kg/m^2, in the case of pure rain water, to 2.4 kg/m^2 in the case of pure cloud water). Finally note that, as expected, the use of a multi-frequency approach allows a better discrimination of different rain rates. For example, the saturation of the PD signals begins at lower rain rates at 30 GHz (lower panel in Figure 3.47), compared with 19 GHz. Thus the sensitivity to small amounts of rain LWP is significantly better at the higher frequency.

3.9.2.2 Downwelling polarisation signal at 85 and 37 GHz from mixed-phase clouds

Troistsky *et al.* (2003) have analysed data acquired by two frequencies (85 and 37 GHz) dual polarised ground-based radiometer looking at a zenith angle of 65° during the Alliance Icing Research Project held in Ottawa in the winter of 1999/2000. Characteristic values of the polarisation difference are about 2 K (with maximum values up to 4.5 K) at 85 GHz while they are 2–6 times less at 37 GHz (with maximum values up to 2.5 K). Both positive and negative PD were observed (although negative values occur more frequently than positive ones): the sign of the polarisation differences and spectral relations at the two analysed frequencies are related to cloud microstructure (ice shape, size and orientation) but they are usually decorrelated from the liquid water path. A typical observation is shown in Figure 3.48.

On the other hand, when LWP lower than 0.4 kg/m^2 are considered the magnitude |PD(37 GHz)| is generally bigger than |PD(85 GHz)| (opposite to the common behaviour). This seems to suggest the presence of very large ice particles (that can cause a reduction of PD(85 GHz) down to zero) and is confirmed by observations by the same authors of |PD(85 GHz)| close to zero in heavy snowfall.

Therefore, when looking at mixed phase non-precipitating clouds, while the intensity of thermal MW radiation provides information about the liquid water path of

Figure 3.47 Resulting T_B and PD for different combinations of cloud and rain LWP at 19 GHz (upper panel) and at 30 GHz (lower panel). Solid, dotted and dash-dotted lines correspond to constant total LWP, constant rain LWP and constant cloud LWP, respectively

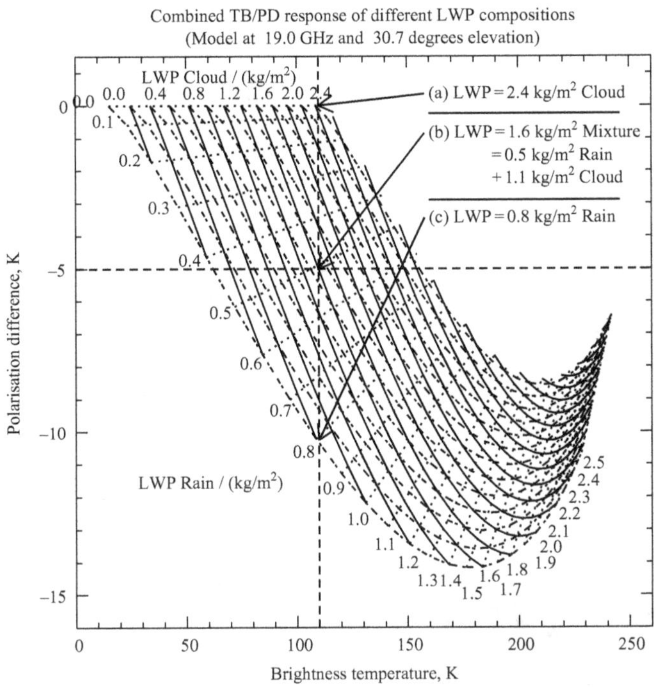
Combined TB/PD response of different LWP compositions
(Model at 19.0 GHz and 30.7 degrees elevation)
LWP Cloud / (kg/m²)
0.0 0.4 0.8 1.2 1.6 2.0 2.4
(a) LWP = 2.4 kg/m² Cloud
(b) LWP = 1.6 kg/m² Mixture
= 0.5 kg/m² Rain
+ 1.1 kg/m² Cloud
(c) LWP = 0.8 kg/m² Rain
LWP Rain / (kg/m²)
Polarisation difference, K
Brightness temperature, K

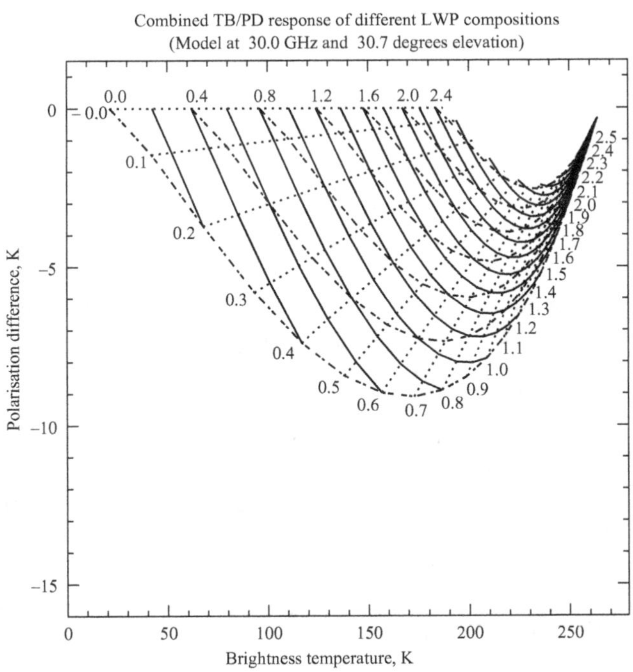
Combined TB/PD response of different LWP compositions
(Model at 30.0 GHz and 30.7 degrees elevation)
0.0 0.4 0.8 1.2 1.6 2.0 2.4
Polarisation difference, K
Brightness temperature, K

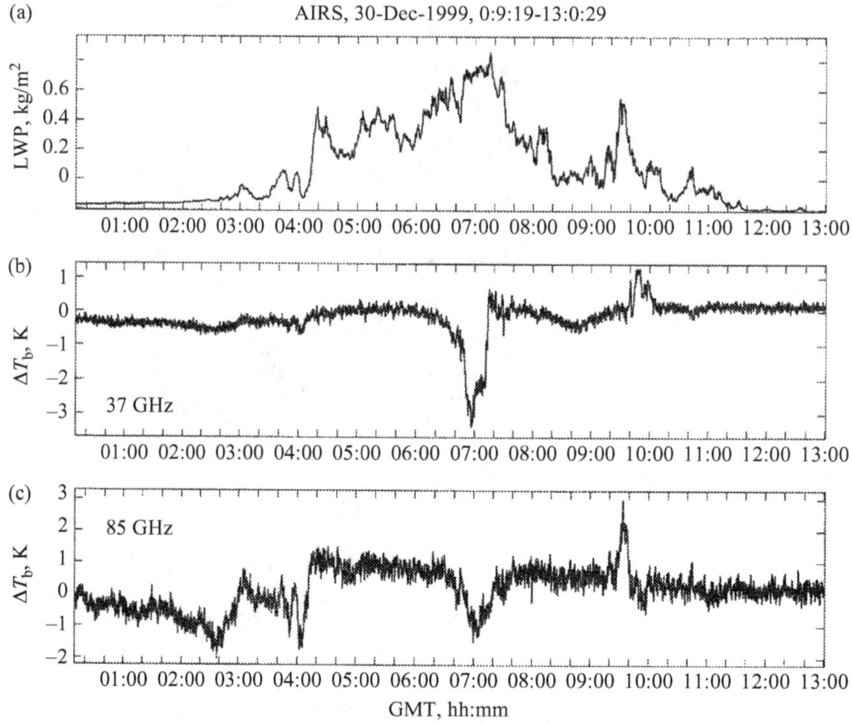

Figure 3.48 *Time history of (a) LWP, (b)PD(37 GHz), (c) PD(85 GHz) measured during AIRS on 30 Dec 1999 in Ottawa. Reprinted from Troistsky et al., 2003*

the cloud, the polarisation signatures are related to the ice water path and to the microphysics of the crystals particles.

3.9.2.3 Upwelling polarisation signal at 85 GHz over cloud systems

Scattering signals are measured by satellites at 85 GHz over convective cloud systems. This information is used to estimate precipitation and cloud ice content over land and ocean.

Prigent *et al.* (2001, 2005) carefully characterised the scattering-related polarisation difference observed at 85 GHz with the special sensor microwave/imager (SSM/I) and with the tropical rainfall measuring mission (TRMM) microwave instrument (TMI).

Figure 3.49 presents scatterplots of the TMI brightness temperature polarisation differences ($T_V - T_H$) versus the brightness temperature at T_V at 85 GHz for July 1998. This is done for cloudy pixels only (cloudiness determined by the International Satellite Cloud Climatology Project (ISCCP) (Rossow and Schiffer, 1999)). The ocean and land cases are treated separately. For each 1 K by 1 K box, the colour indicates the mean cloud top temperature obtained by ISCCP. Contours delimit regions

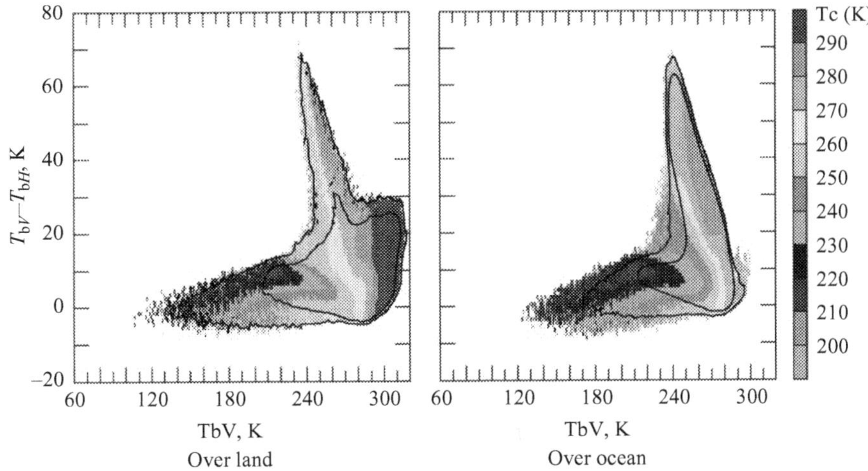

Figure 3.49 Scatterplots of the TMI brightness temperature polarisation differences ($T_V - T_H$) versus the brightness temperatures in the vertical polarisation (T_V), at 85 GHz for July 1998 over the tropics for cloudy pixels only. See text for more details.

where the pixel population in the boxes is larger than 0.001 and 0.0001 per cent of the total population for the month. The emission and scattering regimes are easily recognisable, especially over the ocean: against a cold background, thin clouds have rather low T_V with large polarisation differences caused by the ocean surface, but as cloud opacity increases, usually associated with a decrease of the cloud top temperature, T_V increases and the polarisation difference decreases. In the scattering regime, most often observed for clouds with colder top temperatures, T_V decreases due to scattering by large ice particles and can achieve very low values, especially over land. Limiting the consideration to 60 K$< T_V <$ 260 K and $(T_V - T_H) <$ 15 K, the mean polarisation difference at first increases up to 7 K for T_V 220 K and then decreases below 0 K for T_V 140 K down to about −2 and −1 K over land and ocean, respectively. Using the spatial standard deviation of the 85 GHz observations, we find that the larger positive polarisation differences are associated with areas with lower spatial standard deviations (i.e. stratiform cloud regions), whereas the negative polarisation differences are dominant in spatially heterogeneous regions (i.e. in deep convective regions). A potential contribution of a polarised signal coming from the surface was ruled out, and the signal was clearly related to the presence of cold-topped, optically thick clouds. Radiative-transfer simulations with ATM (see Section 2.6) demonstrated that the presence of large spherical particles can only generate positive polarisation differences of the order of 2 K. Randomly oriented non-spherical particles cannot produce positive polarisation differences as large as seen.

Significant polarisation differences can only be explained by the presence of mostly oriented non-spherical particles, as shown by the radiative-transfer simulation. ATM (see Section 2.6) incorporates the possibility of simulating the presence of

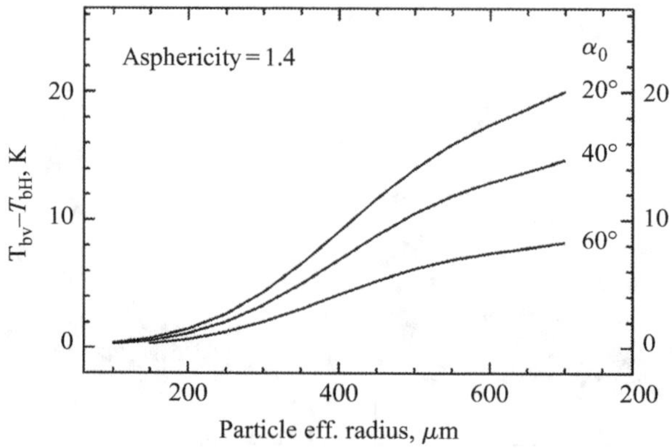

Figure 3.50 Sensitivity of the 85 GHz polarisation difference to the orientation of the oblate particles as a function of particle size. The orientation of the particles is random within α_0 from the horizontal axis

prolate and oblate particles that are randomly oriented within an angle α_0 with respect to the vertical or the horizontal.

In a first simulation set, oblate spheroids that are mostly horizontally oriented (Figure 3.50) are considered. Large positive polarisation differences are obtained that can explain the satellite observations.

In the more stratiform region of the cloud system, forces on falling non-spherical particles are expected to align them horizontally, especially the largest precipitation-sized particles (the forces being size dependent). This is consistent with the occurrence of the positive polarisation ($T_V - T_H > 0$) observed in the most homogeneous part of the cloud structure.

In a second simulation set, prolate particles essentially vertically oriented are examined (Figure 3.51). A significant negative polarisation difference is then obtained. In the deep convective cores of the cloud systems, vertically oriented non-spherical particles related to the electrical processes are present. Using the lightning imaging sensor also on board TRMM, the relationship between the presence of the negative polarisation difference and the strong electrical activity in the cloud is shown in Prigent *et al.* (2005).

3.10 Recommendations and outlook to future developments

In this chapter we have reviewed the entire flow chart necessary to analyse and understand the MW radiative processes in clouds and precipitation. Obviously the descriptions cannot be exhaustive and reflect the authors' own bias. After a review of the principal methodologies of single scattering computations and of appropriate VRTE solutions, an overview of the main issues related to the identification and exploitation of passive MW signatures of clouds and precipitation has been carried

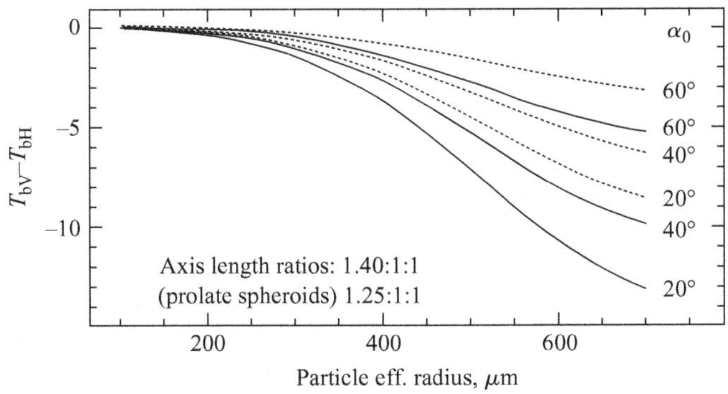

Figure 3.51 Sensitivity of the 85 GHz polarisation difference to the orientation of the non-spherical ice particles as a function of particle size. The orientation of the particles is random within α_0 from the vertical axis

out. As a concluding remark, we can formulate some general recommendations on the discussed topics.

- *Development of new techniques and databases for single scattering properties.* An improvement in the specification of optical properties (phase function, extinction matrix, absorption vector) for ice and mixed phase hydrometeors is urgently needed especially in the submillimetre region. This includes both the development and refinement of theoretical and numerical techniques for solving Maxwell equations and the creation of an accessible database for optical properties of ice crystal-snowflakes-melting particles, that can be used as input for the RT computations.
- *Improvement of the accuracy and of the generality of radiative-transfer codes.* The expanding computational capabilities will allow us to extensively test and use 3D fully polarised codes for sensitivities studies. It is important to cross-validate different codes and to produce benchmark results in the most general RT context. Making codes and benchmark results available to the scientific community in a well documented manner is highly recommended.
- *Assessment of RT errors.* When dealing with approximate solutions of the VRTE (like Eddington or two-stream approximation codes) it is always necessary to evaluate carefully the region of validity of the assumptions underlying the codes. In particular when the plane-parallel scalar approximation is used, a better characterisation of potential 3D effects and of the impact of the polarisation-coupling effect needs to be investigated on a case by case basis.
- *Realism in the specification of cloud scenarios.* The specification of the size, shape and orientation distribution of hydrometeors including their three-dimensional spatial distribution in clouds and precipitation needs a constant feedback and validation process by in situ or other remote sensing observations. This will finally lead to better agreement and consistency between observed and CRM-simulated MW T_Bs.

- *Exploitation of the submillimetre frequency range by spaceborne radiometers.* The region between 200 and 1500 GHz has been shown to be potentially rich with applications (including limb sounding). Therefore we recommend further studies involving upper tropospheric ice clouds, with focus on their characterisation in terms of ice water path and characteristic size. The potential of geostationary MW missions has to be assessed by identifying the capabilities of the submillimetre sounding channels in retrieving the cloud structure as well.
- *Insight into polarisation signatures.* Polarisation measurements have the potential to provide information on the predominant shapes and fall orientations of particles, allowing for a rough habit classification. Further studies are, however, necessary to improve and understand the use of the polarisation signal in retrievals both from space and ground-based radiometers. In this context, ground-based polarimetric measurements below 200 GHz would represent a tremendous help in validating physically-based radiative-transfer models and rain-liquid water content discrimination techniques.

References

ADLER, R. F., YEH, H. M., PRASAD, N., TAO, W., and SIMPSON, J.: 'Microwave simulations of a tropical rainfall system with a three dimensional cloud model', *J. Appl. Meteorol.*, 1991, **30**, pp. 924–53

ARABEY, E. N.: 'Radiosonde data as means for revealing cloud layers', *Meteor. Gidrol.*, 1975, **6**, pp. 32–37

BATTAGLIA, A., and KUMMEROW, C.: 'Forward MonteCarlo computations of polarized microwave radiation'. 13th international conference on *Clouds and precipitation*, Reno Area, Nevada, USA, 2000, pp. 256–59

BATTAGLIA, A., and MANTOVANI, S.: 'Forward MonteCarlo computations of fully polarized microwave radiation in non isotropic media', *J. Quant. Spectrosc. Radiat. Transfer*, 2005, **95**(3), pp. 285–308

BATTAGLIA, A., KUMMEROW, C., SHIN, D.-B., and WILLIAMS, C.: 'Toward characterizing the effect of radar bright bands on microwave brightness temperatures', *J. Atmos. Ocean Technol.*, 2003, **20**(6), pp. 856–71

BATTAGLIA, A., PRODI, F., PORCU, F., and SHIN, D.-B.: 'Measuring Precipitation from space: EURAINSAT and the future'. LEVIZZANI, V., BAUER, P., and TURK, F.J. (Eds): '3D effects in MW radiative transport inside precipitating clouds: Modeling and applications' (Kluwer Academic Publishers, 2005)

BATTAN, L. J.: 'Radarmeteorology' (Zanichelli, Bologna, 1973)

BAUER, P.: 'Over-ocean rainfall retrieval from multi-sensor data of tropical rainfall measuring mission (TRMM): Design and evaluation of inversion databases', *J. Atmos. Ocean Technol.*, 2001, **18**, pp. 1315–30

BAUER, P., and MUGNAI, A.: 'Precipitation profile retrieval using temperature-sounding microwave observations', *J. Geophys. Res.*, 2003, **108**(D23), pp. 4730–43

BAUER, P., SCHANZ, L., and ROBERTI, L.: 'Correction of three dimensional effects for passive microwave remote sensing of convective clouds', *J. Appl. Meteorol.*, 1998, **37**, pp. 1619–32

BAUER, P., KHAIN, A., POKROVSKY, A. *et al.*: 'Combined cloud-radiative transfer modeling of stratiform rainfall', *J. Atmos. Sci.*, 2000, **57**, pp. 1082–104

BAUER, P., BUROSE, D., and SCHULZ, J.: 'Rain detection over land surfaces using passive microwave satellite data', *Meteorologische Zeitschrift*, 2002, **11**(1), pp. 37–48

BENNARTZ, R., and PETTY, G.W.: 'The sensitivity of microwave remote sensing observations of precipitation to ice particle size distributions', *J. Appl. Meteorol.*, 2001, **40**, pp. 345–64

BENNARTZ, R., and BAUER, P.: 'Sensitivity of microwave radiances at 85–183 GHz to precipitating ice particles', *Radio Sci.*, 2003, **38**(4), doi:10.1029/2002RS002626

BOHREN, C.F., and HUFFMAN, D.R.: 'Absorption and scattering of light by small particles' (John Wiley & Sons, New York, 1983)

CHANDRASEKHAR, S.: 'Radiative transfer' (Dover Publications, New York, 1960)

CHERNYKH, I.V., and ESKRIDGE, R.E.: 'Determination of cloud amount and level from radiosonde soundings', *J. Appl. Meteorol.*, 1996, **35**, pp. 1362–69

CHEVALLIER, F., and BAUER, P.: 'Model rain and clouds over oceans: comparison with ssm/i observations', *Monthly Weather Rev.*, 2003, **131**, pp. 1240–55

CHUANG, C., and BEARD, K. V.: 'A numerical model for the equilibrium shape of electrified raindrops', *J. Atmos. Sci.*, 1990, **47**, pp. 1374–89

CREWELL, S., and LÖHNERT, U.: 'Accuracy of cloud liquid water path from ground-based microwave radiometry. Part II. Sensor accuracy and synergy', *Radio Sci.*, 2003, **38**(3), p. 8042

CREWELL, S., CZEKALA, H., LÖHNERT, U., SIMMER, C., ROSE, T., and ZIMMERMANN, R.: 'MICCY – a 22 channel ground-based microwave radiometer for atmospheric research', *Radio Sci.*, 2001, **36**, pp. 621–38

CREWELL, S., DRUSCH, M., MEIJGAARD, E. V., and LAMMEREN, A. V.: 'Cloud observations and modelling within the European BALTEX cloud liquid water network', *Boreal Environ. Res.*, 2002, **7**, pp. 235–45

CREWELL, S., SIMMER, C., BLOEMINK, H. *et al.*: 'The BALTEX bridge campaign: An integrated approach for a better understanding of clouds', *Bull. Amer. Met. Soc.*, 2004, **85**(10), pp. 1565–84, doi: 10.1175/BAMS-85-10-1565

CZEKALA, H.: 'Effects of ice particle shape and orientation on polarized microwave radiation for off-nadir problems', *Geophys. Res. Lett.*, 1998, **25**, pp. 1669–72

CZEKALA, H.: 'Microwave radiative transfer calculations with multiple scattering by nonspherical hydrometeors'. Ph.D. thesis, Rheinische Friedrich-Wilhems-Universität Bonn, Auf dem Hügel 20, 53121 Bonn, 1999

CZEKALA, H., and SIMMER, C.: 'Microwave radiative transfer with nonspherical precipitating hydrometeors', *J. Quant. Spectrosc. Radiat. Transfer*, 1998, **60**, pp. 365–74

CZEKALA, H., and SIMMER, C.: 'On precipitation induced polarization of microwave radiation measured from space', *Meteorologische Zeitschrift*, 2002, **11**, pp. 49–60

CZEKALA, H., HAVEMANN, S., SCHMIDT, K., ROTHER, T., and SIMMER, C.: 'Comparison of microwave radiative transfer calculations obtained with three different approximations of hydrometeor shape', *J. Quant. Spectrosc. Radiat. Transfer*, 1999, **63**, pp. 545–58

CZEKALA, H., CREWELL, S., SIMMER, C., and THIELE, A.: 'Discrimination of cloud and rain liquid water path by groundbased polarized microwave radiometry', *Geophys. Res. Lett.*, 2001a, **28**(2), pp. 267–70

CZEKALA, H., CREWELL, S., HORNBOSTEL, A., SCHROTH, A., SIMMER, C., and THIELE, A.: 'Interpretation of polarization features in ground based microwave observations as caused by horizontally aligned oblate rain drops', *J. Appl. Meteorol.*, 2001b, **40**, pp. 1918–32

d'AURIA, G., MARZANO, F. S., PIERDICCA, N., NOSSAI, R. P., BASILI, P., and CIOTTI, P.: 'Remotely sensing cloud properties from microwave radiometric observations by using a modeled cloud database', *Radio Sci.*, 1998, **33**, pp. 369–92

DAVIS, C. P., EMDE, C., and HARWOOD, R. S.: 'A 3D polarized reversed monte carlo radiative transfer model for mm and sub-mm passive remote sensing in cloudy atmospheres', *IEEE Trans. on Geosci. Remote Sens.*, 2005, **43**, pp. 1096–101

DEETER, M. N., and EVANS, K. F.: 'A hybrid eddington-single scattering radiative transfer model for computing radiances from thermally emitting atmospheres', *J. Quant. Spectrosc. Radiat. Transfer*, 1998, **60**(4), pp. 635–48

DI MICHELE, S., MARZANO, F. S., MUGNAI, A., TASSA, A., BAUER, P., and BAPTISTA, J. P. V. P.: 'The bayesian algorithm for microwave-based precipitation retrieval (BAMPR): Description and application to TMI data', *IEEE Trans. Geosci. Remote Sens.*, 2005, **43**(4), pp. 778–791

DRAINE, B. T., and FLATAU, P. J.: 'Discrete-dipole approximations for scattering calculations', *J. Opt. Soc. Am. A*, 2000a, **11**, pp. 1491–99

DRAINE, B. T., and FLATAU, P. J.: 'Light scattering by nonspherical particles: theory, measurements, and applications', MISHCHENKO, M. I., HOVENIER, J. W., and TRAVIS, L. D. (Eds): in 'The discrete dipole approximation for light scattering by irregular targets' (Academic Press, San Diego, 2000b)

DRAINE, B. T., and FLATAU, P. J.: 'Userguide for the discrete dipole approximation code DDSCAT 6.1', 2004. Available on line at http://arxiv.org/abs/astro-ph/

DUDHIA, J.: 'A nonhydrostatic version of the penn state-ncar mesoscale model: Validation tests and simulation of an Atlantic cyclone and cold front', *Mon. Weather Rev.*, 1993, **121**, pp. 1493–513

EMDE, C.: 'A polarized discrete ordinate scattering model for radiative transfer simulations in spherical atmospheres with thermal source'. Ph.D. thesis, University of Bremen, 2005

EMDE, C., BUEHLER, S. A., ERIKSSON, P., and SREEREKHA, T. R.: 'The effect of cirrus clouds on microwave limb radiances', *Atmos. Res.*, 2004a, **72**, pp. 383–401

EMDE, C., BUEHLER, S. A., DAVIS, C., ERIKSSON, P., SREEREKHA, T. R., and TEICHMANN, C.: 'A polarized discrete ordinate scattering model for simulations of limb and nadir longwave measurements in 1D/3D spherical atmospheres', *J. Geophys. Res.*, 2004b, **109**(D24), D24207, doi:10.1029/2004JD005140

EVANS, K. F.: 'The spherical harmonics discrete ordinate method for three-dimensional atmospheric radiative transfer', *J. Atmos. Sci.*, 1998, **55**, pp. 429–46

EVANS, K. F., and STEPHENS, G. L.: 'A new polarized atmospheric radiative transfer model', *J. Quant. Spectrosc. Radiat. Transfer*, 1991, **46**, pp. 413–23

EVANS, K. F., and STEPHENS, G. L.: 'Microwave radiative transfer through clouds composed of realistically shaped ice crystals. I. Single scattering properties', *J. Atmos. Sci.*, 1995a, **52**, pp. 2041–57

EVANS, K. F., and STEPHENS, G. L.: 'Microwave radiative transfer through clouds composed of realistically shaped ice crystals. II. Remote sensing of ice clouds', *J. Atmos. Sci.*, 1995b, **52**, pp. 2058–72

EVANS, K. F., and VIVEKANANDAN, J.: 'Multiparameter radar and microwave radiative transfer modeling of nonspherical atmospheric ice particles', *IEEE Trans. Geosci. Remote Sens.*, 1990, **28**, pp. 423–37

EVANS, K. F., TURK, J., WONG, T., and STEPHENS, G.: 'A Bayesan approach to microwave precipitation profile retrieval', *J. Appl. Meteorol.*, 1995, **34**, pp. 260–79

EVANS, K. F., WALTER, S. J., HEYMSFIELD, A. J., and DEETER, M. N.: 'Modeling of submillimiter passive remote sensing of cirrus clouds', *J. Appl. Meteorol.*, 1998, **37**, pp. 184–205

EVANS, K. F., EVANS, A. H., NOLT, I. G., and MARSHALL, B. T.: 'The prospect for remote sensing of cirrus clouds with a submillimetre-wave spectrometer', *J. Appl. Meteorol.*, 1999, **38**, pp. 514–25

EVANS, K. F., WALTER, S. J., HEYMSFIELD, A. J., and McFARQUHAR, G. M.: 'Submillimiter-wave cloud ice radiometer: simulations of retrieval algorithm performance', *J. Geophys. Res.*, 2002, **107**(D3), 10.1029/2001JD000709

FABRY, F., and SZYRMER, K.: 'Modeling of the melting layer. Part II: Electromagnetic', *J. Atmos. Sci.*, 1999, **56**, pp. 3593–600

GASIEWSKI, A.: 'Atmospheric remote sensing by microwave radiometry' (Wiley & Sons, New York, 1993), Chap. 4

GOODY, R. M., and YOUNG, Y. L.: 'Atmospheric radiation: theoretical basis' (Oxford University Press, New York, 1989)

GRODY, N. C.: 'Atmospheric remote sensing by microwave radiometry', in JANSSEN, M. A. (Ed.): 'Remote sensing of the atmosphere from satellites using microwave radiometry' (John Wiley & Sons Inc., New York, 1993)

HAFERMAN, J. L.: 'A polarized multi-dimensional discrete-ordinates radiative transfer model for remote sensing applications'. Ph.D. thesis, University of Iowa, 1995

HAFERMAN, J.: 'Light scattering by nonspherical particles: Theory, measurements, and applications in MISHCHENKO, M. I., Hovenier, J. W., and TRAVIS, L. D. (Eds.): 'Microwave scattering by precipitation' (Academic Press, San Diego, CA, 2000)

HAFERMAN, J. L., KRAJEWSKI, W. F., SMITH, T. F., and SANCHEZ, A.: 'Radiative transfer for a three-dimensional raining cloud', *Appl. Opt.*, 1993, **32**, pp. 2795–802

HAFERMAN, J. L., SMITH, T. F., and KRAJEWSKI, W. F.: 'A multi-dimensional discrete ordinates method for polarized radiative transfer. i. validation for randomly oriented axisymmetric particles', *J. Quant. Spectrosc. Radiat. Transfer*, 1997, **58**, pp. 379–98

HEYMSFIELD, A.: 'Ice crystal terminal velocities', *J. Atmos. Sci.*, 1978, **29**, pp. 1348–57

HEYMSFIELD, G. M., and FULTON, R.: 'Passive microwave and infrared structure of mesoscale convective systems', *Meteorol. Atmos. Phys.*, 1994, **54**, pp. 123–39

HEYNEY, L. G., and GREENSTEIN, J. L.: 'Diffuse radiation in the galaxy', *Astrophys. J.*, 1941, **93**, pp. 70–83

HONG, Y., HAFERMAN, J. L., OLSON, W. S., and KUMMEROW, C. D.: 'Microwave brightness temperatures from tilted convective systems', *J. Appl. Meteorol.*, 2000, **39**, pp. 983–98

HOUSE, L., and AVERY, L. W.: 'The Monte Carlo tecnique applied to radiative transfer', *J. Quant. Spectrosc. Radiat. Transfer*, 1969, **9**, pp. 1579–91

HOUZE, R. A.: 'Cloud dynamics' (International Geophysics Series. Academic Press, 1993) p. 573

HOVENIER, J. W., VAN DER MEE, C., and DOMKE, H.: 'Transfer of polarized light in planetary atmospheres-basic concepts and practical methods' (Kluwer Academic Publishers, Dordrecht, The Netherlands, 2004)

ISHIMARU, A.: 'Wave propagation and scattering in random media' (Academic Press, New York, NY, 1978)

ISHIMARU, A., and CHEUNG, R.-T.: 'Multiple scattering effects on wave propagation due to rain', *Ann. Telecommun.*, 1980, **35**, pp. 373–79

ISSIG, C.: 'Ein spektrales Wolkenmodell mit integriertem Strahlungsübertragungsmodell zur Unterstützung von Niederschlagsalgorithmen'. Ph.D. thesis, Meteorological Institute, University of Bonn, 1997

JANSSEN, M. A.: 'Atmospheric remote sensing by microwave radiometry' (John Wiley & Sons, 1996)

JOSEPH, J. H., WISCOMBE, W. J., and WEINMAN, J. A.: 'The Delta Eddington approximation for radiative flux transfer', *J. Atmos. Sci.*, 1976, **33**, pp. 2452–59

KAHNERT, F. M.: 'Numerical methods in electromagnetic scattering theory', *J. Quant. Spectrosc. Radiat. Transfer*, 2003, **79**, pp. 775–824

KARSTENS, U., SIMMER, C., and RUPRECHT, E.: 'Remote sensing of cloud liquid water', *Meteorol. Atmos. Phys.*, 1994, **54**, pp. 157–71

KATSUMATA, M., UYEDA, H., IWANAMI, K., and LIU, G.: 'The response of 36 and 89 GHz microwave channels to convective snow clouds over ocean: Observation and modeling', *J. Appl. Meteorol.*, 2000, **39**, pp. 2322–35

KEY, J.: 'Streamer user's, Version 3.0' (Cooperative Institute for Meteorological Satellite Studies, University of Wisconsin, 2002) p. 96

KHLEBTSOV, N. G.: 'Orientational averaging of light-scattering observables in the T-matrix approach', *Appl. Opt.*, 1992, **31**(25), pp. 5359–65

KING, M., and HARSHVARDHAN, H.: 'Comparative accuracy of selected multiple scattering approximations', *J. Atmos. Sci.*, 1986, **43**, pp. 784–801

KUIK, F., HAAN, J. F. D., and HOVENIER, Q. W.: 'Benchmark results for single scattering by spheroids', *J. Quant. Spectrosc. Radiat. Transfer*, 1992, **47**, pp. 477–89

KUMMEROW, C.: 'On the accuracy of the Eddington approximation for radiative transfer in the microwave frequencies', *J. Geophys. Res.*, 1993, **98**, pp. 2757–65

KUMMEROW, C.: 'Beamfilling errors in passive microwave rainfall retrievals', *J. Appl. Meteorol.*, 1998, **37**, pp. 356–70

KUMMEROW, C., and GIGLIO, L.: 'A passive microwave technique for estimating rainfall and vertical structure information from space. Part I: algorithm description', *J. Appl. Meteorol.*, 1994, **33**, pp. 3–18

KUMMEROW, C., and WEINMAN, J. A.: 'Determining microwave brightness temperatures from precipitating horizontally finite and vertically structured clouds', *J. Geophys. Res.*, 1988, **93**(D4), pp. 3720–28

KUMMEROW, C., OLSON, W. S., and GIGLIO, L.: 'A simplified scheme for obtaining precipitation and vertical hydrometeor profiles from passive microwave sensors', *IEEE Trans. Geosci. Remote Sens.*, 1996, **34**, pp. 1213–32

KUMMEROW, C., HONG, Y., OLSON, W. S. *et al.*: 'The evolution of the goddard profiling algorithm (gprof) for rainfall estimation from passive microwave sensors', *J. Appl. Meteorol.*, 2001, **40**, pp. 1801–20

LAFORE, J.: 'The Meso-NH atmospheric simulation system, Part I: adiabatic formulation and control simulations', *Ann. Geophys.*, 1998, **16**, pp. 90–109

LEITERER, U., DIER, H., NAGEL, D. *et al.*: 'Correction method for RS80-A humicap humidity profiles and their validation by lidar backscattering profiles in tropical cirrus clouds', *J. Atmos. Ocean Technol.*, 2005, **22**, pp. 18–29

LENOBLE, J. E.: 'Radiative transfer in scattering and absorbing atmospheres: Standard computational procedures' (Deepak, Hampton, VA, 1985)

LILJEGREN, J. C.: 'Microwave radiometry and remote sensing of the Earth's surface and atmosphere', in PAMPALONI, P., and PALOSCIA, S. (Eds.): 'Automatic self-calibration of ARM microwave radiometers' (Utrecht, VSP Press, 2000) pp. 433–41

LIOU, K. N.: 'An introduction to atmospheric radiation' (Academic Press, 2nd edn, 2002)

LIU, Q.: 'Approximation of single scattering properties of ice and snow particles for high microwave frequencies', *J. Atmos. Sci.*, 2004, **61**, pp. 2441–56

LIU, G., and CURRY, J. A. 'Determination of ice water path and mass median particle size using multichannel microwave measurements', *J. Appl. Meteorol.*, 2000, **39**, pp. 1318–29

LIU, Q., and SIMMER, C.: 'Polarization and intensity in microwave radiative transfer', *Beitr. Phys. Atmosph.*, 1996a, **69**, pp. 535–45

LIU, Q., and SIMMER, C.: 'Three-dimensional radiative transfer effects of clouds in the microwave spectral range', *J. Geophys. Res.*, 1996b, **101**(D2), pp. 4289–98

LIU, Q., and WENG, F.: 'A microwave polarimetric two stream radiative transfer model', *J. Atmos. Sci.*, 2002, **59**, pp. 2396–402

LÖHNERT, U., and CREWELL, S.: 'Accuracy of cloud liquid water path from ground-based microwave radiometry. Part I. Dependency on cloud model statistics', *Radio Sci.*, 2003, **38**(3)

LÖHNERT, U., CREWELL, S., MACKE, A., and SIMMER, C.: 'Profiling cloud liquid water by combining active and passive microwave measurements with cloud model statistics', *J. Atmos. Ocean Technol.*, 2001, **18**, pp. 1354–66

LÖHNERT, U., CREWELL, S., and SIMMER, C.: 'An integrated approach towards retrieving physically consistent profiles of temperature, humidity and cloud liquid water', *J. Appl. Meteorol.*, 2004, **43**(9), pp. 1295–307

LYNN, B. H., KHAIN, A. P., DUDHIA, J., ROSENFELD, D., POKROVSKY, A., and SEIFERT, A.: 'Spectral (bin) microphysics coupled with a mesoscale model (MM5). Part I: model description and first results', *Monthly Weather Rev.*, 2005, **133**(1), pp. 44–58

MACKOWSKI, D. W., and MISHCHENKO, M. I.: 'Calculation of the T matrix and the scattering matrix for ensembles of spheres', *J. Opt. Soc. Am. A*, 1996, **13**, pp. 2266–78

MARCHUK, G. G., MIKHAILOV, M., NAZARALIEV, M., DARBINJAN, R., KARGIN, B., and ELEPOV, B.: 'The monte carlo methods in atmospheric optics' (Springer-Verlag, Berlin, 1980)

MARSHALL, J. S., and PALMER, W. M.: 'The distribution of raindrop with size', *J. Meteorol.*, 1948, **5**, pp. 165–6

MARTIN, L., LÜDI, A., and MÄTZLER, C.: 'Tropospheric monitoring with ASMUWARA'. Proceedings of the 6th international symposium on *Tropospheric profiling (ISTP): needs and technologies*, Leipzig, Germany, Sep. 14–20, 2003, available from http://istp2003.tropos.de:8085/

MARZANO, F. S., and ROBERTI, L.: 'Numerical investigation of intense rainfall effects on coherent and incoherent slant-path propagation at K band and above', *IEEE Trans. Antennas Propagat.*, 2003, **41**(5), pp. 965–77

MARZANO, F., MUGNAI, A., PANEGROSSI, G., PIERDICCA, N., SMITH, E., and TURK, J.: 'Bayesian estimation of precipitating cloud parameters from combined measurements of spaceborne microwave radiometer and radar', *IEEE Trans. Geosci. Remote Sens.*, 1999a, **37**, pp. 596–613

MARZANO, F. S., FIONDA, E., and CIOTTI, P.: 'Simulation of radiometric and attenuation measurements along earth-satellite links in the 10- to 50-GHz band through horizontally-finite convective raincells', *Radio Sci.*, 1999b, **34**, pp. 841–58

MARZANO, F. S., TURK, J., CIOTTI, P., MICHELE, S. D., and PIERDICCA, N.: 'Potential of combined spaceborne microwave and infrared radiometry for near real-time rainfall attenuation monitoring along earth-satellite links', *Int. J. Satell. Commun.*, 2001, **19**(4), pp. 385–412

MÄTZLER, C.: 'MATLAB functions for mie scattering and absorption – Version 2'. Technical Report 2002-11, Universität Bern, 2002

McKEE, T. B., and COX, S. K.: 'Scattering of visible radiation by finite clouds', *J. Atmos. Sci.*, 1974, **31**, pp. 1885–92

MISHCHENKO, M. I.: 'Extinction and polarization of transmitted light by partially aligned nonspherical grains', *Astrophys. J.*, 1991a, **367**, pp. 561–74

MISHCHENKO, M. I.: 'Light scattering by randomly oriented axially symmetric', *J. Opt. Soc. Am. A*, 1991b, **8**(6), pp. 871–81

MISHCHENKO, M. I.: 'Light scattering by size-shape distributions of randomly oriented axially symmetric particles of a size comparable to a wavelength', *Appl. Opt.*, 1993, **32**(24), pp. 4652–66

MISHCHENKO, M. I.: 'Calculation of the amplitude matrix for a nonspherical particle in a fixed orientation', *Appl. Opt.*, 2000, **39**(6), pp. 1026–31

MISHCHENKO, M. I., and TRAVIS, L. D.: 'Capabilities and limitations of a current fortran implementation of the T-matrix method for randomly oriented rotationally symmetric scatterers', *J. Quant. Spectrosc. Radiat. Transfer*, 1998, **60**(3), pp. 309–24

MISHCHENKO, M. I., LACIS, A. A., and TRAVIS, L. D.: 'Errors induced by the neglect of polarization in radiance calculations for rayleigh-scattering atmospheres', *J. Quant. Spectrosc. Radiat. Transfer*, 1994, **51**, pp. 491–510

MISHCHENKO, M. I., TRAVIS, L. D., and MACKOWSKI, D. W.: 'T-matrix computations of light scattering by nonspherical particles: a review', *J. Quant. Spectrosc. Radiat. Transfer*, 1996, **55**(5), pp. 535–75

MISHCHENKO, M. I., HOVENIER, J. W., and TRAVIS, L. D. (Eds): 'Light scattering by nonspherical particles' (Academic Press, New York, 2000)

MISHCHENKO, M. I., TRAVIS, L. D., and LACIS, A. A.: 'Scattering, absorption and emission of light by small particles' (Cambridge University Press, 2002)

MISHCHENKO, M. I., HOVENIER, J. W., and MACKOWSKI, D. W.: 'Single scattering by a small volume element', *J. Opt. Soc. Am. A*, 2004, **21**, pp. 71–87

MUGNAI, A.: 'Contribution to the global precipitation measurements by the proposed european GPM (EGPM) satellite', *Geophys. Res. Abst.*, 2003, **5**, p. 12470

MUGNAI, A., SMITH, E. A., and TRIPOLI, G. J.: 'Foundations for statistical-physical precipitation retrieval from passive microwave satellite measurements. Part I: Emission source and generalized weighting function properties of a time dependent cloud-radiation model', *J. Appl. Meteorol.*, 1993, **32**, pp. 17–39

MUGNAI, A., MICHELE, S. D., SMITH, E. A. *et al.*: 'Snowfall measurements by the proposed European GPM Mission', in LEVIZZANI, V., BAUER, P., and TURK, F. J. (Eds.): 'Measuring precipitation from space: EURAINSAT and the future' (Kluwer Academic, Dordrecht, 2005a)

MUGNAI, A., BAORDO, F., BIZZARRI, B. *et al.*: 'Studies and simulation of the radiometer'. Technical report, (EGPM-Ice Final REPORT, ESA, ESTEC, Netherlands, 2005b)

MUINONEN, K.: 'Coherent backscattering of light by complex random media of spherical scatterers: numerical solution', *Waves Random Media*, 2004, **14**, pp. 365–88

OLSON, W. S., BAUER, P., VILTARD, N. F. *et al.*: 'A melting layer model for passive/active microwave remote sensing applications. Part I: Model formulation and comparison with observations', *J. Appl. Meteorol.*, 2001a, **48**(7), pp. 1145–63

OLSON, W. S., BAUER, P., KUMMEROW, C. D., HONG, Y., and TAO, W.: 'A melting layer model for passive/active microwave remote sensing applications. Part II: Simulation of TRMM observations', *J. Appl. Meteorol.*, 2001b, **48**(7), pp. 1164–79

PANEGROSSI, G., DIETRICH, S., and MARZANO, F. S.: 'Use of cloud model microphysics for passive microwave-based precipitation retrieval', *J. Atmos. Sci.*, 1998, **55**, pp. 1644–73

PETTY, G. W.: 'Physical and microwave radiative properties of precipitating clouds. Part II: A parametric 1D rain-cloud model for use in microwave radiative transfer simulations', *J. Appl. Meteorol.*, 2001a, **40**, pp. 2115–29

PETTY, G. W.: 'Physical and microwave radiative properties of precipitating clouds. Part II: Principal component analysis of observed multichannel microwave radiances in tropical stratiform rainfall', *J. Appl. Meteorol.*, 2001b, **40**, pp. 2105–14

PLATNICK, S.: 'Vertical photon transport in cloud remote sensing problems', *J. Geophys. Res.*, 2000, **105**, pp. 22919–35

PRIGENT, C., PARDO, J., MISHCHENKO, M. I., and ROSSOW, W. B.: 'Microwave polarized scattering signatures in clouds: SSM/I observations interpreted with radiative transfer simulations', *J. Geophys. Res.*, 2001, **106**, pp. 28243–58

PRIGENT, C., DEFER, E., PARDO, J., PEARL, C., ROSSOW, W. B., and PINTY, J.-P.: 'Relations of polarized scattering signatures observed by the TRMM microwave instrument with electrical processes in cloud systems', *Geophys. Res. Lett.*, 2005, **32**

PRUPPACHER, H. R. and KLETT, J. D.: 'Microphysics of Clouds and Precipitation' (Kluver Academic, Dordrecht, 1997)

PURCELL, E. M., and PENNYPACKER, C. R.: 'Scattering and absorption of light by non spherical dielectric grains', *Astrophys. J.*, 1973, **186**, pp. 705–14

ROBERTI, L.: 'Monte Carlo radiative transfer in the microwave and in the visible: Biasing tecniques', *Appl. Opt.*, 1997, **36**(30), pp. 7929–38

ROBERTI, L., and KUMMEROW, C.: 'Monte Carlo calculations of polarized microwave radiation emerging from cloud structures', *J. Geophys. Res.*, 1999, **104**, pp. 2093–104

ROBERTI, L., HAFERMAN, J., and KUMMEROW, C.: 'Microwave radiative transfer through horizontally inhomogeneous precipitating clouds', *J. Geophys. Res.*, 1994, **99**(16), pp. 707–16

ROGERS, R. R.: 'A short course in cloud physics' (Pergamon Press, New York, NY, 1979)

ROGERS, R. R., and YAU, M. K.: 'A short course in cloud physics' (Butterworth-Heinemann, Woburn, MA, USA, 1989) p. 290

ROSE, T., CREWELL, S., LÖHNERT, U., and SIMMER, C.: 'A network suitable microwave radiometer for operational monitoring of the cloudy atmosphere', specify Atmos. Res. 2005: Special issure: CLIWA-NET: Observation and modelling of Liquid Water Clouds, **75**(3), 183–200, 2005, doi:10.1016/j.atmoves.2004.12.005

ROSSOW, W. B., and SCHIFFER, R. A.: 'Advances in understanding clouds from ISCCP', *Bull. Amer. Met. Soc.*, 1999, **80**, pp. 2261–87

SKOFRONICK-JACKSON, G. M. and WANG, J. R.: 'The estimation of hydrometeor profiles from wideband microwave observations', *J. Appl. Meteorol.*, 2000, **39**, pp. 1645–56

SKOFRONICK-JACKSON, G. M., GASIEWSKI, A. J., and WANG, J. R.: 'Influence of microphysical cloud parametrization on microwave brightness temperatures', *IEEE Trans. Geosci Remote Sens.*, 2002, **40**(1), pp. 187–96

SKOFRONICK-JACKSON, G., WNAG, J. R., HEYMSFIELD, G. M. *et al.*: 'Combined radiometer-radar microphysical profile estimations with emphasis on high frequency brightness temperature observations', *J. Appl. Meteorol.*, 2003, **42**, pp. 476–87

SKOFRONICK-JACKSON, G., KIM, M.-J., WEINMAN, J. A., and CHANG, D.-E.: 'A physical model to determine snowfall over land by microwave radiometry' *IEEE Trans. Geosci Remote Sens.*, 2004

SMITH, E. A., and MUGNAI, A.: 'Radiative transfer to space through a precipitating cloud at multiple microwave frequencies. iii. Influence of large ice particles', *J. Meteorol Soc. Jpn.*, 1989, **67**, pp. 739–55

SMITH, E. A., MUGNAI, A., COOPER, H. J., TRIPOLI, G. J., and XIANG, X.: 'Foundations for statistical-physical precipitation retrieval from passive microwave satellite measurements. Part I: Brightness temperature properties of a time dependent cloud-radiation model', *J. Appl. Meteorol.*, 1992, **31**, pp. 506–31

SMITH, E. A., XIANG, X., MUGNAI, A., and TRIPOLI, G. J.: 'Design of an inversion-based precipitation profile retrieval algorithm using an explicit cloud model for initial guess microphysics', *Meteorol. Atmos. Phys.*, 1994, **54**, pp. 53–78

SMITH, E. A., BAUER, P., MARZANO, F. S. *et al.*: 'Intercomparison of microwave radiative transfer models for precipitating clouds', *IEEE Trans. Geosci. Remote Sens.*, 2002, **40**, pp. 541–49

SOBOLEV, V. V.: 'Light scattering in planetary atmospheres' (Pergamon, New York, 1975)

SOLHEIM, F. A. J. G., WESTWATER, E. R., HAN, Y., KEIHM, S., MARSH, K., and WARE, R.: 'Radiometric profiling of temperature, water vapor and cloud liquid water using various inversion methods', *Radio Sci.*, 1998, **33**, pp. 393–404

SPENCER, R. W., GOODMAN, H. M., and HOOD, R. E.: 'Precipitation retrieval over land and ocean with the SSM/I: identification and characteristics of the scattering signal', *J. Atmos. Ocean Technol.*, 1989, **6**, pp. 254–73

STAMNES, K., and SWANSON, R. A.: 'A new look at the dicrete ordinate method for radiative transfer calculations in anisotropically scattering atmospheres', *J. Atmos. Sci.*, 1981, **38**, pp. 387–99

STAMNES, K., TSAY, S., WISCOMBE, W., and JAYAWEERA, K.: 'Numerically stable algorithm for discrete-ordinate-method radiative transfer in multiple scattering and emitting layered media', *Appl. Opt.*, 1988, **27**, pp. 2502–9

STEPHENS, G.: 'Remote sensing of the lower atmosphere: An introduction' (Oxford University Press, 1994)

TANG, C., and AYDIN, K. 'Scattering from ice crystals at 94 and 220 GHz millimeter wave frequencies', *IEEE Trans. Geosci. Remote Sens.*, 1995, **33**, pp. 93–9

TAO, W. K., and SIMPSON, J.: 'Goddard cumulus ensemble model. Part I: Model description', *Terrest. Atmos. Oceanic Sci.*, 1993, **4**, pp. 35–72

TASSA, A., MICHELE, S. D., D'ACUNZO, E. *et al.*: 'Microwave radiometry and remote sensing of the earths surface and atmosphere' (VSP Int. Sci. Pub., Utrecht (NL), 1999) Chap. x, pp. 371–7

TASSA, A., MICHELE, S. D., MUGNAI, A., MARZANO, F. S., and BAPTISTA, P. P.: 'Cloud-model based Bayesian technique for precipitation profile retrieval from TRMM Microwave Imager', *Radio Sci.*, 2003, **38**(4), doi:10.1029/2002RS002674

THOMAS, G. E., and STAMNES, K.: 'Radiative transfer in the atmosphere and ocean' (Cambridge University Press, 1999)

TRIPOLI, G. J.: 'A nonhydrostatic model designed to simulate scale interaction', *Mon. Wea. Rev.*, 1992, **120**, pp. 1342–59

TROITSKY, A. V., OSHARIN, A. M., KOROLEV, A. V., and STRAPP, J. W.: 'Polarization of thermal microwave atmospheric radiation due to scattering by ice particles in clouds', *J. Atmos. Sci.*, 2003, **60**, pp. 1608–20

TSANG, L., KONG, J. A., and SHIN, R. T.: 'Theory of microwave remote sensing' (Wiley-Interscience, 1985)

TSANG, L., KONG, J. A., and DING, K.: 'Scattering of electromagnetic waves: Theories and applications' (Wiley-Interscience, 2000)

ULABY, F. T., MOORE, R. K., and FUNG, A. K.: 'Microwave remote sensing active and passive' (Artech House, Boston, MA, 1986)

VAN DE HULST, H. C.: 'Light scattering by small particles' (Wiley-Interscience, New York, 1981)

VAN MEIJGAARD, E., and CREWELL, S.: 'Comparison of model predicted liquid water path with ground-based measurements during CLIWA-NET', *Atmos. Res.*, Special issue: CLIWA-NET: Observation and modelling of liquid water clouds, 2005, **75**(3), pp. 201–26, doi: 10.1016/j.atmosres.2004.120006

VILTARD, N., KUMMEROW, C. D., OLSON, W. S., and HONG, Y.: 'Combined use of the radar and radiometer of TRMM to estimate the influence of drop size distribution on rain retrievals', *J. Appl. Meteorol.*, 2000, **39**

WARNER, J.: 'The water content of cumuliform clouds', *Tellus*, 1955, **7**, pp. 449–57

WATERMAN, P. C.: 'Symmetry, unitarity, and geometry in electromagnetic scattering', *Phys. Rev. D*, 1971, **3**(4), pp. 825–39

WAUBEN, W. M., and HOVENIER, J. W.: 'Polarized radiation of an atmosphere containing randomly-oriented spheroids', *J. Quant. Spectrosc. Radiat. Transfer*, 1992, **47**, pp. 491–504

WEINMAN, J. A., and DAVIES, R.: 'Thermal microwave radiances from horizontally finite clouds of hydrometeors', *J. Geophys. Res.*, 1978, **83**(6), pp. 3099–107

WEINMAN, J. A., and GUETTER, P. J.: 'Determination of rainfall distributions from microwave radiation measured by the Nimbus 6 ESMR', *J. Appl. Meteorol.*, 1977, **16**, pp. 437–42

WENG, F., and GRODY, N. C.: 'Retrieval of ice cloud parameters using a microwave imaging radiometer', *J. Atmos. Sci.*, 2000, **57**, pp. 1069–81

WESTWATER, E. R.: 'The accuracy of water vapor and cloud liquid determination by dual-frequency ground-based microwave radiometry', *Radio Sci.*, 1978, **13**, pp. 667–85

WESTWATER, E. R., CREWELL, S., and MÄTZLER, C. 'A review of surface-based microwave and millimeter-wave radiometric remote sensing of the troposphere', *Radio Sci. Bull.*, 2004, 3010, pp. 59–80

WIEDNER, M., PRIGENT, C., PARDO, J. R. *et al.*: 'Modeling of passive microwave response in convective situations using output from mesoscale models: comparison with TRMM/TMI satellite observations', *J. Geophys. Res.*, 2004, **109**(D06214), doi:10.1029/2003JD004280

WILHEIT, T. T., CHANG, A. T. C., and CHIU, L. S. 'Retrieval of monthly rainfall indices from microwave radiometric measurements using probability distribution function', *J. Atmos. Ocean Technol.*, 1991, **8**, pp. 118–36

WISCOMBE, W. J., and JOSEPH, J. H.: 'The range of validity of the Eddington approximation', *Icarus*, 1977, **32**, pp. 362–77

YANG, P., and LIOU, K. N.: 'Single scattering properties of complex ice crystals in terrestrial atmosphere', *Contr. Atmos. Phys.*, 1998, **71**, pp. 223–48

Chapter 4

Surface emission

Jean-Pierre Wigneron, Søren Andersen, Jacqueline Boutin, Jean-Christophe Calvet, Kun-Shan Chen, Patricia de Rosnay, Emmanuel P. Dinnat, Stephen J. English, Maria José Escorihuela, Paolo Ferrazzoli, Leila Guerriero, Martti Hallikainen, Georg Heygster, Thomas R.H. Holmes, Brian K. Hornbuckle, Adrian Jupp, Yann H. Kerr, Christian Mätzler, Thierry Pellarin, Catherine Prigent, Jouni Pulliainen, Kauzar Saleh Contell, Mike Schwank, Jiancheng Shi, Rasmus T. Tonboe, Leif Toudal Pedersen, Adriaan van de Griend, Philippe Waldteufel, and Andreas Wiesmann

4.1 Introduction, purpose and scope *Jean-Pierre Wigneron*

This chapter describes research activities made over surfaces, including ocean, bare soil (mainly in arid regions), vegetation-, snow- and ice-covered areas. Even though an exhaustive review of all studies could not be considered in the framework of the book, the chapter includes the most significant developments which have been made recently in the radiative-transfer modelling of the microwave surface emission.

Most of the contributions included in this chapter are reviews dedicated to a specific field of research:

- physical aspects in the radiative-transfer modelling over covered surfaces (by Mätzler);
- modelling of the ocean microwave signature at L-band (by Dinnat and Boutin) and at higher frequencies, including polarimetric effects (by English and Jupp);
- modelling of the soil microwave emission (by Wigneron *et al.*);

- accounting for the effects of the vegetation canopy, with emphasis on (1) the analysis of the herbaceous vegetation structure (by Wigneron and Hornbuckle), (2) the so-called b-factor relating the vegetation attenuation to the canopy water content (by Van de Griend) and (3) forest emission (by Ferrazzoli and Guerriero);
- modelling of the microwave emission from snow (by Mätzler *et al.*) and sea ice (by Tonboe *et al.*).

In addition to these reviews some other contributions present results on more specific aspects. A contribution by Mätzler proposes simple analytical equations to model the topographic effects, which play a key role in the land surface emission and which are rarely accounted for to date. Over bare soil surfaces, two contributions correspond to the description of the original modelling approaches of (1) the roughness effects (the air to soil transition model) by Schwank and Mätzler and (2) the effective soil temperature by de Rosnay *et al.* Two contributions by Prigent analyse issues in modelling of the soil and vegetation microwave signatures which can be revealed by the analysis of spaceborne microwave observations (mainly SSM/I). Some puzzling measurements, which could not be explained by the current knowledge, are key information to show in which directions new developments should be made to improve the modelling. Section 4.14 by Wigneron *et al.* describes the L-MEB model. This model is the result of an extensive review of the current knowledge of the microwave emission of various land covers (herbaceous and woody vegetation, frozen and unfrozen bare soil, snow, etc.). The model was developed to simulate global maps of brightness temperature over the continental areas.

As in case of L-MEB, most of the time, the main motivation for the modelling developments presented in this chapter was the possibility of improving retrieval methods in view of monitoring surface characteristics from spaceborne observations.

Over the ocean, the contribution by English and Jupp reviews studies carried out with the objective of retrieving instantaneous wind vectors from existing passive microwave observations (SSM/I and polarimetric Windsat instrument launched in 2002 mainly). However, a majority of contributions were written in the framework of near future spatial missions based on instruments operating at L-band (SMOS, HYDROS and Aquarius):

- over the ocean, two missions will attempt to globally monitor sea surface salinity: the ESA mission SMOS (soil moisture and ocean salinity) that will provide dual-polarisation and multi-angular observations, and the NASA mission Aquarius. These missions should be launched, respectively, in 2007 and 2009;
- over the land surfaces, two missions are designed to monitor surface soil moisture: the ESA SMOS and NASA HYDROS missions.

Considering the SMOS system, the possibility to profit by the multi-angular and dual-polarisation capability of the instrument is a new challenge for the retrieval process. This aspect leads to new developments investigating the sensitivity of the model parameters to the configuration system in terms of incidence angle, polarisation and frequency (studies by Wigneron *et al.*, Van de Griend, and Ferrazzoli and Guerriero over vegetated surfaces). Also the retrievals based on spaceborne observations will require simple models that can be used on a global scale. This was the main motivation for the contributions investigating new parametrisations based on simple analytical equations to account for relief, roughness, effective soil temperature and vegetation effects.

Finally, recent developments in the modelling of snow over the land surfaces, and of ice over the ocean are considered in Sections 4.15 and 4.16, respectively. Improvement in the modelling of these effects is a major issue in the understanding of the spatial observations made by current instruments (SSM/I, AMSU, etc.). Also, these studies are essential to investigate which information could be potentially obtained from the near future instruments (even though the instruments were not developed specifically for monitoring these covers).

4.2 Comparison of emission models for covered surfaces

Christian Mätzler

Different approaches used to model microwave emission from surfaces covered by a layer of incoherent volume scatterers are compared, including a zero-order, a single-scattering and a multiple-scattering model. For optically thin layers, all models can be tuned to converge to the same limit. However, differences occur in optically thick situations. If scattering is dominant, the zero-order model is generally closer to the multiple-scattering than the single-scattering solution. The opposite is true if absorption dominates. Furthermore, applications of one-dimensional models set stringent limits on the horizontal inhomogeneity of the cover.

4.2.1 Introduction

Microwave emission of the earth's surface is often dominated or at least influenced by surface covers, consisting of living and dead vegetation with different structure, loose soil, sand or dust, snow, foam and clouds, all above a solid ground or water surface. If the cover is sufficiently deep and random, its effect can be described by incoherent scatterers, which reflect sky radiation, emit thermal radiation and transmit radiation from the underlying surface. It is instructive to compare different approaches with model microwave emission of such systems (Figure 4.1). Note that coherent approaches are needed if the cover thickness is of the order of or less than a wavelength; see for instance the air-to-soil transition model (Section 4.7).

The purpose of this chapter is to use analytical expressions for the comparison of thermal microwave radiation as functions of temperature, layer thickness, absorption and scattering coefficient. For an observer from above, a growing cover is an increased screening of the underlying surface by scattering, absorption and emission. Here the discussion is limited to soft covers by which is meant that the transition from the atmosphere to the canopy is gentle enough to avoid reflections. Reflection is accounted for at the bottom surface, and volume interactions are allowed inside the cover. For simplicity polarisation effects have been neglected, and the angular dependence has been treated in a simplified way. The models can be extended

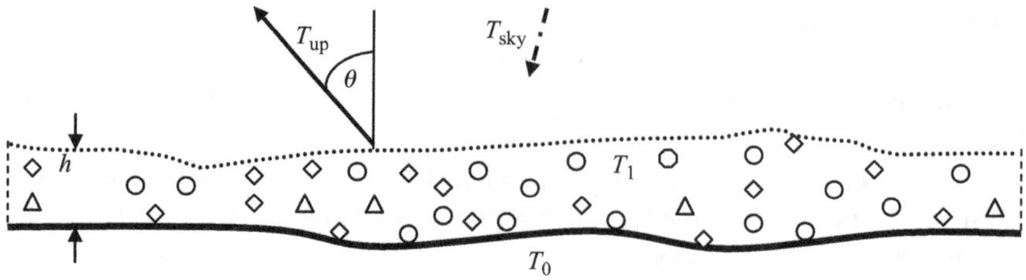

Figure 4.1 *Geometry of thermal radiation from the earth's surface at temperature T_0 covered by a soft canopy at temperature T_1 of height h, being illuminated by T_{sky}*

to include better angular dependence, polarisation, hard covers and multi-layer systems.

For radiative transfer of sun light, the described situation has been called the planet problem, treatments of which were given by Van De Hulst (1948), Chandrasekhar (1960), and by modern textbooks such as Thomas and Stamnes (1999). Through Kirchhoff's Law the methods can be applied to thermal radiation, leading to models of the type to be discussed in Sections 4.2.3–4.2.5. Although the models are one dimensional, nature is usually not. The applicability of the models to real situations with lateral variations will be studied in Section 4.2.6.

4.2.2 Zero-order scattering model

The zero-order scattering model of Kirdiashev *et al.* (1979), also called the 'ω–τ model' (e.g. Wigneron *et al.*, 1995) has often been used in passive microwave radiative transfer mainly to describe vegetation-covered soil. Zero order means that the scattering phase function is set to zero in the radiative-transfer equation (Chanzy and Wigneron, 2000). This model gives a lower bound of the emissivity. Energy conservation is recovered by a proper sky term fulfilling Kirchhoff's Law. The upwelling brightness temperature T_{up} at an angle θ from zenith, including single interactions between canopy and bottom surface, is given by

$$T_{up}(\theta) = (1 - s_0)L^* T_0 + (1 + L^* s_0)(1 - L^*)(1 - \omega^*)T_1 \\ + [\omega^*(1 - L^*)(1 + s_0 L^*) + s_0 L^{*2}]T_{sky} \tag{4.1}$$

where T_0 and T_1 are soil and vegetation temperature, respectively, T_{sky} is the downwelling sky brightness temperature, s_0 is the reflectivity at the bottom surface and ω^* is the effective single scattering albedo

$$\omega^* = \gamma_s^* / \gamma_e^*; \qquad \gamma_e{}^* = \gamma_a + \gamma_s^* \tag{4.2}$$

where γ_a is the absorption, γ_s the scattering and γ_e the extinction coefficient, respectively (effective values denoted by *), and L^* is the effective cover attenuation

$$L^* = \exp(-\tau^*/\mu); \qquad \tau^* g = g\gamma_e^* h \tag{4.3}$$

$\mu = \cos\theta$ and θ is the propagation angle of the radiation within the layer, and τ^* is the effective zenith opacity. Equation (4.1) corresponds to Equation (3) of Chanzy and Wigneron (2000), but with the additional contribution by sky radiation. The use of effective values (with *) allows a certain correction of the simplification made by not including scattered radiation from areas other than the sky. For this purpose the contribution to extinction from scattering in an appropriate forward hemisphere has to be removed (similar to the Delta–Eddington approximation of Joseph *et al.*, 1976: see also Section 3.7.1).

4.2.2.1 Relationship to other models

1. With regard to the HUT snow emission model (Pulliainen *et al.*, 1999), it can be shown that for negligible reflection at the air–snow interface (s_1=0), this model is equivalent to the one of Kirdiashev *et al.* (1979) if the effective extinction coefficient is represented by

$$\gamma_e^* = \gamma_e - q(\gamma_e - \gamma_a); \qquad 0 < q < 1 \quad (q = 0.5 \text{ for isotropic scattering}) \tag{4.4}$$

   where q is the fraction of forward-scattered radiation (Ishimaru, 1978). In the HUT model q is regarded as a fitting parameter. Basically, q is a function of frequency and snow grain size and can be estimated from theoretical considerations or by empirical data analysis. The definition of γ_e^*, given by (4.4), is equivalent to the one of Chanzy and Wigneron (2000). Finally, it is noted that the HUT model includes polarisation and non-zero values of s_1.
2. The emission model of Ulaby and Stiles (1980), see also (1)–(5) in Johnsen and Heygster (2000), again for negligible reflection at the cover surface ($s_1 = 0$), is similar to the model of Kirdiashev *et al.* (1979); however, the factors $(1+s_0L^*)$ in Equation (4.1) are replaced by 1, and the term s_0L^{*2} is ignored. Thus reflections of downwelling radiation at the bottom surface are ignored.

4.2.3 Single-isotropic-scattering model

Single-isotropic scattering is considered in a layer above a specular surface with reflectivity s_0. This situation was described by Chandrasekhar (1960) and, in a simplified way, also by Tsang *et al.* (1982). To develop an expression for a partly reflecting bottom surface, two extreme cases are considered, a surface consisting of a black body, and a fully reflective surface. The general case follows by a linear combination of the situations shown in Figures 4.2 and 4.3.

4.2.3.1 Non-reflective bottom surface

First, since $s_0 = 0$, the upwelling brightness temperature is

$$T_{up,0} = t_0T_0 + (1 - r_0 - t_0)T_1 + r_0T_{sky} \tag{4.5}$$

The one-way layer transmissivity t_0 and the layer reflectivity r_0 are given by the special function F_2 related to exponential integrals, see Chandrasekhar (1960) and

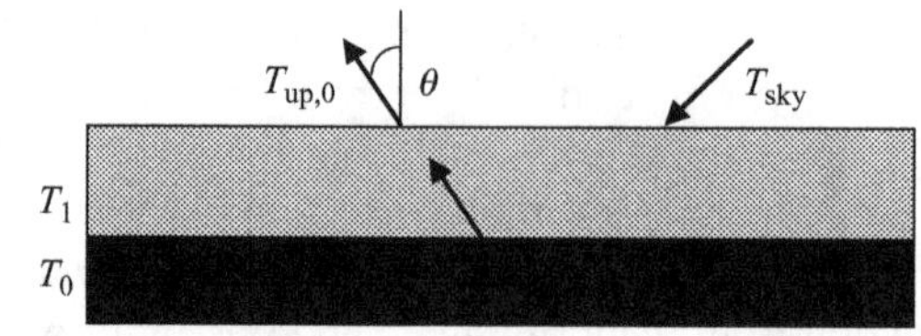

Figure 4.2 Radiation from a scattering layer of optical thickness τ, temperature T_1 above a non-reflective (black) bottom surface at temperature T_0

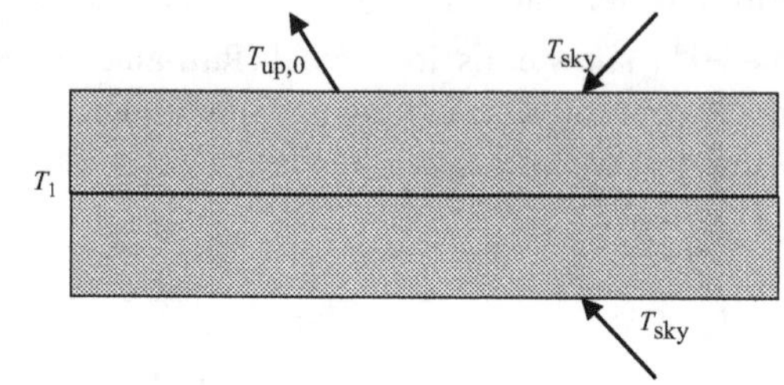

Figure 4.3 Emission from a scattering layer at temperature T_1 above a totally reflective bottom surface. The layer is mirrored and appears twice with illumination by T_{sky} from above and from below

Tsang *et al.* (1982)

$$t_0 = \left[1 + \frac{\omega F_2(\tau, \mu)}{2\mu}\right] L; \qquad L = e^{-\tau/\mu}; \qquad r_0 = \frac{\omega F_2(\tau, -\mu)}{2\mu} \tag{4.6}$$

where $\omega = \gamma_s/\gamma_e$ is the single-scattering albedo, and $\tau = \gamma_e h$ is the opacity of the layer. The two terms in t_0 are transmissivities of the direct beam and of the scattered radiation, respectively, whereas the reflectivity r_0 is due to single scattering of sky radiation (Figure 4.2).

4.2.3.2 Totally reflective bottom surface

If $s_0 = 1$, the upwelling brightness temperature $T_{up,1}$ is given by the radiation from the layer and from its mirror image, illuminated by T_{sky} from both sides (Figure 4.3)

$$T_{up,1} = (1 - r_1 - t_1)T_1 + (r_1 + t_1)T_{sky} \tag{4.7}$$

where r_1 and t_1 are now the reflectivity and transmissivity for a layer of the double thickness

$$t_1 = \left[1 + \frac{\omega F_2(2\tau, \mu)}{2\mu}\right] L^2; \qquad r_1 = \frac{\omega F_2(2\tau, -\mu)}{2\mu} \tag{4.8}$$

4.2.3.3 Partly reflective bottom surface

In the general case of a partly reflective bottom surface, the upwelling radiation is a weighted sum of $T_{up,0}$ and $T_{up,1}$:

$$T_{up} = (1 - s_0)T_{up,0} + s_0 T_{up,1} \tag{4.9}$$

4.2.3.4 Optically thick layer

For an optically thick cover ($\tau \to \infty$), there is zero transmission and the upwelling radiation becomes

$$T_{up} = (1 - R_\infty)T_1 + R_\infty T_{sky} \tag{4.10}$$

where the single-scattering reflectivity R_∞ of the optically thick half space simplifies to

$$R_\infty = 0.5\omega[1 - \mu \cdot \log(1 + 1/\mu)] \tag{4.11}$$

and where log is the natural logarithm. Note that, due to the neglect of multiple scattering, the maximum value of R_∞ is 0.5; this value is obtained for $\omega = 1$, $\mu \to 0$.

4.2.3.5 Relationship to other single-scattering models

Slighty different expressions for the reflectivity and transmissivity are obtained if the surface is a Lambert scatterer. This situation was described by Chandrasekhar (1960) as the planetary problem in section 72, and also by Van De Hulst (1948). The models include corrections up to second-order scattering. A separation of the radiation field into orders of scattering is described in Chapter 7 of Thomas and Stamnes (1999).

4.2.4 Multiple-scattering model in two-stream approach

There are two different approaches of two-flux models: a simpler one only considers up- and downwelling fluxes in the scattering layer; a more sophisticated one distinguishes between directed and diffuse beams (therefore also called the four-flux model). Meador and Weaver (1980) showed that different formulations of the models are principally the same, and thus lead to the same type of solution. Here we consider the approach of the snow emission model MEMLS (Wiesmann and Mätzler, 1999). It can be adapted to the situation of Figure 4.1, by considering a single layer ($n = 1$) on top of a soil surface and by setting s_1=0 as in the HUT model. For a canopy with low density, the effective refractive index can be approximated by 1; then there is no refraction and no reflection at the cover surface. Furthermore, there is no trapped radiation within the cover because total reflection is absent. With these simplifications the observable brightness temperature T_{up} is given by

$$T_{up} = t_2(1 - s_0)T_0 + r_2 T_{sky} + (1 - r_2 - t_2)T_1 + t_2 s_0 \times \frac{r_2\,(1 - s_0)\,T_0 + t_2 T_{sky} + (1 - r_2 - t_2)T_1}{1 - r_2 s_0} \tag{4.12}$$

where the reflectivity r_2 and transmissivity t_2 of the scattering layer appear as symmetrical functions of two very different quantities (r_∞ and t_F)

$$r_2 = r_\infty(1 - t_F^2)(1 - r_\infty^2 t_F^2)^{-1} \tag{4.13}$$

$$t_2 = t_F(1 - r_\infty^2)(1 - r_\infty^2 t_F^2)^{-1} \tag{4.14}$$

For a layer of height h, the transmissivity t_F of a flux through the slab is

$$t_F = \exp(-\gamma_F h/\mu) \tag{4.15}$$

where again $\mu = \cos\theta$, r_∞ is the reflectivity at infinite thickness

$$r_\infty = \gamma_b(\gamma_a + \gamma_b + \gamma_F)^{-1} \tag{4.16}$$

and the flux-attenuation coefficient γ_F (eigenvalue of the two-stream model) is

$$\gamma_F = \sqrt{\gamma_a(\gamma_a + 2\gamma_b)} \tag{4.17}$$

where γ_a is again the absorption coefficient and γ_b is coefficient for scattering in the backward hemisphere, i.e. it corresponds to γ_s^*. The first term of Equation (4.12) represents the direct contribution from the subsurface, the second term is the directly reflected sky component, the third term is the cover emission and the last one is the interaction term between cover and subsurface. In case of a non-vanishing cover-surface reflectivity, the appropriate formulae (7) and (10) of Wiesmann and Mätzler (1999) with $s_1 > 0$ should be used instead of Equation (4.12) above, and for a multi-layer medium, the model becomes the full MEMLS, including vertical and horizontal polarisation.

4.2.5 Comparison

4.2.5.1 General remarks

All models describe T_{up} as linear combinations of the three temperatures T_0, T_1 and T_{sky}

$$T_{up} = tT_0 + eT_1 + rT_{sky} = e_{sys}T + rT_{sky} \tag{4.18}$$

where t is the effective transmissivity for radiation from the underlying surface, e the layer emissivity, $e_{sys} = e + t$ and r are the emissivity and reflectivity of the system consisting of the bottom surface and cover and T is the effective temperature of this system, i.e. a weighted mean value:

$$T = \frac{tT_0 + eT_1}{e + t} \tag{4.19}$$

Often the temperature of the bottom surface and of the layer are the same; then $T = T_0 = T_1$. The parameters, e, r, t and e_{sys}, are related by Kirchhoff's Law, as described in Chapter 1:

$$e + r + t = 1, \quad \text{or} \quad e_{sys} = e + t = 1 - r \tag{4.20}$$

Table 4.1 Model parameters, describing microwave radiation according to (4.18) of a layer on top of a surface with reflectivity s_0. For isotropic scattering $\gamma_s^* = \gamma_b = 0.5\gamma_s$

Parameter	Zero-order scattering	Single-isotropic scattering	Multiple-scattering
Absorption coefficient	γ_a	γ_a	γ_a
Scattering coefficient	γ_s^*	γ_s	γ_b
t	$L^*(1-s_0)$	$t_0(1-s_0)$	$t_2\dfrac{1-s_0}{1-r_2s_0}$
e	$(1-\omega^*)(1-L^*)$ $\times(1+s_0L^*)$	$(1-s_0)(1-r_0-t_0)$ $+s_0(1-r_1-t_1)$	$(1-r_2-t_2)$ $\times\dfrac{1+s_0(t_2-r_2)}{1-r_2s_0}$
r	$\omega^*(1-L^*)$ $\times(1+s_0L^*)+s_0L^{*2}$	$(1-s_0)r_0$ $+s_0(r_1+t_1)$	$\dfrac{r_2+s_0(t_2^2-r_2^2)}{1-r_2s_0}$

Expressions for e, r and t of the three models are shown in Table 4.1 together with the corresponding absorption and scattering coefficients. Note that $\gamma_s^* = \gamma_b$, and for isotropic scattering $q = 0.5$ in (2.4), thus $\gamma_s^* = \gamma_b = 0.5\gamma_s$. Furthermore Kirchhoff's Law (4.20) is satisfied in all cases.

Apart from the incidence angle θ (with $\mu = \cos\theta$) and the three temperatures T_0, T_1 and T_{sky}, there are three parameters in each model to determine e, r and t, namely ω^*, τ^* and s_0 in the zero-order model, ω, τ (determining r_0, t_0,r_1, t_1) and s_0 in the single-scattering model and r_2, t_2 and s_0 in the multiple-scattering model; these model parameters are derived from four physical parameters (absorption coefficient, scattering coefficient, h and s_0). Of the three output parameters e, r, t only two are independent, since their sum equals 1. Multiple interactions lead to products of s_0r_2 in Table 4.1 and to terms including r_∞^2 in (4.13) and (4.14).

The coefficient t is composed of the product of the cover transmissivity (L^*, t_0, t_2) with the emissivity $(1 - s_0)$ of the bottom surface. The additional factor arising for multiple scattering is due to interactions between cover and surface.

The coefficient r can best be understood for the two cases, (1) no surface scattering ($s_0 = 0$) and (2) no volume scattering. In case (1) we get the pure volume-scattering solution for $r \in (\omega^*, r_0 , r_2)$ and for case (2), we simply get (since $\gamma_s = 0$ and $t_1 \equiv t_0^2$) the two-way attenuated surface reflectivity: $r = s_0L^{*2} \equiv s_0t_0^2 \equiv s_0t_2^2 \equiv s_0 \cdot \exp(-2\gamma_a h/\mu)$.

This comparison has not been aimed at studies of the angular dependence. Of course, part of it is given by the angular variation of s_0. The treatment of the cover has been limited to isotropic volume scattering. Furthermore, the angular behaviour

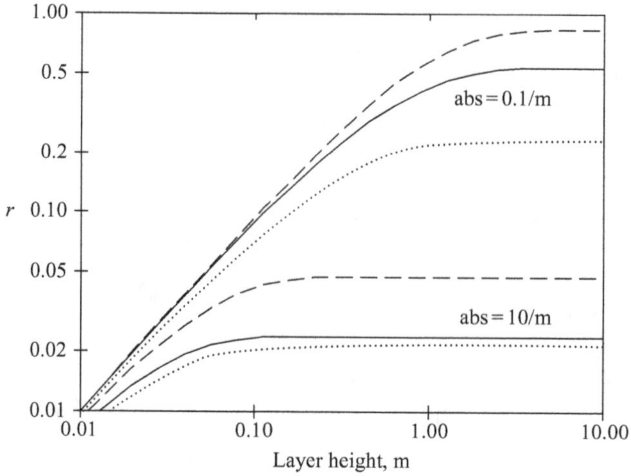

Figure 4.4 *Reflectivity r for $s_0 = 0$, $\theta = 60°$, $\gamma_s = 1$/m versus h for two values of the absorption coefficient ($\gamma_a = 0.1$/m, 10/m) in the zero-order (dashed), single- (dotted) and multiple-scattering (solid line) model, respectively*

has been simplified in the two-flux model by assuming that the same $\mu = \cos\theta$ applies to represent the cosines of the observation direction (direct beam) and of the effective angle θ_{eff} (called mean inclination by Thomas and Stamnes, 1999) applicable to the radiation of the up- and downwelling fluxes. Accurate results are obtained if the two angles are equal. This happens in optically thin layers, and also for all optical depths if the observation angle is near θ_{eff}. Fortunately for the conically scanning radiometers often used on satellites, θ is constant and close to θ_{eff}. Improved angular dependence is obtained by including a direct beam to the two-flux model (Meador and Weaver, 1980). In the discussion to follow, we will concentrate on the behaviour of the system reflectivity $r = 1 - e_{sys}$. This is the key parameter of interest here.

4.2.5.2 Optically thin cover with isotropic scattering

Although not quite apparent, it can be shown that for an optically thin cover all models converge to the same result with $L^* = t_0 = t_2 \Rightarrow 1$, and for vanishing surface reflection ($s_0 = 0$) and for isotropic scattering we find: $r \Rightarrow 0.5\gamma_s h/\mu = \gamma_s^* h/\mu = \gamma_b h/\mu$. The reflectivity linearly increases with h. This is shown in Figure 4.4 ($s_0 = 0$, $\gamma_s^* = \gamma_b = 1$/m) for small h. Above a certain thickness, the models start to differ with the smallest r for single scattering, and largest r for the zero-order model. The screening effect of the soil reflectivity by the canopy is shown in the semi-log representation of Figure 4.5 for $s_0 = 0.3$. The values of r are quite similar in all models up to a thickness of about 3 cm in the high absorption case and up to about 20 cm for low absorption. The multiple-scattering solution is between the two other models. If scattering is dominant ($\gamma_a = 0.1$/m), the zero-order model is closer to the

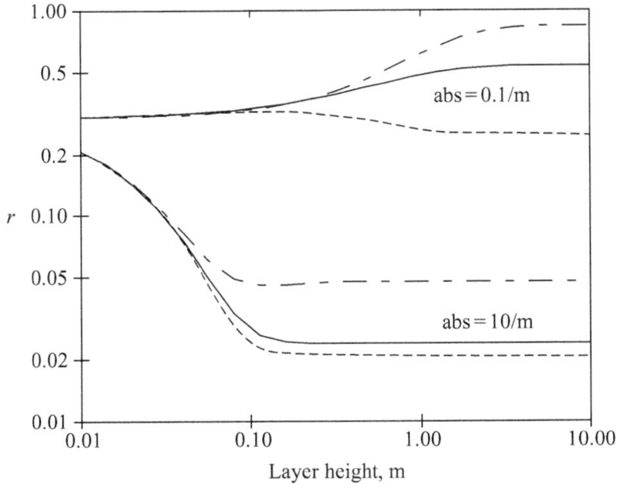

Figure 4.5 Same as Figure 4.4, but for $s_0 = 0.3$

multiple-scattering solution than the single-scattering model. The opposite is true if absorption dominates ($\gamma_a = 10$/m).

4.2.5.3 Optically thick cover with isotropic scattering

Differences between the models are especially apparent if a deep cover is considered, more exactly if $t = 0$. Then the system reflectivity r of Table 4.1 reduces to ω^* in the zero-order model, to R_∞ in the single-scattering approximation and to r_∞ in the two-flux model, respectively. That there is a significant difference between ω^*, R_∞ and r_∞ is apparent from Figure 4.6 showing r versus the ratio $x = \gamma_b/\gamma_a = \gamma_s^*/\gamma_a = 0.5\gamma_s/\gamma_a$. This quantity gives the number of effective scattering events before the wave is absorbed. Note that since $t = 0$, this ratio is the only variable of r for the zero-order and the two-flux model. In the case of the single-scattering model, there is also a dependence on θ as expressed in Figure 4.6 by the two values $\mu = 0.5$ and 1, respectively.

For small x the difference in r is a factor of 2 between zero-order and multiple scattering. The discrepancy indicates a significant deficiency of the zero-order model. Nevertheless, in both models r converges to unity for $x \Rightarrow \infty$. However, the convergence is different. In the zero-order model we have a rapid convergence

$$e = 1 - r = 1 - \omega^* = 1 - \frac{x}{x+1} = \frac{1}{x+1} \Rightarrow x^{-1}; \qquad x >> 1 \qquad (4.21)$$

whereas in the multiple-scattering model the convergence is random-walk like with

$$e = 1 - r = 1 - r_\infty = 1 - \frac{x}{x + 1 + \sqrt{1 + 2x}} \Rightarrow \sqrt{\frac{2}{x}}; \qquad x >> 1 \qquad (4.22)$$

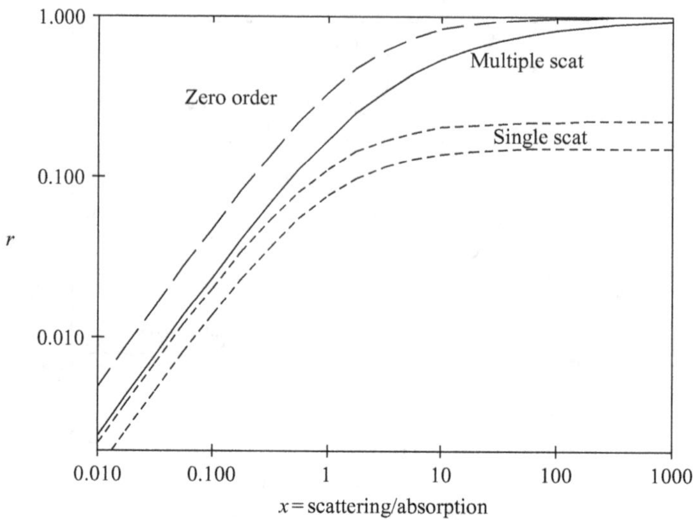

Figure 4.6 Reflectivity r of a half space ($t = 0$) versus ratio, $x = 0.5\gamma_s/\gamma_a$, in zero-order (dashed), single-scattering (dotted, upper curve for $\mu = 0.5$, lower curve for $\mu = 1$) and multiple-scattering (solid). For isotropic scattering assumed here, we have $\gamma_s^* = \gamma_b = 0.5\gamma_s$

Equation (4.22) is closer to reality. The behaviour has been widely discussed in the literature, e.g. by Bohren and Barkstrom (1974), Bohren (1987), and it can be compared with the exact theory of Chandrasekhar (1960) given by

$$e = \sqrt{1-\omega} \cdot H(\mu, \omega) \tag{4.23}$$

where $H(\mu, \omega)$ is a monotonous function as shown in Figure 4.7. Numerical values were presented in table 11 of Chandrasekhar (1960). Since ω is related to x by

$$1 - \omega = \frac{1}{1+2x} \tag{4.24}$$

the convergence $1 - \omega \Rightarrow 0$ is linked with $x \Rightarrow \infty$. Then for the average value $\mu = 0.5$, H is very close to 2. Inserting this value in (4.23) we find that the exact model converges to the limiting emissivity of Equation (4.22). Higher emissivities result for steeper incidence ($\mu > 0.5$), and vice versa. As a final remark it is noted that according to (4.23) isotropic scattering does not show Lambertian behaviour which would imply an angle-independent emissivity.

4.2.6 *Effects of lateral inhomogeneity*

The availability of analytic expressions to relate the physical parameters of the scene to the radiative parameters allows us to perform sensitivity studies by analytical means. This is a powerful tool in remote sensing and in earth-system science. An example is the study of the influence of lateral inhomogeneity. Kokhanovsky (2003)

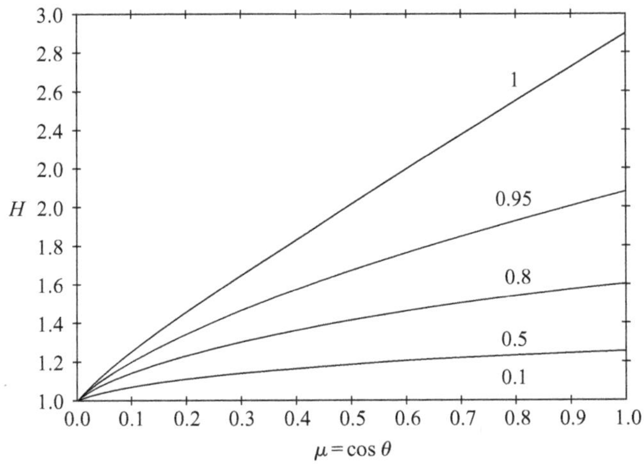

Figure 4.7 Chandrasekhar's H-function for isotropic scattering versus μ and for values of ω ranging from 0.1 to 1

used the method to investigate the influence of cloud inhomogeneity on the radiative characteristics of the atmosphere. For this purpose the models should be three dimensional. Under certain conditions the one-dimensional radiative transfer models can be expanded to three dimensions, if a smoothness condition for the lateral inhomogeneity is fulfilled for the parameter vector $\mathbf{P} = (h, \gamma_a, \gamma_b, s_0)$ over the lateral distance, $L \cong h \cdot \tan\theta + \delta$, over which a ray disperses when traversing the layer. Here δ is the lateral diffusion length due to scattering which may be estimated from the scattering coefficient or from the penetration depth. One-dimensional models can be applied to three-dimensional canopies if locally the canopy appears as a plane-parallel medium with small lateral variation. This smoothness condition can be expressed by

$$L \left| \frac{\partial P_j}{\partial l} \right| << P_j \tag{4.25}$$

where P_j is a component of $\mathbf{P}$, and l is the lateral displacement. If (4.25) is fulfilled, Equation (4.18) is applicable to all surface elements with extent L, and by averaging the results of T_{up} over a given antenna footprint of size A, the pixel brightness temperature at the bottom of the atmosphere

$$\overline{T_{up}} = \frac{1}{A} \int_A T_{up} \, dA \tag{4.26}$$

is obtained. The exponential attenuation introduces non-linearities of T_{up} with regard to height, biomass, water content, etc. Linearity between $\overline{T_{up}}$ and pixel-averaged P_j only results for optically thin canopies. Very often the opacity is large; then it may not be possible to retrieve average surface parameters, even if the smoothness condition (4.25) is fulfilled. Therefore we need a second condition to test the homogeneity within a given footprint area $A = a^2$, where a is the lateral size at the surface. The

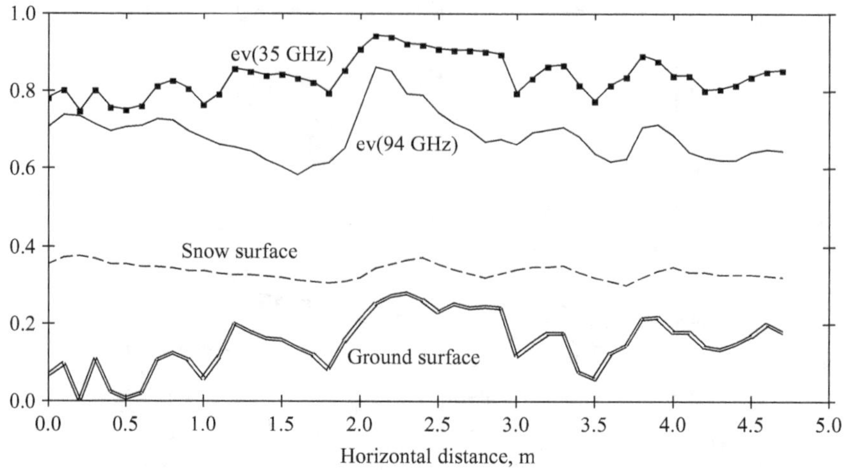

Figure 4.8 *System emissivity at vertical polarisation, $\theta = 50°$, 35 GHz (pointed solid line) and 94 GHz (solid line) along a horizontal distance of a 4-layer snowcover in the tundra of Alaska. The surface heights of ground (double) and snowpack (dashed line) are shown (in m). Snow data from Sturm and Holmgren (1994), microwave simulation with MEMLS*

non-linearity can be handled by the observations if

$$a \left| \frac{\partial P_j}{\partial l} \right| << P_j \tag{4.27}$$

An example of an inhomogeneous cover with thicknesses ranging from 10 to 40 cm is the tundra snowcover presented by the horizontal profile over several meters as seen in Figure 4.8. The snow data, published by Sturm and Holmgren (1994), showed four different snow layers. The system emissivity e_{sys} was computed with MEMLS. The cover is optically thin at 35 GHz; thus the results are correlated with total layer thickness, here dominated by the ground height (see Figure 4.9 (left)). On the other hand, the snow is optically thick at 94 GHz, thus there is no correlation between emissivity and thickness; however a correlation with the height of the snow surface can be seen (see Figure 4.9 (right)). This is due to a modulation by hoar layers present within the snowpack (Sturm and Holmgren, 1994). Let us test the applicability of the one-dimensional model for the parameter, snow depth $h(l)$, assuming $L \cong h$. Then the first smoothness condition (4.25) requires that $\partial h/\partial l << 1$. As can be seen from Figure 4.8, typical derivatives of snow depth are 0.1 to 0.2 which means that the requirement is just fulfilled. Indeed, snowpacks are good examples of layered media. However, in order to cope with the second smoothness condition, a must be of the order of 1 m or less. This is in large contrast to the typical resolution of radiometers on board of satellites.

More delicate is the situation for vegetation canopies. Indeed, trees and most plants show predominant vertical structure with large inhomogeneity on the submetre scale

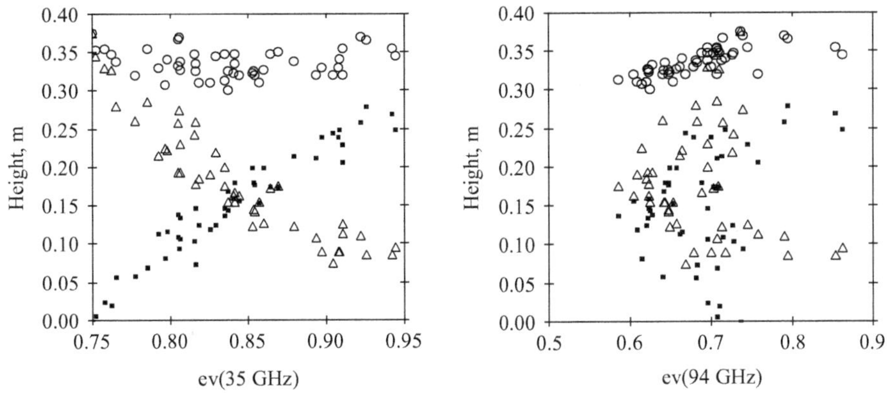

Figure 4.9 *Correlation between system emissivity at vertical polarisation, $\theta = 50°$, 35 GHz (left) and 94 GHz (right) and height of the ground surface (points), height of the snow surface* (○) *and snowpack thickness* (△)

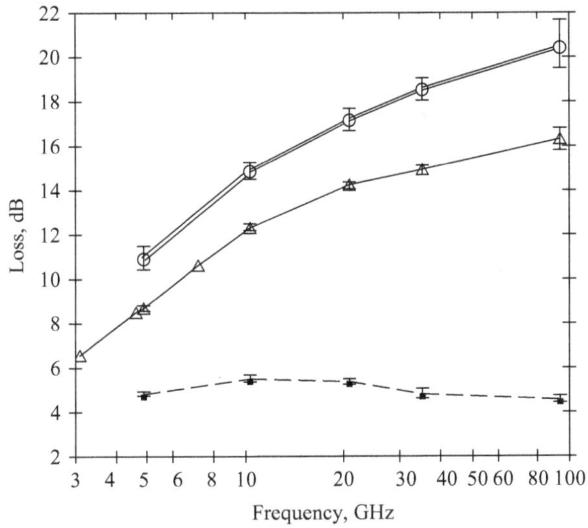

Figure 4.10 *Transmission loss (dB) through the beech with leaves in 1987* (△) *and 1988* (○) *and without leaves (points) during winter (from Mätzler, 1994). Error bars show the standard variation over time*

whereas the height can reach tens of metres. In radiometer experiments of a beech tree, the effects of the heterogeneity were studied by Mätzler (1994). Downwelling microwave radiation below the tree was observed at frequencies from 3 to 94 GHz with antenna beam widths of 10°, leading to beam diameters from 0.5 to about 5 m within the canopy. Figure 4.10 shows the spectra of transmission loss through the foliated (two summer seasons) and defoliated tree (winter in between). An objective analysis showed that the mean opacity of the leaves was about 50 per cent larger than the value for a homogeneous vegetation layer with the same transmission loss. Therefore the inherent heterogeneity questions the applicability of one-dimensional radiative-transfer models to forest canopies.

4.2.7 Conclusions

Zero-order, single- and multiple-scattering models for microwave emission from covered surfaces were presented and compared. Apart from three temperatures, the system is characterised by three independent model parameters to be derived from four physical parameters (cover height, absorption coefficient, scattering coefficient and soil-surface reflectivity). The conclusions are:

- all models consider the same situation, and all can be represented by Equation (4.18) where only the values of e, r, t are model dependent;
- physically all three parameters, (e, r, t), should cover the full range from 0 to 1, with the condition $e + r + t = 1$. This condition is fulfilled in all models, but in the single-scattering model r is limited to $r \leq 0.5$, showing that single scattering cannot describe optically thick situations with significant volume scattering;
- differences between the models arise for appreciable optical depth due to volume scattering. Therefore any physical interpretation in terms of scattering parameters should be done carefully. This is important if radiometric data are used to derive scattering and absorption coefficients, e.g. single-scattering albedo and optical depth;
- multiple-scattering models are physically more correct; they should be given preference in physical interpretation if the opacity is not small. Noticeable differences occur even for small values of the ratio, scattering to absorption coefficient. Furthermore, since forward scattering is ineffective in reflecting radiation, suitable modifications of the scattering and extinction coefficients are needed;
- due to the neglect of the scattering source function, the zero-order model generally leads to a smaller emissivity, i.e. larger reflectivity, than the multiple-scattering model. The opposite is true for the single-scattering model;
- if scattering is dominant, the zero-order model is generally closer to the multiple-scattering solution than the single-scattering solution. The opposite is true if absorption dominates;
- a serious problem for the application of one-dimensional radiative-transfer models is the inhomogeneity of natural covers. Two different smoothness conditions can be used to test the applicability.

4.3 Relief effects for microwave radiometry *Christian Mätzler*

4.3.1 Introduction

The signal of a microwave radiometer observing a land surface from space is composed of surface and atmospheric contributions, both of which depend on the relief.

First, the path through the atmosphere between the surface and the sensor depends on the altitude of the emitting surface, thus leading to a height-dependent atmospheric influence. It can be taken into account by atmospheric radiative-transfer models if the elevation of the surface and the atmospheric state are known. The effect is described in Section 4.3.2. Next, the variable topography of landscapes, consisting of peaks, slopes, ridges and valleys interact radiatively, not only with the atmosphere, but also with each other, leading to the tendency to enhance the effective emission. Deviations from the standard hemispheric emission of a horizontal surface occur by elevated horizons. This is a large-scale roughness effect. The interactions not only depend on topography and emissivity, but also on the bistatic scattering behaviour. The effects are discussed in Section 4.3.3. A simple analytical expression for the radiation enhancement in landscapes of Lambertian surfaces is expressed by an effective reflectivity (4.57) being smaller than the surface reflectivity. A realistic situation is illustrated by simulated data of southern Norway.

4.3.2 Flat horizon

The classical geometry of remote sensing of the terrestrial surface is a pair of half spaces separated by a horizontal surface. In this situation there is no shadow of any kind. Relief effects are determined by the dependence of the emitted radiation on surface altitude h. Blackbody radiation with a brightness temperature equal to the physical temperature T_0 is produced in the lower half space, part of which is transmitted, and thus emits T_e into the upper half space where it can be sensed by a radiometer. The p-polarised emitted brightness temperature T_{ep} of a half space with reflectivity r_p (or emissivity e_p) is given by

$$T_{ep} = e_p T_0 = (1 - r_p) T_0 \tag{4.28}$$

The upwelling radiation T_{up} just above the surface is the sum of the radiation emitted and the fraction of sky radiation reflected by the lower half space.

$$T_{up} = (1 - r_p) T_0 + r_p T_{sky,eff} \tag{4.29}$$

The reflectivity r_p is an integral of the bistatic scattering coefficient σ_0^{bi} over the upper hemisphere, and the effective sky radiation $T_{sky,eff}$ is a weighted integral of the incident radiation $T_{sky}(\mu', \phi', h)$ (Peake, 1959; Ulaby *et al.*, 1981: see also Equation (3.36))

$$r_p(\mu, \phi, h) = \frac{1}{4\pi\mu} \int \sigma_0^{bi}(\mu, \phi, \mu', \phi')\, d\Omega' \tag{4.30}$$

$$T_{sky,eff}(\mu, \phi, h) = \frac{1}{4\pi\mu \cdot r_p(\mu)} \int \sigma_0^{bi}(\mu, \phi, \mu', \phi') T_{sky}(\mu', \phi', h)\, d\Omega' \tag{4.31}$$

where $\mu' = \cos\theta'$, $\mu = \cos\theta$ are the cosines of the incidence angles before and after surface scattering and ϕ' and ϕ are the corresponding azimuth angles. The definition of σ_0^{bi} is identical with the S-function of (Chandrasekhar, 1960), but deviates from the bidirectional reflection distribution function (BRDF) often used at optical wavelengths.

The brightness temperature T_{bp} at satellite level is the attenuated upwelling radiation plus the radiation emitted by the atmosphere (atmospheric scattering being neglected):

$$T_{\mathrm{bp}} = tT_{\mathrm{up}} + (1-t)T_{\mathrm{a}} \tag{4.32}$$

where $t = t(h, \mu, \phi)$ is the atmospheric transmissivity at the surface height h in the observation direction (μ, ϕ), $(1-t)$ is the atmospheric emissivity and T_{a} is the temperature of the atmosphere. The above equation applies to the polarisation and direction corresponding to T_{up}. In the case of an inhomogeneous atmosphere T_{a} is a weighted air temperature $T_{\mathrm{a,up}}$ for upwelling radiation. For a plane-parallel atmosphere the azimuthal dependence disappears, and t is given by

$$t(h, \mu) = e^{-\tau_h/\mu} \tag{4.33}$$

where the zenith opacity τ_h is the vertical path integral of the absorption coefficient $\alpha(z)$ through the atmosphere, starting at the surface.

To be more specific, we assume the reflectivity r_{p} at polarisation $p = h, v$ in (4.29) to be composed of a specular, polarised component r_{sp} and of a diffuse, unpolarised component r_{d} with bistatic scattering coefficients $\sigma_{0\mathrm{sp}}^{\mathrm{bi}}$ and $\sigma_{\mathrm{d}}^{\mathrm{bi}}$, respectively.

$$\sigma_0^{\mathrm{bi}} = \sigma_{0\mathrm{sp}}^{\mathrm{bi}} + \sigma_{0\mathrm{d}}^{\mathrm{bi}} \tag{4.34}$$

The specular component

$$\sigma_{0\mathrm{sp}}^{\mathrm{bi}} = 4\pi\mu \cdot r_{\mathrm{sp}}(\mu)\delta(\mu - \mu')\delta(\phi - \phi') \tag{4.35}$$

contains delta functions in azimuth and μ to indicate that scattering is limited to the specular direction. The integration of (4.35) in Equation (4.30) simply gives $r_{\mathrm{sp}}(\mu)$. The diffuse component r_{d} is assumed to be described by Lambert scattering. For this component the diffuse bistatic scattering coefficient is independent of azimuth and given by

$$\sigma_{0\mathrm{d}}^{\mathrm{bi}} = 4r_{\mathrm{d}} \cdot \mu\mu' \tag{4.36}$$

With Equations (4.29) to (4.31) and (4.34) to (4.36), T_{up} can be written as

$$T_{\mathrm{up}} = e_{\mathrm{p}}(\mu)T_0 + r_{\mathrm{sp}}(\mu)T_{\mathrm{sky}}(\mu, h) + r_{\mathrm{d}}T_{\mathrm{d}} \tag{4.37}$$

where $r_{\mathrm{d}}T_{\mathrm{d}}$ is the diffusely scattered sky radiation with T_{d} given by

$$T_{\mathrm{d}} = \frac{1}{\pi}\int_0^{2\pi}\int_0^1 T_{\mathrm{sky}}(\mu', \phi', h)\mu' \, \mathrm{d}\mu' \, \mathrm{d}\phi' = 2\int_0^1 T_{\mathrm{sky}}(\mu', h)\mu' \, \mathrm{d}\mu' \tag{4.38}$$

and the integral is taken over the sky hemisphere. As a result of Lambert scattering T_{d} only depends on h or on the atmospheric opacity τ_h at the surface. For a plane-parallel, isothermal atmosphere at temperature T_{a} with a cosmic background T_{c} we have

$$T_{\mathrm{sky}}(\mu, h) = T_c e^{-\tau_h/\mu} + T_{\mathrm{a}}(1 - e^{-\tau_h/\mu}) \tag{4.39}$$

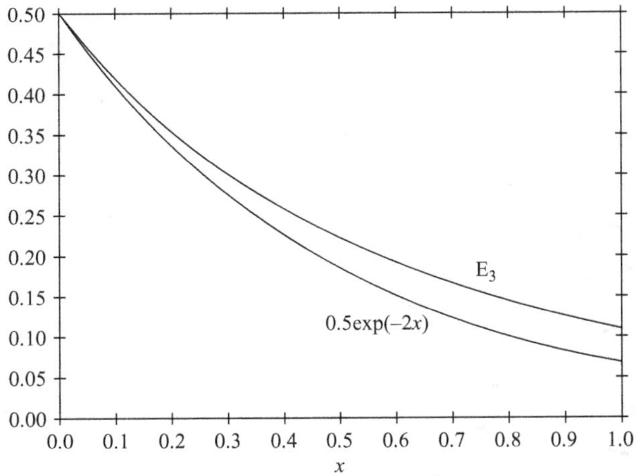

Figure 4.11 Comparison of $E_3(x)$ with the exponential function $e^{-2x}/2$

Equation (4.39) can be used also for a non-isothermal atmosphere if T_a is an effective air temperature $T_{a,down}$ for downwelling radiation (see, e.g. Mätzler, 1992; Ingold *etal.*, 1998). Integration of (4.38), using (4.39) gives

$$T_d = T_a - (T_a - T_c)t_d \tag{4.40}$$

where the diffuse atmospheric transmissivity is given by

$$t_d = 2E_3(\tau_h) = e^{-\tau_h/\mu_d} \tag{4.41}$$

Here E_3 is the exponential integral of order 3 (Chandrasekhar, 1960; Abramowitz and Stegun, 1974), and the last expression in (4.41) in accordance with (4.33) defines the cosine of an effective incidence angle $\mu_d = \cos\theta_d$ of sky radiation for Lambert scattering. For small arguments $E_3(x) \cong 0.5e^{-2x}$, but for larger x, E_3 decreases more slowly than the exponential function as shown in Figure 4.11. Therefore, in the case of an optically thin atmosphere ($\tau_h < 0.2$), we have $\mu_d \cong 0.5$ which means $\theta_d \cong 60°$, and for larger τ_h, θ_d slowly decreases, e.g. to 49° for $\tau_h = 1$ (Mätzler, 1987). Now we get

$$T_{up} = T_0 e_p(\mu) + T_a r_p(\mu) - (T_a - T_c)[r_{sp}(\mu)t(h,\mu) + r_d t_d] \tag{4.42}$$

A simplification occurs if $\mu = \mu_d$, i.e. ($\theta = \theta_d$); then $t = t_d$, and the bracket expression in (4.42) is simply $r_p \cdot t_d$. Since θ_d is often between 50 and 60°, and since this is also the case for the incidence angle θ of sensors with conical scanning (SMMR, SSM/I, AMSR), the simplified expression has been used frequently. The difference to (4.42) is negligible if at least one of the following conditions applies (plane-parallel atmosphere):

- the atmosphere is sufficiently transparent ($t_d \cong 1$),
- $\theta \cong \theta_d$,
- $r_d \cong 0$.

Figure 4.12 Landscape of sandstones with surface facets oriented in different directions. Photo: Courtesy of Bernd Nicolaisen

However, for vertically scanning sensors such as AMSU, $\cos\theta$ changes from 0.5 to 1 and thus clearly requires the distinction between diffuse and specular reflection.

Inserting T_{up} in Equation (4.32) leads to the height-dependent radiation T_{bp} at satellite level. For the antenna temperature, see Equation (4.51) below.

4.3.3 *Terrain with tilted surfaces*

In addition to the altitude effects on atmospheric radiation, there are effects due to tilted surfaces. On the one hand, the local incidence angle of a tilted surface depends on the orientation of the surface with respect to the view direction of the sensor, and on the other hand, the tilted surfaces imply variable and elevated horizons as shown in Figure 4.12. Below these horizons the incident sky radiation is replaced by radiation from the elevated landscape; this is a shadowing effect. If the scale of the relief is sufficiently large with respect to the sensing wavelength, both effects can be described by geometrical optics. A facet model is indicated, see e.g. Schanda (1986). Here we allow, in accordance to Section 4.3.2, that the facets have a partly specular and a partly Lambertian component.

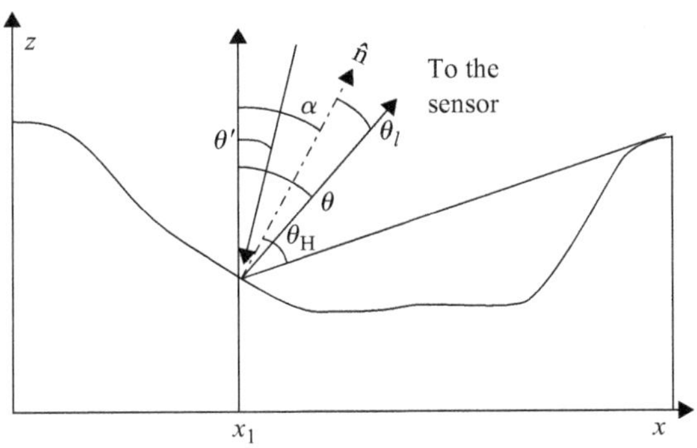

Figure 4.13 *Horizontal profile through a landscape showing a facet at position x_1 with local and global angles of incidence and surface normal tilted by α, and the local angle θ_H of incidence from the horizon*

4.3.3.1 Local and global angles

Reflection at and emission from a surface facet in direction $\hat{\mathbf{k}}$ with cartesian (x, y, z) components $(\sin\theta, 0, \cos\theta)$ towards the radiometer can be treated as in Section 4.3.2. Here however, the surface normal $\hat{\mathbf{n}}$ used to define the plane of incidence deviates from the vertical $\hat{\mathbf{z}}$ direction by a tilt angle α, oriented by an azimuth angle β with respect to the global plane of incidence (Figures 4.12 and 4.13). The transformation from the global to the local plane affects both the scattering geometry and the polarisation. The local angle of incidence θ_l is given by

$$\cos\theta_l = \hat{\mathbf{n}} \cdot \hat{\mathbf{k}} = \cos\theta\cos\alpha + \sin\theta\sin\alpha\cos\beta \tag{4.43}$$

This quantity is positive for all visible surface facets. Furthermore the orientation of the linear polarisation is rotated by an angle φ, given by

$$\sin\varphi = \sin\beta\sin\theta_l \tag{4.44}$$

The local reflectivities r_{vl} and r_{hl} can be represented in the global reference system, taking into account the polarisation rotation:

$$r_v(\theta) = r_{vl}(\theta_l)\cos^2\varphi + r_{hl}(\theta_l)\sin^2\varphi \tag{4.45}$$

$$r_h(\theta) = r_{hl}(\theta_l)\cos^2\varphi + r_{vl}(\theta_l)\sin^2\varphi \tag{4.46}$$

This transformation only acts on the specular components r_{sp}, since r_d is independent of incidence angle and polarisation. The incident radiation that is specularly reflected to the radiometer comes from a direction $-\hat{\mathbf{k}}'$ where the unit vector $\hat{\mathbf{k}}' = (\sin\theta'\cos\phi', \sin\theta'\sin\phi', \cos\theta')$ is defined by the polar angles (θ', ϕ') given by

$$\cos\theta' = 2\cos\alpha\cos\theta_l - \cos\theta \tag{4.47}$$

$$\cot\phi' = \cot\beta - \frac{\sin\theta}{2\sin\alpha\sin\beta\cos\theta_l} \tag{4.48}$$

These equations can be found from the relationship between the unit vectors $\hat{\mathbf{k}}, \hat{\mathbf{k}}', \hat{\mathbf{n}}$:

$$\hat{\mathbf{k}} + \hat{\mathbf{k}}' = 2\hat{\mathbf{n}} \cos\theta_l \tag{4.49}$$

Note that the incoming radiation is from the sky only if $\hat{\mathbf{k}}'$ points above the horizon; otherwise the radiation originates from the surrounding landscape. This situation will be discussed below.

4.3.3.2 Projection effect

Tilted surface facets appear enhanced or reduced in size, or are even hidden from a given view direction, depending on incidence angle, slope and orientation of the facets. The contribution to the radiation depends on the solid angle $\Omega = A\cos\theta_l/R^2$ under which a facet appears, where A is the true surface area of the facet and R is the distance to the radiometer antenna. A surface area is usually represented on a map by its projection A_h on a horizontal plane. If we use this projected area, then we get

$$\Omega = \frac{A_h \cos\theta_l}{R^2 \cos\alpha} \tag{4.50}$$

Now, the total signal at a given polarisation collected by the radiometer antenna – also called antenna temperature – is a beam-weighted sum over the radiation from all visible facets (numbered from $j = 1$ to n) within the antenna footprint:

$$T_{\mathrm{b,beam}} = \frac{1}{4\pi} \sum_{j=1}^{n} G_j T_{\mathrm{bp}}(A_j)\Omega_j \tag{4.51}$$

where G_j is the antenna gain in the direction to facet j with the normalisation $\sum_{j=1}^{n} G_j \Omega_j = 4\pi$.

4.3.3.3 Shadowing effect

A further effect to be taken into account is due to shadowing which occurs if there is an elevated horizon, leading to the interaction of radiation between different surface facets, i.e. radiation emitted by some elevated part of the relief is incident on another part of the surface. This shadow radiation replaces the radiation from the hidden sky. It means that the incident radiation is composed of a sky term and of a terrestrial term. The remainder of this section is devoted to the estimation of this effect. The accurate computation of the effective reflectivity can be very difficult for facets with general bidirectional scattering. In situations with a complex relief, ray-tracing techniques may be needed to include multiple specular reflections and emissions between facets. However, the assumption of a Lambertian landscape (facet reflectivity r_{d}) simplifies the computation. Then the emission is unpolarised and independent of direction. This allows us to tilt the coordinate system at the facet using the surface normal as the local z-axis (in Figure 4.13). A quantity of interest is the local angle of incidence θ_{H} of the horizon for a given surface facet. There the sky radiation is limited to local incidence angles θ_l smaller than θ_{H}. For larger values, the incident radiation emanates from the surrounding landscape at the brightness temperature T_{sr}. Now, the total upwelling brightness temperature $T_{\text{up,relief}}$ is the sum of the radiation from a flat horizon, $T_{\text{up,flat}}$ of Section 4.3.2, plus the increase ΔT_{up}, i.e.

$$T_{\text{up,relief}} = T_{\text{up,flat}} + \Delta T_{\text{up}} \tag{4.52}$$

The increase describes the enhancement resulting from T_{sr} being larger than the hidden T_{sky}:

$$\Delta T_{\text{up}} = \frac{r_{\text{d}}}{\pi} \int_0^{2\pi} \int_0^{\cos\theta_{\text{H}}} \{T_{\text{sr}}(\phi, x, \mu) - T_{\text{sky}}(\phi, \mu)\}\mu \, \text{d}\mu \text{d}\phi \tag{4.53}$$

Here ϕ is the azimuth angle at the local surface and $\mu = \cos\theta_l$. The choice of the local frame of reference allows simpler boundaries of integration than in an earlier treatment (Mätzler and Standley, 2000). The lower limit is $\cos(\pi/2) = 0$, and the upper limit is $\cos^2\theta_{\text{H}}$ which depends on ϕ.

First, an upper limit $\Delta T_{\text{up,max}}$ can be estimated by assuming that the elevated surface is a black body at constant temperature ($T_h = T_0$) whereas the flat surface facet is a non-black Lambert scatterer at the same physical temperature. With this simplification we get

$$\Delta T_{\text{up,max}} = \frac{r_{\text{d}}(T_0 - T_{\text{sky}})}{2\pi} \int_0^{2\pi} \cos^2\theta_{\text{H}} \, \text{d}\phi = c r_{\text{d}}[T_0 - T_{\text{sky}}] \tag{4.54}$$

where

$$c = \overline{\cos^2\theta_{\text{H}}} \tag{4.55}$$

is the local, azimuthal average of $\cos^2\theta_{\text{H}}$. It is straightforward to compute $\Delta T_{\text{up,max}}$ from a digital elevation model (DEM). In a hilly terrain, a typical angle θ_{H} may be 70°, leading to $\cos^2\theta_{\text{H}} = 0.12$. In order to produce a noticeable brightness increase, the reflectivity r_{d} of the facet should be sufficiently large. For $r_{\text{d}} = 0.2$ and for $T_0 - T_{\text{sky}} = 260$ K, the correction term $\Delta T_{\text{up,max}} = 6$ K. The effect gets larger as the ruggedness of the terrain increases. In a valley with steep slopes c may exceed 0.5.

Now we assume constant surface reflectivity $r = r_{\text{d}}$ and temperature T_0, negligible sky brightness ($T_{\text{sky}} = 0$) and a constant value of c for all interacting surface elements. The upwelling brightness temperature consists of radiation emitted by the facet eT_0 and scattering of incident radiation from the surroundings: $cr_{\text{d}} \cdot eT_0$. In addition there is radiation emitted from other facets scattered at the surroundings which again scatters to the facet. This interactive process goes on, leading to a geometrical series

$$T_{\text{up,relief}} = (1 - r_{\text{d}})T_0[1 + cr_{\text{d}} + (cr_{\text{d}})^2 + (cr_{\text{d}})^3 + \cdots] = T_0 \frac{1 - r_{\text{d}}}{1 - cr_{\text{d}}} \tag{4.56}$$

where e has been replaced by $1 - r_{\text{d}}$. The effective reflectivity r_{eff} can now be evaluated from

$$r_{\text{eff}} = 1 - \frac{T_{\text{up,relief}}}{T_0} = r_{\text{d}} \frac{1 - c}{1 - cr_{\text{d}}} \tag{4.57}$$

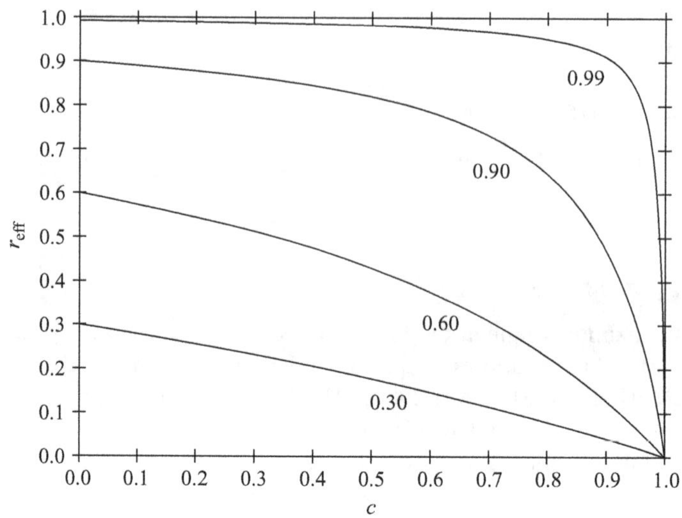

Figure 4.14 *Effective reflectivity of the landscape versus c for different values of the facet reflectivity* r_d

This effective reflectivity is also the valid expression for the effective reflection of the incident sky radiation. Thus reintroducing sky radiation, the upwelling brightness temperature becomes

$$T_{up,relief} = (1 - r_{eff})T_0 + r_{eff} T_{sky} \tag{4.58}$$

where r_{eff} is given by the right-most expression in Equation (4.57). Remember that specular components have not been included here. For rays without multiple interactions with the landscape, they can be added by Equations (4.42) to (4.51). Equation (4.57) and Figure 4.14 show that the relief effect is a decrease of the reflectivity and thus an increase of the emissivity. In another view, this is a large-scale roughness effect for thermal radiation. Multiple surface interactions make the landscape more black-body like. For small values of r_d the decrease of r_{eff} is linear with increasing c, but for large r_d the decrease is limited more and more to a narrow interval near $c \to 1$, e.g. to $0.9 < c < 1$ for $r_d = 0.99$ (Figure 4.14). This behaviour indicates that highly reflecting surfaces cannot easily be forced to emit thermal radiation. This is only possible if the radiation is trapped by the landscape, requiring a large number of bounces before it can leave the surface (e.g. in a deep canyon).

4.3.4 An example

As an illustration we consider the rugged terrain of southern Norway (Mätzler and Standley, 2000). The relief is shown by the DEM in Figure 4.15, and Figure 4.16 shows the computed values of c whose range extends from 0 to about 0.3 with extremes up to 0.4. Roughly speaking, Figure 4.16 looks like a negative of Figure 4.15 because of an anticorrelation between h and c. On a closer look, there are significant differences.

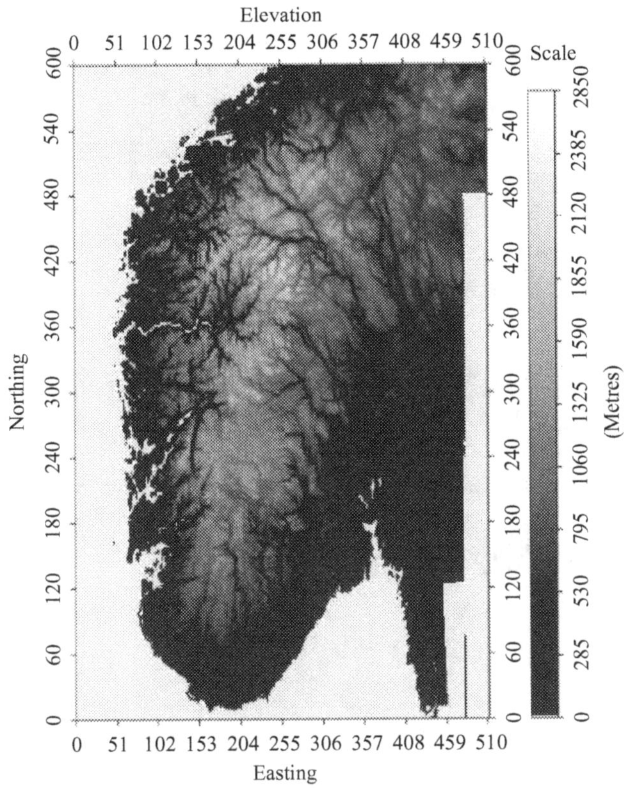

Figure 4.15 Digital elevation model of southern Norway, resolution 1 km (horizontal) 100 m (vertical); from Mätzler and Standley (2000)

The main valleys and fiords seen in Figure 4.15 almost disappear in Figure 4.16 where the brightest areas are found in small and narrow valleys, e.g. in the upper left part of the image. The extended mountain areas in the southern part are practically unaffected.

4.3.5 Conclusions

Relief effects of microwave radiation have been considered for the upwelling brightness temperature at the surface and above the atmosphere. Formulas were derived for surfaces whose reflectivity can be described by specular and Lambertian scattering behaviour. At frequencies of relevant atmospheric attenuation, effects result from the variable atmospheric contributions due to their dependence on the altitude h of the emitting surface. Tilted surfaces lead to projection effects with differences between local and global angles of incidence, polarisation rotation and to radiation from elevated horizons. For Lambert surfaces only the latter effect applies; it can be estimated from the parameter c defined by Equation (4.55). The main effect is an

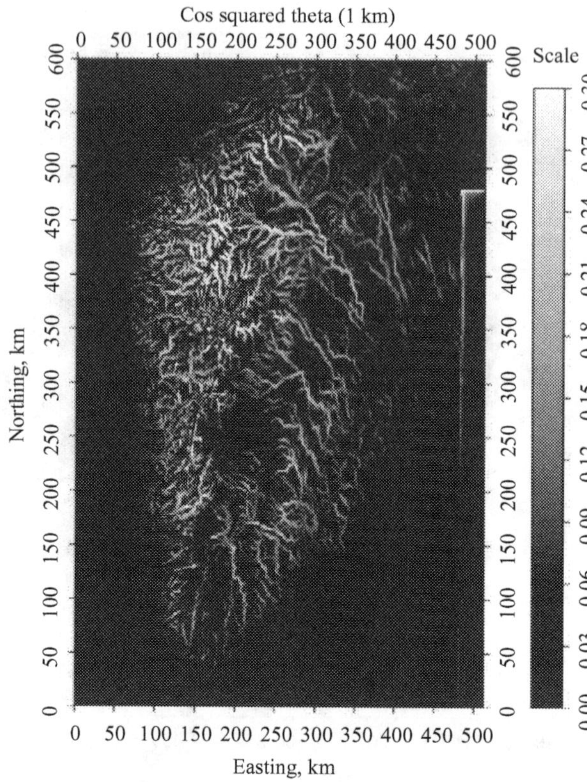

Figure 4.16 *Values of* $c = \overline{\cos^2 \theta_H}$ *in southern Norway; from Mätzler and Standley (2000)*

increase ΔT_{up} of T_{up} or equivalently a decrease of the effective reflectivity (4.57); it can be regarded as a correction for the large-scale ($\gg$wavelength) roughness of the relief. As an example the situation of southern Norway was shown. The formulas can also be applied if a specular component exists, as long as the specularly reflected rays escape to the sky. However, ray-tracing methods are required to model multiple specular reflections.

4.4 Ocean emissivity models *Stephen J. English and Adrian Jupp*

4.4.1 Existing observations used in near surface wind analysis

Numerical weather prediction (NWP) requires an initial 3D analysis of atmospheric temperature, humidity and wind. At present the user requirement for temperature and humidity is met primarily through satellite data assimilation, in particular data from the Advanced Microwave Sounding Unit (AMSU) and Special Sensor Microwave Imager (SSM/I) although *in situ* measurements (radiosonde temperature and humidity and SYNOP surface pressure) retain some value. By contrast radiosonde wind information remains very important, and radiosonde impact on short range wind forecasts remains larger than the impact of satellite data.

Dual polarised high incidence angle passive microwave data can provide surface windspeed (ocean-only) information by taking advantage of the lack of sensitivity of vertical polarisation to surface windspeed between 10 and 100 GHz, whereas horizontal polarisation increases rapidly as windspeed increases. Data from the single-look passive SSM/I microwave radiometer, which measures vertically and horizontally polarised brightness temperatures, is used operationally at many NWP centres to provide valuable ocean wind speed information over a relatively wide swath (1400 km), but cannot be used to derive directional information. A comparison of SSM/I derived wind speeds with buoy measurements has shown that a directional signal in measured brightness temperatures contributes to errors in wind speed retrievals (Wentz, 1992). Wentz went on to demonstrate that monthly averaged wind vectors could be derived from SSM/I data. These results led to a revival of interest in the retrieval of instantaneous wind vectors using a spaceborne polarimetric passive microwave radiometer, culminating in 2002 with the launch of the Windsat instrument on Coriolis. In recent years most effort on improving sea surface emissivity models has been on handling the azimuthal variation (due to roughness and foam) and the polarimetric phase signal. These developments are discussed in more detail in the next section.

This section summarises the main conclusions of a review of theory and models for polarimetric radiometry provided for and funded by the European Space Agency. The full report containing consideration of more general polarimetric applications and use of wind observations in NWP is given by English *et al.* (2003).

4.4.2 The effects of changes in surface windspeed on ocean surface emissivity

The microwave emission from an undisturbed ocean surface is highly polarised. The hydrodynamic processes associated with wind forcing alter the emissivity characteristics, tending to increase the emissivity while reducing the degree of polarisation. The relationship between wind forcing and the emissivity has been long established using geometric optics (e.g. Wilheit 1979). Using such models it is possible to analyse wind speed (e.g. Phalippou 1996). To be able to obtain wind direction from ocean surface radiance measurements these models need to be extended to be fully polarimetric. Alternatively an empirical approach can be taken which, combined with some physical insight, can provide statistically reliable retrievals, but with a dependence on the training set that may impose undesirable features into the retrieval.

Wind forcing of the ocean excites gravity-capillary and capillary waves (wavelengths of the order of those of microwaves), which tend to increase the surface emissivity. These small-scale features do not appear to be significant at wind speeds below 2 m/s. Larger scale waves are also generated, tilting the gravity-capillary and capillary waves and further modifying the emissivity characteristics. In addition, features such as swells which are generated non-locally may also influence measured brightness temperatures, and will introduce errors to wind retrievals based on the microwave emission/scattering theory.

In addition to capillary waves and the large-scale waves, there are two other important ocean surface features directly associated with local wind forcing which may have an impact on the surface emissivity. At wind speeds above approximately 7 m/s, whitecaps are generated as the surface begins to break. These also increase the unpolarised emissivity. Foam, which forms when the surface breaks and air bubbles form and rise to the surface, also dramatically increases the emissivity, but at wind speeds below 10–12 m/s, its contribution to microwave brightness temperatures is negligible (Trokhimovski and Irisov, 1995). The foam polarimetric signal has been studied by Rose *et al.* (2002) and although research in this area is still ongoing, it is possible that a very precise treatment of foam distribution on wave facets, taking full account of the wave 'lifecycle', will be required to accurately represent the true azimuthal variation of emissivity.

The nature of the wind-roughened surface leads to a directional variation of the surface emission. This is due to the fact that the amplitude of small-scale waves is much greater in the upwind-downwind direction in comparison to the crosswind direction. In addition, the small-scale waves are not uniformly distributed over the surface, but tend to cluster on the downwind side of the large scale wave slopes. The forward side of breaking waves exhibits higher brightness temperatures due to increased whitecap and foam coverage. All of the above considerations imply that observed microwave brightness temperatures may be dependent on the relative angle between the wind direction and the instrument viewing direction (known as the azimuth angle).

4.4.3 *The Stokes vector formulation applied to polarimetric radiometry*

One may consider radiation emerging from the ocean surface as having an amplitude A and some random phase ϕ. The horizontally and vertically polarised components of the electric field, E_V and E_h, may then be described as follows:

$$E_h = B\cos(\omega t)$$

$$E_V = C\cos(\omega t + \delta)$$

The values of A, B, C and δ are virtually impossible to measure directly, and so the Stokes vector was developed to cast them in a measurable form. The Stokes vector provides a full description of the polarisation state of the radiation emitted from the ocean surface, and may be defined in terms of A, B, C and δ or the electric field components as follows:

$$T_B = \begin{pmatrix} T_h \\ T_V \\ U \\ V \end{pmatrix} = \begin{pmatrix} A^2 + B^2 \\ A^2 + C^2 \\ 2BC\ \cos\delta \\ 2BC\ \sin\delta \end{pmatrix} = \begin{pmatrix} T_h \\ T_V \\ 2T_{45^\circ} - T_V - T_h \\ 2T_r - T_V - T_h \end{pmatrix} = \frac{\lambda^2}{kz} \begin{pmatrix} \langle E_h E_h^* \rangle \\ \langle E_V E_V^* \rangle \\ 2\mathbf{Re}\langle E_V E_h^* \rangle \\ 2\mathbf{Im}\langle E_V E_h^* \rangle \end{pmatrix}$$

where λ is the radiation wavelength, k is Boltzmann's constant and z is the impedance. T_h, T_V, T_{45° and T_r represent the horizontal, vertical, 45° linear and right-hand circular polarisations, respectively.

This vector characterises the phase correlation between the vertically and horizontally polarised fields. The form above is known as the modified Stokes vector. Alternatively, the brightness Stokes vector, which contains equivalent information, is given by:

$$\overline{T_B} = \begin{pmatrix} I \\ Q \\ U \\ V \end{pmatrix} = \begin{pmatrix} T_V + T_h \\ T_V - T_h \\ 2T_{45^\circ} - T_V - T_h \\ 2T_r - T_V - T_h \end{pmatrix} = \frac{\lambda^2}{kz} \begin{pmatrix} \langle E_V E_V^* \rangle + \langle E_h E_h^* \rangle \\ \langle E_V E_V^* \rangle - \langle E_h E_h^* \rangle \\ 2\mathbf{Re}\langle E_V E_h^* \rangle \\ 2\mathbf{Im}\langle E_V E_h^* \rangle \end{pmatrix}$$

This second form proves to be more convenient for describing the surface emission.

4.4.4 *A theoretical basis for polarimetric wind direction signals*

Ocean surface emission has typically been modelled by a two-scale approach which approximates the oçean surface as a series of independent small surface facets with capillary waves riding on them, tilted by underlying longer waves. The total emission is then calculated from the sum of the emission over the tilted surface facets. Traditionally, the approach has not predicted the observed dependence of brightness temperatures with the relative angle between the wind direction and the instrument look direction (azimuth angle), and yields zero values for U and V, the third and fourth Stokes parameters. This is because the anisotropic distribution of the small-scale surface features, as discussed in Section 4.4.2, is not taken into account. A two-scale model which takes the anisotropies into account has been developed by Yueh (1997), and shows reasonable agreement with the results from airborne experiments.

Yueh *et al.* (1994) present a full Stokes vector small-scale model of the ocean surface emission at nadir angles, which provides a useful insight into the possible nature of direction signals, and is summarised in this section.

The approach derives near-nadir polarimetric wind direction signals entirely from the consideration of the small-scale surface waves, with tilting by the large-scale waves not taken into account. This neglecting of tilting seems intuitively reasonable at nadir given the nature of the surface anisotropies.

Provided that the rms height of the gravity capillary waves is small compared to the wavelength of the emitted radiation, and the rms slope of the small-scale waves is much less than 1, perturbation theory may be used to estimate the emission from the small-scale surfaces. If the assumption is made that the sea surface is reflection symmetric with respect to the upwind look direction ($\phi = 0$), it can be shown that the Stokes parameters I and Q are even functions of ϕ, whereas U and V are odd. The Stokes parameters are approximated as Fourier series, to second harmonics, of the azimuth angle, ϕ:

$$I = I_0 + I_1 \cos\phi + I_2 \mathbf{cos}\, 2\phi$$

$$Q = Q_0 + Q_1 \cos\phi + Q_2 \mathbf{cos}\, 2\phi$$

$$U = U_1 \sin\phi + U_2 \mathbf{sin}\, 2\phi$$

$$V = V_1 \sin\phi + V_2 \mathbf{sin}\, 2\phi$$

Since it has been assumed that there is no upwind/downwind asymmetry, the Stokes vector $\overline{T}_{\mathrm{B}}$ has the property $\overline{T}_{\mathrm{B}}(\phi) = \overline{T}_{\mathrm{B}}(\phi + \pi)$. The first harmonics must therefore be equal to zero, i.e.

$$I = I_0 + I_2 \cos 2\phi$$

$$Q = Q_0 + Q_2 \cos 2\phi$$

$$U = U_2 \sin 2\phi$$

$$V = V_2 \sin 2\phi$$

If one takes the vertical polarisation channel of an instrument antenna to be at an angle ϕ relative to the wind direction, then the horizontal channel is at an angle $\pi/2 - \phi$. The vertically and horizontally polarised electric fields at a general angle ϕ may then be related to E_{v0} and E_{h0} and their values at $\phi = 0$:

$$E_{\mathrm{V}}(\phi) = E_{\mathrm{v0}} \cos\phi + E_{\mathrm{h0}} \sin\phi$$

$$E_{\mathrm{h}}(\phi) = E_{\mathrm{h0}} \cos\phi - E_{\mathrm{v0}} \sin\phi$$

From the original definition of the Stokes parameters in terms of the electric field components, the theoretical directional dependence of the Stokes parameters at nadir may be derived to yield:

$$I(\phi) = T_{\mathrm{V0}} + T_{\mathrm{h0}}$$

$$Q(\phi) = (T_{\mathrm{V0}} - T_{\mathrm{h0}}) \cos 2\phi$$

$$U(\phi) = -(T_{\mathrm{V0}} - T_{\mathrm{h0}}) \sin 2\phi$$

$$V(\phi) = 0$$

Note that because U is phase shifted from Q by 45°, a measurement of Q and U is effectively equivalent to taking two measurements of Q at azimuth angles separated by 45°.

The effective equivalence of a U measurement to taking a second measurement of Q with the azimuth angle shifted by 45° means that it could be possible to retrieve a wind direction pair separated by 180° using a single look passive polarimetric radiometer. Using a similar two look instrument, the possibility of retrieving a unique wind direction will be increased still further.

Comparisons between values of the Stokes parameters predicted using this version of the small perturbation model and Monte Carlo simulations has demonstrated reasonably good agreement. However, predicted values of the Stokes parameters are dependent upon an empirical spectral density function which describes the nature of the surface wave spectra. Several of these functions have been suggested, and each is sensitive to the values of parameters contained within the function. Yueh *et al.* (1994) use a corrected version of the model devised by Durden and Vesecky (1985), which showed reasonable agreement with aircraft experiments, and perform a limited analysis of the model sensitivity to these parameters. However, the issue has yet to be explored thoroughly, and it must be borne in mind that aircraft measurements are very different in nature to those made by a polar orbiting satellite.

Although the nature of directional signature appears to have been reasonably well predicted by the small perturbation method, the predicted peak to peak magnitudes have not.

Yueh *et al.* (1994) also examine the theoretical sensitivity of the Fourier components of the Stokes parameters to incidence angle. The results are once again based solely upon the small-scale surfaces, and so should only be used as a very rough guide given the likely increased importance of the large-scale slopes on the directional signature. Even without taking the large-scale waves into account, the results suggest that I will tend to develop a $\cos 2\phi$ dependency, due to the I_2 component becoming non-zero. As well as a directional dependence, the first three Stokes parameters are also shown to exhibit a windspeed dependence, as can be expected given the windspeed dependence of the vertically and horizontally polarised brightness temperatures observed in other spaceborne microwave radiometer data (e.g. SSM/I). All three parameters are found to exhibit an increase in magnitude with increasing windspeed, with I and Q exhibiting greater increases with incidence angle for angles up to 40°.

In order to provide satisfactory coverage, a spaceborne polarimetric radiometer must operate at a relatively large incidence angle, and modelling of sea-surface brightness temperatures will therefore need to take into account large scale waves (e.g. Yueh, 1997).

4.4.5 Models available for polarimetric radiometry

Pampaloni *et al.* (1996) have provided a physical two-scale model solving the small-scale scattering for anisotropic ocean surfaces based primarily on Yueh *et al.* (1994) providing a sound physical model for comparison with faster models. For data assimilation a very fast model is required. For example, the proposed Windsat instrument will provide 8 000 000 channel brightness temperatures per 6 h assimilation cycle. These must be processed within a few minutes within the operational schedule. Therefore, fast models are required for all aspects of the radiative transfer, if data is to be optimally assimilated. Speed can be achieved in three different ways: simplification, empirical regression and physical regression. It is unlikely that the ocean surface scattering model can be simplified sufficiently without losing accuracy, so the favoured approach is a regression-based model. There is a lack of good quality observational data for an empirical regression model (as is done for scatterometers, e.g. CMOD5, SASS). Therefore regression against the output of a physical model is preferred.

Liu and Weng (2003) have developed a fast model based on the model of St. Germain and Poe (1998) which has been validated against aircraft data. This model will be compared with the output of the Pampaloni model. The difference can be taken as an estimate of the likely maximum error in the physical model. An initial comparison of the Liu and Pampaloni models is shown in Figure 4.17. There is a difference in the amplitude of the signal; it is already known that the size of this variation depends on sea-surface temperature and windspeed. For the specific case of Figure 4.17 the Liu model amplitude at 8.7 m/s is about 60 per cent larger than

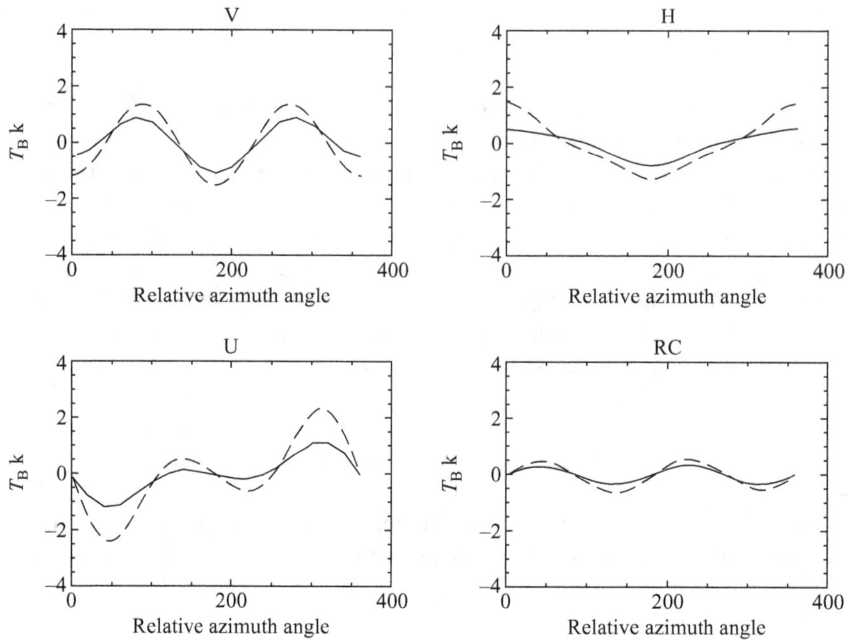

Figure 4.17 *Comparison of azimuthal brightness temperature dependence from the fast model of Liu (dashed) and the model of Pampaloni (continuous). Frequency = 37 GHz, windspeed = 8.7 m/s, sea surface temperature = 293 K, view angle = 55°*

that predicted by the Pampaloni model, and is closer to the amplitude predicted at 12.7 m/s by the Pampaloni model, as seen in Figure 4.18.

The Liu model needs to be incorporated in a fast radiative-transfer model to allow testing within a data assimilation environment, including writing of gradient code. Fortunately the Satellite Application Facility for Numerical Weather Prediction has undertaken to include this code in the RTTOV (Saunders *et al.*, 2002) model which is widely used at NWP centres (e.g. Met Office, ECMWF, Meteo-France). RTTOV uses an emissivity model known as Fastem. Fastem-1 was described in the COST-712 book (Matzler, 2000) and simply employed a linear regression to compute a term which allowed for modification of the surface emissivity due to roughness, whilst retaining a physical model for the emissivity of a flat smooth surface. Fastem-2 (Deblonde and English, 2000) modified this to allow a formulation of the apparent surface temperature which is closer to that in the geometric optics model. Fastem-3 simply added the polarimetric terms from the model of Liu and Weng (2003) to Fastem-2. This was included in RTTOV-8, which is described in Section 2.7 of this book.

By assuming a linear framework it will be possible to determine how accurate the surface emissivity model must be to obtain useful information on wind direction

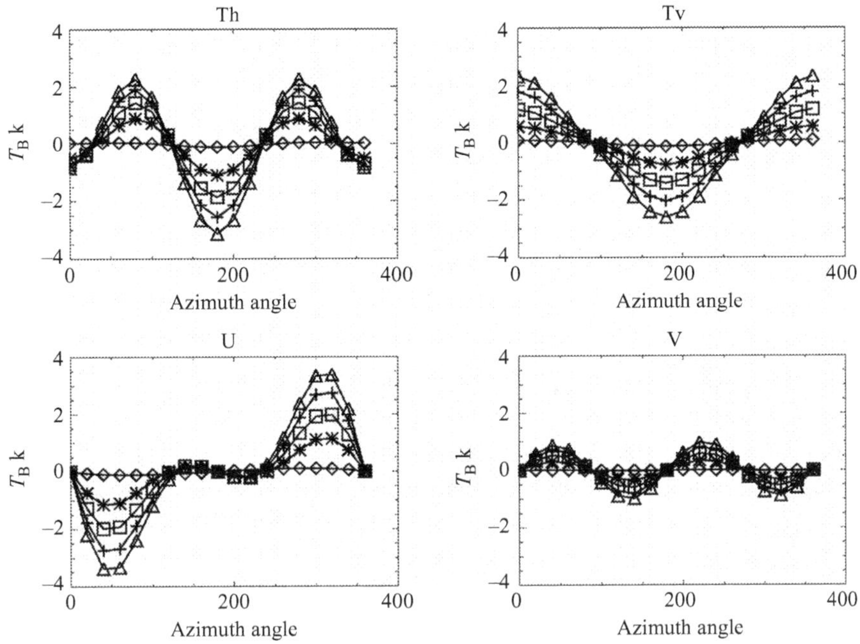

Figure 4.18 *The variation of brightness temperature with azimuth angle predicted by the Pampaloni model for the four elements of the Stokes vector for a range of windspeeds (2.8 m/s = diamonds, 8.7 m/s = asterix, 12.7 m/s = squares, 16.3 m/s = crosses, 19.7 m/s = triangles)*

within the data assimilation environment. Currently it is considered that the total 'observation error' (combination of measurement and radiative-transfer error) must be ~0.1 K (Van Woert *et al.*, 1996; Skou, 2000).

4.5 Modelling the emission at 1.4 GHz for global sea-surface salinity measurements *Emmanuel P. Dinnat and Jacqueline Boutin*

4.5.1 Introduction

Ocean salinity is a key parameter of oceanic circulation, which plays an important role in the climate system. Together with temperature, salinity determines the density of the water masses which drives the thermohaline circulation. Sea-surface salinity (SSS) is also a tracer of the water cycle, as it is affected by precipitation and evaporation. However, until now there has been a lack of SSS measurements over large areas of the global ocean and a lack of long-term SSS monitoring because it is yet not possible to measure it by satellite. The European Space Agency's (ESA) mission soil moisture and ocean salinity (SMOS), to be launched in 2007, is the first attempt to globally monitor

SSS by satellite (Kerr *et al.*, 1998). The SMOS mission will acquire multi-angular and multi-polarisation L-band (1.4 GHz) brightness temperature (T_b) measurements by means of a 2D synthetic aperture antenna. In 2009, the NASA mission Aquarius will measure SSS by means of an L-band radiometer and scatterometer combination (Koblinsky *et al.*, 2003).

The threshold accuracy in SSS required for oceanic studies, as recommended by the Global Ocean Data Assimilation Experiment (GODAE) group, is 1 practical salinity unit (psu) on salinity averaged over $200 \times 200\,\text{km}^2$ and 10 days, the optimum accuracy being 0.1 psu. Decreasing microwave measurement frequency induces increasing T_b sensitivity to SSS but also decreasing spatial resolution. Therefore, the choice of $\lambda_0 = 21$ cm for SSS measurement results from a trade-off between accuracy and resolution constraints. Moreover, this wavelength offers two advantages: it is protected from anthropogenic emission, and atmospheric water vapour influence on radiative transfer is small.

In this chapter we describe models for sea surface T_b , atmospheric radiative transfer and cosmic and galactic contributions at 21 cm in the frame of SSS remote sensing. As the largest incidence angles for SMOS and Aquarius measurements will be less than 60°, results for larger incidence angles are not discussed.

The brightness temperature (T_b, i.e. radiative power arriving to the antenna) at an altitude z, an incidence angle θ and an azimuth angle ϕ results from various contributions (see detailed description in Section 4.3). When observing the ocean at L-band, the emitted sea surface brightness temperature ($T_{b\,\text{sea}}$) is the largest contributing term. Its modelling and its sensitivity to oceanic parameters are described in Section 4.5.2. Atmospheric contributions are described by the atmospheric optical thickness ($\tau_\theta(a,b)$), at an incidence angle θ and between altitudes a and b, and the down- and upward atmospheric brightness temperatures ($T_{b\,\text{atm}\downarrow}(z,\theta)$ and $T_{b\,\text{atm}\uparrow}(z,\theta)$, respectively), at an incidence angle θ, for the atmospheric layer between the tropopause and altitude z, and between sea surface and altitude z, respectively. Atmospheric T_b's and optical thickness are assumed azimuthally isotropic (no dependence on ϕ). The radiative-transfer model and atmospheric parameters' influence on T_b are given in Section 4.5.3. The contribution of brightness temperature emitted by extra-terrestrial sources is discussed in Section 4.5.4.

Some sources contribute directly to the measured signal ($T_{b\,\text{sea}}$, $T_{b\,\text{atm}\uparrow}$) and some other ($T_{b\,\text{atm}\downarrow}$, extra-terrestrial sources) are reflected on the sea surface before reaching the instrument, and thus are affected by the surface reflection coefficient r. For these latter sources, only the reflection occurring along the specular direction is accounted for. This hypothesis is too simplistic for some of the indirect sources, such as the sun (see Sections 4.5.3 and 4.5.4). Modelling of sea-surface reflectivity, r, is part of $T_{b\,\text{sea}}$ modelling and is therefore discussed in Section 4.5.2.

4.5.2 Sea-surface brightness temperature

By definition, the emitted brightness temperature $T_{b\,\text{sea}}$ in the direction (θ,ϕ) is

$$T_{b\,\text{sea}}(\theta,\phi) = \text{SST} \cdot e(\theta,\phi) \tag{4.59}$$

where e is sea-surface emissivity and SST is sea-surface temperature.[1] Assuming local thermodynamic equilibrium, the Kirchhoff Law applies and $e = a = 1 - r$, where a and r are sea-surface absorptivity and reflectivity, respectively. When the electromagnetic radiation is polarised, the reflectivity is a four element vector, and the emitted brightness temperature is expressed through the so-called modified Stokes vector (see Chapter 1, Section 1.4) as

$$\boldsymbol{T}_{\mathbf{b\,sea}} = \begin{bmatrix} T_\mathrm{v} \\ T_\mathrm{h} \\ T_3 \\ T_4 \end{bmatrix} = \mathrm{SST} \times \left(\begin{bmatrix} 1 \\ 1 \\ 0 \\ 0 \end{bmatrix} - \boldsymbol{r} \right) \tag{4.60}$$

where T_v and T_h are $T_{\mathrm{b\,sea}}$ in vertical and horizontal polarisations (hereafter V- and H-pol), respectively, related to modified first and second Stokes parameters by $T_1 = T_\mathrm{v} + T_\mathrm{h}$ and $T_2 = T_\mathrm{v} - T_\mathrm{h}$, respectively, and T_3 and T_4 are related to third (U) and fourth (V) Stokes parameters, respectively. The first element of $\boldsymbol{r}$ is the reflectivity in vertical polarisation, accounting for incident waves polarised vertically or horizontally that are reflected as vertically polarised waves. The second element of $\boldsymbol{r}$, similar to the first one, is the H-pol reflectivity accounting for all incident waves reflected as H-pol waves. The third Stokes parameter being a difference of energy between waves linearly polarised at $\pm 45°$ off the H-pol vector, it is directly proportional to the third element of $\boldsymbol{r}$ (contrary to the first and second Stokes parameters that are proportional to $1 - r$) that is the difference of the reflectivities along these $\pm 45°$ linear polarisations. Similarly, the fourth Stokes parameter represents the amount of circular polarisation and is directly proportional to the fourth element of $\boldsymbol{r}$.

Due to the sea surface not being flat, diffraction induced by sea waves slightly modifies the reflectivity from Fresnel's equations. Consequently, $\boldsymbol{r}$ depends not only on incidence angle θ and sea water dielectric constant (ε) (and in turn on SST and SSS) but also on ϕ and shape of the surface, i.e. the roughness. A model for computing reflectivity of the ocean surface is given in Section 4.5.2.2.

$T_{\mathrm{b\,sea}}$ can be formally partitioned into two terms,

$$T_{\mathrm{b\,sea}} = T_{\mathrm{b\,flat}}(\varepsilon, \theta) + T_{\mathrm{b\,wind}}(\varepsilon, \theta, \phi, \boldsymbol{U}) \tag{4.61}$$

where $T_{\mathrm{b\,flat}}$ is $T_{\mathrm{b\,sea}}$ for a flat sea surface and $T_{\mathrm{b\,wind}}$ a correcting term due to wind $(\boldsymbol{U})$ (note that roughness is not necessarily related to local and instant wind only, and that wind effects also imply its duration and action distance, as well as the presence of swell). Actually, most of the influence of SST and SSS is due to $T_{\mathrm{b\,flat}}$, even though $T_{\mathrm{b\,wind}}$ depends slightly on SST (Dinnat *et al.*, 2003a). Therefore, $T_{\mathrm{b\,sea}}$ sensitivity to these parameters is described in 4.5.2.1, with Section 4.5.2.2 being focused on surface roughness influence.

[1] At L-band, penetration depth of electromagnetic waves in sea water is of the order of 1 cm. Very little is known about the temperature gradient in the first centimetre, and it has to be considered that SST for L-band radiometry might not correspond exactly to SST derived from instruments at higher frequencies, or *in situ* methods at slightly different reference depths.

Additional phenomena are likely to cause noticeable modification of $T_{b\,sea}$. The first one is foam, that appears above a threshold wind speed, and whose permittivity largely differs from the one of sea water (see Section 4.5.2.3). The second one, still to be investigated for L-band radiometry, is the presence of surface slicks of natural or non-natural origin. Slicks are known to damp roughness at specific scales, and their permittivity different from that of sea water might change $T_{b\,sea}$.

4.5.2.1 Sea water dielectric constant and brightness temperature for a flat surface

For a flat sea surface, reflectivity is derived from Fresnel's equations, and hence depends on θ and ε only (see for example Swift and MacIntosh (1983)). Until 2002, Ho *et al.* (1974) measurements of real (ε') and imaginary (ε'') parts of sea water permittivity were the only ones performed at L-band. They showed that permittivities for NaCl solution and sea water exhibit significant differences (confirmed later on by Ellison *et al.*, 1996, 1998). Therefore only measurements made on synthetic or real sea water, and subsequent models, should be considered. Re-analysing Ho *et al.* (1974) measurements at 1.43 GHz, and Ho and Hall (1973) measurements at 2.65 GHz, Klein and Swift (1977) derived a model for relative sea water permittivity (ε_r) based on the Debye (1929) equation (see Chapter 5, Section 5.2).

A straightforward analysis of Debye's and Fresnel's equations shows that, in most of the open ocean conditions (i.e. SSS >30 psu), $T_{b\,flat}$ is more sensitive to ε'' than to ε', by a few percent in cold waters and by a factor of $\simeq$4.6 in warm waters (note that sensitivity ratio is only slightly affected by incidence angle). Moreover, contrary to what occurs at higher frequencies, ε'' is mostly due to the conductivity component (75 per cent at 0°C and 95 per cent at 30°C), with less absorption being caused by dipolar polarisation (see Chapter 5, Section 5.2 for absorption terms description). Consequently, stringent accuracy is required on ionic conductivity (σ_i) model for any low frequency ε model. Accuracy on ε measurements required to achieve a 0.1 K accuracy on T_b is of the order of 0.2 per cent (Dinnat *et al.*, 2002).

Variation of $T_{b\,flat}$ with θ (Figure 4.19) is obviously complementary to the one of the Fresnel reflectivity. It is small close to nadir, and increases with increasing incidence angle. When θ gets larger, T_v increases as well (except for $\theta > 84°$) and T_h gets smaller. The incidence angle is the cause for, by far, the largest variation in T_b. Variation of $T_{b\,flat}$ with SSS is monotonic, whereas sensitivity to SST changes sign around 15°C, being close to zero for temperate waters (Figure 4.20). As the usual uncertainty on SST products is of the order of 1°C; a rough estimation shows that SST will be in most cases (i.e. for temperate waters) a very small source of error, and a small one for extreme cases (in very cold or warm waters). Actually, the most noticeable effect of SST is the large increase of $T_{b\,flat}$ sensitivity to SSS from cold to warm waters (by a factor of four). The error budget derived from these sensitivities is discussed in Section 4.5.5. $T_{b\,sea}$ sensitivity given in Figure 4.20 will vary only slightly with surface roughness. The order of magnitude is thus representative of all conditions, provided that the permittivity model is accurate enough. However, the accuracy of this model is questionable.

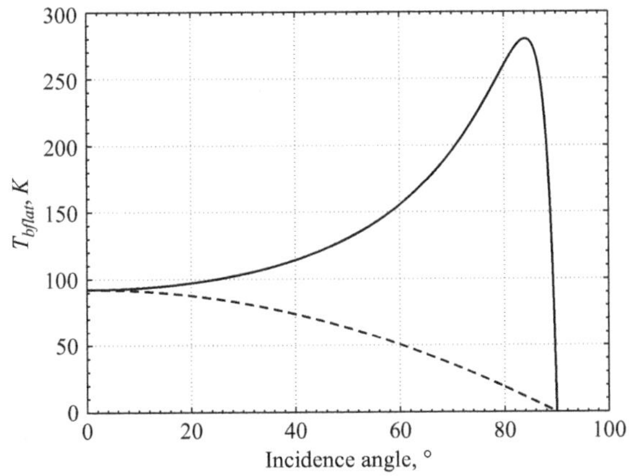

Figure 4.19 *Emitted brightness temperature for a flat sea surface and SSS = 35 psu, SST = 18°C (according to permittivity model by Klein and Swift, 1977)*

Klein and Swift (1977) revealed a bias in Ho *et al.* (1974) measurements which questions the claimed accuracy of 0.2 and 0.4 per cent on ε' and ε'' measurements. Moreover, few of the measurements were performed in the range of salinities encountered in open ocean (30–40 psu) and, even though SSS influence is close to linear, model accuracy in this particular range might be decreased. Until recently, no other ε measurements at L-band were available to allow direct comparison. Dinnat *et al.* (2002b, 2003a) compared the Klein and Swift (1977) model with models by Stogryn *et al.* (1995), Ellison *et al.* (1996, 1998) and Stogryn *et al.* (1997) that were derived from ε measurements at frequencies mostly between 3 and 20 GHz. Models are extrapolated at L-band using the Debye's equation (see Yueh *et al.*, 2001 for $T_{b\,sea}$ sensitivity plots derived from the Ellison *et al.*, 1998 model). Discrepancies between ε-models can be larger than 4 per cent, which is considerably larger than the required 0.2 per cent accuracy. The largest discrepancies are between the Klein and Swift (1977) and the Ellison *et al.* (1996) models; they concern mostly the influence of SST, and are largely variable between −2°C and 7°C (Figure 4.20). They result in $T_{b\,flat}$ discrepancies at nadir from 0.55 K, at 0°C, to 1.15 K, at 10°C (~1 K for SST >10°C). Note that whereas absolute discrepancy in T_b depends on θ because of r dependence, uncertainty in inverted SSS is quite independent of θ as $T_{b\,sea}$ sensitivity to SSS evolves in the same proportions as $T_{b\,sea}$ uncertainty. An explanation for the large ε discrepancies is the too simple equation used by Ellison *et al.* (1996) to fit σ_i measurements (Stogryn, 1997), leading to an error of 8 per cent on σ_i significantly larger than the 1 per cent claimed accuracy. Stogryn *et al.* (1995) and Stogryn (1997) models result in differences with respect to the Klein and Swift (1977) model of less than 1 K in T_b.

Two experiments were conducted in 2002 to improve the ε-model, one consisting in T_b measurements over a saltwater pond (Wilson *et al.*, 2004) and the other in

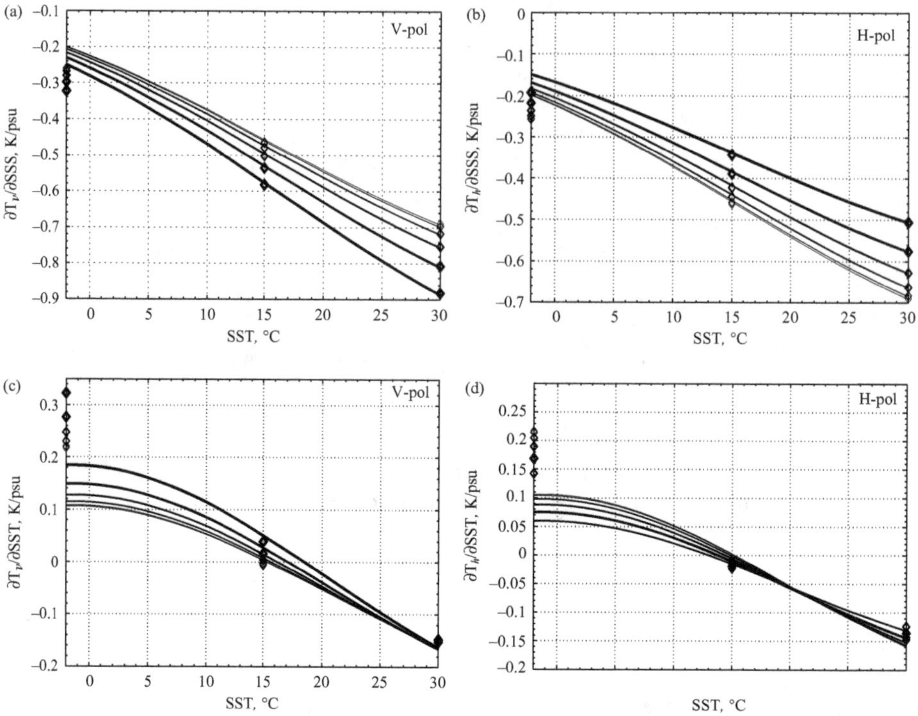

Figure 4.20 *Curves are sensitivity of $T_{b\,flat}$ to (a, b) SSS (in K/psu) and (c, d) SST (in K/ °C) at various incidence angles (the line gets thicker with larger θ that goes from 0° to 50° by 10° steps) using the model by Klein and Swift (1977) at SSS = 35 psu (results vary slightly with SSS), for vertical (a, c) and horizontal (b, d) polarisations. The diamonds are the same, but use the model by Ellison et al. (1996)*

new ε measurements (Blanch and Aguasca, 2003). T_b measurements suggest a good agreement with Klein and Swift (1977). However, they have been performed for SSTs larger than 8°C and corrected for a bias to better agree with models. Therefore, measurements allow assessment of relative variation with SST only, and not the validation of absolute $T_{b\,\mathrm{flat}}$. Whereas the largest relative differences (i.e. difference on T_b sensitivity to SST and SSS) between models occur at cold SSTs, no measurements are available in this range. Therefore, it is difficult to validate the Klein and Swift (1977) model from these measurements, even though they confirm that the model predicts realistic sensitivity to SST for temperate and warm seas. Moreover, first results from the new permittivity measurements show a relative good agreement with Klein and Swift (1977) regarding ε', but a significant discrepancy on ε'' (Blanch and Aguasca, 2003) to which T_b is very sensitive. Measurements are still underway at various institutes, and no definitive conclusions will be drawn before high accuracy measurements are performed. However, precise knowledge of sea water permittivity is still a critical open issue, in particular for cold waters where uncertainty is large and requirements are stringent.

In the following section, we use the Klein and Swift (1977) model for the sea water permittivity, because it relies on L-band measurements and no other model has been shown to be more reliable.

4.5.2.2 Wave spectrum and brightness temperature induced by surface roughness

In the previous section, T_b for a flat surface was discussed. However, the sea surface is never flat, with roughness at very different scales being created by local and instantaneous wind and/or distant wind (inducing swell), as well as by wave interactions (see Section 4.5.2.2.3).

Roughness at the sea surface diffracts impinging electromagnetic waves and consequently modifies r from the Fresnel's equation. We discuss two widely-used approximative models, the two-scale model and the so-called one-scale small slope approximation. A simpler approach based on geometric optics (GO) (Stogryn, 1967; Prigent and Abba, 1990) is discarded for use at low frequency. Indeed, whereas at high frequencies (i.e. in the millimetre domain) waves shorter than λ_0 are negligible, and all ocean waves can be considered as large-scale, $T_{b\,sea}$ simulation at $\lambda_0 = 21$ cm showed that a significant signal is induced by small scales, and that a large part of the roughness-induced signal is not predicted by GO (Dinnat *et al.*, 2002b). Noticeably, GO predicts a very small roughness effect on $T_{b\,sea}$ at nadir and moderate incidence angles, in contradiction with observations from Hollinger (1971), Swift (1974), Webster and Wilheit (1976), Camps *et al.* (2004) and Etcheto *et al.* (2004), as well as those from Blume *et al.* (1977) at 2.65 GHz. Note that it is very unlikely that the observed T_b variations, correlated with wind speed variations, are due to foam only, because they were observed also at small wind speed U, and the trend was close to linear in U (in the limit of measurement precision).

4.5.2.2.1 *The two-scale or composite model*

In the two-scale model, surface is modelled as the superimposition of small waves upon large waves, roughness scales being parted into small and large scales by a cutoff wavelength λ_c. Small scales are sea waves whose height is small compared to λ_0 and large scales are waves whose curvature radius is large compared to λ_0. To derive $T_{b\,sea}$, one combines both scales (Yueh, 1997) by integrating contributions from all large waves over the slope domain (S_x, S_y) and weighting contributions by the slope probability density function (pdf) of the large waves $(P(S_x, S_y))$. It follows that

$$\boldsymbol{T}_{\mathbf{b\,sea}}(\theta,\phi) = \iint \boldsymbol{T}_{\mathbf{b,l}}(\theta_l,\phi_l)\,(1 - S'_x \tan\theta)\,P(S_x,S_y)\,\mathrm{d}S_x\,\mathrm{d}S_y \tag{4.62}$$

where P is Gaussian, and its width depends on the mean square slope (MSS) of the large-scale waves, and S'_x is the large wave slope in the radiometer direction. Local brightness temperature for a large wave ($\boldsymbol{T}_{\mathbf{b,l}}$) differs from $T_{b\,flat}$ because (1) incidence and azimuth angles (θ, ϕ) are modified by the large wave's tilting, resulting in local incidence and azimuth angles (θ_l, ϕ_l), and (2) diffracting small-scale roughness is present on the large wave. Hence, $\boldsymbol{T}_{\mathbf{b,l}}$ is expressed as

$$\boldsymbol{T}_{\mathbf{b,l}}(\theta_l, \phi_l) = \mathrm{SST} \cdot (1 - \boldsymbol{r}_{\mathbf{ss}}(\theta_l, \phi_l)) \tag{4.63}$$

where $\boldsymbol{r}_{\mathbf{ss}} = \boldsymbol{r}_{\mathbf{c}} + \boldsymbol{r}_{\mathbf{i}}$ is the reflectivity of the small-scale roughness covered surface, separated into coherent ($\boldsymbol{r}_{\mathbf{c}}$) and incoherent ($\boldsymbol{r}_{\mathbf{i}}$) components. The incoherent term, that accounts for waves impinging from non-specular directions and scattered toward the radiometer, according to the first-order small-perturbation method (SPM1 – Rice, 1951), is written as

$$\boldsymbol{r}_{\mathbf{i}}(\theta_l, \varphi_l) = \int_0^{\pi/2} \sin\theta_\alpha \, d\theta_\alpha \int_0^{2\pi} \frac{\cos\theta_\alpha}{4\pi\cos\theta_l} \cdot \begin{bmatrix} \gamma_{\mathrm{vvvv}} + \gamma_{\mathrm{vhvh}} \\ \gamma_{\mathrm{hhhh}} + \gamma_{\mathrm{hvhv}} \\ 2\mathrm{Re}\{\gamma_{\mathrm{vhhh}} + \gamma_{\mathrm{vvhv}}\} \\ 2\mathrm{Im}\{\gamma_{\mathrm{vhhh}} + \gamma_{\mathrm{vvhv}}\} \end{bmatrix} d\varphi_\alpha \tag{4.64}$$

where the γ functions are the bistatic scattering coefficient, dependent on the sea surface power spectrum of small-scale roughness (Ψ_{ss}).

The coherent term $\boldsymbol{r}_{\mathbf{c}}$, that expresses the reflectivity in the specular direction, is modelled using either a so-called effective Fresnel coefficient or second-order SPM (SPM2, Yueh *et al.*, 1988). Effective Fresnel coefficients are derived from coefficients for a flat surface ($|r_{\mathrm{vv}}^{(0)}|^2$, $|r_{\mathrm{hh}}^{(0)}|^2$) multiplied by the scattering loss factor $\exp(-4\beta^2\sigma_{\mathrm{h,ss}}^2)$, with $\sigma_{\mathrm{h,ss}}$ the rms height of small-scale roughness and $\beta = 2\pi\cos\theta_l/\lambda_0$. The scattering loss factor accounts for the fact that part of the radiative power impinging from the specular direction, with respect to the instrument observation direction, is scattered toward non-specular directions (i.e. not toward the instrument). Coefficients derived from SPM2 are (Yueh, 1997)

$$\begin{aligned} \boldsymbol{r}_{\mathbf{c}}(\theta_i, \varphi_i) = & \begin{bmatrix} \left|r_{\mathrm{vv}}^{(0)}\right|^2 \\ \left|r_{\mathrm{hh}}^{(0)}\right|^2 \\ 0 \\ 0 \end{bmatrix} + \int_0^{2\pi} d\varphi_\alpha \int_0^{\infty} k_0^2 k_{\rho\alpha} \Psi_{\mathrm{ss}}\left(\vec{k}_{\mathrm{c}}\right) \\ & \times \begin{bmatrix} 2\mathrm{Re}\left\{r_{\mathrm{vv}}^{(0)*} g_{\mathrm{vv}}^{(2)}\right\} \\ 2\mathrm{Re}\left\{r_{\mathrm{hh}}^{(0)*} g_{\mathrm{hh}}^{(2)}\right\} \\ 2\mathrm{Re}\left\{\left(r_{\mathrm{hh}}^{(0)*} - r_{\mathrm{vv}}^{(0)*}\right) g_{\mathrm{vh}}^{(2)}\right\} \\ 2\mathrm{Im}\left\{\left(r_{\mathrm{hh}}^{(0)*} + r_{\mathrm{vv}}^{(0)*}\right) g_{\mathrm{vh}}^{(2)}\right\} \end{bmatrix} dk_{\rho\alpha} \end{aligned} \tag{4.65}$$

Both SPM1 and SPM2 lead to comparable results at moderate incidence angles (Boutin *et al.*, 2002).

From (4.64) and (4.65), Johnson and Zhang (1999) introduce the unified equation

$$\boldsymbol{r}_{\mathbf{ss}} = \begin{bmatrix} \left| r_{\mathrm{vv}}^{(0)} \right|^2 \\ \left| r_{\mathrm{hh}}^{(0)} \right|^2 \\ 0 \\ 0 \end{bmatrix} + \underbrace{\int_{k_c}^{\infty} k_0^2 k_\rho' \, \mathrm{d}k_\rho' \int_0^{2\pi} \Psi(k_\rho', \phi') \cdot \begin{bmatrix} g_v \\ g_h \\ g_3 \\ g_4 \end{bmatrix} \mathrm{d}\phi'}_{\delta \boldsymbol{r}_{\mathbf{ss}}} \tag{4.66}$$

where Ψ is the surface power spectrum, $k_c = 2\pi/\lambda_c$ is the cutoff wavenumber, with λ_c chosen typically between $3\lambda_0$ and $5\lambda_0$, and g_p functions ($p = v, h, 3$ or 4) account for both coherent and incoherent contributions to $\delta \boldsymbol{r}_{\mathbf{ss}}$, the correction to Fresnel reflectivity induced by small-scale waves.

Expanding physical quantities in a Fourier series with respect to azimuth direction, and under the assumption of even symmetry for surface roughness, one has

$$\delta \boldsymbol{r}_{\mathbf{ss}} = \delta \boldsymbol{r}_{\mathbf{ss,0}} + \delta \boldsymbol{r}_{\mathbf{ss,2}} \, f(2\phi_0) \tag{4.67}$$

$$C(k, \phi) = C_0(k) + C_2(k) \cos(2\phi_0) \tag{4.68}$$

$$g_p = g_{p,0} + g_{p,2} \, f(2\phi_0) \tag{4.69}$$

where f is the cosine function for T_{v} and T_{h} and the sine function for T_3 and T_4, $C(k, \phi) = k^4 \, \Psi(k, \phi)$ is the 2D surface curvature spectrum. Therefore, the omni-directional component ($\delta \boldsymbol{r}_{\mathbf{ss,0}}$) and second harmonic amplitude ($\delta \boldsymbol{r}_{\mathbf{ss,2}}$) result from weighted integrals of the respective harmonics of the curvature spectrum

$$\delta \boldsymbol{r}_{\mathbf{ss,0}} = \int_{k_c/k_0}^{\infty} C_0(k_\rho') \cdot \begin{bmatrix} g'_{\mathrm{v},0} \\ g'_{\mathrm{h},0} \\ g'_{3,0} \\ g'_{4,0} \end{bmatrix} \mathrm{d}\xi \quad \text{and}$$

$$\delta \boldsymbol{r}_{\mathbf{ss,2}} = \int_{k_c/k_0}^{\infty} C_2(k_\rho') \cdot \begin{bmatrix} g'_{v,2} \\ g'_{h,2} \\ g'_{3,2} \\ g'_{4,2} \end{bmatrix} \mathrm{d}\xi \tag{4.70}$$

where $\xi = k_{\rho'}/k_0$ and $g'_{p,n} = g_{p,n}/\xi$ are scattering weighting functions given by Johnson and Zhang (1999) and depending on bistatic coefficients for the first and second-order SPM given in Yueh (1997).

Dinnat and Drinkwater (2004) assessed the relative influence of the various scales on $T_{\mathrm{b\,sea}}$ from weighting functions in (4.70). Similar to the radar case, there is a specific range of wavelengths (i.e. typically around λ_0) that contributes most to $T_{\mathrm{b\,sea}}$, particularly when θ is small. However, significant additional contributions also arise from various scales especially at large θ. Therefore, good knowledge of roughness is required over a wide range of scales (typically from 1 m to 2 cm).

4.5.2.2.2 *The one-scale small-slope approximation*

The small-slope approximation (SSA) developed by Voronovich (1994) fills the gap left by the high- and low-frequency asymptotic approximations that are SPM and GO, as it is applicable to all ocean waves, at any scale. Composite models require us to determine an arbitrary cutoff wavelength that fulfils GO and SPM requirements at the same time, which is problematic. SSA does not require us to determine any cutoff wavelength. However, Irisov (1997) showed that emissivity calculations using SPM and SSA lead to identical results, leading Johnson and Zhang (1999) to apply the so-called SPM/SSA over the whole range of ocean wave scales. $T_{b\,sea}$ is derived accounting for all scales from (4.70) by extending the limit k_c to 0 and removing any further computation of the large-wave effect such as the one performed in (4.62). One noticeable difference between composite and SPM/SSA models, is that the first one allows accounting for hydrodynamic modulation (i.e. upwind/downwind asymmetry) through an empirical term in (4.62) whereas the second one does not, as the surface spectrum cannot render for phase correlation. This phenomenon is small for small θ but could be noticeable (± 0.4 K at 8 m/s) at large θ (60°).

Dinnat *et al.* (2003) assessed the influence of varying k_c from $2\pi/(3\lambda_0)$ to $2\pi/\infty$ (i.e. in the latter case, the model corresponds to SPM/SSA): the maximum sensitivity of $T_{b\,sea}$ to k_c is between $2\pi/(3\lambda_0)$ and $2\pi/(10\lambda_0)$. Changing k_c for θ less than 30° changes the omnidirectional $T_{b\,sea}$ by less than 0.1 K (up to 0.3 K at 40°). For the dependence on U, the sensitivity to k_c increases mostly in the first metres per second. $T_{b\,sea}$ dependence on U becomes quite insensitive to k_c for U larger than 3 m/s. Although the cutoff wavenumber influence is not negligible, it is of the second order with respect to roughness modelling uncertainties (see Section 4.5.2.2.3).

4.5.2.2.3 *Sea-surface roughness model*

Sea-surface roughness is described using a 2D surface power spectrum $\Psi(k_{\rho'}, \phi')$, i.e. a Fourier transform of the autocorrelation function of sea-surface height, that appears in (4.65), (4.66) and implicitly in (4.64), or using a 2D curvature spectrum $C(k, \phi)$ that appears in (4.70). Ψ is also used in the composite model to compute large-scale MSS that defines P in (4.62). There exist very different wave spectrum models (e.g. Durden and Vesecky, 1985; Donelan and Pierson, 1987; Apel, 1994; Yueh, 1997; Elfouhaily *et al.*, 1997; Lemaire, 1998; Kudryavtsev *et al.*, 1999...). In the following, we focus on DV and ECKV models as they were widely used to simulate $T_{b\,sea}$ at L-band.

The Durden and Vesecky (1985) (hereafter DV) model is a semi-analytic spectrum, that relies on work by Pierson and Moskowitz (1964) for gravity wave ranges, on Phillips (1977) for general form in equilibrium range, and that is fitted to HH-pol radar data at 13.9 GHz in order to account for deviation from the Phillips spectrum. The model is tuned to agree with Cox and Munk (1954) (hereafter CM) measured MSS. Yueh (1997) (hereafter DV2) model is the DV model multiplied by a factor of two to account for possible underestimation of MSS measured by Cox and Munk (1954), as suggested by Donelan and Pierson (1987) and Apel (1994), and to better fit data at 19.65 and 37 GHz. It should be noted however that, if needed, the multiplying

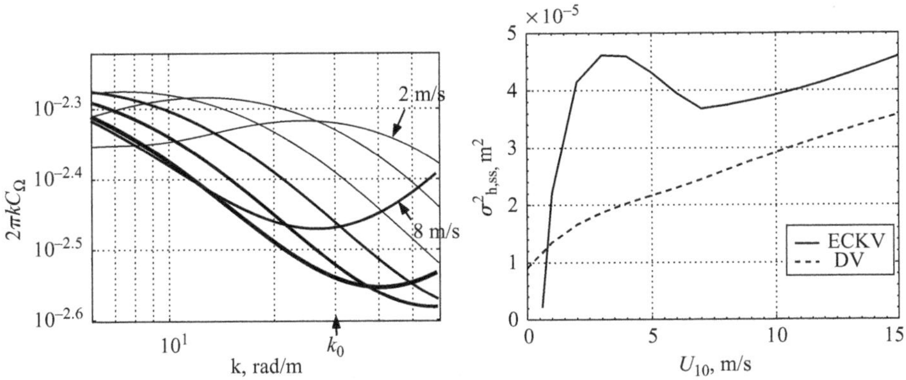

Figure 4.21 *Left: curvature spectrum from Elfouhaily et al. (1997) model versus wavenumber, around $k_0 = 30$ rad/m (i.e. 21 cm). Wind speeds are from 2 to 8 m/s, with 1 m/s steps (the thicker the line, the larger the wind speed). Right: small-scale height variance versus 10 metre height wind speed for Elfouhaily et al. (1997) (solid line) and Durden and Vesecky (1985) (dashed line) models*

factor is quite uncertain. Elfouhaily *et al.* (1997) (hereafter ECKV) model is quite singular as it was developed independently of radar data (i.e. not adjusted to fit radar measurements), and is therefore not bounded to any electromagnetic theory.

Spectrum models differ in shape, as well as in amplitude (Elfouhaily, 1997; Johnson and Zhang, 1999; Dinnat *et al.*, 2003a). It is particularly noticeable on the spreading function, DV (as well as DV2) model predicting no azimuthal anisotropy for large waves, contrary to ECKV model. When looking at wind influence on spectra, a particularity of the ECKV model arises (Figure 4.21): increasing wind speed between 3 and 7 m/s results in decreasing curvature for decimetric waves, contrary to DV (and DV2) model that predicts increasing curvature at all scales. Consequently, $\sigma_{h,ss}$ decreases with increasing U in that specific range when it is derived from ECKV model, whereas it monotonically increases when derived from the DV model. This behaviour will have important consequences on $T_{b\,sea}$ sensitivity to wind speed.

4.5.2.2.4 *Models results and comparison with data*

Sensitivity to roughness and model uncertainty studies are reported in Dinnat *et al.* (2003a). The influence of roughness depends slightly on SST and SSS . Using an SST wrong by 30°C for estimating roughness effect would induce an error on retrieved SSS of the order of 0.5 psu (less than 0.1 psu for an error of 7 psu on SSS). Therefore, $T_{b\,wind}$ dependence on SST should not be neglected in the case of SSS retrieval in the context of largely variable SST, as for example, for global ocean measurements where there is a risk of introducing regional biases.

Whatever the sea-surface spectrum model is, wind increases T_v and T_h for most incidence angles, by the order of a few tenths of a Kelvin per metre per second (see

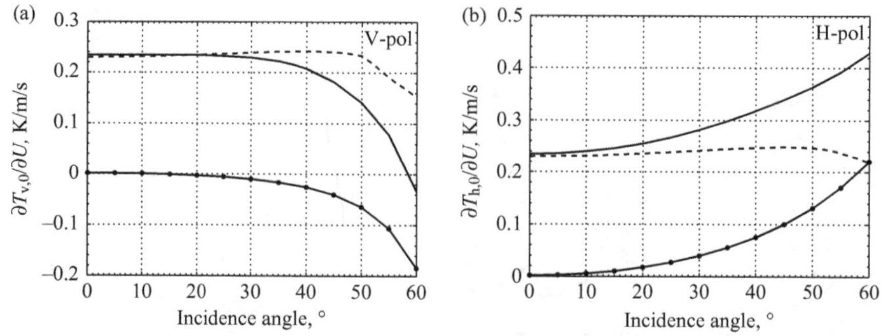

Figure 4.22 *Sensitivity of omnidirectional $T_{b\,sea}$ to wind speed for (a) V-pol and (b) H-pol and SST = 15°C, SSS = 36 psu, and $U = 8$ m/s. Note that this sensitivity depends on U, particularly for small wind speeds. Respective contribution of (dashed line) small and (pointed line) large scales to (solid line) total signal are illustrated. Model for sea spectrum is twice Durden and Vesecky (1985)*

Figure 4.22 derived from the DV2 spectrum). Sensitivity of T_v varies very slightly with θ up to 30°, where it starts to decrease to reach 0 close to 55°, and to become negative above. The sensitivity of T_h increases with the incidence angle. These large sensitivity differences between V- and H-pol are due to the combination of large- and small-scale effects. Figure 4.22 reports $T_{b\,wind}$ simulated for each of the two scales separately. Whereas the effect of small scales is an increase of roughly the same amount in T_v and T_h up to 50°, the effect of large scales is opposite in V- and H-pol, and very dependent on θ. Therefore, the increase in T_h sensitivity with θ results from the addition of both scale effects, whereas the decrease in T_v sensitivity at large θ translates the increasing relative importance of large-scale effects that counteract small scales, $\theta \sim 55°$ being the incidence angle at which both scale effects cancel each other.

Sensitivity reported in Figure 4.22 is derived from the DV2 spectrum model. It is fairly constant for U between 8 and 15 m/s, therefore $T_{b\,wind}$ can be approximated by a linear function of U in this range. At larger U, sensitivity increases slightly, except at large incidence angles. For smaller U, sensitivity is slightly smaller (note that spectrum is very uncertain for U less than 2 m/s). Although linearity cannot be assumed for large incidence angle (i.e. $\theta > 50°$), it is a reasonable approximation in a wide range of U for moderate θ. The Kudryavtsev *et al.* (1999) spectrum model leads also to linear $T_{b\,wind}$ dependence for U larger than 2 m/s (at smaller U, sensitivity is very large due to a very steep increase of $T_{b\,wind}$ with U). The ECKV model leads to linear $T_{b\,wind}$ dependence only above 8 m/s: up to this limit, sensitivity is largely variable, due partly to a 'plateau' feature where $T_{b\,sea}$ is almost insensitive to U (Dinnat *et al.*, 2003a). The plateau results from the peculiar behaviour of the curvature spectrum in the decimetre wave domain illustrated in Figure 4.21. The more the signal depends on small scales (i.e. the smaller θ), the more the plateau is pronounced. This plateau does not appear at high frequencies, where this model has been validated with data.

Therefore, linear approximation is non-relevant for the ECKV-model over the whole range of wind speeds.

Upwind/downwind $T_{\mathrm{b\,sea}}$ asymmetry is uncertain at L-band: using the empirical model for roughness asymmetry developed by Yueh (1997) from high frequency radiometric data, it is estimated to be up to ± 0.4 K at $\theta = 60^\circ$ and $U = 8$ m/s, and very small at moderate incidence angles.

Upwind/crosswind asymmetry is very dependent on the spectrum model, as discussed in Section 4.5.2.2.3. The ECKV model predicts asymmetry three times larger than the DV2 model (and hence six times larger than DV), but still at most ± 0.3 K for $U = 10$ m/s.

In view of the uncertainty of the wind influence, the WISE (Camps *et al.*, 2004) and EuroSTARRS (López-Baeza *et al.*, 2003) campaigns were conducted on behalf of ESA in the Atlantic ocean and Mediterranean sea to improve knowledge of sea-surface emissivity and its dependence on roughness. Model/data comparisons are reported in Camps *et al.* (2003), Gabarró *et al.* (2004a,b), Camps *et al.* (2004) and Etcheto *et al.* (2004). Both Camps *et al.* (2004) and Etcheto *et al.* (2004) find the sensitivity of WISE and EuroSTARRS T_{b} measurements to wind speed significantly larger than the one predicted by the DV model, and very close to the one predicted by the DV2 model. Camps *et al.* (2004) derive a nadir extrapolated sensitivity of $\sim$0.23 K/(m/s), and sensitivities of -0.2 and $+0.5$ K/(m/s) at 65° incidence angle in V- and H-pol, respectively; Etcheto *et al.* (2004) derived $T_{\mathrm{v\,wind}}$ sensitivities of 0.27 and 0.25 ± 0.2 K/(m/s) at 21.5° and 38.4° incidence angles, respectively. For U less than 8 m/s, no L-band data ever suggest that the plateau predicted using ECKV spectrum model would exist. Camps *et al.* (2004) suggest that multiplying the ECKV model by a factor of two, similarly to what was done for the DV2 model, would result in simulations in agreement with measured T_{b} sensitivity. However, multiplying by a factor of two the ECKV model results in significantly larger $T_{\mathrm{b\,wind}}$ than those predicted by DV2, and than those measured. This surprising statement certainly results from the computation of $T_{\mathrm{b\,wind}}$ sensitivity assuming a linear behaviour of T_{b} versus U over the whole range of U for the ECKV model (and consequently for ECKV $\times$ 2), which is clearly not the case. This leads to irrelevant slopes as they are largely affected by the considered wind speed range.

Although being quite non-physical, the DV2 model appears to be more appropriate than ECKV over the global wind speed range. Above 8 m/s (i.e. beyond the plateau), it is difficult to separate DV2 and ECKV models because their trends are similar and campaign datasets are noisy.

Gabarró *et al.* (2004a) proposed an original empirical approach accounting for possible decorrelation of large- and small-scale roughness. The following empirical relationship was derived from WISE 2001 data

$$T_{\mathrm{v\,wind}} = 0.12\left(1 - \frac{\theta}{40^\circ}\right)U_{10} + 0.59\left(1 - \frac{\theta}{50^\circ}\right)H_{\mathrm{s}} \tag{4.71}$$

$$T_{\mathrm{h\,wind}} = 0.12\left(1 + \frac{\theta}{24^\circ}\right)U_{10} + 0.59\left(1 - \frac{\theta}{50^\circ}\right)H_{\mathrm{s}} \tag{4.72}$$

where H_s is the significant wave height, related mostly to large-scale waves, and U_{10} is the local and instant 10 metre height wind speed. This model better fits data, and permits better retrieval of SSS with respect to any other model. However, as specified in the paper, this good performance might be a consequence of the increasing degrees of freedom of the inversion process. Although it is argued that this degree of freedom has a physical sense, one may question this sense. Indeed, WISE measurements indicate that H_s and U_{10} are largely correlated (typically the correlation coefficient is 0.7 and $H_s/U_{10} \sim 3.1$). According to (4.71) and (4.72), two features in contradiction with the theoretical models occur: (1) at nadir, large waves contribute as much to the signal as small scales and (2) the effect of large waves is of the same sign at both polarisations (i.e. increasing $T_{b\,\mathrm{sea}}$). This is contrary to any prediction of GO, SPM and two-scale models (see Figure 4.22). However, one has to consider that conditions (sea state around the Casablanca oil platform where T_b and sea state measurements were performed, limited antenna footprint) during WISE campaigns are very peculiar. Moreover, H_s, being largely correlated to U, could stand for an implicit non-linear T_b sensitivity. Although limited to the case of WISE campaign up to now, this work emphasises the likely need for knowing the history of the wind, or knowing the roughness at various scales (through other instruments such as the altimeter or from wave models).

4.5.2.2.5 Influence of wave development

Simulations for a fully developed sea assume that wind has blown over a large enough distance (fetch) and a long enough time (duration) for the sea to be in equilibrium with the local wind. This assumption is difficult to verify in general, and proved to be wrong during WISE campaigns (Dinnat *et al.* 2003b; Miranda *et al.*, 2003) inverse wave ages (Ω_w) being sometimes larger than 1.6. Miranda *et al.* (2003) assessed the effect of wave development comparing $T_{b\,\mathrm{sea}}$ simulation derived from the fully developed ECKV spectrum model and from the same spectrum model forced to reproduce actual spectrum energy (measured by a swell buoy during the WISE 2001 campaign) using an effective wind speed as input. Due to limitations on wave scales measured by the buoy, the influence of sea state deviation from a fully developed sea was tested accounting for large scales only. Dinnat *et al.* (2003b) estimated wave-age from the same sea state measurements acquired during the same campaign, and used it as an input parameter for the ECKV model that accounts for the wave development effect on small as well as on large scales (using an aerodynamic roughness scale model by Donelan *et al.*, 1993).

The wave spectrum of a young or mature sea (i.e. not yet fully developed) is characterised by two main features: a smaller gravity peak (in terms of wavelength and amplitude) and larger slopes for short waves, as illustrated in Figure 4.23 (note that the curve at 8 m/s is for illustration of the relative influences of wave development and wind speed variation). Both studies lead to similar results regarding the influence of wave development caused by the large scales only: when the sea is not fully developed, $T_{b\,\mathrm{sea}}$ is weakly affected at nadir (because large waves do not induce significant signals here) but is changed by the order of ± 0.1 K at 50° incidence angle, sign depending on polarisation, as large-scale roughness increases T_h but decreases

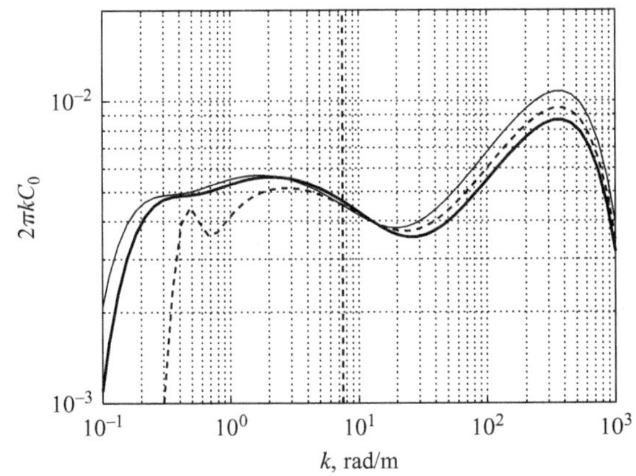

Figure 4.23 *Curvature spectrum versus wavenumber for (thicker solid line) a fully developed sea and (dashed line) a mature sea (i.e. inverse wave age $\Omega_W = 1.6$), at a wind speed of 7 m/s. The thinner solid line is also for a fully developed sea, but for a wind speed increased by 1 m/s (i.e. 8 m/s) with respect to thick solid line. It illustrates effects of wave development relative to a variation in wind speed. Spectrum model is by Elfouhaily et al. (1997) and roughness scale model by Donelan et al. (1993)*

T_V (see Figure 4.22). However, the overall effect is quite different when accounting for the small scale influence: when the sea is young, $T_{b\,sea}$ is increased by the order of 0.1 K at nadir, and by slightly more (less) at 50° in H-pol (V-pol), as small scales increases $T_{b\,sea}$ at both polarisations and as this effect is quite constant over incidence angle (see Figure 4.22). Therefore, the overall effect of non-fully-developed waves is to induce a slight increase of $T_{b\,sea}$ at nadir for both polarisations (meaning that this effect cannot always be minimised using T_I, as suggested in Miranda *et al.* (2003) who based this argument on results for large scale effect only), and slight increase and decrease for V- and H-pol, respectively, at larger incidence angles.

Although not negligible, the influence of wave development should be small and of second order with respect to wind effect and uncertainty on its modelling, as illustrated in Figure 4.23 and in Section 4.5.2.2.3.

4.5.2.3 Foam influence at L-band

When sea waves break, air bubbles are injected under the surface and formation of various very different types of whitecap (in terms of surface coverage, layer thickness, bubble size, lifetime) will succeed each other (Monahan and Lu, 1990).

It is known, at least since the work by Williams (1969), that whitecaps noticeably increase the emissivity of the sea surface because of their large emissivity (close to 1 at high frequencies), even though they often fill only a few per cent of the area seen by a radiometer. Foam effect on surface emitted T_b was observed at frequencies larger than 1.4 GHz by Nordberg *et al.* (1971), Stogryn (1972), and down to L-band by Webster and Wilheit (1976). Multi-frequency observations suggest a decrease of foam influence when frequency gets smaller. This is coherent with the Droppleman (1970) theoretical work, and explained by an increasing ratio λ_0/foam layer thickness. However, foam structure (layer thickness, bubble size distribution), electromagnetic properties and surface coverage are still subject to investigation (not mentioning foam classification and influence of wave development), and it is a particularly complex task to decouple roughness and foam effects on *in situ* T_b measurements.

Two different types of models exist for foam: emissivity models, that describe intrinsic $T_{b\ \mathrm{foam}}$, and coverage models, that describe what fraction of the sea surface is covered by foam (F_r), mostly as a function of wind. The resulting influence of foam on surface emitted T_b is obtained combining both parts, according to

$$T_b = (1 - F_r) \cdot T_{b\,\mathrm{sea}} + F_r \cdot T_{b\,\mathrm{foam}} \tag{4.73}$$

Droppleman (1970) proposed an emissivity model for foam, considering it as a mixture of air and water whose permittivity results from the combination of the components and their relative volume. Stogryn (1972) proposed an empirical spectral model, relying on measurements between 13.4 and 37 GHz, and suggested that it can be extrapolated down to 3 GHz. Boutin *et al.* (2002) showed that both models give similar results (i.e. $T_{b\,\mathrm{foam}}$ larger than 210 K at nadir) assuming a foam thickness larger than 10 mm and an air volume fraction close to 0.94. Reul and Chapron (2002) performed a review of recent foam emissivity models and an extensive sensitivity study including effects of foam layer thickness, mean bubble radius, air fraction and packing coefficient.

There is a wealth of models for foam coverage, one of the most complete of them accounting for wind speed, atmospheric stability and the various types of foam (Monahan and Lu, 1990). Recently, Reul and Chapron (2003) proposed accounting for the foam layer thickness distribution (i.e. the fact that thickness depends on the scale of the breaking wave and hence is heterogeneous). This model agrees with empirical laws for both crest- and static-foam when using realistic values for the persistence time parameter.

Radiometric data acquired during the recent WISE and EuroSTARRS campaigns are in contradiction with the large foam emissivity predicted by the models of Droppleman (1970), Stogryn (1972). Indeed, these measurements do not show any noticeable non-linear behaviour for wind speeds above 10 m/s where wave breaking becomes significant (Etcheto *et al.*, 2004). Moreover, simulations using the Stogryn (1972) model for foam emissivity and the model by Weill *et al.* (2003) or Monahan and Lu (1990) (active foam only) for coverage overestimate significantly WISE measurements (Dinnat *et al.*, 2002a; Etcheto *et al.*, 2004).

The problem of comparing $T_{b\,\mathrm{sea}}$ measurements and simulations is the difficulty in identifying the respective roles of the various sources of discrepancies that are foam coverage, emissivity and roughness. To overcome this problem, Villarino *et al.* (2003) derived $T_{b\,\mathrm{foam}}$ from WISE data when 100 per cent of the ocean surface in the radiometer field of view was foam covered. They emphasised the need for a dedicated experiment with artificially generated foam, an experiment that took place later on (Camps *et al.*, 2005) and whose first results are in agreement with Reul and Chapron (2003) on the fact that $T_{b\,\mathrm{foam}}$ is largely smaller (~130 K at nadir) than the one predicted by the Droppleman (1970) and Stogryn (1972) models. These results suggest a small effect of foam at wind speeds below 12 m/s wind speed. However, it does not rule out the likely need for a foam parametrisation at very large wind speeds.

4.5.3 Effects of the atmosphere

In order to derive $T_{\mathrm{b\,sea}}$ from T_{b}, one must correct for the effects of the atmosphere on the signal. They are of various types: the effect in the troposphere is absorption/thermal emission, scattering being neglected at this wavelength, and effects in the ionosphere are absorption/emission and Faraday rotation.

4.5.3.1 Tropospheric effects

Troposphere effects at L-band are studied in Yueh *et al.* (2001), Delahaye *et al.* (2002), Dinnat (2003). Atmospheric contribution terms introduced in Section 4.5.1 (see also Section 4.3) are modelled by

$$\tau_\theta(z_1, z_2) = \int_{L(z_1,\theta)}^{L(z_2,\theta)} \gamma_e(z')\,\mathrm{d}L(z',\theta) \tag{4.74}$$

$$T_{\mathrm{b\,atm}\downarrow}(0,\theta) = \int_{L(z,\theta)}^{L(z_{\mathrm{lim}},\theta)} \gamma_{\mathrm{e}}(z')T(z')\,e^{-\tau(z',z)}\,\mathrm{d}L(z',\theta) \tag{4.75}$$

$$T_{\mathrm{b\,atm}\uparrow}(z,\theta) = \int_{0}^{L(z,\theta)} \gamma_{\mathrm{e}}(z')T(z')\,e^{-\tau(z',z)}\,\mathrm{d}L(z',\theta) \tag{4.76}$$

where $L(z',\theta)$ is the path length at incidence angle θ and altitude z', z_{lim} is the upper limit of the atmosphere, $T(z)$ is the adiabatic vertical profile of atmospheric temperature and γ_{e} the extinction coefficient (here identified with the absorption coefficient γ_{a}) in the absence of liquid water derived from Liebe *et al.* (1993) (note that it depends on T, relative humidity (R_{h}) and pressure (p) vertical profiles).

Optical thickness is very small, of the order of 0.01 neper (except for grazing incidence angle), atmosphere is quasi-transparent, and because of larger values at the bottom of the profiles for T and γ_{e}, most of the atmospheric contribution is due to the first kilometres above sea surface. Therefore, atmospheric effects are mostly sensitive to conditions (i.e. T, p and R_{h}) near sea-surface level, more than to the vertical profile. Up- and downwelling contributions are similar and from 4.6 to 6.5 K for θ from 0 to 60°. However, final downwelling contributions will be smaller because of reflection at the surface. Attenuation is insignificant on reflected downwelling radiation, but from 0.7 to 2 K on $T_{\mathrm{b\,sea}}$.

One remarkable feature of L-band atmospheric effects is their slight dependence on atmospheric parameters. Peculiarly, γ_{e} sensitivity to R_{h} is very small (H_2O absorptivity is 20 times smaller than that of O_2) compared to what occurs at larger frequencies. Atmospheric T_{b} sensitivities to R_{h}, T and p are less than 0.001 K/per cent, 0.004 K/°C and 0.007 K/mb, respectively. Regarding wind effect, as $T_{\mathrm{b\,atm}\downarrow}$ depends on incidence angle and is reflected on surface before reaching the radiometer, wind should *a priori* modify this part of atmospheric contribution. Actually, contrary to what occurs at higher frequencies where roughness signature at small incidence angles is due to modification of the atmospheric contribution (Prigent and Abba, 1990), incidence angle modulation by large waves does not induce significant variability at L-band (Dinnat, 2003). Scattering of the downwelling atmospheric radiation by the short waves has not been assessed, but is likely to be small as a modification of reflectivity by the short waves is of the order of a few per cent and $T_{\mathrm{b\,atm}\downarrow}$ is of the order of a few Kelvins only.

4.5.3.2 Ionospheric effects

The Faraday rotation induced by the ionospheric electrons modifies polarised L-band radiometric temperatures. This is especially important for ocean spaceborne radiometric missions, since such missions require a very high measuring accuracy. According to the magnetoionic theory and to the quasi-longitudinal approximation, and assuming a plane parallel ionosphere, the Faraday rotation angle (in degrees) is

$$F_{\mathrm{a}} \simeq 6900 \times B \times (\cos L_{\mathrm{a}} / \cos E) \times \mathrm{VTEC} \tag{4.77}$$

where B is the representative magnitude of the earth's magnetic flux density (teslas), L_{a} is the angle between the vector $\boldsymbol{B}$ and the line of sight, E is the angle between nadir and the line of sight and VTEC is the vertical total electron content below the instrument in TEC units, where 1 TECu is 10^{16} electrons per m^2 (Waldteufel *et al.*, 2004).

Global maps for Faraday rotation angle, ionospheric emission and attenuation were simulated at L-band by Le Vine and Abraham (2002) and Waldteufel *et al.* (2004), using models for generating the VTEC and $\boldsymbol{B}$. Ionospheric T_{b} and attenuation are small at 6 a.m., up to 0.03 and 0.004 K, respectively, at nadir. On the other hand, the Faraday rotation angle is large (it can be larger than the 10°) and is highly variable, resulting in T_{b} variations that can be of several Kelvins, and thus the Faraday rotation correction is a critical issue.

Polarisation rotation can either be estimated using ancillary data (VTEC and magnetic field), deduced from radiometric measurements or circumvented using invariant quantities in the SSS retrieval scheme. When fully polarimetric information is available, Yueh (2000) proposes to use the measured third Stokes parameter, assumed to be zero for sea-surface emission, to estimate the rotation angle. Skou (2003) proposed a method applicable when dual polarisation only is available, using a ratio between V- and H-pol. Waldteufel *et al.* (2004) showed that, in the particular case of the 2D interferometer, VTEC can be retrieved accurately (less than 0.5 TECu). Finally, one of the retrieval scenarios assessed by Boutin *et al.* (2004) showed that using the first Stokes parameter T_{I}, by definition insensitive to polarisation rotation, increases SSS error by only 10 per cent, even though there are 50 per cent less independent T_{b} estimations with respect to T_{v} and T_{h}. This is partly explained by the larger T_{I} sensitivity to SSS.

4.5.4 *Extra-terrestrial sources*

At L-band, radiation from celestial sources is strong and spatially variable; they have been reviewed by Delahaye *et al.* (2002), Le Vine and Abraham (2004), and associated corrections needed to interpret L-band radiometric measurements have been thoroughly described by Le Vine and Abraham (2004). Radiation originates from three types of sources. The hydrogen line emission corresponds to a hyperfine atomic transition in neutral hydrogen: the radiation is maximum around the plane of the galaxy, most of the time less than 2 K. The cosmic background is a remnant signal of the origin of the universe and is almost constant in space and time (2.7 K). In addition to the almost constant cosmic background, a very variable continuum radiation (up to more than 10 K) is due to emissions from discrete radiosources. As in the case of atmospheric emission, the cosmic background adds a contribution to the radiometric temperature that depends on the incidence angle linked to the reflection of the signal on the sea surface. The two other types of sources add a signal that varies according to the incidence angle and to the geometry of the measurement.

The galactic signal has to be thoroughly modelled and corrected before data can be interpreted. This requires us to know the galactic signal with an accuracy better than 0.1 K, to very well monitor the attitude of the radiometer and to correctly model the antenna pattern of the radiometer. Actually Le Vine and Abraham (2004) have shown that the integration of the galactic signal over the antenna pattern may modify the galactic correction by more than 1 K in the case of a strong and variable galactic signal.

The sun is a potential critical noise source for L-band radiometry because of its very large T_b , ranging from 100 000 to 500 000 K depending on solar activity (Delahaye *et al.*, 2002). Its influence on measurements can be direct, through antenna side-lobes, or indirect through reflection at the sea surface.

Reflections can be minimised by setting the satellite orbit crossing time at 6 a.m./p.m. at the equator, but not completely ruled out as the sun will still be reflected from the grazing angle at higher latitude. Actually, the reflection problem does not come from specular reflection – that can be avoided or easily detected – but from roughness-induced reflection (i.e. specular reflection on steep waves or scattering by small-scale roughness). This problem is clearly illustrated by Swift (1974), where Figure 11 shows an easily identified sun effect of more than 360 K in the specular direction, but also a very large residual effect even when the sun image is not in the main beam.

Yueh *et al.* (2001) assessed the sun effect through the side-lobes of the antenna pattern in the case of a real aperture antenna. A more complex effect occurs in the case of an interferometric radiometer, because the sun image can also be present in the aliased field of view. This complex effect is currently being studied in the frame of the SMOS mission preparation.

The moon is less problematic than the sun, as its T_b is of 275 K. Therefore, its non-specular effects can be neglected but its reflected image has to be corrected for.

4.5.5 Perspectives

Using $T_{b\,sea}$ sensitivities to SST, SSS and U derived from a two-scale model, Boutin *et al.* (2004) assessed the influence of uncertainties on U, SST and measured T_b on the accuracy of global retrieved SSS in the frame of the SMOS mission. SST uncertainty is of secondary order, uncertainties on U and T_b inducing at least 94 per cent of the uncertainty on SSS, radiometric noise being dominant close to the edges of the swath. It is concluded that random noise on U (~2 m/s), SST (1°C) and T_b (from 1 to several Kelvins depending on pixel position in SMOS field of view) should not hamper the SSS retrieval within a precision better than 0.1 psu over 200 × 200 km^2 and 10 day average. However, requirements are stringent for the absolute accuracy on T_b (0.07 and 0.03 K for warm and cold seas, respectively) and U (0.3 and 0.15 m/s for warm and cold seas, respectively). A comparison of various retrieval scenarios showed that estimation of U at high temporal resolution is required, and that the use of the first Stokes parameter T_I decreases SSS precision by 10 per cent only, even though it decreases the number of independent measurements by a factor of two. These results were obtained by assuming a perfect knowledge of the emissivity model; one can expect more issues than obtaining good ancillary data, as some critical aspects of the models are not yet well characterised.

Some of the modelling issues are likely to be solved in the near future thanks to ongoing and future experiments (e.g. permittivity model, galactic emissivity), but roughness modelling will certainly remain a critical issue even after the satellite is launched. Given the complexity of roughness modelling and the stringent accuracy required on it, empirical approaches will be necessary.

The Aquarius mission is a bit different from SMOS as it will dispose of scatterometer measurements, simultaneous to those of the radiometer, that will allow us to estimate properties of the surface roughness. Yueh *et al.* (2001) demonstrated that an empirical correction of roughness effect on T_b using radar measurements leads to a better retrieved SSS than not accounting for roughness (but the use of the emissivity model combined with wind speed estimation to correct for roughness effect was not assessed). However, Dinnat and Drinkwater (2004) showed that a scatterometer and a radiometer are in general not sensitive to the same roughness scales in the same relative proportions. Consequently, the relation between measurements from both instruments is not straightforward and is likely to depend on other parameters than wind speed (e.g. wave development, presence of slicks, ...). It is therefore likely that the Aquarius mission also will benefit from external data (or from a combination of various radar measurements) to estimate the large-scale sea state.

Finally, sun glint and foam are potential critical issues for both missions, particularly in the case of large wind speeds.

4.6 Modelling the soil microwave emission *Jean-Pierre Wigneron, Jiancheng Shi, Maria José Escorihuela and Kun-Shan Chen*

4.6.1 Introduction

In modelling the bare surface microwave emission signals, they are typically described as $1 - R_p^e$ where R_p^e is the surface effective reflectivity for polarisation p. Two main types of approaches can be roughly distinguished to model surface effective reflectivity:

- physical model approaches: the model is based on electromagnetic equations of the surface scattering effects. The model input parameters for the surface roughness characteristics: the standard deviation in surface height (S_D), the autocorrelation function and the associated correlation length (L_C) are the most commonly used parameters. The surface effective reflectivity is obtained by integrating the bistatic coefficient over the hemisphere.

– semi-empirical approaches: the model is based on simple equations derived originally from physical considerations. The form of the equations and the values of the model parameters are generally obtained following the best-fit approaches, namely by looking empirically for equations and values providing a good fit between measurements and simulations.

Recent developments in the use of these two types of approaches will be discussed in the following two sections.

In recognising of the complexity of natural surfaces, it should be noted that these two types of approaches cannot easily be compared. To explain this remark, let us have a quick look at the general structure of soil: there are generally strong heterogeneities in the three-dimensional spatial variations of the soil characteristics (structure, moisture content, ...). For instance for ploughed soils, after dry and sunny conditions following rainfall events, large emerging clods dry out more rapidly than hollows within the fields. These effects may lead to a large spatial variability in the soil moisture content at the soil surface and within soil at a spatial scale of about 1 m. There are also non-uniformities in the soil moisture in the soil clods, at a lower spatial scale of a few centimetres (the surface of the clods drying more quickly than inside). These heterogeneities produce significant dielectric discontinuities at the soil surface and within the soil volume. On the other hand, the physical models assume isotropic roughness properties and homogenous dielectric half-space.

At low frequencies, the thickness of the soil layer whose dielectric properties contribute to the soil microwave emission exceeds 10 cm in dry conditions. Therefore, the soil microwave emission results from effects due to both the 'geometric' roughness effects (namely the geometry of the interface between soil and atmosphere) and the 'dielectric' roughness (namely the large dielectric discontinuities at the soil surface and within the soil volume that are due to soil heterogeneities in terms of moisture, texture, structure, chemical properties, etc.).

The surface scattering model accounts only for the geometric 'roughness' (roughness resulting from variation of surface heights). This kind of approach does not account for volume effects which may be significant in soils. Therefore, the use of this type of approach should probably be limited to rather wet soils, for which surface scattering effects are dominant. Conversely, the model parameters used in semi-empirical approaches are derived from observations and implicitly account for both the 'geometric' and 'dielectric' roughness effects. Both surface and volume effects are accounted for in this latter type of approach. The weakness of the semi-empirical approaches, however, is that they were derived typically from limited field observations.

An intermediate modelling approach between physical and semi-empirical approaches was recently developed and was denoted as an air-to-soil transition model. This new model describes the influence of the topsoil structure on the L-band radiation as an impedance matching between the dielectric constants of the top-soil and air. This approach was developed from physical considerations. However, as an accurate estimation of all model input parameters may be uneasy, this model may also require the use of best-fit parameters. This model is described in Section 4.7.

4.6.2 Physical modelling approaches

4.6.2.1 General

Several physically-based models have been developed to account for the effect of surface roughness on the observed brightness temperature T_B (Fung *et al.*, 1994). Two main types of approaches can be distinguished:

1. *Numerical simulations:* These methods solve directly the Maxwell equations and allow the computation of the surface reflectivity without any approximations. Three-dimensional (3D) method-of-moment Monte Carlo simulations of soil emission have been recently developed using fast computation methods such as sparse matrix canonical grid and physical-based two grids (Li *et al.*, 2000; Tsang *et al.*, 2001). These methods require ensemble average of several simulation realisations. Each realisation requires about 10 h CPU time on a workstation and these methods have been mainly used for 'punctual' validation of simplified analytical models (e.g. Chen *et al.*, 2003).

2. *Analytical model:* By making several assumptions, analytical equations based on a physical description of wave scattering from random rough surfaces have been developed to model the bistatic scattering coefficient. These models are generally driven by surface characteristics derived from measurements of surface height profiles.

For analytical models, the traditional models are the small perturbation method (SPM) and the Kirchhoff approximation (KA) (that includes the physical optics (PO) models and the geometric optics (GO) in the high-frequency limit) applicable, respectively, for slightly rough surfaces and surfaces with large curvature. Much effort has been devoted to broadening the range of validity of these approximations. The small slope approximation (SSA) (Broschat, 1993) and the integral equation model (IEM) (Fung *et al.*, 1994) overlap these two methods. In particular, the IEM has bridged the gap between these two models successfully and was shown to be in agreement with SPM and KA, respectively, in the low- and high-frequency region. The KA and SPM models, namely, can be considered as special cases of IEM in each frequency region.

4.6.2.2 IEM and AIEM models

The IEM was developed to obtain an explicit analytical formula easy to compute, while maintaining reasonable accuracy under several assumptions. The main assumptions were as follows (Chen *et al.*, 2003):

- Common approximations for the local angle in the Fresnel coefficients are formed by either using the incident angle for slightly rough surfaces or by using the specular angle for the rough surface.

- Edge diffraction terms are excluded.
- To compute the complementary field coefficients, the absolute phase terms in the surface Green's function and its gradient are eliminated.

Recently some developments have been carried out to extend the original IEM model proposed by Fung *et al.* and the above assumptions were partially removed.

4.6.2.2.1 Computing complementary field coefficients

First the complementary field coefficients were re-derived, leading to more complete expressions of the multiple scattering (Chen *et al.*, 2000) and single scattering terms (Chen *et al.*, 2003). Let us explicit these improvements. To compute the far-zone scattered field in the medium above the rough surface, in accordance with the Stratton–Chu integral, an estimate of the tangential surface fields is written as the sum of the Kirchhoff surface fields and the complementary surface fields. The complementary fields were expressed as a function of field coefficients (F_{qp}) that can be computed from spectral representation of the Green's function G and of its gradients (∇G). In the original IEM, phase terms used in the derivation of G and ∇G were assumed to be cancelled out.

In a first improvement (Chen *et al.*, 2000), modifications were made to include all the phase terms in G and ∇G for multiple scattering while single scattering terms were left unchanged. However, the computation of surface emissivity requires integrating all the scattering power over the half-space and energy conservation requires accurate scattering calculations (Chen *et al.*, 2003). In order to extend the model to handle the surface emission problem, all the terms in Green's function and its gradient in the surface fields, should be kept for the calculation of both single and multiple scattering.

Therefore, in a second step, the modifications were applied also for the single scattering term (Chen *et al.*, 2003). The major task lies in the ensemble average over the scattering field, and hence many more mathematical manipulations have to be carried out. Chen *et al.* (2003) expanded the expression of the bistatic coefficients in three terms: the Kirchhoff, cross, and complementary terms (the Kirchhoff term remaining the same as in the previous version of the IEM). The cross term consists of eight terms, while the complementary term contains 64 terms. The new IEM version was called advanced IEM (AIEM). The authors concluded that AIEM contains many more terms compared with the original version and a little more computation time is required. The payoff is much higher accuracy which is critical for emissivity calculations and AIEM remains in algebraic form for ease of numerical implementation.

4.6.2.2.2 Transition model for the reflection coefficient

To compute the bistatic scattering coefficient of randomly dielectric rough surfaces, it is usually assumed that the Fresnel reflection coefficients could be approximately evaluated at either the incident angle (θ_{i}) or the specular angle (θ_{sp}) in the low- and high-frequency range, respectively. However, these two considerations are only applicable to their respective regions of validity. A physical-based transition function that naturally connects these two approximations was proposed by Wu *et al.* (2001)

and was included in the new version of the IEM model (Fung and Chen, 2004; Wu *et al.*, 2004). We explicit this improvement in the following.

The IEM model for single scattering can be written using three terms, namely, the Kirchhoff term (σ^{k}_{qp}), the cross term (σ^{kc}_{qp}) and the complementary term (σ^{c}_{qp}) (Wu *et al.*, 2001):

$$\sigma^{\circ}_{qp} = \sigma^{k}_{qp} + \sigma^{kc}_{qp} + \sigma^{c}_{qp} \tag{4.78}$$

The Kirchhoff term accounts for the large scale surface roughness effects, while the other two terms correct for the Kirchhoff approximations and account for the effects of the smaller structure of the surface. As the frequency increases from low to high, surface roughness scales become largely relative to the wavelength, and the scattering coefficient is gradually dominated by the Kirchhoff term. When the incident frequency is sufficiently high, the single scattering coefficient given by the IEM using the specular angle reduces mathematically to the standard geometric optics model (GOM), which is the Kirchhoff model in the high-frequency limit and is independent of the frequency. Incidence angle also plays a role in the weight of the three terms in Equation (4.78): as the incident angle increases, the effective roughness increases (the small-scale components on the surface were found to be more important, e.g. Wu *et al.*, 2004), and contributions from the complementary terms become more significant.

The Kirchhoff and complementary field coefficients (f_{qp} and F_{qp}) which are used to compute the three terms in Equation (4.78) are a function of the Fresnel reflection coefficients. In the low-frequency region, the local incident angle in the Fresnel reflection coefficients was replaced by the incident angle (note that this is always the case to compute the complementary field coefficients). Conversely, in the high-frequency region, the local angle was replaced by the specular angle. The separation of the two regions leads to discontinuity in the angular response of the reflection coefficients and thus, unsatisfied scattering behaviour.

The authors proposed to join the two approximations by introducing a transition function (Υ_p), so that the scattering model could have a wider region of validity. Using Υ_p, their approximation to the Fresnel coefficient was expressed in the form:

$$R_p(T) = R_p(\theta_i) + [R_p(\theta_{sp}) - R_p(\theta_i)]\Upsilon_p \tag{4.79}$$

Υ_p was expressed as a function of a parameter S_{qp} defined as the ratio of the complementary term to the total scattering coefficients:

$$S_{qp} = \sigma^{c}_{qp}/\sigma^{\circ}_{qp} \tag{4.80}$$

Υ_p lies between 0 (i.e. R_p (T) = $R_p(\theta_i)$ in the low frequency region) and 1 (i.e. R_p (T) = $R_p(\theta_{sp})$ in the high frequency region). Validation of the proposed form for the transition function was tested against numerical simulations with moment method (Chen *et al.*, 2001) and experimental data (Fung and Chen, 2004).

4.6.2.3 AIEM simulations of soil emission

AIEM has been used mostly for simulations of the backscattering coefficients (Chen *et al.*, 2000; Fung *et al.*, 2004; Wu *et al.*, 2004). However, it has also been used in the field of microwave radiometry and as discussed above, the IEM model has been improved recently with the view of modelling the emission of rough surfaces (Chen *et al.*, 2003). These advances have allowed more accurate calculations of surface reflectivity for simulating soil emission at L-band (Liou *et al.*, 2001; Shi *et al.*, 2002).

Based on energy conservation law, the surface reflectivity was generally written as the sum of two components: non-coherent ($R_{\mathrm{p}}^{\mathrm{non}}$) and the coherent components ($R_{\mathrm{p}}^{\mathrm{coh}}$). $R_{\mathrm{p}}^{\mathrm{non}}$ was obtained by integrating the bistatic scattering coefficient σ, simulated with AIEM, over the upper hemisphere:

$$R_{\mathrm{p}}^{\mathrm{non}} = 1/4\pi \cos(\theta) \int_{\mathrm{hemisph.}} (\sigma_{\mathrm{pp}}(\theta,\theta_{\mathrm{s}},\varphi_{\mathrm{s}}-\varphi) + \sigma_{\mathrm{pq}}(\theta,\theta_{\mathrm{s}},\varphi_{\mathrm{s}}-\varphi)) \sin\theta_{\mathrm{s}}\, d\theta_{\mathrm{s}}\, d\varphi_{\mathrm{s}} \quad (4.81)$$

the subscript p or q is for polarisation and s indicates the direction of the scattered radiation.

$R_{\mathrm{p}}^{\mathrm{coh}}$ was written as a function of the Fresnel reflectivity $R_{\mathrm{p}}^{*}(\theta)$:

$$R_{\mathrm{p}}^{\mathrm{coh}}(\theta) = R_{\mathrm{p}}^{*}(\theta) \exp[-(4\pi S_{\mathrm{D}} \cos\theta/\lambda)^2] \quad (4.82)$$

where S_{D} is the standard deviation of the surface height.

The AIEM was coded only for correlation functions $\rho(\zeta)$ that are in the form of:

$$\rho(\zeta) = \exp(-(\zeta/l_{\mathrm{C}})^N) \quad (4.83)$$

where l_{C} is the correlation length. The exponent N is equal to 1 or 2 for, respectively, an exponential and Gaussian function that are the most commonly used correlation functions.

Experimental profiles of the surface heights that are fitted to Equation (4.83) show that natural surfaces are generally between the Gauss and exponential functions and are closer to the exponential function (in Shi *et al.* (1997) over an agricultural area; 76 per cent of the measured profiles could be described by an exponential function with $n \leq 1.4$).

Two studies (Liou *et al.*, 2001; Shi *et al.*, 2002) have evaluated AIEM simulations of soil emission at L-band for both Gaussian ($n = 2$) and exponential ($n = 1$) autocorrelation functions. An important and rather new result was obtained. Both studies showed that roughness effects differ strongly at different incidence angles and polarisations. In particular, the effects of roughness at rather large incidence angles ($\theta \approx 50°$) were found to be very different in magnitude and direction for the two H and V polarisations. As the roughness effects increase (e.g. kS_{D} increases) the emission was found to increase at H polarisation. However, at V polarisation, the emission was found to decrease so that the emission of rough surfaces was generally lower than that of flat surfaces.

This last result is very different from the general idea that emissivity increases with increasing roughness effects, an idea which is also reflected by simulations of simplified semi-empirical models (Mo and Schmugge, 1987; Wigneron *et al.*, 2001).

Shi *et al.* (2002) explained this result by comparing the relative magnitude of both the coherent and non-coherent component of surface reflectivity while surface roughness changes. As surface roughness kS_D increases, R_p^{coh} decreases while R_p^{non} increases. At H polarisation the rate of the decrease in R_p^{coh} is larger than the rate of increase in R_p^{non}. Inverse results were obtained at V polarisation and at rather large angles. For this configuration, due to the Brewster angle in the coherent component, the dominant effect is the increase in R_p^{non} with increasing roughness effects. This result impacts strongly the roughness effects on the ratio of measurements at H and V polarisation (R_V/R_H). Based on simulated datasets using the AIEM model, Shi *et al.* (2002) developed a new parametrised AIEM model of the surface reflectivity that accounted for these new results.

As noted in the introduction, as surface scattering models only account for the geometric 'roughness' and do not account for volume effects, the use of this type of approach should probably be limited to rather wet soils. Over rather dry soils, emission originates from soil layers at depth, and volume scattering effects due to 'dielectric' roughness may be significant. For instance, in the active field, reasonable agreement between AIEM and backscattering measurements over bare soil surfaces was found to be limited to rather wet soil conditions, in a recent study (Zribi *et al.*, 2005). As discussed in the following section, ground-based observations of soil emission seem to indicate that the 'dielectric' roughness effects enhance the microwave emission of soil in dry conditions.

Recently, based on AIEM, Shi *et al.* (2005) developed a parametric model which is based on a roughness parameter (QP) depending on polarisation P and frequency (larger than about 3 GHz). It was found that the dependance of QP on the soil roughness characteristics could be accurately accounted for by the ratio of the surface rms height and correlation length (S_D/L_c).

4.6.3 A semi-empirical parametrisation of the soil emission at L-band

Simplified semi-empirical approaches have also been developed. Most of these approaches are based on two roughness parameters: the roughness height h_S and a polarisation mixing parameter Q_S that can be retrieved from brightness temperature measurements (Choudhury *et al.*, 1979, Wang *et al.*, 1981). In these studies, a modified Fresnel reflection formulation was used to calculate the soil reflectivity $R_P(\theta)$. For most applications, this simple formulation was found to be adequate.

The polarised soil reflectivity $R_P(\theta)$ is given by:

$$R_P(\theta) = [(1 - Q_S)R_P^*(\theta) + Q_S R_Q^*(\theta)] \exp(-h_S \cos^{N_s}(\theta)) \tag{4.84}$$

where p and q stand for the polarisation v or h, $R_P^*(\theta)$ is the polarised specular reflectivity; N_S is an exponent. The specular reflectivity $R_P^*(\theta)$ can be simply computed from the soil dielectric permittivity ε_S and the incidence angle θ, using the Fresnel equations.

These modelling approaches are generally found to be very useful in most soil moisture retrieval studies. However the dependence of the model roughness parameters (h_S and Q_S) on the surface roughness characteristics (S, $L_C, \ldots$) is not well known. Several studies provided values for h_S and Q_S obtained from microwave observations over bare fields (Choudhury *et al.*, 1979; Wang *et al.*, 1981; Mo *et al.*, 1987). However, the experimental datasets that were used in these studies were usually not large enough to allow the development of parametrisations with a large range of validity. More recently, Wegmüller and Mätzler (1999) developed another parametrisation based on a dataset obtained over a large frequency range (2–94 GHz). However, the L-band was not included in this frequency range and the obtained accuracy (from 0.06 at V-polarisation to 0.095 at H-polarisation in terms of microwave surface emissivity, i.e. about 20–30 K in terms of brightness temperature) is generally not adequate for soil moisture studies.

A simple parametrisation of the soil roughness effects was developed at L-band as a function of soil variables (Wigneron *et al.*, 2001). This parametrisation is valid over the whole range of soil roughness conditions encountered in agricultural fields. The parametrisation was developed from a large dataset (PORTOS-93) obtained during a three-month experiment covering seven surface roughness conditions (representing the total range which can be expected on agricultural fields) and a large range in soil moisture and temperature conditions (Laguerre *et al.*, 1994).

A detailed analysis of the PORTOS-93 dataset showed that both Q_S and N_S could be set equal to zero. This result is generally in good agreement with most of the studies based on a large experimental dataset including multi-angular measurements. After two initial studies considering that $N_S = 2$ (Choudhury *et al.*, 1979; Wang *et al.*, 1981), Wang *et al.* (1983) considered in a more detailed study that the dataset used in (Wang *et al.*, 1981) was not adequate to test the θ dependence of surface roughness and that the $\cos^2(\theta)$ dependence (i.e. $N_S = 2$) was much too strong. In the study of Wang *et al.* (1983), $N_S = 0$ was found to be consistent with measurements over θ range of 10–60° at three frequencies (1.4, 5 and 10.7 GHz). Mo and Schmugge (1987) also considered that the $\cos^2(\theta)$ dependence should not be used for rough surfaces at L- and C-bands. In a recent study over a rather smooth soil, Escorihuela *et al.* (2004) showed that values of N_S should be distinguished for both polarisations: $N_S \approx 1$ at H polarisation and $N_S \approx 0$ at V polarisation. Generalisation of this result for other roughness conditions should be made. Note also that at higher frequencies (23–90 GHz), the value of N_S was found to vary between 0 for a rather smooth soil and 0.5 for a rather rough soil (Prigent *et al.*, 2000).

Considering that $Q_S = 0$ at L-band is also in agreement with most of the published studies based on a large experimental dataset. Wang *et al.* (1983) have found that the frequency dependence of Q_S was strong and very small values for Q_S were obtained at L-band (three values of Q_S were computed for three soil types: 0, 0.01 and 0.12). Two other detailed studies at L- and C-band (Mo and Schmugge, 1987), and over a large frequency range (Wegmüller and Mätzler, 1999), have also considered that $Q_S = 0$. Njoku and Entekhabi (1996) noted that the assumption that $Q_S = 0$ can be made at low frequencies. Note that the value of Q_S was found to increase with frequency and it was a non-zero parameter in the studies of Wang *et al.* (1983) and Prigent *et al.* (2000) at higher frequencies.

Therefore, from (4.84) and by setting both Q_S and N_S equal to zero, the semi-empirical modelling of the L-band surface microwave emissivity could be simplified as:

$$e_P = 1 - R_P(\theta) = 1 - [R_P^*(\theta)\exp(-h_S)] \tag{4.85}$$

The roughness parameter (h_S) was retrieved from the PORTOS-93 experimental data over a range of incidence angles of 0–40°. The retrieved parameter h_S can be considered an effective parameter that accounts for (1) 'geometric roughness' effects, in relation to spatial variations in the soil surface height, and (2) 'dielectric roughness' effects in relation to the variation of the dielectric constant at the soil surface and within the soil which can be caused by non-uniformities in the soil characteristics (moisture content, bulk density, ...). A simple analytical formula was developed to express the parameter h_S as a function of averaged geophysical variables, estimated from ground-based measurements.

As in the study of Mo and Schmugge (1987), it was found that the best geophysical parameters to model h_S were the slope parameter ($m = S_D/L_C$) and the surface soil moisture w_S.

The formula providing the best results, can be written as

$$h_S = A(w_S)^B \cdot (S_D/L_C)^C \tag{4.86}$$

with $A = 0.5761$, $B = -0.3475$ and $C = 0.4230$.

This best fit formulation is illustrated in Figure 4.24. In this figure, retrieved and simulated values of the roughness parameter h_S are plotted versus the slope parameter $m = (S_D/L_C)$ for different conditions in the soil moisture content w_S. The best-fit regression curve (4.86) is shown (—) for three values of w_S ($w_S = 0.03$, 0.15, 0.35 m^3/m^3). Retrieved values of h_S are illustrated by different symbols according to the field soil moisture. It can be seen that the value of h_S may decrease by a factor of two from dry to wet soils.

The dependence of the h_S parameter on soil moisture could be explained by an effect of volume scattering: as the soil dries out, emission results from the deeper layers within soil. Possibly, the spatial fluctuations of the dielectric constant within the soil volume are rather strong during drying out, producing an important 'dielectric' roughness effect. Accurate 3D description of the soil layer and physical models accounting for both volume and surface scattering effects would be required to analyse this empirical result.

If we do not consider the dependence of h_S on w_S, the following best-fit formulation was obtained:

$$h_S = A(S_D/L_C)^C \tag{4.87}$$

with $A = 1.3972$ and $C = 0.5879$. These two parametrisations were successfully applied to reproduce the measured angular distributions of polarised brightness temperature at L-band over the seven plots: The range of the average rms error and bias were, respectively, about 8–10 K and 0–2 K.

Note that recent results obtained from the analysis of long term (three-years) radiometric measurements acquired over a fallow and a bare soil (SMOSREX experiment),

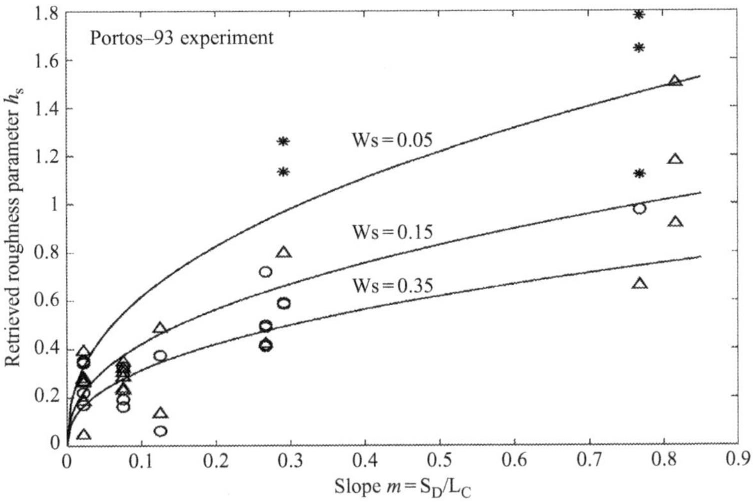

Figure 4.24 *Retrieved and simulated values of the roughness parameter h_S versus the slope parameter $m = (S_D/L_C)$ for different conditions in the soil moisture content w_S. The best-fit regression curve (4.86) is shown (—) for three values of w_S (w_S = 0.03, 0.15, 0.35 m^3/m^3). Retrieved values of h_S are illustrated by different symbols according to the soil moisture content: '*' for dry soils (w_S < 0.07 m^3/m^3); 'Δ' for intermediate moisture content (0.07 $m^3/m^3 \leq w_S \leq$ 0.25 m^3/m^3); 'o' for wet soils (w_S > 0.25 m^3/m^3)*

confirm the general soil moisture dependence of h_S. Preliminary results showed a linear dependence (instead of exponential as given by (4.86)) was preferrable and are illustrated in Figure 4.25 (Escorihuela *et al.*, 2004).

4.6.4 Conclusion

In the literature, two main types of methods used to model soil emission could be distinguished: semi-empirical and theoretical approaches.

Semi-empirical methods are calibrated directly from experimental datasets and thus they can account for actual soil roughness conditions: including both 'geometric' and 'dielectric' roughness as defined in the introduction, row effects, volume scattering, etc. The measured difference between the emissivity of dry soil and that of wet soil (e_P(DRY) − e_P(WET)) was larger than that predicted by classical semi-empirical parametrisations (e.g. Wang *et al.*, 1981). To correct for this discrepancy, the dependence of the h_S roughness parameter on soil moisture was accounted for in two recent studies. This dependence may account for possible enhanced 'dielectric' roughness effects due to volume scattering for rather dry soil conditions.

However, only a few experimental datasets are available to date and this is not enough to develop accurate methods for a large range of roughness conditions (namely the methods are probably site specific). Also, the obtained accuracy still remains

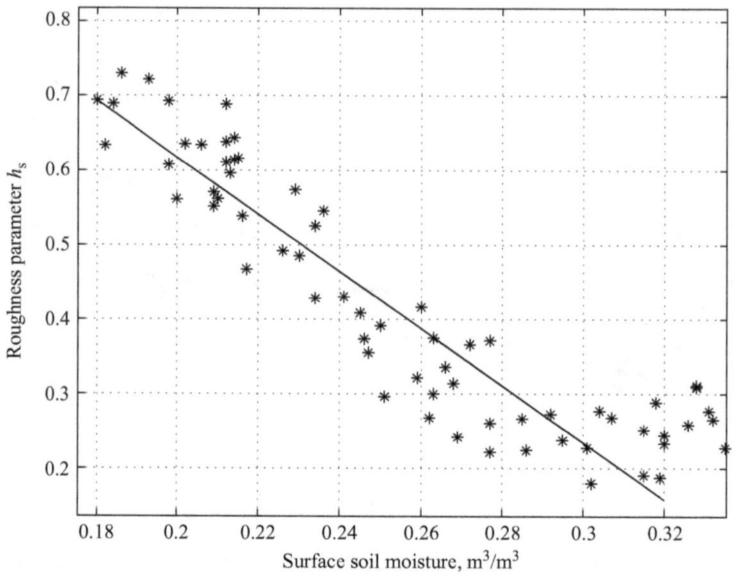

Figure 4.25 Dependence of the roughness parameter h_S on soil moisture – SMOSREX experiment (Escorihuela et al., 2004)

relatively low (about 10 K in terms of rms error on T_B) for soil moisture studies at L-band.

Conversely, physical models allow simulating soil emission for a large range of conditions in terms of soil moisture and geometric roughness. Very interesting features of the soil microwave signature, that could not be revealed by the analysis of the few available experimental datasets, were obtained by Shi *et al.* (2002) and Liou *et al.* (2001) for instance.

These models could simulate very accurately the emission of a variety of rough surfaces. The main problem however is to know whether these surfaces correspond actually to what is seen by the radiometer.

For instance and as mentioned in the introduction, these methods can only simulate 'geometric' effects and cannot account for the heterogeneities in the dielectric constant at the soil surface and within the soil volume. The use of these methods may thus be limited to rather wet soil conditions. Moreover, these models simulate 'ideal' surfaces with homogeneous roughness properties and specific parametrisation of the autocorrelation function. To date these models cannot account for more complex geometric roughness features, such as regular or irregular row patterns that can be found over agricultural sites.

Only, numerical simulation methods, that solve directly the Maxwell equations, allow the computation of soil emission without any approximations. However, it is still very difficult to implement these methods which are time consuming and require a complex description of the 3D soil characteristics.

The most promising approaches are probably combined methods based on the use of theoretical simulations to develop accurate and simple analytical equations that can be fitted and/or compared to experimental datasets. Such approaches were found promising in the estimation of soil moisture by Shi *et al.* (2002, 2003) and should be further evaluated. This will require long term experimental datasets allowing analyses of soil emission for a variety of conditions in terms of moisture, temperature and surface geometry. An example of this kind of 'combined' methods is given in the following section. Even though it has some strong physical basis, the air to soil-transmission model has a simple form that be easily parametrised from comparison with field observations.

4.7 Air-to-soil transition model *Mike Schwank and Christian Mätzler*

4.7.1 Introduction

Within the last three years, the L-band radiometer ELBARA has been applied for bare soil experiments (Schneeberger, Stamm *et al.*, 2004), (Schneeberger, Schwank *et al.*, 2004), a grass land experiment (De Jeu, Holmes *et al.*, 2003), a cold season experiment (Schwank, Stähli *et al.*, 2004) and a measurement campaign on a field site with growing clover-grass vegetation (Schwank *et al.*, 2005). In these ground truth field experiments, ELBARA was mounted on a tower approximately 7 m above ground directed at a fixed observation angle. The radiometer observations were compared with simultaneously measured ground truth dielectric constant and temperature data. The remotely sensed and *in situ* measured data were synchronous and therefore suitable for testing radiative-transfer models by comparing predicted and measured data. The soil reflectivities derived from the radiometer and from the ground truth data were inconsistent when using the Fresnel formula without correcting for soil roughness. To examine possible reasons for the observed discrepancy, other physical soil reflectivity models accounting for the topsoil structure were tested. The models of Wegmüller and Mätzler (1999) and Wigneron *et al.* (2001) are empirical models and thus not suitable to examine causes for the discrepancy.

However, the physically-based surface reflectivity model of Shi *et al.* (2002) was tested in Schneeberger, Schwank *et al.* (2004). This model is a parametrisation of simulated reflectivity data derived from an integral equation model (IEM) (Fung, 1994) applied to a wide range of soil water content and roughness properties. In Schneeberger, Schwank *et al.* (2004) we found that even this physically-based model is not capable of explaining the observed discrepancy between the ground truth and remotely sensed data.

The problem of tackling the roughness effect at L-band leads to the development of a new model for describing the influence of the topsoil structure on the L-band radiation as an impedance matching between the dielectric constants of soil and air. The concept was first proposed by Hüppi (1987), and a model based on similar ideas is described in Cherny and Raizer (1998), in connection with the microwave emission of the ocean, affected by small-scale roughness.

In Section 4.7.2 we discuss possible reasons for the failure of surface scattering models. A detailed description of the air-to-soil transition model is presented in Section 4.7.3. Finally, the model is verified (Section 4.7.4.1) and applied for comparing radiometer measurements with *in situ* measured ground truth data (Section 4.7.4.2).

4.7.2 *Scope of roughness models for L-band observations*

The Fraunhofer criterion (Ulaby, Moore *et al.*, 1982) is a common way of testing whether a surface may be considered electromagnetically smooth in the microwave range:

$$\sigma < \frac{\lambda}{32 \cos \vartheta} \tag{4.88}$$

This criterion relates the observation wavelength λ to the standard deviation σ of the surface height distribution (roughness) and to the observation angle ϑ (relative to the vertical direction). According to the Fraunhofer criterion a surface can be treated as smooth if expression (4.88) is true.

Evaluating the right side of criterion (4.88) for the experimental conditions presented later ($\vartheta = 55°$, $\lambda = 210\,\text{mm}$) yields 11.4 mm. The photogrammetrically determined values (Schneeberger and Willneff, 2003) of σ measured in $30 \times 30\,\text{cm}^2$ areas of the radiometer footprint were between 6 and 9.2 mm during the whole observation period. Hence, the Fraunhofer criterion is fulfilled. However, from our experimental dataset we have clear evidence that criterion (4.88) does not generally hold for the L-band observation of soils (Schneeberger, Schwank *et al.*, 2004). The Fraunhofer criterion only considers the vertical roughness dimension σ assuming that the horizontal wavelength Λ of the soil features is considerably larger than the observation wavelength λ. This implies that the validity of the Fraunhofer criterion is restricted by the Bragg limit:

$$\Lambda > \frac{\lambda}{2 \sin \vartheta} \tag{4.89}$$

For the L-band observation of soil surfaces comprising crumbs, stones, cracks and organic debris, the Bragg limit is not necessarily fulfilled for all the spectral components. Therefore, the Fraunhofer criterion is not sufficient to decide (1) whether roughness has to be considered for L-band observations and (2) whether surface scattering models are applicable to fully describe the effect of small-scale soil roughness. Furthermore, surface Bragg scattering as used in the small perturbation model (Ulaby, Moore *et al.*, 1982) is not capable of solving this problem.

4.7.3 *Model description*

In contrast to Bragg scattering an impedance matching approach between the dielectric constants of air and soil has the potential to describe the effect of roughness components with short wavelengths Λ because it is not restricted by the Bragg limit. This is a consequence of the impedance match taking place over distances significantly smaller than one wavelength. The air-to-soil transition model is best understood in the quasi-static limit where the mean field is homogeneous and extends over a region much larger than the local soil heterogeneity. In this case the mean field determines

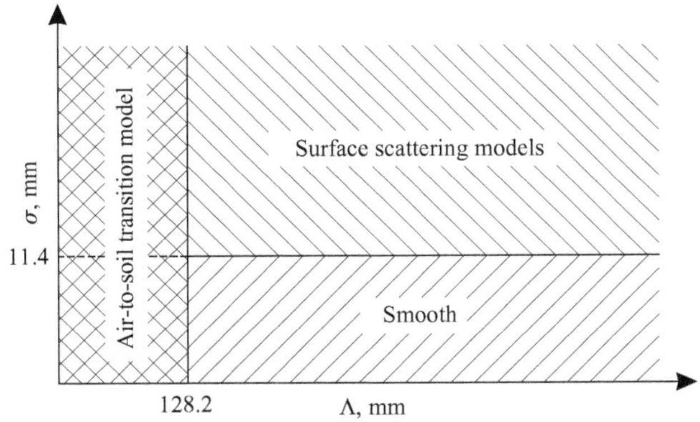

Figure 4.26 *Scope of roughness models depending on the roughness parameters σ and Λ according to the Fraunhofer criterion (4.88) and the Bragg limit (4.89) for the experimental conditions presented in Section 4.7.4 ($\vartheta = 55°$, $\lambda = 210$ mm)*

the behaviour of the wave, and this mean field is influenced by the effective dielectric constant (Sihvola and Kong, 1988). Since its value changes with height, there is a softer transition from the air to the soil than in a case of an abrupt surface, as for example at the calm sea. Figure 4.26 distinguishes between the spectral regions of the air-to-soil transition model and the scattering models for the situation $\vartheta = 55°$, $\lambda = 210$ mm.

The air-to-soil transition model postulates a transition layer of thickness h related to the soil roughness σ (Equation (4.95)). If the transition layer leads to a round-trip phase difference $\Delta\phi$ between partially reflected waves, a distinct reduction of the soil reflectivity R at L-band can be explained. As an example consider a homogeneous transition layer with $h = 8.75$ mm (corresponding to $\sigma \approx 2.17$ mm), $\lambda = 210$ mm and $\varepsilon = 9$ (corresponding to a volumetric soil water content of $\approx 0.10\ \mathrm{m^3/m^3}$) in-between two semi-infinite media with $\varepsilon_{air} = 1$ and $\varepsilon_{soil} = 81$ representing the dielectric constants of air and soil material, respectively. For vertical incidence this results in $\Delta\phi = 2h\varepsilon^{1/2} \cdot 2\pi/\lambda = \pi/2$ and consequently $R = 0$. This simple example illustrates the situation of ideal impedance matching as realised in anti-reflection coatings.

Our model makes use of the fact that a large fraction of the structure in the topsoil is smaller than the observation wavelength λ and, in addition, below the Bragg limit. As a consequence, the structured topsoil appears as a transition layer of thickness h with volume fraction ν_h^{soil} of the soil material which increases with depth. We describe this transition layer with a depth-dependent two-phase mixing model considering the two phases 'air' and 'soil material' (Figure 4.27). In this way, mixing effects caused by cracks, stones, organic debris and topographic features are included. As depicted in Figure 4.27, a more specific interpretation of the transition layer thickness h is the peak-to-peak roughness of the soil surface. Thereby, h is the distance between the deepest depression and the highest peak at the surface in an area

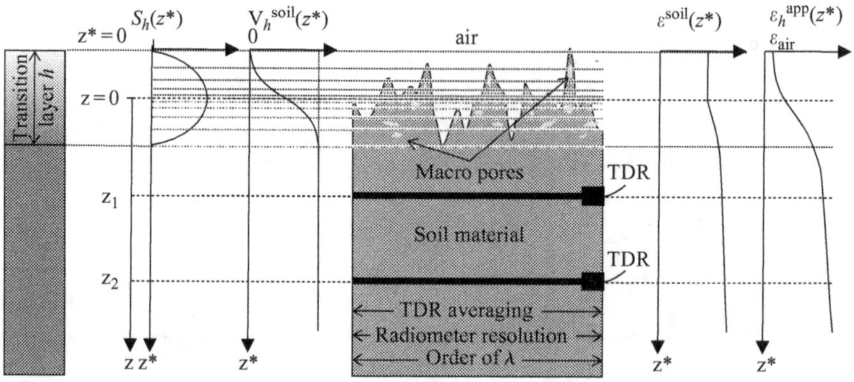

Figure 4.27 *Illustration of the ideas implemented in the air-to-soil transition model. $S_h(z^*)$ is the surface height distribution and $\nu_h^{soil}(z^*)$ its cumulated value corresponding to the volume fraction of the soil material; $z = 0$ is the average surface height of a soil area A of the order $\lambda \times \lambda$; $z^* = 0$ and $z^* = h$ refer to the highest peak and the deepest depression within A. The TDR and temperature probes are located at the depths z_1 and z_2. $\varepsilon^{soil}(z^*)$ is the dielectric constant of soil material (including soil water) and $\varepsilon_h^{app}(z^*)$ is the apparent dielectric constant due to the mixture of air- and soil-phase in the transition layer of thickness h*

corresponding to the resolution limit of the radiometer observation. At L-band this area is of the order of $A = \lambda \times \lambda = 21 \times 21\ \text{cm}^2$. The interpretation of h as peak-to-peak roughness is applicable under the assumption that the proportional air-to-soil transition is dominated by topographical effects.

4.7.3.1 Formulation of the model

The apparent dielectric profile $\varepsilon_h^{app}(z^*)$ relevant for the L-band emission of the soil is described by a dielectric two-phase mixing model (Sihvola, 1999) for the phases 'air' and 'soil'. Inserting the dielectric constants $\varepsilon^{soil}(z^*)$ and ε_{air} of the soil- and air-phase together with their corresponding volume fractions $\nu_h^{soil}(z^*)$ and $1 - \nu_h^{soil}(z^*)$ into the mixing formula yields

$$\varepsilon_h^{app}\left(z^*\right) = \left[\nu_h^{soil}\left(z^*\right) \cdot \varepsilon^{soil}\left(z^*\right)^{\beta} + \left[1 - \nu_h^{soil}\left(z^*\right)\right] \cdot \varepsilon_{air}^{\beta}\right]^{1/\beta} \tag{4.90}$$

The exponent β is set to $\beta = 0.5$ representing the refractive mixing model (Birchak, Gardner *et al.*, 1974).

The transition layer thickness h is interpreted as the topographic roughness parameter and consequently the depth dependence of the volume fraction $\nu_h^{soil}(z^*)$ is related to the height distribution $S_h(z^*)$ of the surface. This becomes obvious when $\nu_h^{soil}(z^*)$

is understood as the cumulative probability density of $S_h(z^*)$:

$$\nu_h^{\text{soil}}\left(z^*\right) = \int_0^{z^*} S_h\left(z'\right) \cdot \mathrm{d}z' \tag{4.91}$$

The depth variable z^* defined to be zero at the highest peak is related to the depth z by $z^* = z + h/2$. We parametrise $S_h(z^*)$ with a normalised quadratic function of z^* comprising h as the roughness parameter (Figure 4.27):

$$S_h\left(z^*\right) = \begin{cases} -\dfrac{6z^{*2}}{h^3} + \dfrac{6z^*}{h^2} & \text{for } 0 \le z^* \le h \\ 0 & \text{for } z^* < 0 \text{ and } z^* > h \end{cases} \tag{4.92}$$

The integral of $S_h(z^*)$ over the depth z^* and hence $\nu_h^{\text{soil}}(z^*)$ is zero above the highest soil peak ($z^* < 0$) and approaches one below the deepest soil depression at $z^* > h$ (Figure 4.27):

$$\nu_h^{\text{soil}}\left(z^*\right) = \begin{cases} 0 & z^* < 0 \\ \dfrac{3z^{*2}}{h^2} - \dfrac{2z^{*3}}{h^3} & 0 \le z^* \le h \\ 1 & z^* > h \end{cases} \tag{4.93}$$

Since $\nu_h^{\text{soil}}(z^*)$ depends on the roughness parameter h, the dielectric profile $\varepsilon_h^{\text{app}}(z^*)$ determining the L-band emission (Equation (4.90) also depends on h). Therefore, we need a relation between h and the experimentally accessible standard deviation σ of the surface height within an area of the order $A = \lambda \times \lambda$:

$$\int_{h/2-\sigma}^{h/2+\sigma} S_h\left(z^*\right) \cdot \mathrm{d}z^* = \frac{1}{\sigma^*\sqrt{2\pi}} \int_{\xi-\sigma^*}^{\xi+\sigma^*} e^{-(z^*-\xi)^2/2\sigma^{*2}} \cdot \mathrm{d}z^* = 0.6827 \tag{4.94}$$

The left side is the probability of finding a surface height within $\pm\sigma$ around the average surface height at $z^* = h/2$ for the quadratic height distribution $S_h(z^*)$ given by Equation (4.92).

This integral is evaluated and equated with the corresponding $\pm\sigma^*$-probability for the Gaussian distribution with standard deviation σ^* centred around ξ. This relates the roughness parameter h to the standard deviation σ of the surface height:

$$\sigma\left(h\right) = 0.2479 \cdot h \tag{4.95}$$

Relation (4.95) states that the experimentally accessible roughness σ is approximately one quarter of the corresponding roughness parameter h.

4.7.3.2 Calculation of soil reflectivity

The procedure for calculating the soil reflectivity $R(h)$ considering roughness effects in terms of the air-to-soil transition model is described here.

As illustrated in Figure 4.28, the calculation of $R(h)$ is based on water content and temperature profiles, $\theta(z_i)$ and $T(z_i)$. These profiles are derived from *in situ* measurements and interpreted as ground truth data. The measurements were performed at five depths z_i relative to the average surface height at $z = 0$ (corresponding to $z^* = h/2$).

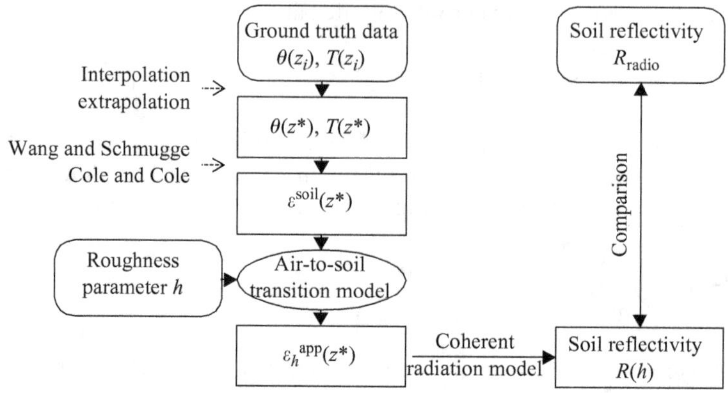

Figure 4.28 *Flow chart illustrating the procedure for calculating the soil reflectivity R(h) using the air-to-soil transition model based on in situ measured ground truth data $\theta(z_i)$ and $T(z_i)$. The comparison with the soil reflectivity, R_{radio} derived from the radiometer measurement is presented in Section 4.7.4*

This interpretation of the measurement depths z_i is consistent with the averaging effect of the horizontally installed TDR (time domain reflectometer) probes. The water content is spatially averaged over the rod-length of the TDR probes which is of the same order as the L-band wavelength λ (Figure 4.27). Therefore, $\theta(z_i)$ is interpreted as the soil-water content profile disregarding the increasing volume fraction of air with decreasing depth due to the small-scale roughness.

Due to the coarse depth resolution the profiles $\theta(z_i)$ and $T(z_i)$, must be interpolated and/or extrapolated to construct the profiles $\theta(z^*)$ and $T(z^*)$ at the required resolution. The depth resolution has to be finer than one-tenth of the observation wavelength λ to allow calculation of the soil reflectivity $R(h)$. Furthermore, the profiles $\theta(z^*)$ and $T(z^*)$ have to be defined for all $z^* \geq 0$. In this work we apply third-order polynomials to interpolate $\theta(z_i)$ and $T(z_i)$ below the uppermost *in situ* probe installed at the depth z_1. The temperature profile $T(z^*)$ was assumed to be constant above z_1 with the value of $T(z_1)$. The extrapolation applied to the data $\theta(z_i)$ representing the profile $\theta(z^*)$ between z_1 and the average soil surface height at $z^* = h/2$ is adapted to the measurement $\theta(z_1)$ observed with the uppermost TDR probe (Equation (4.98)). From $z^* = h/2$ to the highest soil peak at $z^* = 0$ the water content profile $\theta(z^*)$ of the soil material is assumed to be constant with the extrapolated value $\theta_{\mathrm{surface}}$ at $z^* = h/2$:

$$\theta(z^*) = \begin{cases} \theta(z^*) & \text{for } z^* > h/2 \\ \theta_{\mathrm{surface}} = \theta(h/2) & \text{for } 0 \leq z^* \leq h/2 \end{cases} \tag{4.96}$$

From the profiles $\theta(z^*)$ and $T(z^*)$ we calculated dielectric profiles $\varepsilon^{\mathrm{soil}}(z^*)$ of the soil material using the empirical model for the complex dielectric permittivity of soils as a function of water content developed by Wang and Schmugge (1980). This model accounts for soil texture and porosity as well as for bound and free water (see Section 5.7.5). The dielectric constant of water was corrected for temperature effects according to the model of Cole and Cole (1941).

In the next step, the profile $\varepsilon^{\text{soil}}(z^*)$ is used to calculate the apparent dielectric profile $\varepsilon_h^{\text{app}}(z^*)$ according to Equation (4.90).

Finally, the soil reflectivity $R(h)$ is calculated from $\varepsilon_h^{\text{app}}(z^*)$ by applying a coherent radiation-transfer model for layered dielectric media. We used a model based on a matrix formulation of the boundary conditions at the layer interfaces derived from Maxwell's equations (Bass, Stryland *et al.*, 1995). The corresponding algorithm was implemented in Mathematica 5 and a Matlab representation is also available from the authors.

The resulting $R(h)$ are compared to the reflectivities R_{radio} derived from the radiometrically measured brightness temperature T_{B}. At L-band, and assuming thermodynamic equilibrium, R_{radio} is extracted from

$$T_{\text{B}} = (1 - R_{\text{radio}})\, T_{\text{A}} + R_{\text{radio}} T_{\text{sky}} \tag{4.97}$$

where T_{A} is the thermodynamic temperature of the observed scene and T_{sky} is the radiation temperature of the sky.

4.7.4 *Comparison between radiometer and ground truth data*

In Section 4.7.4.1 we discuss the validation of the proposed air-to-soil transition model in terms of investigating the sensitivity of calculated soil reflectivities to the extrapolation applied to the data $\theta(z_i)$ and $T(z_i)$. Section 4.7.4.2 demonstrates an application to compare time series of radiometrically measured reflectivity data with reflectivities derived from the ground truth data. These reflectivities are calculated following the procedure presented in Section 4.7.3.2 (Figure 4.28).

The database used consists of horizontally polarised L-band measurements of T_{B} conducted on a bare arable field near Zürich (Switzerland). A detailed description of the experimental conditions and the instruments is reported in Schneeberger, Stamm *et al.* (2004) and Mätzler, Weber *et al.* (2003). The L-band radiometer was mounted on a 7 m high tower at the observation angle (relative to the vertical direction) of $\vartheta = 55°$, resulting in an elliptic footprint of approximately $9 \times 8\,\text{m}^2$. Soil profiles were measured *in situ* within the main target area of the radiometer. We installed 15 two-rod TDR probes at five depths ($z_1 = 2$ cm, $z_2 = 7$ cm, $z_3 = 15$ cm, $z_4 = 25$ cm, $z_5 = 45$ cm). The rods are 25 cm in length each, 0.5 cm in diameter and separated by 3 cm.

4.7.4.1 Model verification

The depth resolution of the ground truth data $\theta(z_i)$ and $T(z_i)$ is limited for practical reasons, especially within the topsoil. Due to the fact that the upper soil region strongly influences the soil reflectivity $R(h)$, it is expected that the chosen extrapolation applied to $\theta(z_i)$ and $T(z_i)$ affects the calculated $R(h)$ considerably. Therefore, we discuss the effect of the most extreme extrapolation approaches applied to $\theta(z_i)$. We will see that independent of the chosen extrapolation the air-to-soil transition model is able to explain the remotely sensed observations. On the other hand, for a flat surface ($h = 0$ mm) the L-band measurements cannot be explained by the ground truth data.

The extrapolation is applied in steps of 1 mm to the depth range $0 \leq z \leq z_1$, corresponding to the topsoil between the uppermost *in situ* probes at $z_1 = 2$ cm and the average surface height at $z^* = h/2$. The temperature profile $T(z^*)$ was assumed to be constant above z_1 and equal to $T(z_1)$. The soil phase water-content $\theta(z^*)$ within the range $h/2 \geq z^* \geq 0$ (between the average surface height and the highest soil peak) was assumed to be constant and equal to θ_{surface} (Equation (4.96)). Consequently, $\varepsilon^{\text{soil}}(z^*)$ is also constant within the depth range $h/2 \geq z^* \geq 0$ (Figure 4.27).

The investigation of the influence of the extrapolation on $R(h)$ is performed by means of two pairs of ground truth profiles $\theta_{\text{wet}}(z_i)$, $T_{\text{wet}}(z_i)$ and $\theta_{\text{dry}}(z_i)$, $T_{\text{dry}}(z_i)$, as measured under wet and dry conditions, respectively. With these representative profiles, the influence of the extrapolation is discussed. The two pairs of profiles were constructed by averaging the measurements over Julian day 157 (which was a wet day) and over Julian day 173 (which was a dry day). In this way the possible effects of a temperature gradient are eliminated.

Dry conditions. These conditions were associated with $\theta(z_1) \approx 0.16\,\text{m}^3/\text{m}^3$. The right graph of Figure 4.29 shows the resulting dependence of calculated reflectivities $R_{\text{dry}}(h)$ for horizontal polarisation on the roughness parameter h and on the extrapolation. The lack of topsoil ground truth data is taken care of by assuming two extreme examples of profile extrapolations as illustrated in the left graph of Figure 4.29: A constant water content $\theta(z_1)$ (dashed line) and an extrapolation to $\theta_{\text{surface}} = 0\,\text{m}^3/\text{m}^3$ representing a fully desiccated soil surface (dotted line). The extrapolation to the desiccated surface was realised by a quadratic function with the same gradient at z_1 as the fitted profile $\theta_{\text{dry}}(z^*)$. As can be seen, $R_{\text{dry}}(h)$ decreases with increasing h. For $h > 35$ mm the soil reflectivities $R_{\text{dry}}(h)$ approach each other for the two extrapolations. This means that the shape of the water content distribution in the topsoil becomes less important with increasing soil roughness. The measured reflectivity $R_{\text{radio,dry}} = 0.32$ for horizontal polarisation is reproduced by the model calculation $R_{\text{dry}}(h_{\text{dry}}^{\text{opt}})$ with the optimised roughness parameter $h_{\text{dry}}^{\text{opt}} \approx 34$ mm independent of the chosen extrapolation. With $h = 25$ mm and $\theta_{\text{surface}} = 0\,\text{m}^3/\text{m}^3$, the deviation between the model and the radiometer observation is $\Delta_{\text{dry}} = 0.022$.

Wet conditions. These conditions were associated with $\theta(z_1) \approx 0.30\,\text{m}^3/\text{m}^3$. The right graph of Figure 4.30 shows the dependence of $R_{\text{wet}}(h)$ for horizontal polarisation on h and on the extrapolation. The lack of ground truth data in the topsoil is taken care of by assuming two extreme extrapolations as illustrated in the left graph of Figure 4.30: A constant water content $\theta(z_1)$ (dashed line) and an extrapolation to $\theta_{\text{surface}} = \theta_{\text{sat}} = 0.5\,\text{m}^3/\text{m}^3$ representing a saturated soil surface (dotted line). The extrapolation to the saturated surface was realised by a corresponding quadratic extrapolation. The decrease of $R_{\text{wet}}(h)$ with increasing h is clearly visible. The difference in $R_{\text{wet}}(h)$ for the two extrapolation approaches also becomes smaller with increasing roughness but is less pronounced compared with the behaviour in dry conditions. At wet conditions, the reflectivity $R_{\text{radio,wet}} = 0.55$ measured at horizontal polarisation is reproduced by the model for the range $12 < h_{\text{wet}}^{\text{opt}} < 21$ mm depending

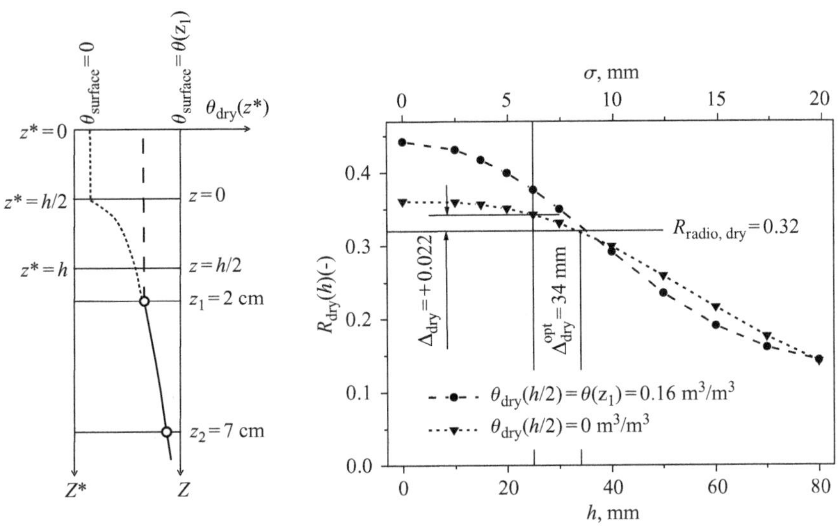

Figure 4.29 *Influence of h on the horizontally polarised soil reflectivity* R_{dry} *(h) for dry conditions for two different assumptions made on the extrapolation of the in situ measurements applied within the top 2 cm (dotted and dashed lines of the left graph). For comparison, the soil reflectivity* $R_{\text{radio,dry}}$ *measured at horizontal polarisation, as well as the smallest deviation* Δ_{dry}*, from the reflectivity* R_{dry} *(h) calculated for h* $= 25$ *mm is indicated*

on the chosen extrapolation. With $h = 25$ mm and assuming $\theta_{\text{surface}} = \theta_{\text{sat}}$ under wet conditions, the deviation between the model and the radiometer observation is $\Delta_{\text{wet}} = -0.031$.

In both wet and dry conditions the air-to-soil transition model can reproduce the radiometrically measured reflectivity from the ground truth data for optimised roughness parameters $h^{\text{opt}}_{\text{wet}}$ and $h^{\text{opt}}_{\text{dry}}$, respectively. On the other hand, a completely flat surface ($h = 0$ mm) cannot reproduce measured values with any reasonable extrapolation. As already mentioned, even existing roughness models are not capable of solving this problem.

Under the assumption that h has to be the same for wet and dry conditions, $R_{\text{radio,wet}}$ and $R_{\text{radio,dry}}$ are best fitted with $h = 25$ mm, extrapolation of the profiles $\theta(z_i)$ to saturation for wet condition and to the desiccated soil surface state at dry condition. This finding is used in the following section to reproduce a time series of reflectivity data R_{radio} measured at horizontal polarisation from simultaneously recorded ground truth measurements. However, requiring h to be constant is questionable due to the fact that the dielectric structure of the topsoil (roughness) may be modified by rainfall events. Therefore, we will also demonstrate the possibility of achieving an even better agreement between R_{radio} and modelled data allowing a water content dependent roughness.

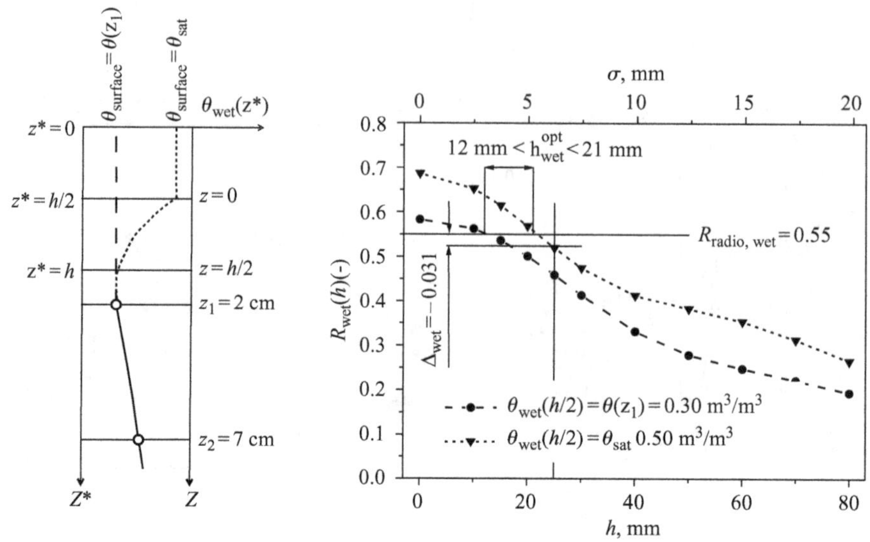

Figure 4.30 *Same as Figure 4.29 but for wet conditions. For comparison, the measured soil reflectivity $R_{radio,wet}$, as well as the smallest deviation Δ_{wet}, from the reflectivity R_{wet} (h) calculated for h = 25 mm is indicated*

4.7.4.2 Fitting a time series of radiometer measurements

Figure 4.31(a) and (b) shows the comparison between horizontally polarised reflectivities R_{radio} (bold black line) and various modelled reflectivity data. The modelled reflectivities are based on the ground truth measurements $\theta(z_i)$ and $T(z_i)$ and are calculated following the procedure described in Section 4.7.3.2 (Figure 4.28).

The dotted line in Figure 4.31(a) illustrates the reflectivity $R^{in\ situ}(h = 0\,\text{mm})$ calculated without considering roughness ($h = 0$ mm) and assuming that the water content above z_1 is constant ($\varepsilon^{soil}(z^*) = \varepsilon_h^{app}(z^*) = \varepsilon(z_1)$ for $z^* = z \le z_1$). The data $R(h = 25\,\text{mm})$ (solid thick grey line) and $R(h = 0\,\text{mm})$ (solid thin black line) were calculated for $h = 25$ mm and $h = 0$ mm, respectively, under the assumption that the extrapolated surface water content $\theta_{surface}$ at $z^* = h/2$ depends on the water content $\theta(z_1)$ as measured with the uppermost TDR probe (Figure 4.31(c)) according to the following relation (all water contents in m^3/m^3):

$$\theta_{surface} = (\theta(z_1)) = \begin{cases} 0 & \theta(z_1) < 0.2 \\ 0.05 \cdot (\theta(z_1) - 0.2) & 0.2 \le \theta(z_1) \le 0.3 \\ \theta_{sat} = 0.5 & \theta(z_1) > 0.3 \end{cases} \tag{4.98}$$

This water content dependent extrapolation is in accordance with the perceptions made in Section 4.7.4.1 (Figures 4.29 and 4.30) where the soil surface was assumed to be desiccated for dry conditions ($\theta_{surface} = 0\,m^3/m^3$ for $\theta(z_1) < 0.2\,m^3/m^3$) and saturated for wet conditions ($\theta_{surface} = \theta_{sat} = 0.5\,m^3/m^3$ for $\theta(z_1) > 0.3\,m^3/m^3$).

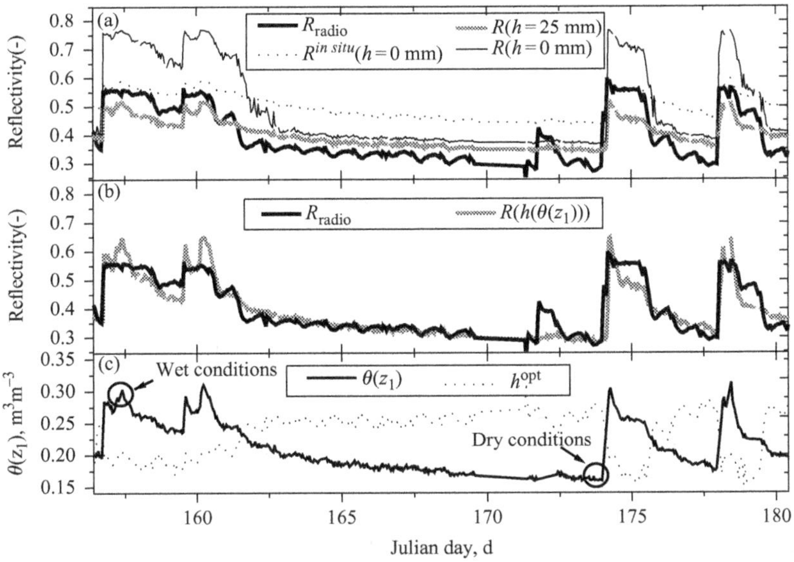

Figure 4.31 *Reflectivity R_{radio} measured with the L-band radiometer at horizontal polarisation compared with calculated reflectivities based on ground truth data $\theta(z_i)$ and $T(z_i)$. The data $R^{in\ situ}$ ($h = 0$ mm) were calculated for horizontal polarisation without considering the air-to-soil transition model. $R(h = 25$ mm) and $R(h = 0$ mm) were calculated by applying the air-to-soil transition model considering the water content dependent extrapolation (Equation (4.98)) with $h = 25$ mm and $h = 0$ mm, respectively. (b) R_{radio} compared with modelled reflectivities considering the water content dependent roughness (Equation (4.98)) and extrapolation (Equation (4.98)). (c) Water content $\theta(z_1)$ was measured with TDR probes installed at $z_1 = 2$ cm and optimises roughness parameter h^{opt}. The data labelled with wet- and dry-conditions refer to the discussion given in Section 4.7.4.1*

As can be seen, for $h = 25$ mm assumed to be the same in wet and dry conditions, $R(h = 25$ mm) is best adapted to R_{radio}. The value $h = 25$ mm corresponds to a standard deviation σ of the surface height of approximately 6.2 mm (Equation (4.95)). This is of the same order as the measured σ within an area of 30×30 cm^2, which was between 6 and 9.2 mm during the whole observation period.

As can be seen in Figure 4.31(a), the modelled data $R(h = 25$ mm) are better adapted to the measured data R_{radio} compared to the data $R^{in\ situ}(h = 0$ mm) and $R(h = 0$ mm). However, the data $R(h = 25$ mm) still show a systematic lower dynamic range compared to R_{radio}, possibly due to the unrealistic assumption that the roughness is independent of the soil water content. Local ponding and subsequent silting due to rainfall was observed. Therefore, an apparent decreasing roughness with increasing water content is supposed, capable of partially explaining the discrepancy between remotely sensed and ground truth measurements.

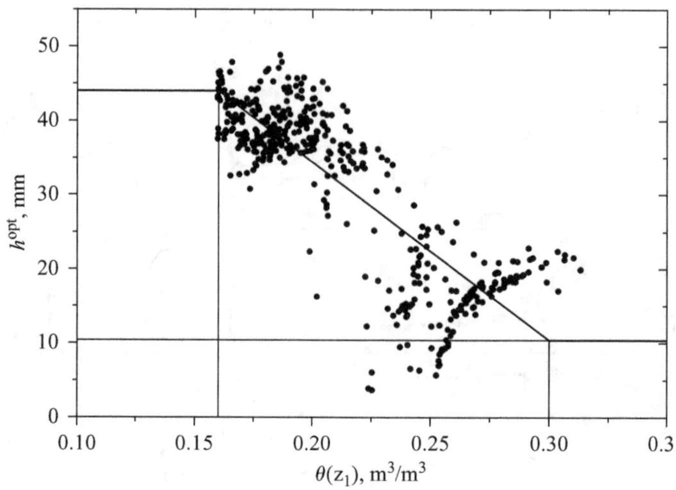

Figure 4.32 *Optimised roughness parameter $h^{opt}(\theta(z_1))$ versus ground truth water content $\theta(z_1)$ so that $R(h^{opt}) = R_{radio}$ for horizontal polarisation. The solid black line represents the approximation given by Equation (4.98)*

This conjecture is supported by investigating the optimal roughness parameters h^{opt} leading to total agreement between R_{radio} and modelled reflectivities $R(h^{opt})$ at any time. We computed h^{opt} as a function of time, revealing an anti-correlation with $\theta(z_1)$ (Figure 4.31(c)). Figure 4.32 shows h^{opt} plotted versus the corresponding $\theta(z_1)$, showing the decrease of h^{opt} with increasing $\theta(z_1)$ as supposed from observations. A similar behaviour was observed in Figure 4.25. However, this could also partly result from the electromagnetic problem of overestimating $\varepsilon_h^{app}(z^*)$ under dry conditions. Using a better adapted representation of the dielectric mixing similar to the Maxwell Garnett formula (Sihvola, 1999) may help to explain this observation. The fitted relation between h^{opt} and $\theta(z_1)$ is given by (all water contents in m^3/m^3):

$$h^{opt}(\theta(z_1)) = \begin{cases} 44 & \theta(z_1) < 0.16 \\ 82.4 - 240 \cdot \theta(z_1) & 0.16 \le \theta(z_1) \le 0.3 \\ 10.4 & \theta(z_1) > 0.3 \end{cases} \tag{4.99}$$

This relation is applied to calculate reflectivities $R(h(\theta(z_1)))$ for horizontal polarisation considering both a water content dependent roughness and a water content dependent extrapolation approach (Equation (4.98)). The comparison between $R(h(\theta(z_1)))$ and R_{radio} is depicted in Figure 4.31(b). The dynamic range of the modelled data is now mostly coincident with the measurements. This illustrates that $R(h(\theta(z_1)))$ represents a further improvement of the model.

4.7.5 Fast model

A multi-parameter regression formula is derived allowing the air-to-soil transition model to be applied with little computational effort. Therefore the model is evaluated for a restricted parameter range and the dielectric profile $\varepsilon^{\text{soil}}(z) = \varepsilon$ of the soil is assumed to be real and constant in depth. The procedure described in Section 4.7.3.2 was applied for calculating reflectivities $R^p(h, \varepsilon, \vartheta)$ for horizontal and vertical polarisation (p = hor, vert) as a function of the peak-to-peak roughness h, uniform soil dielectric constant ε and observation angle ϑ. For investigating the reflectivity dependence from these parameters, we define the relative reflectivity $\mathbf{R}^p(\Sigma, \varepsilon, \vartheta)(p = \text{hor, vert})$ normalised to the specular reflectivity $R_{\text{specular}}(\varepsilon, \vartheta)$ (= Fresnel-reflectivity) as

$$\mathbf{R}^p(\Sigma, \varepsilon, \vartheta) = \frac{R^p(\Sigma, \varepsilon, \vartheta)}{R^p_{\text{specular}}(\varepsilon, \vartheta)} \quad \text{with } p = \text{hor, vert} \tag{4.100}$$

The relative roughness $\Sigma \equiv \sigma/\lambda$ is defined as the standard deviation $\sigma = 0.2479 \cdot h$ (Equation (4.95)) of the soil surface height divided by the observation wavelength λ.

Figure 4.33 shows $\mathbf{R}^p(\Sigma, \varepsilon, \vartheta)$ as a function of Σ calculated for the angles $\vartheta = 0°$ and $55°$ each with $\varepsilon = 8$ and 20 corresponding to soil water contents of $\theta_{\text{dry}} = 0.16\,\text{m}^3/\text{m}^3$ and $\theta_{\text{wet}} = 0.30\,\text{m}^3/\text{m}^3$ (compare Section 4.7.3.1). The range of the relative roughness Σ is confined to $0 \leq \Sigma \leq 0.062$ corresponding to the range $0 \leq H \leq 1/4$ of the relative peak-to-peak roughness $H = h/\lambda$. Within this parameter range, $\mathbf{R}^p(\Sigma, \varepsilon, \vartheta)$ decreases with increasing Σ, whereas the decline is more pronounced for $\varepsilon = 20$ (wet conditions) than for $\varepsilon = 8$ (dry conditions).

The relative reflectivity $\mathbf{R}^p(\Sigma, \varepsilon, \vartheta)$ is approximated by a multiple regression $\Re^p(\Sigma, \varepsilon, \vartheta)$ (p = hor, vert) valid within the parameter range $0 \leq \Sigma \leq 0.062, 8 \leq \varepsilon \leq 20$ and $0 \leq \vartheta \leq 1$ (= 55°) outlined in Figure 4.33. The decrease of $\mathbf{R}^p(\Sigma, \varepsilon, \vartheta)$ with increasing Σ is approximated by Gaussian functions with fitted widths $a^p(\varepsilon, \vartheta)$ for all ε and ϑ within the parameter range and for horizontal and vertical polarisation (p = hor, vert):

$$\Re^p(\Sigma, \varepsilon, \vartheta) = \exp\left(-\frac{\Sigma^2}{2a^p(\varepsilon, \vartheta)^2}\right) \tag{4.101}$$

Applying a further fit involving the variable ε and the parameters $k_1^p(\vartheta)$, $k_2^p(\vartheta)$ (p = hor, vert) allows approximating the widths $a^p(\varepsilon, \vartheta)$ by

$$a^p(\varepsilon, \vartheta) = k_1^p(\vartheta) + \frac{k_2^p(\vartheta)}{\varepsilon} \tag{4.102}$$

This approach accounts for the more pronounced decline of $\mathbf{R}^p(\Sigma, \varepsilon, \vartheta)$ at higher ε which requires $a^p(\varepsilon, \vartheta)$ to decrease with increasing ε. The fit parameters $k_1^p(\vartheta)$ and $k_2^p(\vartheta)$ calculated for the angles ϑ are represented by the

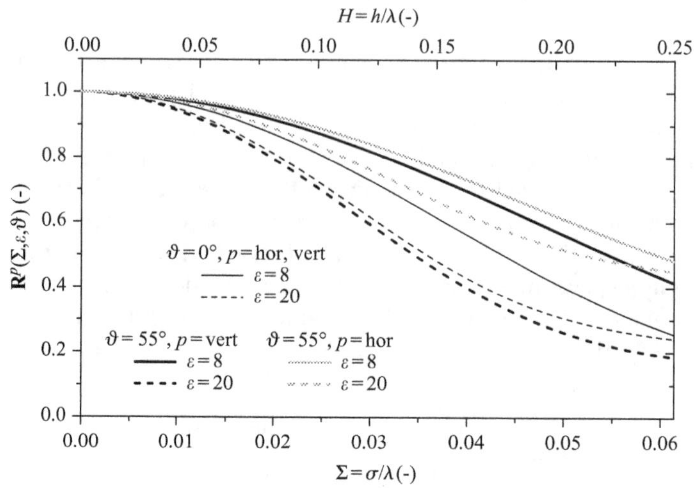

Figure 4.33 *Relative reflectivity* $\mathbf{R}^p(\Sigma, \varepsilon, \vartheta)$ *calculated using the air-to-soil transition model*

following fits:

$$\begin{aligned}
k_1^{\text{hor}}(\vartheta) &= 0.0283663 + \vartheta^2 0.00813961 + \vartheta^4 0.00259038 + \vartheta^8 0.000858079 \\
k_2^{\text{hor}}(\vartheta) &= 0.0714485 + \vartheta^2 0.019949 + \vartheta^4 0.00432687 + \vartheta^8 0.00182565 \\
k_1^{\text{vert}}(\vartheta) &= 0.0284217 - \vartheta^2 0.00773684 + \vartheta^4 0.00381838 - \vartheta^8 0.00713309 \\
k_2^{\text{vert}}(\vartheta) &= 0.0702825 + \vartheta^2 0.0955159 - \vartheta^4 0.104509 + \vartheta^8 0.194922
\end{aligned} \tag{4.103}$$

The multiple regression $\Re^p(\Sigma, \varepsilon, \vartheta)$ given by the Equations (4.101) to (4.103) approximate the modelled relative reflectivity $\mathbf{R}^p(\Sigma, \varepsilon, \vartheta)$ within the restricted parameter range with standard deviations of0.0159444 and 0.0139947 for horizontal and vertical polarisation.

4.7.6 Summary

A new model was developed to explain the effect of dielectric small-scale features (in the horizontal sense) within the topsoil on the L-band radiation. This air-to-soil transition model makes use of the fact that structure significantly smaller than the observation wavelength can influence the impedance mismatch between air and soil. This allows describing the small-scale topography and the volume fraction ν_h^{soil} of soil material as a surface height distribution and its cumulated distribution, both comprising roughness parameter h. From this, the dielectric profile $\varepsilon_h^{\text{app}}$, relevant for the L-band emission, was calculated by means of a dielectric two-phase mixing model applied to the phases 'air' and 'soil material' with their corresponding volume fractions. The resulting profile $\varepsilon_h^{\text{app}}$ was utilised to calculate the reflectivity $R(h)$

by applying a coherent radiation-transfer model for layered dielectric media. $R(h)$ was interpreted as the roughness dependent soil reflectivity R_{radio} extracted from the measured L-band radiation at horizontal polarisation.

For the case of a homogeneous dielectric soil profile a multiple parameter fit approximating the modelled reflectivities was calculated. The resulting formula allows for calculating soil reflectivities considering the air-to-soil transition model with little calculation effort as required in inverse modelling approaches.

4.8 Microwave emissivity in arid regions: What can we learn from satellite observations? *Catherine Prigent*

4.8.1 Introduction

In arid and semi-arid regions, the passive microwave signal emerging from the surface is primarily sensitive to the surface roughness (at small and large topographic scales), to the dielectric properties of the medium (related to the soil composition, texture and humidity) and potentially to volume scattering below the surface. These interactions are complex to model, being dependent on a large number of highly variable and difficult to access parameters. Efforts have been made to better understand the mechanisms responsible for the microwave emission of arid surfaces, from both theoretical analysis and field experiments. Model developments include complex simulations of bare soil with integral electromagnetic models (e.g. Shi, 2002). Ground or airborne-based measurements have been performed (e.g. Calvet *et al.*, 1996) to help develop the models. However, even assuming that a perfect land surface emissivity model exists, would the inputs it will require (e.g. soil composition, texture, humidity and surface roughness) be available on a global basis with a resolution compatible with the satellite one and with the required accuracy?

Very few groups have so far examined the problem of microwave land surface emissivities at a global scale. Global land surface emissivity maps were produced at special sensor microwave/imager (SSM/I) frequencies by Prigent *et al.* (1997, 1998), directly from the satellite observations, by removing the contribution of the atmosphere, clouds, rain and the surface temperature using ancillary data. The emissivities are estimated for SSM/I observation conditions, i.e. for a 53° zenith angle at 19.35, 22.235, 37.0 and 85.5 GHz for both orthogonal polarisations. Advanced microwave sounder unit (AMSU) emissivities have also been calculated, up to 150 GHz (Karbou *et al.*, 2005; Prigent *et al.*, 2005c). On the other hand, Weng *et al.* (2001) and Pellerin *et al.* (2003) chose to develop a global model to estimate the emissivity for the various surface conditions encountered over the continents, using different radiative transfer solutions depending on the surface characteristics. Model inputs are provided by a land surface model, such as the one in the Global Data Assimilation System of the National Center for Environmental Prediction (NCEP) (Weng *et al.*, 2001). Land surface emissivity maps obtained from direct calculation based on SSM/I satellite observations are first presented and compared with model calculations. The next section describes three specific lessons that have been learnt from the direct emissivity calculations on a large scale.

4.8.2 Direct estimates of emissivities from satellite observations and comparison with model calculations in arid regions

The SSM/I instruments on board the DMSP polar orbiters observe the earth twice daily at 19.35, 22.235, 37.0 and 85.5 GHz with both vertical and horizontal polarisations, with the exception of 22 GHz which is vertical polarisation only. The observing incident angle is close to 53°, and the elliptical fields of view decrease in size proportionally with frequency, from 43×69 to $13 \times 15\,\text{km}^2$. Microwave emissivity calculations are performed by subtracting the contributions from the atmosphere, clouds and rain using ancillary data from the International Satellite Cloud Climatology Project (ISCCP) (Rossow and Schiffer, 1999) and the National Centers for Environmental Prediction (NCEP analyses) (Kalnay *et al.*, 1996). The method is fully described by Prigent *et al.* (1997, 1998) and summarised here.

Over a flat lossy surface, the integrated radiative-transfer equation in the Rayleigh–Jeans approximation, for a non-scattering plane-parallel atmosphere, can be expressed in terms of brightness temperature for a given polarisation state p:

$$T_{\text{bp}} = T_{\text{surf}} \times \epsilon_{\text{p}} \times e^{-\tau(0,H)/\mu} + T_{\text{atm}}^{\downarrow} \times (1 - \epsilon_{\text{p}}) \times e^{-\tau(0,H)/\mu} + T_{\text{atm}}^{\uparrow} \tag{4.104}$$

with $T_{\text{atm}}^{\downarrow} = \int_H^0 T(z)\alpha(z)e^{-\tau(z,0)/\mu}\,dz$ and $T_{\text{atm}}^{\uparrow} = \int_0^H T(z)\alpha(z)e^{-\tau(z,H)/\mu}\,dz$. T_{bp} is the brightness temperature measured by the satellite for polarisation state p, T_{surf} is the surface 'skin' temperature, ϵ_{p} is the surface emissivity for polarisation state p, $\mu = \cos(\theta)$, θ being the incidence angle on the surface, $\alpha(z)$ is the atmospheric absorption by gases at altitude z, $T(z)$ is the atmospheric temperature at altitude z, $\tau(z_0, z_1) = \int_{z_0}^{z_1} \alpha(z)\,dz$ is the atmospheric extinction from z_0 to z_1 and H is the orbiter height.
This leads to

$$\epsilon_{\text{p}} = \frac{T_{\text{bp}} - T_{atm}^{\uparrow} - T_{\text{atm}}^{\downarrow} \times e^{-\tau(0,H)/\mu}}{e^{-\tau(0,H)/\mu} \times (T_{\text{surf}} - T_{\text{atm}}^{\downarrow})} \tag{4.105}$$

At frequencies above ~10 GHz, the radiation is expected to emanate from only a thin surface layer of bare soil with the penetration depth of the order of the wavelength with the following consequences: If there is no volume scattering, the surface temperature is the skin temperature and for flat surfaces the reflection is quasi-specular. However, volume scattering can be involved in the cases of very dry sand since the microwave radiation can arise from below the surface. When the terrain is rough on scales between the radiation wavelength and the size of the field of view, the surface acts as a set of scattering facets with a complex distribution of orientations. In these cases (4.104) and (4.105) involve some 'effective' emissivity and temperature, aggregated over the depth of penetration and the field of view of the satellite instrument. The method consists solving the radiative-transfer Equation (4.105) for the 'effective' surface

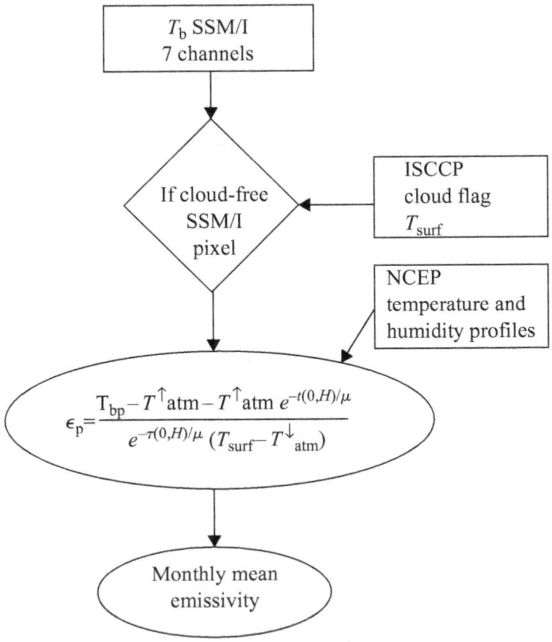

Figure 4.34 Schematic presentation of the emissivity calculation process

emissivity for each channel using ancillary data to specify the atmospheric and other surface parameters.

Cloud-free SSM/I observations are first isolated using collocated visible/infrared satellite observations (ISCCP data). The cloud-free atmospheric contribution is then calculated using a radiative-transfer model and an estimate of the local atmospheric temperature–humidity profile from NCEP reanalysis. Finally, with the surface skin temperature derived from IR observations (ISCCP estimate), the surface emissivity is calculated for all seven SSM/I channels. The standard deviation of day-to-day variations of the emissivities retrieved over a month are typically 0.013. An emissivity error analysis was conducted (Prigent *et al.*, 1997) and concluded in a 1 per cent accuracy for the retrieved emissivities. Monthly mean values are calculated with a spatial resolution of 0.25° at the equator. For the sake of simplicity in the following, the term surface emissivity will be used instead of 'effective' surface emissivity. Figure 4.34 summarises the different steps in this emissivity calculation.

Figure 4.35 shows a global monthly-mean emissivity map at 37 GHz horizontal polarisation for July 1993. Desert areas are clearly noticeable and are characterised by large emissivity differences and low emissivities at horizontal polarisation. Tropical rain forests display high emissivities at both polarisations and negligible emissivity polarisation differences. Water surfaces (lakes, rivers and wetlands) show high emissivity polarisation differences at 37 GHz. The major river systems (e.g. Congo and Amazon) and wetlands (e.g. Pantanal in South America) appear clearly.

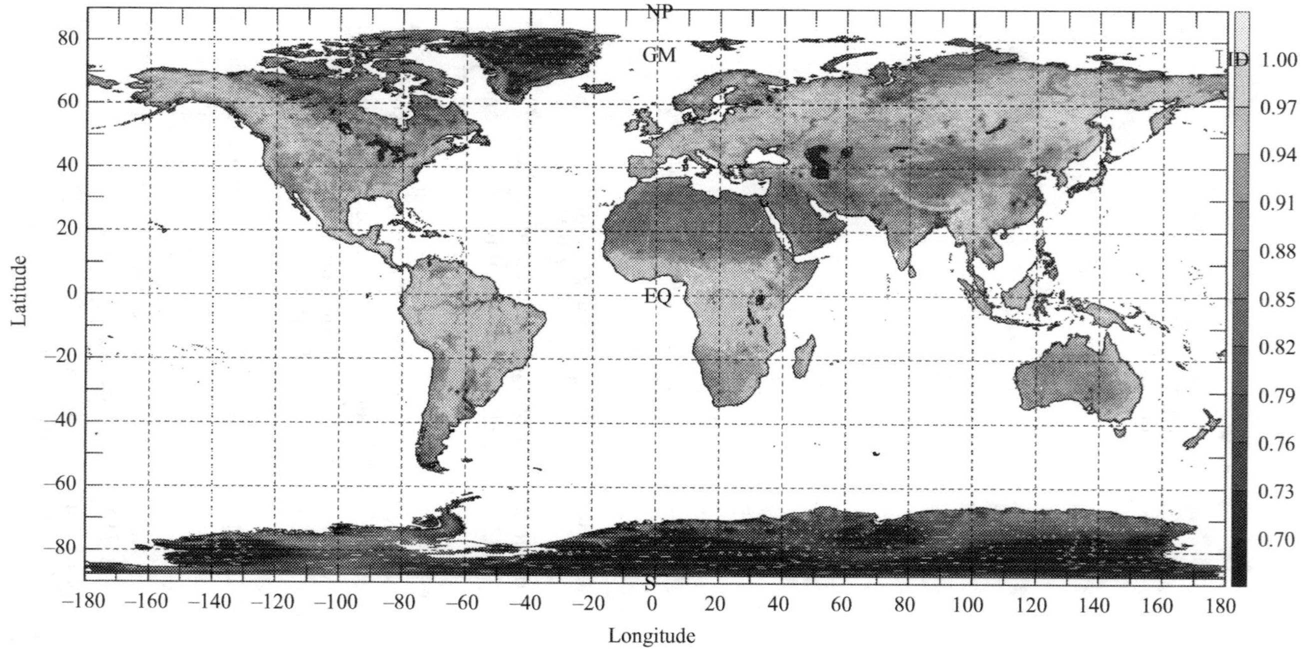

Figure 4.35 *Monthly-mean emissivity map calculated from SSM/I observations at 37 GHz (horizontal polarisation) at 53° incidence for July 1993*

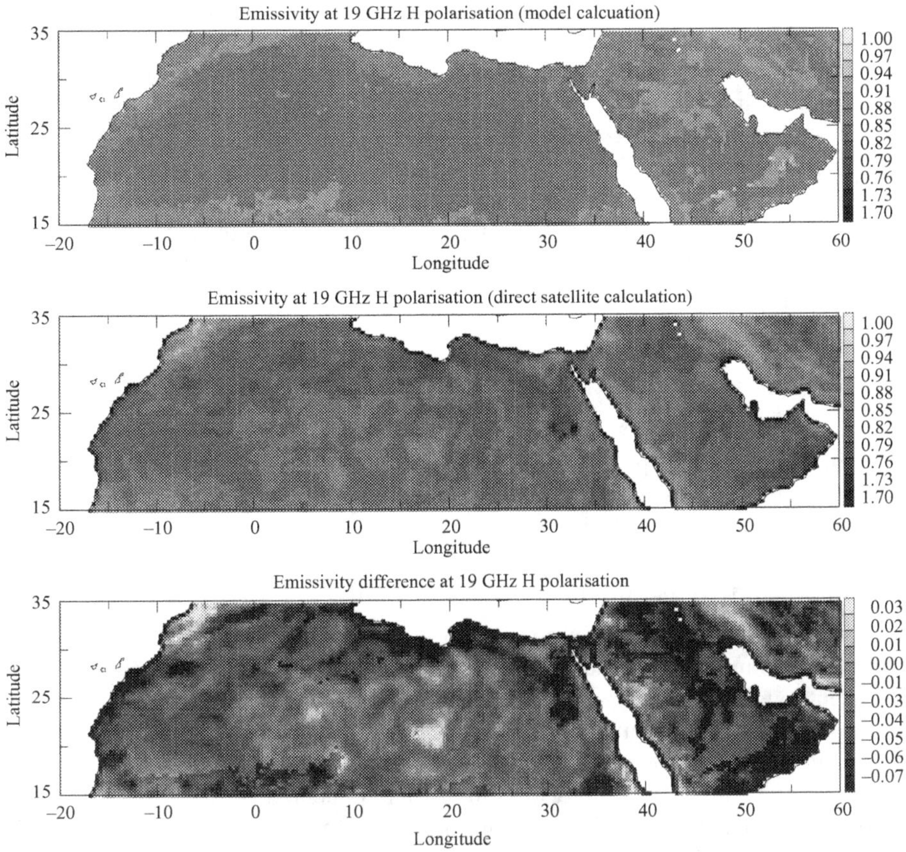

Figure 4.36 *Top panel: model calculation using the Weng et al. (2001) model and ECMWF outputs for the same month. Middle panel: satellite derived emissivities at 19 GHz horizontal polarisation at 53° incidence for July 2002. Lower panel: the difference between the emissivity direct satellite estimate and the model*

The following will focus on the arid regions. Figure 4.36 presents the satellite derived emissivities at 19 GHz horizontal polarisation (upper panel), for July 2002. The emissivity is rather low except in high topography regions (e.g. Atlas, Tibesti, and Air mountains). A specific feature characterised by low emissivities can also be noticed in Oman and Egypt. The direct emissivity calculations are compared with the model calculation (middle panel) for the same frequency and conditions for July 2003 for the North African arid region. Both calculations (direct and model) have been performed at ECMWF in collaboration with Dr. Chevallier. Model calculations are based on the Weng *et al.* (2001) emissivity model using the ECMWF forecast model relevant surface variables (soil temperature and soil moisture, essentially). The Weng *et al.* model provides a very uniform emissivity for this whole arid region, with very limited geographical structures, even in regions of high topography. Models dedicated to relief effects, like the one developed by Matzler and Standley (2000), would perform better. The emissivity differences (Figure 4.36, lower panel) clearly show that the model does not capture the emissivity variability that is observed by the satellite. It must be underlined that emissivity differences of ~0.03 translate into brightness temperature differences of the order of 10 K meaning that the use of such an emissivity model will inevitably yield detrimental effects in the assimilation of these passive microwave observations in these arid regions. Comparison between the directly calculated emissivities and the model ones shows fewer differences above vegetated areas (Prigent *et al.*, 2005c).

4.8.3 Lessons learnt from direct calculations of emissivities from satellite observations

We will focus on three specific features observed from direct emissivity calculations that cannot be easily taken into account in model calculation. The third feature is still not fully explained and is the object of ongoing research.

4.8.3.1 Emissivity sensitivity to surface roughness

As already observed on the emissivity maps (Figures 4.35 and 4.36), surface roughness has a strong effect on the emissivities, especially in arid and semi-arid regions. Different roughness scales are encountered within a satellite field of view, from the small-scale roughness related to surface irregularities, small compared to the wavelength, to the large-scale topographic effects. The effect of topography is to locally modify the viewing angle with respect to the horizontal and vertical polarisations as defined for a flat surface and to increase the effective emissivity (see Section 4.3). Within a field of view, different surface slopes related to topography tend to mix the vertical and horizontal polarisations, as defined, relative to the mean flat surface. Thus with increasing topographic roughness, the reflection properties of the surface approach a lambertian behaviour.

The histograms of the emissivities are presented at 19 and 85 GHz, for the North African desert, for selected ranges of terrain roughness (Figure 4.37). The roughness at a point is measured as the standard deviation of the elevation of the terrain, on a 90×90 km square centred on this point.

The difference of polarisation decreases with increasing terrain roughness. The lower the frequency, the higher the sensitivity to the topography. In areas of low topographic roughness, the polarisation difference is smaller at 85 GHz because the higher the frequency, the larger the effect of the small-scale roughness of the soil surface.

4.8.3.2 Microwave penetration depth in arid regions

As already mentioned, the standard deviation in the microwave emissivities within a month is typically about 0.013 for all the SSM/I frequencies. However, examination revealed that the standard deviations over specific arid areas are much higher than in

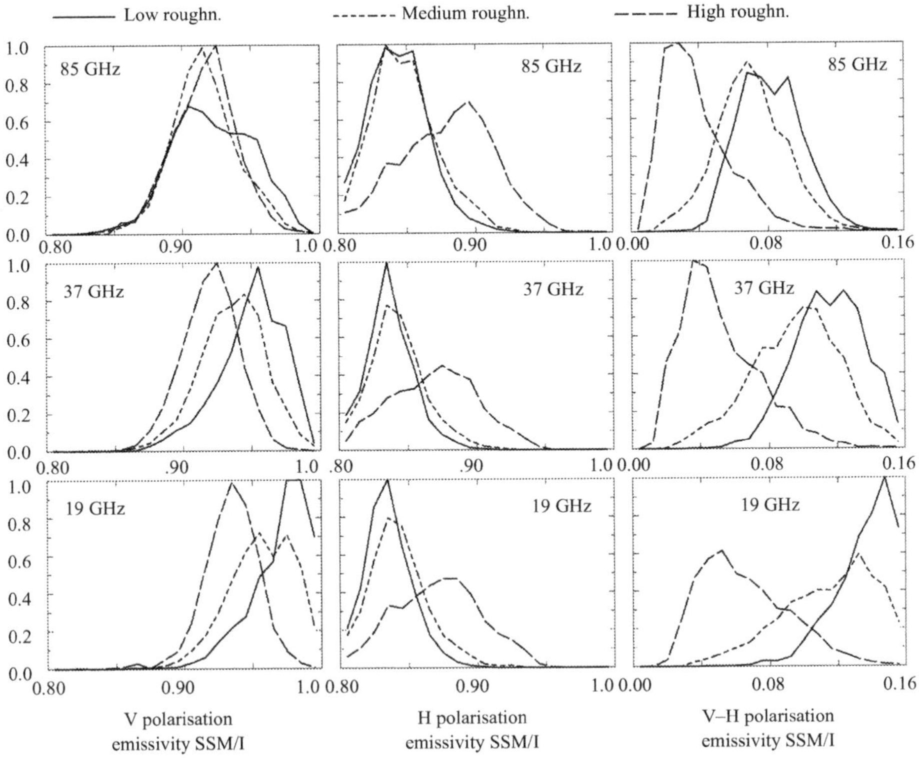

Figure 4.37 Histograms of the emissivities at 19, 37 and 85 GHz, for the North African desert, for selected ranges of terrain roughness

surrounding areas. These areas have been examined more closely in order to explain this feature (Prigent *et al.*, 1999).

Time series of IR surface temperatures and microwave brightness temperatures have been analysed, using the ISCCP surface skin temperature and the SSM/I observations derived from the F11 and F10 satellites in 1992 and 1993. In the arid areas associated with high emissivity standard deviations, these time series show that for a given diurnal change in surface temperature, the corresponding change in the microwave response is smaller in magnitude than expected and the discrepancy between the expected and observed changes increases with wavelength. Figure 4.38 compares the time series of microwave brightness temperatures and surface skin temperatures at two locations, one situated at 25° N and 12.5° E in an area of high emissivity standard deviations and a second location at 28° N and 17° E in an area where the emissivity standard deviation over a month is low. It is suggested that the observed behaviour with SSM/I over the desert can be explained by microwave emissions from different layers in the soil, depending on the soil type and on the microwave radiation frequency. The soil subsurface temperature varies with time and

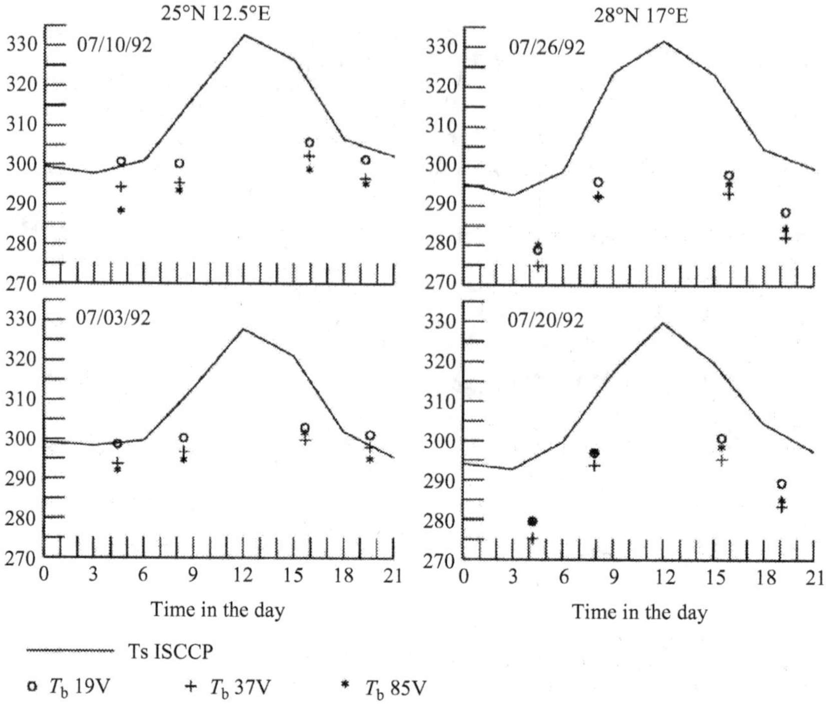

Figure 4.38 Time series of microwave brightness temperatures and surface skin temperatures at two locations (25° N and 12.5° E; 28° N and 17° E)

depth such that with increasing depth in the soil, the amplitude of the diurnal temperature variation decreases and the phase lags as compared to the diurnal variation of the surface skin temperature.

To examine the hypothesis, a one-dimensional, time-dependent heat conduction model is used to account for the transfer of heat in the soil over time and depth. On the basis of the eight-times daily estimates of the surface temperature from ISCCP, the diurnal cycle of the surface skin temperature is represented by the first two terms of a Fourier series fit. Then, using a simple Fresnel model for each individual location, the month-long time series of the seven SSM/I brightness temperatures and the surface temperature match-ups are used to obtain best-fit estimates of the effective emissivity and a penetration depth parameter. Calculations are performed and the Sahara and the Arabian Peninsula resulting penetration depths are shown in Figure 4.39 (a colour version of these maps is presented in Prigent *et al.* (1999)), assuming a thermal diffusivity of 0.002 cm^2/s for dry quartz sand (Ingersoll *et al.*, 1948). The microwave radiation is estimated to come from soil layers down to depths of at least five wavelengths in some locations. Regions where the microwave radiation comes from deeper soil layers also have large microwave emissivity polarisation differences, i.e. are rather flat. As expected, the penetration depth decreases with increasing frequency. The

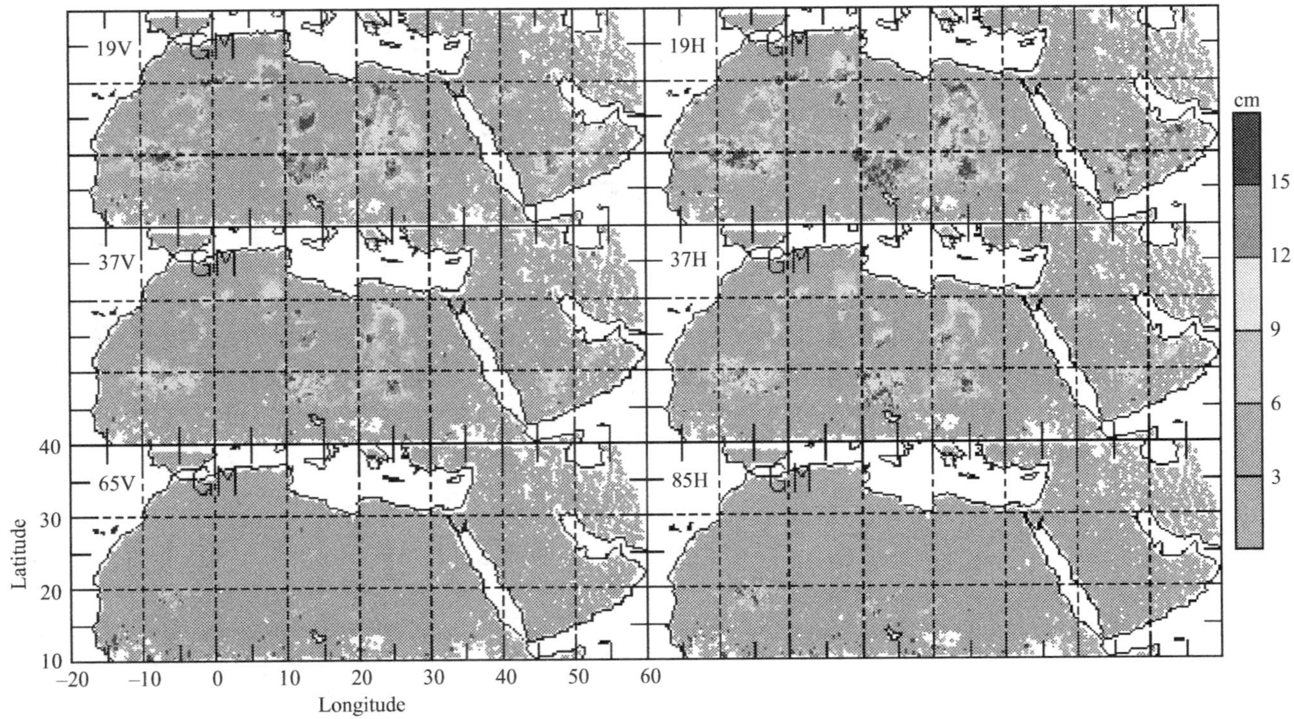

Figure 4.39 Estimated penetration depths at 19, 37 and 85 GHz, assuming a thermal diffusivity of 0.002 cm²/s

ratio between the penetration depth at the various frequencies is higher than the wavelength ratio, following Matzler's results that the frequency dependence of the penetration depth is weaker than expected (Matzler, 1998). The regions characterised by large penetration depth also have large visible reflectances, suggesting that these areas correspond to sand dune fields. In order to verify it, the Food and Agriculture Organization (FAO, 1977) soil maps have been analysed in coincidence with the retrieved penetration depth. The correspondences between the sand dune locations and the regions of high penetration are striking and have been further analysed in Prigent *et al.* (1999).

This study clearly shows that the penetration depth in very dry sand can be significant and it is possible to quantify this penetration depth for frequencies between 19 and 85 GHz. Penetration depths as large as five wavelengths in some locations in sand deserts are observed. This specific feature should be taken into account in the emissivity models of these regions.

4.8.3.3 Emissivity sensitivity to geological structure

Surprisingly low emissivities are observed in arid regions over the south of Arabia and in the north-east of Egypt around the Nile river in Figures 4.34 and 4.35. They are very stable over long time periods (several years of data have been examined) and are observable over a large frequency range (SSM/I and AMSU A and B observations have been analysed, i.e. from 19 to 190 GHz). Low emissivities are usually associated with the presence of standing water: Compared with the other materials, water has high dielectric values that translate into low emissivities. The regions of concern in the south of Arabia and Egypt are particularly arid areas with very limited precipitation all year long and their low emissivity cannot be explained by the presence of standing water. It could also be noted that these regions do not correspond to sandy deserts where large penetration depths have been observed (see the above section).

In order to understand the surprising signatures observed with passive microwave measurements, other satellite observations covering the same regions with comparable spatial resolution have been checked. The visible (0.58–0.68 μm) and near infrared (0.73–1.1 μm) reflectances measured by the Advanced Very High Resolution Radiometer do not show any specific structures in these regions: The reflectances are rather high, similar to what is observed in other arid areas. Nothing special has been observed, either, with the ERS scatterometer that measures the backscattering signal at 5.25 GHz in the vertical polarisation.

A geological analysis has been conducted using available geological maps of these areas. Examination of the geological maps reveals that the regions of low microwave emissivities coincide with the outcrops from the paleogene. In Figure 4.40, in Egypt, the contours of the paleogene outcrops from the tertiary era (Egyptian Geological Survey and Mining Authority, 1981) have been superimposed on the microwave emissivities at 19 GHz horizontal polarisation. An amazing correspondence is observed between the low emissivities and the sedimentary outcrops. The sedimentary outcrops that correspond to the passive microwave low emissivities are dominated by carbonated rocks. Dielectric measurements of a large range of rocks and sand have

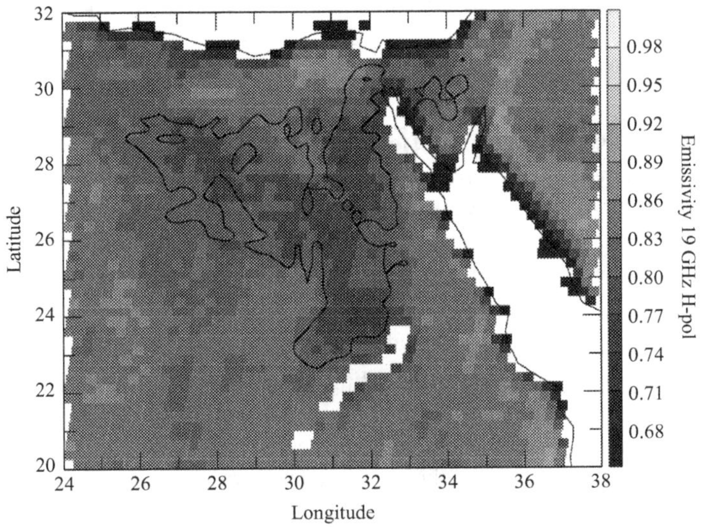

Figure 4.40 Microwave emissivity at 19 GHz horizontal polarisation over Egypt (July 1992). Contours indicate the sedimentary outcrops from the tertiary era

been measured at Physique des Interactions Ondes-Matières (PIOM) in Bordeaux (France), at the Paris Observatory and at the Institute for Applied Physics in Bern. Samples from the low emissivity regions have been collected and measured. The satellite observations are interpreted in terms of dielectric properties: the carbonate rocks have high permittivities (higher than 8 as compared to typical values of 5 for other rocks or 2.5 for silicate sands) and this explains the observed emissives (Prigent *et al.*, 2005a). In this case, the direct calculation of the emissivities from the satellite observations makes it possible to evidence an unexpected sensitivity of the passive microwave observations to the geological properties.

4.8.4 Conclusion

Emissivities have been directly calculated from satellite observations (SSM/I and AMSU) and several lessons have been learnt from these maps. Modelling studies are very important for understanding the complex interaction between the land surface and the microwave radiation. However, they show limitations when used on a large scale, due to the models themselves and to the uncertainty in the input parameters. Local radiometric measurements can help tune the models for specific conditions but they have difficulties capturing and representing the larger scale components. Analysis of the satellite derived emissivity is complementary to these studies. Over arid areas, the satellite derived emissivities have shown an unexpected variability that is still not fully explained. Comparison of microwave observations on a global scale for different modes (passive and active), angles and polarisations, along with observations at other frequencies (from the visible to the far infrared) will help disentangle the complex interaction between the arid surface and the radiation.

4.9 Parametrisations of the effective temperature for L-band radiometry. Inter-comparison and long term validation with SMOSREX field experiment *Patricia de Rosnay, Jean-Pierre Wigneron, Thomas Holmes and Jean-Christophe Calvet*

4.9.1 Introduction

Measurements of passive microwave brightness temperature quantify the intensity of the soil microwave radiation. According to the Rayleigh–Jeans approximation, in the microwave domain the emitted energy from the soil is proportional to the thermodynamic temperature, and the brightness temperature T_b is expressed as a function of the emissivity, e, and the effective temperature, T_{eff}, as:

$$T_b = eT_{eff} \tag{4.106}$$

The whole soil layer contributes to the soil thermal emission. From the point of origin to the soil surface, the intensity is attenuated by the intervening soil, whose absorption is related to the moisture content. The net intensity at the soil surface is, therefore, a superposition of the intensities emitted at various depths within the soil (Choudhury *et al.*, 1982).

The effective temperature is used to normalise the measured brightness temperature and obtain values of surface emissivity. An accurate computation of the effective temperature is thus critical for obtaining relevant values of the soil emissivity from brightness temperature measurements. It follows that the soil moisture can be retrieved from the estimates of the soil emissivity.

The theoretical formulation of the effective temperature requires fine vertical profile information on both soil moisture and soil temperature. The required information is not available from remote sensing and only a few test sites provide a sufficiently fine measurement of the vertical profiles in the soil. To estimate the effective temperature with minimum soil profile information, several parametrisations have been developed. The simplest method normalises the brightness temperature by a surface temperature that represents only a small part of the soil profile, for example using thermal infrared remotely sensed observations (Jackson, 1993), air temperature (Van de Griend and Owe, 1994) or relating the 37 GHz (V) brightness temperatures to the $T_{soil(5\,cm)}$ (Owe *et al.*, 2001; De Jeu, 2003). But this approximation is only valid at frequencies higher than 10 GHz, while at lower frequencies the penetration depth is larger and deep temperatures influence the effective temperature (Choudhury *et al.*, 1982).

To account for the effects of the soil temperature profile, Choudhury *et al.* (1982) have proposed a parametrisation of the effective temperature for microwave radiometry as a linear function of the soil temperature at two depths. Wigneron *et al.* (2001) improved this formulation for L-band radiometry, by taking into account the influence of the soil moisture layer contributing to the emission. For instance, the dryer the soil, the deeper the contribution.

These effective temperature parametrisations where tested and validated with field experiment datasets. But all the passive microwave remote sensing experiments conducted in the past were limited to short time periods ranging from days to weeks. Parametrisations of the effective temperature were shown to be relevant for the studied periods, but their suitability at longer temporal scales, including inter-seasonal and inter-annual variability with large variations in the soil and temperature conditions, remained unknown.

In the context of the first long term field experiment SMOSREX (surface monitoring of soil reservoir experiment) for soil moisture remote sensing (de Rosnay *et al.*, 2005), Holmes *et al.* (2005) proposed a new parametrisation of the effective temperature based on the relation between the attenuation of the emitted energy and the dielectric constant at the soil surface. This parametrisation is as simple as that of Wigneron *et al.* (2001) in terms of the required information. The former authors explored the new opportunity provided by the long term SMOSREX field experiment dataset to analyse the long term features and characteristics of the L-band effective temperature for a large range of soil moisture and temperature profiles. Based on this dataset, they provided an inter-comparison and validation of the different parametrisations of the effective temperature.

This section reviews the main parametrisations of the effective temperature which have been proposed in the literature by Choudhury *et al.* (1982), Wigneron *et al.* (2001) and Holmes *et al.* (2005). Based on the SMOSREX dataset, the paper investigates the effective temperature dependency to soil moisture and presents an inter-comparison and validation of the effective temperature models at the inter-annual time scale.

4.9.2 SMOSREX experimental dataset

SMOSREX is a long term field experiment for soil moisture and land surface process remote sensing (de Rosnay *et al.*, 2005). The SMOSREX test site is located in southern France, near the town of Toulouse, at 43°23′N 1°17′E and 188 m altitude. SMOSREX is equipped with an L-band radiometer (LEWIS L-band radiometer for estimating water in soils) which was installed at the top of a 15 m structure (Lemaître, *et al.*). The L-band surface brightness temperature has been monitored since January 2003 for two plots, one which is kept as a bare soil, while the other is a natural grass. An automatic scanning mode provides L-band brightness temperatures at five incidence angles (20, 30, 40, 50, 60°) on both plots. Two thermal infrared radiometers are installed on the structure and measure the surface skin temperature for both fallow and bare soil areas. Simultaneous ground measurements are performed for meteorological and soil conditions. Soil moisture profiles are measured for the fallow and the bare soil area by Delta T Device theta-probes at 5, 10, 20, 30, 40, 50, 60, 70, 80 and 90 cm depths. Soil temperatures are measured at 1, 5, 20, 50 and 90 cm by PT100 sensors. The experiment, including ground and remote sensing measurements of soil conditions, is scheduled to last three years from January 2003 to December 2005. The

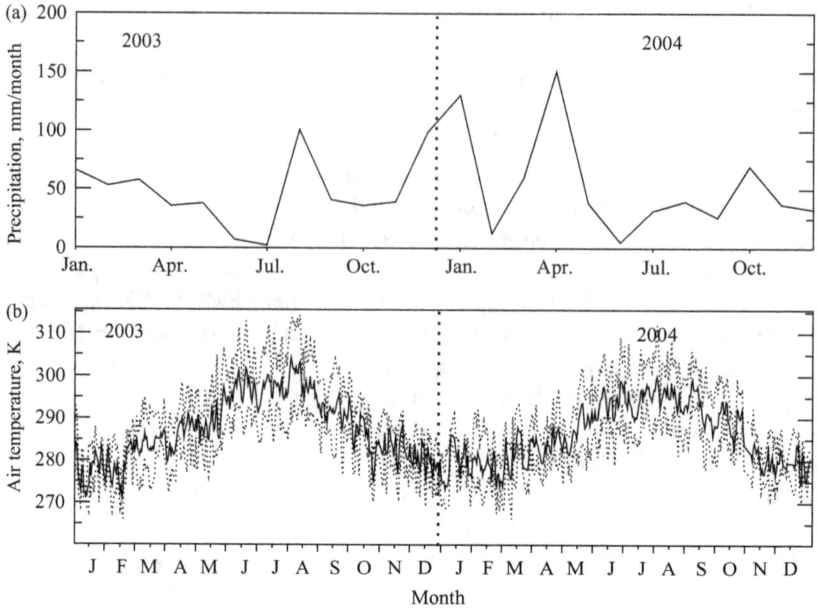

Figure 4.41 *Annual cycles for 2003 and 2004 of monthly precipitation (mm/month) and daily air temperature (K). For air temperature, dotted lines show the daily maximum and minimum value and the black line depicts the daily mean temperature*

long term component of SMOSREX is innovative in the field of soil moisture remote sensing research. It opens new possibilities for analysing the microwave emission in contrasting conditions. Based on the 2003 and 2004 datasets, this paper focuses on the long term component of the effective temperature.

Figure 4.41 depicts the annual cycles in 2003 and 2004 of both precipitation and air temperature. It shows that the annual cycle is well contrasted with dry and warm conditions in the summer and cold wet winter conditions. The cumulated precipitation is 574 mm in 2003 and 634 mm in 2004. Although there are only small differences in the cumulative precipitation, the inter-annual variability is strong between the two studied years. Very dry conditions during spring and the first part of summer, associated with particularly warm temperatures, followed by very wet conditions at the end of summer and autumn, characterised 2003. In contrast, the spring in 2004 is particularly wet with 150 mm of rainfall in April, but autumn is much dryer. For both years, several soil freezing events were monitored. These different features in the annual cycle of meteorological conditions allows the consideration of a large range of soil moisture/temperature combinations for the two-year period.

4.9.3 Theoretical formulation of the effective temperature

The effective temperature is controlled by the soil dielectric and temperature profiles. From radiative-transfer theory (Chapter 1, Section 1.4.5) the effective mean temperature, T_{eff}, can be expressed for a bare soil surface as:

$$T_{\mathrm{eff}} = \int_0^{\infty} T_{\mathrm{s}}(z) \cdot W(z) \cdot \mathrm{d}z \tag{4.107}$$

where z is the vertical axis (positive downward), $T_{\mathrm{S}}(z)$ the soil thermodynamic temperature at depth z and $W(z)$ is a temperature weighting function of the contribution of each soil layer to the effective temperature (Ulaby *et al.*, 1986). $W(z)$ is defined as

$$W(z) = \alpha(z) \cdot \exp\left[- \int_0^{z} \alpha(z') \cdot \mathrm{d}z' \right] \tag{4.108}$$

where

$$\alpha(z) = (4\pi/\lambda) \cdot \epsilon''(z)/2(\epsilon'(z))^{0.5} \tag{4.109}$$

$\alpha(z)$ is an attenuation coefficient related to the soil dielectric constant, λ is the observation wavelength and ϵ' and ϵ'' are the real and imaginary parts of the soil dielectric constant. The shape of the weighting function is determined only by the soil dielectric profile. For wet soils, the weighting function decreases rapidly with depth, indicating a shallow contributing soil profile. For dryer soils, the weighting function declines more slowly and the contributing profile is deeper (Raju *et al.*, 1995; Chan *et al.*, 1997).

Using Equations (4.107–4.109), the theoretical effective temperature, T_{eff}, was computed for the two-year 2003–2004 SMOSREX dataset using measurements of soil moisture and temperature profiles. Despite uncertainties in the measured soil moisture and temperature profiles, the theoretical approach is considered to provide a good approximation of the effective temperature. It is used in the following as the 'true' effective temperature.

The influence of the soil moisture on the effective temperature is illustrated in Figure 4.42 which shows the time variations of the effective temperature and of the soil temperature at different depths for three days in March, 2003 and three days in July, 2003. The soil is rather wet in March: values of the volumetric water content are higher than 30 per cent for 17 days, between 25 and 30 per cent for eight days and between 23 and 25 per cent for six days. The effective temperature is close to the near surface soil temperature at a 5 cm depth. This indicates that the soil contribution to the brightness temperature originates mainly from a rather shallow surface layer. In contrast, the soil dryness in July (volumetric soil moisture lower than 12 per cent the whole month and lower than 10 per cent for 18 days) is associated with a much deeper soil contribution to the soil microwave emission. For this dry period, the amplitude of the effective temperature diurnal cycle is close to that of the 20 cm temperature, while its phase remains correlated to the 5 cm temperature.

The effective temperature is in better agreement with the soil temperature at a depth that depends mainly on the soil moisture conditions. This aspect is analysed, at the monthly scale for 2003 and 2004, based on the minimum rmse (root mean square error). This depth varies from 3.4 cm in January 2003 (rmse 0.13 and square correlation 0.998) to 16.9 cm in July 2003 (rmse 0.99, correlation 0.967). The depth which gives the best fit of soil temperature with effective temperature is different from the thermal sensing depth which quantitatively describes the soil depth

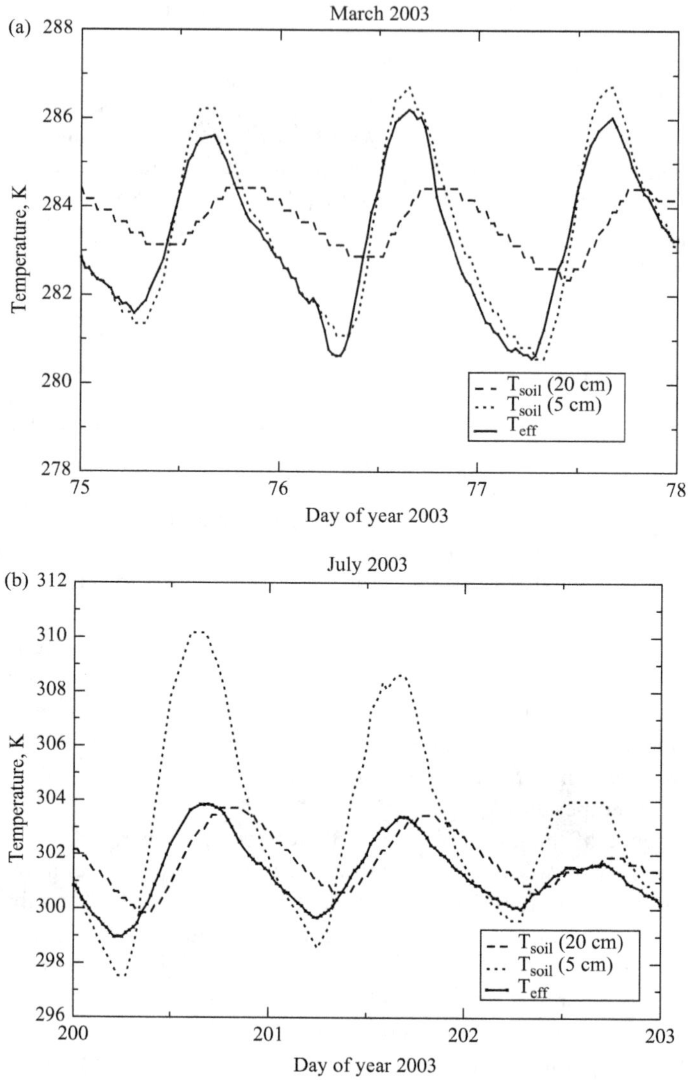

Figure 4.42 Temporal evolution of the soil temperature at 5 and 20 cm and effective temperature for three days in March 2003 (wet conditions) and three days in July 2003 (dry conditions). Diurnal cycle of the effective temperature indicates a much deeper contribution in July, when the soil is dry, than in March

that contributes to the effective temperature. The thermal sensing depth is defined as the top soil depth below which the integrated contribution to the total emission is divided by 1/exp(1). Accordingly, the thermal sensing depth contributes to 63 per cent (e.g. 1 − 1/exp(1)) to the effective temperature. Calculated at a monthly scale, for the two-year SMOSREX dataset, it varies from 6.5 cm in January 2003 to 29.09 cm in July 2003.

This seasonal variability of the temperature sensing depth observed at the SMOSREX field site is expected to be also important at larger spatial scales in regions characterised by a contrasted annual cycle in terms of the meteorological conditions. For long term remote sensing applications, it is critical to use effective temperature parametrisations that are able to capture these seasonal variations.

4.9.4 Simple parametrisations of the effective temperature

Because the calculation of the effective temperature given by Equations (4.107–4.109) requires detailed information on the temperature and water content profiles, it is necessary to use a simpler parametrisation for application studies on a larger scale.

4.9.5 Choudhury et al., 1982

Based on the theoretical formulation and a 15-day field experiment, Choudhury *et al.* (1982) developed a simple parametrisation of the effective temperature for passive microwave radiometry:

$$T_{\mathrm{eff}} = T_{\mathrm{deep}} - (T_{\mathrm{deep}} - T_{\mathrm{surf}})C \tag{4.110}$$

where T_{deep} represents a soil temperature at a depth ranging from 50 to 100 cm and T_{surf} is the surface soil temperature between 0 and 5 cm. In this parametrisation, C is a fitting parameter that varies with frequency and soil moisture. For simplicity, Choudhury *et al.* assumed a constant value of C at a given wavelength. They fitted the value of C for their field experiment at different wavelengths between 2.8 and 49 cm, and showed that C varies from 0.802 at a wavelength of 2.8 cm to 0.084 at 49 cm: the larger the wavelength, the deeper the contribution. This formulation is largely used to approximate effective temperature with a minimum amount of soil profile information.

4.9.6 Wigneron et al., 2001

To take into account the sensitivity of the fitting parameter C to the soil moisture, Wigneron *et al.* (2001) proposed a parametrisation for low-frequency microwave radiometry where :

$$T_{\mathrm{eff}} = T_{\mathrm{deep}} - (T_{\mathrm{deep}} - T_{\mathrm{surf}})C(w) \tag{4.111}$$

where $C(w)$ is a parametric function of the surface soil moisture, w_{surf}, given as:

$$C(w) = (w_{surf}/w_0)^b \tag{4.112}$$

w_0 and b are the constant fitting parameters. This formulation, which was developed for L-band radiometry, takes into account the sensing depth variation with the soil moisture content. For dryer soils, $C(w_{\mathrm{surf}})$ is lower and T_{eff} is closer to T_{deep} than for wet soils. Calibration of w_0 and b was performed for the INRA (Institut National de la

Recherche Agronomique) Avignon test site using several options with respect to the choice of the variables T_{surf} and T_{deep}. Best results were obtained when $T_{surf} = T_{2-4\,cm}$ and $T_{deep} = T_{50\,cm}$. The three month dataset used for the calibration (April 20–July 10 1993) allowed the consideration of a seasonal variability in soil moisture and temperature profiles. This parametrisation proved to be robust at this seasonal scale.

4.9.7 Holmes et al., 2005

Based on the SMOSREX long term field experiment, Holmes *et al.* (2005) showed that Wigneron's parametrisation was able to approximate the effective temperature at the annual scale. But the calibrated parameters were shown to vary significantly depending on the season. An analytical computation of the values of the fitting function, C, revealed that C is more clearly related to the ratio between the imaginary and real parts of the surface dielectric constant than to the surface soil moisture. This result is consistent with the formulation of the effective temperature (Equation (4.107)) where the attenuation coefficient is expressed as a function of soil dielectric constant (Equation (4.109)). Accordingly, Holmes *et al.* (2005) proposed a new parametrisation of the effective temperature where the fitting parametric function depends on the ratio between the imaginary and real parts of the surface dielectric constant (namely ϵ'' and ϵ'):

$$T_{eff} = T_{deep} - (T_{deep} - T_{surf})C(\epsilon) \tag{4.113}$$

with $C(\epsilon)$

$$C(\epsilon) = [(\epsilon''/\epsilon')/\epsilon_0]^b \tag{4.114}$$

where ϵ_0 and b are fitting parameters. Holmes *et al.* (2005) used the model developed by Wang and Schmugge (1980) to compute the real and imaginary parts of the dielectric constant from the value of the surface soil moisture and information on soil texture. They showed that their parametrisation (1) is slightly improved compared to Wigneron's method in terms of rms error and (2) provides a description of the variations of sampling depth at the inter-seasonal temporal scale which is in better agreement with the theoretical computations.

4.9.8 Inter-comparisons

Table 4.2 summarises the different parametrisations of the effective temperature that were considered and inter-compared in this study. All of them are based on the following expression:

$$T_{eff} = T_{deep} - (T_{deep} - T_{surf})C \tag{4.115}$$

Depending on the parametrisation, the parametric fitting function is defined either as a constant, or a function of the surface soil moisture or a dielectric constant (Choudhury *et al.*, 1982; Wigneron *et al.*, 2001; Holmes *et al.*, 2005). These three parametrisations

Table 4.2 Different parametrisations of L-band effective temperature

Reference	Weighting function C	Parameters	Required information
Choudhury *et al.* (1982)	$C = \text{constant}$	1: C	T_{deep}, T_{surf}
Wigneron *et al.* (2001)	$C = (w_{surf}/w_0)^b$	2: w_0, b	T_{deep}, T_{surf}
Holmes *et al.* (2005)	$C = ((\epsilon''/\epsilon')/\epsilon_0)^b$	2: ϵ_0, b	T_{deep}, T_{surf}

will be referred as [Ch], [Wig], [Hol] in the following. The most simple parametrisation of the effective temperature, not listed in Table 4.2, considers $T_{eff} = T_{surf}$. It is a particular case ($C = 1$) of [Ch], and its validity is restricted only to shallow contributions of the soil temperature profile to the brightness temperature. A multi-frequency analysis conducted by Choudhury *et al.* (1982) showed that it is not suitable for frequencies lower than 10 GHz.

The values of the coefficients used in these parametrisations were calibrated using the values of the theoretical effective temperature computed using Equations (4.107–4.109) during the whole period of the measurements at the SMOSREX site. This calibration was carried out for different choices of T_{surf}: Using the infrared temperature or the 1 or 5 cm soil temperature. Best results were obtained using $T_{surf} = T_{5\,cm}$ and $T_{deep} = T_{50\,cm}$. All of the results reported hereafter considered these two soil depths in the parametrisation of the effective temperature. The calibration was performed separately for three time periods in order to analyse the stability and robustness of the parametrisation at annual and inter-annual temporal scales: For the single year 2003, the single year 2004 and for the whole dataset (years 2003–2004). For each parametrisation, the calibration was also tested for 2004 with the optimised parameters obtained in 2003.

For each parametrisation, Table 4.3 indicates the best fit parameters at different temporal scales and the averaged error (rmse) between the parametrised and the theoretical effective temperature. It also indicates, for each time period, the maximum error and the occurrence of errors larger than 1 K. This provides a quantitative assessment of the percentage of situations where the model is not able to reproduce effective temperature with a 1 K accuracy. Choudhury's parametrisation is suitable for short time periods (Choudhury *et al.*, 1982; Schneeberger *et al.*, 2004), but it is clear from this table that it is not able to reproduce the time variations in the effective temperature for inter-seasonal and inter-annual scales. This is illustrated in Figure 4.43 for selected time periods during 2003 and 2004. For intermediate and more uniform conditions of moisture and temperature vertical gradients, Choudhury's parametrisation is in good agreement with the theoretical formulation (black line). But as soon as the soil profiles differ significantly from the average, such as for freezing or warm and dry conditions, the use of a constant C parameter does not allow representation of the temporal variations of the effective temperature with a constant value of the parameter C. This is indicated by the high frequency of errors larger than 1 K which

Table 4.3 *Results of the calibration of the three parametrisations. The best fitting parameters are indicated for each parametrisation as well as the associated rmse and square correlation coefficients compared with the theoretical value of effective temperature. Maximum error (Emax) on the simulated effective temperature and occurrence (Freq in %) of errors larger than 1 K are indicated for each case. The calibration was performed for different periods: 2003, 2004, 2003–2004. The case '2003 applied to 2004' is not a calibration but an evaluation of the best 2003 fitting parameters to the year 2004. This test addresses the robustness of each parametrisation for temporal extrapolation. Maximum error and occurrence of errors larger than 1 K are indicated for each case*

Parametrisation	Period	Parameters		rmse (K)	R^2	Emax (K)	Freq (%)
Choudhury		C					
et al. (1982)	2003	0.62		0.897	0.996	3.29	21
	2004	0.71		0.970	0.993	4.10	18
	2003 applied to 2004	0.62		1.920	0.974	5.30	51
	2003 and 2004	0.68		0.953	0.994	3.69	39
Wigneron		w_0	b				
et al. (2001)	2003	0.36	0.70	0.478	0.999	2.38	4
	2004	0.32	0.58	0.592	0.997	2.64	4
	2003 applied to 2004	0.36	0.70	0.734	0.996	2.22	6
	2003 and 2004	0.33	0.63	0.573	0.998	3.11	10
Holmes		ϵ_0	b				
et al. (2005)	2003	0.08	0.95	0.412	0.999	2.5	2
	2004	0.08	0.81	0.482	0.998	1.67	2
	2003 applied to 2004	0.08	0.95	0.515	0.998	2.00	4
	2003 and 2004	0.08	0.87	0.458	0.999	2.43	6

varies between 21 and 39 per cent with [Ch] for the different time periods considered. The maximum model errors occur in these situations and range between 3.3–5.3 K.

The Wigneron's model (dashed line) is in better agreement with theoretical effective temperature as shown in Table 4.3 and Figure 4.43. For 2003, the rmse decreases from 0.897 K for [Ch] to 0.478 K for [Wig] and from 0.97 to 0.592 K for 2004. In particular, [Wig] drastically improves the case '2003 applied to 2004' (rmse 1.92 with [Ch] and 0.734 with [Wig]), thereby showing it is more temporally stable. The simulation of [Wig] is in better agreement with the theoretical formulation than [Ch] for a large range of surface conditions, including rather extreme conditions of the soil profile. This is pointed out by lower values of maximum errors with [Wig] than [Ch]. At the annual scale the occurrence of errors larger than 1 K is reduced from 21 per cent with [Ch] to 4 per cent with [Wig] in 2003, and from 18 to 4 per cent

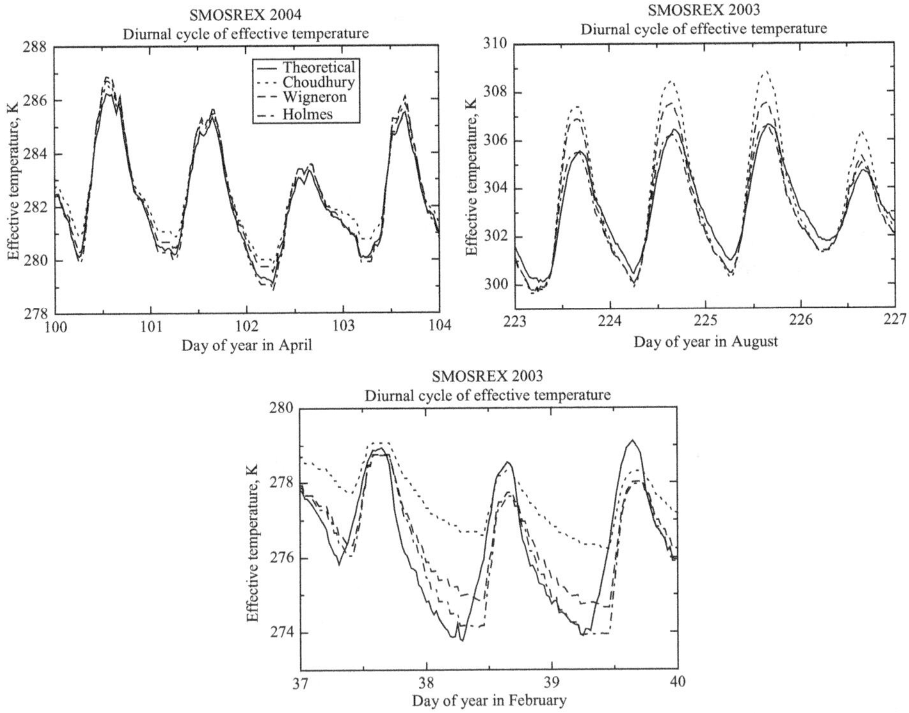

Figure 4.43 *Diurnal cycles of theoretical and parametrised effective temperature for a selected wet period in April 2004 (top), a very dry and warm period in August 2003 and a soil freezing period in February 2003 (bottom). For each parametrisation, the best fit parameters given in Table 4.3 at the annual scale are used for the considered year (top: 2004, middle and bottom: 2003). Black line represents the effective temperature computed from theoretical microwave radiometry, dotted line uses Choudhury's parametrisation, the dashed line corresponds to Wigneron's model and the dashed–dotted line uses the Holmes parametrisation*

in 2004. This illustrates that a 1 K accuracy in ensured in 79 and 82 per cent of the situations with [Ch] for 2003 and 2004, against 96 per cent with [Wig] for both years. This improvement shows that it is very important to take into account the role of the surface soil moisture in the effective temperature parametrisation.

Table 4.3 indicates that these results are further improved using the Holmes *et al.* (2005) model. The rmse are lower than those obtained with [Wig] at any temporal scales, from month (not shown) to annual and inter-annual scales. At the inter-annual scale (2003–2004), the rmse decreases from 0.953 with [Ch] to 0.573 with [Wig], to 0.458 with [Hol]. Maximum error decreases from 3.69 with [Ch] to 3.11 K with [Wig] and 2.43 K with [Hol] and the occurrence of errors larger than 1 K is 39, 10 and 6 per cent for the three models, respectively. The 1 K accuracy is thus ensured at this inter-annual scale in 61 per cent of the situations with [Ch], 90 per cent with [Wig] and 94 per cent with [Hol]. The stability of the calibrated parameters for [Hol] (see Table 4.3) is particularly noteworthy. With only one parameter ($\epsilon_0 = 0.08$ is shown to be almost constant at any time scale) the parametrisation is able to describe the variations of the effective temperature encountered at different temporal scales. These results indicate that the simple [Hol] parametrisation is able to take into account the main processes that govern the thermal sampling depth. Specific short time scale events, such as the very warm period that occurred in August 2003 or the soil freezing, are also well represented. This is clearly illustrated by Figure 4.43: For intermediate soil moisture conditions, such as in April 2004 (top of Figure 4.43), the three parametrisations provide similar results which are in good agreement with the theoretical formulation, although minimum values of the diurnal cycle are not very well represented with the most simple parametrisation. The differences between the three parametrisations are more significant for February 2003 and August 2003, which represent rather extreme conditions. In these conditions, the relevance of the model is strongly dependent on the approach used to account for the effects of soil moisture in the computation of the effective temperature. The [Hol] model, which represents this effect through the modifications of the dielectric constant, is closer to the theoretical effective temperature than the two other parametrisations.

These results underscore that (1) the soil moisture effect on the thermal sampling depth is critical. It affects the relationship between the effective temperature and the surface and deep soil temperatures. This explains the improvements of the Wigneron's parametrisation compared to that of Choudhury. (2) According to the theoretical formulation of the attenuation coefficient, the soil moisture influences the thermal sampling depth through the soil dielectric profile. This explains why Holmes's parametrisation is still better than that of Wigneron.

Figure 4.44 shows the scatter plot of the parametrised versus theoretical effective temperature for the case '2003 applied to 2004'. Each parametrisation uses its 2003 best fit parameters for 2004. The scatter plot includes all the point measurements of 2004. The figure shows the ability of each parametrisation to extrapolate the effective temperature from one year to another. The top left scatter plot shows the soil temperature at different depths compared against the theoretical effective temperature. It shows that using a surface temperature as an effective temperature is not suitable at the annual scale. The surface soil temperature at a 1 cm depth shows a much higher scatter for high temperatures than the 5 and 20 cm temperatures. This is due to the fact that higher temperatures occur in summer in dry conditions and are associated with a deeper contribution to the soil emission. Accordingly, the deeper temperature (20 cm) is in better agreement with the effective temperature (less scatter) for high temperatures. In winter, higher soil moisture leads to shallower contribution to the soil thermal emission and the surface soil temperature is closer to the effective temperature. This explains the lower scatter with a 5 cm temperature than with a 20 cm temperature for colder soil conditions.

The strong improvement between [Ch] (top right) and [Wig] (bottom left) is clearly shown. It illustrates the results of Table 4.3 where the ratio between [Ch] and [Wig] is 2.6 for the rmse and 8.5 for the 1 K error frequency for the case 2003 applied

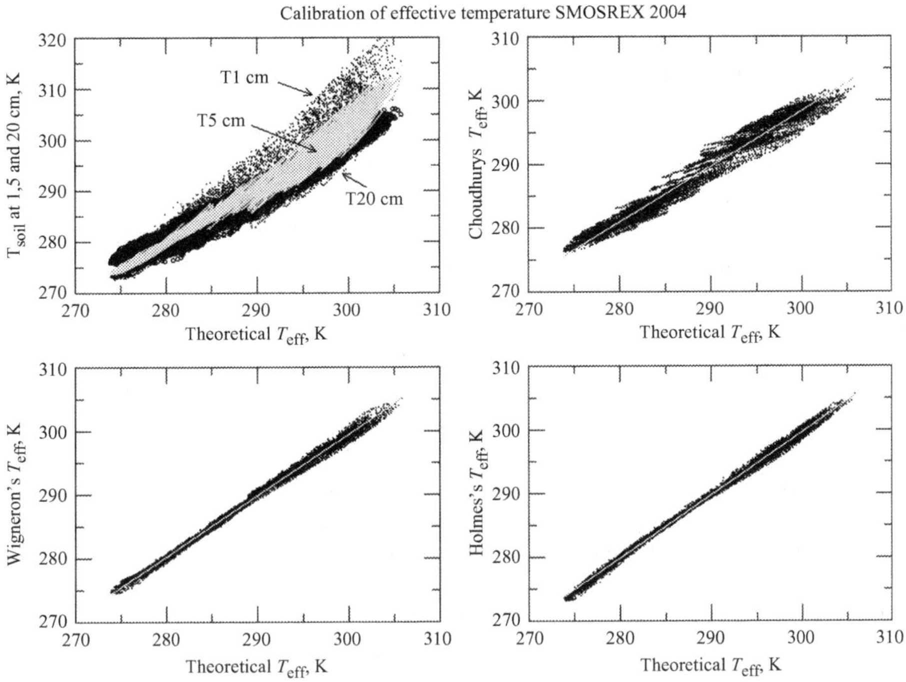

Figure 4.44 Scatter plot of surface (top left) and parametrised (bottom and top right) versus theoretical effective temperature for all point measurements in 2004. For each parametrisation, the 2004 effective temperature is computed from the optimised parameters obtained in 2003

to 2004. The improvement from [Wig] to [Hol] (Figure 4.44 bottom left and right) is also clear at the annual scale (the rmse drops from 0.734 to 0.515 and the 1 K error frequency decreases from 6 to 4 per cent).

4.9.9 Conclusion

This section reviews the parametrisations of the effective temperature for L-band radiometry. Three parametrisations, Choudhury *et al.* (1982), Wigneron *et al.* (2001) and Holmes *et al.* (2005) are described and calibrated using the SMOSREX dataset. The calibration is made at different temporal scales for which the parametrisations are validated and inter-compared. The SMOSREX field experiment provides two-year long time series of soil moisture and temperature profiles and L-band surface brightness temperature measurements. It provides a new opportunity to understand the role of long term soil moisture variation in the computation of the effective temperature.

The analysis confirms that the simple model of Choudhury [Ch] is suitable for short time periods (weeks). This parametrisation considers a constant weighting factor, C, which expresses the effective temperature as a linear function of soil temperature at two depths. This is an implicit assumption of a constant thermal sampling depth. For longer temporal scales (season to years), large variations in the soil moisture profile lead to temporal variations in the soil depth that contribute to the soil thermal emission. This effect, which is not taken into account in [Ch], may explain the poor results at the annual scale (see Table 4.3) and the large scatter obtained in Figure 4.44.

Wigneron *et al.* (2001) improved the Choudhury parametrisation by introducing a soil moisture dependency in the weighting factor C. This parametrisation allows us to account for the seasonal, annual and inter-annual variations of the thermal sampling depth in relation to soil moisture conditions. The rms errors and the 1 K error occurrence, given in Table 4.3, are significantly reduced at any temporal scale when compared to [Ch].

In their parametrisation, Holmes *et al.* (2005) went further in terms of the representation of the effect of the soil moisture on the thermal sampling depth. In order to better approximate the theoretical formulation of the effective temperature, where attenuation depends on the dielectric profile, they expressed the weighting factor C as a function of the surface dielectric constant. Their parametrisation is still very simple in terms of the required information and number of parameters. It is shown to be in better agreement with the theoretical effective temperature for all the temporal scales considered in this study. Low values of the maximum errors and low frequency of errors larger than 1 K, given in Table 4.3, underline its robustness for specific conditions such as soil freezing or very dry and warm conditions. This is illustrated at the diurnal scale for a few contrasted situations, shown in Figure 4.43. The good temporal stability of the best fit parameters shown in Table 4.3 confirms that the process that governs the sensing depth variation is closely related to the variations in the dielectric profile. The introduction of this information in the C fitting function ensures this simple parametrisation is able to capture the variations of the effective temperature for a large range of surface conditions. Further work, in other soil moisture and temperature conditions, corresponding to different climate conditions, is now necessary to generalise these long term inter-comparison results to larger spatial scales.

4.10 Modelling the effect of the vegetation structure – evaluating the sensitivity of the vegetation model parameters to the canopy geometry and to the configuration parameters (frequency, polarisation and incidence angle) *Jean-Pierre Wigneron and Brian K. Hornbuckle*

4.10.1 Introduction

Vegetation structure, which varies with vegetation type and phenological stage, affects surface microwave emission. For crops with scattering elements that have either (1) a non-uniform orientation and size distribution and/or (2) a row structure, attenuation properties may also be non-isotropic. As a consequence, vegetation model parameters, such as the optical depth (τ) and single scattering albedo (ω) in the zero-order model, may depend strongly on the configuration parameters of frequency (f), polarisation

(P) and incidence angle (θ). Retrieval of soil moisture and vegetation biomass using multi-channel observations composed of unique combinations of the triplet f, P and θ has been found to be a promising approach (Wigneron *et al.*, 2004).

A few studies have investigated the f, P and θ dependence of microwave emission. Some analyses have shown that τ may depend on P and θ over wheat crops (Ulaby and Wilson, 1985; A. Van de Griend *et al.*, 1996). Numerical and experimental results have also shown that the b_{P} parameter (see Section 4.12) may depend on the vegetation moisture content (Mg) (Le Vine and Karam, 1996; Wigneron *et al.*, 1996), which varies during the crop cycle. In most studies, ω is assumed to be independent of time and incidence angle and very few studies have provided detailed and quantitative results on these aspects. Recently, several studies have been carried out to investigate the dependence of vegetation model parameters on:

- Polarisation and incidence angle at L-band for wheat (Pardé *et al.*, 2003) and for a variety of canopy types (Wigneron *et al.*, 2004; Schwank *et al.*, 2005);
- Frequency at H polarisation (Van de Griend and Wigneron, 2004);
- Incidence and azimuth angle over corn at L-band (Hornbuckle *et al.*, 2003);
- Incidence angle at C-band at large spatial scales (Owe *et al.*, 2001).

The most significant results of these different studies are reviewed in the following. In the first section, a brief discussion of coherent scattering effects within a vegetation canopy is given. Coherent effects often explain the sensitivity of the measured signal to the vegetation structure.

4.10.2 Coherent effects

Several studies investigating the effect of the vegetation structure have been carried out in the field of active (radar) measurements. They will not be discussed in detail here, but they generally indicate a strong dependence of the measurements on the angular configuration, both in incidence (θ) and azimuth (Φ), for crops. These effects have been explained by considering a fully-phase coherent scattering model (Yueh *et al.*, 1992; Lin and Sarabandi, 1999; Stiles and Sarabandi, 2000). For many dense crops with row structure, microwave backscatter cannot be modelled simply as the incoherent addition of power from each element of the plant as done in radiative-transfer (RT) models used frequently in microwave radiometry. A coherent model that considers scattered fields, and not only scattered intensity, is required. Phase interference that occurs at the scale of plant components such as leaves, stems and fruits, and at the scale of plant distribution (e.g. row structure) may explain the sensitivity of the measurement to the vegetation architecture. Yueh *et al.* (1992) noted that information pertaining to the structure and biomass of vegetation is embedded within the f, P and θ responses of the backscatter signal.

Coherency effects are important at low frequencies (L- and P-bands) where the wavelength is of the order or greater than the typical size of the vegetation elements. For instance, very significant coherent effects were shown over wheat at L-band (Stiles *et al.*, 2000). The strong sensitivity of the radar measurements to the azimuth angle could be explained by two main terms: the coherent plant scattering terms and the coherent response due to row structure. The magnitude of the first term was much greater than the incoherent scattering power (the difference was ~5 dB). The second term could explain the strong $\sin(x)/x$ pattern of the dependence of backscatter on azimuth angle. The results clearly show that phase incoherent approaches such as RT models cannot correctly represent coherent scattering effects from densely-vegetated crops.

In the field of passive microwaves, coherency effects are probably less marked since the signal measured by a radiometer originates as a random thermally-generated microwave emission from the soil and vegetation. However, coherent effects should also be considered. For example, Hornbuckle and England (2004) found that the sensitivity of L-band brightness to soil moisture through dense corn was much higher than that predicted by the zero-order RT model. They hypothesised that radiation scattered by the vegetation and reflected by the soil surface, similar to the enhancement of backscatter in scattering canopies, increased the soil moisture sensitivity. In the following sections, even though coherency effects are neglected in the zero-order RT model used for the retrievals, it is likely that the retrieved values of the vegetation parameters implicitly account for these effects. These values should thus be considered as effective values that are used to fit simple RT simulations to observations often resulting from complex coherent scattering mechanisms.

4.10.3 Characterising attenuation by a wheat crop (Pardé et al., 2003)

Pardé *et al.* (2003) computed attenuation effects, parametrised by τ, from the brightness temperatures measured during an experiment over a wheat field (the PORTOS-1993 at the INRA Avignon test site). The dependence of τ as a function of incidence angle θ and for both polarisations is shown in Figure 4.45 for several time periods during crop development. It can be seen that at H polarisation τ is almost constant, while it increases with viewing angle at V polarisation. For vertical-stem dominated vegetation, a simple formulation based on a correction parameter (C_{pol}), derived from a physical analysis of the attenuation by vertical stems (Ulaby and Wilson, 1985) was proposed by Wigneron *et al.* (1995) to model the dependence of τ_{p} on view angle and polarisation. The C_{pol} formulation is given by the equations:

$$\tau_{\mathrm{V}}/\tau_{\mathrm{H}} = \sin^2\theta C_{\mathrm{pol}} + \cos^2\theta \tag{4.116}$$

$$\tau_{\mathrm{H}} = \tau_{\mathrm{H}}(\theta = 0) = \text{constant} \tag{4.117}$$

In this approximation, τ for H polarisation (τ_{h}) is assumed to be independent of θ. Simulations of τ using the C_{pol} parametrisation were compared with the values of τ retrieved for each view angle and both polarisations. This comparison is illustrated in Figure 4.45. The use of C_{pol} was found to be very efficient, since (1) it led to very accurate simulations of brightness temperature (Tb), and (2) it is a very convenient approach since only two parameters (C_{pol} and $\tau_{\mathrm{H}}(\theta = 0)$) are required to simulate τ as a function of view angle and polarisation. The values of the two parameters C_{pol} and $\tau_{\mathrm{H}}(\theta = 0)$ were computed for each day, by minimising the rms error between measured and simulated Tb. The time variations in the values of C_{pol} are plotted in

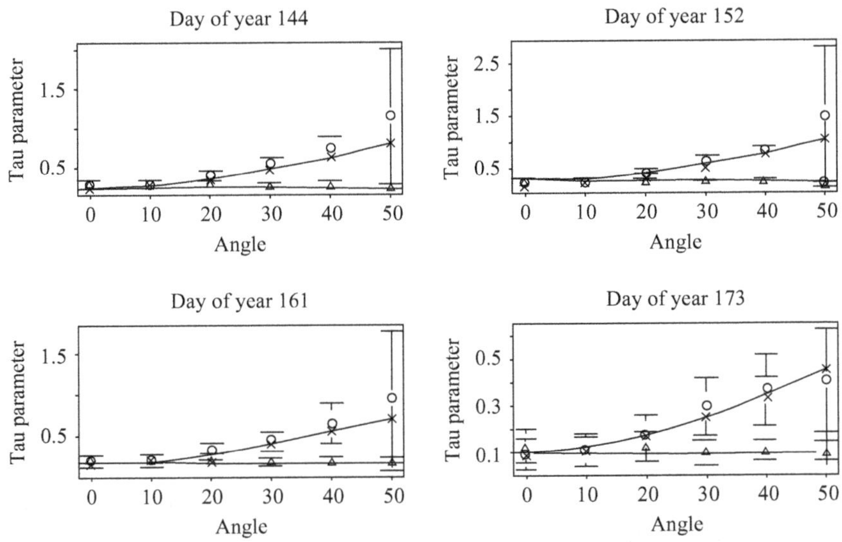

Figure 4.45 *Retrievals of τ as a function of θ and polarisation (V (o), H (Δ)) and computed values using the C_{pol} formulation (__) (PORTOS-93 experiment). (Reprinted from Pardé et al. (2003), © 2003, with permission from Elsevier)*

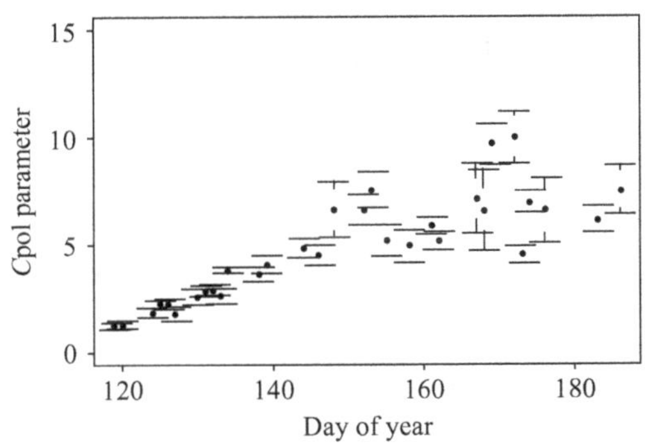

Figure 4.46 *Time variations in the retrieved C_{pol} parameter during the wheat cycle (PORTOS-93 experiment). (Reprinted from Pardé et al., (2003), © 2003, with permission from Elsevier)*

Figure 4.46. Note that C_{pol} increases almost linearly during wheat growth until the vegetation is well-developed. This increase can be related to the growth of the vertical stems which have high attenuation properties at V polarisation. At the beginning of the crop development (~DoY 120), stems are very short and the vegetation is leafy. Associated values of C_{pol} are close to one, which correspond to the case of no dependence of τ on view angle and polarisation. C_{pol} increases simultaneously

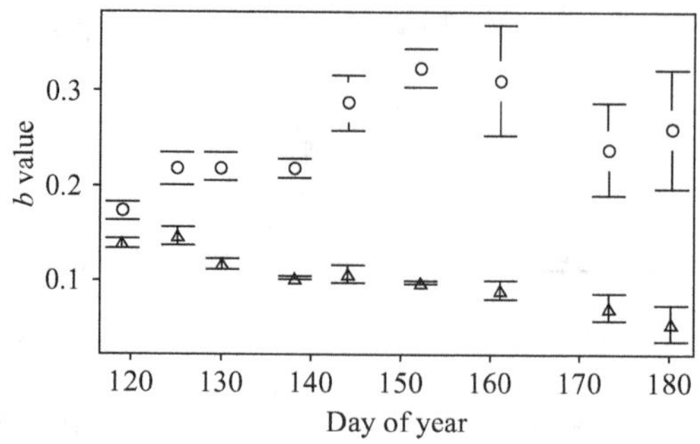

Figure 4.47 *Time variations in the retrieved b_P parameter during the wheat cycle (PORTOS-93 experiment) (V(o), H(Δ) at $\theta = 40°$). (Reprinted from Pardé et al. (2003), © 2003, with permission from Elsevier)*

with the development of the stems. In Wigneron *et al.* (2004), high correlation was found between C_{pol} and vegetation height (H_C) ($R^2 = 0.92$, $C_{pol} = 8.431 H_C$ (m) – 0.795, for $H_C > 0.25$ m). The time variation of b_P, computed by dividing τ by vegetation water content (VWC, in kg per square metre), is plotted in Figure 4.47 at $\theta = 40°$. The large values of b_P initially ($b_P = 0.15$) correspond primarily to leafy vegetation at the beginning of crop growth. As the structure of the vegetation began to be dominated by the vertical stems, the b_P values decreased at H polarisation, while they increased at V polarisation. During senescence, the b_P values decreased at both V and particularly H polarisation (to approximately 0.03).

4.10.4 *Characterising scattering and attenuation by crops at L-band (Wigneron et al., 2004)*

Wigneron *et al.* (2004) analysed several experimental datasets of multi-angular and dual-polarisation measurements:

- At the BARC test site (Beltsville Agricultural Research Center) over soybean, corn, alfalfa and grass (Wang *et al.*, 1982) with a duration of approximately one month;
- At the INRA Avignon test site (Institut National de la Recherche Agronomique): with the PORTOS instrument over soybean, sorghum and wheat and the EMIRAD instrument over corn (Wigneron *et al.*, 1993, 1995; Pardé *et al.*, 2003).

The study investigated first the dependence of ω on the configuration parameters θ and P over time. Two main results were illustrated over corn but were found to be also valid for the other crop types. First, the best modelling results were generally obtained by considering that ω changes with time at both polarisations. This time dependence is in relation to the changing vegetation structure, which is very important over crop fields at the beginning of the vegetation development (Figure 4.48). This time dependence might be much lower for natural vegetation covers.

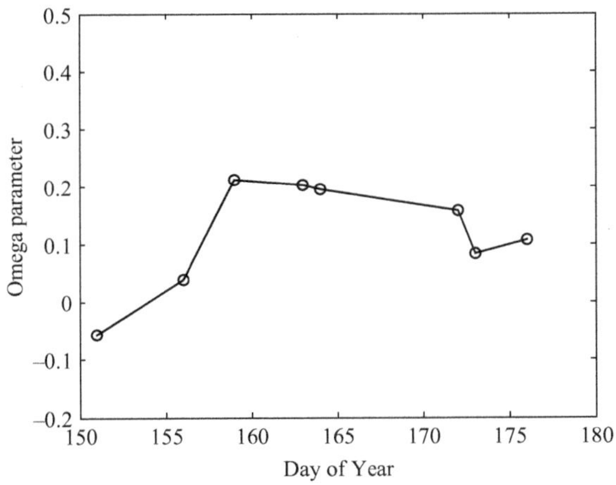

Figure 4.48 Retrieved time dependence of ω over a corn field (EMIRAD'2001 experiment; Wigneron et al., 2004; reproduced by permission of IEEE, ©2004)

Second, even though dependences on time, incidence angle or polarisation were incorporated into the model, the resulting simulations were found to be very close and for most applications it seems that ω can be considered to be a constant.

Retrieved values of ω were found to be close to zero for soybean and alfalfa. These rather low values of ω could be probably related to low multiple scattering effects within the canopy at L-band. Low values of ω were also obtained for crops with scattering patterns with a strong peak in the forward direction, such as a wheat canopy. Conversely, for corn, and to a lower extent for grass, ω was found to be significantly larger than zero ($\omega = 0.06$ for BARC, $\omega \approx 0.13$ for EMIRAD). The ω parameter was then set to a constant value for each crop and the b_P parameter was retrieved as a function of time, θ and P. The main results of the study were the following:

- The b_P values appear to be primarily determined by crop type. At H polarisation, the average values of b_P were about 0.5 for grassland, 0.15–0.25 for soybean and 0.1–0.15 for corn and wheat. Large values of the b_P parameter have been estimated previously over prairies (Jackson and Schmugge, 1991) which may be related to the high values of the volume fraction of vegetation and possibly to thatch (dead vegetation matter at the soil surface). More detailed results will be given in Section 4.12.
- The dependence of b_P on polarisation was significant for several crops. In particular, a marked effect can be noted over wheat and to a slighter extent over corn. For these two types of vegetation, the polarisation dependence was likely due to the vegetation structure being dominated by vertical stems since $b(\mathrm{v}) > b(\mathrm{h})$.
- The dependence of b_P on incidence angle θ was generally low at H polarisation. This is illustrated for wheat in Figure 4.45 and for corn in Figure 4.49. At V

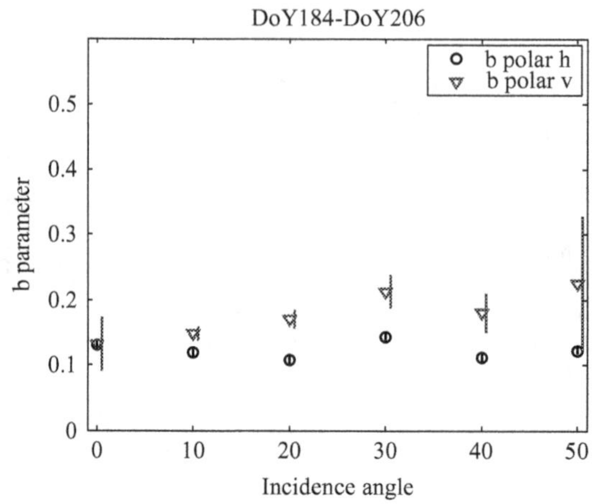

Figure 4.49 *Dependence of* b_P *on incidence angle* θ *and polarisation for* $\Phi = 90$ *degrees (viewing direction is* $\perp$ *to the rows) over corn (EMIRAD'2001 experiment, Wigneron et al., 2004; reproduced by permission of IEEE, ©2004)*

polarisation, a clear increasing and decreasing trend could be noted for wheat and herbaceous canopies, respectively.

– Excluding the early phases of vegetation development and senescence, the dependence of b_P on time was found to be relatively low. A small decreasing trend was noted for herbaceous canopies. Wheat is an outlier: b_P decreases at H polarisation and increases at V polarisation, reflecting stem growth.

4.10.5 *Anisotropy in relation to the row structure of a corn field at L-band (Hornbuckle et al., 2003)*

Although several studies have investigated emission from vegetation planted in rows, no clear conclusion on the effect of row structure has been obtained. It was noted in Section 4.10.1 that radar measurements of wheat were found to be sensitive to incidence and azimuth angle (Stiles *et al.*, 2000). This result could be explained by coherent modelling (Stiles and Sarabandi, 2000). In the field of passive microwaves, Hornbuckle *et al.* (2003) recently investigated anisotropy (incidence and azimuth angle dependence) in 1.4 GHz brightness induced by a corn canopy. Several clear conclusions were made.

First, both polarisations of 1.4 GHz brightness are isotropic in azimuth during most of the growing season. However, when the canopy is senescent, brightness is a strong function of row direction. In that case, brightness is highest parallel to the rows at H-pol and perpendicular to the rows at V-pol. At senescence, brightness temperature varies from 5 to 10 K in azimuth, depending on the incidence angle (Figure 4.50). To explain this result, the authors examined the changes in water content of the leaves and stems over the course of the growing season. They concluded that leaves play

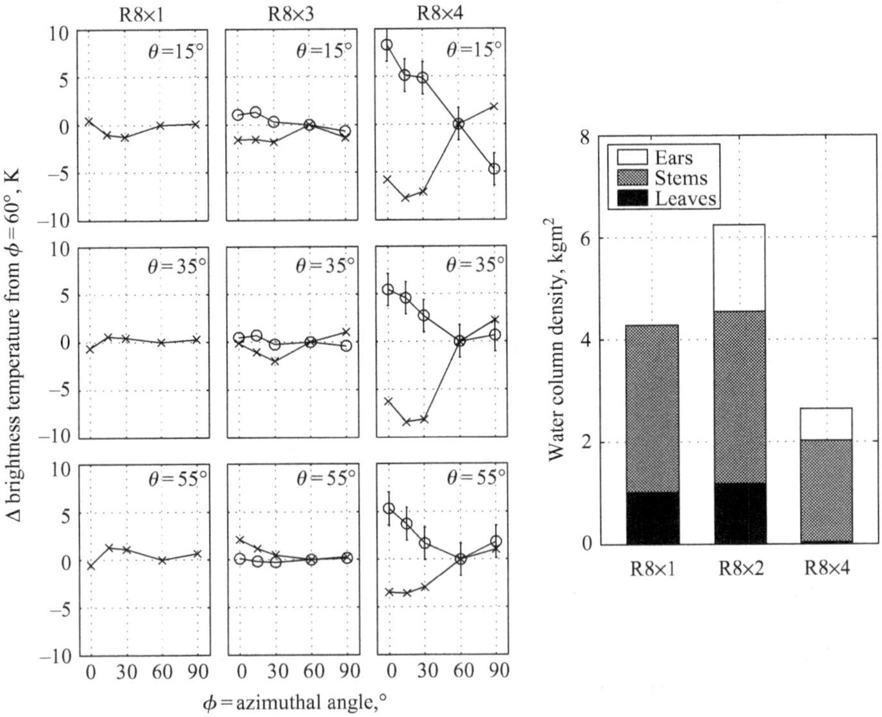

Figure 4.50 *At left, change in V-pol (x) and H-pol (o) brightness temperature as a function of azimuth angle (ϕ) with respect to row direction. $\phi = 0$ degree is parallel to the row direction. Experiment R8 × 1 occurred in early July, R8 × 3 in mid-August, and R8 × 4 in early October. At right, vegetation water content (in kg per square metre) during the experiments, distributed among the primary canopy components*

an important role: instead of acting simply as a 'cloud' of moisture, they scatter microwave radiation at 1.4 GHz and mask the internal, stem-dominated structure of a corn canopy. At senescence when they lose their water, the leaves become essentially invisible and the internal structure is exposed. In that case, the row arrangement of moist stems was the dominant effect in the crop microwave signature and was found to be anisotropic in azimuth as expected.

On the other hand, 1.4 GHz brightness was found to be anisotropic in elevation: an isotropic zero-order RT model could not reproduce observed changes in brightness with θ. The observed decrease in brightness with θ is likely due to scatter darkening in the canopy, since the electrical size of the stems and ears of field corn are significant at 1.4 GHz (Ulaby *et al.*, 1987). Scatter darkening increased with incidence angle. At H-pol, brightness temperatures modelled assuming no scattering overestimated measured brightness temperature by about 1–2 K at 35° and about 7–9 K at 55° for corn at maximum biomass. The variation of scatter darkening with incidence angle was greater at V-pol than at H-pol. To model this effect with the zero-order model at both polarisations, the volume scattering coefficient and, consequently, the extinction coefficient and the single-scattering albedo, were allowed to be a function of incidence angle. The change was very small between 15 and 35 at H polarisation (Figure 4.51). The decrease in the V-pol extinction coefficient during crop growth is a consequence of the replacement of water by vegetation matter within the canopy volume as the corn matures. The rather high values of the single scattering albedo (~0.06) are in general agreement with values obtained by Wigneron *et al.* (2004).

For the zero-order model to retain its physical significance, the single scattering albedo must be small in magnitude. Hornbuckle *et al.* (2003) found that this was the case for corn. Observed scatter darkening, particularly at large incidence angles, was not that high. Therefore, provided the zero-order parametrisation is formulated to account for observed variations of extinction and scattering with incidence angle, the zero-order radiative-transfer solution can correctly model 1.4 GHz brightness. Such anisotropic effects should be less drastic in many types of vegetation.

4.10.6 Anisotropy at large spatial scale (Owe et al., 2001)

Owe *et al.* (2001) evaluated vegetation anisotropy at a large spatial scale (pixel size ~150 km) using satellite observations from SMMR at 6.6 GHz. Even though this was only a small aspect of the whole study, it is one of the first times such an attempt has been made. The authors noted that it is reasonable to assume that vegetation elements are randomly oriented for most crops and naturally occurring vegetation. Consequently, it is likely that optical depth (τ) is polarisation-independent on a satellite scale. Validation of this hypothesis was attempted by retrieving optical depth when soil moisture was assumed to be at saturation after very large storm events. Over a footprint covered by a range of canopy types and densities, the authors obtained a relationship between optical depths at V and H polarisation close to the 1:1 line (Figure 4.52). However, this preliminary result was obtained over a very large pixel which was much larger than the typical size of footprints (~50 km) of future microwave sensors. There was also significant scatter in this relationship. Additional studies investigating anisotropy at large scales are thus required for modelling the observations that will be acquired from future spaceborne sensors.

4.10.7 Conclusions

Several results have recently improved our knowledge of the sensitivity of vegetation model parameters to canopy geometry and to configuration parameters (frequency, polarisation and incidence angle). This knowledge can be very useful in retrieval studies. For instance, the C_{pol} parameter (presented in Section 4.10.3) was used to account for the dependence of optical depth on polarisation and incidence angle. Pardé *et al.* (2004) showed that retrieving this parameter simultaneously with soil moisture (using a method referred to as the N-parameter retrieval approach) could improve the accuracy of soil moisture retrievals over a variety of vegetation canopies. An accurate estimation of the single scattering albedo ω for each kind of canopy type was also found to be useful for soil moisture retrieval applications.

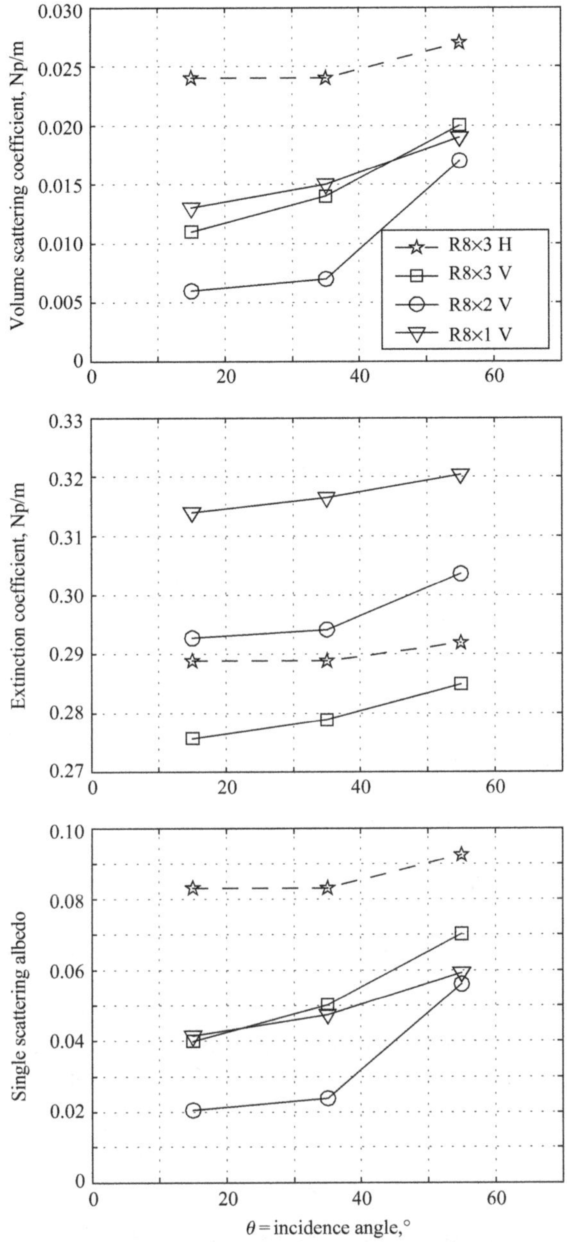

Figure 4.51 *Variation of the volume scattering coefficient Ks, the extinction coefficient Ke, and the single scattering albedo ω, with incidence angle θ. Experiment R8 × 1 occurred in early July, R8 × 2 in mid-July, and R8 × 3 in mid-August. (Hornbuckle et al., 2003; reproduced by permission of IEEE, ©2003)*

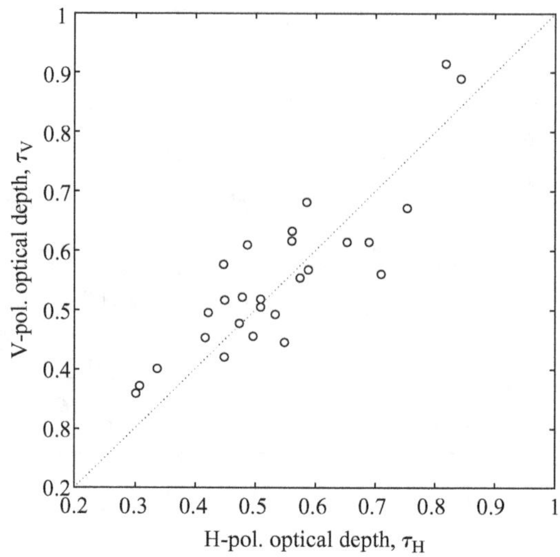

Figure 4.52 *Relationship between the calculated optical depths at V and H polarisation over a large agricultural area from 6.6 GHz SMMR observations (Owe et al., 2001; reproduced by permission of IEEE, ©2001)*

On a satellite scale as discussed by Owe *et al.* (2001), footprints normally include a variety of canopy types, and it is likely that this mixing should result in a low dependence of the vegetation model parameters on polarisation and incidence angle. Nevertheless, further investigations (experimental and perhaps synthetic datasets) are required to confirm this rather intuitive result. Moreover, footprint size may soon approach 30 km as future spaceborne sensors are developed. At this scale, the footprint could be considered as nearly homogeneous over many large boreal and tropical forests and agricultural areas, resulting in a rather low 'mixing' effect.

Most of the results discussed in this chapter pertain to agricultural crops. Other aspects peculiar to the natural environment should also be investigated in the future. These include the effect of litter (thatch) and necromass (dead vegetation material included within the vegetation layer) in areas which are never or rarely ploughed (prairies, fallow and natural covers) and the specific effects of the architecture trees and understory in forests. Some first results relating to this last topic are discussed in Section 4.13.

4.11 Passive microwave emissivity in vegetated regions as directly calculated from satellite observations *Catherine Prigent*

4.11.1 Introduction

Satellite microwave sensors have shown some ability to characterise the vegetation at spatial resolutions compatible with climatological applications. However, these instruments have, to date, triggered less interest than their visible and near infrared counterparts which offer a much better spatial resolution, and the NDVI derived from the visible and near infrared reflectances is still preferred for vegetation applications. Most studies at microwave frequencies have focused on the use of simple indices like the microwave vegetation index which is based on the polarisation difference at 37 GHz (e.g. Choudhury and Tucker, 1987; Justice *et al.*, 1989). However, as noted by several authors (e.g. Tucker, 1989; Kerr and Njoku, 1993), atmospheric effects, especially cloud cover, are responsible for a large part of the 37 GHz polarisation difference casting doubt on the interpretation of simple indices solely in terms of surface properties.

The microwave land surface emissivities directly calculated from the SSM/I observations between 19 and 85 GHz by subtracting the atmospheric effect (see Section 4.8 above and Prigent *et al.* (1997, 1998)) showed promising correspondences between geographical and seasonal patterns of emissivities and the vegetation (Prigent *et al.*, 2001a).

First, the sensitivity of the large-scale emissivities to the vegetation density and phenology is examined, as compared to the NDVI. Second, an unexpected sensitivity of the emissivities to the underlying surface is evidenced casting doubt on our understanding of the vegetation interaction with the microwave signal.

4.11.2 Sensitivity to vegetation density and phenology: Comparison with the NDVI

A monthly mean map of the microwave emissivities (polarisation difference at 37 GHz) is presented in Figure 4.53 top panel (colour versions of the maps in Figure 4.53 are available in Prigent *et al.* (2001a)). For comparison, the NDVI monthly mean dataset derived from AVHRR is also shown (NASA and NOAA Earth Observing System Pathfinder project (James and Kalluri, 1994) (Figure 4.53, middle panel)). Green vegetation has a specific reflectance curve in the visible and near-infrared, and the NDVI has been found to be correlated with the fraction of photosynthetically active radiation absorbed by green vegetation. In addition, the Matthews (1983) vegetation classification is presented (Figure 4.53, lower panel). The latter was compiled from a large number of published sources. At a 1° spatial resolution the vegetation classification distinguishes a large number of vegetation types, typically grouped to 30 classes of natural vegetation that are further grouped and combined with cultivation intensity maps to provide ten classes of decreasing vegetation density. Vegetation absorbs, emits and scatters microwave radiation. Radiative properties of vegetation are mainly controlled by the dielectric properties of vegetation components, their density and the relative dimension of vegetation components with respect to the wavelength. Dielectric properties of vegetation are closely related to their water content. Increasing vegetation density usually reduces the emissivity polarisation difference. There are obvious qualitative correspondences between the two types of satellite observations and the vegetation classification. Desert areas (V10) are characterised by low NDVI and large emissivity differences and are clearly noticeable. Tropical rain forests (V1) display negligible emissivity polarisation differences.

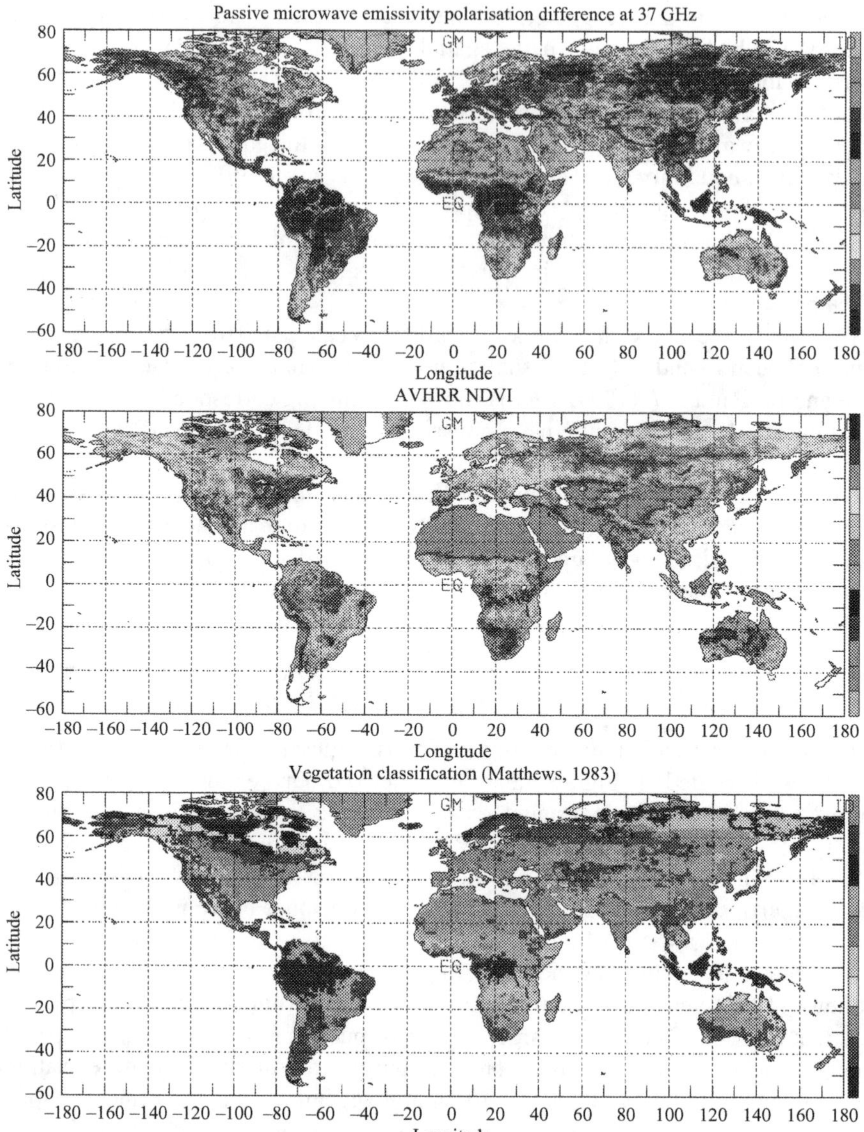

Figure 4.53 *Top panel: Monthly mean emissivity polarisation differences (vertical–horizontal) at 37 GHz from SSM/I for August 1992. Middle panel: NDVI(AVHRR) from the Pathfinder for the same month. Lower panel: Simplified version of the Matthews (1983) vegetation classification (ten classes of decreasing vegetation density)*

However, the tropical rain forest in Africa shows rather low NDVI values compared with the rain forest in South America. Grassland (V7) and woodland (V2, V3, V5) are also distinguished in Africa and South America. The sharp gradient southwards of 15°N in Africa, observable on all the maps, corresponds to the Sahelian transition between the arid shrub grassland and the more humid grassland with tree cover. For a closer examination of the sensitivity of the microwave observations and NDVI to the vegetation, Figure 4.54 shows histograms of the different microwave observations and NDVI for four major vegetation types in the northern hemisphere where the cultivation intensity is $\leq$20 per cent (as separated by Matthews' classification). For a given observation the histograms are normalised to have the same area, giving an estimate of the probability distribution function. Vegetation information is similar at all microwave frequencies between 19 and 85 GHz. Separating vegetation via individual SSM/I channels (vertical or horizontal polarisation) cannot be done. The emissivity in horizontal polarisation is lower for grassland than for tree-covered areas. With decreasing vegetation density the contribution of bare soil surfaces within a pixel increases, reducing emissivity in the horizontal polarisation. The emissivity in vertical polarisation shows little variation. The emissivity polarisation difference helps distinguish rain forest from forest/woodland and from grassland. Nevertheless, discrimination between forest and woodland is not possible. It appears that it is easier to distinguish rain forest and other forests with passive microwave than with the NDVI. Other authors also observed saturation of the NDVI response for high green-leaf density (e.g. Tucker *et al.*, 1985). Correspondences between the passive microwave emissivities and the NDVI translate into a correlation of 0.72, at a global scale over snow free continental surfaces during one year. Figure 4.55 shows the scatterplot of the NDVI observations versus the emissivity polarisation difference for 1993 and 1994, over the regions described in the Global Soil Moisture Data Bank (Robock *et al.*, 2000). For comparison, for the same locations and time period, the scatterplot of the soil moisture *in situ* measurements versus the passive microwave signal is also presented. The relationship between the microwave emissivities and the NDVI is much stronger than with the soil moisture (see Prigent *et al.* (2005b) for a detailed analysis of the sensitivity of the emissivities to soil moisture). In order to analyse the sensitivity of the microwave emissivities to the vegetation phenology, the emissivity maps at all SSM/I frequencies have been examined for a full year. For instance, in Africa, the emissivity gradient southwards of 15°N that corresponds to the Sahelian transition is broader in winter than it is in summer. In the grassland area in Southern Africa, vegetation density declines during the southern hemisphere winter: The emissivities are lower. For a detailed study of the sensitivity of the emissivities to the vegetation phenology, see Prigent *et al.* (2001a).

4.11.3 Puzzling observations in densely vegetated areas

Radiative-transfer theory predicts that absorption/emission contribution from the vegetation increases with frequency (Wigneron *et al.*, 1993; Karam, 1995). As a consequence, in very dense forested regions, the microwave satellite observations are

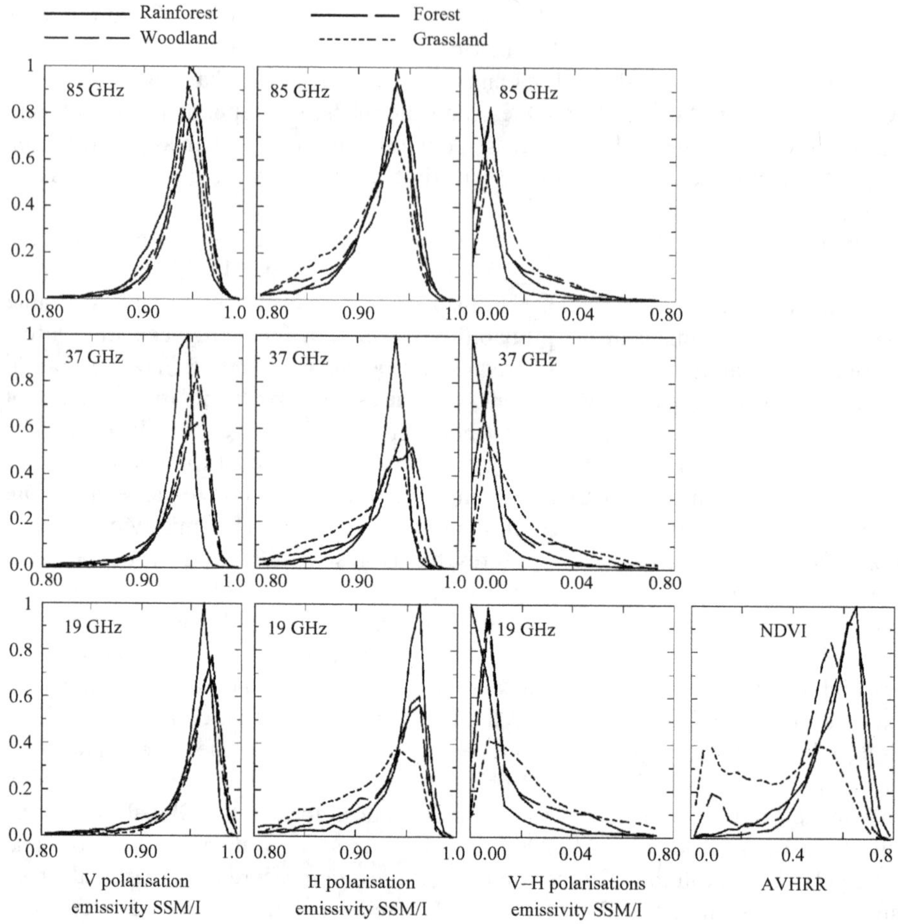

Figure 4.54 Histograms of the different satellite observations for four major vegetation types, for the northern hemisphere in July 1992. For a given observation, histograms are normalised to have the same area. Evergreen and deciduous forests have been grouped, as well as the evergreen and deciduous woodlands. Only pixels with cultivation intensity ≤ 20% are considered

not expected to be sensitive to the underlying surface and with increasing frequency, the sensitivity to the underlying surface should decrease.

Figure 4.56 shows the response of three instruments in the Amazonian rain forest every other month for a year over the Amazon river (a colour version of this figure is presented in Prigent *et al.* (2001a)). The SSM/I at 37 GHz, the ERS scatterometer signal at 5.25 GHz and the AVHRR NDVI responses are presented. The passive microwave observations clearly detect the Amazon river and its associated wetlands,

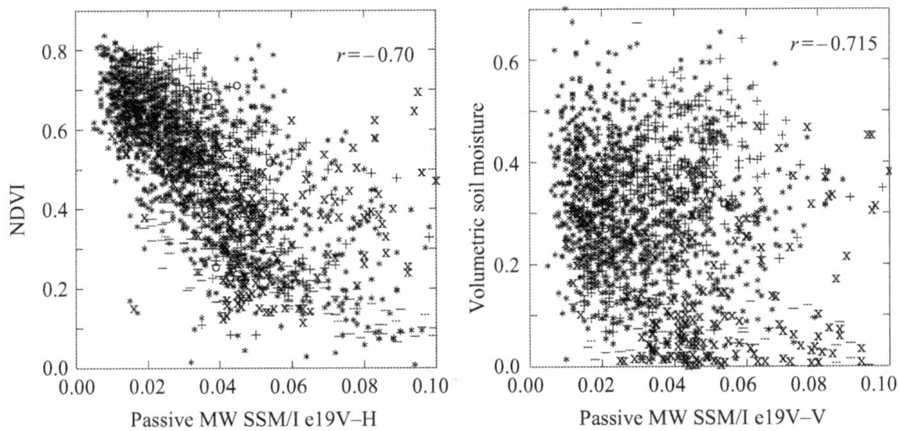

Figure 4.55 *Left: Scatterplots of the NDVI observations versus the microwave emissivities at 19 GHz (polarisation difference), over the Global Soil Moisture Data Bank regions, for 1993 and 1994. Right: Similar scatterplot with the in situ soil moisture measurements over the same regions and time period. The linear correlation coefficients are indicated*

with lower emissivities (at both H and V polarisations) and larger polarisation differences above standing water. From summer to the next spring, there is a significant increase of the areas of low emissivities, likely related to an increase in the wetland extent. Similar results are obtained at 19 and 85 GHz (not shown) as at 37 GHz, although absorption by the vegetation canopy is expected to increase significantly with frequency. The active microwave instrument at 5.25 GHz does not show any sensitivity to the presence of the river and surrounding inundations. With increasing frequency, one expects higher attenuation and scattering by vegetation and as a consequence a lower sensitivity to soil properties. Although operating at a lower frequency than SSM/I, the ERS-1 scatterometer shows much less sensitivity to inundated areas (note that the scatterometer and SSM/I at 19 GHz have very similar spatial resolution). The visible/near infrared signals (here with a 8 km spatial resolution) do not detect the presence of the river either. Similar observations have been made in various environments, with always a much more limited sensitivity of the active microwave at low frequency to the standing water than with SSM/I observations. This feature has been exploited to estimate the extent and seasonality of the wetlands at a global scale from a combination of these satellite observations (Prigent *et al.*, 2001b).

These observed signatures are difficult to interpret. Different factors might interfere with the radiation (rain deposition on the leaves, open spaces between the trees, role of multiple reflections between the canopy and the standing water below, to name a few). Efforts to understand the complex physical interactions between the soil surface and the canopy have not yet succeeded. Further investigations have to be conducted.

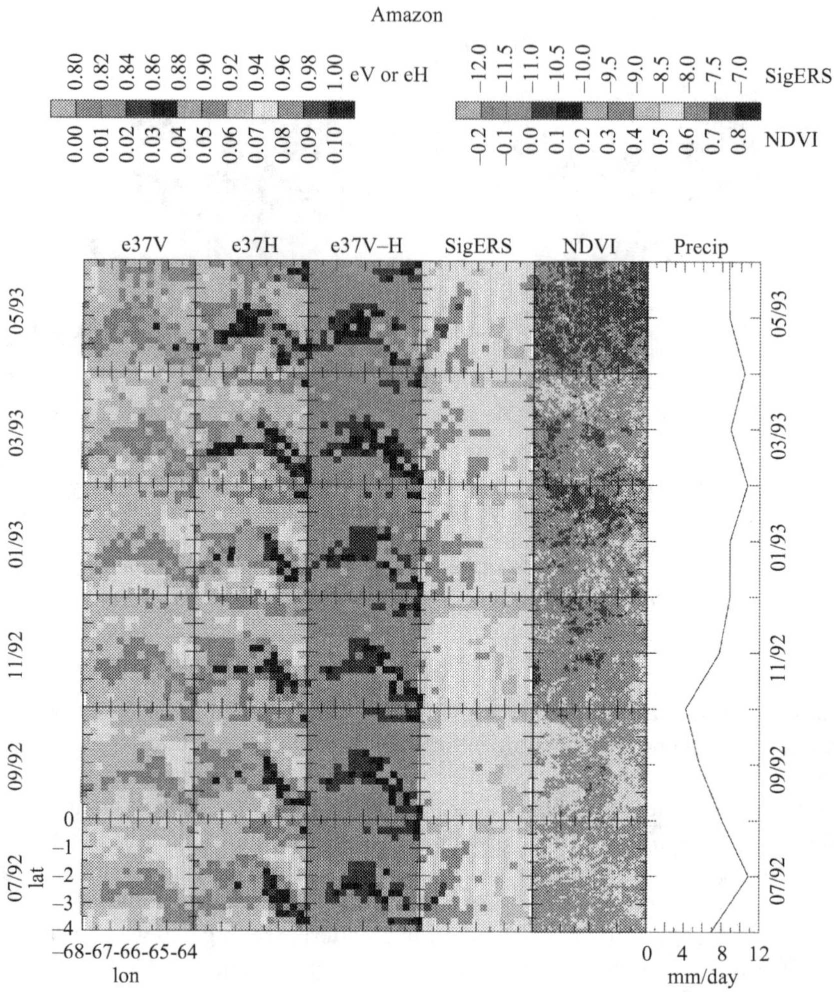

Figure 4.56 *Response of various instruments over wetlands in an area in the Amazon rain forest, for every other month between July 1992 and June 1993. From left to right: The SSM/I 37 GHz emissivities vertical, horizontal polarisations and their difference, the ERS scatterometer response linearised at 45° incidence and the AVHRR NDVI response. The precipitation is also indicated as given by the Global Precipitation Climatology Project (Huffman, 1996)*

4.11.4 Conclusion

Passive microwave emissivities are sensitive to vegetation density, although it would be difficult to accurately discriminate vegetation types from passive microwave observations only. Microwave observations respond to the absorption/emission by vegetation elements including woody parts and are not directly sensitive to the photosynthesis activity. These measurements can therefore complement the visible/near-infrared observations by indicating structural density, the NDVI being primarily sensitive to the photosynthetically active part of the plant. In addition, passive microwave can help separate rather dense vegetation for which the NDVI saturates.

Contrary to the expectation, passive microwave observations are very sensitive to the underlying surface even in very densely vegetated areas. Joint analysis of the emissivity and backscattering model responses, compared with passive and active satellite observations, will help better understand the vegetation and soil interaction with the microwave radiation.

4.12 The *b*-factor relating vegetation optical depth to vegetation water content *Adriaan Van de Griend*

4.12.1 Introduction

The transparency of the canopy (also denoted transmissivity) is expressed in terms of the (nadir) optical depth τ_o according to

$$t_\mu = e^{-(\tau_o/\cos\theta)} \tag{4.118}$$

where θ is the incidence angle from nadir. Although, for several reasons (see e.g. Ulaby *et al.*, 1986) microwave observations are seldom taken at nadir, the nadir optical depth (usually simply denoted 'optical depth') has been introduced to correct for the slant path when looking off-nadir. This, however, assumes validity of the cosine rule of Equation (4.118), which may be violated in reality (see below). For this reason, it should be common practice to identify the original angle of incidence from which the nadir optical depth has been derived or $\tau_{o(\theta)}$.

For a certain vegetation type and frequency, optical depth is usually directly related to the vegetation water content, W (kg/m^2). Jackson and O'Neill (1990) reported the following linear relationship between τ_o and canopy water content (W)

$$\tau_o = b \cdot W \tag{4.119}$$

which has been referred to in the literature many times. Here, b is a regression coefficient which is frequency and polarisation dependent and more or less characteristic for a certain type of canopy. The relationship has also been studied on a theoretical basis by Le Vine and Karam (1996) using a discrete scatter model. They concluded that the linear assumption is reasonable for $1.4\,\text{GHz} < f < 10\,\text{GHz}$, which is the frequency range relevant for soil moisture studies.

Knowledge of the parameter b and its dependence on incidence angle, frequency and polarisation for different canopy types is important for the development of multi-frequency inversion models, e.g. L-band SMOS (Kerr *et al.*, 2001; Berger *et al.*, 2003) and C-band AMSR-E (Njoku *et al.*, 2000). Such models, especially if multi-angle observations are involved, allow multiple parameter retrievals, e.g. ε (soil dielectric constant), τ and T_{eff} (effective physical temperature of the emitting body). For example, based on combined L- and C-band observations, Wigneron *et al.* (1995) demonstrated that the retrieval accuracy for soil moisture and optical depth is very sensitive to the ratio of optical depths at 5 and 1.4 GHz. Knowledge of the b-factor is also important for forward modelling of brightness temperatures in assimilation schemes of atmospheric weather prediction models and global circulation models using auxiliary data on wet vegetation biomass.

In order to get a better insight into the frequency dependence of the parameter b, Van de Griend and Wigneron (2004a) analysed all available data on b and their corresponding frequencies that they could find in the literature. This chapter builds highly on this latter paper.

4.12.2 The b-factor and its theoretical dependence of wavelength

The dependence of the parameter b on wavelength (and thus also on frequency) follows from theoretical considerations. If multiple scattering is ignored ($\gamma_{\mathrm{s}} \cong 0$) then absorption ($\gamma_{\mathrm{a}}$) is the dominant process, which can be defined in terms of the 'effective' canopy dielectric constant, $\varepsilon_{\mathrm{can}}$, according to (Ulaby, 1986):

$$\gamma_{\mathrm{a}} = 2k_{\mathrm{o}}\mathrm{Im}[\sqrt{(\varepsilon_{\mathrm{can}}/\varepsilon_{\mathrm{o}})}] \tag{4.120}$$

Here, $k_{\mathrm{o}}$and ε_{o} are the wave number ($k_{\mathrm{o}} = 2\pi/\lambda_{\mathrm{o}}$) and the dielectric constant in free space, respectively. For different canopy architectures, the parameter $\varepsilon_{\mathrm{can}}$ may be calculated using canopy mixing models (see e.g. Van de Griend and Wigneron, 2004b). If we ignore extinction as a result of scattering ($\gamma_{\mathrm{e}} \cong \gamma_{\mathrm{a}}$), then, with

$$\tau_{\mathrm{o}(\theta)} = \int_0^d \gamma_{\mathrm{e}}\mathrm{d}z = \gamma_{\mathrm{e}}d \tag{4.121}$$

the optical depth at nadir, for a canopy with thickness d, can be written as

$$\tau_{\mathrm{o}(\theta=0)} = \gamma_{\mathrm{e}}d = \frac{4\pi d}{\lambda_o}[\mathrm{Im}\sqrt{\varepsilon_{\mathrm{can}}/\varepsilon_{\mathrm{o}}}] \tag{4.122}$$

Since $\varepsilon_{\mathrm{can}}$ is determined by the dielectric properties of the plant–air mixture and the dielectric constant of leaves for frequencies $<$10 GHz is only slightly dependent on frequency (see e.g. El-Rayes and Ulaby, 1987), it follows from Equations 4.119 and 4.122 that b is approximately proportional to $1/\lambda$. For higher frequencies, however, a significant frequency dependence due to Debye relaxation (Mätzler, 1994) has been discerned from measurements based on a quasi-optical method using radiometers up to 94 GHz (Mätzler and Sume, 1989).

The general decrease of b with wavelength was confirmed amongst others by Kirdyashev *et al.* (1979), Chukhlantsev and Shutko (1988), Pampaloni and Paloscia (1986) and by Jackson and Schmugge (1991). However, Kirdyashev *et al.* (1979) suggested

$$b = c/\sqrt{\lambda} \tag{4.123}$$

for fine-leafed vegetation.

Theoretical and experimental evidence has shown that scattering becomes dominant for $\lambda < \sim 5$ cm (corresponding to $f > \sim 6$ GHz), although the precise boundary depends on the dimensions of canopy elements and spaces between elements. Since multiple scattering is ignored in the zero-order τ-ω model (for the τ-ω model see the contribution in Section 4.2), Jackson and Schmugge (1991) argued that both populations should be regarded separately and suggested

$$b = c/\sqrt{\lambda} \quad (\text{for } \lambda > 5\,\text{cm}) \tag{4.124}$$

$$b = c/\lambda \quad (\text{for } \lambda < 5\,\text{cm}) \tag{4.125}$$

Ulaby and Wilson (1985) measured a wide range of wavelengths for the same canopy (wheat and sorghum, each at $\lambda = 2.8$, 6.5 and 18.2 cm) and, based on those data, Jackson and Schmugge (1991) suggested for these crops

$$b = c/\lambda^x \tag{4.126}$$

which is basically a generalisation of the relationship between b and λ in which x is a coefficient, assumed to be specific for a certain canopy type. Based on their analyses, Jackson and Schmugge (1991) concluded that in terms of Equation (4.126), most (agricultural) vegetation can be categorised as either leaf dominated (soybean, alfalfa) or stem dominated (corn, cereal, wheat) and that grass behaves differently from all other crops. At L-band, they found a rather constant value of b (~ 0.15 at 1.4 GHz) regardless of canopy type, except for grass. At higher frequencies, several experiments demonstrated an increasing dependence of b on canopy structure (Wegmüller *et al.*, 1995; Le Vine and Karam, 1996; Njoku and Li, 1999).

4.12.3 Comparison of b-factors from different sources

Direct comparison of b-values from different sources should be done with care because their derivation may be based on different methods and approaches. Van de Griend and Wigneron (2004a) identified the following five major points of concern which should be recognised when comparing b-values from different sources. These are:

First. Most investigations assumed the cosine correction for the slant path of angular measurements (Equation (4.118)) to be valid. At H-pol this assumption seems to be reasonable for most agricultural crops (Ulaby *et al.*, 1986). However, at V-pol both experimental evidence (Ulaby *et al.*, 1987; Tavakoli *et al.*, 1991) and theoretical considerations (Ulaby *et al.*, 1986) indicate a strong angular dependence of the optical depth for canopies with vertical stalks. This angular dependence – for an overview see Van de Griend and Wigneron (2004a) – implies that b-values at V-pol are also angular dependent. This has been confirmed from multi-angle measurements over wheat (Wigneron *et al.*, 1996) and therefore a comparison of b-values on an angular basis would be required. Analysis of an extended dataset obtained over a wheat field (Pardé *et al.*, 2003) indicated the same angular dependence of the parameter b for V-pol, whereas the angular dependence at H-pol turned out to be smaller or even absent. A more recent study by Wigneron *et al.* (2004) showed that the *b*-factor (and thus also τ_0) is, except for wheat at H polarisation, a function of view angle and different for different crop types (corn, grass, alfalfa, soybean and wheat). The angular dependence at H polarisation, however, was found to be relatively small compared to V polarisation, which is in line with several earlier studies as summarised by Van de Griend and Wigneron (2004b). The number of experiments on V-pol measurements found in the literature is, however, rather limited.

Second. For agricultural crops, the azimuth angle with respect to the row direction could play an important role (see e.g. Tavakoli *et al.*, 1991). Most experiments, however, were performed with look directions perpendicular to the row direction, whereas observations from space will include a mixture of azimuth angles with respect to row directions. Note that results obtained over corn (Section 4.10) indicated a very low sensitivity of the measurements to azimuth at L-band and for green vegetation.

Third. Different approaches and models have been used to derive τ (and the single scattering albedo ω) from the experimental data. The approaches and models have been critically reviewed recently (Van de Griend and Wigneron, 2004b) and the reader is referred to this paper for details. The most frequently used model is the zero-order radiative-transfer model, also known as the τ-ω model, in which both τ and ω describe the radiative-transfer properties of the canopy. This model ignores multiple scattering effects, which is only acceptable for frequencies smaller than approximately 6 GHz. At higher frequencies, multiple scattering may become substantial, depending on the type of canopy, the dimension of the canopy elements and the spaces between the canopy elements, all relative to the wavelength.

Fourth. It may be difficult to decompose both parameters τ and ω in a retrieval algorithm (Burke *et al.*, 1999).

Fifth. Several investigators assumed a fixed value for ω or even assumed ω to be zero. An erroneous value of ω may have a substantial effect on the retrieval of τ, unless the canopy transmissivity is high. Therefore, the variations in the data presented in the literature may in part result from the different approaches and models used and from the different assumptions made.

4.12.4 Functional behaviour of the b-factor

The behaviour of the *b*-factor as a function of canopy type and frequency has been studied using data which were either directly available or could be derived from data presented in the literature. All these data have been tabulated by Van de Griend and Wigneron (2004a). Part of these data have been brought together already in an earlier study by Jackson and Schmugge (1991). The data comprise a total of 13 vegetation types and cover five frequencies between 1.4 GHz (L-band) and 37 GHz. Most of the data found in the literature are derived *b*-values for specific crop types, whereas in a few cases canopy water content and τ_0 were given, from which the corresponding *b*-values were calculated. The various data collection procedures can be found in Van de Griend and Wigneron (2004a).

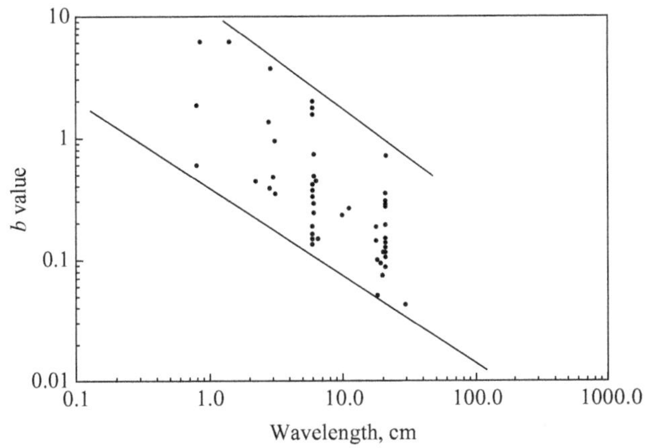

Figure 4.57 *Plot of the parameter b versus wavelength (from Table 4.1 of Van de Griend and Wigneron, 2004a) (Reproduced by permission of IEEE © 2004)*

The calculation of b from Equation (4.119) assumes angular independence of τ_o. Many experiments reported in the literature measured only at one or a few angles and reported either a single value of τ_o (derived from measurements taken at one incidence angle) or an average value (derived from measurements at several angles). Despite the evidence of angular dependence of $\tau_{o(\theta)}$ being a real issue, as argued above, Van de Griend and Wigneron either assumed no angular dependence or took the average in the case of multiple measurements of $\tau_{o(\theta)}$. Also, only observations over green vegetation (measurements over senescent canopies are not considered) were included in this study.

Equation (4.126) can be linearised according to

$$\log b = \log c - x \log \lambda \tag{4.127}$$

Figure 4.57 shows a double-log plot of all available data and demonstrates quite some scatter. This scatter is mainly the result of different canopies having different regression lines. However, at a certain wavelength, the variation within one canopy type is in some cases larger than the variation between the canopy types themselves. The variation within one canopy type may be the result of different measurement procedures and analysis methods and is possibly also due to different growth stages which may substantially affect absorption and scattering. The dynamic range in b-values increases significantly with decreasing wavelength and is smallest at $\lambda = 21$ cm as notified earlier by Jackson and Schmugge (1991). Averaged over all measurements we find an average value of $b_C = 0.57 \pm 0.47$ (b at C-band) and $b_L = 0.20 \pm 0.12$ (b at L-band).

The limited number of wavelengths and the fact that many experiments were done for only one or two wavelengths impede a standard statistical analysis. The cluster

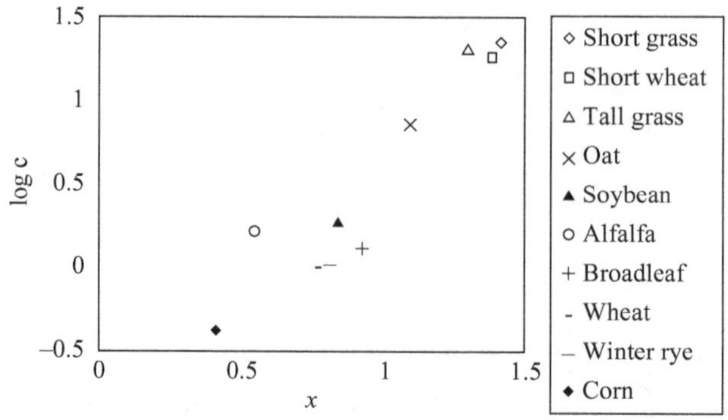

Figure 4.58 Plot of log c versus x for different canopy types; log b = log c − x log λ (from: Van De Griend and Wigneron, 2004a) (Reproduced by permission of IEEE © 2004)

of data seems to be bounded by two lines of which the lower one crosses the horizontal axis approximately at $\lambda = 200$ cm, which would mean that, if theoretical considerations allow such an extrapolation, the canopy would be almost completely transparent ($b \cong 0.01$) at this and higher wavelengths. The data cluster does not allow a confirmation of the relationship between b and λ to be different for $\lambda > \sim 5$ cm and $\lambda < \sim 5$ cm, as suggested by Jackson and Schmugge (1991) on the basis of theoretical considerations. Validation of theoretical statements may require data gathered in a more consistent manner, which is definitely not the case here.

In order to classify the different canopies, two different methods were used, i.e. (1) on the basis of the regression lines in the $\log b$ versus $\log \lambda$ domain, and (2) on the basis of the ratio b_C/b_L (with the subscripts C and L denoting C- and L-band, respectively).

1. *Distinction on the basis of regression lines.* Only a few experiments provided b-values for more than two frequencies of the same canopy type. Therefore, all data for each canopy type were grouped together in order to analyse the regression equation (Equation (4.127)) for each canopy separately. Figure 4.58 shows a plot of *log c* versus x (Equation (4.127)) and demonstrates a sequence from more or less vertically oriented, but leaf dominated canopies such as grass and short wheat to more randomly oriented and leaf dominated crops such as oat, followed by soybean, alfalfa and broadleaf, and stem dominated crops such as wheat and winter rye, to end with corn having the most pronounced vertical structure. Sorghum, cereals and sweet have only one case each and cannot be included in this analysis.

Based on visual interpretation, two groups/clusters and three individual canopy types were distinguished. The clusters are

cluster A: short grass; short wheat; tall grass ($\log c \cong 1.3$; $x \cong 1.4$);
cluster B: soybean; broadleaf; wheat; winter rye ($0 < \log c < 0.3$; $0.7 < x < 1$).

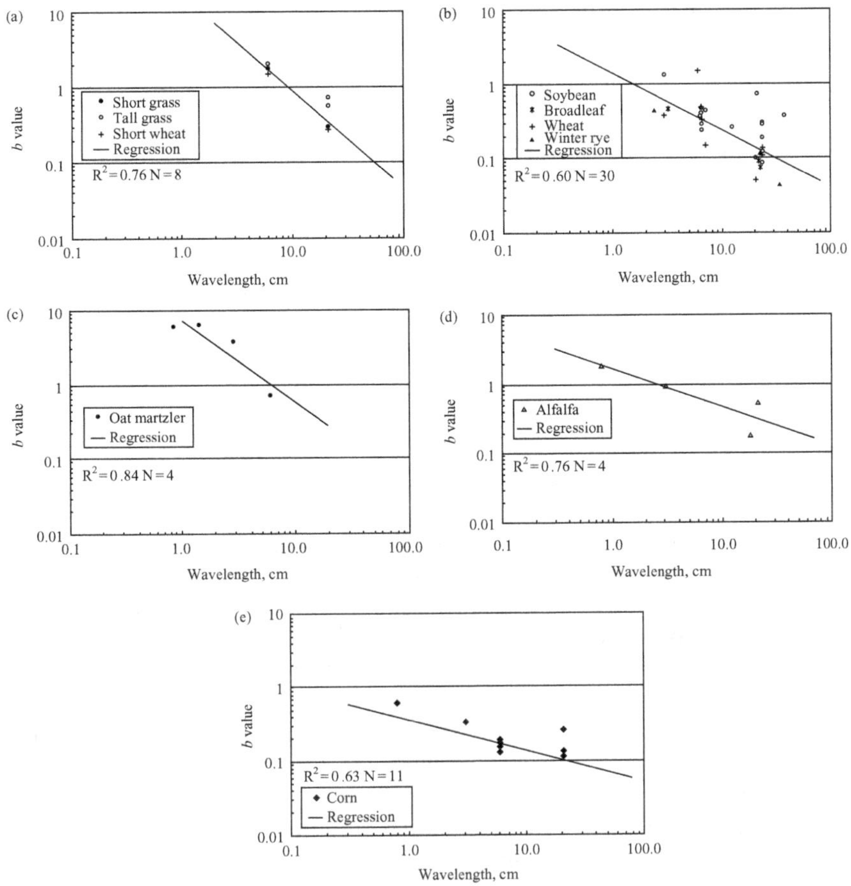

Figure 4.59 *Plots of log b versus log λ for the clusters/canopy types in Figure 4.58 (from: Van De Griend and Wigneron, 2004a) (Reproduced by permission of IEEE © 2004)*

The individual canopy types are: oat ($\log c \cong 0.9$, $x \cong 1.1$); alfalfa ($\log c \cong 0.2$; $x \cong 0.6$) and corn ($\log c \cong -0.4$; $x \cong 0.4$). Within cluster B, wheat and winter rye seem to behave very similarly.

Figure 4.59(a–e) shows the double-log plots of the clusters and individual canopy types identified in Figure 4.58, all demonstrating the double-log linear relationship. As expected, the broad cluster B (Figure 4.59(b)) shows the largest spreading and seems to be composed of canopies with a somewhat different behavior. Wheat, rye and broadleaf are located predominantly under the regression line, whereas soybean tends to have somewhat higher b-values. The graph for corn (Figure 4.59(e)) shows somewhat deviating values from the regression line for $\lambda = 20$ cm. This deviation was also simulated by Le Vine and Karam (1996) using the discrete scatter model mentioned earlier in the introduction. They calculated a hump around

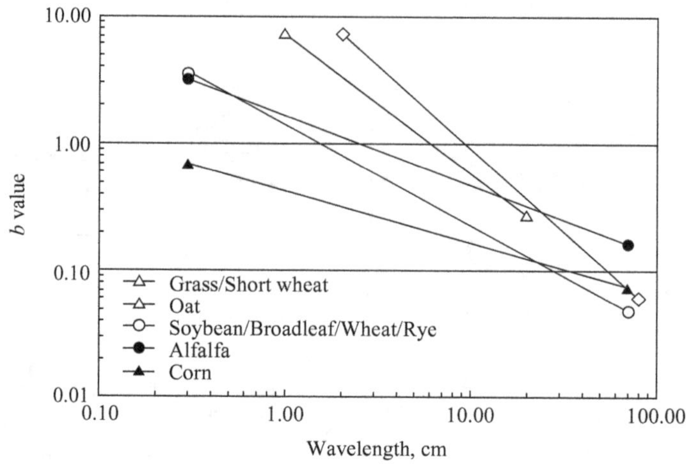

Figure 4.60 *Regression lines* $\log b = \log c - x \log \lambda$ *for the identified groups in Figure 4.59 (from: Van De Griend and Wigneron, 2004a) (Reproduced by permission of IEEE © 2004)*

1.8 GHz ($\lambda = 17$ cm) which they ascribed to the effect of stalks. This hump was not simulated for soybean which has a more random orientation.

Figure 4.60 shows the regression lines of the different canopy types/clusters in one graph. The family of regression lines, which corresponds to the spreading of points in the $\log c - x$ domain of Figure 4.58, suggests that there is a broad range rather than a sharp boundary between dominating canopy structures such as 'erectophyl' and 'planophyl'. Based on the relative contribution of stems, leaves and shoots, different canopies seem to create a spectrum of possible combinations of $\log c$ and x. Of course, differences in growth stage may also have an effect, as reported in several experiments, but it is unlikely that the growth stage has a dominant effect on the findings presented here.

2. *Distinction on the basis of* b_C/b_L. For the synergistic analysis of L-band in combination with C-band, knowledge of the relationship between b_C and b_L is important. Notwithstanding the limited available data for several canopies, calculation of the ratio b_C/b_L results in basically the same distinction as found on the basis of regression lines. Short grass, short wheat, tall grass and oat all have high ratios in the range $4.2 < b_C/b_L < 5.9$. The canopy types with more than one experiment, i.e. soybean and wheat, are positioned somewhere in the middle, $2.9 < b_C/b_L < 3.5$, whereas corn has the smallest value, $b_C/b_L \cong 1.36$, with a relatively small standard deviation compared with the other canopy types.

4.12.5 Summary and conclusions

Analysis of b-values as a function of wavelength (λ) and canopy type confirms the validity of the general equation $b = c/\lambda^x$ proposed by Jackson and Schmugge (1991), in which the coefficient b relates optical depth (τ) to canopy water content (W) according to $\tau = b \cdot W$. The large spreading of the b-values, especially at wavelengths smaller than 21 cm, together with the limited number of wavelengths, and the fact that many experiments described in the literature were done for only one or two wavelengths, impede a standard statistical analysis such as cluster-analysis or principal component analysis. However, a simple regression analysis reflects a logical gradual change from leaf dominated 'erectophyl' canopies like grass and young wheat, to more randomly oriented leaf dominated crops such as alfalfa and soybean towards stem dominated crops like wheat and corn. More or less the same distinction can be made on the basis of ratios of b at C-band to b at L-band, i.e b_C/b_L.

Jackson and Schmugge (1991) reported that the variation in b is much smaller at L-band than at C-band. However, also at L-band, there seems to be a range of b-values rather than a clear distinction between dominant canopy structures such as 'erectophyl' and 'planophyl'. The different canopies seem to create a spectrum of radiative-transfer properties based on the relative contribution of stems, leaves and shoots in combination with their respective orientations.

It must be realised that, in many cases, canopy measurements have been done under different environmental conditions and using different instruments and methods of analysis. Also, as reported in several studies, the growth stage may have a substantial effect on the relationship between the b-factor and canopy water content. Notwithstanding these 'restrictions' the different behavior of canopies seems to be logically related to canopy structure. However, because of the rather limited number of experiments for most canopies, the results found are far from final.

Most data presented in the literature were collected over homogeneous (agricultural) canopies and, therefore, there is a strong need for more experiments especially on natural mixed canopies such as found in the savannas, for example. Derivation and validation of the relationships described in this section over inhomogeneous 'real world' environments, however, is not an easy task. For example, over sparse vegetation, similar to the Sahelian environment, Chanzy *et al.* (1994) have shown the inability to assign a single value of b to this type of ecosystem at a given frequency.

Knowledge of the ratios b_C/b_L for different canopies is important for synergistic analysis using both C-band (AMSR-E) and L-band (SMOS), of which the latter is planned for launch in 2007 (Kerr *et al.*, 2001; Berger *et al.*, 2003). This allows a correction for the effect of vegetation on the emitted microwave radiation in multi-frequency inversion algorithms. It also allows for global simulation of brightness temperatures at multiple frequencies in support of global surface–atmosphere assimilation schemes and direct comparison of simulated brightness temperatures with observed brightness temperatures from space.

4.13 Modelling forest emission *Paolo Ferrazzoli and Leila Guerriero*

4.13.1 Summary

The main steps of forest emissivity models based on radiative-transfer theory are summarised. These include: subdivision of forest medium into discrete elements, modelling the permittivity of elements, computation of bistatic scattering cross sections and extinction cross sections and, finally, combination of contributions.

The main results obtained by applying physical models and comparing model outputs with experimental data are reviewed. Some main aspects are considered: (1) trends versus biomass or age; (2) wave penetration to the soil; (3) contributions of various forest components to overall emissivity; (4) absolute emissivity values predicted for dense forests.

Some considerations about the potential and limits for applications, as well as the needs of future validations, are written.

4.13.2 Introduction

Basically, emission from forests is determined by processes similar to those of emission from other vegetated surfaces, such as agricultural fields or herbaceous vegetation. In all cases some fundamental contributions, such as soil emission attenuated by vegetation, vegetation emission and interaction, are present. Nevertheless, there are some important properties of forests which make it necessary to develop specific models. First of all, forest elements, i.e. trunks, branches, twigs and leaves (or needles), show a wide variability of dimensions and shapes. Differently from crops, non-vertical elements, such as branches, may also have large dimensions, thus producing strong scattering even at low frequencies. As a consequence of this, multiple interactions among elements cannot be neglected, so that single scattering models often fail to represent scattering and emission. Moreover, a simple dependence of emissivity from biomass cannot be established, since most of the biomass is contained in trunks, while simulations do not indicate trunks to be the main contributors to the emissivity. Finally, there are specific problems in modelling soil emission, since the soil is often covered by litter and understory vegetation, and the dynamic range of its moisture is reduced by rain interception from the forest crown.

Several microwave forest models have been developed for the active case (see, e.g. Richards, 1987; Durden *et al.*, 1989; Ulaby *et al.*, 1990; Chauan *et al.*, 1991; Karam *et al.*, 1992; Ferrazzoli and Guerriero, 1995; Karam *et al.*, 1995; Lang, 2004), while only a few papers describe forest models for the passive case (Ferrazzoli and Guerriero, 1996; Karam, 1997, Macelloni *et al.*, 2001; Ferrazzoli *et al.*, 2002). Probably, this is due to the poor spatial resolution of spaceborne passive systems, making them less suitable for some applications, such as biomass monitoring. Anyhow, the interest in the problem is increasing, since new space radiometer missions are planned and several pixels will be covered, at least partially, by forests.

This section is focused on the description of theories and algorithms adopted in forest emission models. A general review of all aspects of forest radiometry, with description of experiments and applications, was recently published by Pampaloni (2004).

Presently available forest emission models are based on a discrete approach. Four main steps may be identified:

- subdivision of forest medium into discrete elements;
- modelling the permittivity of elements;

- computation of bistatic scattering cross sections and extinction cross sections;
- combination of contributions.

The first two steps, and partially the third one, are common to passive and active models. The last step must be specifically developed for the computation of emissivity. The basic modelling steps will be described in Section 4.13.3, while some main features of simulated emissivities will be summarised in Section 4.13.4.

4.13.3 Basic modelling steps

4.13.3.1 Subdivision into discrete elements

In the development of a discrete model, the first step consists of identifying the elements composing the forest structure and assigning them realistic locations within the canopy. Then, a simplified geometrical shape is selected in order to describe their scattering and absorption properties by means of electromagnetic theories. A general sketch is depicted in Figure 4.61.

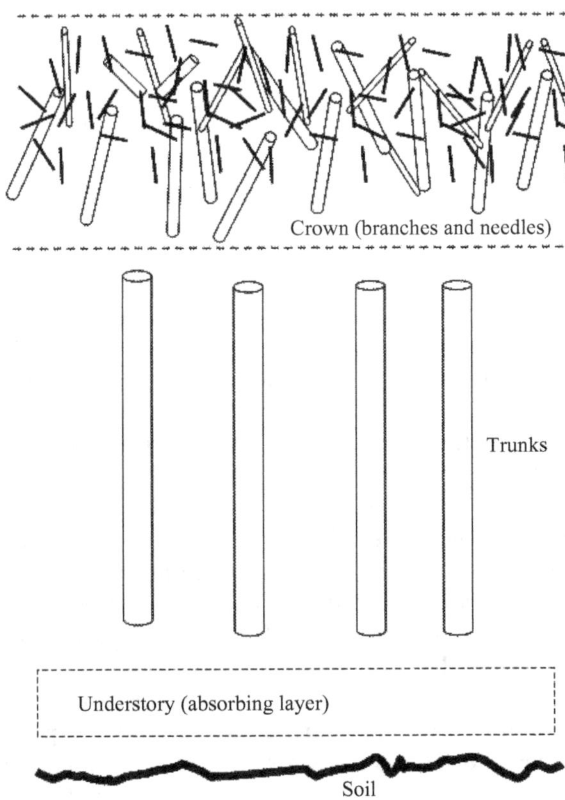

Figure 4.61 Sketch of the structure adopted to model a coniferous forest

The soil is described as a homogeneous half-space with rough interface, producing surface scattering. This approximation is adopted in most discrete models. In the case of forests, soils are often covered by litter and/or understory vegetation. These elements may appreciably modify soil emission properties, but it is difficult to have detailed information on their structure. A reasonable assumption is to represent them as absorbing layers.

In most passive and active models, trunks, branches and twigs are represented as dielectric cylinders. Trunks are near vertical. Their inclination angle may influence backscattering contribution due to soil–trunk double bounce, but it is much less important for emissivity simulations, so that orientation may be assumed to be vertical, in passive models. For branches and twigs, realistic distributions of dimensions and orientations must be given. These distributions depend on tree species and may be derived by detailed measurements or allometric equations, if available (e.g. Kasischke *et al.*, 1994; Saleh *et al.*, 2004b). In general, the diameter of a trunk section decreases with the distance from the soil, and the diameter of a branch section decreases with the distance from origin at the trunk base. Moreover, branches are often bent. These problems may be partially overcome by subdividing trunks into sections with different diameters, and branches into 'grow units' with different diameters and inclination (Saleh *et al.*, 2004b).

Leaves of coniferous trees may be described as thin dielectric cylinders (needles). Leaves of deciduous trees are generally described as dielectric discs. For long leaves and/or leaves with complex shape, each one may be subdivided into several circular discs, as suggested in Chapter 11 of Fung (1994).

4.13.3.2 Modelling the permittivity of elements

For soil permittivity, most of the scattering models adopt either the semi-empirical formula proposed by Dobson *et al.* (1985) and refined by Ulaby *et al.* (1986) or the empirical formula derived by Hallikainen *et al.* (1985). The latter expresses the real and imaginary parts of soil relative dielectric constant (ϵ_s) as a second-order polynomial in the volumetric soil moisture content (m_v). The polynomial coefficients depend on the soil texture (i.e. percentage of clay and sand) and on the frequency, in the range 1.4–18 GHz.

As far as vegetation is concerned, most discrete models compute permittivity of vegetation elements using the semiempirical model proposed by Ulaby and El-Rayes (1987). This formula was developed by fitting experimental data collected by the same scientists (El-Rayes and Ulaby, 1987). Measurements were done over several vegetation samples, in a frequency range of 0.2 to 20 GHz. The proposed permittivity model is physically based in that it includes significant parameters, such as volumetric moisture, free-water fraction, bound-water fraction and free-water ionic conductivity. The values of these parameters were simply derived by fitting experimental data.

An extended formulation of vegetation permittivity model was derived by Mätzler (1994a). In this paper, a new empirical formula was derived by joining experimental data published by El-Rayes and Ulaby (1987) with new permittivity estimates, based on radiometric measurements over various kinds of leaves. This work is important in that it includes a significant amount of measurements and this mainly for vegetation in its natural condition, with high moisture content. Moreover, measurements covered an extended frequency range, up to 94 GHz. The formula proposed by Mätzler (1994a) related the leaf permittivity to dry matter fraction (i.e. the complement of gravimetric moisture) and was validated in a range of gravimetric moistures from 0.5 to 0.9.

More recently, Franchois *et al.* (1998) published results of permittivity measurements for various components of two tree species, i.e. fir and spruce, in the 1–10 GHz range. Results indicate that wood is not isotropic and that there is an appreciable variation of permittivity with depth within the trunk. Until now, these experimental efforts have not been used to test and refine permittivity models.

4.13.3.3 Simulating scattering and extinction of elements

For soil, since it is the lowest component and is characterised by a homogeneous half-space, only upward scattering needs to be estimated. This may be subdivided into two components: a coherent specular reflection, reduced by roughness, and an incoherent bistatic scattering component, related to roughness. Specular reflection may be computed by means of the Fresnel coefficients corrected by the well-established roughness factor proposed by Beckmann and Spizzichino (1963), requiring only surface height standard deviation as input. For bistatic scattering, both height standard deviation and correlation length are required as input. Two classical theories, based on small perturbation and geometrical optics approximations, are valid in the low-frequency and high-frequency limits, respectively (Ulaby *et al.*, 1986). In order to bridge the frequency gap and to improve the general accuracy, the more advanced integral equation model (IEM) was developed by Fung (1994). The bistatic version, which is required for emissivity computation, was made available by Fung (1994).

Trunk height is large with respect to diameter and, at microwave frequencies, it is also large with respect to the wavelength. Therefore, extinction and bistatic scattering cross sections may be computed by means of the 'infinite length' approximation (Karam and Fung, 1988). For high trunks and/or at the higher frequencies, the bistatic scattering pattern becomes more and more peaked in the forward direction. This means that losses are essentially due to absorption, while a correct characterisation of bistatic scattering angular pattern would require very narrow angular steps and, hence, heavy computational efforts. In this case, trunks may be modelled as pure absorbers. The validity of this approximation was demonstrated in Ferrazzoli and Guerriero (1996b), where a close formulation of absorption cross section was also given.

The same 'infinite length' approximation may be generally adopted for branches also. Moreover, larger branches also, at the higher frequencies, may be modelled as absorbers.

For needles, if the diameter is small with respect to the wavelength, the Rayleigh–Gans approximation may be applied (Schiffer and Thielheim, 1979; Eom and Fung, 1986). In this case, the cylinder is generated by a prolate ellipsoid, with two axes much smaller than the third. The extinction cross section must be expressed as the sum of absorption and scattering cross sections, since the optical theorem is not valid

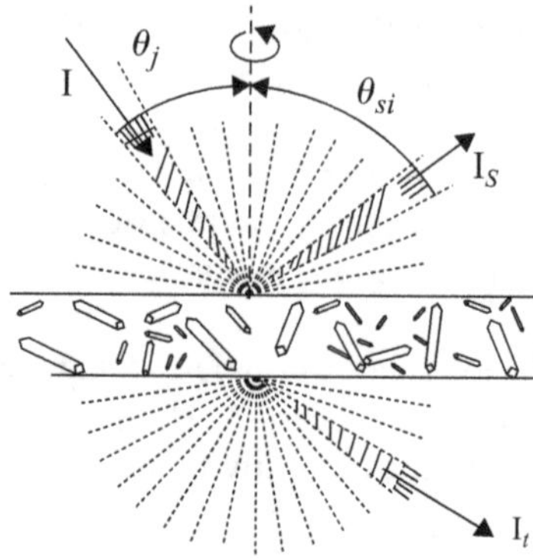

Figure 4.62 Sketch of a scatterer sublayer

with this approximation. At the higher frequencies, and/or for longer needles, the 'infinite length' approximation may be adopted for needles also.

Deciduous leaves may be described as circular or elliptic discs. If the disc thickness is small with respect to the wavelength, the Rayleigh–Gans theory may be applied. In this case, the disc is generated by an oblate ellipsoid with one axis much smaller than the others. The extinction cross section must be expressed as the sum of absorption and scattering cross sections. In the high-frequency limit, if the disc diameter is high with respect to both the wavelength and the thickness, the bistatic scattering and extinction cross sections may be computed using the physical optics theory (Le Vine *et al.*, 1983). In this case, the extinction cross section may be computed by using the optical theorem (Karam and Fung, 1982).

4.13.3.4 Combination of contributions

In the available models, contributions of the various sources of emission/absorption and scattering are combined by using, basically, the radiative-transfer theory. Nevertheless, this theory was implemented using different numerical procedures, as it will be illustrated below.

4.13.3.4.1 Multiple scattering approach based on the matrix doubling algorithm

Ferrazzoli and Guerriero (1996a) consider three main forest components, i.e. crown, trunk and soil.

The crown layer, filled with scatterers of different kinds, is subdivided into N optically thin elementary sublayers (Figure 4.62). The structure is assumed to be symmetrical in azimuth. For each sublayer, both upper and lower half-spaces are subdivided into discrete intervals of incidence and scattering off-normal angles, θ and θ_s, respectively. The interval amplitude is $\Delta\theta$. The incidence and scattering azimuth angles are denoted by ϕ and ϕ_s, respectively. Due to the azimuthal symmetry assumption, the scattering depends on the difference $\phi_s - \phi$. The sublayer behaviour is characterised by the scatter matrices **S** and the transmission matrices **T**. Each element of an **S** (or **T**) matrix represents the ratio I_s/I (or I_t/I) between the specific intensity I_s (or I_t) scattered into an upper (lower) angular interval of θ_s and the specific intensity I incoming from an upper angular interval of θ. The dependence on $\phi_s - \phi$ is expressed in Fourier series, so that the number of **S** and **T** matrices is equal to the number of the series terms. Since the emissivity has to be simulated, averaging over azimuth angles is performed. As a consequence of this, computation of the first terms (0th order terms) of the Fourier series is sufficient, as demonstrated in chapter 8 of Fung (1994). The elements of the 0th order **S** and **T** matrices can be computed as a function of bistatic scattering cross section of elements filling the sublayer, according to

$$S_{ijpq0} = \int_0^{2\pi} \frac{n\Delta z \Delta\theta \sin\theta_j}{4\pi \cos\theta_{si}} \cdot \sigma_{ijpq}(\phi_s - \phi) \mathrm{d}(\phi_s - \phi) \tag{4.128}$$

$n\Delta z$ is the number of elements filling the sublayer per unit of underlying surface. $\sigma_{ijpq}(\phi_s - \phi)$ is the bistatic scattering cross-section averaged within the jth and ith angular intervals of the upper half-space. p and q are scattering and incidence polarisations, respectively. Similarly, the elements of the $\mathbf{T}_0$ matrix can be computed by

$$T_{ijpq0} = \int_0^{2\pi} \frac{n\Delta z \Delta\theta \sin\theta_j}{4\pi \cos\theta_{si}} \cdot \sigma'_{ijpq}(\phi_s - \phi) \mathrm{d}(\phi_s - \phi) \tag{4.129}$$

$\sigma'_{ijpq}(\phi_s - \phi)$ is the bistatic scattering cross-section averaged within the jth and ith angular intervals of the lower half-space. Equations (4.128) and (4.129) are valid for equal scatterers. Since the sublayer is filled with scatterers of different shape, dimensions and orientation, averaging must be done also among different shapes, dimensions and orientations.

In order to correctly include both the attenuation and the scattering effects, the downward scattered power, expressed by the 0th order **T** matrix, must be added to the fraction of undisturbed power. This is accomplished by adding to the diagonal elements of the matrix a quantity equal to one minus the fraction of power travelling within each angular interval at each polarisation which undergoes extinction in the considered sublayer.

The contributions of two adjacent thin sublayers are then combined through the matrix 'doubling' algorithm (chapter 8 of Fung, 1994). For a pair of sublayers, characterised by matrices $\mathbf{S}_{10}$, $\mathbf{T}_{10}$ and $\mathbf{S}_{20}$, $\mathbf{T}_{20}$, respectively, the matrices of the combination are given by

$$\mathbf{S}_0 = \mathbf{S}_{10} + \mathbf{T}_{10}\mathbf{S}_{20}(\mathbf{1} - \mathbf{S}_{10}\mathbf{S}_{20})^{-1}\mathbf{T}_{10} \tag{4.130}$$

and, analogously,

$$\mathbf{T}_0 = \mathbf{T}_{20}(\mathbf{1} - \mathbf{S}_{10}\mathbf{S}_{20})^{-1}\mathbf{T}_{10} \tag{4.131}$$

By reiterating this procedure, the N sublayers are successively combined, and the scattering and transmission matrices of the whole crown are computed.

The scattering properties of the soil are expressed by the dimensionless bistatic scattering coefficient $\sigma^o_{\mathrm{BS}gpq}(\theta, \theta_s, \phi_s - \phi)$, which is employed to obtain a 0th order

$\mathbf{S_g}$ matrix. They may be defined similarly to the 0th order $\mathbf{S}$ matrix of the crown elementary layer. The formula is

$$S_{gijpq0} = \int_0^{2\pi} \frac{\Delta\theta \sin\theta_j}{4\pi \cos\theta_{si}} \cdot \sigma^o_{\mathrm{BS}gijpq}(\phi_s - \phi)\mathrm{d}(\phi_s - \phi) \tag{4.132}$$

The models of scattering from rough soil surfaces require as input data the soil moisture and the height standard deviation and correlation length of the surface. Details of this subject are available in Sections 4.3, 4.6 and 4.7. Ferrazzoli *et al.* (2002) refined the soil model by including an absorbing layer representative of the understory.

Trunk absorption is described by a transmission matrix $\mathbf{T_t}$ whose diagonal elements are related to absorption cross sections.

Crown, soil and trunk contributions are combined through matrix operations. In this way the scatter matrix $\mathbf{S}_{T0}$, describing the upper half-space scattering of the whole forest medium, is obtained. The polarised emissivity may be related to the bistatic scattering coefficient of the whole medium $\sigma^o_{\mathrm{BS}pq}(\theta, \theta_s, \phi_s - \phi)$ by the energy conservation law:

$$e_q(\theta) = 1 - \frac{1}{4\pi} \int_0^{2\pi} \int_0^{\pi/2} \sum_{p=1}^{2} \frac{\sigma^o_{\mathrm{BS}pq}(\theta, \theta_s, \phi_s - \phi)}{\cos\theta} \sin\theta_s \mathrm{d}\theta_s \mathrm{d}(\phi_s - \phi) \tag{4.133}$$

Its discretised version, which is suitable for this numerical approach, is given by

$$e_{jq} = 1 - \sum_{i=1}^{2N_\theta} \sum_{p=1}^{2} \frac{\cos\theta_{si} \sin\theta_{si}}{\cos\theta_j \sin\theta_j} S_{Tijpq0} \tag{4.134}$$

S_{Tijpq0} is an element of the scatter matrix $\mathbf{S}_{T0}$, describing the upper half-space scattering of the whole medium.

Results reported by Ferrazzoli and Guerriero (1996a) were obtained by assuming the crown to be uniform along the vertical direction, so that it was subdivided into equal sublayers. However, the algorithm may be applied also to the case of a composite crown without specific difficulties.

4.13.3.4.2 Multiple scattering approach based on iterative computations

Also the algorithm developed by Karam (1997) is based on the main subdivision into crown and trunk and a further subdivision into thin layers. In each layer, according to the radiative-transfer theory, three processes take place: thermal emission, extinction and scattering. The basic idea of this model is to apply a recurrence formulation. Each layer is assumed to produce thermal emission, while extinction and scattering effects are applied to the radiation computed in the previous iteration. The procedure starts from the 0th order iteration, in which emitted radiation is computed by considering only two main contributions, i.e. soil emission attenuated by vegetation and vegetation thermal emission. Scattering is neglected in this 0th order iteration, so that the absorption coefficient is assumed to be equal to the extinction coefficient. Then, in the first-order iteration, a new term is included, which is given by previously computed 0th order radiation subject to scattering and extinction.

Only 0th and first-order iterations are described in Karam (1997), although the procedure may be extended to any number of iterations. This model is able to simulate emitted radiation also in case of a non-uniform temperature profile.

4.13.4 Results

This section shows some main results obtained by applying physical models described in the previous section and comparing model outputs with experimental data. Some main aspects will be considered: (1) trends versus biomass or age; (2) wave penetration to the soil; (3) contributions of various forest components to overall emissivity; (4) emissivity values predicted for dense forests.

4.13.4.1 Trends of emissivity versus forest biomass or forest age

Some trends of emissivity versus biomass were simulated and presented by Ferrazzoli and Guerriero (1996a). The work was performed for both coniferous and deciduous forests and for both vertical and horizontal polarisation, and covered various combinations of frequency and angle. At the time of publication, radiometric measurements over forests, as well as ground truth, were few and sparse. Therefore, inputs to the model were given on the basis of reasonable assumptions, and comparisons with experimental data were not direct. Anyhow, simulations indicated that, especially at L-band and horizontal polarisation, the emissivity gently increases with increasing forest biomass. Saturation effects were estimated to be less severe than with radar systems. Also Ferrazzoli *et al.* (2002) simulated emissivity trends versus forest age at L- and C-band. In this case, inputs to the model were through detailed information about Les Landes forest. Results are shown in Figures 4.63 and 4.64. The general properties of the trends were confirmed by these new simulations, although it was pointed out that properties may be substantially modified by soil conditions and the presence of understory vegetation. At C-band, results were in general agreement with measurements carried out over Les Landes forest using the PORTOS radiometer (Wigneron *et al.*, 1997).

Some experimental results published in the literature confirm the presence of a gentle trend of emissivity versus biomass at L-band. Results of measurements performed by the ESTAR radiometer at L-band horizontal polarisation were reported by Lang *et al.* (2000). A sensible increase in the brightness temperature was detected with increasing biomass, and no evident saturation effects were observed.

In the summer of 1999, a European radiometric campaign was conducted over six Italian forests which were covered by deciduous trees (Macelloni *et al.*, 2001). The L-band data showed a gradual increase with both tree height and woody volume, also in these measurements.

The brightness temperatures of snow-covered forests with different biomass were measured in Finland in the framework of EMAC-95 campaign (Kroupis *et al.*, 1999). Forest transmissivities, retrieved by means of an empirical model, showed a gradual decrease with increasing biomass even at high frequencies (18 GHz and above). These

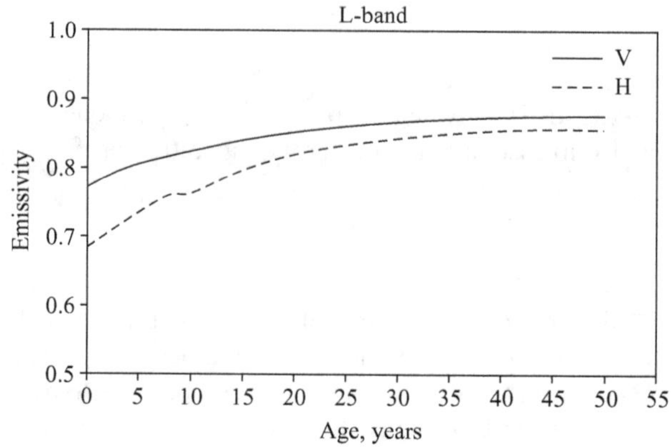

Figure 4.63 Trends of emissivity as a function of forest age for a coniferous forest. L-band. Observation angle: 35°. Volumetric soil moisture: 20 per cent. Both polarisations

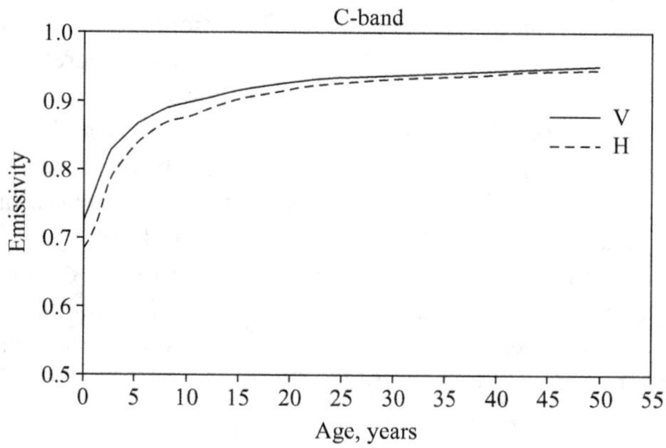

Figure 4.64 Trends of emissivity as a function of forest age for a coniferous forest. C-band. Observation angle: 35°. Volumetric soil moisture: 20 per cent. Both polarisations

results, however, may be partially due to the low temperatures, affecting the dielectric properties of vegetation elements.

In November 2001, measurements were carried out at 1.4 GHz (L-band), vertical polarisation, over Les Landes forest, in the framework of the Eurostarrs campaign (Saleh *et al.*, 2004a). In this case, the dynamic range of emissivity versus biomass was low, and emissivity values under young forests were slightly higher than under old forests. This result may be explained by some particular properties of the measurement campaign. First of all, only vertical polarisation, for which soil emission may be close to vegetation emission, was used. Moreover, the soil was dry (with a volumetric

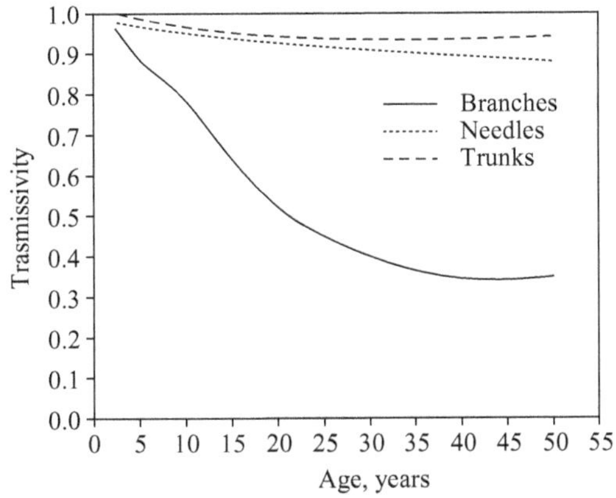

Figure 4.65 *Trends of transmissivity of forest components as a function of age for a coniferous forest. L-band. Observation angle: 35°. Horizontal polarisation*

moisture ranging from ~10 per cent in young forests and ~15 per cent in old forests), and the understory vegetation was dense.

4.13.4.2 Wave penetration to the soil

The problem of wave penetration into the forest canopy receives great attention, due to its importance for applications, such as soil moisture retrieval. It is commonly believed that forest canopy attenuation is high, even at low frequencies, so that penetration is very low.

Simulations published by Ferrazzoli and Guerriero (1996a) indicate crown attenuation to range between 2 dB, for young forests, and about 10 dB, for old forests. Ferrazzoli *et al.* (2002) estimated attenuation due to different forest components by using input data of the Les Landes forest. Results are shown in Figure 4.65 for L-band, horizontal polarisation and an angle of 35°. Similar values were obtained at vertical polarisation. It was found that the overall L-band transmissivity of an old forest was about 0.3, at both polarisations, for an angle of 35°.

Simulation results were also used to calibrate a first-order model. It was found that the 'equivalent' albedo to be assigned to the first-order model, in order to predict emissivity values with minimum rms difference with respect to values predicted by the physical model, was slightly higher than 0.1.

Simulation results of Karam (1997) predicted, at L-band, an emissivity of the order of 0.7 at horizontal polarisation and 0.85 at vertical polarisation. These figures are typical of an emission with an appreciable soil component. However, it must be considered that they were obtained for a walnut canopy and not for a forest.

The simulation results mentioned above would indicate the presence of some penetration. Some experimental data confirm this. In Lang *et al.* (2001) it is shown that a certain sensitivity to soil conditions is maintained even when the soil is overlaid by

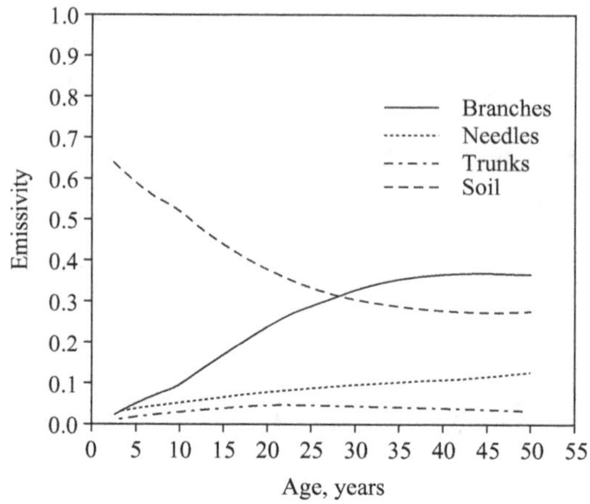

Figure 4.66 Trends of emissivity of forest components as a function of age for a coniferous forest. L-band. Observation angle: 35°. Volumetric soil moisture: 20 per cent. Horizontal polarisation

significant woody biomass. Macelloni *et al.* (2003) found that the seasonal variation of emissivity is low for forests with a woody volume higher than 300 m^3/ha, but is appreciable at L-band and with 200 m^3/ha. This result may be related to the seasonal cycle of leaves, but could be also due to seasonal variations of soil moisture and understory.

This problem needs further investigations. A multi-temporal experiment, with continuous registration of emissivity by a radiometer located on top of a forest, would be the best way to verify the model results.

4.13.4.3 Relative importance of forest components

Ferrazzoli *et al.* (2002) simulated the single contributions of trunks, branches, needles and understory to overall emissivity and overall transmissivity. Simulations were performed at L-band, both vertical and horizontal polarisation and an observation angle of 35°. Results for the horizontal polarisation are shown in Figure 4.66. It was found that branches play a dominant role, with respect to other components. Needle contribution was underestimated, since recently published results of detailed measurements (Saleh *et al.*, 2004b) indicate needle dimensions of Les Landes forest to be higher than values adopted by Ferrazzoli *et al.* (2002). This leads to an overall needle contribution comparable to one of the branches.

Macelloni *et al.* (2001) simulated the emissivity of a deciduous forest by using a first-order scattering model. Branches were subdivided into categories, according to their dimensions. It was found that the dominant contribution was due to branches with larger dimensions.

Ferrazzoli and Guerriero (1996a) simulated the frequency trends of branch attenuation and leaf attenuation of deciduous forests. Branch attenuation dominates at low frequencies but shows a limited increase with increasing frequency. On the contrary, leaf attenuation shows a marked increase with frequency. These results agree with experimental data published by Mätzler (1994b), which indicated crown attenuation to increase with frequency in the summertime (with leaves), while such an evident frequency trend was not observed in the wintertime (without leaves).

4.13.4.4 Absolute emissivity values

Ferrazzoli *et al.* (2002) predicted that the emissivity of coniferous old forests at L-band ranges between 0.85 and 0.9, with vertical polarisation values slightly higher than horizontal polarisation ones. These values were obtained by using a preliminary forest structure model. Recent refinements of the forest model, based on measurements at Les Landes (Saleh *et al.*, 2004b), led to use higher values of needle dimensions, to subdivide branches into growth units and to compute detailed branch orientation distributions for each category of branch diameter. By running the model with these refined inputs, simulated emissivities of old forests were about 0.9 at horizontal polarisation and 0.92 at vertical polarisation. The latter values are close to experimental values collected during the Eurostarrs campaign (Saleh *et al.*, 2004a).

Values slightly lower than 0.9 were predicted also for deciduous forests (Ferrazzoli and Guerriero, 1996a). In this case, running the model with detailed input data is more difficult, since allometric equations are scarcely available.

4.13.5 Concluding remarks

Although most forest modelling efforts have been directed towards active systems, some passive models, based on sound electromagnetic theories, are now available. Simulations show that model outputs are appreciably affected by geometrical properties of vegetation elements. Therefore, physical models must be accompanied by detailed structural models. Also, permittivity models should be tested and refined, on the basis of recent and future measurements.

In the models described in this Section, the forest medium has been supposed to be symmetrical in azimuth. This assumption is acceptable for dense forests. Specific correction must be applied to simulate the emission from single trees or single rows or sparse forests, particularly when the resolution of the instrument is comparable with the distance between trees. This problem has been addressed also in Section 4.2.

Overall, models need to receive further tests versus experimental data, since presently available experimental datasets are still limited.

As far as model use for applications is concerned, a first effort was presented by Ferrazzoli *et al.* (2002), where the outputs of the physical model calibrated a simple first-order model. The latter, in turn, was included into a global earth emission model, in order to estimate the potential of future satellite missions, such as SMOS, in soil moisture monitoring at global scale (Pellarin *et al.*, 2003). This subject is described also in Section 4.14. As far as forest biomass is concerned, both model simulations and some experimental data indicate that passive systems at L-band have good potential, provided a multi-polarisation system is adopted. However, the sensitivity is reduced if the soil is dry and/or the understory is dense. Anyhow, these assumptions need to be verified over extensive experimental datasets.

4.14 L-MEB: a simple model at L-band for the continental areas – application to the simulation of a half-degree resolution and global scale dataset *Jean-Pierre Wigneron, Thierry Pellarin, Jean-Christophe Calvet, Patricia de Rosnay, Kauzar Saleh, Philippe Waldteufel and Yann Kerr*

4.14.1 Introduction

L-band (1–2 GHz) microwave radiometry is the most relevant remote sensing technique to monitor soil moisture over land surfaces at the global scale. A synthetic multi-angular brightness temperature dataset over land surfaces was simulated at 1.4 GHz, at a half-degree resolution and at a global scale (Pellarin *et al.*, 2003a). This dataset was built in order to develop and validate methods to retrieve soil moisture for near-future 1.4 GHz space missions.

Brightness temperatures were computed using a simple model (L-MEB, L-band microwave emission of the biosphere) based on radiative-transfer equations. The L-MEB model is the result of an extensive review of the current knowledge of the microwave emission of various land covers (herbaceous and woody vegetation, frozen and unfrozen bare soil, snow, etc.) at L-band considering that the model should be simple enough to be compatible with the simulation of a half-degree resolution and global scale dataset. This model was parametrised for simulating L-band observations (in the 1–2 GHz range), but the model equations remain valid in a low-frequency range (about 1 to 10 GHz) and thus including the L-, C- and X-bands.

The soil and vegetation characteristics needed to initialise the L-MEB model were derived from existing land cover maps. Continuous simulations from a land-surface scheme for 1987 and 1988 provided time series of the main variables driving the L-MEB model: soil temperature at the surface and at depth, surface soil moisture, proportion of frozen surface soil moisture and snow cover characteristics (depth, density, grain size, liquid water content).

The different components of the emission model are described in the following sections. These sections present the general formulation of T_B for a composite pixel and the microwave emission modules for soil, vegetation-covered surfaces, open water, snow-covered surfaces and atmospheric effects. For vegetation, soil and snow, many of these formulations are developed and discussed in other sections.

4.14.2 Composite pixel emission

Four main surface types, and the associated values of their cover fraction within each pixel, were considered in the simulations: (1) bare soil surfaces, (2) low vegetation canopies (either grassland or crop; mixed grassland/crop landscapes were not considered), (3) forest canopies (a distinction was made between tropical, temperate broadleaf and coniferous forests) and (4) open water surfaces (lakes, rivers, etc.).

The brightness temperature of the mixed pixel was written as a linear combination of each cover fraction:

$$Tb_{P} = f_{B}Tb_{P\text{-}B} + f_{F}Tb_{P\text{-}F} + f_{H}Tb_{P\text{-}H} + f_{W}Tb_{P\text{-}W} \tag{4.135}$$

where f_X and $Tb_{P\text{-}X}$ are the cover fraction and the p-polarised brightness temperature of the different surface types (X = B for bare soil, X = F for forests, X = H for herbaceous vegetation-covered surfaces, X = W for open water surfaces). The modelling of $Tb_{P\text{-}X}$ also included the reflection of the downwelling sky radiation Tb_{SKY-D} by the surface.

Tb_P at the surface level, i.e. as observed immediately above the canopy, is described by (4.135). As described in the literature (Ulaby *et al.*, 1986) upward atmospheric emission ($Tb_{SKY\text{-}U}$) and atmospheric attenuation (parametrised by the atmospheric optical thickness τ_{ATM}) had to be taken into account for simulating top-of-the-atmosphere (TOA) brightness temperature:

$$Tb_{P\_SPACE} = Tb_P \exp(-\tau_{ATM}) + Tb_{SKY\text{-}U} \tag{4.136}$$

where $Tb_{P\text{-}SPACE}$ is the brightness temperature as seen by a spaceborne instrument. A simplified approach was developed in this study to compute τ_{ATM} and Tb_{SKY-U}.

The modelling of $Tb_{P\text{-}X}$ for each kind of surface type (soil, vegetation, water, forest) including the effect of snow and atmosphere is described in the following sections.

4.14.3 Soil emission

The bare soil brightness temperature $Tb_{SOIL\text{-}P}$ (for polarisation p) can be expressed as Ulaby *et al.* (1986)

$$Tb_{SOIL\text{-}P} = e_P T_S^E = (1 - r_P) T_S^E \tag{4.137}$$

where, e_P is the soil microwave emissivity, r_P is the reflectivity at the soil/atmosphere interface and T_S^E is the soil effective temperature which accounts for the fact that the soil emission is originating from a soil layer in which soil temperature and moisture content generally vary with depth. The soil reflectivity r_P is mainly a function of soil moisture, through the soil dielectric permittivity ε_S, and of surface roughness effects.

Computation of the soil effective temperature (T_S^E) was done following the approach of Wigneron *et al.* (2001) that accounts for surface soil moisture and which was improved by de Rosnay (see Section 4.9 this chapter). Modelling of surface roughness effects was based on the semi-empirical approach by Wigneron *et al.* (2001) and is also described in section 4.6.

Only modelling of the soil dielectric permittivity (ε_S) will be described here. This parameter is mainly driven by the soil moisture content, but soil textural and structural properties also influence ε_S. Several models have been developed to relate ε_S to soil parameters (soil moisture, bulk density, proportion of sand and clay, etc.) in the lower microwave frequency range (1–20 GHz). In L-MEB, the model of Dobson *et al.* (1985) was used to compute ε_S in the general case. For dry sandy soils, the approach proposed by Mätzler (1998) was used. The Mätzler model accounts for the specific dielectric behavior of dry desert sand: a Debye relaxation model is used to fit experimental measurements in the 1–10 GHz frequency range, and the soil dielectric permittivity ε_S is computed as

$$\varepsilon_S = \varepsilon_\infty + (\varepsilon_0 - \varepsilon_\infty)/(1 - i(f/f_0)) + i a'' \tag{4.138}$$

where f is the frequency, $\varepsilon_0 = 2.79$, $\varepsilon_\infty = 2.53$, $f_0 = 0.27$ GHz, and $a'' = 0.002$.

Dry sandy soils were defined in this study as soils with a volumetric soil moisture, w_S, lower than 0.02 m^3/m^3 and a sand fraction higher than 90 per cent. Note that the use of two different modelling approaches (Dobson *et al.* and Mätzler) may introduce some discontinuities in the modelling of ε_S. A more general modelling approach of ε_S should be developed to avoid these non-physical discontinuities.

Soil freezing affects ε_S considerably. Measurements of soil dielectric constant at L-band have been obtained by Hallikainen *et al.* (1984, 1985) in wet, frozen and unfrozen conditions. According to these observations, the real part of ε_S (ε'_S) is close to 5 for frozen soils. This value is relatively independent of soil texture, temperature and frequency. For very low temperatures (close to -50°C) and for specific soil textures, ε'_S may decrease down to 4. The imaginary part of ε_S (ε''_S) of a frozen soil was found to be lower than 1 for most soil types. The values of ε''_S depend significantly on temperature. For very low soil temperatures (lower than -50°C), ε''_S can be close to zero. However, for temperatures between -0.5 and -10°C, ε''_S is close to 0.5 for most soil types. Similar conclusions were obtained by Mätzler (personal communication). C. Mätzler noted that for frozen soil (a sandy loam) very constant L-band ε_S values were measured by the PAMIR system. With decreasing temperature this value seems to decrease to about 4 at -6°C to -10°C. The freezing temperature was found to be very close to 0°C and larger than -1°C. Due to ions and salts in the soil there is still some liquid water present in the frozen soil.

As no detailed model is available to parametrise accurately the possible effect of water inclusion in frozen soil on the soil dielectric constant, the permittivity of frozen soil (ε_{SF}) was set equal to a constant:

$$\varepsilon_{SF} = 5 + 0.5i \tag{4.139}$$

A rapid evaluation of this formula was done. A simple sensitivity study (Figure 4.67) showed, that if the real part of ε is set equal to 5 ($\varepsilon' = 5$), the surface emissivity at both polarisations is almost insensitive to the variation of the imaginary part of ε in the range (0.01–1). On the contrary, it can be seen that the emissivity is strongly sensitive to the real part of ε (when $\mathrm{Re}(\varepsilon)$ varies from 4 to 5; an increase in emissivity of about 10 K can be seen at H polarisation at incidence angle $\theta = 40°$).

In this study, two soil variables describe the status of water within the soil: the volumetric fraction of ice (X_I) and of liquid water (X_M). It was assumed that these fractions were also good descriptors of the frozen and unfrozen soil fractions. Using these variables, a simple linear mixing model was used to compute the dielectric constant of partially frozen soils (ε_{SP}):

$$\varepsilon_{SP} = X_I/(X_I + X_M)\varepsilon_{SF} + X_M/(X_I + X_M)\varepsilon_S \tag{4.140}$$

where ε_{SF} is given by (4.138), and ε_S is computed using the Dobson *et al.* or Mätzler models.

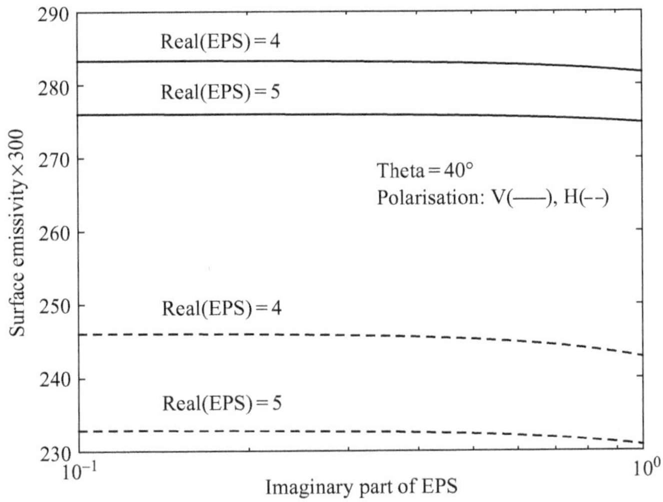

Figure 4.67 Sensitivity of the emissivity to the real and imaginary part of the frozen soil dielectric constant ε_{SF}

4.14.4 *Vegetation emission*

In the (0.3–2 GHz) frequency range, the emission of a vegetation canopy is well described using a simple radiative-transfer formulation, hereafter referred to as the τ–ω model (Mo *et al.*, 1982; Brunfeldt *et al.*, 1984; Ulaby *et al.*, 1986). This model is a zero-order solution of the radiative-transfer (RT) equations as it assumes that the phase matrix term can be neglected (Chanzy and Wigneron, 2000; see Section 4.2). Even at L-band (scattering effects are considered to be very low at low frequencies) the elimination of the scattering source term is a strong hypothesis. However, the τ–ω model has usually been found to be an accurate approach to model the L-band emission from a vegetation canopy in numerous studies (Mo et al., 1982; Brunfeldt *et al.*, 1984, 1986; Pampaloni *et al.*, 1986; Jackson and Schmugge, 1991; Wigneron *et al.*, 1995; Van de Griend *et al.*, 1996), and it is also a tractable tool for inversion processes (Van De Griend *et al.*, 1993; Wigneron *et al.*, 2003). Detailed equations of the τ–ω model are given in the literature (e.g. Mo *et al.*, 1982) and will not be reproduced here.

Using this model, the main task was to assign values to the vegetation model parameters (mainly τ and ω) as a function of the vegetation characteristics (canopy type, phenological stage, water content, biomass, etc.).

The vegetation optical thickness τ is generally found to be linearly related to the total vegetation water content VW_C (kg/m^2), using the so-called b parameter (Jackson and Schmugge, 1991):

$$\tau = b\mathrm{VWC} \tag{4.141}$$

Table 4.4 L-MEB vegetation parameters of the τ–ω model for two classes of herbaceous vegetation (grasslands and crops) and three classes of woody vegetation (rainforests, deciduous forests and coniferous forests)

Surface type	ω	b	VW_C ($\tau = b \cdot VW_C$)
Grasslands	0.05	0.2	0.5 LAI ($kg\,m^{-2}$)
Crops	0.05	0.15	0.5 LAI ($kg\,m^{-2}$)
Rainforests	0.15	0.33	6 $kg\,m^{-2}$ (branches)
Deciduous forests	0.15	0.33	4 $kg\,m^{-2}$ (branches)
Coniferous forests	0.15	0.33	3 $kg\,m^{-2}$ (branches)

At 1.4 GHz, a value of $b = 0.12+/-0.03$ was found to be representative of most agricultural crops. Numerous authors have investigated the parametrisation of b as a function of cover types, temperature and vegetation biomass or water content (see Section 4.12). A summary of the values of the vegetation parameters used in L-MEB is given in Table 4.4. For forests, the application of a zero-order RT theory is not straightforward, since branch scattering is appreciable even at L-band. However, an equivalent τ–ω formulation was derived by Ferrazzoli *et al.* (2002) from a detailed analysis of forest microwave signatures at L-band. See previous section for a description of forest emissivity.

For herbaceous vegetation, VW_C was linearly related to the leaf area index (LAI, m^2/m^2) (see Table 4.4). Using these relationships, VW_C values could be estimated from global LAI data given by existing maps derived from remote sensing observations. An analysis made over agricultural crops from INRA datasets showed that the VW_C/ LAI ratio is about 0.5 kg/m^2 when the vegetation is well developed. This value was close to that obtained for herbaceous vegetation over a fallow ($VW_C/LAI \approx 0.42\,kg/m^2$ in Figure 4.68). This value was also representative of the range of values obtained for soybean ($VW_C/LAI \approx 0.27$) at early vegetation stage and for corn ($VW_C/LAI \approx 1.04$) by Anderson *et al.* (2004) during the SMEX02 experiment in Iowa. For all these different datasets, the use of a linear functional relationship between LAI and VWC was found to be best suited.

For future versions of L-MEB, direct relationships between vegetation optical depth τ and LAI, which is an output variable in some soil–vegetation–atmosphere–transfers (SVAT) models with interactive vegetation (Wigneron *et al.*, 2002) could be developed for a variety of vegetation types. This would allow us to avoid the use of the intermediary vegetation variable VWC which was usually estimated from LAI data (as in the first version of L-MEB) or from vegetation indices, such as NDVI or NDWI, computed from optical remote sensing observations (Jackson *et al.*, 2004).

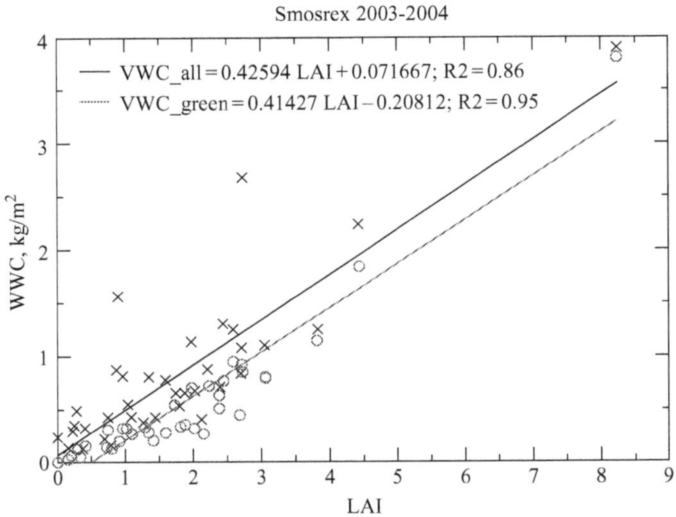

Figure 4.68 *Relationship between leaf area index (LAI) and the vegetation water content (VWC, kg/m^2) over a fallow. One relationship (×) is based on the water content of all vegetation elements (green and litter), the other (○) is based on the water content of green vegetation only (P. De Rosnay, SMOSREX experiment, 2003–2004)*

Note that for herbaceous vegetation, two different relationships could be computed in Figure 4.68, including or not the water content of litter. The effect of litter on the measured microwave observations is very significant as discussed in previous studies (Jackson *et al.*, 1991; Wigneron *et al.*, 2004) and was probably implicitly accounted for by the use of very high values of the b parameter ($b \approx 0.2$–0.4). Current studies based on the SMOSREX dataset attempt to evaluate the combined effects of (1) green vegetation, (2) water interception by the aerial vegetation layer and (3) litter, on the canopy extinction parameter τ (Saleh *et al.*, 2004). Some first results, illustrated in Figure 4.69, showed that τ can be approximately linearly related to LAI in 'dry conditions', i.e. when no intercepted water is present in the canopy. In 'wet conditions', combined effects of water intercepted by the aerial vegetation and by the litter may be very significant and increase τ by a factor of 2 or more (results not shown here). In dry conditions, the ratio τ/LAI computed over the SMOSREX experiment was about 0.062 (see Figure 4.69). Note that the values computed for all these ratios are coherent. For instance, using $\tau/\text{LAI} \approx 0.062$ and $\text{VW}_\text{C}/\text{LAI} \approx 0.5\ \text{kg/m}^2$ as in L-MEB corresponds to a value of the parameter $b \approx 0.125$ ($b = (\tau/\text{LAI})\cdot(\text{LAI}/\text{VWC})$), which is a value commonly used for crops.

4.14.5 The emission of water bodies

It was assumed that the emissivity of calm, unfrozen lakes could be represented by the Fresnel formula (as that used for smooth bare soil) applied with the permittivity of

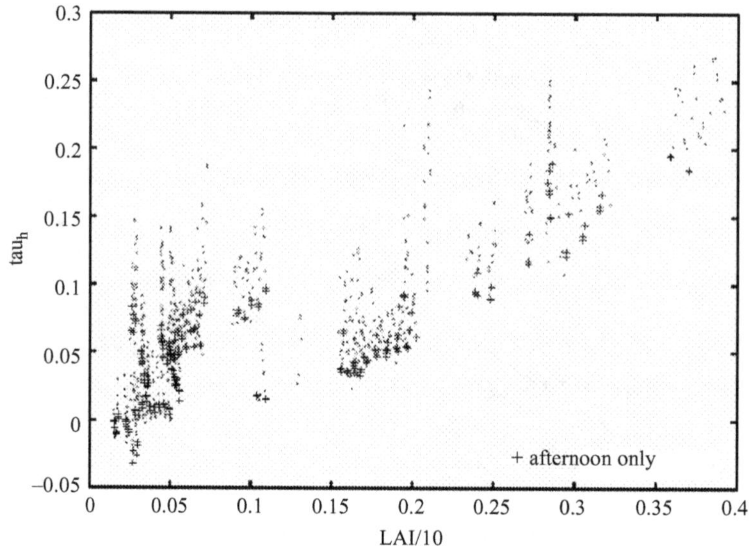

Figure 4.69 Relationship between leaf area index (LAI) and retrieved optical depth (τ) over a fallow. Obtained values of τ correspond to dry conditions: there is no intercepted water in vegetation and litter is rather dry (Saleh et al., 2004; SMOSREX experiment, 2003–2004). Crosses (+) correspond to measurements acquired during the afternoon

pure water. As no modelling approach could be found in the literature to account for roughness effects (in windy conditions, roughness conditions are very different from those over the ocean), the water bodies were assumed to have a smooth surface. Permittivity of pure ice was used instead of permittivity of saline water for temperatures below −0.5°C.

4.14.6 Snow-covered surfaces

The zero-order HUT (Helsinki University of Technology) model, developed by Pulliainen *et al.* (2001), was used to model snow emission (see Section 4.15). In the presence of vegetation, only two cases were considered in this study: (1) over herbaceous covers, it was assumed that the snow cover overlays the vegetation canopy, (2) over forest covers, it was assumed that the snow cover overlays the soil surface (the snow located on the forest branches was assumed to be transparent). For snow overlaying herbaceous vegetation canopies, a simple two-step modelling approach was applied:

1. First, the soil/vegetation emission (T_{B1}) was computed using the τ–ω model.
2. Second, it was considered that this emission is that of the ground which is overlaid by snow.

The HUT model was used to compute the snow cover emission, T_{B1} being substituted to the soil emission.

Using this approach, the continuity of the simulations is maintained when a snow mantle appears or disappears in relation to the climatic conditions. Also, it would be difficult to demonstrate that more complex approaches would be more accurate. If the snow is dry, the total emission is that of the vegetation cover. If the snow is rather wet, the total emission is dominated by the snow layer contribution. For intermediate cases, the total emission is due to the contributions of soil, vegetation and snow. For snow under forest canopies, the snow layer was assumed to overlay the ground, below the trees. A two-step modelling approach quite similar to that used in the case of low canopies was implemented.

These approaches do not really account for the emission by a multi-layer medium (in theory, a four-layer medium: soil/vegetation/snow/atmosphere or soil/snow/forest/atmosphere should be considered). However, they are very simple and fit well to global simulations, for which a simplified characterisation of the surface variables has to be used.

4.14.7 Influence of the atmosphere at L-band

The atmosphere is relatively transparent to L-band microwave radiation. However, atmospheric constituents such as clouds, water vapour or precipitation slightly affect the land surface emission as measured by spaceborne observation systems, and TOA brightness temperatures must be corrected for atmospheric effects. From the radiative-transfer equations, the downwelling atmospheric emission, $Tb_{\text{SKY-D}}(\theta)$, can be expressed as a function of the atmospheric optical thickness (τ_{ATM}) and the equivalent temperature of the atmosphere (T_{ATMeq}):

$$Tb_{\text{SKY-D}}(\theta) = T_{\text{ATMeq}}(1 - \exp(-\tau_{\text{ATM}}/\cos(\theta))) + T_{\text{cosmos}}\exp(-\tau_{\text{ATM}}/\cos(\theta)) \tag{4.142}$$

where $T_{\text{cosmos}} = 2.7\,\text{K}$. Since these values are relatively little affected by weather, it is expected that a simple parametrisation, using commonly available atmospheric variables, may be used instead of a full radiative-transfer scheme requiring a lot of input data (radio soundings, etc.).

A study was carried out using a one-day global numerical weather forecast of the European Centre for Medium-range Weather Forecasts (ECMWF). Twelve atmospheric and surface fields were derived from this archive: altitude, surface (screen level) air temperature and specific humidity, integrated water vapour, rain rate, etc.

The 3D fields, together with the available information on rain and clouds, were processed by the earth vegetation atmosphere (EVA) model proposed by Kerr and Njoku (1990), and global estimates of τ and T_{ATMeq} were obtained. A statistical analysis was performed to underscore the influence of each of the twelve fields on τ_{ATM} and T_{ATMeq}. The most explicative fields for optical thickness τ_{ATM} were found to be surface altitude (Z) and surface temperature ($T_{2\text{m}}$) with values of square correlation coefficient R^2 of 82 and 17 per cent, respectively. It was found that the other fields have a much lower effect on τ_{ATM}. The equivalent atmospheric temperature, T_{ATMeq}, is strongly related to the surface air temperature ($R^2 = 95.7$ per cent) and less to the water content of the atmosphere ($R^2 \sim 60$ per cent).

In order to quantify the influence of water on the atmospheric absorption at L-band, τ_{ATM} and T_{ATMeq} were computed as previously, but the water content of the atmosphere was forced to be equal to zero (this case is referred to as dry conditions). Z and T_{2m} were relevant to explain the variations of the obtained values of τ_{ATM} and T_{ATMe} in dry conditions. A simple, statistical parametrisation of τ_{ATM} and T_{ATMeq} that used only the surface altitude, Z, and surface temperature, T_{2m}, that accounted respectively for oxygen absorption and air temperature T_{2m} was obtained:

$$\tau_{ATM} = \exp[-3.9262 - 0.2211\mathbf{Z}(\mathrm{km}) - 0.00369\mathbf{T}_{2m}(\mathrm{K})] \tag{4.143}$$

$$T_{ATMeq} = \exp[4.9274 + 0.002195\mathbf{T}_{2m}(\mathrm{K})] \tag{4.144}$$

The average value at nadir of the downwelling atmospheric emission (first term in Equation (4.142)) is 1.42 K. The rms error resulting from the use of (4.143) and (4.144) to estimate that term is only 0.06 K, despite the fact that the water content of atmosphere was negelected.

4.14.8 Global half-degree maps of synthetic L-band brightness temperatures

Global half-degree maps of synthetic L-band T_B were simulated with L-MEB at 0600 and 1800 LST, each day, from 1 January 1987 to 31 December 1988, for incidence angles of 0, 20, 30, 40 and 50°, at vertical and horizontal polarisations (V and H, respectively). T_B values ranged between about 140 and 300 K. To illustrate the obtained global synthetic dataset, the annual amplitude of T_B and the polarisation difference D_P is presented in Figure 4.70. The L-band T_B observations are unlikely to be useful over regions where the annual amplitude of both T_B and D_P are small. This occurred mainly over rainforests in Africa and South America. Conversely, very large amplitude could be seen over desertic areas (Sahara, Arabic Peninsula, centre of Australia, etc.) where T_B is strongly sensitive to soil moisture and in areas dominated by crops (western part of Europe, centre of the USA, south-eastern part of China, etc.) where T_B is strongly sensitive to both soil moisture and to the vegetation development. Over boreal forests, there is a low amplitude in the polarisation difference D_P, but the seasonal change in the L-band emission is significant. These global daily maps have been used in two studies to develop and evaluate methods to retrieve soil moisture from near-future L-band space mission using statistical (Pellarin *et al.*, 2003b) or forward model inversion (Pellarin *et al.*, 2003c) methods.

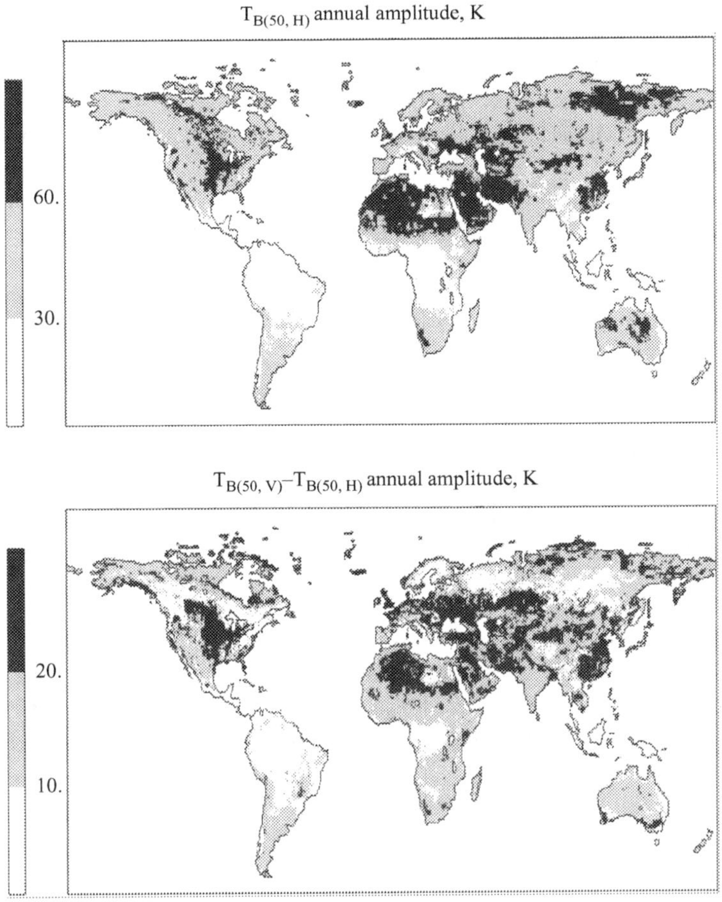

Figure 4.70 *Global half-degree maps of the annual amplitude of (top) the synthetic L-band H-polarised brightness temperatures and (bottom) the polarisation difference, at an incidence angle of 50°, for 1988 at 0600 LST. The three considered classes correspond to (top) 0–30 K, 30–60 K and more than 60 K and (bottom) 0–10 K, 10–20 K and more than 20 K (Pellarin et al., 2003a; reproduced by permission of IEEE, ©2003)*

4.15 Microwave emission of snow *Christian Mätzler, Andreas Wiesmann, Jouni Pulliainen and Martti Hallikainen*

4.15.1 Passive microwave remote sensing of snow

4.15.1.1 Discovering microwave signatures of snowpacks and firn

Passive microwave signatures of seasonal snowpacks are well known by their strong contrast between wet and dry snow, by the decreasing emissivity for dry snow with increasing frequency (10–100 GHz range) and by their potential to give snow–volume information in the case of dry snow. Signatures have been observed for over 30 years mainly by ground-based studies (e.g. Kennedy and Sakomoto, 1966; Hofer and Schanda, 1978; Stiles and Ulaby, 1980; Ulaby and Stiles, 1980; Mätzler, 1987, 1994, Wiesmann *et al.*, 1998), and also from aircraft and indirectly from satellite observations (e.g. McFarland *et al.*, 1987, Josberger *et al.*, 1993), and an early review of the potential and limitations for remote sensing was published by Bernier (1987). A study, covering frequencies up to 220 GHz (Tait *et al.*, 1999), showed that volume-scattering signatures of snowpacks continue throughout the millimetre wavelength range.

Different microwave signatures were observed for the perennial snow and firn on polar ice sheets (Gurvich *et al.*, 1973; Gloersen *et al.*, 1974; Zhang *et al.*, 1989; Rott, 1989; Rott *et al.*, 1993). Lowest emissivities are often observed at low microwave frequencies around 10 GHz. This property is a consequence of the large depth of scattering layers coupled with the increase of grain size with increasing depth to which low-frequency radiation can better penetrate, typically from 10 m at 10 GHz to at most 1 m at 35 GHz (Mätzler, 1987; Rott, 1989). Penetration depth versus frequency simulated by the microwave emission model of layered snowpacks (MEMLS, Wiesmann and Mätzler, 1999) for different types of dry and wet snow are shown in Figures 4.71(a) and (b), respectively (from Mätzler 2002a).

4.15.1.2 Search for inversion algorithms and why physical modelling is needed

With varying success, different researchers developed algorithms, based on spectral features, to retrieve the water equivalent of snowpacks for radiometer data from satellites (e.g. Hallikainen and Jolma, 1986; Chang *et al.*, 1987; Wang *et al.*, 1992; Goodison and Walker, 1995; Tait and Armstrong, 1996; De Sève *et al.*, 1997, Armstrong and Brodzik, 2001). Although high correlations were found by algorithms for snow depth under special conditions, the usefulness of these empirical models was limited to specific geographic regions and special snowpack histories (Rango *et al.*, 1979; Foster *et al.*, 1980; Künzi *et al.*, 1982). Rott and Nagler (1995) showed that different snow types (wet snow, snow with crusts, dry snow) can well be delineated, but that a universal algorithm for snow water equivalent does not exist. Mätzler (1994) found from surface-based experiments that scattering by different types of grains and layering of dry snow can dominate the signal variation. Snowpacks are inhomogeneous and strongly layered, as shown in Figure 4.72.

Tait and Armstrong (1996), being aware of special snowpack signatures, optimised snow-depth algorithms by excluding certain data (types of snowpacks, special regions and weather conditions). These experiences tell us that a more physical approach, including all snow parameters, would be more adequate than empirical algorithms for single parameters. The information gain accompanied with a physical analysis is manifold because all snow properties provide geophysical information, not only the water equivalent. Parameters to be derived from the snow structure include the optical albedo and the diffusivity for mass and heat transfer within the snowpack. Furthermore, in a physical treatment it is possible to include the relief (Section 4.3 and

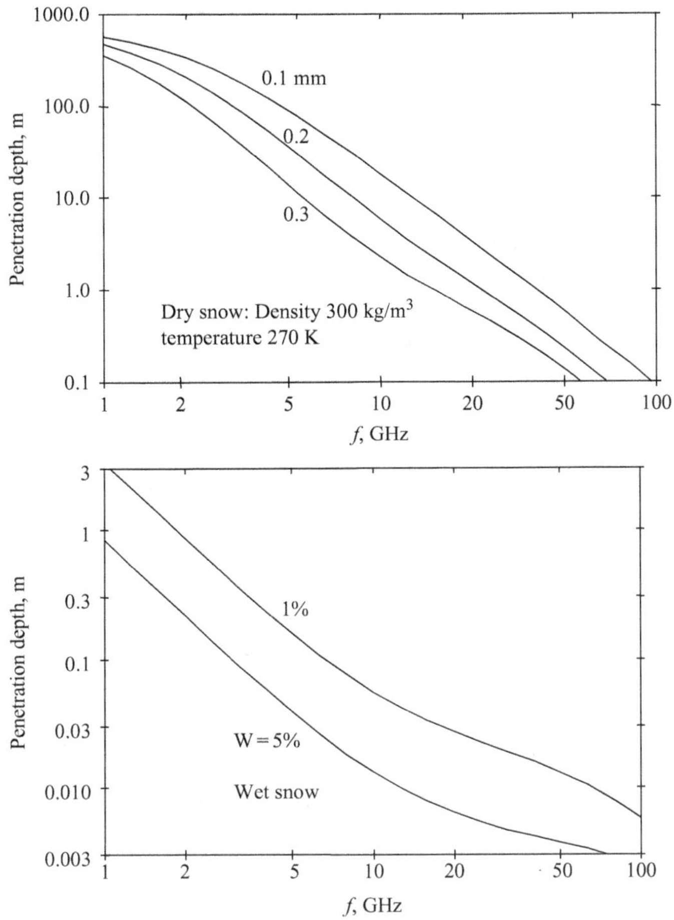

Figure 4.71 *Penetration depth d_p in snow ($T = 270\,K$, $\rho = 300\,kg/m^3$) versus frequency, computed with MEMLS. (a) Dry snow for 3 different correlation lengths p_{ex} (0.1 mm is typical for fresh and fine-grained snow, and 0.3 mm for depth hoar). (b) Wet snow (with $p_{ex} = 0.3$ mm) for liquid water content W= 1 per cent and 5 per cent by volume, respectively*

Mätzler and Standley, 2000) and the background radiation. Experiments showed that even the ground emission can be measured through dry, alpine snowpacks, and that the data are useful for permafrost mapping by microwave radiometry at frequencies near 10 GHz (Gubler and Hauck, 1998).

4.15.2 Modelling efforts for seasonal snow and ice sheets

In order to understand the influence of snowpack properties on the microwave signal and to explain snow signatures, theoretical investigations have been performed

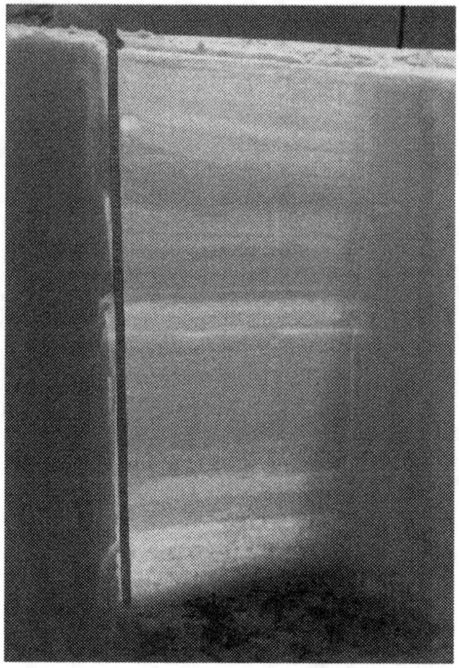

Figure 4.72 Layers of a 1m deep snowpack at Weissfluhjoch in January 1994 observed in transparent light. Bright regions represent snow with a small specific surface (light snow, ice layers, depth hoar) and dark regions represent high specific surface (compacted snow with small grain size)

through both electromagnetic wave theory (Tsang, 1987; Stogryn, 1986; Surdyk and Fily, 1995) and radiative transfer (e.g. Chang *et al.*, 1976; Tsang and Kong, 1977). Validating the results of modelling efforts has been difficult, mainly because the snow structure cannot be quantified easily. If it is described by grain size we need realistic size distributions. Such distributions are often hypothetical because of the required assumptions on grain shape (e.g. Jin, 1995). An alternative is to use the Born approximation or the strong fluctuation theory (SFT) where the structure is quantified by the autocorrelation function whose main parameter is the correlation length p_c, defined by the slope of the autocorrelation function at zero displacement. The quantity can be determined from snow sections (e.g. Reber *et al.*, 1987). This alternative has an important advantage by the link of p_c with the specific surface s and thus with the optical grain size D_o by

$$p_c = \frac{4f(1-f)}{s} = \frac{2D_o}{3}(1-f) \tag{4.145}$$

where f is the volume fraction of ice (Grenfell and Warren, 1999; Mätzler, 2002b). Note that p_c and D_o are mainly determined by the minor axis of grains and thus are often much smaller than the grain size. The SFT model of Stogryn (1986) was tested by Weise (1996a) in the implementation of Surdyk (1992), using Weise's experimental data. The test showed that the model is inaccurate, probably because multiple scattering is ignored in SFT. Indeed, West *et al.* (1993) showed, on the basis of the dense-medium radiative-transfer theory applied to backscattering from snow, that multiple scattering by snow grains is important.

The development of snowpack emission models is further complicated by the fact that snow profiles show strong vertical inhomogeneity. Since the penetration depth of microwaves and the scattering coefficient depend on frequency and snow properties, a general model should account for realistic snow profiles of all sensitive parameters. A successful model might then be used for the development of remote sensing tools for observing processes on and within the snowpack. An important internal process, modifying heat and vapour transfer and being responsible for the release of avalanches, is temperature-gradient metamorphism, often taking place near the bottom of the dry snowpack. This process is based on large water-vapour gradients; it leads to the formation of cohesionless depth hoar (Armstrong, 1977, 1985). Microwave emission of depth-hoar layers is very special due to the enhanced volume-scattering by the large crystals (Weise, 1996a,b). On the other hand, snowpack stratification has important implications for physical processes, such as avalanche formation by weak layers (Föhn *et al.*, 1998; Fierz, 1998), reduced vertical diffusion of water vapour and heat transfer and percolation of liquid water in the case of wet snow (Arons and Colbeck, 1995, and references therein). Snow layers also mark meteorological events in the history of snowpacks, and thus they contain information about past weather conditions. Specific microwave signatures related to stratification are expressed, for instance, by polarisation features of microwave emission, and, sometimes, by special spectral properties. Since the earliest radiometer observations, interest has been shown in the layering of polar ice sheets (Gurvich *et al.*, 1973; Rott *et al.*, 1993; Surdyk and Fily, 1995; Steffen *et al.*, 1999), and layer effects have also been found in snow-covered sea ice (e.g. Mätzler *et al.*, 1984). Microwave emission of layered media, such as firn, was proposed by several authors (e.g. Gurvich *et al.*, 1973; Djermakoye and Kong, 1979; West *et al.*, 1996). Other models were related to volume scattering by the granular structure (e.g. Zwally, 1977; Boyarskii and Tikhonov, 2000).

Driven by the need for realistic microwave emission models, combining soil, vegetation, snowpacks and atmosphere, ESA launched a study to combine theoretical and experimental investigations in 1995. Two different snowpack emission models evolved from this initiative (Pulliainen *et al.*, 1998): A single-layer model, developed at the Helsinki University of Technology, called the 'HUT model', and the microwave emission model of layered snowpacks (MEMLS), developed at the University of Bern. Whereas the first model is a semi-empirical, zero-order-scattering model suitable for rapid emission computation, the second one is a multiple-scattering model allowing many layers to simulate snow cover emission including most of the known physical effects. A comparison of the two models with regard to the aspect of zero-order versus multiple scattering is given in Section 4.2. Both models are described below.

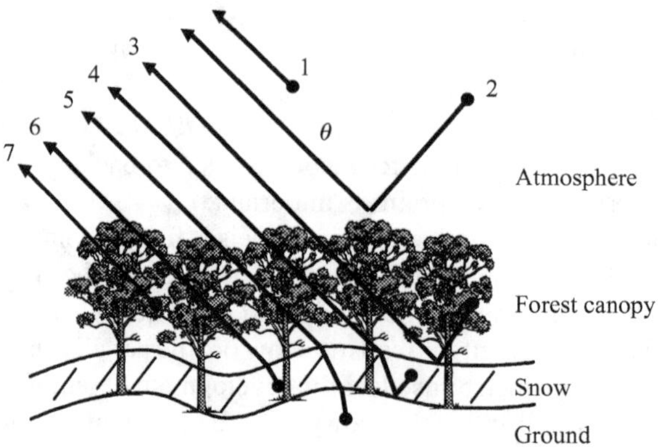

Figure 4.73 *Schematic presentation of the main emission contributions included in the HUT snow emission model. The contributions are*

(1) upward emitted atmospheric radiation;
(2) downward emitted reflected atmospheric contribution;
(3) downward emitted reflected forest canopy emission contribution;
(4) downward emitted reflected snowpack emission contribution;
(5) upward emitted soil emission contribution;
(6) upward emitted snowpack emission contribution;
(7) upward emitted forest canopy emission contribution

4.15.2.1 The HUT model

The HUT model describes the snowpack as a single homogeneous layer and uses an empirical formulation for the snow extinction coefficient and an empirical forward scattering factor. It assumes that the scattering is mostly concentrated in the forward direction. The snow extinction coefficient is modelled as a function of snow grain size and frequency (Hallikainen *et al.*, 1987), whereas the forward scattering factor is approximated to be a constant coefficient. The effects of forest canopy, soil surface roughness and atmosphere on the brightness temperature are considered by using empirical/semi-empirical formulae (Pulliainen *et al.*, 1993; Kurvonen and Hallikainen, 1997; Kruopis *et al.*, 1999; Wegmüller and Mätzler, 1999). A schematic presentation of the main emission contributions is shown in Figure 4.73.

The inclusion of vegetation and atmospheric emission contributions in modelling enables the use of the HUT model for interpreting spaceborne radiometer observations. Comparisons of the HUT emission model predictions with independent experimental snowpack emission data and spaceborne SSM/I observations show good agreement in general (Pulliainen *et al.*, 1999). An example of a comparison with spaceborne SSM/I observations is shown in Figure 4.74. The HUT snow emission model is simple and thus useful for direct retrievals of the water equivalent in simple snowpacks. Care has to be taken at high frequencies (e.g. 90 GHz) where

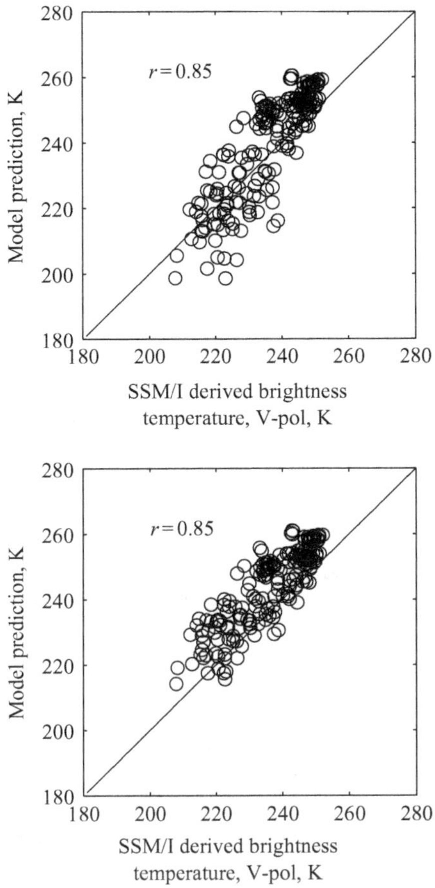

Figure 4.74 Comparison of SSM/I observations with the HUT snow emission model predictions. The observations and predictions for all SSM/I frequencies are combined. In-situ snow water equivalent, snow density, forest/land cover and air temperature information are used as model input. Snow grain size is treated with a fixed value of 1.1 mm. Top: Comparison using balloon sounding-based in-situ atmosphere profiles. Bottom: Comparison using the statistical atmosphere model (Pulliainen et al., 1993)

the penetration depth is often smaller than the snow depth. The single-layer model is unable to account for vertical gradients of snowpack properties.

4.15.2.2 The microwave emission model of layered snowpacks (MEMLS)

MEMLS is based on a six-flux radiative-transfer model, using a correlation-function approach to quantify snow structure, including multiple scattering both by stratification and by snow grains, refraction and radiation trapping by total internal reflection

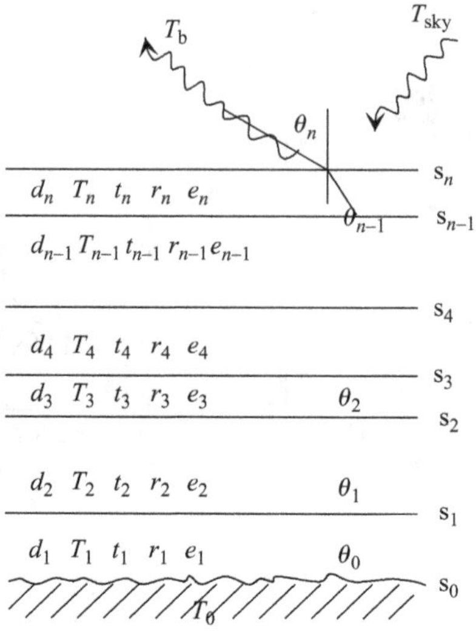

Figure 4.75 Geometry and radiative-transfer parameters of the n-layer snowpack assumed in MEMLS; T_b is the scene brightness (model output) observed in direction θ_n and T_{sky} is the sky brightness

and a combination of coherent and incoherent superposition of interface reflections (Wiesmann and Mätzler, 1999). Simplifications were applied in the treatment of the distinction between directional and diffuse radiation and in the mixing of H and V polarisation by volume scattering. The geometry of the snowpack, consisting of n ($n \geq 1$) layers, is illustrated in Figure 4.75. Each layer j is characterised by primary and secondary model parameters. The primary parameters are layer thickness d_j, temperature T_j, density ρ_j, volumetric liquid-water content W_j and correlation length. Secondary parameters are dielectric constant, refractive index, refracted propagation angle θ_{j-1}, absorption and scattering coefficient, γ_a and γ_s, respectively. They lead to parameters characterising the radiative transfer in and between the layers: interface reflectivity s_j, transmissivity t_j, reflectivity r_j and emissivity e_j of each layer j, with $e_j + r_j + t_j = 1$. Finally the transfer of radiation through the multi-layer pack, including the refraction, is computed with a matrix method. MEMLS resulted after detailed experimental studies of single snow slabs by Weise (1996a,b) and Wiesmann *et al.* (1998); they found an empirical relationship between γ_s, frequency ν and the exponential correlation length p_{ex}, the e-folding length of an exponential autocorrelation function, $A(x) = \exp(-x/p_{ex})$, fitted to the observed one. Experiments showed a slight difference to the correlation length p_c of Equation (4.145), namely $p_{ex} \cong 0.75 p_c$ (Mätzler 2002b), indicating that $A(x)$ is not exactly exponential. The empirical function determined for γ_s was implemented in MEMLS together with the physical expression for γ_a, related to the complex dielectric constant. Excellent agreement between measured and computed γ_a was found. For γ_s the experimental result is similar to Rayleigh scattering ($\gamma_s \sim \nu^4 p_{ex}^3$) with regard to the correlation length as size parameter, namely proportional to $p_{ex}^{2.5}$. However, the exponent of ν is smaller than 4, about 2.5. A limitation of the empirical fits for γ_s is in the applicable range of frequency (5–100 GHz), density (0.1–0.4 g/cm^3) and p_{ex} (0.05–0.3 mm), i.e. the typical situations found in dry winter snow.

To extend MEMLS to coarse-grained snow, a physical model of volume scattering was developed and validated by Mätzler and Wiesmann (1999), based on the improved Born approximation (Mätzler, 1998). With this addition, MEMLS obtained a complete physical basis. The extended model is devoid of free parameters. The validation included radiometric snow-samples of Wiesmann *et al.* (1998) on black-body and on metal plate, respectively, and coarse-grained crusts growing and decaying during melt-and-refreeze cycles. In all cases studied, i.e. for correlation lengths up to 0.7 mm, the model showed good agreement with the observations.

4.15.2.3 Combination of microwave and physical snow cover models

A handicap of physical emission models is the large number of input parameters. The quality of the simulations depends on the accuracy of these parameters, especially if the involved medium is complex. Ground-truth measurements have been used to provide the input. Often they are incomplete, and the correlation length is difficult to measure. A library of snow sections including all snow parameters for comparison was developed by Wiesmann *et al.* (1998), and a short version of the results was presented by Mätzler (2002b). Data from this library can serve as a guideline for the parameter selection. Another solution is to derive the parameters from physical snow cover models. They describe the snowpack from standard meteorological observations. These models have been developed to predict the mechanical, thermal and optical snow properties for various applications, such as climate modelling, avalanche forecasting and hydrology. The combination of such models with radiative-transfer models is a necessary step in the closure problem of surface effects in numerical-weather prediction and data assimilation. In addition the combination is needed in the validation of both types of models. In operational tasks, the input data needed for realistic radiative-transfer computations can only be provided by such combinations. Several efforts have been made: Shih *et al.* (1997) published a model to calculate backscattering from layered snowpacks, based on dense medium radiative transfer in combination with SNTHERM, a snow physical model to be explained below. Galantowicz and England (1997) presented a passive microwave model in combination with a soil–vegetation–atmosphere scheme based on the physical snow model of Anderson (1975). For an area in Colorado, Wilson *et al.* (1999) combined a single-layer dry-snow emission model with a hydrological model derived from SNTHERM. Wiesmann *et al.* (2000) developed and tested interfaces to combine MEMLS with CROCUS and SNTHERM. SNTHERM is a one-dimensional mass and energy balance model, developed at the US Army Cold Regions Research and Engineering Laboratory (CRREL) for predicting temperature profiles within the snow and frozen soil. The model is formulated to describe a snow cover over soil with high vertical and temporal resolution. CROCUS is a one-dimensional numerical energy and mass evolution model of snow covers developed at Météo France. Its main objective is

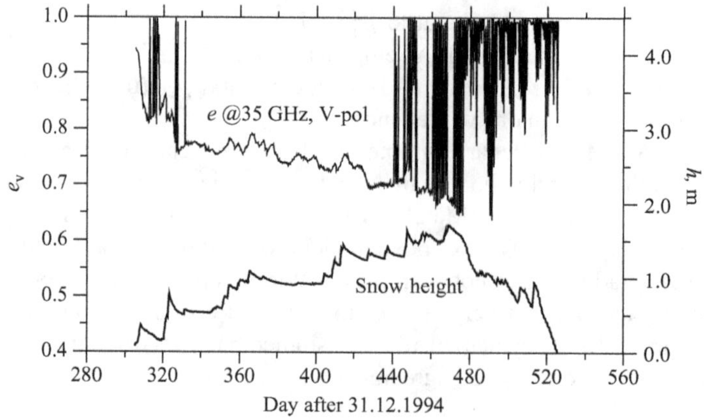

Figure 4.76 Variation of snow height h and snowpack emissivity e_V at 35 GHz, vertical polarisation, during the snow season 1995/96 at the test site, Weissfluhjoch-Davos. The emissivity was simulated by the combined use of CROCUS and MEMLS from standard meteorological observations

operational avalanche forecasting. Both models use meteorological data versus time as input. The output of both models can almost directly be used as input in MEMLS. Only the different descriptions of snow structure had to be adapted to the correlation length used in MEMLS. An example of a combined simulation at 35 GHz is shown in Figure 4.76. The strong emissivity fluctuations found early and late in the snow season are due to melt–refreeze cycles. During the dry snow phase an anticorrelation of the emissivity with snow height can be recognised in the long-term behaviour; the short-term variations, however, are uncorrelated.

The results obtained with both, CROCUS and SNTHERM, indicate reasonable agreement between the combined model and the observations. Discrepancies are observed in the polarisation difference, $T_V - T_h$, which is often underestimated. This is due to the fact that the physical snowpack models cannot simulate the crusts (SNTHERM), or that their density is underestimated (CROCUS). This means that there is a need for improving the physical snowpack models.

4.15.2.4 Outlook to applications and further improvements

The HUT model can be used for snowpack emission computation, including contributions from soil, forest canopies and the atmosphere. It has the potential to retrieve information on snow water equivalent (SWE) for dry snow, extent of dry snow cover, onset of snow melt and information on snow grain size and density.

The investigations carried out show promising SWE retrieval accuracies as the HUT model is directly applied to SSM/I observations using a multi-parameter constrained iterative inversion procedure. For example, the area SWE of the 50 000 km^2-sized River Kemijoki drainage area was estimated with an overall RMSE of 24 mm for a single winter. The corresponding coefficient of determination (r^2)

was 0.92 with 30 reference observations for comparison. These results were obtained under dry snow conditions (period from 1 November 1993 to 31 March 1994) without using any training reference data.

MEMLS is useful in all applications where realistic simulations of snow covers are needed. This applies to data assimilation in numerical-weather prediction and to sensitivity studies of snow parameters on the microwave signal. It is also helpful for designing new radiometer systems. Furthermore MEMLS can become a key for new developments in snow physics. Some steps of future work are outlined here

- to assess the sensitivity of MEMLS to special parameter variations, weighting functions, defined by $w_{ij} = \partial e_p / \partial x_{ij}$, where e_p is the emissivity at polarisation p and x_{ij} is the ith physical parameter of layer j, can be computed.
- the assumption of an isotropic and exponential correlation function is a simplification, but so far, it has been a reasonable assumption. By including snow-structure anisotropy (observed, for instance, in new snow deposited during calm wind condition, or in surface hoar), the model can be improved, especially its polarisation behaviour at $\nu > 30$ GHz.
- multiple versus single layer: It has been argued that single-layer snow emission models are more practical and faster than multi-layer models. This statement may be tested by simulations using MEMLS for different values of n. MEMLS is adaptive with respect to n. But it must be pointed out that layering is an inherent property of snowpacks. Solutions may come from fast-model approximations to results of MEMLS simulations.
- a combination of SAFRAN/CROCUS (Durand *et al.*, 1993) with MEMLS gives the possibility of simulating brightness temperature maps. SAFRAN (Système d'Analyse Fournissant des Renseignements Atmosphériques à la Neige) is a sophisticated meteorological model, calculating the input variables for snow cover simulations in a model topography over large areas (about 1000 km^2). Comparison of such maps with observations makes it possible to validate forecasts on a large scale. On a local scale it may help to assess snowpack and background properties.
- meteorological information is useful to simulate the emissivity development providing a tool for reanalysing historical satellite data. This application is needed for model validation, and furthermore, opens the way to climate studies using different observational systems.
- with accurate knowledge of the surface emissivity, it is possible to determine atmospheric parameters from space-based observations (Karbou *et al.*, 2004).
- Due to the high sensitivity of microwave emission on internal snow-structure parameters, MEMLS can help to improve the physical models, especially the understanding of snow metamorphosis and layer structure. As a feedback of such studies, MEMLS also will be further improved.

Although MEMLS has been developed for snow on land, by suitable scaling of the dielectric and geometrical parameters it was extended to sea-ice, as shown in Section 4.16.

4.15.3 Recommended emission models

The two described emission models for snowpacks, HUT and MEMLS, were developed, tested and validated in a combined effort. They were successful in their fields of application. Therefore both models can be recommended. The specifications and present limitations are summarised here.

4.15.3.1 Physical model: MEMLS

Described by Pulliainen *et al.* (1998), Wiesmann and Mätzler (1999), extended by Mätzler and Wiesmann (1999) and by Wiesmann *et al.* (2000) (see page 383 for inline table).

4.15.3.2 Semi-empirical model: HUT Model

Described by Pulliainen *et al.* (1999) and Pulliainen *et al.* (1998) (see page 384 for inline table).

4.16 Sea ice emission modelling *Rasmus T. Tonboe, Georg Heygster, Leif Toudal Pedersen and Søren Andersen*

4.16.1 Introduction

The strategic importance of sea ice emission models is to relate physical properties of the target to brightness temperatures (T_b) at different microwave frequencies (ν) and polarisations and further to interpolate between observations at different places and frequencies. Because the radiances received at the satellite generally contain contributions from both atmosphere and surface, the surface emission model relationships are needed for the retrieval of both surface and atmospheric parameters. While there are physical microwave models describing the atmosphere, the open ocean and partly the land covering a wide range of microwave frequencies and polarisations, such models are generally missing for ice-covered polar regions. The importance of sea ice emission models is illustrated in the applications of this chapter. The aim here is to give a short overview of the quantitative influence of different microphysical parameters in sea ice and the snow on top of it at 50° incidence angle and how these link to emissivity, brightness temperature, dielectric properties and scattering exemplified by using a sea ice version of MEMLS. We further present three examples of sea ice emissivity modelling applications: sensitivity of emissivity to snow parameters at sounding frequencies, sea ice concentration estimate sensitivity to ice surface emissivity and investigation of new sensors (i.e. SMOS) potential for sea ice mapping.

A. Sea ice microwave emission measurements. Formation environment, ice cover distribution, snow cover, ice surface roughness, ice density, ice salinity and inclusion content are important properties for the understanding and interpretation of sea ice microwave data. Emission models provide a link between these physical properties and microwave measurements and thereby to the understanding of measurements.

Purpose of the model	MEMLS computes transmissivity t_p, emissivity e_p and brightness temperature T_p at polarisation p (h,v) of a snow-covered surface under given illumination by T_{sky}
Applications, constraints	Seasonal and perennial snowpacks of any thickness with defined stratigraphy Assimilation of T_p for NWP, avalanche forecasts, hydrology, climatology, simulation of T_p for sensitivity studies, as source for construction of simple models and inversion algorithms for snow parameters and ground temperature. Constraints to be assessed
Model variables and ranges	Frequency range: 5–100 GHz, incidence angle θ: 0–60°, polarisation: linear h and v
Input parameters, including applicable ranges, options for selection and potential sources	Sky brightness temperature T_{sky} (unlimited), ground temperature T_0, ground–snow interface reflectivity s_0 (0 to 1), for each of the n ($\geq$1) snow layers the following model parameters: Correlation length p_{ex}: 0 to 0.7 mm, temperature T: 100 to 273.15 K, snow density ρ: 50 to 500 kg/m^3, (model accepts up to 917 kg/m^3) volumetric liquid-water content W: 0 to about 0.15. Input parameters can be generated by physical snowpack models and meteorological information (Wiesmann *et al.*, 2000)
Output parameters	Standard: Upwelling brightness temperature T_p above the surface; Optional: Scene emissivity (0 to 1), snowpack transmissivity (0 to 1)
Expected errors	Validation made for various snow conditions (Wiesmann *et al.*, 1998; Wiesmann and Mätzler, 1999; Mätzler and Wiesmann, 1999) Present estimate: for T_p: $\leq$10 K, for e_p: $\leq$0.03
Accuracy requirements	Useful: for T_p: 10 K, for e_p: 0.03; further assessment needed; Adequate: for T_p: 3 K, for e_p: 0.01
Further developments	For special parameter ranges it is possible to develop fast models by fitting simple functions to the results of simulations Weighting functions, $w_{ij} = \partial e_p / \partial x_{ij}$, where x_{ij} is the ith physical parameter of layer j, can be computed in order to assess the sensitivity of the model to special parameter variations Isotropic correlation functions other than exponential can be included in the improved Born approximation. More work is needed to account for anisotropic snow structures Presently the main uncertainty is due to errors of the input parameters. More accurate data on correlation length from physical snowpack models are needed

Purpose of the model	Computes brightness temperature and emissivity of snow-covered terrain, optionally including the effect of sparse coniferous forest canopy (under winter conditions) and atmosphere
Applications, constraints	Applicable for iterative inversion: snow water equivalent retrieval/monitoring in boreal forest zone (tested and demonstrated). Assimilation to hydrological models to be done. Potential areas of application: hydrology, meteorology, climatology, hydropower production, flood prediction. Constraints to be assessed
Model variables and ranges	Frequency: 5–100 GHz, nadir angle 20° to 70° (restricted by the soil emission model), polarisation: linear h and v
Input parameters (including applicable ranges, options for selection and potential sources)	Sky brightness temperature (unlimited), ground temperature, snow temperature, ground–snow interface reflectivity, effective snow grain size (up to 1.6 mm), snow density (0.1–0.4 g/cm^3). Optionally snow wetness (up to a few percentage units); calculation of ground–snow reflectivity, e.g. using the model by Wegmüller and Mätzler (1999); calculation of atmospheric effects, e.g. using the statistical model by Pulliainen *et al.* (1993)
Output parameters	Brightness temperature and emissivity of snow covered terrain (above the snow surface). Optionally: airborne or spaceborne observed scene brightness temperature or emissivity including the effect of forest canopy and atmosphere
Expected errors, validation and quality checks	Validation performed against experimentally controlled conditions, measurements and SSM/I observations. Further error analysis to be done
Accuracy ranges	To be assessed
Scopes for further development	Currently one empirical parameter (q) included. The behaviour of q as function of frequency and snow grain size to be modelled using theoretical approaches Development of operative snow parameter retrieval methods based on the HUT model

Measurements are acquired from surface-based installations, aircraft or satellite. Polar orbiting satellites carrying microwave radiometers cover the globe daily with measurements and, in particular, the polar regions, several times a day. Reported sea ice measurements at different microwave frequencies cover the range of 1.4 to 183 GHz; for example, the aircraft 1.4 GHz (L-band) measurements acquired recently over sea ice during the summer melt (Klein *et al.*, 2004). Eppler *et al.* (1992), and the references herein, describe measured sea ice emissivities between 4.9 and 94 GHz for different ice types. Hewison and English (1999) and Selbach (2003) report millimetre wavelength measurements between 89 and 183 GHz in the Baltic and Arctic Ocean, respectively. Our focus is on the frequencies: 1.4, 7, 10, 19, 23, 37, 50, 89, 157 and 183 GHz, which represent channels on present and future satellite sensors (e.g. CMIS, SMOS, SSM/I, AMSR, AMSU-B). The T_b measured at these frequencies has a different sensitivity to natural variability of snow and sea ice microphysical properties. The general appearances of the three predominant ice types in the Arctic, new-ice, first-year ice and multi-year ice types, are distinguishable using microwave remote sensing because of the differences in dielectric and scattering properties. It is therefore convenient in this investigation to distinguish between these three major classes of sea ice.

Thin ice represented in the following by nilas belongs to the category new-ice (<10 cm). It forms under quiescent conditions. It has a smooth surface with little or no snow cover and relatively high salinity, S, of 14–16 ppt. The density is about 920 kg/m^3 (Tucker *et al.*, 1991). Under continued growth, nilas will form young ice and eventually first-year ice.

The level first-year ice (0.3–2 m) surface salinity is stable above 6 ppt during winter. In fact, during winter, the level first-year ice microwave signatures change primarily due to snow cover related processes (Eppler *et al.*, 1992). Snow cover properties directly affect T_b, e.g. liquid water content (W), grain size, density (ρ) etc., and indirectly, by the thermodynamic control of the snow cover on the ice e.g. brine volume (Barber *et al.*, 1995).

Multi-year ice has survived the seasonal melt. The melt processes recrystallise the snow and ice, flush brine from the ice interior and create surface topography i.e. melt ponds and hummocks. The hummocks are porous with large air bubbles and voids; frozen melt ponds are less porous. Salinity increases downwards with about 0.0–0.1 ppt above sea level and 3.0–3.5 ppt below (Weeks and Ackley, 1986). It differs from other ice types by a large number density of voids in the upper part of the ice, its low salinity and thickness (often several metres).

During summer (May–August), about half of the Arctic ice cover melts (its area shrinks from about $15 \times 10^6\,\mathrm{km}^2$ to $8.5 \times 10^6\,\mathrm{km}^2$) (Parkinson and Cavalieri, 1989). Melting sea ice undergoes a significant and complex transformation described in connection with microwave remote sensing by e.g. Garrity (1992) and Gogineni *et al.* (1992). These processes are understood relatively poorly in terms of microwave remote sensing signatures (Carsey *et al.*, 1992). The summer season is therefore beyond the scope of the present text, which focuses on winter conditions.

B. Satellite radiometer field-of-view, sea ice heterogeneity and modelling. Sea ice and snow cover on sea ice is heterogeneous within the satellite field-of-view e.g. from 5 to 50 km for spaceborne radiometers (Eppler *et al.*, 1992). Direct comparison between point measurements (and surface-based sensors) on the ice and satellite data is therefore not feasible (Tonboe and Andersen, 2004). Satellite measurements are an ensemble average of the different emissivities present within the field-of-view and the antenna gain pattern. Models compute emissivities of laterally homogeneous structures of snow and sea ice. Therefore, a comparison between emission models and brightness temperature measurements is possible only on local scales where snow and sea ice are laterally homogenous. These comparisons can be done with laboratory measurements (e.g. Barber *et al.*, 1998) or with surface-based radiometer measurements on the ice (e.g. Mätzler *et al.*, 1984). In this study, the model results are not compared directly with radiometer measurements.

C. Sea ice emission modelling. The permittivity (ε) of snow is determined by its density and liquid water content (Ulaby *et al.*, 1986). The permittivity of snow affects the reflection, transmission and the absorption coefficients. Scattering in the snow pack is detectable for frequencies higher than 10 GHz (Barber *et al.*, 1998) and is important for coarse snow grains or high frequency ($\nu > 20$ GHz) (Mätzler, 1987). The permittivity of sea ice is largely given by the brine and air-inclusion volume (Shokr, 1998). Important for scattering in sea ice is the size and number density of brine pockets or air bubbles (Winebrenner *et al.*, 1992).

Important radiative processes in a homogeneous snow cover can be described using simple emission models (Ulaby and Stiles, 1980; Mätzler, 1987). However, snow cover on land, ice or sea ice is a layered medium, and therefore the simple models fail to simulate observations by not accounting for important reflections between layers (Mätzler *et al.*, 1984; Surdyk and Fily, 1993). Winebrenner *et al.* (1992) provide a review of different types of emission models that exist for sea ice. The microwave emission model for layered snow-packs (MEMLS) is a 'model suitable for simulations of all kinds of physical effects' (Mätzler *et al.*, 2000, p. 107), and it has been tested and validated for snow cover on land with satisfactory results. Here MEMLS is extended to include emission from sea ice. The sea ice model and modifications are described in the next section.

4.16.2 Extension of MEMLS to sea ice emission

MEMLS, described in Wiesmann and Mätzler (1999), uses the physical snow quantities and structure as input i.e. sequence of layers (j), density (ρ), exponential correlation length (p_{ec}), thermometric temperature (T) and volumetric liquid water content (W). In order to apply this model to compute the emission of both snow and sea ice it is necessary to include modules that compute the dielectric properties and scattering of sea ice. Small liquid brine inclusions, also called brine pockets, dominate scattering in nilas and first-year ice. In multi-year ice, the voids and air bubbles in the upper ice are the primary scatterers (Nghiem *et al.*, 1995). The permittivity of liquid brine is an order of magnitude larger than the permittivity of solid ice, and the permittivity of sea ice is therefore primarily a function of brine volume (Ulaby *et al.*, 1986). The permittivity of brine is computed using appendix E in Ulaby *et al.* (1986).

The permittivity of sea ice is computed using Polder–Van Santen mixing formulae described in e.g. Sihvola and Kong (1988). It is a function of pure ice permittivity (using Mätzler, 1998b), inclusion shape and orientation, volume and the brine pocket permittivity. The formulation in Mätzler (1998a) for a media with spherical inclusions is used.

MEMLS is valid for snow cover in the range 5–100 GHz. The primary limitation is the estimation of the scattering coefficient using empirical relations, which fit scattering in natural snow cover. For the use of MEMLS outside of this frequency range, and for sea ice, it is necessary to compute the scattering coefficient using theoretical relations (Mätzler and Wiesmann, 1999).

The scattering in sea ice is therefore computed using the improved Born approximation (Mätzler, 1998). 'The scattering [using the improved Born approximation] increases by a power law of the microwave frequency times the correlation length with a power of approximately 2.5. Above a certain frequency or above a certain correlation length, the increase will saturate in a similar way as Mie scattering does for spheres.' (Mätzler and Wiesmann, 1999, p. 317). It is further noted by Mätzler and Wiesmann that the improved Born approximation fits observations for snow grains which are large compared with the wavelength. It is therefore assumed, in this study, that the improved Born approximation is valid also at high frequency (157 and 183 GHz). Scatterers are exclusively air bubbles and voids in multi-year ice and brine pockets in first-year ice. The scattering coefficient is, in general, a function of the permittivity of pure ice, the permittivity of brine or air, the permittivity of the sea ice mixture, volume of brine or air, microwave frequency and the correlation length of scatterers. The exponential correlation length (p_{ec}) is a measure of scatterer size and distribution, see e.g. Mätzler (2002).

Natural snow cover consists of both rounded and oblate grains. Recrystallised snow grains have cup-like forms (Mätzler, 2002). Congelation first-year ice has vertically oriented needle shaped brine pockets while frazil first-year ice has randomly oriented elongated brine pockets (Shokr, 1998). The upper centimetres of first-year ice typically consist of frazil and deeper ice of congelation ice. Frazil layers may also appear at depth (Weeks and Ackley, 1986). In this study, both the snow and sea ice scattering and dielectric properties are assumed isotropic within a given layer. This assumption significantly simplifies the *a priori* input to the model and the interpretation of the results, although it is in conflict with the description of the natural types of snow and ice above. Scattering anisotropy is a secondary effect that should be included in case-specific studies; however, here we are interested in the primary properties affecting the emissivity in a more general description.

These additional modules (sea ice scattering and dielectrics) are hence included in the model computation scheme. MEMLS is coded in MATLAB. With a few minor changes, this sea ice version of MEMLS is now running in GNU Octave (www.octave.org).

Table 4.5 Nilas profile used as input to MEMLS

No. j	Layer depth z_j (m)	Density ρ (kg/m^3)	Exponential correlation length p_{ec} (mm)	Salinity S (ppt)	Permittivity ε
1	0.0–0.10	920	0.15	14.0	$3.88 + 0.07i^1$ $3.55 + 0.27i^2$ $3.30 + 0.12i^3$

Note: The permittivity of the sea ice is frequency dependent. Permittivity at frequencies 1.4, 37 and 183 GHz covers the range for both the permittivity real part and the loss factor. The permittivity, ε, at these frequencies is marked by the superscript numbers: 1. ε at 1.4 GHz; 2. ε at 37 GHz and 3. ε at 183 GHz

Table 4.6 First-year ice profile used as input to MEMLS

No. j	Layer depth z_j (m)	Density ρ (kg/m^3)	Exponential correlation length p_{ec} (mm)	Salinity S (ppt)	Permittivity ε
1	0.00	260	0.07	0	$1.43 + 0.00i$
2	0.07	410	0.10	0	$1.76 + 0.00i$
3	0.12	300	0.14	0	$1.56 + 0.00i$
4	0.13–1.15	920	0.15	7	$3.62 + 0.04i^1$ $3.45 + 0.17i^2$ $3.26 + 0.09i^3$

Note: The permittivity of the sea ice is frequency dependent. Permittivity at frequencies 1.4, 37 and 183 GHz covers the range for both the permittivity real part and the loss factor. The permittivity, ε, at these frequencies is marked by the superscript numbers: 1. ε at 1.4 GHz; 2. ε at 37 GHz and 3. ε at 183 GHz. The small frequency dependent snow loss factor is rounded to 0.00

4.16.3 Sea ice emission modelling experiments using MEMLS

A. The sea ice profiles: nilas, first-year and multi-year ice. The constructed profiles of nilas, first-year ice and multi-year ice with typical properties for these ice types are used as input to the model. Tables 4.5–4.7 present the input parameters. For all three profiles the skin temperature (synonymous with the upper 1 cm) is $-15°$C (258.15 K). The ice/water interface temperature is $-1.8°$C (271.35 K). The temperature profile (1 cm increments) in snow and ice is linear with the thermal conductivity in the snow and ice of 0.3 W/mK and 2.1 W/mK, respectively. During simulations with liquid water in the snow, the snow temperature is constant $-1°$C and the snow/ice interface temperature is $-5°$C. The multi-year ice profile is adapted from Mätzler *et al.*

Table 4.7 Multi-year ice profile used as input to MEMLS

No. j	Depth z_j (m)	Density ρ (kg/m^3)	Exponential correlation length p_{ec} (mm)	Salinity S (ppt)	Permittivity ε
1	0.0	260	0.07	0	1.43 + 0.00i
2	0.07	450	0.10	0	1.85 + 0.00i
3	0.08	300	0.10	0	1.52 + 0.00i
4	0.10	450	0.10	0	1.85 + 0.00i
5	0.11	300	0.10	0	1.52 + 0.00i
6	0.16	450	0.10	0	1.85 + 0.00i
7	0.17	300	0.14	0	1.52 + 0.00i
8	0.19	750	0.25	0.5	3.20 + 0.00i[1] 3.20 + 0.01i[2] 3.18 + 0.01i[3]
9	0.29–2.49	900	0.20	2.5	3.31 + 0.01i[1] 3.26 + 0.05i[2] 3.21 + 0.03i[3]

Note: The permittivity of the sea ice is frequency dependent. Permittivity at frequencies 1.4, 37 and 183 GHz covers the range for both the permittivity real part and the loss factor. The permittivity, ε, at these frequencies is marked by the superscript numbers: 1. ε at 1.4 GHz; 2. ε at 37 GHz and 3. ε at 183 GHz. The small frequency dependent snow loss factor is rounded to 0.00

(1984), and the first-year ice profile is constructed to fit satellite T_b measurements (Tonboe and Andersen, 2004).

B. Penetration depth. Assessment of the penetration depth is important when interpreting microwave data: is the T_b at a certain microwave frequency a result of snow, ice or even deeper ice or water interface radiative processes?

The microwave penetration is strongly reduced in sea ice due to the presence of liquid brine compared with snow or freshwater ice. The penetration depth, δ_p, is defined as (Hallikainen and Winebrenner, 1992),

$$\frac{P(\delta_p)}{P(0+)} = \frac{1}{e} \tag{4.146}$$

where $P(\delta_p)$ is the transmitted power at depth δ_p and $P(0+)$ is the transmitted power just beneath the surface. δ_p is a function of the extinction coefficient, i.e. both scattering and absorption, and the transmission loss.

Figures 4.77–4.79 show the penetration depth at 50° incidence angle in nilas, first-year and multi-year ice profiles presented in Tables 4.5–4.7.

C. Depth of icy crust and scattering layer in the first-year ice snow cover. Both scattering and reflections at layers in the snow and ice are processes influencing the emitted brightness temperature at most frequencies. For the natural range of

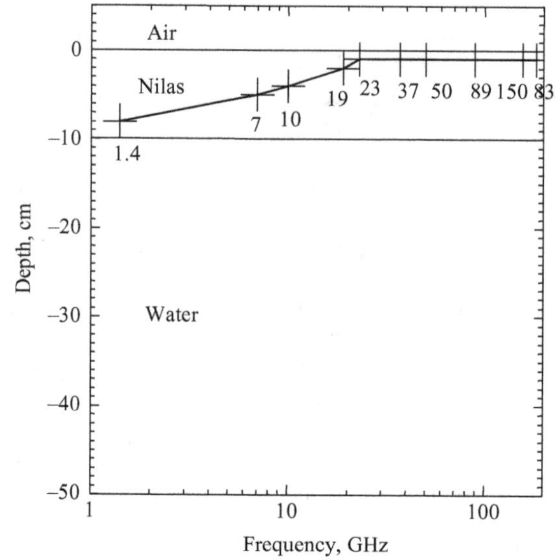

Figure 4.77 *Simulated penetration depth, δ_p, at 50° incidence angle in the nilas ice profile described in Table 4.5, computed at 1.4, 7, 10, 19, 23, 37, 50, 89, 157 and 183 GHz marked with crosses and numbers. The δ_p is similar for V and H polarisations*

snow grain sizes, scattering is not important at 1.4 GHz. For higher frequency, it is increasingly important to about 89 GHz and for even higher frequency (157 and 183 GHz), scattering tends to saturate for coarse grains. While the influence of grain size and density in unlayered typical sea ice structures has been investigated by Fuhrhop *et al.* (1997), reflections between layers (snow crusts, air/snow and snow/ice interfaces) are also significant for T_b and the polarisation difference ($T_v - T_h$) at frequencies between 1.4 and 183 GHz, but have not yet been examined in detail. This is the goal of the present study. However, the importance of layering decreases as a function of frequency, due to the shallow penetration and relative significance of scattering even for small scatterers at high frequency (89, 157 and 183 GHz). In the following model experiment, T_b at 89 GHz is affected by both scattering and dense layer depth. Therefore, 89 GHz is used as an example in Figure 4.80.

MEMLS is initialised with a 13 cm new-snow cover ($\rho = 260\,\text{kg/m}^3$, $p_{ec} = 0.07\,\text{mm}$) on thick first-year ice. The depth of two 1 cm thick layers is varied between the top and bottom of the snow pack at 1 cm intervals: a dense layer ($\rho = 500\,\text{kg/m}^3$) and a scattering layer ($p_{ec} = 0.4\,\text{mm}$). The dense layer is a crust, and the scattering layer is recrystallised snow. At frequencies lower than 89 GHz (1.4–50 GHz) the position of the scattering layer is largely insignificant, while the position of the dense layer is significant at all frequencies (1.4–183 GHz). This model experiment indicates

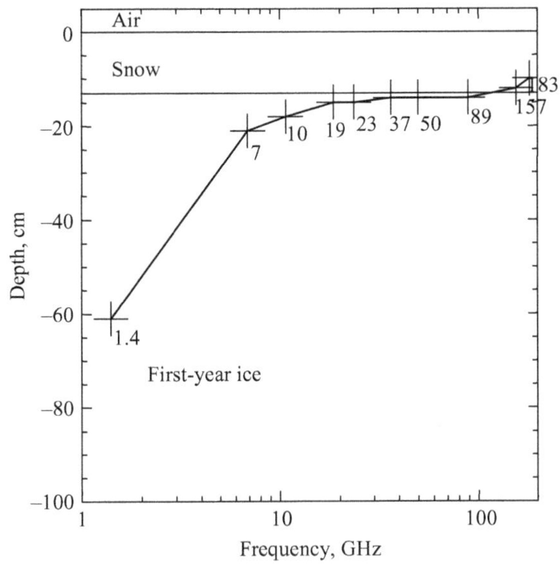

Figure 4.78 *Simulated penetration depth, δ_p, at 50° incidence angle in the first-year ice profile described in Table 4.6, computed at 1.4, 7, 10, 19, 23, 37, 50, 89, 157 and 183 GHz marked with crosses and numbers. The δ_p is similar for V and H polarisations*

that dense layers in the snow pack are most significant when they modify the large dielectric contrasts in the system, i.e. air/snow and snow/ice interface, and less important at intermediate depths. The depth of scattering layers is important at frequencies where extinction due to scattering in the background snow pack is significant.

4.16.4 Parametrisation of sea ice emissivity for atmospheric retrieval

The assimilation of atmospheric parameters derived from microwave satellite data e.g. AMSU has a significant impact on both global (ECWMF) and regional (HIRLAM) weather prediction models (Prigent *et al.*, 2004). The principle is to separate the atmospheric emissivity from the surface emissivity in the T_b measurements from space, thus the atmospheric part is parametrised in terms of temperature or water vapour. The surface emissivity, which is high for sea ice compared with open water, can be determined at frequencies where the atmosphere is largely transparent in the 'atmospheric windows'. This allows referring the emissivities at the sounding frequencies to those at the window frequencies by interpolation using emission models. Temperature sounding uses frequencies around 50 GHz, and humidity sounding uses frequency ≥85 GHz. Surface emissivity estimation at sounding frequencies will become even more important with SSMIS (Wessel *et al.*, 2004) because it combines observations at the window and at the sounding frequencies on one sensor observing at all frequencies with a constant incidence angle.

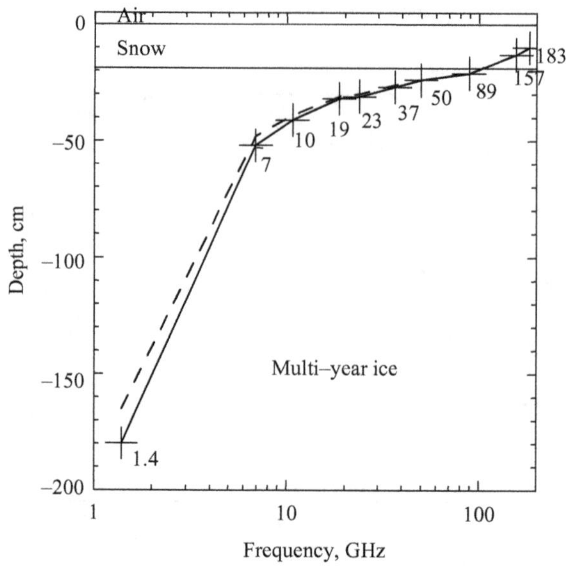

Figure 4.79 *Simulated penetration depth, δ_p, at 50° incidence angle in the multi-year ice profile described in Table 4.7, computed at 1.4, 7, 10, 19, 23, 37, 50, 89, 157 and 183 GHz marked with crosses and numbers. In this particular case, the δ_p is different for v (solid line) and h (dashed line) polarisations*

Figure 4.81 shows the simulated correlation between nilas, first-year and multi-year ice emitted T_v at the six microwave frequencies (23–183 GHz) using all parameter pairs described in Table 4.8.

The many off-diagonal points at the 37/50 GHz (Figure 4.81b) and 89/157 GHz (Figure 4.81c) plots indicate that simple linear models are inadequate to relate T_b at window frequencies to sounding frequencies.

In Figures 4.81(c) and 4.81(d) the off-diagonal points are associated with the upper snow density/bottom snow layer correlation length parameter pair for both first-year and multi-year ice. This is the most important parameter combination among the pairs in Table 4.8 for the first-year and multi-year ice T_v sensitivity at medium and high frequencies. The T_v variability is similar when varying more than two parameters simultaneously, e.g. first-year ice upper and mid snow density and bottom snow correlation length and salinity (not shown).

4.16.5 Sensitivity of sea ice concentration estimates to surface emissivity

Current ice concentration algorithms use satellite microwave T_b between 6 and 89 GHz as input. The mean accuracies of some of the more common algorithms, used for SSM/I data, such as NASA Team (Cavalieri *et al.*, 1984) and Bootstrap (Comiso, 1986) are reported to be 1–6 per cent in winter (Steffen and Schweiger, 1991; Emery *et al.*, 1994; Belchansky and Douglas, 2002). At high ice concentrations, even

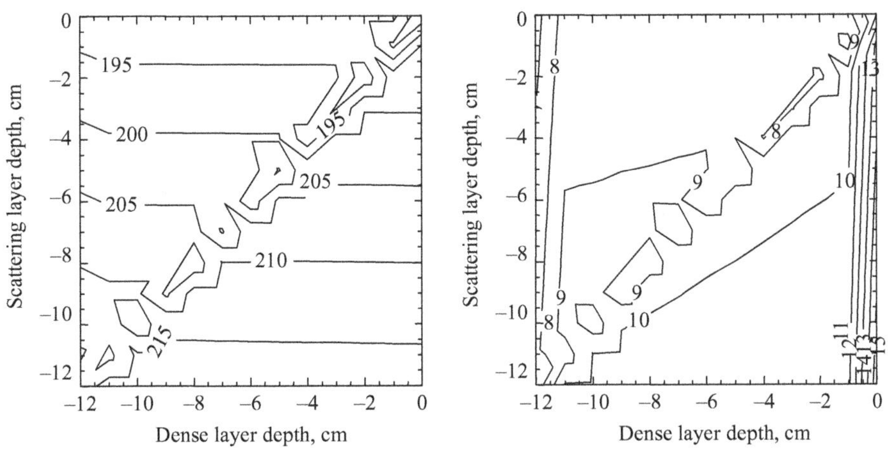

Figure 4.80 *The simulated T_v (89 GHz) (left) and $T_v - T_h$ in Kelvin (right) sensitivity to the depth of both a dense ($\rho = 500\,kg/m^3$) and a scattering layer ($p_{ec} = 0.4\,mm$) of 1 cm thickness each in a 13 cm first-year ice snowpack ($\rho = 260\,kg/m^3$ and $p_{ec} = 0.07\,mm$). The scattering layer has the density of the background ($\rho = 260\,kg/m^3$) and the dense layer the correlation length of the background ($p_{ec} = 0.07\,mm$). The depth is measured from the top of the snow pack (at 0 cm) to the top of the dense (x-axis) and scattering (y-axis) layer, respectively. The irregular values along the main diagonal correspond to the overlap between the dense and scattering layers. This part of the diagram is unphysical, hence ignored in the discussion*

small changes in the sea ice concentration have a significant impact on energy fluxes between the ocean and the atmosphere. Therefore, ice concentration is an important ice cover parameter to estimate accurately (Steffen and Schweiger, 1991).

The computed ice concentration accuracy is further degraded by particular atmospheric constituents like cloud liquid water, where NASA Team ice concentration erroneously can increase by about 10 per cent (Oelke, 1997). Changes in the surface emissivity can depress the computed ice concentration by 20 per cent (Tonboe *et al.*, 2003). Toudal (1994) argued that variations in surface emissivity is a very important error source in sea ice concentration retrievals over consolidated ice from satellite passive microwave measurements.

The observed sensitivity of the different ice concentration algorithms using SSM/I satellite data e.g. NASA Team, Bootstrap and Near 90 GHz (Svendsen *et al.*, 1987) to the atmosphere or surface brightness temperature is different as shown in Figure 4.82 for a case in Baffin Bay. Coincident SAR images show that the real ice concentration in this area is stable above 94 per cent (Tonboe *et al.*, 2003).

The sea ice emissivity normally varies during winter because of ice growth, snowfall, diurnal cycling and snow/ice metamorphism. Warm air outbreaks (Figure 4.82) over the consolidated sea ice pack in Baffin Bay and Fram Strait/Arctic Ocean during

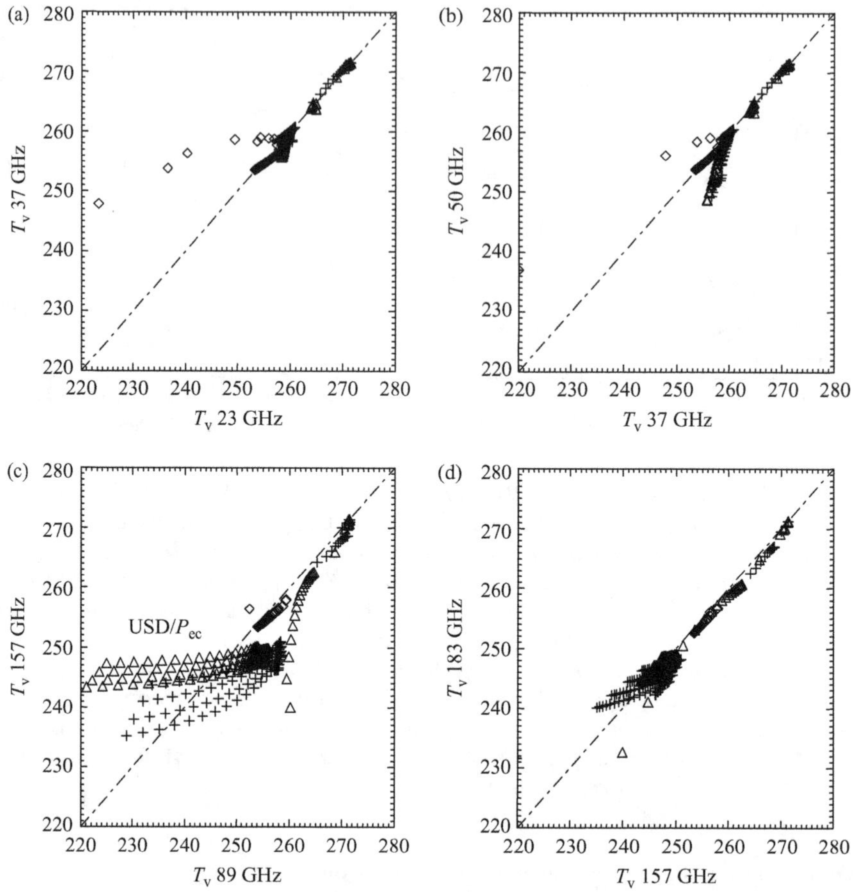

Figure 4.81 *The simulated correlation between window and sounding frequencies for nilas (◇), first-year ice (+) and multi-year ice (Δ) emitted T_v at the 6 frequencies (23, 37, 50, 89, 157 and 183 GHz) using all parameter pairs in Table 4.8 (every 5th data point). The off-diagonal points marked by USD/p_{ec} in 6(c) are associated with the upper snow density (USD)/bottom snow layer correlation length (p_{ec}) parameter pair*

winter offer a possibility to investigate the sensitivity of the ice concentration estimate to changes in the snow and ice cover emissivity in the course of days (Tonboe *et al.*, 2003). While the actual ice concentrations remain close to 100 per cent during and after the advection of warm air followed by snowfall and possibly rain, the formation of depth hoar and icy layers in the snow pack and in general the metamorphism accelerates as described in e.g. Drinkwater *et al.* (1995) and Garrity (1992). These winter warm air events may influence large areas (about 360 000 km^2) and on a weekly timescale smaller areas (15 000 km^2) along the ice edge (Tonboe *et al.*, 2003; Voss *et al.*, 2003).

Table 4.8 The parameters, which are varied in sensitivity studies

Ice type	Parameter	Depth [cm]	Range
Nilas/young ice	Salinity	Entire ice column	2.5–18 ppt
	Ice thickness	–	3–31.5 cm
First-year ice	Upper snow density	0–7	260–450 kg/m^3
	Mid snow density	7–12	260–450 kg/m^3
	Bottom snow p_{ec}	12–13	0.07–0.45 mm
	Upper ice salinity	13–19	2.5–18 ppt
	Snow liquid water	Entire snow column	0–14%
	Snow thickness	–	1.5–31.5 cm
Multi-year ice	Upper snow density	0–4	260–450 kg/m^3
	Mid snow density	4–7	260–450 kg/m^3
	Bottom snow p_{ec}	17–19	0.07–0.45 mm
	Upper ice density	19–29	750–900 kg/m^3
	Snow liquid water	Entire snow column	0–14%
	Snow thickness	–	1.5–31.5 cm

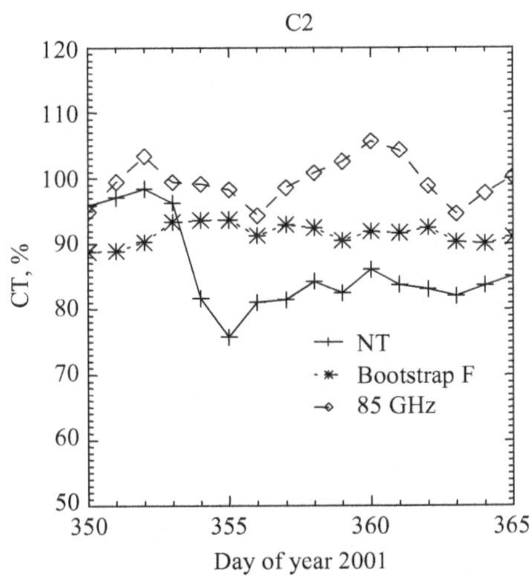

Figure 4.82 The average ice concentration, observed by satellite in a small area in Baffin Bay (62.6° W, 71.5° N) using NASA Team (+), Bootstrap in frequency mode (∗) and Near 90 GHz (◇) algorithms, respectively, December 2001. The variability is a function of ice concentration and both surface and atmospheric emissivity. The measured air temperature on a nearby meteorological station on the Greenland coast (Upernavik) increases to above 0° C on day 353/354 and the freeze-up begins on day 357 (from Tonboe et al., 2003)

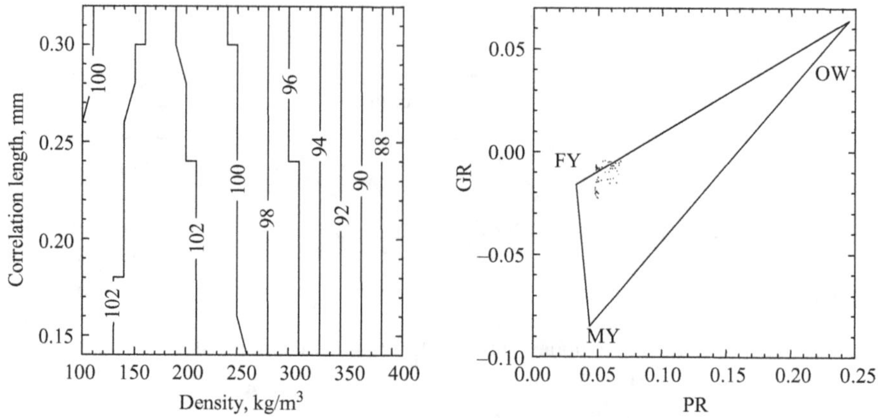

Figure 4.83 Simulated sensitivity at ground level of the NASA Team algorithm ice concentration (%) to variations in the upper snow layer density (layering) and bottom snow layer correlation length (grain size) in the snow cover on 100% first-year ice (left). Simulated data placed in the PR_{19} *and* $GR_{19v/37v}$ *space with the approximate NASA Team retrieval triangle in the background (right). Tie-points for open water (OW), first-year ice (FY) and multi-year ice (MY) based on Andersen (1999) (from Tonboe and Andersen, 2004)*

The changes are persistent but recover in the course of months, e.g. the NASA Team ice concentration estimate is sensitive to layering in the snow pack and, in particular, to the air/snow dielectric contrast (i.e. density of the snow surface). Surface crusts with relatively high permittivity form during temporary warming (this strongly affects the NASA Team ice concentration); however, new snow on top of the crust will decrease the dielectric contrast again thus reducing its influence on the brightness temperature (see Section 4.16.3). The NASA Team total ice concentration is robust to the simulated variations in the bottom snow layer correlation length.

The algorithm sensitivity to surface emissivity and thermometric temperature of the target depends on the selected polarisations and frequencies (Emery *et al.*, 1994; Toudal, 1994; Comiso *et al.*, 1997). Emission models can be used to compute and analyse the sensitivity of the retrieved ice concentration to the microphysical properties of the snow and ice and to select and develop algorithms with low sensitivity to variations in the surface emissivity. Figure 4.83 shows the modelled sensitivity of the NASA Team ice concentration to the density of the upper snow layer and correlation length of the bottom 1 cm snow layer.

Tonboe and Andersen (2004) made a model sensitivity study using MEMLS of nine common ice concentration algorithms where both the upper snow density and the bottom snow correlation length of a first-year ice profile (Table 4.6) were varied as described in Table 4.8. The model simulations showed

(1) the combination of T_v (19 GHz) and T_v (37 GHz) as used in the Comiso Bootstrap algorithm in frequency mode, in the NORSEX (Svendsen *et al.*, 1983) and in

the Cal-val (Ramseier, 1991) algorithms has low sensitivity to the simulated emissivity changes;

(2) algorithms also using T_h like Bristol (Smith, 1996), NASA Team and Bootstrap in polarisation mode are sensitive to the simulated density of the upper snow layer.

They further found that the simulated high sensitivity of the 89 GHz polarisation difference ($T_v - T_h$) to both density and the correlation length makes the algorithms using the high-frequency channel like Near 90 GHz and TUD (Pedersen, 1998) very sensitive to the simulated snow cover emissivity.

Mätzler *et al.* (1984) reported results from a field experiment which showed that the high-frequency (94 GHz) polarisation difference appeared insensitive to ice lenses at 7 cm depth in the snow. This observation was explained by strong attenuation and isotropic scattering in the upper snow layer. They further noted that '. . . at 94 GHz the surface reflectivity will lead primarily to polarisation effects . . .' (Mätzler *et al.*, 1984, p. 335). The shallow penetration of the near 90 GHz microwaves (confined to the upper centimetres of the snow cover) means that different ice types with snow cover have similar radiative signatures. Furthermore, ice lenses below the upper centimetres of snow are only vaguely affecting the near 90 GHz emissivity. In other words, the near 90 GHz emissivity is largely insensitive to ice type and ice layers within the snow pack, which are affecting and complicating the interpretation of emissivity at e.g. 19 and 37 GHz. These observations lead to the development of the Near 90 GHz ice concentration algorithm that exploits the higher spatial resolution at this frequency (Svendsen *et al.*, 1987; Pedersen, 1998; Kaleschke *et al.*, 2001).

However, the near 90 GHz polarisation difference is indeed sensitive to snow–ice surface emissivity. The model experiment indicates that the polarisation difference (at 89 GHz) increases as a function of air–snow dielectric contrast/reflectivity (upper snow layer density). The air–snow density contrast was not explicitly the theme of the laboratory experiment of Barber *et al.* (1998). Even so, they note that as the snow cover permittivity increases: (1) e_h (90 GHz) decreases, (2) e_v is stable. This confirms our model experiments.

4.16.6 New sensors: L-band sea ice radiometry with SMOS

Before the launch of the European Soil Moisture and Ocean Salinity mission (SMOS) in 2007 it is largely unknown what the benefit of L-band (1.4 GHz) radiometer measurements will be for sea ice mapping. The SMOS mission is described in Kerr *et al.* (2001). L-band radiometer measurements have been acquired over sea ice only during summer in order to investigate the influence of melt ponds (melt pond 2000, US campaign, Klein *et al.*, 2004). Models need to be used to investigate the winter sea ice mapping potential for L-band radiometers. The primary limitation to the valid frequency range in MEMLS is the estimation of scattering. MEMLS is expected to be valid at L-band since scattering at this frequency is insignificant.

We notice that the simulated radiation at L-band penetrates to near the ice/water interface in nilas (Figure 4.77). To investigate this further Figure 4.84 shows the evolution of the emissivity at 1.4 GHz and 7 GHz for a growing new young ice sheet

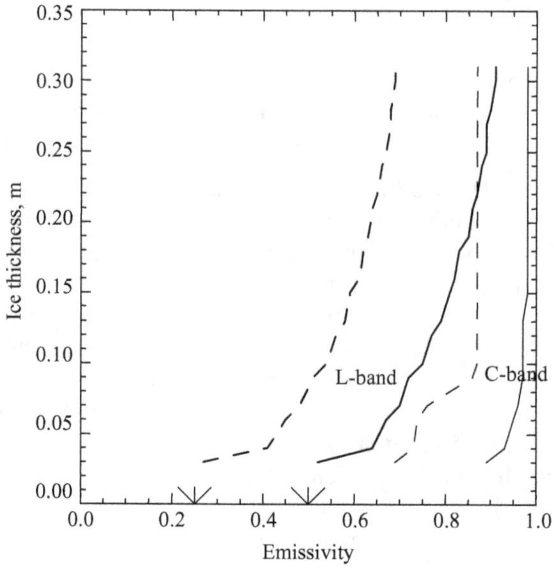

Figure 4.84 *Simulated 1.4 GHz (L-band) and 7 GHz (C-band) emissivity from a growing nilas/young ice sheet (3–31 cm) without snow cover.* e_v *(full line) and* e_h *(dashed line). The water emissivity is assumed similar at the two frequencies:* $e_v = 0.50$ *and* $e_h = 0.25$*, marked by arrowheads on the x-axis*

(3–31 cm) without snow cover. It is assumed that the thin ice is in the form of nilas and young ice (complete coverage of the ocean surface) and not a heterogeneous ice cover such as pancake ice consisting of irregular floes in a frazil shush matrix. The model simulations show that the 1.4 GHz emissivity is affected by the water emissivity throughout the range of thickness while at 7 GHz the emissivity relatively quickly rises from typical water to ice emissivities and reaches saturation at about 20 cm thickness.

Sea ice salinity increases with the growth rate (Weeks and Ackley, 1986) which, in turn, decreases with thickness because the ice layer insulates and reduces cooling of the ice/water interface. Hence, as ice grows from the initial thin nilas (3 cm) to thick young ice (30 cm) it experiences a decrease in total salinity. In addition, the brine volume (i.e. the permittivity) in the upper ice decreases as a function of ice thickness and the resulting reduction in upper ice temperature. Figure 4.85 shows the simulation of the emitted T_v (1.4 GHz) as a function of the total thickness and of the salinity (assumed homogeneous here) of a thin ice sheet. Interference effects are insignificant for the emitted nilas T_v. The signal variation is primarily a function of ice surface permittivity. The results for thin saline ice T_v (1.4 GHz) and thick less saline ice T_v are similar. Also, the simulated $T_v - T_h$ (1.4 GHz) is ambiguous with respect to salinity and thickness. Mapping of thin ice thickness using L-band is therefore difficult. In any case, formation of thin ice (nilas)

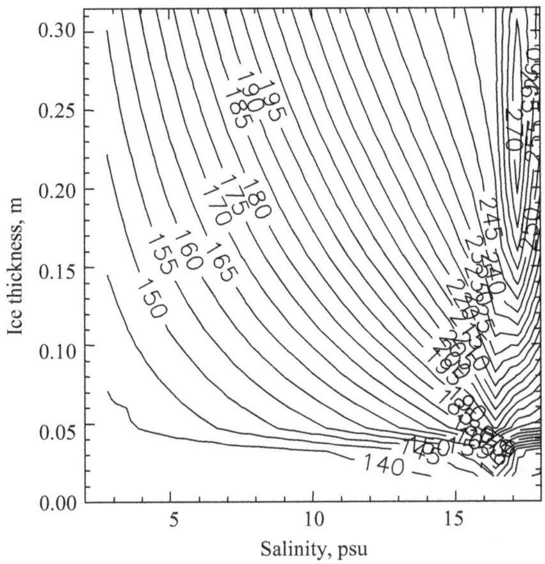

Figure 4.85 *Simulation of the emitted T_V (1.4 GHz) sensitivity in Kelvin to ice thickness (3–31.5 cm) and salinity (2.5–18 ppt) in nilas/ young ice*

on the scale of the SMOS resolution (50 km) is rare, and in most cases, new ice areas are typically mixed with older ice (young-, first- and multi-year ice) and open water.

4.16.7 Conclusions

MEMLS has been set up for sea ice and the usefulness of sea ice emission modelling demonstrated with three different applications, i.e. surface emissivity and atmospheric parametrisation, sensitivity of the ice concentration estimate to the variable emissivity and future sensors. The specific conclusions are

- the penetration depth at 157 and 183 GHz in first- and multi-year ice is confined to the snow layer and 89 GHz penetrates to the ice surface. Penetration depth is considerable at 1.4 GHz with 60 cm depth in first-year and 180 cm in multi-year ice. On the one hand, frequencies higher than 19 GHz only penetrate the nilas surface. On the other hand, radiation at 1.4 GHz penetrates to near the nilas/water interface;
- the position of icy crusts in the snow layer influences brightness temperature when the large dielectric contrasts of the system are modified by the crust, i.e. if the crust is near the air/snow or near the snow/ice interface. The scattering layer depth is significant at high frequency (89, 157 and 183 GHz);
- simple emission models to interpolate between window and sounding frequencies are needed to separate surface and atmospheric contribution to radiation measured at the satellite emission for atmospheric parametrisation. Since scattering

in the snow is important at sounding frequencies, an adequate scattering description in the model seems to be paramount for obtaining the correct level of T_V or e_V;

- MEMLS has successfully been used in the pursuit of ice concentration algorithms with low sensitivity to surface emissivity. In the case presented, T_V combinations at 19 and 37 GHz seemed to be most robust.
- 1.4 GHz brightness temperature is sensitive to both thin ice thickness and salinity. The ambiguity cannot be lifted by additionally considering $T_V - T_h$.

4.16.8 Open challenges

Sea ice emission modelling is still young compared with modelling of land and ocean. Several issues remain:

One of the most important problems remaining for sea ice emission modellers is to develop computationally fast and relatively simple emissivity models for sea ice to estimate the surface emissivity from window to sounding frequencies. This is required for atmospheric parametrisation (temperature and water vapour), and the results are significant for assimilation of microwave radiometer data into weather prediction models operating globally or regionally at high latitudes.

Snow mapping on sea ice is an important issue not only for the parametrisation of fluxes but also for other microwave remote sensing applications. As shown earlier, the snow cover on first-year ice has a significant impact on, for example, the ice concentration estimate. Models for snow parametrisation using inversion techniques must be relatively simple (computationally fast) and rely only on the most important snow parameters (e.g. Hewison and English, 1999). The development of such models could take departure in all-round complex models like MEMLS to arrive at models such as 'the HUT snow microwave emission model' (Pulliainen *et al.*, 1999), successfully used for model inversion for land snow cover parametrisation. Toudal (1994) demonstrated earlier how forward models of ice emissivity could be used in inversion algorithms. Results from MEMLS will provide a very important step towards such forward models.

Wiesmann *et al.* (2000) describe a coupled physical and emissivity snow model that successfully simulates the effects of snow depth on microwave parameters. Physical sea ice models coupled to emission models similar to the approach of Wiesmann *et al.* (2000) would make it possible to seek brightness temperature combinations, which are unlikely to be influenced by ice emissivity for use in ice concentration algorithms. Realistic description of sea ice metamorphosis and the snow and ice profile are necessary prerequisites for reliable model results. Both models should further treat the physics of saline snow.

The validation of sea ice emission models is the overriding issue in all of the above future applications. Future model developments should include a substantial validation component. Logistics is a larger problem on sea ice than on land, and conditions are not as homogeneous as in the ocean, but validation is a task worth undertaking since it is the key to the above sea ice emission model applications.

References

ABRAMOWITZ, M., and STEGUN, I.: 'Handbook of mathematical functions' (Dover, New York, NY, 1974)

ARMSTRONG, R. L.: 'Continuous monitoring of metamorphic changes of the internal snow structure as a tool in avalanche studies', *J. Glaciol.*, 1977, **19**, pp. 325–34

ARMSTRONG, R. L.: 'Metamorphism in subfreezing, seasonal snow cover: the role of thermal and vapor pressure conditions'. PhD thesis, University of Colorado, Faculty of Graduate School, Department of Geography, 1985

ARMSTRONG, R. L., and BRODZIK, M. J.: 'Recent northern hemisphere snow extent: a comparison of data derived from visible and microwave sensors', *Geophys. Res. Lett.*, 2001, **28**(19), pp. 3673–6

ANDERSON, E. A.: 'A point energy and mass balance model of a snow cover'. PhD thesis, Department of Civil Engineering, Common on Graduate Students of Stanford University, 1975

ANDERSON, M. C., NEALE, C. M. U., LI, F. *et al.*: 'Upscaling ground observations of vegetation water content, canopy height and leaf area index during SMEX02 using aircraft and Landsat imagery', *Remote Sens. Environ.*, **92**, 2004, pp. 447–64

APEL, J. R.: 'An improved model of the ocean surface wave vector spectrum and its effects on radar backscatter', *J. Geophys. Res.*, 1994, **99**, pp. 16,269–91

ARONS, E. M., and COLBECK, S. C.: 'Geometry of heat and mass transfer in dry snow: a review of theory and experiment', *Rev. Geophys.*, 1995, **33**, pp. 463–93

BARBER, D. G., REDDAN, S. P., and LEDREW, E. F.: 'Statistical characterisation of the geophysical and electrical properties of snow on landfast first-year sea ice', *J. Geophys. Res.*, 1995, **100**(C2), pp. 2673–86

BARBER, D. G., FUNG, A. K., GRENFELL, T. C. *et al.*: 'The role of snow on microwave emission and scattering over first-year sea ice', *IEEE Trans. Geosci. Remote Sens.*, 1998, **36**(5), pp. 1750–63

BASS, M., VAN STRYLAND, E. W., WILLIAMS, D. R., and WOLFE, W. L.: 'Optical properties of films and coatings, Part 11: Handbook of optics' (McGraw-Hill, Inc., New York, San Francisco, Washington, DC, vol. I, 1995) pp. 42.9–42.14

BECKMANN, P., and SPIZZICHINO, A.: 'The scattering of electromagnetic waves from rough surfaces' (Pergamon Press, Oxford, UK, 1963)

BELCHANSKY, G. I., and DOUGLAS, D. C.: 'Seasonal comparisons of sea ice concentration estimates derived from SSM/I, OKEAN, and RADARSAT data', *Remote Sens. Environ.*, 2002, **81**, pp. 67–81

BERGER, M., KERR, Y., FONT, J. *et al.*: 'Measuring soil moisture with ESA's SMOS mission; advancing the science', *ESA-Bull.*, 2003, **115**, pp. 40–5

BERNIER P.Y.: 'Microwave remote sensing of snowpack properties: potential and limitations', *Nordic Hydrol.*, 1987, **18**, pp. 1–20

BIRCHAK, J. R., GARDNER, C. G., HIPP, J. E. *et al.*: 'High dielectric constant microwave probes for sensing soil moisture'. Proceedings of the

Institute of Electrical and Electronics Engineers, January 1974, vol. 62, pp. 93–8

BLANCH, S., and AGUASCA, A.: 'Sea water dielectric permittivity models: Review and impact on the brightness temperature at L-band'. *First results* Workshop, EuroSTARRS/WISE/LOSAC Campaigns, ESA SP-525, pp. 137–41, ESTEC/European Space Agency, 2003

BLUME, H.-J. C., LOVE, A. W., MELLE, M. J. V., and HO, W. W.: 'Radiometric observations of sea temperature at 2.65 GHz over Cheasapeake Bay', *IEEE Trans. Antennas and Propagat.*, 1977, **AP-25**, pp. 121–8

BOHREN, C.F.: 'Multiple scattering of light and some of its observable consequences', *Am. J. Phys.*, 1987, **55**, pp. 524–33

BOHREN, C.F., and BARKSTROM, B.R.: 'Theory of the optical properties of snow', *J. Geophys. Res.*, 1974, **79**, pp. 4527–35

BOUTIN, J., OBLIGIS, E., and DINNAT, E.: 'Influence of surface roughness on Tb simulated in L-band by Yueh-LODYC emissivity model and by UCL model – Analyse of the differences', in 'Scientific requirements and impact of space observation of ocean salinity for modeling and climate studies: final report'. NERSC technical report n°.214 under contract n°.14273/00/NL/DC, European Space Agency, 2002

BOUTIN, J., WALDTEUFEL, P., MARTIN, N., CAUDAL, G., and DINNAT, E.: 'Salinity retrieved from SMOS measurements over global ocean: Imprecisions due to surface roughness and temperature uncertainties', *J. Atm. Ocean. Technol.*, 2004, **21**, pp. 1432–47

BOUTTIER, F., and KELLY, G.: 'Observing system experiments in the ECMWF 4D-var data assimilation system', *Q.J.R. Meteorol. Soc.*, 2001, **127**, pp. 1469–88

BOYARSKII, D. A., and TIKHONOV, V. V.: 'The influence of stratigraphy on microwave radiation from natural snow cover', *J. Electromag. Waves Appl.*, 2000, **14**, pp. 1265–85

BROSCHAT, S. L.: 'The small slope approximation reflection coefficient for scattering from a Pierson–Moskowitz sea surface', *IEEE Trans. Geosci. Remote Sens.*, 1993, **31**, pp. 1112–14

BRUNFELDT, D. R., and ULABY, F. T.: 'Measured microwave emission and scattering in vegetation canopies', *IEEE Trans. Geosci. Remote Sens.*, 1984, **22**, pp. 520–4

BRUNFELDT, D. R., and ULABY, F. T.: 'Microwave emission from row crops', *IEEE Trans. Geosci. Remote Sens.*, 1986, **24**, pp. 353–9

BURKE, E. J., WIGNERON, J.-P., and GURNEY, R. J.: 'The comparison of two models that determine the effects of a vegetation canopy on passive microwave emission', *Hydrol. and Earth Sys. Sci.*, 1999, **3**(3), pp. 439–44

CALVET, J.-C., CHANZY, A., and WIGNERON, J.-P.: 'Surface temperature and soil moisture retrieval in the Sahel from airborne multifrequency microwave radiometry', *IEEE Trans. Geosci. Remote Sens.*, 1996, **34**, pp. 588–600

CAMPS, A. *et al.*: 'L-band sea surface emissivity: Preliminary results of the WISE-2000 campaign and its application to salinity retrieval in the SMOS mission', *Radio Sci.*, 2003, **38**(4), 8078, doi: 10.1029/2002 RS002729

CAMPS, A., CORBELLA , I., VALL-LORSERA, M. *et al.*: 'The WISE 2000 and 2001 field experiments in support of the SMOS mission: Sea surface L-band brightness temperature observations and their application to sea surface salinity retrieval', *IEEE Trans. Geosci. Remote Sens.*, 2004, **42**, pp. 804–23

CAMPS, A., VALL-LOSSERA, N., VILLARINO, R., REUL, N., CHAPRON, B., CORBELLA, I., DUTT, N., TORRES, F., MIRANDA, J., SABIA, R., MONEMIS, A., and RODRIGUEZ, R.: 'The emissivity of foam-covered water surface at L-band: theoretical modelling and experimental results from the FROG 2003 field experiment'. *IEEE Trans. Geosci. Remote Sens.*, 2005, **43**(5), pp. 925–37

CARSEY, F. D., BARRY, R. G., ROTHROCK, D. A., and WEEKS, W. F.: 'Status and future directions for sea ice remote sensing', in CARSEY, F. D. (Ed.): 'Microwave remote sensing of sea ice, geophysical monograph 68' (American Geophysical Union, Washington, DC, 1992), pp. 444–6

CAVALIERI, D. J., GLOERSEN, P., and CAMBELL, W. J.: 'Determination of sea ice parameters with the NIMBUS 7 SMMR', *J. Geophys. Res.*, 1984, **89**(D4), pp. 5355–69

CHANDRASEKHAR, S.: 'Radiative transfer' (Dover, New York, NY, 1960, reprint of first edition)

CHANG, A. T. C., GLOERSEN, P., and SCHMUGGE, T.: 'Microwave emission from snow and glacier ice', *J. Glaciol.*, 1976, **16**(1), pp. 23–39

CHANG, A. T. C., FOSTER, J. L., and HALL, D. K.: 'Nimbus-7 SMMR derived global snow cover parameters', *Ann. Glaciol.*, 1987, **9**, pp. 39–44

CHANZY, A., and WIGNERON, J.P.: 'Microwave emission from soil and vegetation', in MÄTZLER, C. (Ed.): 'Radiative transfer models for microwave radiometry' (COST Action 712 'Application of microwave radiometry to atmospheric research and monitoring', Final Report of Project 1, 2000), pp. 89–103

CHANZY, A., HABOUDANE, D., WIGNERON, J.-P., CALVET, J.-C., and GROSJEAN, O., 'Radiométrie micro-onde sur divers types de couverts végétaux: Influence de l'humidité du sol'. Proceedings of the sixth international symposium on *Physical measurements and signatures in remote sensing*, Val d'Isère, France, Jan. 17–21, pp. 505–12, 1994

CHANZY, A., RAJU, S., and WIGNERON: 'Estimation of soil microwave effective temperature at l and c bands', *IEEE Trans. Geosci. Remote Sens.*, 1997, **35**, pp. 570–80

CHAUHAN, N. S., LANG, R. H., and RANSON, K. J.: 'Radar modeling of a boreal forest', *IEEE Trans. Geosci. Remote Sens.*, 1991, **29**, pp. 627–38

CHEN, K. S., WU, T. D., and FUNG, A. K.: 'A note on the multiple scattering in an IEM model', *IEEE Trans. Geosci. Remote Sens.*, 2000, **38**, pp. 249–56

CHEN, K. S., TZONG-DAR WU, LEUNG TSANG, QIN LI, JIANCHENG SHI, and FUNG, A. K.: 'Emission of rough surfaces calculated by the integral equation method with comparison to three-dimensional moment method simulations', *IEEE Trans. Geosci. Remote Sens.*, January 2003, **41**(1), pp. 90–101

CHERNY, I. V., and RAIZER, V. YU.: '*Passive microwave remote sensing of oceans*' (John Wiley and Sons, 1998)

CHOUDHURY, B. J., and TUCKER, C. J.: 'Monitoring global vegetation using Nimbus-7 37 GHz data. Some empirical relations', *Int. J. Remote Sens.*, 1987, **8**, pp. 1085–90

CHOUDHURY, B. J., SCHMUGGE, T. J., CHANG, A., and NEWTON, R.W.: 'Effect of surface roughness on the microwave emission from soils', *J. Geophys. Res.*, 1979, **84**, pp. 5699–706

CHOUDHURY, B., SCHMUGGE, T., and MO, T.: 'A parameterization of effective soil temperature for microwave emission', *J. Geophys. Res.*, pp. 1301–4

CHUKHLANTSEV, A. A., and SHUTKO, A. M.: 'An account of the effect of vegetation during remote microwave radiometric sounding of terrestrial deposits', *Remote Sens. Earth Space*, 1998, **2**, pp. 67–72 (Engl. Transl.)

COLE, K. S., and COLE, R. H.: 'Dispersion and absorption in dielectrics', *J. Chem. Phys.*, 1941, **9**, 341–51

COMISO, J.C.: 'Characteristics of arctic winter sea ice from satellite multispectral microwave observations', *J. Geophys. Res.*, 1986, **91**(C1), pp. 975–94

COMISO, J.C., CAVALIERI, D. J., PARKINSON, C. L., and GLOERSEN, P.: 'Passive microwave algorithms for sea ice concentration: A comparison of two techniques', *Remote Sens. Environ.*, **60**, pp. 357–84

COX, C., and MUNK, W.: 'Measurement of the roughness of the sea surface from photographs of the sun's glitter', *J. Opt. Soc. Am.*, 1954, **44**, pp. 838–50

DE JEU, R.: 'Retrieval of land surface parameters using passive microwave remote sensing'. PhD thesis, Faculty of Earth and Life Sciences, Vrije Universiteit Amsterdam, the Netherlands, 2003, 122pp

DE ROSNAY, P., KERR, Y., CALVET, J., ESCORIHUELA, L. F. M., and WIGNERON, J.: 'SMOSREX: surface monitoring of the soil reservoir experiment'. Proceedings of *MICRORAD'04*, Rome (Italy), 24–27 February 2004

DE ROSNAY, P., CALVET, S.-C., KERR, Y. *et al.*: 'SMOREX: a long term field campaign experiment for soil moisture and land surface processes remote sensing' Submitted to Remote Sensing of Environment, 2005

DE SÈVE, D., BERNIER, M., FORTIN, J.-P., and WALKER, A.: 'Preliminary analysis of snow microwave radiometry using the SSM/I passive-microwave data: the case of La Grande River watershed (Quebec)', *Ann. Glaciol.*, 1997, **25**, pp. 353–61

DEBLONDE, G., and ENGLISH, S. J.: 'Evaluation of the Fastem2 fast microwave ocean surface emissivity model'. Proceedings of the international *TOVS* study conference, Budapest, 20–26 September 2000, LEMARSHALL, J. F., and JASPER, J. D. (Eds): (Published by Bureau of Meteorology Research Centre, PO Box 1289 K, GPO Melbourne, Vic 3001, Australia) pp. 67–78

DEBYE, P.: 'Polar molecules' (Chemical Catalogue Company, New York, 1929)

DELAHAYE, J.-Y., GOLÉ, P., and WALDTEUFEL, P.: 'Calibration error of L-band sky-looking ground-based radiometers', *Radio Sci.*, 2002, **37**, pp. 11-1–11-11

DINNAT, E. P.: 'De la détermination de la salinité de surface des océans à partir de mesures radiométriques hyperfréquences en bande L'. PhD thesis, Université Pierre et Marie Curie, Paris, France, 2003

DINNAT, E. P., and DRINKWATER, M. R.: 'Optimizing the active/passive synergy in the frame of sea surface salinity retrieval from microwave measurements at L-band'. Proceedings of *MicroRad'04*, Rome, Italy, 2004.

DINNAT, E., ETCHETO, J., BOUTIN, J. *et al.*: 'Sea state influence on L-band emissivity in various fetch conditions'. *Geoscience Remote Sensing* Symposium on *IGARSS'02, 2002 IEEE Int.*, 2002a, **6**, pp. 3632–34 t

DINNAT, E. P., BOUTIN, J., CAUDAL, G., ETCHETO, J., and WALDTEUFEL, P.: 'Influence of sea surface emissivity model parameters at L-band for the estimation of salinity', *Int. J. Remote Sens.*, 2002b, **23**, pp. 5117–22

DINNAT, E. P., BOUTIN, J., CAUDAL, G., and ETCHETO, J.: 'Issues concerning the sea emissivity modeling at L-band for retrieving surface salinity', *Radio Sci.*, 2003a, **38**, pp. 25-1–25-11

DINNAT, E. P., BOUTIN, J., CAUDAL, G., ETCHETO, J., and CONTARDO, S.: 'On the use of EuroSTARRS and WISE data for validating L-band emissivity models'. First Results Workshop, EuroSTARRS/WISE/LOSAC Campaigns, ESA SP-525, pp. 117–24, ESTEC/European Space Agency, 2003b

DJERMAKOYE, B., and KONG, J. A.: 'Radiative-transfer theory for the remote sensing of layered random media', *J. Appl. Phys.*, 1979, **50**, pp. 6600–4

DOBSON, M. C., ULABY, F. T., HALLIKAINEN, M. T., and EL-RAYES, M. A.: 'Microwave dielectric behavior of wet soil – Part II: Dielectric mixing models', *IEEE Trans. Geosci. Remote Sens.*, 1985, **23**, pp. 35–45

DONELAN, M. A., and PIERSON, Jr., W. J.: 'Radar scattering and equilibrium ranges in wind-generated waves with applications to scatterometry', *J. Geophys. Res.*, 1987, **92**, pp. 4971–5029

DONELAN, M. A., DOBSON, F. W., SMITH, S. D., and ANDERSON, R. J.: 'On the dependence of sea surface roughness on wave development', *J. Phys. Ocean.*, 1993, **23**, pp. 2143–9

DRINKWATER, M. R., HOSSEINMOSTAFA, R., and GOGINENI, P.: 'C-band backscatter measurements of winter sea ice in the Weddell Sea, Antarctica', *Int. J. Remote Sens.*, 1995, **16**(17), pp. 3365–89

DROPPLEMAN, J. D.: 'Apparent microwave emissivity of sea foam', *J. Geophys. Res.*, 1970, **75**, pp. 696–7

DURAND, Y., BRUN, E., MÉRINDOL, L., GUYOMARC'H, G., LESAFFRE, B., and MARTIN, E.: 'A meteorological estimation of relevant parameters for snow models', *Ann. Glaciol.*, 1993, **18**, pp. 65–71

DURDEN, S. L., and VESECKY, J. F.: 'A physical radar cross-section model for a wind-driven sea with swell', *IEEE J. Ocean. Eng.*, 1985, **OE-10**, pp. 445–51

DURDEN, S. L., VAN ZYL, J., and ZEBKER, H. A.: 'Model and observation of the radar polarization signature of forested areas', *IEEE Trans. Geosci. Remote Sens.*, 1989, **27**, pp. 290–301

EL-RAYES, M. A., and ULABY, F. T.: 'Microwave dielectric spectrum of vegetation – Part I: Experimental observations', *IEEE Trans. Geosci. Remote Sens.*, 1987, **25**, pp. 541–49

ELFOUHAILY, T., CHAPRON, B., KATSAROS, K., and VANDEMARK, D.: 'A unified directional spectrum for long and short wind-driven waves', *J. Geophys. Res.*, 1997, **102**, pp. 15,781–96

ELLISON, W. J., BALANA, A., and DELBOS, G.: 'Study and measurement of the dielectric properties of sea water'. Technical Report 11197/94/NL/CN, ESTEC/European Space Agency, 1996

ELLISON, W., BALANA, A., DELBOS, G. *et al.*: 'New permittivity measurements of sea water', *Radio Sci.*, 1998, **33**, pp. 639–48

EMERY, W. J., FOWLER, C., and MASLANIK, J.: 'Arctic sea ice concentrations from special sensor microwave imager and advanced very high resolution radiometer satellite data', *J. Geophys. Res.*, **99**(C9), pp. 18329–42

ENGLISH, S. J., JUPP, A., CANDY, B., BEBBINGTON, D., and HOLT, A.: 'Capability and requirements for polarimetric radiometry in numerical weather prediction for surface wind analysis'. ESA report contract no. CAO/CB/cb/02.1025, 2003

EOM, H. J., and FUNG, A. K.: 'A scatter model for vegetation up to Ku-band', *Remote Sens. Environ.*, 1984, **15**, pp. 185–200

EPPLER, D. T., FARMER, L. D., LOHANICK, A. W. *et al.*: 'Passive microwave signatures of sea ice', in CARSEY, F. D. (Ed.): 'Microwave remote sensing of sea ice, Geophysical monograph 68' (American Geophysical Union, Washington, DC, 1992), pp. 47–71

ESCORIHUELA, M. J., KERR, Y., WIGNERON, J.-P., ROSNAY, P., CALVET, J.-C., and LEMAÎTRE, F.: 5th SMOS Workshop, ESA-ESRIN, Frascati, Italy, November 29–December 1, 2004b

ETCHETO, J., DINNAT, E., BOUTIN, J. *et al.*: 'Wind speed effect on L-band brightness temperature inferred from EuroSTARRS and WISE 2001 field experiments', *IEEE Trans. Geosci. Remote Sens.*, 2004, **42**, pp. 2206–13

FERRAZZOLI, P., and GUERRIERO, L.: 'Radar sensitivity to tree geometry and woody volume: A model analysis', *IEEE Trans. Geosci. Remote Sens.*, 1995, **33**, pp. 360–71

FERRAZZOLI, P., and GUERRIERO, L.: 'Passive microwave remote sensing of forests: A model investigation', *IEEE Trans. Geosci. Remote Sens.*, 1996a, **34**, pp. 433–43

FERRAZZOLI, P., and GUERRIERO, L.: 'Emissivity of vegetation: Theory and computational aspects', *J. Electromagn. Waves Applicat.*, 1996b, **10**, pp. 609–28

FERRAZZOLI, P., GUERRIERO, L., and WIGNERON, J.-P.: 'Simulating L-band emission of forests in view of future satellite applications', *IEEE Trans. Geosci. Remote Sens.*, 2002, **40**, pp. 2700–8

FIERZ, C.: 'Field observations and modelling of weak-layer evolution', *Ann. Glaciol.*, 1998, **26**, pp. 7–13

FÖHN, P. M., CAMPONOVO, C., and KRÜSI, C.: 'Mechanical and structural properties of weak snow layers measured in situ', *Ann. Glaciol.*, 1998, **26**, pp. 1–6

Food and Agriculture Organization of the United Nations – United Nations Educational, Scientific and Cultural Organization, Soil map of the world, scale 1:5M, UNESCO, Paris, 1977

FOSTER, J. L., RANGO, A., HALL, D. K., CHANG, A. T. C., ALLISON, L. J., and DIESEN, B. C.: 'Snowpack monitoring in North America and Eurasia using passive microwave satellite data', *Rem. Sens. Environ.*, **10**, pp. 285–98

FRANCHOIS, A., PIÑEIRO, Y., and LANG, R. H.: 'Microwave permittivity measurements of two conifers', *IEEE Trans. Geosci. Remote Sens.*, 1998, **36**, pp. 1384–94

FUHRHOP, R., SIMMER, C., SCHRADER, M., HEYGSTER, G., JOHNSEN, K.-P., and SCHÜSSEL, P.: 'Study of passive remote sensing of the atmosphere and surface ice', (Berichte aus dem Institut für Meereskunde an der Christian-Albrechts Universität), *Kiel*, 1997, **297**, p. 241

FUNG, A. K.: 'Microwave scattering and emission models and their applications' (Artech House, Norwood, USA, 1994)

FUNG, A. K., and CHEN, K. S.: 'An update on the IEM surface backscattering model', *IEEE Trans. Geosci. Remote Sens. Lett.*, April 2004, **1**(2)

GABARRÓ, C., FONT, J., CAMPS, A., VALL-LLOSSERA, M., and JULIÀ, A.: 'A new empirical model of sea surface microwave emissivity for salinity remote sensing', *Geophys. Res. Lett.*, 2004a, **31**, p. L01,309

GABARRÓ, C., VALL-LLOSSERA, M., FONT, J., and CAMPS, A.: 'Determination of sea surface salinity and wind speed by L-band microwave radiometry from a fixed platform', *Int. J. Remote Sens.*, 2004b, **25**, pp. 111–28

GALANTOWICZ, J. F., and ENGLAND, A. W.: 'Seasonal snowpack radiobrightness interpretation using a SVAT-linked emission model', *J. Geophys. Res.*, 1997, **102**(D18), pp. 21,933–46

GARRITY, C.: 'Characterisation of snow on floating ice and case studies of brightness temperature changes during the onset of melt', in CARSEY, F. D. (Ed.): 'Microwave remote sensing of sea ice, Geophysical monograph 68' (American Geophysical Union, Washington, DC), 1992, pp. 313–28

Geological Map of Egypt Egyptian Geological Survey and Mining Authority, Geological Map of Egypt, 1:2,000,000, 1981

GLOERSEN, P., WILHEIT, T. T., CHANG, A. T. C., NORDBERG, W., and CAMPBELL, W. J.: 'Microwave maps of the polar ice of the earth', *Bull. Am. Meteo. Soc.*, 1974, **55**, pp. 1442–8

GOODISON, B. E., and WALKER, A. E.: 'Canadian development and use of snow cover information from passive microwave satellite data', in CHOUDHURY, B. J., KERR, Y. H., NJOKU, E. G., and PAMPALONI, P. (Eds.): 'Passive microwave remote sensing of land–atmosphere interactions' (VSP, Utrecht, The Netherlands, 1995), pp. 245–262

GOGINENI, S. P., MOORE, R. K., GRENFELL, T. C., BARBER, D. G., DIGBY, S., and DRINKWATER, M.: 'The effects of freeze-up and melt processes on microwave signatures', in CARSEY, F. D. (Ed.): 'Microwave remote sensing of sea ice, Geophysical monograph 68' (American Geophysical Union, Washington, DC, 1992), pp. 329–41

GRENFELL, T., and WARREN, S. G.: 'Representation of a nonspherical ice particle by a collection of independent spheres for scattering and absorption of radiation', *J. Geophys. Res.*, 1999, **104**, 31,697–709

GUBLER, H., and HAUCK, C.: 'Preliminary measurements of BTS using a microwave radiometer, permafrost and climate in Europe (PACE)'. Project of the European Commission, ENV4-CT97-0492, WP 2, 1998

GURVICH, A. S., KALININ, V. I., and MATVEYEV, D. T.: 'Influence of the internal structure of glaciers and their thermal radio emission', *Atm. Oceanic Phys.*, 1973, **12**, pp. 713–17

HALLIKAINEN, M. T.: 'Microwave dielectric behaviour of wet soil – Part I: Empirical models and experimental observations', *IEEE Trans. Geosci. Remote Sens.*, 1985, **23**, pp. 25–33

HALLIKAINEN, M. T., and JOLMA, P. A.: 'Retrieval of the water equivalent of snow cover in Finland by satellite microwave radiometry', *IEEE Trans. Geosci. Remote Sens.*, 1986, **GE-24**, pp. 855–62

HALLIKAINEN, M., ULABY, F., DOBSON, M., and EL-RAYES, M.: 'Dielectric measurements of soils in the 3- to 37-GHz band between −50°C and 23°C'. IGARSS'84, Strasbourg, France, 1984, pp. 163–8

HALLIKAINEN, M., ULABY, F., DOBSON, M., EL-RAYES, M., and WU, L. K.: 'Microwave dielectric behavior of wet soil – Part I: Empirical models and experimental observations', *IEEE Trans. Geosci. Remote Sens.*, 1985, **GE-23**(1), pp. 25–34

HALLIKAINEN, M., ULABY, F., and DEVENTER, T.: 'Extinction behavior of dry snow in the 18- to 90-GHz range', *IEEE Trans. Geosci. Remote Sens.*, 1987, **GE-25**, pp. 737–45

HEWISON, T. J., and ENGLISH, S. J.: 'Airborne retrievals of snow and ice surface emissivity at millimeter wavelengths', *IEEE Trans. Geosci. Remote Sens.*, 1999, **37**(4), pp. 1871–79

HO, W. W., and HALL, W. F.: 'Measurements of the dielectric properties of seawater and NaCl solutions at 2.65 GHz', *J. Geophys. Res.*, 1973, **78**, pp. 6301–15

HO, W. W., LOVE, A. W., and MELLE, M. J. V.: 'Measurements of the dielectric properties of sea water at 1.43 GHz'. Technical Report CR-2458, NASA, 1974

HOFER, R., and SCHANDA, E.: 'Signatures of snow in the 5–94 GHz range', *Radio Sci.*, 1978, **13**, pp. 365–9

HOLLINGER, J. P.: 'Passive microwave measurements of sea surface roughness', *IEEE Trans. Geosci. Electron.*, 1971, **GE-9**, pp. 165–9

HOLMES, T., DE ROSNAY, P., DE JEU, R., WIGNERON, J.-P., *et al.*: 'A new parameterization of the effective temperature for L-band radiometry', *Geophys. Res. Lett.*, 2005, submitted

HORNBUCKLE, B. K., and ENGLAND, A. W.: 'Radiometric sensitivity to soil moisture at 1.4GHz through a corn crop at maximum biomass', *Water Resources Res.*, 2004, 40, W10204, doi: 10.1029/2003WR002931

HORNBUCKLE, B. K., ENGLAND, A. W., DE ROO, R. D., FISCHMAN, M. A., and BOPRIE, D. L.: 'Vegetation canopy anisotropy at 1.4 GHz', *IEEE Trans. Geosci. Remote Sens.*, 2003, **41**, pp. 2211–22

HUFFMAN, G. J.: GPCP version 1 combined precipitation dataset, SSAI, NASA Lab. for Atmos., Goddard Space Flight Cent., Greenbelt, Md., 1996

HÜPPI, R. A.: 'RASM: A microwave-scatterometer to measure microwave signatures of soil, vegetation, and snow'. PhD thesis, University of Bern, Switzerland, 1987

INGERSOLL, L., ZOBEL, O. J., and INGERSOLL, A. C.: 'Heat conduction' (McGraw-Hill, New York, 1948)

INGOLD, T., PETER, R., and KÄMPFER, N.: 'Weighted mean tropospheric temperature and transmittance determination at millimeterwave frequencies for groundbased applications', *Radio Sci.*, 1998, **33**(4), pp. 905–18

IRISOV, V. G.: 'Small-slope expansion for thermal and reflected radiation from a rough surface', *Waves Random Med.*, 1997, **7**, pp. 1–10

ISHIMARU, A.: 'Wave propagation and scattering in random media', Vol. 2 (Academic Press, New York, 1978), Equation (14-73)

JACKSON, T.: 'Measuring surface soil moisture using passive microwave remote sensing', *Hydrol. Process.*, 1993, **7**, 139–52 –76

JACKSON, T. J., and O'NEILL, P. E.: 'Attenuation of soil microwave emissivity by corn and soybeans at 1.4 and 5 GHz', *IEEE Trans. Geosci. Remote Sens.*, 1990, **28**, pp. 978–80

JACKSON, T. J., and SCHMUGGE, T. J.: 'Vegetation effects on the microwave emission of soils', *Remote Sens. Environ.*, 1991, **36**, pp. 203–12

JACKSON, T. J., CHEN, D., COSH, M. *et al.*: 'Vegetation water content mapping using Landsat data derived normalised difference water index for corn and soybeans', *Remote Sens. Environ.*, 2004, **92**, pp. 475–82

JAMES, M. E., and KALLURI, S. N. V.: 'The Pathfinder AVHRR land data set: An improved coarse resolution data set for terrestrial monitoring', *Int. J. Remote Sens.*, **15**, 1994, pp. 3347–64

DE JEU, R., HOLMES, T. *et al.*: 'Deriving land surface parameters from three different vegetated sites with the ELBARA 1.4 GHz passive microwave radiometer'. Proceedings of SPIE series: 'Remote sensing for agriculture, ecosystems and hydrology V', vol. 5232, Barcelona, Spain, 2003

JIN, Y. Q.: 'Size parameters of snow grains in scattering and emission models', in CHOUDHURY, B. J., KERR, Y. H., NJOKU, E. G., and PAMPALONI, P. (Eds.): 'Passive microwave remote sensing of land-atmosphere interactions' (VSP, Utrecht, The Netherlands, 1995, pp. 273–83), pp. 245–62

JOHNSEN, K. P., and HEYGSTER, G.: 'Emissivity modelling of sea and lake ice', in MÄTZLER, C. (Ed.): 'Radiative transfer models for microwave radiometry' (COST Action 712 'Application of microwave radiometry to atmospheric research and monitoring', Final Report of Project 1, 2000), p. 149

JOHNSON, J. T., and ZHANG, M.: 'Theoretical study of the small slope approximation for ocean polarimetric thermal emission', *IEEE Trans. Geosci. Remote Sens.*, 1999, **37**, pp. 2305–16

JOSBERGER, E. G., CAMPBELL, W. J., GLOERSEN, P., CHANG, A. T. C., and RANGO, A.: 'Snow conditions and hydrology of the upper Colorado river basin from satellite passive microwave observations', *Ann. Glaciol.*, 1993, **17**, pp. 322–6

JOSEPH, J.H., WISCOMBE, W.J., and WEINMANN, J.A.: 'The delta–Eddington approximation for radiative flux transfer', *J. Atmos. Sci.*, 1976, **33**, pp. 2452–9

JUSTICE, C. O., TOWNSHEND, J. R., and CHOUDHURY, B. J.: 'Comparison of AVHRR and SMMR data for monitoring vegetation phenology on a continental scale', *Int. J. Remote Sens.*, 1989, **10**, pp. 1607–32

KALESCHKE, L., LÜPKES, C., VIHMA, T. *et al.*: 'Sea ice remote sensing for mesoscale ocean-atmosphere interaction analysis', *Canad. J. Remote Sens.*, 2001, **27**(5), pp. 526–37

KALNAY, E., *et al.*: 'The NCEP/NCAR 40-year reanalysis project', 1996, **77**, pp. 437–70

KARAM, M. A.: 'A physical model for microwave radiometry of vegetation', *IEEE Trans. Geosci. Remote Sens.*, 1997, **35**, pp. 1045–58

KARAM, M. A., and FUNG, A. K.: 'Vector forward scattering theorem', *Radio Sci.*, 1982, **17**, pp. 752–56

KARAM, M. A., and FUNG, A. K.: 'Electromagnetic scattering from a layer of finite length, randomly oriented, dielectric, circular cylinders over a rough interface with application to vegetation', *Int. J. Remote Sens.*, 1988, **9**, pp. 1109–34

KARAM, M. A., FUNG, A. K., LANG, R. H., and CHAUHAN, N. S.: 'A microwave scattering model for layered vegetation', *IEEE Trans. Geosci. Remote Sens.*, 1992, **30**, pp. 767–84

KARAM, M. A., AMAR, F., FUNG, A. K. *et al.*: 'A microwave polarimetric scattering model for forest canopies based on vector radiative transfer theory', *Remote Sens. Environ.*, 1995, **53**, 16–30

KARBOU, F., PRIGENT, C., EYMARD, L., and PARDO, J.: 'Microwave land emissivity calculations using AMSU measurements', *IEEE Trans. Geosci. Remote Sens.*, 2005, **43**, 948–59

KASISCHKE, E. S., CHRISTENSEN, N. L., and HANEY, E. M.: 'Modeling of geometric properties of loblolly pine tree stand characteristics for use in radar backscatter studies', *IEEE Trans. Geosci. Remote Sens.*, 1994, **32**, pp. 800–22

KENNEDY, J. M., and SAKAMOTO, R. T.: 'Passive microwave determination of snow wetness factors'. Proceedings of the 4th international symposium on *Remote sensing of environment* (Ann Arbor, Michigan, 1966), pp. 161–71

KERR, Y. H., and NJOKU, E. G.: 'A semiempirical model for interpreting microwave emission from semiarid land surfaces as seen from space', *IEEE Trans. Geosci. Remote Sens.*, 1990, **28**, pp. 384–93

KERR, Y. H., and NJOKU, E. G.: 'On the use of passive microwave at 37 GHz in remote sensing of vegetation', *Int. J. Remote Sens.*, 1993, **14**, pp. 1931–43

KERR, Y. H. *et al.*: 'SMOS-MIRAS on RAMSES: Radiometry applied to soil moisture and salinity measurements'. Full proposal, a.o. earth explorer opportunity missions, ESA, 1998

KERR, Y. H., WALDTEUFEL, P., WIGNERON, J.-P., MARTINUZZI, J.-M., FONT, J., and BERGER, M.: 'Soil moisture retrieval from space: The soil moisture and ocean salinity (SMOS) mission', *IEEE Trans. Geosci. Remote Sens.*, 2001, **39**(8), pp. 1729–35

KIRDIASHEV, K. P., CHUCKLANTSEV, A. A., and SHUTKO, A.: 'Microwave radiation of the earth's surface in the presence of vegetation cover', *Radioteknica*, 1979, **24**, pp. 256–64

KLEIN, L. A., and SWIFT, C. T.: 'An improved model for the dielectric constant of sea water at microwave frequencies', *IEEE Trans. Antennas Propagat.*, 1977, **AP-25**, pp. 104–11

KLEIN, M., GASIEWSKI, A. J., CAVALIERI, D. J., and MARKUS, T.: 'Meltpond2000 polarimetric scanning radiometer sea ice brightness temperatures' (National Snow and Ice Data Center, Boulder, CO, USA, Digital Media, 2004)

KOBLINSKY, C., HILDEBRAND, P., LEVINE, D. *et al.*: 'Sea surface salinity from space: Science goals and measurement approach', *Radio Sci.*, 2003, **38**, 29 MAR pp. 1–7

KOKHANOVSKY, A. K.: 'The influence of horizontal inhomogeneity on radiative characteristics of clouds: an asymptotic case study', *IEEE Trans. Geosci. Remote Sens.*, 2003, **41**, pp. 817–25

KRUOPIS, N., PRAKS, J., ARSLAN, A. N., ALASALMI, H. M., KOSKINEN, J. T., and HALLIKAINEN, M. T.: 'Passive microwave measurements of snow-covered forest areas in EMAC'95', *IEEE Trans. Geosci. Remote Sens.*, 1999, **37**, pp. 2699–705

KUDRYAVTSEV, V. N., MAKIN, V. K., and CHAPRON, B.: 'Coupled sea surface-atmosphere model – 2. Spectrum of short wind waves', *J. Geophys. Res.*, 1999, **104**, pp. 7625–39

KÜNZI, K. F., PATIL, S., and ROTT, H.: 'Snowcover parameters retrieved from Nimbus-7 scanning multichannel microwave radiometer (SMMR) data', *IEEE Trans. Geosci. Remote Sens.*, 1982, **GE-20**, pp. 452–67

KURVONEN, L., and HALLIKAINEN, M.: 'Influence of land-cover category on brightness temperature of snow', *IEEE Trans. Geosci. Remote Sens.*, 1997, **GE-35**, pp. 367–77

LAGUERRE, L., CALVET, J.-C., KERR, Y., CHANZY, A., WIGNERON, J.-P., and RAJU, S.: 'Influence of surface roughness on the microwave emission from bare soils for surface soil moisture algorithms'. Proceedings of *Progress in electromagnetics research* symposium (PIERS'94), Noordwiik, pp. 2127–31, 1994

LANG, R. H.: 'Scattering from a layer of discrete random medium over a random interface: Application to microwave backscattering from forests', *Waves Random Med.*, 2004, **14**, pp. S359–S391

LANG, R. H., DE MATTHAEIS, P., LE VINE, D.M., BIDWELL, S., HAKEM, M.,and CHAUHAN, N.: 'L-band radiometer measurements of coniferous forests'. Proceedings of 2000 international *Geoscience and remote sensing* symposium, 2000

LANG, R. H., UTKU, C., DE MATTHAEIS, P., CHAUHAN, N., and LE VINE, D. M.: 'ESTAR model brightness temperatures over forests: Effects of soil moisture'. Proceedings of 2001 international *Geoscience and remote sensing* symposium, 2001

LE VINE, D. M., and ABRAHAM, S.: 'The effect of the ionosphere on remote sensing of sea surface salinity of space: Absorption and emission at L-band', *IEEE Trans. Geosci. Remote Sens.*, 2002, **40**, p. 771

LE VINE, D. M., and ABRAHAM, S.: 'Galactic noise and passive microwave remote sensing from space at L-band', *IEEE Trans. Geosci. Remote Sens.*, 2004, **42**, pp. 119–29

LE VINE, D. M., and KARAM, M. A.: 'Dependency of attenuation in a vegetation canopy on frequency and plant water content', *IEEE Trans. Geosci. Remote Sensing*, 1996, **34**(5), pp. 1090–94

LE VINE, D. M., MENEGHINI, R., LANG, R. H., and SEKER, S. S.: 'Scattering from arbitrarily oriented dielectric disks in the physical optics regime', *J. Opt. Soc. Am.*, 1983, **73**, pp. 1255–62

LEMAIRE, D.: 'Non-fully developed sea state characteristics from real aperture radar remote sensing'. PhD thesis, Université Catholique de Louvain, 1998

LEMAÎTRE, F., POUSSIÈRE, J., KERR, Y. *et al.*: 'Design and test of the ground based L-band radiometer for estimating water in soils (LEWIS)', *IEEE Trans. Geosci. Remote Sens.*, 2004, **42**(8), 1666

LI, Q., TSANG, L., SHI, J. C., and CHAN, C. H.: 'Application of physics based two-grid method and sparse matrix canonical grid method for numerical simulations of emissivities of soils with rough surfaces', *IEEE Trans. Geosci. Remote Sens.*, July 2000, **38**, pp. 1635–43

LIEBE, H. J., HUFFORD, G. A., and COTTON, M. G.: 'Propagation modeling of moist air and suspended water/ice particles at frequencies below 1000 GHz'. AGARD 52nd specialists meeting of the electromagnetic wave propagation panel (Palma de Mallorca, Spain, 1993), pp. 3-1–3-10

LIN, YI-CHENG, and SARABANDI, K.: 'A Monte Carlo coherent scattering model for forest canopies using fractal-generated trees', *IEEE Trans. Geosci. Remote Sens.*, 1999, **37**(1), pp. 440–51

LIOU, Y.-A., CHEN, K.-S., and WU, T.-D.: 'Reanalysis of L-band brightness predicted by the LSP/R model-for prairie grassland: Incorporation of rough surface scattering', *IEEE Trans. Geosci. Remote Sens.*, 2001, **39**(1), pp. 129–35

LIU, Q. H., and WENG, F. Z.: 'Retrieval of sea surface wind vector from simulated satellite microwave polarimetric measurements', *Radio Sci.*, 2003, 38(4), 8078, doi: 10.1029/2002RS002729

LÓPEZ-BAEZA, E. *et al.*: 'EuroSTARRS: Campaign description', in First Results Workshop, EuroSTARRS/WISE/LOSAC Campaigns, ESA SP-525, pp. 37–47, ESTEC/European Space Agency, 2003

MACELLONI, G., PALOSCIA, S., PAMPALONI, P., and RUISI, R.: 'Airborne multi-frequency L- to Ka-band radiometric measurements over forests', *IEEE Trans. Geosci. Remote Sens.*, 2001, **39**, pp. 2507–13

MACELLONI, G., PALOSCIA, S., PAMPALONI, P., RUISI, R., and SANTI, E.: 'Microwave radiometric features of mediterranean forests: Seasonal variations'. Proceedings of 2003 international *Geoscience and remote sensing* symposium, 2003

MATTHEWS, E.: 'Global vegetation and land use: New high-resolution data bases for climate studies', *J. Clim. Appl. Meteorol.*, 1983, **22**, 474–86

MÄTZLER, C.: 'Seasonal evolution of microwave radiation from an oat field', *Remote Sens. Environ.*, 1990, **31**, pp. 161–73

MÄTZLER, C.: 'Applications of the interaction of microwaves with the natural snow cover', *Remote Sens. Rev.*, 1987, **2**(2), pp. 259–387

MÄTZLER, C.: 'Groundbased observations of atmospheric radiation at 5, 10, 21, 35, and 94 GHz', *Radio Sci.*, 1992, **27**(3), pp. 403–15

MÄTZLER, C.: 'Microwave (1–100 GHz) dielectric model of leaves', *IEEE Trans. Geosci. Remote Sens.*, 1994a, **32**, pp. 947–9

MÄTZLER, C.: 'Microwave transmissivity of a forest canopy: Experiments made with a beech', *Remote Sens. Environ.*, 1994b, **48**, pp. 172–80

MÄTZLER, C.: 'Passive microwave signatures of landscapes in winter', *Meteorol. Atm. Phys.*, 1994c, **54**, pp. 241–60

MATZLER, C.: 'Microwave permittivity of dry sand', *IEEE Trans. Geosci. Remote Sens.*, 1998, **36**, 317–9

MÄTZLER, C.: 'Improved Born approximation for scattering of radiation in a granular medium', *J. Appl. Phys.*, 1998a, **83**(11), pp. 6111–7

MÄTZLER, C.: 'Microwave properties of ice and snow', in SCHMITT, B. *et al.* (Eds.): 'Solar system ices' (Kluwer Academic Publishers, 1998b), pp. 241–57

MÄTZLER, C.: 'Relation between grain-size and correlation length of snow', *J. Glaciol.*, 2002, **48**(162), pp. 461–66

MÄTZLER, C. (Ed.): 'Radiative transfer models for microwave radiometry'. COST Action 712 'Application of microwave radiometry to atmospheric research and monitoring', Metrology, Final Report of Project 1, European Commission, Directorate General for Research, EUR 19543, ISBN 92-828-9842-3 (2000)

MÄTZLER, C.: 'COST Action 712: Microwave Radiometry', in MARZANO, F. S., and VISCONTI, G. (Eds.): 'Remote sensing of atmosphere and ocean from space: Models, instruments and techniques', *Adv. Global Change Res.*, 2002a, **13**, pp. 231–46 (Kluwer Acad. Publ., Dordrecht)

MÄTZLER, C.: 'Relation between grain size and correlation length of snow', *J. Glaciol.*, 2002b, **48**(162), pp. 461–6

MÄTZLER, C., and STANDLEY, A.: 'Relief effects for passive microwave remote sensing', *J. Geophys. Res.*, 2000, **21**, pp. 2403–12

MÄTZLER, C., and SUME, A.: 'Microwave radiometry of leaves' in PAMPALONI, P. (Ed.): 'Microwave radiometry and remote sensing applications' (VSP Utrecht, 1989), pp. 133–48

MÄTZLER, C., and WIESMANN, A.: 'Extension of the microwave emission model of layered snowpacks to coarse-grained snow', *Remote Sens. Environ.*, 1999, **70**, pp. 317–25

MÄTZLER, C., RAMSEIER, R. O., and SVENDSEN, E.: 'Polarisation effects in sea ice signatures', *IEEE J. Ocean. Eng.*, 1984, **OE-9**(5), pp. 333–8

MÄTZLER, C., WIESMANN, A., PULLIAINEN, J., and HALLIKAINEN, M.: 'Development of microwave emission models of snowpacks', in MÄTZLER, C.

(Ed.): 'Radiative transfer models for microwave radiometry' (European Commission COST Action 712, final report, 2000), pp. 105–16

MÄTZLER, CH., WEBER, D. *et al.*: 'ELBARA, the ETH L-band radiometer for soil moisture research'. Proceedings of the international *Geoscience and remote sensing* symposium, Toulouse, France, 2003

MCDONALD, M., and WHITT, M. C. DOBSON: 'Michigan microwave canopy scattering model', *Int. J. Remote Sens.*, 1990, **11**, pp. 1223–53

MCFARLAND, M. J., WILKE, G. D., and HARDER, P. H.: 'Nimbus 7 SMMR investigation of snowpack properties in the Northern Great Plains for the Winter 1978-1979', *IEEE Trans. Geosci. Remote Sens.*, 1987, **GE-25**, pp. 35–45

MEADOR, W.E., and WEAVER, W.R.: 'Two-stream approximations to radiative transfer in planetary atmospheres: A unified description of existing methods and a new improvement', *J. Atm. Sci.*, 1980, **37**, pp. 630–43

MIRANDA, J., VALL-LLOSSERA, M., CAMPS, A., DUFFO, N., CORBELLA, I., and ETCHETO, J.: 'Sea state effect on the sea surface emissivity at L-band', *IEEE Trans. Geosci. Remote Sens.*, 2003, **41**, pp. 2307–15

MO, T., CHOUDHURY, B. J., SCHMUGGE, T. J., WANG, J. R., and JACKSON, T. J.: 'A model for microwave emission from vegetation-covered fields', *J. Geophys. Res.*, 1982, **87**, pp. 11.229–37

MO, T., and SCHMUGGE, T. J.: 'A parameterization of the effect of surface roughness on microwave emission', *IEEE Trans. Geosci. Remote Sens.*, 1987, **25**, pp. 47–54

MONAHAN, E. C., and LU, M.: 'Acoustically relevant bubble assemblages and their dependence on meteorological parameters', *IEEE J. Ocean. Eng.*, 1990, **15**, pp. 340–9

NGHIEM, S. V., KWOK, R., YUEH, S. H., and DRINKWATER, M. R.: 'Polarimetric signatures of sea ice – 1. Theoretical model', *J. Geophys. Res.*, 1995, **100**(C7), pp. 13665–79

NJOKU, E. G., and ENTEKHABI, D.: 'Passive microwave remote sensing of soil moisture', *J. Hydrol.*, 1996, **184**, pp. 101–29

NJOKU, E. G., and LI, L.: 'Retrieval of land surface parameters using passive microwave measurements at 6–18 GHz', *IEEE Trans. Geosci. Remote Sens.*, 1999, **37**(1), pp. 79–93

NJOKU, E. T., KOIKE, T., JACKSON, T. J., and PALOSCIA, S.: 'Retrieval of soil moisture from AMSR data', in PAMPALONI, P., and PALOSCIA, S. (Eds.): 'Microwave radiometry and remote sensing of the earth's surface and atmosphere' (VSP Publications, Utrecht, The Netherlands, 2000), pp. 525–33

NORDBERG, W., CONAWAY, J., ROSS, D. B., and WILHEIT, T.: 'Measurements of microwave emission from a foam-covered, wind-driven sea', *J. Atm. Sci.*, **28**, 1971, pp. 429–35

OELKE, C.: 'Atmospheric signatures in sea ice concentration estimates from passive microwaves: Modelled and observed', *Int. J. Remote Sens.*, 1997, **18**(5), pp. 1113–36

OWE, M., DE JEU, R., and WALKER, J.: 'A methodology for surface soil moisture and vegetation optical depth retrieval using the microwave polarization difference index', *IEEE Trans. Geosci. Remote Sens.*, 2001, **39**(8), pp. 1643–54

PAMPALONI, P.: 'Microwave radiometry of forests', *Waves Rand. Media*, 2004, **14**, pp. S275–98

PAMPALONI, P., and PALOSCIA, S.: 'Microwave emission and plant water content: A comparison between field measurements and theory', *IEEE Trans. Geosci. Remote Sens.*, 1986, **24**, pp. 900–5

PARDÉ, M., WIGNERON, J.-P., CHANZY, A., WALDTEUFEL, P., KERR, Y., and HUET, S.: 'Retrieving surface soil moisture over a wheat field: Comparison of different methods', *Remote Sens. Environ.*, 2003, **87**, pp. 334–44

PARDÉ, M., WIGNERON, J.-P., CHANZY, A. *et al.*: 'N-parameter retrievals from L-band microwave measurements over a variety of agricultural crops', *IEEE Trans. Geosci. Remote Sens.*, 2004, **42**(6), pp. 1168–78

PARKINSON, C. L., and CAVALIERI, D. J.: 'Arctic sea ice 1973–1987: Seasonal, regional, and interannual variability', *J. Geophys. Res.*, 1989, **94**(C10), pp. 14499–523

PEAKE, W. H.: 'Interaction of electromagnetic waves with some natural surfaces', *IRE Trans. Antennas Propag.*, 1959, **AP-7**, pp. S324–9 (special supplement)

PEDERSEN, L. T.: in SANDVEN *et al.* 'IMSI report no. 8. Development of new satellite ice data products'. NERSC technical report no. 145, Nansen Environmental and Remote Sensing Center, Bergen, Norway, 1998

PELLARIN, T., CALVET, J.-C., and WIGNERON, J.-P.: 'Surface soil moisture retrieval from L-band radiometry: A global regression study', *IEEE Trans. Geosci. Remote Sens.*, September 2003, **41**(9), pp. 2037–51

PELLARIN, T., WIGNERON, J.-P., CALVET, J.-C. *et al.*: 'Two-year global simulation of L-band brightness temperatures over land', *IEEE Trans. Geosci. Remote Sens.*, September 2003a, **41**(9), pp. 2135–9

PELLARIN, T., WIGNERON, J.-P., CALVET, J.-C., and WALDTEUFEL, P.: 'Global soil moisture retrieval from a synthetic L-band brightness temperature data set', *J. Geophys. Res.*, 2003b, **108**(D12), p. 4364, doi:10.1029/2002JD003086

PHILLIPS, O. M.: 'The dynamics of the upper ocean' (Cambridge University Press, New York, 1977)

PIERSON, W. J., Jr., and MOSKOWITZ, L.: 'A proposed spectral form for fully developed wind seas based on the similarity theory of S. A. Kitaigorodskii', *J. Geophys. Res.*, 1964, **69**, pp. 5181–90

PRIGENT, C., and ABBA, P.: 'Sea surface equivalent brightness temperature at millimeter wavelengths', *Ann. Geophys.*, 1990, **8**, pp. 627–34

PRIGENT C., ROSSOW, W. B., and MATTHEWS, E.: 'Microwave land surface emissivities estimated from SSM/I observations', *J. Geophys. Res.*, 1997, **102**, 21,867-21

PRIGENT, C., ROSSOW, W. B., and MATTHEWS, E.: 'Global maps of microwave land surface emissivities: Potential for land surface characterization', *Radio Sci.*, 1998, **33**, 745–51

PRIGENT, C., ROSSOW, W. B., MATTHEWS, E., and MARTICORENA, B.: 'Microwave radiometric signatures of different surface types in deserts', *J. Geophys. Res.*, 1999, **104**, 12,147–58.

PRIGENT, C., WIGNERON, J.-P., ROSSOW, W. B., and PARDO-CARRION, J. R.: 'Frequency and angular variations of land surface microwave emissivities: Can we estimate SSM/T and AMSU emissivities from SSM/I emissivities ?', *IEEE Trans. Geosci. Remote Sens.*, 2000, **38**, pp. 2373–86

PRIGENT, C., AIRES, F., ROSSOW, W. B., and MATTHEWS, E.: 'Joint characterization of vegetation by satellite observations from visible to microwave wavelength: A sensitivity analysis', *J. Geophys. Res.*, 2001a, **106**, pp. 20665–85

PRIGENT, C., MATTHEWS, E., AIRES, F., and ROSSOW, W. B.: 'Remote sensing of global wetland dynamics with multiple satellite data sets', *Geophys. Res. Let.*, 2001b, **28**, pp. 4631–4

PRIGENT, C., MUNIER, J. M., THOMAS, B., and RUFFIE, G.: 'Microwave signatures over carbonate sedimentary platforms in arid areas: potential geographical applications of passive microwave observations, *Geophys. Res. Lett.*, 2005a, **32**, doi:10.1029/2005GL024691

PRIGENT, C., AIRES, F., ROSSOW, W. B., and ROBOCK, A.: 'Sensitivity of satellite microwave and infrared observations to soil moisture at a global scale: Relationship of satellite observations to in situ soil moisture measurements', *J. Geophys. Res.*, 2005a, **110**, p. D07110, doi:10.1029/2004JD005087

PRIGENT, C., CHEVALLIER, F., KARBOU, F., BAUER, P., and KELLY, G.: 'AMSU-A land surface emissivity estimation for numerical weather prediction assimilation schemes', *J. Appl. Meteorol.*, 2005b, **44**(4), pp. 416–26.

PULLIAINEN, J., and HALLIKAINEN, M.: 'Retrieval of regional snow water equivalent from space-borne passive microwave observations', *Remote Sens. Environ.*, 2001, **75**, pp. 76–85

PULLIAINEN, J., KÄRNÄ, J-P., and HALLIKAINEN, M.: 'Development of geophysical retrieval algorithms for the MIMR', *IEEE Trans. Geosci. Remote Sens.*, 1993, **GE-31**, pp. 268–77

PULLIAINEN, J., TIGERSTEDT, K., HUINING, W. *et al.*: 'Retrieval of geophysical parameters with integrated modeling of land surfaces and atmosphere (models/inversion algorithms)'. Final report, ESA/ESTEC, Noordwijk, Netherlands, contract no. 11706/95/NL/NB(SC), Oct. 1998

PULLIAINEN, J., GRANDELL, J., and HALLIKAINEN, M.: 'HUT snow emission model and its applicability to snow water equivalent retrieval', *IEEE Trans. Geosci. Remote Sens.*, 1999, **37**, pp. 1378–90

RAJU, S., CHANZY, A., WIGNERON, J., CALVET, J., KERR, Y., and LAGUERRE, L.: 'Soil moisture and temperature profile effect on microwave emission at low frequencies', *Remote Sens. Environ.*, 1995, **54**, pp. 85–97

RAMSEIER, R. O.: 'Sea ice validation', in HOLLINGER, J. P. (Ed.): 'DMSP special sensor microwave/imager calibration/validation – Final report volume II' (Naval Research Laboratory, Washington, DC, 1991)

RANGO, A., CHANG, A. T. C., and FOSTER, J. L.: 'The utilization of spaceborne microwave radiometers for monitoring snowpack properties', *Nord. Hydrol.*, 1979, **10**, pp. 25–40

REBER, B., MÄTZLER, C., and SCHANDA, E.: 'Microwave signatures of snow crusts: Modelling and measurements', *Int. J. Remote Sens.*, **8**, pp. 1649–65

REUL, N., and CHAPRON, B.: 'WP1300, Effects of foam on the emissivity of sea surface at L-band'. Technical Report 14273/00/NL/DC, ESTEC/European Space Agency, 2002

REUL, N., and CHAPRON, B.: 'A model of sea-foam thickness distribution for passive microwave remote sensing applications', *J. Geophys. Res.*, 2003, **108**, pp. 19-1–19-14

RICE, S. O.: 'Reflection of electromagnetic waves from slightly rough surfaces', *Commun. Pure Appl. Math.*, 1951, **4**, pp. 351–78

RICHARDS, J. A., SUN, G. Q., and SIMONETT, D. S.: 'L-band radar backscattering modeling of forest stands', *IEEE Trans. Geosci. Remote Sens.*, 1987, **25**, pp. 487–98

ROBOCK, A., VINNIKOV, K. Y., SRINIVASAN, G. *et al.*: 'The global soil moisture data bank', *Bull. Am. Meteorol. Soc.*, 2000, **81**, 1281–99

ROTT, H.: 'Multispectral microwave signatures of the Antarctic ice sheet', in PAMPALONI, P. (Ed.): 'Microwave radiometry and remote sensing applications' (VSP, Utrecht, the Netherlands, 1989), pp. 89–101

ROTT, H., and NAGLER, T., 'Intercomparison of snow retrieval algorithms by means of spaceborne microwave radiometry', in CHOUDHURY, B. J., KERR, Y. H., NJOKU, E. G., and PAMPALONI, P. (Eds.): 'Passive microwave remote sensing of land-atmosphere interactions' (VSP, Utrecht, The Netherlands, 1995), pp. 245–62

ROTT, H., STURM, K., and MILLER, H., 'Active and passive microwave signatures of Antarctic firn by means of field measurements and satellite data', *Ann. Glaciol.*, 1993, **17**, pp. 337–43

ROSE, L. A., ASHER, W. E., REISING, S. C. *et al.*: 'Radiometric measurements of the microwave emissivity of foam', *IEEE Trans. Geosci. Remote Sens.*, 2002, **40**(12), pp. 2619–25

ROSSOW, W. B., and SCHIFFER, R. A.: 'Advances in understanding clouds from ISCCP', *Am. Meteorol. Soc.*, 1999, **80**, 2261–87.

SALEH, K., WIGNERON, J. P., CALVET, J. C. *et al.*: 'The EuroSTARRS airborne campaign in support of the SMOS mission: First results over land surfaces', *Int. J. Remote Sens.*, 2004a, **25**, pp. 177–94

SALEH, K., GUERRIERO, L., DELLA VECCHIA, A., FERRAZZOLI, P., and WIGNERON, J.-P.: 'A radiative model to simulate forest emission at L-band: Sensitivity of brightness temperature to forest components'. Proceedings of 2004 international *Geoscience and remote sensing* symposium, 2004b

SALEH, K., WIGNERON, J.-P., CALVET, J.-C. *et al.*: 'Statistical methods to estimate soil moisture from Lband radiometry: Application to the SMOSREX experiment over a fallow site'. Proceedings of IEEE international *Geoscience*

Remote Sensing symposium (IGARSS'04), 20–24 Sept., Anchorage, Alaska, USA, 2004c

SCHANDA, E.: (Springer-Verlag, Berlin, Heidelberg, 1986, 1st edition)

SCHIFFER, R., and THIELHEIM, K. O.: 'Light scattering by dielectric needles and disks', *J. Appl. Phys.*, 1979, **50**, pp. 2476–83

SCHNEEBERGER, K., and WILLNEFF, J.: 'Digital surface models for the estimation of hydraulic properties of soils', Videometrics VIII (EI14), Proceedings of SPIE, Vol. 5013, January 20–24, 2003, Santa Clara, California.

SCHNEEBERGER, K., SCHWANK, M., STAMM, CH., DE ROSNAY, P., MÄTZLER, CH., and FLÜHLER, H.: 'Topsoil structure influencing soil water retrieval by microwave radiometry', *Vadose Zone J.*, 2004, **3**, pp. 1169–79

SCHNEEBERGER, K., STAMM, CH., MÄTZLER, CH., and FLÜHLER, H.: 'Ground-based dual-frequency radiometry of bare soil at high temporal resolution', *IEEE Trans. Geosci. Remote Sens.*, 2004b, **42**(3), pp. 588–95

SCHWANK, M., STÄHLI, M. *et al.*: 'Microwave L-band emission of freezing soil', *IEEE Trans. Geosci. Remote Sens.*, 2004, **42**(6), pp. 1252–61

SCHWANK, M., MÄTZLER, C., GUGLIELMETTI, M., and FLÜHLER, H.: 'L-band radiometer measurements of soil water under growing clover grass', *IEEE Trans. Geosci. Remote Sensing*, 2005, **43**(10), pp. 2225–37

SELBACH, N.: 'Determination of total water vapour and surface emissivity of sea ice at 89GHz, 157GHz and 183GHz in the Arctic winter', PhD thesis, *Berichte aus dem Institut für Umweltphysik*. Berlin: Logos-Verlag, **21**, 2003, p. 191

SHI, J., JIANG, L., ZHANG, L., CHEN, K.-S., WIGNERON, J.-P., and CHANZY, A.: 'A parameterized multifrequency-polarization surface emission model', *IEEE Trans. Geosc. Remote Sens.*, 2005, **43**, pp. 2831–41

SHI, J., WANG, J., HSU, A., O 'NEILL, P., and ENGMAN, E. T.: 'Estimation of bare surface soil moisture and surface roughness parameters using L-band SAR image data', *IEEE Trans. Geosci. Remote Sens.*, Sept. 1997, **35**, pp. 1254–66

SHI, J., CHEN, K. S. *et al.*: 'A parameterized surface reflectivity model and estimation of bare-surface soil moisture with L-Band radiometer', *IEEE Trans. Geosci. Remote Sens.*, 2002a, **40**(12), pp. 2674–86

SHI, J., NJOKU, E. G., CHEN, K. S., JACKSON, T., and O'NEILL, P.: 'Estimation of soil moisture change with PALS's L-band radiometer'. *Geoscience remote sensing* symposium, 2002b, IGARSS '02; *IEEE Int.*, 2002, **3**, pp. 24–28; pp. 1393–5

SHIH, S.-E., DING, K.-H., KONG, J. A., and YANG, Y. E.: 'Modeling of millimeter wave backscatter of time-varying snowcover', *PIER*, 1997, **16**, pp. 305–30

SHOKR, M. E.: 'Field observations and model calculations of dielectric properties of Arctic sea ice in the microwave C-band', *IEEE Trans. Geosci. Remote Sens.*, 1998, **36**(2), pp. 463–78

SIHVOLA, A.: 'Electromagnetic mixing formulas and applications' (The Institution of Electrical Engineers, London), 1999

SIHVOLA, A., and KONG, J. A.: 'Effective permittivity of dielectric mixtures', *IEEE Trans. Geosci. Remote Sens.*, 1988, **26**(4), pp. 420–9

SKOU, N.: 'Retrieval of wind direction over the ocean from polarimetric radiometer data'. in EYMARD, L., and MÄTZLER, C. (Eds.) COST712, Report 2 of projects 2&3, Workshop Amsterdam, 18–19 Jan. 2000, pp. 20–32

SKOU, N.: 'Faraday rotation and L band oceanographic measurements', *Radio Sci.*, 2003, **38**(4), pp. MAR 24-1

SMITH, D.M.: 'Extraction of winter sea ice concentration in the Greenland and Barents Seas from SSM/I data', *Int. J. Remote Sens.*, 1996, **17**(13), pp. 2625–46

St. GERMAIN, K., and POE, G.: 'Polarimetric emission model of the sea at microwave frequencies, Part II: Comparison with measurements' (Naval Research Laboratory Report, Washington, DC, 1998)

STEFFEN, K., and SCHWEIGER, A.: 'NASA team algorithm for sea ice concentration retrieval from defence meteorological satellite program special sensor microwave imager: Comparison with landsat satellite imagery', *J. Geophys. Res.*, 1991, **96**(C12), pp. 21971–87

STEFFEN, K., ABDALATI, W., and SHERJAL, I.: 'Faceted crystal formation in the NE-Greenland low accumulation region', *J. Glaciol.*, 1999, **45**, pp. 63–8

STILES, J. M., and SARABANDI, K.: 'Electromagnetic scattering from grassland. I. A fully phase-coherent scattering model', *IEEE Trans. Geosci. Remote Sens.*, 2000, **38**(1), pp. 339–48

STILES, W.H., and ULABY, F. T.: 'The active and passive microwave response to snow parameters, 1. Wetness', *J. Geophys. Res.*, 1980, **85**, pp. 1037–44

STILES, J. M., SARABANDI, K., and ULABY, K. T.: 'Electromagnetic scattering fron grassland – Part II: Measurement and modeling results', *IEEE Trans. Geosci. Remote Sens.*, 2000, **38**, pp. 349–56

STOGRYN, A.: 'The apparent temperature of the sea at microwave frequencies', *IEEE Trans. Antennas Propagat.*, 1967, **77**, pp. 1658–66

STOGRYN, A.: 'The emissivity of sea foam at microwave frequencies', *J. Geophys. Res.*, 1972, **77**, pp. 1658–66

STOGRYN, A.: 'Equations for the permittivity of sea water' Technical Report, GenCorp Aerojet, 1997

STOGRYN, A.: 'A study of the microwave brightness temperature of snow from the point of view of strong fluctuation theory', *IEEE Trans. Geosci. Remote Sens.*, 1986, **GE-24**(2), pp. 220–31

STOGRYN, A. P., BULL, H. T., RUBAYI, K., and IRAVANCHY, S.: 'The microwave dielectric properties of sea and fresh water'. Technical Report, GenCorp Aerojet, 1995

STURM, M., and HOLMGREN, J.: 'Effects of microtopography on texture, temperature and heat flow in Arctic and sub-Arctic snow', *Ann.. Glaciol.*, 1994, **19**, pp. 63–68

SURDYK, S.: 'Etudes des signatures spectrales micro-ondes obtenues par télédétection sur la calotte polaire Antarctique'. Thèse de doctorat de l'institut national polytechnique de Grenoble (1992)

SURDYK, S., and FILY, M.: 'Comparison of the passive microwave spectral signature of the Antarctic ice sheet with ground traverse data', *Ann. Glaciol.*, 1993, **17**, pp. 161–6

SURDYK, S., and FILY, M.: 'Results of a stratified emissivity model based on the wave approach: Application to the Antarctic ice sheet', *J. Geophys. Res.*, 1995, **100**, pp. 8837–48

SVENDSEN, E., KLOSTER, K., FARELLY, B. *et al.*: 'Norwegian remote sensing experiment: Evaluation of the Nimbus 7 scanning multichannel microwave radiometer for sea ice research', *J. Geophys. Res.*, 1983, **88**(C5), pp. 2781–91

SWIFT, C. T.: 'Microwave radiometer measurements of the Cape Cod Canal', *Radio Sci.*, 1974, **9**, pp. 641–53

SWIFT, C. T., and MCINTOSH, R. E.: 'Considerations for microwave remote sensing of ocean-surface salinity,' *IEEE Trans. Geosci. Remote Sens.*, 1983, **GE-21**, pp. 480–91

TAIT, A. B., and ARMSTRONG, R. L.: 'Validation of SMMR satellite-derived snow depth with ground-based measurements', *Int. J. Remote Sens.*, 1996, **17**, pp. 657–65

TAIT, A., HALL, D., FOSTER, J., CHANG, A., and KLEIN, A.: 'Detection of snow cover using millimeter-wave imaging radiometer (MIR) data', *Rem. Sens. Environ.*, 1999, **68**, pp. 53–60

TAVAKOLI, A., SARABANDI, K., and ULABY, F. T.: 'Microwave propagation through cultural vegetation canopies'. Technical report 026511-3-T, JPL-contract, C-958437 (NASA) (Electrical Engineering Department, The University of Michigan, 1991)

THOMAS, G.E., and STAMNES, K.: 'Radiative transfer in the atmosphere and ocean'. Cambridge Atmospheric Space Science Library Series (Cambridge, UK, 1999)

TONBOE, R. T., and ANDERSEN, S.: 'Modelled radiometer algorithm ice concentration sensitivity to variations of the Arctic sea ice snow cover'. Danish Meteorological Institute scientific report 04-03, 2004, p. 34

TONBOE, R. T., ANDERSEN, S., and TOUDAL, L.: 'Anomalous winter sea ice backscatter and brightness temperatures'. Danish Meteorological Institute Scientific Report 03-13, 2003, p. 59

TOUDAL, L.: 'Merging microwave radiometer data and meteorological data for improved sea ice concentrations', *EARSeL Adv. Remote Sens.*, 1994, **3**(2) XII, pp. 81–9

TROKHIMOVSKI, YU. G., and IRISOV, V. G.: 'Wind speed and direction measurements using microwave polarimetric radiometers'. NOAA technical memorandum ERL ETL-250, US Department of Commerce, 1995

TSANG, L.: 'Passive remote sensing of dense nontenuous media', *J. Electrom. Waves Appl.*, 1987, **1**(2), pp. 159–73

TSANG, L., and KONG, J. A.: 'Theory for thermal microwave emission from a bounded medium containing spherical scatterers', *J. Appl. Phys.*, 1997, **48**(8), pp. 3593–99

TSANG, L., BLANCHARD , A., NEWTON, R.W., and KONG, J.A.: 'A simple relation between active and passive microwave remote sensing measurements of earth terrain', *IEEE Trans.*, 1982, **GE-20**, pp. 482–5

TSANG, L., KONG, J. A., DING, K. H., and AO, C. O.: 'Scattering of electromagnetic waves' (Wiley, New York, 2001, Vol. 2, Numerical Simulations)

TUCKER, C. J.: 'Comparing SMMR and AVHRR data for drought monitoring', *Int. J. Remote Sens.*, 1989, **10**, pp. 1663–72

TUCKER, C. J., TOWNSHEND, J. R. G., and GOFF, T. E.: 'African land-cover classification using satellite data', *Science*, 1985, **227**, pp. 369–75

TUCKER, W. B., GRENFELD, T. C., ONSTOTT, R. G. *et al.*: 'Microwave and physical properties of sea ice in the winter marginal ice zone', *J. Geophys. Res.*, 1991, **96**(C3), pp. 4573–87

ULABY F. T., and STILES, W. H.: 'The active and passive microwave response to snow parameters', *J. Geophys. Res.*, 1980, **85**, pp. 1045–9

ULABY, F. T., and WILSON, E. A.: 'Microwave attenuation properties of vegetation canopies', *IEEE Trans. Geosci. Remote Sens.*, 1985, **23**, pp. 746–53

ULABY, F. T., MOORE, R., and FUNG, A. K.: 'Microwave remote sensing, active and passive, Vol I, Microwave remote sensing fundamentals and radiometry (1981), Vol II, Radar remote sensing and surface scattering and emission theory (1982), Vol III, From theory to applications' (1986), (Artech House Inc., Noorwood, MA)

ULABY, F. T., RAZANI, M., and DOBSON, M. C.: 'Effects of vegetation cover on the microwave radiometric sensitivity to soil moisture', *IEEE Trans. Geosci. Remote Sens.*, 1983, **21**, pp. 51–61

ULABY, F. T., TAVAKOLI, A., and SENIOR, T. B. A.: 'Microwave propagation constant for a vegetation canopy with vertical stalks', *IEEE Trans. Geosci. Remote Sens.*, 1987, **GE-25**(6), pp. 714–25

VAN DE GRIEND, A. A., and OWE, M.: 'Determination of microwave vegetation optical depth and single scattering albedo from large scale soil moisture and Nimbus/SMMR satellite observations', *Int. J. Remote Sens.*, 1993, **14**(10), pp. 1875–86

VAN DE GRIEND, A., and OWE, M.: 'Microwave vegetation optical depth and single scattering albedo from large scale soil moisture and NIMBUS/SMMR Satellite observations', *Meteor. Atm. Phys.*, 1994, **54**, pp. 225–39

VAN DE GRIEND, A. A., and WIGNERON, J.-P.: 'The *b*-factor as a function of frequency and canopy type at H-polarization', *IEEE Trans. Geosci. Remote Sens.*, 2004a, **42**(4), pp. 786–94

VAN DE GRIEND, A. A., and WIGNERON, J.-P.: 'On the measurement of microwave vegetation properties: Some guidelines for a protocol', *IEEE Trans. Geosci. Remote Sens.*, 2004b, **42**(10), pp. 2277–89

VAN DE GRIEND, A. A., OWE, M., DE RUITER, J., and GOUWELEEUW, B. T.: 'Measurement and behavior of dual-polarization vegetation optical depth and single scattering albedo at 1.4- and 5-GHz microwave frequencies', *IEEE Trans. Geosci. Remote Sens.*, 1996, **34**(4), pp. 957–65

VAN DE HULST, H.C.: 'Scattering in a planetary atmosphere', *Astrophys. J.*, 1948, **107**, pp. 220–46

VAN WOERT *et al.*: 'Ocean winds from space . . . the next generation'. Office of Naval Research report no. 32296-13, Arlington, Virginia, 1996, 61 pp

VILLARINO, R., CAMPS, A., VALL-LLOSSERA, M., MIRANDA, J., and ARENAS, J.: 'Sea foam and sea state effects on the instantaneous brightness temperatures at L-band', in First Results Workshop, EuroSTARRS/WISE/LOSAC Campaigns, ESA SP-525, pp. 95–103, ESTEC/European Space Agency, 2003

VORONOVICH, A. G.: 'Wave scattering from rough surfaces' (Springer-Verlag, Berlin, 1994)

VOSS, S., HEYGSTER, G., and EZRATY, R.: 'Improving the sea ice type discrimination by the simultaneous use of SSM/I and scatterometer data', *Polar Res.*, 2003, **22**(1), pp. 35–42

WALDTEUFEL, P., FLOURY, N., DINNAT, E., and CAUDAL, G.: 'Ionospheric effects for L-Band 2-D interferometric radiometry', *IEEE Trans. Geosci. Remote Sens.*, 2004, **42**, 105–18

WANG, J. R., and CHOUDHURY, B. J.: 'Remote sensing of soil moisture content over bare field at 1.4 GHz frequency', *J. Geophys. Res.*, 1981, **86**, pp. 5277–82

WANG, J. R., and SCHMUGGE, T.: 'An empirical model for the complex dielectric permittivity of soils as a function of water content', *IEEE Trans. Geosci. Remote Sens.*, 1980, **GE-18**(4), pp. 288–95

WANG, J. R., MCMURTREY, J. E., ENGMAN, E. T. *et al.*: 'Radiometric measurements over bare and vegetated fields at 1.4-GHz and 5-GHz frequencies', *Remote Sens. Environ.*, 1982, **12**, pp. 295–311

WANG, J. R., O'NEILL, P. E., JACKSON, T. J., and ENGMAN, E. T.: 'Multifrequency measurements of the effects of soil moisture, soil texture, and surface roughness', *IEEE Trans. Geosci. Remote Sens.*, 1983, **21**, pp. 44–51

WANG, J. R., CHANG, A. T. C., and SHARMA, A. K.: 'On the estimation of snow depth from microwave radiometric instruments', *IEEE Trans. Geosci. Remote Sens.*, 1992, **30**(4), pp. 785–92

WEBSTER, W. J., Jr., and WILHEIT, T. T.: 'Spectral characteristics of the microwave emission from a wind-driven foam-covered sea', *J. Geophys. Res.*, 1976, **81**, pp. 3095–9

WEEKS, W. F., and ACKLEY, S. F.: 'The growth, structure and properties of sea ice', in UNTERSTEINER, N. (Ed.): 'The geophysics of sea ice' (New York: Plenum, 1986), pp. 9–164

WEGMÜLLER, U., and MÄTZLER, C.: 'Rough bare soil reflectivity model', *IEEE Trans. Geosci. Remote Sens.*, 1999, **37**, pp. 1391–95

WEGMÜLLER, U., MÄTZLER, C., and NJOKU, N. E.: 'Canopy opacity models', in CHOUDHURY *et al.* (Eds.): 'Passive microwave remote sensing of land-atmosphere interactions' (VSP, Utrecht, The Netherlands, 1995), pp. 375–87

WEISE, T.: 'Radiometric and structural measurements of snow'. PhD thesis, Inst. of Appl. Phys., University of Bern, Bern, Switzerland, 1996a

WEISE, T.: 'Radiometric data (11–94 GHz) and autocorrelation functions of dry snow samples'. IAP-Report 96-2, Inst. of Appl. Phys., Univ. of Bern, 1996b

WEILL, A., BESNARD, T., CONTARDO, S., ETCHETO, J., and BOUTIN, J.: 'WISE results from whitecapping and stereophotogrammetry: A tentative of superposition of topography and foam surface to retrieve active and fossil foam', in First Results Workshop, EuroSTARRS/WISE/LOSAC Campaigns, ESA SP-525, pp. 89–94, ESTEC/European Space Agency, 2003

WENG, F., YAN, B., and GRODY, N. C.: 'A microwave land emissivity model', *J. Geophys. Res.*, 2001, **106**, 20115–23

WENTZ, F. J.: 'Measurement of oceanic wind vector using satellite microwave radiometers', *IEEE Trans. Geosci. Remote Sens.*, 1992, **30**(5), pp. 960–72

WESSEL, J., CORNELIUS, J., FARLEY, R. W. *et al.*: 'First observations from DMSP SSMIS'. Proceedings of the 14th AMS conference on *Satellite meteorology and oceanography*, 2004

WEST, R., TSANG, L., and WINEBRENNER, D. P.: 'Dense medium radiative transfer theory for two scattering layers with a Rayleigh distribution of particle sizes', *IEEE Trans. Geosci. Remote Sens.*, 1993, **31**, pp. 426–36

WEST, R. D, WINEBRENNER, D. P., TSANG, L., and ROTT, H., 'Microwave emission from density-stratified Antarctic firn at 6 cm wavelength', *J. Glaciol.*, 1996, **42**, pp. 63–76

WIESMANN, A., and MÄTZLER, C.: 'Microwave emission model of layered snowpacks', *Remote Sens. Environ.*, 1999, **70**, pp. 307–16.

WIESMANN, A., and MÄTZLER, C.: 'Documentation for MEMLS 98.1, microwave emission model of layered snowpacks'. IAP-report 98-2, Inst. of Appl. Phys., Univ. of Bern, 1999

WIESMANN, A., MÄTZLER, C., and WEISE, T.: 'Radiometric and structural measurements of snow samples', *Radio Sci.*, 1998, **33**(2), pp. 273–89

WIGNERON, J.-P., CALVET, J.-C., KERR, Y. H., CHANZY, A., and LOPES, A.: 'Microwave emission of vegetation: Sensitivity to leaf characteristics', *IEEE Trans. Geosci. Remote Sens.*, 1993, **31**, pp. 716–26

WIGNERON, J.-P., CHANZY, A., CALVET, J.-C., and BRUGUIER, N.: 'A simple algorithm to retrieve soil moisture and vegetation biomass using passive microwave measurements over crop fields', *Remote Sens. Environ.*, 1995, **51**, pp. 331–41

WIGNERON, J.-P., CALVET, J.-C., and KERR, Y.: 'Monitoring water interception by crop fields from passive microwave observations', *Agri. Forest Meteor.*, 1996, **80**, pp. 177–94

WIGNERON, J.-P., COMBAL, B., WEGMÜLLER, U., and MÄTZLER, C.: 'Estimation of microwave parameters of crops from radiometric measurements', *Int. J. Remote Sens.*, 1996, **17**(14), pp. 2875–80

WIGNERON, J. P., GUYON, D., CALVET, J.-C., COURRIER, G., and BRUGUIER, N.: 'Monitoring coniferous forest characteristics using a multifrequency (5-90 GHz) microwave radiometer', *Remote Sens. Environ.*, 1997, **60**, pp. 299–310

WIGNERON, J.-P., WALDTEUFEL, P., CHANZY, A., CALVET, J.-C., and KERR, Y.: 'Two-D microwave interferometer retrieval capabilities of over land surfaces (SMOS Mission)', *Remote Sens. Environ.*, 2000, **73**, pp. 270–82

WIGNERON, J.-P., LAGUERRE, L., and KERR, Y.: 'A simple parameterization of the L-band microwave emission from rough agricultural soils', *IEEE Trans. Geosci. Remote Sens.*, 2001, **39**(8), pp. 1697–1707

WIGNERON, J.-P., CHANZY, A., CALVET, J.-C., OLIOSO, A., and KERR, Y.: 'Modeling approaches to assimilating L-band passive microwave observations over land surfaces', *J. Geophys. Res.*, 2002, **107**, p. D14 (DOI 10.1029/2001JD000958)

WIGNERON, J.-P., CALVET, J.-C., PELLARIN, T., VAN DE GRIEND, A., BERGER, M., and FERRAZZOLI, P.: 'Retrieving near surface soil moisture from microwave radiometric observations: Current status and future plans', *Remote Sens. Environ.*, 2003, **85**, pp. 489–506

WIGNERON, J.-P., PARDÉ, M., WALDTEUFEL, P. *et al.*: 'Characterizing the dependence of vegetation parameters on crop type, view angle and polarization at L-Band', *IEEE Trans. Geosci. Remote Sens.*, 2004, **42**(2), pp. 416–25

WILLIAMS, G.: 'Microwave radiometry of the ocean and the possibility of marine wind velocity determination from satellite observations', *J. Geophys. Res.*, 1969, **74**, pp. 4591–4

WILSON, L. T., TSANG, L., HWANG, J. N., and CHEN, C. T.: 'Mapping snow water equivalent by combining a spatially distributed snow hydrology model with passive microwave remote-sensing data', *IEEE Trans. Geosci. Remote Sens.*, 1999, **37**, pp. 690–704

WILSON, W., YUEH, S., DINARDO, S., and LI, F.: 'High-stability L-band radiometer measurements of saltwater', *IEEE Trans. Geosci. Remote Sens.*, 2004, **42**, pp. 1829–35

WINEBRENNER, D. P., BREDOW, J., FUNG, A. K. *et al.*: 'Microwave sea ice signature modelling', in CARSEY, F. D. (Ed.): 'Microwave remote sensing of sea ice, geophysical monograph 68' (American Geophysical Union, Washington, DC, 1992), pp. 137–75

WU, T. D., and CHEN, K. S.: 'A reappraisal of the validity of the IEM model for backscattering from rough surfaces', *IEEE Trans. Geosci. Remote Sens.*, April 2004, **42**, pp. 743–53

WU, T. D., CHEN, K. S., SHI, J., and FUNG, A. K.: 'A transition model for the reflection coefficient in surface scattering', *IEEE Trans. Geosci. Remote Sens.*, Sept. 2001, **39**, pp. 2040–50

YUEH, S. H.: 'Modeling of wind direction signals in polarimetric sea surface brightness temperatures', *IEEE Trans. Geosci. Remote Sens.*, 1997, **35**(6), pp. 1400–18

YUEH, S. H.: 'Estimates of Faraday rotation with passive microwave polarimetry for microwave sensing of Earth surfaces', *IEEE Trans. Geosci. Remote Sens.*, 2000, **38**, pp. 2434–8

YUEH, H. A., SHIN, R. T., and KONG, J. A.: 'Scattering of electromagnetic waves from a periodic surface with random roughness', *J. Appl. Phys.*, 1988, **64**, pp. 1657–70

YUEH S. H., KONG, J. A., JAO, J. J., SHIN, R. T., and LE TOAN, T.: 'Branching model for vegetation', *IEEE Trans. Geosci. Remote Sens.*, 1992, **30**, pp. 390–402

YUEH, S. H., WEST, R., WILSON, W. J., LI, F. K., NJOKU, E. G., and RAHMAT-SAMII, Y.:'Error sources and feasibility for microwave remote sensing of ocean surface salinity,' *IEEE Trans. Geosci. Remote Sens.*, 2001, **39**, pp. 1049–60

YUEH, S. H., KONG, J. A., KWOK, R., LI, F. K., NGHIEM, S. V., and WILSON, W. J.: 'Polarimetric passive remote sensing of ocean wind vectors', *Radio Sci.*, **29**(4), pp. 799–814

ZHANG, H., TOUDAL PEDERSEN, L., and GUDMANDSEN, P.: 'Microwave brightness temperatures of the Greenland Ice Sheet', *Adv. Space Res.*, 1989, **9**, pp. (1)277–(1)287

ZRIBI, M., BAGHDADI, N., HOLAH, N., FAFIN, O., and GUÉRIN, C.: 'Evaluation of a rough soil surface description with ASAR-ENVISAT radar data', *Remote Sens. Environ.*, 2005, **95**(1), pp. 67–76

ZWALLY, H. T.: 'Microwave emissivity and accumulation rate of polar firn', *J. Glaciol.*, 1977, **18**, pp. 195–215

Chapter 5

Dielectric properties of natural media

Christian Mätzler, William Ellison, Bertrand Thomas, Ari Sihvola, and Mike Schwank

Dielectric, refractive and possibly magnetic properties of the media to be studied with microwave radiometry are important in several ways. First, the absorption coefficient follows from the imaginary part of the refractive index. The refractive index is a simple function of the dielectric constant and of the magnetic permeability. Second, the wave impedance and the scattering properties are related to the dielectric and magnetic properties. And finally, the ray path of the radiation depends on the refractive index in the propagating medium. In short, dielectric and magnetic properties of the media are basic parameters for the topics of this book. Uncertainties of radiative-transfer computations have often been limited by the insufficient accuracy of dielectric properties. The purpose of this chapter is to review the present knowledge in this field. A simplification to be made is the assumption that the magnetic permeability can be approximated by the value found in vacuum. It means that the small diamagnetic and paramagnetic effects can be ignored, and that ferromagneticity is not relevant to remote sensing. As an example water is slightly diamagnetic with a relative magnetic permeability of $\mu = 0.999991$ (Gupta, 1986). In our approximation, this quantity is set to 1. The remaining electromagnetic key parameter is the complex dielectric constant.

The dielectric constant of a substance depends both upon the electromagnetic field which interacts with the matter and upon the physical properties of the substance itself. For the fields considered in this book, ε depends only upon the frequency, ν, of the field. As for its dependence upon the matter, it will be a function of bulk properties such as temperature, pressure etc., a function of the physical constituents and of the molecular and atomic structure of the substance.

5.1 Introduction to dielectric properties *Christian Mätzler*

5.1.1 Outline

This section introduces electromagnetic properties of homogeneous matter. More detailed treatments can be found in the literature on electromagnetic waves and fields. Apart from general textbooks on electrodynamics, there exist a number of books on dielectric media. The first to be mentioned is '*Dielectrics and Waves*' by Von Hippel (1954) who introduced and compared different approaches from microscopic and macroscopic views and from the physics and electrical engineering standpoints, using historical, theoretical and experimental aspects. Less well known, but excellent, too, is the revised two-volume treatise on the '*Theory of Electric Polarisation*' of Böttcher *et al.* (1973) and Böttcher and Bordewijk (1978). Of special importance are the dielectric properties of water and of aqueous solutions with the concept of relaxation phenomena. The leading scientist was Peter Debye who wrote a treatise on polar molecules (Debye, 1929). He received the Nobel Prize in 1936 for his achievements on molecular studies. Progress has been rather slow in comparison with solid-state physics, because polar liquids are extremely complex due to the nature of their structure and due to the difficulty of modelling the internal electric field. Advances through theory and experiments are continuing, for instance, in the discipline called 'dielectric relaxation spectroscopy'. A classic textbook on the dielectric constant of water is the one by Hasted (1973). Apart from basic relations to be presented below, we will concentrate on theoretical and experimental results needed to obtain accurate numerical models. Homogeneous components are discussed in Sections 5.2–5.4. Section 5.5 is devoted to mixing formulas of heterogeneous media in terms of an effective dielectric constant and of other effective-medium properties (Sihvola, 1999). The topic is important for remote sensing because many natural media are inherently heterogeneous, but may still be treated as if they were homogeneous. The effective-medium theories allow significant simplifications without losing the physical basis.

5.1.2 Dielectric constant and refractive index in a homogeneous medium

Although homogeneity is not exactly met in a world of atoms and molecules or in granular media, in a physical sense the homogeneity requirement only means that any structural feature of the medium must be much finer than the wavelength λ of a sensing wave. The microscopic behaviour is reflected by bulk properties, and they are contained in the constitutive relations between the electromagnetic fields. These relations apply to linear media and are shown below. For thermal microwave radiation of the terrestrial environment, all relevant media behave linearly. The fields, i.e. the electric field $\mathbf{E}$, the displacement field $\mathbf{D}$, the magnetic field $\mathbf{H}$, and the magnetic induction field $\mathbf{B}$, are governed by Maxwell's equations. For homogeneous and isotropic media (away from sources), and for time-harmonic fields with time factor $\exp(-i\omega t)$, where t is time and ω is the angular frequency, Maxwell's equations can be written as

$$\nabla \cdot \mathbf{H} = \frac{\partial \mathbf{D}}{\partial t} + \mathbf{j} = -i\omega\mathbf{D}; \quad \text{where } \nabla \cdot \mathbf{D} = \rho_e = 0 \tag{5.1}$$

$$\nabla \cdot \mathbf{E} = -\frac{\partial \mathbf{B}}{\partial t} = i\omega\mathbf{B}; \quad \text{where } \nabla \cdot \mathbf{B} = 0 \tag{5.2}$$

with the constitutive relations

$$\mathbf{D} = \varepsilon\varepsilon_0\mathbf{E} \quad \text{and} \quad \mathbf{B} = \mu\mu_0\mathbf{H} \tag{5.3}$$

where ∇ is the nabla operator, ε is the (relative) dielectric constant (also called relative electric permittivity), $\varepsilon_0 = 8.854 \cdot 10^{-12}$ As/V/m the vacuum permittivity, μ the relative magnetic permeability and $\mu_0 = 4\pi \cdot 10^{-7}$ Vs/A/m the vacuum permeability.

For anisotropic media, ε and μ have to be replaced by tensors, $\boldsymbol{\varepsilon}$ and $\boldsymbol{\mu}$, respectively. Furthermore, in chiral or bi-isotropic media, also the electric and magnetic fields are linearly related (Sihvola, 1999). For simplicity let us assume isotropic media obeying (5.3). The different fields, except for a selected one, can be eliminated from (5.1) to (5.3), leading to the vector-wave equation in the unbounded homogenous medium; for **E** it is

$$c^2\Delta\mathbf{E} = -\omega^2\mathbf{E} \tag{5.4}$$

Here Δ is the Laplace operator, and c is the phase velocity in the medium, determined by

$$c = \frac{c_0}{n}; \qquad c_0 = \frac{1}{\sqrt{\varepsilon_0\mu_0}} \tag{5.5}$$

where c_0 is the speed of light in vacuum. The refractive index n follows from

$$n = \sqrt{\varepsilon\mu} \tag{5.6}$$

or simply $n = \sqrt{\varepsilon}$, for the frequent situation of non-magnetic materials ($\mu = 1$). For a plane wave $\mathbf{E} \propto \exp(ikx - i\omega t)$ propagating along path x with wave number k, the Laplace-operated electric field is $\Delta\mathbf{E} = \partial^2\mathbf{E}/\partial x^2 = -k^2\mathbf{E}$. Insertion of this expression in the wave equation leads to the dispersion relation

$$k = n\frac{\omega}{c_0} \tag{5.7}$$

Since the x direction is arbitrary, the dispersion relation holds for all directions, and it is independent of polarisation. In a medium with losses, the quantities are complex ($k = k' + ik''$, $n = n' + in''$, $\varepsilon = \varepsilon' + i\varepsilon''$), and they depend on frequency. The current density **j** excited in a conducting medium by the electric field **E** is given by Ohm's law $\mathbf{j} = \sigma\mathbf{E}$ where σ is the conductivity. In our notation **j** is expressed as a displacement current, $-i\omega\mathbf{D}$, in Equation (5.3), with the complex dielectric constant $\varepsilon = \varepsilon' + i\varepsilon''$ whose imaginary part ε'' is given by

$$\varepsilon'' = \frac{\sigma}{\varepsilon_0\omega} \tag{5.8}$$

Therefore, conductivity and imaginary permittivity are different representations of the same effect. In the present time convention (see text to Equation (5.1)), the imaginary parts of n, ε and μ are positive for lossy media (negative values of the opposite convention often used by electrical engineers are obtained by the transformation to $\mathrm{j} = -\mathrm{i}$ for the imaginary unit). The losses are responsible for wave attenuation, and if scattering is absent, the attenuation is due to absorption. Since the wave intensity is proportional to $|\mathbf{E}|^2$, its spatial variation is an exponential decay $\propto \exp(-2k''x)$, and k'' is the imaginary part of k. The damping coefficient $2k''$ is the absorption coefficient γ_a, and by (5.7), it is related to the imaginary refractive index n''.

$$\gamma_a = 2k'' = \frac{2n''\omega}{c_0} \tag{5.9}$$

Finally, Equation (5.6) gives the relationship with the dielectric constant. For $\mu = 1$ and for small or moderate dielectric loss ($\varepsilon'' \ll \varepsilon'$) we get

$$\gamma_a \cong \frac{\varepsilon''\omega}{\sqrt{\varepsilon'}c_0} = \frac{2\pi\varepsilon''}{\sqrt{\varepsilon'}\lambda_0} \tag{5.10}$$

where λ_0 is the vacuum wavelength. For high loss, the exact expression (5.6) has to be used.

5.1.3 *Kramers–Kronig relations*

The real $\varepsilon'(\omega)$ and imaginary $\varepsilon''(\omega)$ spectra of $\varepsilon(\omega)$ are not independent functions; they are related by an integral transform, called after Hilbert, and the corresponding expressions are called Kramers–Kronig relations (e.g. Böttcher and Bordewijk (1978) or Kong (1985)). The transform is a consequence of linearity and causality, the latter meaning that a response cannot precede its action. In dielectric terms, the displacement field (or the corresponding polarisation field $\mathbf{P} = \mathbf{D} - \mathbf{E}$) is caused by an exciting electric field. Causality restricts the Fourier transform of the temporal behaviour. It turns out that $\varepsilon(\omega)$ is an analytical function, meaning that $\varepsilon'(\omega)$ and $\varepsilon''(\omega)$ are a Hilbert-transform pair:

$$\varepsilon''(\omega) = -\mathrm{Hi}(\varepsilon') = -\frac{1}{\pi}\int_{-\infty}^{\infty} \frac{\varepsilon'(x) - \varepsilon_\infty}{x - \omega}\mathrm{d}x \tag{5.11}$$

$$\varepsilon'(\omega) - \varepsilon_\infty = \mathrm{Hi}(\varepsilon'') = \frac{1}{\pi}\int_{-\infty}^{\infty} \frac{\varepsilon''(x)}{x - \omega}\mathrm{d}x \tag{5.12}$$

The divergence at $\omega = x$ is allowed for by taking the Cauchy principal value of the integral. Furthermore ε' is a symmetric function of ω, whereas ε'' is antisymmetric. In Equations (5.11) and (5.12) ε_∞ is the dielectric constant at infinite frequency, or more realistically a constant value well above the spectral range of dispersion. Note that except for constant values, $\mathrm{Hi}(\mathrm{Hi}(y)) = -y$, and that the Hilbert transform of a constant is zero. A frequency-independent dielectric constant cannot be obtained from the Hilbert transform. Therefore ε_∞ has been subtracted in (5.12). The properties of the Hilbert transform, including a table of transformed functions, are presented in Bracewell (1965). A useful relationship between the Fourier and Hilbert transforms follows from the fact that $\mathrm{Hi}(\sin x) = \cos x$ and $\mathrm{Hi}(\cos x) = -\sin x$: Since the Fourier transform can be understood as a superposition of sine and cosine transforms, this relationship means that the cosine coefficients of the Hilbert-transformed Fourier transform (i.e. of $\varepsilon' - \varepsilon_\infty$) are the sine coefficients of the untransformed function (i.e. of ε''), and vice versa, except for the negative sign in (5.11). Due to symmetry the sine coefficients of $\varepsilon'(\omega) - \varepsilon_\infty$ are zero, and the same is true for the cosine coefficients of $\varepsilon''(\omega)$.

The key applications of the Kramers–Kronig relations are tests and subsequent improvements of dielectric models. Another application is the computation of the susceptibility spectrum, $\varepsilon'(\omega) - \varepsilon_\infty$, from the absorption spectrum, measured over a sufficiently wide spectral range, or vice versa.

5.2 Freshwater and seawater *William Ellison*

5.2.1 Introduction

The principal interaction of an electromagnetic wave at microwave frequencies with an aqueous solution is via relaxation processes of associations of the polar water molecule and, as we approach the far infrared, atomic resonances can occur. An excellent collection of articles concerning theoretical, experimental and sensing applications of water–microwave interactions can be found in the book edited by Kraszewski (1996).

Purewater is the substance for which the most independently measured permittivity data exists, both in frequency and in temperature. In the review article, Ellison *et al.* (1996a), over 1000 values are reproduced for the frequency range 0–100 GHz and the temperature range 0–100°C. For the frequency range 100–1000 GHz only 130 measurements had been published, mainly for temperatures between 20 and 30°C. There are still large gaps in our knowledge. For example, there is little experimental data for frequencies greater than 30 GHz at temperatures other than 20, 25 and 30°C. For super-cooled water, there is only one dataset at a single frequency of 9.61 GHz.

Until recently there was no useful experimental data for the permittivity of seawater. To date there are only two independent datasets for the frequency range 3–20 GHz and one dataset for the frequency range 30–100 GHz. This poses a serious problem for applications to radiometry, where it is vital to have accurate interpolated permittivity data for purewater and seawater in relatively small steps over the temperature range −20 to 30°C and over the frequency range 1–1000 GHz.

The object of this section is to take a step towards the construction of such an interpolation function. In Section 5.2.2 we describe an appropriate theoretical model for purewater and for seawater permittivity as a function of frequency, then in Sections 5.2.3 and 5.2.4 we discuss the various interpolation functions that have been proposed. Section 5.2.5 describes a new interpolation function, the construction of which uses all the available experimental data for purewater and seawater. It covers the frequency range 0–1000 GHz, the temperature range 0–30°C, the salinity range 0–40‰ and represents all the available measured data to within the limits of experimental uncertainty.

5.2.2 Theoretical considerations

In principle, it ought to be possible to calculate ε for a given material in terms of its composition and molecular structure. A considerable effort has been devoted to this problem (see, e.g. Fröhlich 1958; Hill *et al.* 1968; Eyring and Jhon 1969; Gerschel 1995). The results for perfect gases and perfect crystals are adequate, but for solids and liquids, the complexity of the problem is such that there are significant differences between measured permittivity values and theoretical values. For polar liquids, such as purewater and seawater, Debye simplified the model by supposing that a polar molecule interacts with an electromagnetic field by rotating within a homogeneous viscous medium and he deduced the relaxation expression:

$$\varepsilon = \varepsilon' + \mathrm{i}\varepsilon'' = \varepsilon_\infty + \frac{\varepsilon_\mathrm{s} - \varepsilon_\infty}{1 - i2\pi\nu\tau} \tag{5.13}$$

$$\text{i.e.} \quad \varepsilon' = \varepsilon_\infty + \frac{\varepsilon_\mathrm{s} - \varepsilon_\infty}{1 + 4\pi^2\nu^2\tau^2} \quad \text{and} \quad \varepsilon'' = \frac{2\pi\nu\tau(\varepsilon_\mathrm{s} - \varepsilon_\infty)}{1 + 4\pi^2\nu^2\tau^2}$$

where ν is the frequency of the electromagnetic field, ε_s the permittivity of the polar liquid in a static electric field, τ the relaxation time of the polar molecule within the viscous medium and ε_∞ is the permittivity of the polar liquid at an 'infinite' frequency.

As permittivity is a function of temperature the Debye parameters are also temperature dependent, and we shall denote them, when relevant, by: $\varepsilon_\mathrm{s}(T)$, $\tau(T)$ and $\varepsilon_\infty(T)$.

Equation (5.13) results from the solution of a first-order differential equation for molecules rotating in a viscous fluid under excitation by a harmonic electric field, assuming that inertial effects are negligible. Under these assumptions ε tends to ε_∞ as ν tends to infinity. In real liquids there are often multiple relaxation or resonance phenomena over very wide frequency ranges, and ε_∞ does not make much physical sense. It is better to consider it simply as a fitting parameter.

For materials with a DC conductivity of σ S/m, we have seen in 5.1.2 that Maxwell's equations imply the addition of the term $\sigma/2\pi\nu\varepsilon_0$ to ε'', where $\varepsilon_0 = 8.85418 \cdot 10^{-12}$ F/m. (In later formulae it will be convenient to have the frequency in GHz, in which case $1/(2\pi\varepsilon_0) = 17.9751$ GHz m/S.)

The Debye expression is certainly a reasonable approximation to ε for many polar liquids, and many attempts have been made to calculate ε_s, ε_∞ and τ in terms of the polar constituents of the liquid and its bulk properties such as temperature, density and conductivity. The results of the calculations do not correspond to the measurements with sufficient accuracy. Experimental work and 'curve fitting' have shown that for many polar liquids there are discrepancies between the measured values of ε and 'best fits' to the parameters in the Debye function. Several empirical relations have been proposed, such as the Cole–Cole relation:

$$\varepsilon - \varepsilon_\infty = \frac{\varepsilon_\mathrm{s} - \varepsilon_\infty}{1 + (-\mathrm{i}2\pi\nu\tau)^\alpha} \quad \text{where } 0 < \alpha \leq 1 \tag{5.14}$$

the Cole–Davidson relation:

$$\varepsilon - \varepsilon_\infty = \frac{\varepsilon_s - \varepsilon_\infty}{(1 - i2\pi\nu\tau)^\beta} \quad \text{where } 0 < \beta \leq 1 \tag{5.15}$$

If the molecules of the liquid in question consist of several types of polar structures, it is natural to extend the Debye reasoning to each of the relaxing structures and propose formulae of the kind:

$$\varepsilon - \varepsilon_\infty = \frac{\Delta_1}{1 - \mathrm{i}2\pi\nu\tau_1} + \frac{\Delta_2}{1 - \mathrm{i}2\pi\nu\tau_2} + \cdots + \frac{\Delta_n}{1 - \mathrm{i}2\pi\nu\tau_n} \tag{5.16}$$

where $\Delta_1 + \cdots + \Delta_n = \varepsilon_s - \varepsilon_\infty$.

Experimentation has found polar liquids for which each of the above expressions gives a better representation for ε than the simple Debye formula. There is a long-standing controversy as to whether the permittivity of purewater is best represented by a simple Debye function or by a Cole–Cole function with the parameter α close to 1 or by a double Debye function. Such considerations, whilst interesting from the point of view of the theory of polar liquids, do not have their place in the context of this book. We are primarily concerned with using an adequate function to interpolate and extrapolate the experimental data. For this purpose the Debye functions suffice.

5.2.3 *Freshwater*

Freshwater and purewater are not the same substance from the point of view of dielectric spectroscopy. Purewater is relatively easy to obtain and it has a very low conductivity. However, it is very difficult to maintain in its pure state as it is a powerful solvent and readily absorbs atmospheric gases. In particular, its conductivity rapidly becomes non-negligible as soon as a few ions are dissolved in the liquid. Freshwater is 'natural' and hence contains dissolved ions and gases. Its conductivity is non-zero (e.g. freshwater, such as domestic tap water, has a conductivity σ between 200 and 1000 μS/m). However, there are relatively few dissolved ions and ε' for freshwater is very close to that of purewater, but the term ε'' will have to be corrected by the addition of a conductivity term:

$$\varepsilon_{\text{freshwater}}'' = \varepsilon_{\text{purewater}}'' + \frac{\sigma}{2\pi\nu\varepsilon_0} \tag{5.17}$$

For low frequencies the correction term can be significantly larger than the first term. In the rest of this sub-section we shall consider the permittivity of purewater.

The permittivity of purewater, at $T = 25$°C and over the frequency range 0–30 GHz, is given very closely (to within less than 1 per cent) by a single Debye function. There are hundreds of published data that enable one to estimate the Debye parameters $\varepsilon_s(25)$, $\tau(25)$, $\varepsilon_\infty(25)$ quite accurately. However, it is not true that the extrapolation of this function to the frequency range 30–100 GHz maintains the 1 per cent accuracy. At 90 GHz, the differences between the extrapolated value and independently measured permittivity values are about 10 per cent. There are theoretical and experimental reasons to suspect that a second relaxation process occurs around 140 GHz in water at 25°C, and that the permittivity data should be represented by the sum of two Debye functions.

The experimental evidence for the second relaxation is tantalisingly elusive. The second relaxation was predicted by Haggis *et al.* (1952), it was used to interpret permittivity measurements of water at 4°C by Grant and Sheppard (1974), and it was inferred from experimental results at 25°C by Barthel *et al.* (1990), then by Kaatze (1993) in water at 20°C. They estimated it to be in the region of 150 GHz. Kindt *et al.* (1996), using a femtosecond Terahertz pulse spectroscopy technique in the frequency range 80–1500 GHz, interpreted their data by evoking a second Debye relaxation in water 'at room temperature'. Rønne *et al.* (1997), using a similar technique over the frequency range 100–2000 GHz, also inferred the existence of this high-frequency process and they studied its temperature variation. Buchner *et al.* (1999) attempted to formulate a theoretical model of the relaxation process.

Some insight is gained from the data for seawater reported by Ellison *et al.* (1997, 2003). This data does detect a second relaxation and the phenomenon will be similar to that which occurs in purewater. At 20°C the second relaxation frequency is about 120 GHz, at 15°C it is about 97 GHz and at 10°C it is about 56 GHz. This means that for $T > 20$°C and for frequencies less than about 70 GHz the influence of the second relaxation will be small. However, for temperatures less than about 10°C the effects of this second relaxation will probably be noticeable for frequencies greater than 20 GHz. Unfortunately there are practically no experimental permittivity measurements for purewater in this interesting region.

The radiometric investigation of cold cloud formations containing super-cooled water requires permittivity estimates of purewater down to about −30°C. The only measured values at a microwave frequency are those of Bertolini *et al.* (1982), at 9.61 GHz and down to −18°C. Estimates for the static permittivity and principal relaxation time have been proposed by Noyel *et al.* (1992, 1994). They measured the permittivity of water–alcohol and water–alkylhydrazene mixtures over the frequency range 1 mHz to 10 MHz and at temperatures down to −135°C (the supposed glass transition temperature, T_∞, of purewater) and fitted the data to a single Debye relaxation. From these measurements they extrapolated the data to zero concentration in order to estimate the dielectric parameters of purewater. They proposed, for T in kelvin and relaxation time in seconds:

$$\varepsilon_s(T) = -22.2 + \frac{29950}{T}, \quad \log_e \tau(T) = A_1 + \frac{2B_1}{T - T_\infty + \sqrt{(T - T_\infty)^2 + C_1 T}} \tag{5.18}$$

where $A_1 = -13.72$, $B_1 = 486.3$, $C_1 = 0.3$, with an estimated precision of about 7 per cent.

For the frequency interval $70 < \nu < 1000$ GHz, which contains a relaxation process, we do not have enough data in the scientific literature to do a serious interpolation at fixed temperatures. For example, at 4°C there are only five data points and at 10°C there are only nine data points! Furthermore, the available data, shown in

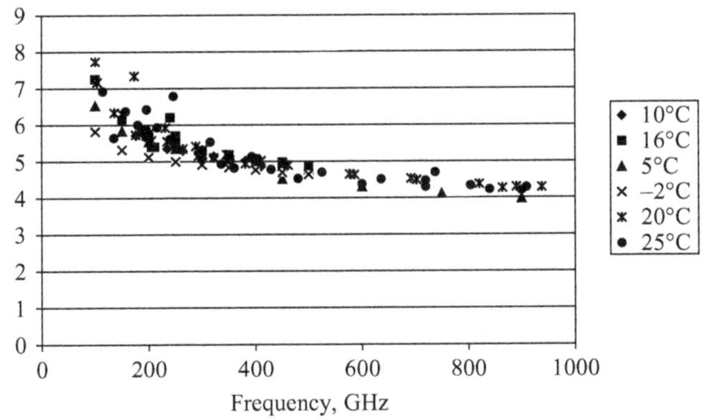

Figure 5.1 *Published data for ε′ of purewater at frequencies above 100 GHz*

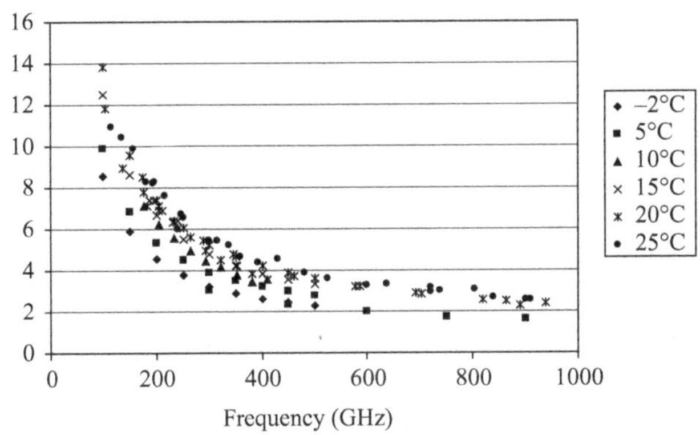

Figure 5.2 *Published data for ε″ for purewater at various temperatures at frequencies above 100 GHz*

Figures 5.1 and 5.2, are not very accurate (author estimates, when given, are usually between 5 and 10 per cent). We observe that for the frequency range 500–1000 GHz the temperature variation in the interval −5 to 30°C is swamped in experimental uncertainty. However, in the frequency range 100–500 GHz the data does indicate a steady temperature variation. This is clearly visible for ε'' (Figure 5.2) and a sympathetic eye sees it for ϵ' (Figure 5.1).

Radiometric studies vitally need an interpolation function in this range and Manabe *et al.* (1987) and Liebe *et al.* (1991) used the existing data to do the 'best possible' at the time. For temperatures T_K satisfying 293 K $< T_K <$ 303 K and a frequency of ν GHz they defined:

$$\theta = 1\text{–}300/T_K, \quad \varepsilon_0 = 77.66 - 103.3\theta, \quad \varepsilon_1 = 0.0671\varepsilon_0,$$

$$\varepsilon_2 = 3.52 + 7.52\theta, \quad \gamma_1 = 20.20 + 146.4\theta + 316\theta^2, \quad \gamma_2 = 39.8\gamma_1$$

and proposed:

$$\varepsilon(T_K, \nu) = \frac{\varepsilon_0 - \varepsilon_1}{1 - i(\nu/\gamma_1)} + \frac{\varepsilon_1 - \varepsilon_2}{1 - i(\nu/\gamma_2)} + \varepsilon_2 \tag{5.19}$$

If this function is extrapolated to $T < 20°C$ no uncertainty estimates are available, other than the gloomy prediction that they are certainly greater than 10 per cent.

The totality of all published permittivity data for purewater up to 1995, together with critical analyses of the experimental techniques used, are assembled in a review article (Ellison *et al.* 1996a). We shall use all this data, together with seawater data, to construct an interpolation function that gives acceptable permittivity values for purewater in the frequency range 0–1000 GHz and the temperature range 0–30°C. This function is described in Section 5.2.5. For super-cooled water the situation is very unsatisfactory. Experimental measurements of the permittivity of water in the intervals −30 to 15°C and 3–100 GHz are necessary. It cannot be emphasised enough that no amount of 'fiddling' with the existing data in the scientific literature, as we shall do later, replaces ignorance with knowledge.

5.2.4 Seawater

Bulk seawater is a complex solution of purewater, inorganic salts and a relatively small percentage of organic matter. The organic matter has no measurable influence upon the permittivity of seawater (Ellison *et al.* 1996b). The global effect of the inorganic salts is quantified by the salinity which, for natural seawater, is usually between 20‰ and 40‰, with a mean value close to 35‰. The permittivity of seawater is a function of its temperature T, its salinity S and ν the frequency of the interacting electromagnetic field. Whilst an essential parameter in radiometric oceanographic studies, seawater permittivity is still an ill-known quantity. Until very recently calculated seawater permittivity values were based upon a proposition of Stogryn (1971). This interpolation function was derived from inaccurate permittivity measurements of saline solutions made by Lane and Saxton in the 1950s. Klein and Swift (1977) proposed an improved formula based upon newer and more accurate data for purewater and for various ionic aqueous solutions. However, the data was limited in frequency (a few spot values less than 10 GHz) and limited in its temperature range (10, 20, 25°C). The only data for natural seawater used in the construction of the model was that of Ho and Hall (1973) and Ho *et al.* (1974). They reported data at 1.43 and 2.653 GHz for a single salinity.

Between 1977 and 2005 only seven publications reported seawater permittivity data:

- Blue (1980),
- Stogryn *et al.* (1996),
- Ellison *et al.* (1996b, 1997, 1998, 2003),
- Meissner and Wentz (2004).

Blue reports data at 97, 103.8, 135.6 and 183.3 GHz for seawater at 20°C. Stogryn *et al.* (1996) gave laboratory permittivity measurements in the frequency range 7–14 GHz at six temperatures and ten representative salinities. Ellison *et al.* (1996b, 1997, 1998, 2003) also reported laboratory measurements of natural seawater, with six representative salinities, over the temperature range −2 to 30°C and for frequencies between 3 and 105 GHz. Meissner and Wentz used radiometric microwave data to deduce seawater permittivity in the frequency range 30–100 GHz. This recent work has resulted in four proposed interpolation functions for the permittivity of seawater as a function of frequency, temperature and salinity. We shall discuss each of them in Section 5.2.4.2.

5.2.4.1 The conductivity of seawater

The conductivity of seawater σ, which is of the order of 2 or 3 S/m, has an important influence upon the measured permittivity, and it is essential to have an accurate description in terms of temperature and salinity. Fortunately, Lewis (1978, 1980), Culkin and Smith (1980) have established an internationally accepted interpolation function for the ionic conductivity of seawater. For a temperature T°C and salinity S‰ the conductivity $\sigma(T, S)$, in siemens per metre, is given by:

$$\sigma(T, S) = \sigma(T, 35) \cdot P(S) \cdot Q(T, S) \tag{5.20}$$

where:

$$\begin{aligned}\sigma(T, 35) = {} & 2.903602 + 8.607 \cdot 10^{-2}T + 4.738817 \cdot 10^{-4}T^2 \\ & - 2.991 \cdot 10^{-6}T^3 + 4.3041 \cdot 10^{-9}T^4\end{aligned}$$

$$P(S) = S\frac{(37.5109 + 5.45216S + 0.014409S^2)}{(1004.75 + 182.283S + S^2)}$$

$$Q(T, S) = 1 + \frac{\alpha_0(T - 15)}{T + \alpha_1}$$

$$\alpha_0 = \frac{6.9431 + 3.2841S - 0.099486S^2}{84.85 + 69.024S + S^2}$$

$$\alpha_1 = 49.843 - 0.2276S + 0.00198S^2$$

This expression will be used in the interpolation function that we propose in Section 5.2.5.

5.2.4.2 The existing seawater interpolation functions

There are four proposed interpolation functions for the permittivity of seawater as a function of frequency, temperature and salinity. Each function was constructed by the respective authors largely based on their own experimental results and in ignorance of the other results, obtained more or less simultaneously. The available numerical data from all this work is gathered together in Appendix C.

Stogryn *et al.* (1996) reported laboratory measurements of the permittivity of seawater for frequencies between 7 and 14 GHz in 1 GHz steps, salinities of 0, 2.09, 3.92, 7.17, 11.2, 15.46, 20.14, 22.47, 31.68, 35.96‰ and temperatures 0, 4, 10,

15, 20, 25, 30°C. They represented the permittivity by the following double Debye function:

$$\varepsilon(\nu, T, S) = \frac{\varepsilon_s(T,S) - \varepsilon_1(T,S)}{1 - i2\pi\nu\tau_1(T,S)} + \frac{\varepsilon_1(T,S) - \varepsilon_\infty(T,S)}{1 - i2\pi\nu\tau_2(T,S)} + \varepsilon_\infty(T,S) + i\frac{17.9751\sigma(T,S)}{\nu}$$

where ν is in GHz, relaxation times in ns and T in °C

$$\varepsilon_s(T,S) = \varepsilon_s(T,0) \cdot a(T,S)$$
$$\tau_1(T,S) = \tau_1(T,0) \cdot b(T,S)$$
$$\varepsilon_1(T,S) = \varepsilon_1(T,0) \cdot c(T,S)$$
$$\tau_2(T,S) = \tau_2(T,0) \cdot d(T,S)$$
$$\varepsilon_\infty(T,S) = \varepsilon_\infty(T,0) \cdot e(T,S)$$

The authors considered that the restricted frequency range was far too low to determine the functions $d(T,S)$ and $e(T,S)$, so they imposed $d(T,S) = 1$ and $e(T,S) = 1$. Using a subset of the literature data for purewater, they gave:

$$\varepsilon_s(T,0) = \frac{3708.6 - 82.168T}{421.854 + T}$$
$$\varepsilon_1(T,0) = 0.0787\varepsilon_s(T,0)$$
$$2\pi\tau_1(T,0) = \frac{255.04 + 0.7246T}{(49.25 + T)(45 + T)}$$
$$2\pi\tau_2(T,0) = 0.00628$$
$$\varepsilon_\infty(T,0) = 4.05 + 0.0186T$$

The functions $a(T,S)$, $b(T,S)$ and $c(T,S)$ were fitted to their experimental data using a least squares fit. The result is:

$$a(T,S) = 1 - \frac{S(0.03838 + 0.00218S)(79.88 + T)}{(12.01 + S)(52.53 + T)}$$
$$b(T,S) = 1 - S\frac{(0.03409 + 0.002817S)}{(7.69 + S)} - ST\frac{(0.00246 + 0.00141T)}{1880 - 7.57T + T^2}$$
$$c(T,S) = a(T,S)$$

The conductivity term $\sigma(T,S)$ is calculated by the function described in 5.2.4.1.

Ellison *et al.* (1996b, 1998) and Guillou *et al.* (1998) reported permittivity measurements of natural seawater. Seawater samples from the N. Atlantic, E. and W. Mediterranean, N. Polar and the Gironde estuary provided representative salinities (23.2, 28, 30.024, 35, 38.024, 38.893). The frequency range was 3–20 GHz in 85 MHz steps and the temperature range was from −2 to 30°C in 1°C steps. The authors estimated that their results were accurate to better than 1 per cent. At each of the temperatures and salinities, the permittivity data could be accurately fitted to a single

Debye function and the corresponding Debye parameters estimated. The parameters were interpolated by polynomials as functions of temperature T and salinity S. The result is

For	a frequency ν in the range	$3 \leq \nu \leq 20$ GHz
	a temperature T in the range	$-2 \leq T \leq 30$°C
	a salinity S in the range	$20‰ \leq S \leq 40‰$

the permittivity $\varepsilon = \varepsilon' + i\varepsilon''$ of seawater is given by

$$\varepsilon'(\nu, T, S) = \varepsilon_\infty(T, S) + \frac{\varepsilon_s(T, S) - \varepsilon_\infty(T, S)}{1 + 4\pi^2\nu^2\tau^2(T, S)}$$

$$\varepsilon''(\nu, T, S) = \frac{(\varepsilon_s(T, S) - \varepsilon_\infty(T, S))2\pi\nu\tau(T, S)}{1 + 4\pi^2\nu^2\tau^2(T, S)} + \frac{17.9751\sigma(T, S)}{\nu}$$

where

$$\varepsilon_s(T, S) = a_1(T) + Sa_2(T)$$

$$\tau(T, S) = (b_1(T) + Sb_2(T)) \cdot 10^{-3} \quad \text{(ns)}$$

$$\sigma(T, S) = c_1(T) + Sc_2(T) \quad \text{(S/m)}$$

$$\varepsilon_\infty(T, S) = d_1(T)$$

with

$$a_1(T) = 81.82 - 6.0503 \cdot 10^{-2}T - 3.1661 \cdot 10^{-2}T^2 + 3.1097 \cdot 10^{-3}T^3 - 1.1791 \cdot 10^{-4}T^4 + 1.4838 \cdot 10^{-6}T^5$$

$$a_2(T) = -0.12544 - 9.4037 \cdot 10^{-3}T + 9.5551 \cdot 10^{-4}T^2 - 9.0888 \cdot 10^{-5}T^3 + 3.6011 \cdot 10^{-6}T^4 - 4.713 \cdot 10^{-8}T^5$$

$$b_1(T) = 17.303 - 0.66651T + 5.1482 \cdot 10^{-2}T^2 + 1.2145 \cdot 10^{-3}T^3 - 5.0325 \cdot 10^{-5}T^4 + 5.8272 \cdot 10^{-7}T^5$$

$$b_2(T) = -6.272 \cdot 10^{-3} + 2.357 \cdot 10^{-4}T + 5.075 \cdot 10^{-4}T^2 - 6.3983 \cdot 10^{-5}T^3 + 2.463 \cdot 10^{-6}T^4 - 3.0676 \cdot 10^{-8}T^5$$

$$c_1(T) = 8.6374 \cdot 10^{-2} + 3.9606 \cdot 10^{-2}T - 4.121 \cdot 10^{-4}T^2$$

$$c_2(T) = 7.7454 \cdot 10^{-2} + 1.687 \cdot 10^{-3}T + 1.937 \cdot 10^{-5}T^2$$

$$d_1(T) = 6.4587 - 4.204 \cdot 10^{-2}T - 6.5881 \cdot 10^{-3}T^2 + 6.4924 \cdot 10^{-4}T^3 - 1.2328 \cdot 10^{-5}T^4 + 5.0433 \cdot 10^{-8}T^5$$

Ellison *et al.* (1997, 2003) reported permittivity data for the range 30–105 GHz, at 31 roughly equally spaced frequencies and for temperatures of −2, 5, 10, 15, 20, 25, 30°C. The estimated uncertainty in the data is 3 per cent. The changes in permittivity due to salinity variations in the range 20‰ to 40‰ are smaller than 3 per cent and so the measurements were made for a unique value of salinity, namely the mean value, 35‰.

For the frequency range 30–105 GHz a single Debye function does not represent the data to within the limits of experimental accuracy. The data could be interpreted either by a Cole–Cole model or by two Debye functions. They chose to use two Debye functions because of the evidence that there is a second relaxation process in the neighbourhood of 150 GHz.

For a frequency ν in the range $30 \le \nu \le 105$ GHz
a temperature T in the range $-2 \le T \le 30°\mathrm{C}$
a salinity S in the range $20‰ \le S \le 40‰$

the permittivity $\varepsilon = \varepsilon' + \mathrm{i}\varepsilon''$ of seawater is given by:

$$\varepsilon(\nu, T, S) = \frac{\Delta_1(T)}{1 - \mathrm{i}2\pi\nu\tau_1(T)} + \frac{\Delta_2(T)}{1 - \mathrm{i}2\pi\nu\tau_2(T)} + \mathrm{i}\frac{17.9751\sigma(T,35)}{\nu} + \varepsilon_\infty(T)$$

where

$$\Delta_1(T) = 68.396 - 0.40643T + 0.022832T^2 - 0.00053061T^3$$

$$\Delta_2(T) = 4.7629 + 0.1541T - 0.033717T^2 + 0.00084428T^3$$

$$\tau_1(T) = (17.535 - 0.61767T + 0.0089481T^2) \cdot 10^{-3} \quad \text{(ns)}$$

$$\tau_2(T) = (3.1842 + 0.019189T - 0.010873T^2 + 0.00025818T^3) \cdot 10^{-3} \quad \text{(ns)}$$

$$\varepsilon_\infty(T) = 5.2214 - 0.022532T + 0.013764T^2 - 0.0023272T^3 + 0.00012962T^4 - 0.0000022464T^5$$

Meissner and Wentz (2004) used the laboratory data of Stogryn *et al.* (1996) for seawater and information obtained from microwave radiometric studies at selected frequencies (7, 10, 18, 24, 37, 85.5 and 89) in the interval 7–100 GHz. Their objective was to cover the temperature range $-20 \le T \le 30°\mathrm{C}$, the salinity range $0 \le S \le 40‰$ and the frequency range $0 \le \nu \le 1000$ GHz with a single interpolation function. Their function is based upon that used by Stogryn *et al.* (1996), namely:

$$\varepsilon(\nu, T, S) = \frac{\varepsilon_s(T,S) - \varepsilon_1(T,S)}{1 - \mathrm{i}\nu/\nu_1(T,S)} + \frac{\varepsilon_1(T,S) - \varepsilon_\infty(T,S)}{1 - \mathrm{i}\nu/\nu_2(T,S)} + \varepsilon_\infty(T,S) + \mathrm{i}\frac{17.9751\sigma(T,S)}{\nu}$$

where

$$\begin{aligned}
\varepsilon_s(T,S) &= \varepsilon_s(T,0)\exp(b_0S + b_1S^2 + b_2TS)\\
\nu_1(T,S) &= \nu_1(T,0)\{1 + S(b_3 + b_4T + b_5T^2)\}\\
\varepsilon_1(T,S) &= \varepsilon_1(T,0)\exp(b_6S + b_7S^2 + b_8TS)\\
\nu_2(T,S) &= \nu_2(T,0)\{1 + S(b_9 + b_{10}T)\}\\
\varepsilon_\infty(T,S) &= \varepsilon_\infty(T,0)\{1 + S(b_{11} + b_{12}T)\}
\end{aligned}$$

and

$$\begin{aligned}
\varepsilon_s(T,0) &= \frac{3708.6 - 82.168T}{421.854 + T}\\
\varepsilon_1(T,0) &= a_0 + a_1T + a_2T^2\\
\nu_1(T,0) &= \frac{45 + T}{a_3 + a_4T + a_5T^2} \quad \text{(GHz)}\\
\varepsilon_\infty(T,0) &= a_6 + a_7T\\
\nu_2(T,0) &= \frac{45 + T}{a_8 + a_9T + a_{10}T^2} \quad \text{(GHz)}
\end{aligned}$$

The coefficients a_i and b_i are given in Table 5.1.

The conductivity term $\sigma(T,S)$ is calculated by the expression (5.20) of Section 5.2.4.1.

Table 5.1 Coefficients used in the Meissner–Wentz interpolation function

i	a_i	b_i
0	+5.723	−0.00356417
1	+0.022379	+0.00000474868
2	−0.00071237	+0.0000115574
3	+5.0478	+0.00239357
4	−0.070315	−0.000031353
5	+0.00060059	+0.000000252477
6	+3.6143	−0.00628908
7	+0.028841	+0.000176032
8	+0.13652	−0.0000922144
9	+0.0014825	−0.0199723
10	+0.00024166	+0.000181176
11		−0.00204265
12		+0.000157883

The form of the interpolation function, and in particular the terms $(45 + T)$, were chosen to comply with the existence of a hypothetical second-order thermodynamical phase transition in water at about -45°C. Experimental evidence is against the existence of such a transition. This point will be further discussed in Section 5.2.5.1.

Meissner and Wentz do not estimate of the precision of the interpolation function. However, when describing the methodology used to exploit the microwave data, they assert that they expect the function should give a more precise estimate for ε'' than for ε' and that they could not make any assessment about the quality of their fit for ε' when $\nu > 30$ GHz.

5.2.4.3 Comparison between the interpolation functions

With the above interpolation functions we are in a classic situation: 'A man who has precisely one wristwatch always knows the "right" time. A man with two watches never knows the "right" time'. Each of the above interpolation functions gives a good representation of the dataset used to construct the function, but there are notable differences when we compare the *different* interpolation functions. This situation is not at all unexpected. Permittivity measurement of aqueous solutions is a difficult art to practice and in any given dataset the possibility, that either the random experimental uncertainty is greater than estimated or that there is some systematic error, always exists. This is certainly true for the permittivity data for purewater and will, no doubt, be confirmed for seawater when more measurements become available. The only way around this difficulty is to combine as many independent datasets as possible in the construction of the interpolation function. We can however make a few useful observations.

First, Ellison *et al.* (1997, 2003) claimed that for frequencies greater than 30 GHz, salinity variations produced a negligible effect upon their measured permittivities. We used the Meissner–Wentz formula to calculate ε at 20‰ and 40‰ at representative temperatures for a frequency of 30 GHz. For higher frequencies the differences are smaller. The results, shown in Table 5.2, support the claims of Ellison *et al.*

Table 5.2 Effect of salinity variations at 30 GHz according the Meissner–Wentz model

	ε'		ε''	
°C	$S = 20$	$S = 40$	$S = 20$	$S = 40$
0	12.14	12.25	23.08	23.81
10	17.00	17.10	28.28	28.65
20	22.86	22.73	32.2	32.35
30	28.89	28.43	34.20	34.26

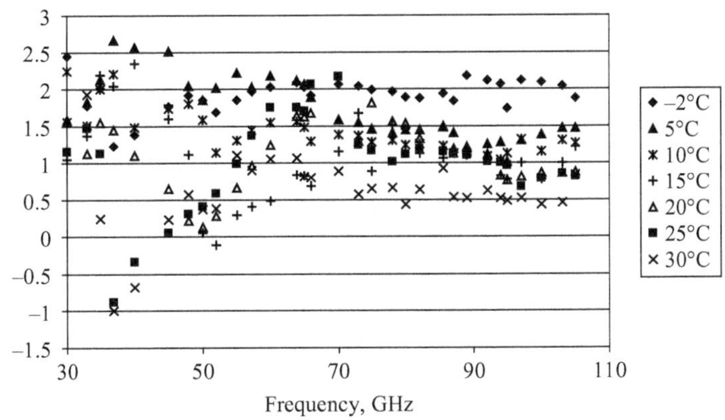

Figure 5.3 *Differences between the experimental data (Table 1A, Appendix C) for ε′ and the Meissner–Wentz interpolation*

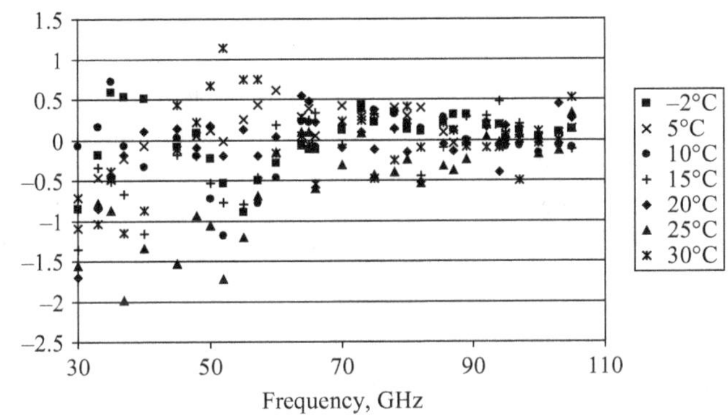

Figure 5.4 *Differences between the experimental data (Table 1A, Appendix C) for ε″ and the Meissner–Wentz interpolation*

Second, we compare the Meissner–Wentz formula, for a salinity of 35‰, with the data in Table 1A in Appendix C. This data was not used in the elaboration of the function and so serves as a test set. The differences between the calculated and measured values of ε' are shown in Figure 5.3 and the differences between the measured and calculated values of ε'' are shown in Figure 5.4.

We observe in Figure 5.3, that at all temperatures and all frequencies, the differences lie systematically between 1 and 2.5. On the other hand, in Figure 5.4, for all temperatures and frequencies, the differences are spread randomly about zero and within a strip slightly larger than ± 0.5.

The data in Figure 5.4 show that the Meissner–Wentz interpolation function for ε'' differs from the measured values of ε'' to within the experimental uncertainty. This suggests that the ε'' data obtained from the radiometric studies, which Meissner and

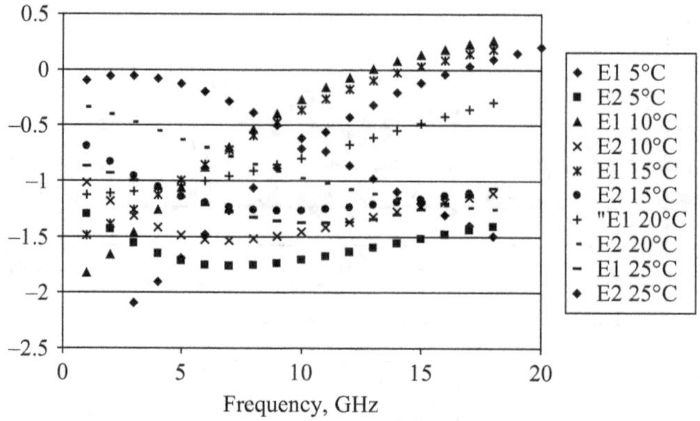

Figure 5.5 Differences between the values of ε′, respectively ε″, calculated using the Meissner–Wentz and the Ellison interpolation formulae, for a salinity of 20‰

Wentz consider to be fairly accurate, correspond very well with the data in Table 1A of Appendix C. These two datasets are totally independent, and their concordance can be interpreted as supporting the claim that the 3 per cent error estimate for ε'', given by Ellison *et al.*, is close to the truth. If this is so, then their error estimate of 3 per cent for ε' in Table 1A is also close to the truth, because the complicated extraction process of the permittivity values from the measured data precludes a systematic error in one of (ε', ε'') and not in the other. The systematic differences between the Meissner–Wentz interpolation for ε' and the experimental data, as shown in Figure 5.3, are probably due to a systematic error in the interpolation formula. This is a reasonable possibility, since Meissner and Wentz remark that they could not make any assessment about the quality of their fit for ε' at frequencies greater than 30 GHz.

Finally, we compare the differences between the Meissner–Wentz interpolation and the Ellison *et al.* interpolation over the frequency range 3–20 GHz, and at typical temperatures between −2 and 30°C. The results for both ε' and ε'' at two different salinities are shown in Figures 5.5 and 5.6. It is clear that the Meissner–Wentz function gives systematically larger values for ε' and for ε'', at almost all temperatures and frequencies, than the Ellison function. The two functions were constructed using two different data sets and each function closely represents its respective data set. Thus, the differences are probably due to systematic differences in the original data.

It is pointless to continue making further comparisons. The proposed interpolation functions represent the best that could be done when the various authors obtained their data, and they have served their purpose. The logical step is to combine all the current data and make a new interpolation function that will tend to reduce some of the systematic errors and pave the way for future improvements when more independent permittivity data becomes available.

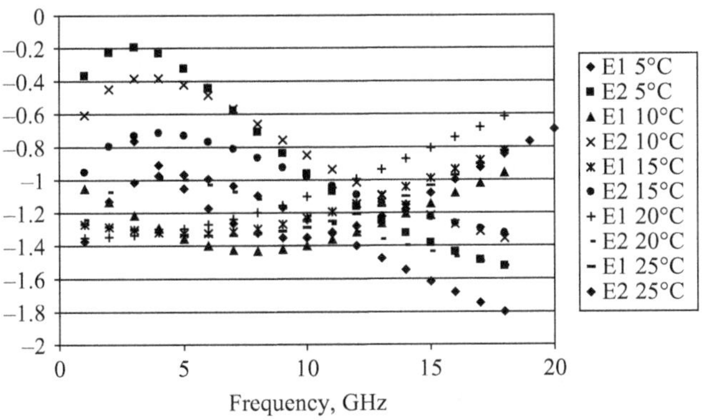

Figure 5.6 *Differences between the values of ε′, respectively ε″, calculated using the Meissner–Wentz and the Ellison interpolation formulae, for a salinity of 40‰*

5.2.5 *A new water interpolation function*

Our objective is to propose an interpolation function for water (including seawater) over the temperature range 0–30°C, the salinity range 0–40‰ and the frequency range 0–500 GHz. There is clear experimental evidence that there are two relaxation frequencies in the range 0–500 GHz and no indication, for the moment, that a third relaxation process influences the permittivity. Several authors (e.g. Mizoguchi *et al.* (1992)) have suggested, on the basis of limited and imprecise experimental data, that the permittivity in the interval 500–2000 GHz will be influenced by processes in the far infrared. But, as we already have little data in the interval 100–500 GHz, we limit our interpolation to the frequency range 0–500 GHz. However, as we shall verify in Section 5.2.5.6, the extrapolation, for purewater, up to 1000 GHz is compatible with the available data.

5.2.5.1 The interpolation function

The basic interpolation function that we use is the 'two Debye' model:

$$\varepsilon(\nu, T, S) = \frac{\varepsilon_s(T,S) - \varepsilon_1(T,S)}{1 - i2\pi\nu\tau_1(T,S)} + \frac{\varepsilon_1(T,S) - \varepsilon_\infty(T,S)}{1 - i2\pi\nu\tau_2(T,S)} + \varepsilon_\infty(T,S) + j\frac{17.9751\sigma(T,S)}{\nu} \tag{5.21}$$

We must now decide upon the mathematical forms for the parameters $\varepsilon_s(T,S)$, $\varepsilon_1(T,S)$, $\tau_1(T,S)$, $\tau_2(T,S)$ and $\varepsilon_\infty(T,S)$. It is reasonable to base our choice upon functions which have a physical sense for purewater and modify them slightly to take into account the factor salinity.

The only Debye parameter for purewater that is known precisely over a wide temperature range is the static dielectric constant $\varepsilon_s(T,0)$. It is given (Ellison *et al.* 1996a),

Table 5.3 Calculated Debye parameters using data over the frequency range 0–30 GHz

°C	ε_∞	τ_1(ps)	N° data values
0	4.499	17.57	12
5	5.669	14.97	12
10	5.412	12.50	20
15	6.079	10.89	15
20	5.295	9.40	58
25	5.474	8.28	133
30	5.825	7.35	38
40	5.319	5.84	18
50	4.641	4.80	20
60	2.452	3.85	9

with an error of ± 0.03 units, over the temperature range $-35° \leq T \leq 100°$C, by

$$\varepsilon_s(T, 0) = A \exp(-bT) \tag{5.22}$$

where $A = 87.85306$ and $b = 0.00456992$. This functional form does have a theoretical basis and the constant b has an interpretation in terms of hydrogen bonding in liquid water.

There is enough experimental data over the frequency range 1–30 GHz to estimate the values of $\tau_1(T, 0)$ and $\varepsilon_\infty(T, 0)$ at ten canonical temperatures (Table 5.3). We can use this data to test several proposed theoretical models that are supposed to represent relaxation times.

Lamkaouchi (1992) compared the 'best fit' parameters of these models with the above relaxation times. As is customary, the temperatures T are now in kelvin.

- The Eyring (1941) model:

  $$\tau(T) = \frac{A}{T} e^{B/T}$$

  The best fit gave $A = 2.3722797 \cdot 10^{-12}$ and $B = 2075.58868$. The mean percentage difference between the model and the above relaxation times was 2.38 per cent.
- The Nabokov (1987) model:

  $$\tau(T) = A \exp\left(\frac{B}{T}\right)^2$$

  The 'best fit' gave $A = 1.697974 \cdot 10^{-13}$ and $B = 588.66747$. The mean percentage difference between the model and the above relaxation times was 0.95 per cent.
- The Hill (1968) model:

  $$\tau(T) = A \exp(B/(T - T_0))$$

The 'best fit' gave $A = 1.12569 \cdot 10^{-13}$ and $B = 719.064347$ and $T_0 = 130.82654$. The mean percentage difference between the model and the above relaxation times was 0.76 per cent. The singularity in the relaxation time at $T = T_0$ should be less than the glass transition temperature for water, which has been estimated to be in the region of 135 K.

– The Speedy (1976) model:

$$\tau(T) = \frac{A}{T}\left(\frac{T}{T_s} - 1\right)^{\gamma}$$

– The 'best fit' gave $A = 4.127754 \cdot 10^{-10}$, $T_s = 219.90747$ and $\gamma = -1.7313$. The mean percentage difference between the model and the above relaxation times was 0.68 per cent. The singularity in the relaxation time at $T = T_s$ corresponds to a hypothetical second-order thermodynamical phase transition that was supposed to occur around 228 K.

The possible existence of the second-order transition was the subject of much speculation. However, experimental measurements of various water mixtures at temperatures down to 135 K by Noyel *et al.* (1992, 1994) seem to indicate that the transition does not, in fact, exist. Consequently, we suggest that the temperature variation of the two relaxation times for purewater be modelled by the functions

$$\tau_1(T, 0) = A\exp(B/(T - T_0)) \tag{5.23}$$

We now revert to temperature T in degrees Celsius. The problem is what to do about $\varepsilon_\infty(T, 0)$? As we explained above, $\varepsilon_\infty(T, 0)$ should be considered a 'fudge factor' which seems to vary slightly with T. We can propose nothing better than taking a linear function of $\varepsilon_\infty(T, 0) = a + bT$.

We have no theoretical model, and very little experimental data, for the changes in the Debye parameters as a function of salinity. Since the perturbations caused by salinity variations are relatively small and apparently 'smooth', we suggest the introduction of first and second-order polynomials in S, which will correspond in some sense to the Taylor series of unknown functions. It is probable that the introduction of new, or more precise, experimental permittivity data in low temperature and high frequency regions will impose the introduction of higher order terms for these polynomials.

Our proposed functions are

$$\varepsilon_s(T, S) = 87.85306\exp\{-0.00456992T - a_1S - a_2S^2 - a_3ST\} \tag{5.24}$$

$$\varepsilon_1(T, S) = a_4\exp\{-a_5T - a_6S - a_7ST\} \tag{5.25}$$

$$\tau_1(T, S) = (a_8 + a_9S)\exp\left(\frac{a_{10}}{T + a_{11}}\right) \quad \text{(ns)} \tag{5.26}$$

$$\tau_2(T, S) = (a_{12} + a_{13}S)\exp\left(\frac{a_{14}}{T + a_{15}}\right) \quad \text{(ns)} \tag{5.27}$$

$$\varepsilon_\infty(T, S) = a_{16} + a_{17}T + a_{18}S \tag{5.28}$$

and $\sigma(T, S)$ in siemens per metre is calculated by Formula (5.20) given in 5.2.4.1.

5.2.5.2 The parameter-fitting algorithm

We must now choose values for the parameters $a_1, \ldots, a_{18}$ so that the above function represents the data 'as closely as possible'. There are several possible interpretations for the requirement 'as closely as possible'. Fitting models with many parameters to noisy experimental data should never be a 'black box' computer program. There are choices to be made, and these choices are subjective. The simplest idea is to minimise the sums:

$$\sum_{i=1}^{N} (\varepsilon'_{\text{exp},i} - \varepsilon'_{\text{calc},i})^2 \quad \text{and} \quad \sum_{i=1}^{N} (\varepsilon''_{\text{exp},i} - \varepsilon''_{\text{calc},i})^2$$

There are large variations in the permittivity values over the range 0–500 GHz (from about 80 to 4 for ε' and from about 40 to 3 for ε'') and such a minimisation would tend to pull the fitting parameters towards a better fit to the numerically large values of ε' and ε'' at the expense of the numerically smaller values. This would tend to give a bad fit at the higher frequencies. Since we have error estimates for the permittivity data it is wiser to look for parameters that minimise the weighted sums:

$$\chi_1^2 = \frac{1}{N}\sum_{i=1}^{N} \frac{1}{\delta\varepsilon'_{\text{exp},i}{}^2} \left(\frac{100(\varepsilon'_{\text{exp},i} - \varepsilon'_{\text{calc},i})}{\varepsilon'_{\text{exp},i}} \right)^2$$

and (5.29)

$$\chi_2^2 = \frac{1}{N}\sum_{i=1}^{N} \frac{1}{\delta\varepsilon'_{\text{exp},i}{}^2} \left(\frac{100(\varepsilon''_{\text{exp},i} - \varepsilon''_{\text{calc},i})}{\varepsilon''_{\text{exp},i}} \right)^2$$

where $\delta\varepsilon'_{\text{exp},i}$ and $\delta\varepsilon''_{\text{exp},i}$ are the author-estimated percentage experimental errors. These latter sums compare the percentage difference between the experimental and the calculated value with the estimated experimental error. It has the advantage of giving equal importance to both numerically large and small values and gives more weight to the more accurate data, but it relies upon the assumption that the error estimates are realistic. Most experimenters try to give reasonable error estimates, but nobody can be certain that there is not some systematic error lurking in the data. The only solution to this problem is to have as much data as possible from independent sources and such systematic errors will gradually emerge.

If, for some choice of the parameters, a term in the above sums is less than one, then the function is within the experimental error and it does not make much sense to continue looking for parameter values that make the term even smaller. Thus, one should look for parameter sets for which the above sums are 'as small as possible' and for which the number of terms less than one is 'as large as possible' and the values of the parameters are 'realistic'. In practice, these requirements are often slightly contradictory!

If one is uniquely interested in interpolating the data a reasonable strategy would be to forget about physically realistic values of the parameters and look first for a set which strictly minimises the sum $\chi_1^2 + \chi_2^2$. However, one should not forget that for seawater we have only two independent datasets for frequencies less than 20 GHz and only one set for frequencies greater than 30 GHz. If any of these sets contain systematic errors, then it would not be a good idea for the interpolation function to try to approximate the data as closely as possible, because the data is wrong!

If one is looking for a physically realistic model, with the secret hope that an extrapolation outside the data range will produce accurate permittivity values, then we suggest looking for parameter sets for which

- the corresponding Debye parameters are physically realistic (what this could mean will be discussed below);
- the number of terms in the sums which are less than one is 'large', but not too large;
- this second condition automatically implies that the sum will be 'small', but one should not seek to make it very small.

The first constraint presupposes that the mathematical model of water permittivity is correct and the latter two constraints are intended to counter the effects of systematic errors and over-optimistic error estimates in the data.

In Section 5.2.5.4 we shall do both types of calculations. The parameters will be determined using a slight modification of the standard Levenberg–Marquart non-linear parameter-fitting algorithm. (See Chapter 15 of the computers 'bible': Press *et al.* (1992), for a description of the method and the basic FORTRAN routines.) A file containing all the experimental data used and a 'quick and dirty' Fortran program are available from the author to readers who wish to do their own numerical experimentation.

5.2.5.3 The datasets

As explained above, it is important to have permittivity data from as many independent sources as possible and to have 'reasonable' error estimates.

More than a 1000 values for the permittivity of purewater in the frequency range 0–100 GHz, 130 values in the frequency range 100–1000 GHz are cited in Ellison *et al.* (1996a) from over 60 different sources, and we do not reproduce them here. Additional measurements made after 1996 are given in Appendix C. Data from Lamkaouchi (1999) are given in Table 14A. Rønne *et al.* (1997) measured the permittivity of purewater at various temperatures over the frequency range 100–2000 GHz. They analysed their data in terms of two Debye relaxations. The relevant parameters are given in Table 15A. Dr. C. Rønne very kindly supplied the original experimental permittivity data and they are given in Tables 16A and 18A.

For the permittivity of seawater we are much less rich in independent measurements. There are only three datasets:

1. Ellison *et al.* (1996b) contains 594 data points at 6.8, 10.65, 18.7 GHz, temperatures −2 to 30°C in 1°C steps and salinities 23.2, 28, 30.025, 38.024, 38.89‰. The data are given in Tables 2A, 3A and 4A of Appendix C. The authors in

fact measured the permittivity of the seawater samples over the frequency range 3–20 GHz in 85 MHz steps and the temperature range −2 to 30°C in 1°C steps. At each temperature and salinity the data were accurately fitted to a single Debye function. The corresponding Debye parameters are reproduced in Tables 5A and 6A of Appendix C. We have used these tables to reconstruct the permittivity values in 1 GHz steps over the interval 3–20 GHz at each temperature and salinity;

2. Ellison *et al.* (1997, 2003) provide 217 data points in the ranges: Frequency 30–105 GHz, salinity 35‰ and temperatures −2, 5, 10, 15, 20, 25, 30°C. The data are reproduced in Table 1A of Appendix C;
3. Stogryn *et al.* (1996) established their interpolation function using data in the frequency range 7–14 GHz in 1 GHz steps, salinities 0, 2.09, 3.92, 7.17, 11.2, 15.46, 20.14, 22.47, 31.68, 35.96‰ and temperatures 0, 4, 10, 15, 20, 25, 30, 37°C. The data is also given in Appendix C.

The first two datasets are independent, even though the same laboratory team reports them, because the measuring apparatus and techniques used were totally different. The first dataset was obtained using a vector network analyser to measure the transmission coefficient of a seawater sample trapped in a coaxial waveguide. The technique is described in Ellison *et al.* (1996b). The data were reproducible to within 1 per cent, but there was a possible source of systematic error in the measured electromagnetic parameters that could not be eliminated. The vector network analyser has to be calibrated before each experimental run in order to fix a reference for the phase measurements. The manufacturer's calibration kit must be used within the temperature range 20–25°C in order to be valid and during an experimental run the measuring cell is at temperatures between −2 and 30°C. Consequently, the phase reference may not be constant and this will induce a systematic error in the phase of the complex transmission coefficient. Also, as we did runs over a large temperature range during 48 h periods, there was an additional source of error due to drift in the analyser's electronics. We finally estimated the uncertainty in the data to be less than 3 per cent.

The second dataset was obtained by a free-air, variable thickness transmission cell method described in Lamkaouchi *et al.* (2003). The apparatus was custom designed and the rather complicated data analysis method separated the permittivity of the sample from variations due to drift from the system electronics. The experimental uncertainty was estimated to be 3 per cent, rather than the usual 10 per cent for lossy liquids in the frequency range 30–100 GHz.

Stogryn *et al.* (1996) used a vector network analyser and a variable thickness transmission cell, but they neither give any experimental details of the techniques nor an estimate for the experimental uncertainty. The sources of experimental and systematic errors are the same as in Ellison (1996b), so we have attributed the same error estimates to the data.

Unfortunately, Meissner and Wentz (2004) do not give any of the permittivity data that they used in constructing their interpolation function.

5.2.5.4 The interpolation parameters

We first look for a 'best fit to data' for $\varepsilon(\nu, T, S)$ in the ranges

$$0 \leq \nu \leq 500\ \text{GHz}, \quad 0 \leq T \leq 30°\text{C} \quad \text{and} \quad 0 \leq S \leq 40‰$$

We used a subset of the data for purewater, restricting it to temperatures less than 40°C and frequencies less than 500 GHz. This gave 1125 values. For seawater permittivity in the frequency range 30–105 GHz we used the data given in Table 1A of Appendix C. For the frequency range 3–20 GHz we used the data in Tables 2A, 3A, 4A, 7A and 8A and with all of Stogryn's data. A total of 5170 data values was used in the data fitting program.

With the parameters given in Table 5.4, both sums χ_1^2 and χ_2^2 are less then 0.4. We found that 4763 calculated values out of 5170 are within the estimated experimental uncertainty. All the seawater data in the frequency range 3–20 GHz are within 3 per cent of the calculated values. This corresponds to the experimental uncertainty. Most of the seawater data in the frequency range 30–105 GHz are within the 3 per cent experimental uncertainty and the values greater than 3 per cent are within 5 per cent of the calculated values. Almost all the remaining terms in the sum, corresponding to purewater, are less than one and the few terms slightly greater than one probably correspond to authors who were a little over-optimistic in their error estimates.

We now look for a 'realistic' choice of parameters. By realistic we mean that the parameters should have a physical interpretation in terms of the model which we have used to describe the permittivity. The temperature coefficient in the static dielectric constant $\varepsilon_S(T, 0)$ has an interpretation in terms of hydrogen bonding between water molecules. A similar interpretation can be given to the temperature coefficient $a(5)$ in the expression for $\varepsilon_1(T, 0)$ and the value given in Table 5.4 is not aberrant. Similarly the value of $\varepsilon_\infty(T, 0)$ given by $a(16)$ and $a(17)$ is not surprising. The only problem lies in the expressions for the temperature dependence of the two relaxation times. We chose to represent them by Hill functions. The functions have a singularity at a certain temperature T_0. In Hill's theory this temperature, when the relaxation times become infinite, should be less than the glass transition for water. If Hill's theory is close to reality, then both relaxation times should become infinite at the same temperature and this temperature should be around −130°C. In Table 5.4 the

Table 5.4 'Best' parameter values for the interpolation function over the ranges: Frequency 0–500 GHz, temperature 0–30°C, salinity 0–40‰

$a(1) = 0.46881323\text{E}{-02}$	$a(7) = 0.42082521\text{E}{-04}$	$a(13) = 0.39569374\text{E}{-06}$
$a(2) = -0.28047338\text{E}{-04}$	$a(8) = 0.25141829\text{E}{-03}$	$a(14) = 0.17748178\text{E}{+04}$
$a(3) = -0.55364942\text{E}{-05}$	$a(9) = -0.22227872\text{E}{-06}$	$a(15) = 0.48000000\text{E}{+03}$
$a(4) = 0.61299826\text{E}{+01}$	$a(10) = 0.49000690\text{E}{+03}$	$a(16) = 0.35813749\text{E}{+01}$
$a(5) = 0.14144372\text{E}{-02}$	$a(11) = 0.11493865\text{E}{+03}$	$a(17) = 0.11364924\text{E}{-01}$
$a(6) = -0.61842225\text{E}{-02}$	$a(12) = 0.12920010\text{E}{-04}$	$a(18) = -0.11830874\text{E}{-01}$

Table 5.5 'Realistic' parameter values for the interpolation function over the ranges: Frequency 0–500 GHz, temperature 0–30° C, salinity 0–40‰

$a(1) = 0.46606917E{-}02$	$a(7) = 0.34414691E{-}04$	$a(13) = 0.38957681E{-}06$
$a(2) = -0.26087876E{-}04$	$a(8) = 0.17667420E{-}03$	$a(14) = 0.30742330E{+}03$
$a(3) = -0.63926782E{-}05$	$a(9) = -0.20491560E{-}06$	$a(15) = 0.12634992E{+}03$
$a(4) = 0.63000075E{+}01$	$a(10) = 0.58366888E{+}03$	$a(16) = 0.37245044E{+}01$
$a(5) = 0.26242021E{-}02$	$a(11) = 0.12634992E{+}03$	$a(17) = 0.92609781E{-}02$
$a(6) = -0.42984155E{-}02$	$a(12) = 0.69227972E{-}04$	$a(18) = -0.26093754E{-}01$

parameters $a(11)$ and $a(15)$ correspond to 'singular' temperatures of -114.9°C and -480°C respectively. These values bear no relation to the glass transition temperature of water.

The question which now arises is: 'Is it possible to have a "good" approximation to the data with $a(11) = a(15)$ and that this common value is about 130 K'. The answer is Yes. The parameters in Table 5.5 have $a(11) = a(15) = 126.35$ and the sums χ_1^2 and χ_2^2 are both less then 0.43. Moreover, 4711 calculated values of ε' are within the experimental error and 4759 calculated values of ε'' are within the experimental error of the data. This is almost as good as the 'best' choice of the parameters (Table 5.4).

Either of the two sets of parameter values can be used to interpolate seawater permittivity; we personally prefer the set given in Table 5.5 because when the function is extrapolated it is likely to be less wrong than the function which uses the parameters in Table 5.4.

5.2.6 Extrapolations

There are no experimental data for seawater at frequencies greater than 105 GHz. However, our function for purewater, when extrapolated up to 1000 GHz and compared with the experimental data, is quite good. The graphs in Figure 5.7 illustrate the comparisons between calculated and experimental values of ε at canonical temperatures for which data exist. With such a good correlation for purewater we would expect that the extrapolation of the function for seawater will be just as close to reality up to 1000 GHz.

What can be said about extrapolation to temperatures less than 0°C? For supercooled water there is only one data set: Bertolini *et al.* (1982), frequency 9.61 GHz for temperatures down to -18°C. Figure 5.8 shows the extrapolated values and the measured permittivity. The comparison for ε' is quite good. Unfortunately, the comparison for ε'' begins to diverge slightly almost as soon as the temperature becomes negative. This is not unexpected, for two reasons. First, the value of ε'' is more sensitive than ε' to the influence of the second, and still ill-defined, relaxation in purewater. We would expect this influence upon extrapolations to be greater for higher frequencies in this temperature range. Second, the data for ε'' have a good chance of being less accurate than supposed. Bertolini *et al.* used a resonating cavity technique to determine the permittivity. With this method, notorious for systematic errors, ε' and

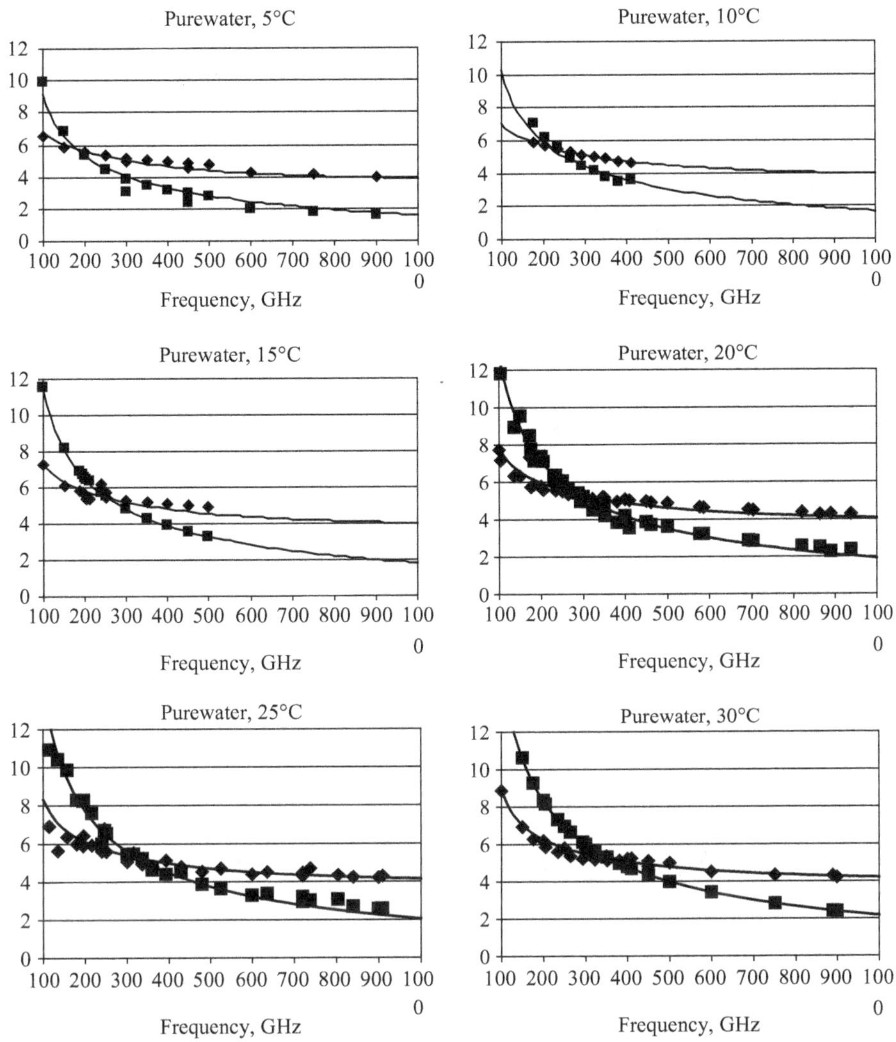

Figure 5.7 Experimental data for ε′ (symbol ♦) and ε″ (symbol ■) and the interpolations

ε'' are determined separately by two kinds of measurements (a resonance frequency shift for ε' and a broadening of a resonance peak for ε'') with reference to a 'known' sample. The first type of measurement is more accurate than the second. One could interpret Figure 5.8 as showing that the extrapolation for ε' is accurate, then infer that the extrapolation for ε'' is probably also just as accurate and that the differences are due to an experimental uncertainty in the data! It is unfortunate that there are no other data for super-cooled water with which to compare the Bertolini *et al.* measurements.

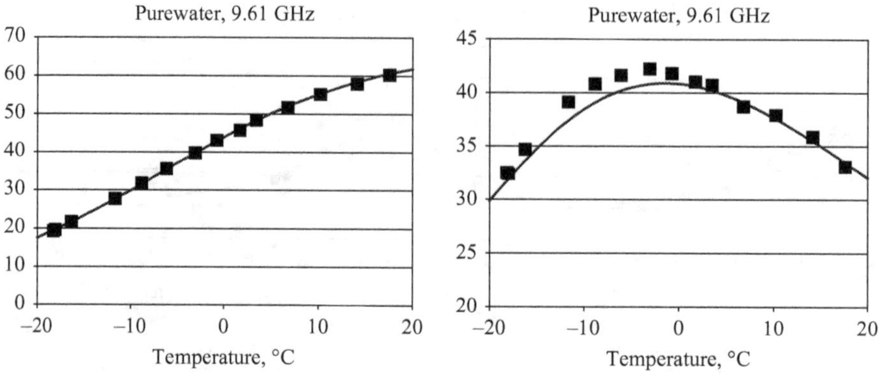

Figure 5.8 *Comparison between calculated values of ε′, ε″ and measured values (Bertolini et al., 1982)*

As far as the radiometric community is concerned, the situation is clear. At the present time there is no choice but to extrapolate the above function in the super-cooled region, with no estimate for the accuracy of the calculated permittivity.

5.2.7 Conclusion

We propose that the permittivity of seawater $\varepsilon(T, S, \nu)$ be represented over the ranges

$$0 \le T \le 30°\text{C}, \quad 0 \le S \le 40‰ \quad \text{and} \quad 0 \le \nu \le 1000 \text{ GHz}$$

by (5.21), where (5.24) to (5.28) and the parameters a_i have the values:

$a_1 = 0.46606917E{-}02$	$a_7 = 0.34414691E{-}04$	$a_{13} = 0.38957681E{-}06$
$a_2 = -0.26087876E{-}04$	$a_8 = 0.17667420E{-}03$	$a_{14} = 0.30742330E{+}03$
$a_3 = -0.63926782E{-}05$	$a_9 = -0.20491560E{-}06$	$a_{15} = 0.12634992E{+}03$
$a_4 = 0.63000075E{+}01$	$a_{10} = 0.58366888E{+}03$	$a_{16} = 0.37245044E{+}01$
$a_5 = 0.26242021E{-}02$	$a_{11} = 0.12634992E{+}03$	$a_{17} = 0.92609781E{-}02$
$a_6 = -0.42984155E{-}02$	$a_{12} = 0.69227972E{-}04$	$a_{18} = -0.26093754E{-}01$

and $\sigma(T, S)$ is calculated by (5.20) as given in Section 5.2.4.1.

This function represents the permittivity of purewater to within 1 per cent over the frequency range 0–20 GHz, to within 3 per cent over the frequency range 30–100 GHz and to within 5 per cent over the frequency range 100–1000 GHz. The permittivity of seawater is represented to within 3 per cent over the frequency range 3–105 GHz. These estimates also correspond to estimates for the experimental uncertainty in the measured data.

There is a huge gap in our knowledge of the permittivity of purewater and seawater for the frequency range 0–100 GHz and the temperature range −20 to +10°C. A second relaxation process occurs in this region and this relaxation renders the extrapolation from existing data probably inaccurate.

The above interpolation function relies heavily upon a unique dataset for the permittivity of seawater in the frequency range 30–100 GHz. It would be comforting to have another dataset for comparison.

There are no experimental results for the permittivity of seawater at frequencies greater than 105 GHz.

The radiometric community should actively encourage experimental measurements in the above mentioned regions. Dielectric spectroscopy is a difficult art to practice, and this work is unlikely to be done spontaneously.

5.3 Microwave dielectric properties of ice *Christian Mätzler*

5.3.1 Introduction

Water ice is one of the most transparent media at microwave frequencies with very low dielectric loss. Early measurements of the imaginary dielectric constant were often contradictory due to the limitation of the available techniques. In his review, Warren (1984) noticed an urgent need for more accurate measurements. New experiments started by Wegmüller (1986) in the frequency range, 2–10 GHz, using a resonator technique, and including a radiometer method up to 94 GHz by Mätzler and Wegmüller (1987). Later, Fujita *et al.* (1993), Matsuoka *et al.* (1993, 1996), Koh (1992, 1997) and Zhang *et al.* (2001) joined the effort with different techniques to clarify the behaviour of the dielectric loss. The progress achieved for pure ice by 1991 led Hufford (1991) to synthesise a model covering the frequency range from 0.1 MHz to 1 THz, and for temperatures ranging from 273 K down to about 200 K (radio frequencies) and to about 100 K (microwave and millimetre-wave range), respectively. Hufford realised that the measurements and the theoretical model of Mishima *et al.* (1983) can be used to formulate the frequency and temperature variation at millimetre and submillimetre wavelengths. By a different weighting of the available data, especially at high frequencies and very low temperature on the one hand, and near 273 K on the other hand, Hufford's formula was adjusted by Mätzler (1998). A similar adjustment was proposed by Jiang and Wu (2004). As an extension of the low-temperature data of Mishima *et al.* (1983), Zhang *et al.* (2001) made first measurements of the complex refractive index of pure ice at 240–270 K in the 250–1000 GHz range, using a time-domain technique based on femtosecond laser pulses. The available dielectric data are used here to improve and to test the dielectric models.

Attention was also paid to the effects of saline and acid impurities (Arcone *et al.*, 1986; Mätzler, 1987; Mätzler and Wegmüller, 1987; Fujita *et al.*, 1992; Corr *et al.*, 1993; Moore and Fujita, 1993; Matsuoka *et al.*, 1993). For a review of dielectric sea-ice properties, see Hallikainen and Winebrenner (1992).

5.3.2 *Dielectric properties of ice: real part*

For ordinary ice (Ih) the real part ε' of the dielectric constant of ice is denoted by ε_i'; it is independent of frequency from 10 MHz to about 300 GHz and only slightly dependent on temperature, the maximum of 3.19 being found at 273 K with the following linear dependence for $T > 240$ K (Mätzler and Wegmüller, 1987)

$$\varepsilon_i' = 3.1884 + 9.1 \cdot 10^{-4}(T - 273\text{ K}); \quad 243 \le T \le 273\text{ K} \tag{5.30}$$

At $T < 240$ K the temperature sensitivity decreases, and a constant value of $\varepsilon_i' \cong 3.10$ is found at $T < 100$ K (Gough 1972). A small anisotropy of single-crystal ice of about 1 per cent was measured at 9.7 GHz (Fujita *et al.* 1993). The frequency variation in the kilohertz to 10 GHz range of ε_i' was discussed by Fujita *et al.* (2000). They admit a possible decrease of ε_i' by 1 per cent with increasing frequency from 2 MHz to 5 GHz. However, by the Kramers–Kronig relations, this hypothesis would imply much higher dielectric loss than observed in the frequency range near 1 GHz, and therefore it must be rejected. Dispersion only starts with the approach of the vibration bands in the infrared, where ε_i' shows a slight increase, starting at about 300 GHz. The behaviour up to 1000 GHz is shown in Figure 5.9 by the data of Zhang *et al.* (2001).

In principle ε_i' depends on ionic and other impurities present in the ice. For low concentration, i.e. below 100 ppm, this dependence is weak and can be neglected. The effect on ε_i' due to salinity in the parts per thousand range of sea ice was reviewed by Hallikainen and Winebrenner (1992). For acid ice, see Moore and Fujita (1993) and Fujita *et al.* (2000).

5.3.3 *Dielectric properties of ice: imaginary part*

5.3.3.1 Purewater ice

Let the imaginary part ε'' of the dielectric constant of ice be denoted by ε_i''. At frequencies ν above 10 kHz and up to 1 GHz, ε_i'' is governed by the high-frequency tail of

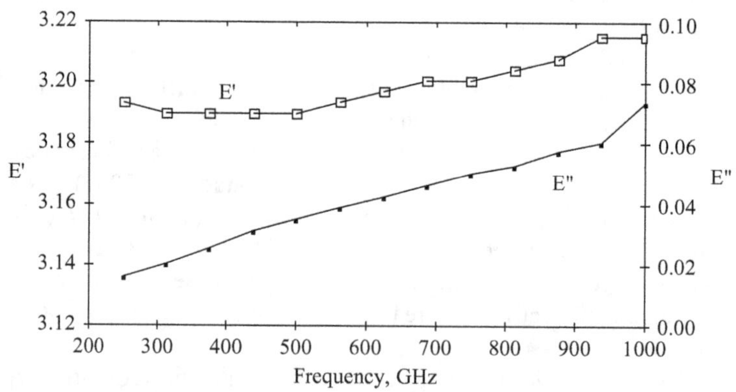

Figure 5.9 *Spectra of ε' and ε'' from 250 to 1000 GHz at $T = 243$ K (from Zhang et al., 2001)*

the orientational Debye relaxation (Auty and Cole, 1952; see the excellent overview of Petrenko 1993, also included in Petrenko and Whitworth 1999) with its relaxation frequency in the hertz to kilohertz range, strongly depending on temperature. Above 100 kHz the relaxation tail can be represented by a $1/\nu$ term, i.e. like a conductive component. This term is quite sensitive to ionic impurities, even at the level of a few parts per million (see e.g. Moore *et al.* 1993). The low-frequency tail of the translational optics vibration bands peaking in the infrared (Mishima *et al.* 1983) is a term proportional to ν. The sum of these two terms leads to a function with a deep minimum in the microwave range. At $T = 250$ K the minimum of $\varepsilon_\mathrm{i}'' \approx 10^{-4}$ is found near $\nu = 1$ GHz. With decreasing temperature the minimum and its frequency decrease, because the change of ε_i'' with changing temperature is mostly due to the decreasing relaxation term. Additional losses due to unexplained relaxation effects in the megahertz range were proposed in the literature; however, they could not be quantified, and thus were not confirmed. Therefore Hufford (1991) explained $\varepsilon_\mathrm{i}''(\nu, T)$ from 100 kHz to 1 THz by the two terms only. The formula was derived from a critical analysis of published data. The mentioned data of Auty and Cole (1952) and of Mishima *et al.* (1983) formed the basis. Unfortunately, these two datasets correspond to non-overlapping temperature ranges. Therefore, additional information was sought to extend Mishima's data at $T > 200$ K. Hufford found suitable measurements of Lamb (1946) at 10 GHz, and of Wegmüller (1986) at frequencies just above the transition region (2–10 GHz, see Figure 5.10). Hufford concluded that ε_i'' can be written as

$$\varepsilon_\mathrm{i}'' = \frac{\alpha}{\nu} + \beta\nu \tag{5.31}$$

where the first term is the unaltered relaxation term of Auty and Cole, and the second term is based on Mishima, corrected by the additional loss observed at temperatures above 200 K. In some papers (Mätzler and Wegmüller 1986; Matsuoka *et al.* 1996) a modified version of this equation includes an exponent C of frequency in the second term. The C value was found to be very close to 1. Here we follow the physically based approach of Mishima *et al.* (1983). Equation (5.31) was already proposed by Walford (1968); only the parameters were inaccurate at that time. Using the modified inverse temperature θ, defined by $\theta = T_0/T - 1$, where $T_0 = 300$ K, Hufford (1991) proposed the following expressions for α:

$$\alpha = (0.00504 + 0.0062\theta) \cdot \exp(-22.1\theta) \quad \text{(GHz)} \tag{5.32}$$

At very low temperatures (below 100 K) and at frequencies above 500 GHz an additional term was found by Mishima *et al.* (1983) due to sound waves interacting with electromagnetic waves by the orientational disorder of the dipole moments. This term increases ε'' with ν^3, and the corresponding term of the absorption coefficient γ_a increases ε'' with ν^4. Here we will include this term as a ν^2 term of β in Equation (5.33). The model of Mishima *et al.* (1983) considered cold water ice ($T < 200$ K); it is expressed by the following β parameter:

$$\beta_\mathrm{M} = \frac{B_1}{T}\frac{\exp(b/T)}{(\exp(b/T) - 1)^2} + B_2\nu^2 \tag{5.33}$$

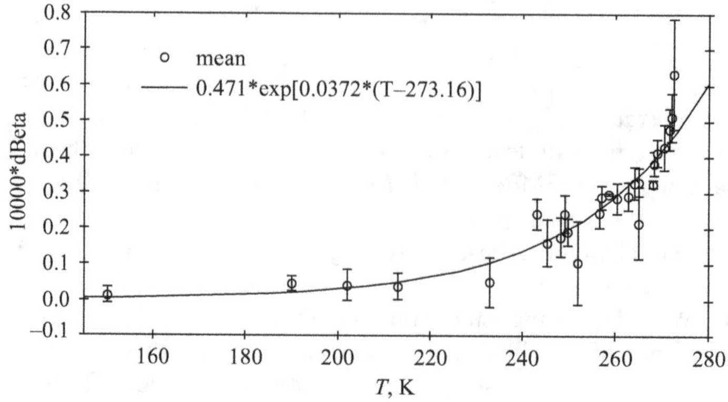

Figure 5.10 *Difference $\Delta\beta$ between the measured β and β_M versus temperature for the 3 to 35 GHz measurements of Mätzler and Wegmüller (1987), the 20–40 GHz measurements of Surdyk and Fujita (1995) and Matsuoka et al. (1996), and for >200 GHz by Mishima et al. (1983) and Zhang et al. (2001). Circles show the mean values of data at different frequencies, and the error bars show the standard deviation. The solid line is a fit to the data, representing the function of Equation (5.35)*

where ν is in GHz, $B_1 = 0.0207$ K/GHz, $b = 335$ K and $B_2 = 1.16 \cdot 10^{-11}/\text{GHz}^3$, independent of temperature. To account for higher temperature ($T > 200$ K), it is proposed to use a correction $\Delta\beta$,

$$\beta = \beta_M + \Delta\beta \tag{5.34}$$

This procedure was previously applied by Hufford (1991) and by Mätzler (1998). The correction is independent of frequency. This is shown in Figure 5.10 where a large number of experimental data at various frequencies of $\Delta\beta$ are shown versus temperature. At frequencies below 10 GHz the data are from Wegmüller (1986), at 20–40 GHz the data are from Mätzler and Wegmüller (1987), Surdyk and Fujita (1995) and from Matsuoka *et al.* (1996), and above 200 GHz the data are from Mishima *et al.* (1983) and from Zhang *et al.* (2001).

Not included for various reasons are the 94 GHz data of Mätzler and Wegmüller (1987), the data of Koh (1992, 1997) and the 5 and 10 GHz data of Matsuoka *et al.* (1996). In comparison with Hufford's analysis, the data of Lamb (1946) near 10 GHz were omitted due to their enhanced temperature dependence towards 273 K, indicating ionic impurities. The same holds for the data of Cumming (1952). The difference $\Delta\beta$ between the measured β and β_M versus temperature is shown in Figure 5.10 together with the fitted curve of Equation (5.35):

$$\Delta\beta = \exp(-9.963 + 0.0372(T - 273.16)); \quad T \text{ in K} \tag{5.35}$$

The standard deviation to the fit is $0.45 \cdot 10^{-5}$, and the multiple R^2 is 0.92. If individual data are used instead of the mean values, the standard deviation increases to $0.68 \cdot 10^{-5}$, and R^2 decreases to 0.84. The function for $\Delta\beta$ in (5.35) is very similar to the one

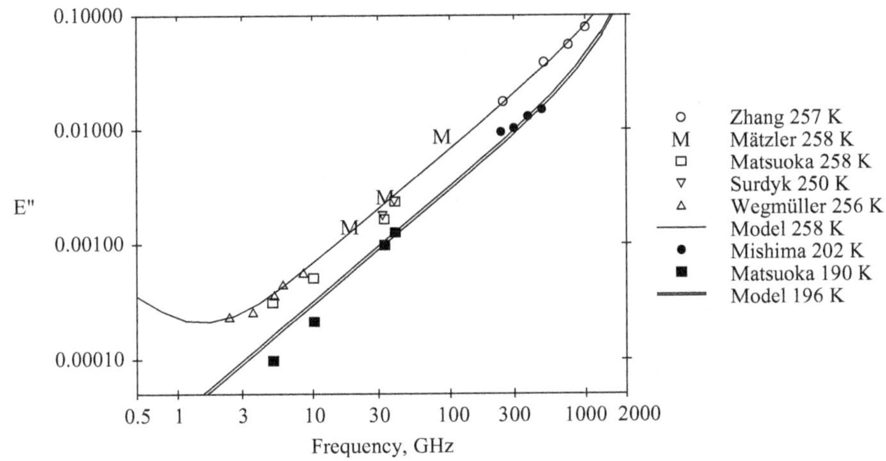

Figure 5.11 *Comparison of measured and computed imaginary dielectric constant of pure ice versus frequency near $T = 258$ K (open symbols and single line) and near 196 K (filled symbols and double line). Reference to measurements is given by the name of the first author and temperature*

proposed by Mätzler (1998), which was only based on the 3–9 GHz data of Wegmüller *et al.* (1986). The constant in the exponent was -10.02 instead of -9.963, and the factor was 0.0364 instead of 0.0372. The change to (5.35) leads to a slightly larger ε_i'' at $T > 200$ K. The difference is less than 1 per cent for $T < 245$ K, and it remains <3 per cent up to 273 K. Also the difference from Hufford's model is small, with two exceptions: First at low temperature and high frequency Hufford's approximation to Mishima's model becomes inaccurate, and second for $T > 260$ K the computed values of ε_i'' are larger by 1, 7 and 12 per cent with respect to the present model at $T = 265$, 270 and 273 K, respectively.

The measured values of ε_i'' versus frequency and computed spectra, using Equations (5.31)–(5.35), are shown in Figure 5.11 for T near 196 K and 258 K. Over most of the spectrum the linear increase of ε_i'' is dominant. At the low-frequency end, the high-frequency tail of the relaxation spectrum is visible at 258 K, and at the high-frequency end of the figure, the effect of the second term in Equation (5.33) becomes apparent at 196 K by the steepening of the curve. The agreement between the model and measurements is in general very good. Exceptions are the 94 GHz data of Mätzler which are thought to be too high, possibly due to scattering at small air bubbles (the 21 and 35 GHz data being much less affected), and the 5 and 10 GHz data of Matsuoka *et al.* (1996) which are obviously smaller than the model. In fact, the $\Delta\beta$ values are negative for these data. In addition to the data points shown in Figure 5.11, additional data of Koh (1992) in the 26–40 GHz range are larger than the model at 258 K, and the data of Koh (1997) at 75–110 GHz are below the model. In spite of these puzzling data, there is plenty of evidence that the model of Equations (5.31)–(5.35) is presently the best choice.

To quantify the uncertainty of ε_i'', it is noted that in the microwave range the error is dominated by the uncertainty of the $\beta\nu$ term in Equation (5.31). By definition, the uncertainty of β is the one of $\Delta\beta$. It is expressed by the standard deviation of the fit to the data shown in Figure 5.10. Since different types of measurements from different sources are included, the standard deviation $\sigma = 0.45 \cdot 10^{-5}$ of $\Delta\beta$ is an objective estimate of the uncertainty. Thus the error $\delta\varepsilon_i''$ of ε_i'' can be expressed as $\delta\varepsilon_i'' = 0.45 \cdot 10^{-5}\nu$, where ν is in GHz. This leads to a relative error $\delta\varepsilon_i''/\varepsilon_i''$ of about 5 per cent at 270 K, increasing to 7.5 per cent at 250 K and to 14 per cent at 200 K. This error is essentially independent of frequency for $\nu > 3$ GHz.

The temperature dependence of ε'' translates to the temperature dependence of the absorption coefficient γ_a. An interesting observation was provided by Woschnagg and Price (2001) who compared their own measurements of γ_a in the visible range with the behaviour in the UV, IR and millimetre wave ranges. Their result was that the relative temperature dependence $\gamma_a^{-1} \cdot d\gamma_a/dT$ of about 0.01/K ($230 < T < 250$ K) is independent of frequency from 30 GHz to the UV range. The temperature dependence of ε'' in Figure 5.11 is slightly larger than this value. If the temperature variation of ε' is considered in the computation of γ_a, then we get a close agreement, namely 0.013/K.

5.3.3.2 Effect of ionic impurities at microwave frequencies

One effect of ionic impurities is to increase the parameter α, i.e. losses are increased at low frequencies. Moore and Fujita (1993) showed that differences exist between saline and acid impurities. If brine is present, additional losses due to the relaxation of liquid water must be included; their maximum is in the microwave range near the effective relaxation frequency ν_0. A comparison of ε'' versus temperature at 3–9 GHz for slightly saline (sea salt, salinity $S_0 = 13$ ppm by weight) and pure ice, respectively, is presented in Figure 5.12 (data from Wegmüller, 1986). In order to model the dielectric data, we found that the inverse values $1/\varepsilon_i''$ are simply related with temperature. The same is true for $1/\Delta\varepsilon_i''$ where $\Delta\varepsilon_i'' = \varepsilon_{is}'' - \varepsilon_i''$ (ε_{is}'' stands for the saline sample and ε_i'' for pure ice) as shown in Figure 5.13, Linear functions of the form

$$1/\Delta\varepsilon_i'' = g_0 + g_1(273.16 - T); \quad T \text{ in K} \tag{5.36}$$

were fitted to the measured values of impure ice and using the model of Equations (5.31)–(5.35) for pure ice. The results are given in Table 5.6.

The fit parameters and fitted functions describing their frequency dependence are shown in Figure 5.14 (left). These functions are

$$g_0 = 1866 \exp(-0.317\nu); \quad g_1 = 72.2 + 6.02\nu \tag{5.37}$$

where ν is in GHz. They can be used for interpolation and possibly for extrapolation to other frequencies. As an example, Figure 5.14 (right) shows spectra of $\Delta\varepsilon_i''$ for an extreme extrapolation to the frequency range, 1–100 GHz. The curve at 270 K shows a peak near 10 GHz. This is reasonable due to the relaxation of brine within the ice. The behaviour is consistent with the dielectric loss of liquid-water films or veins, whose relaxation frequency ν_0 is given by the value of bulk water, i.e. $\nu_0 = 9$ GHz near 0°C. With decreasing temperature, more and more brine freezes,

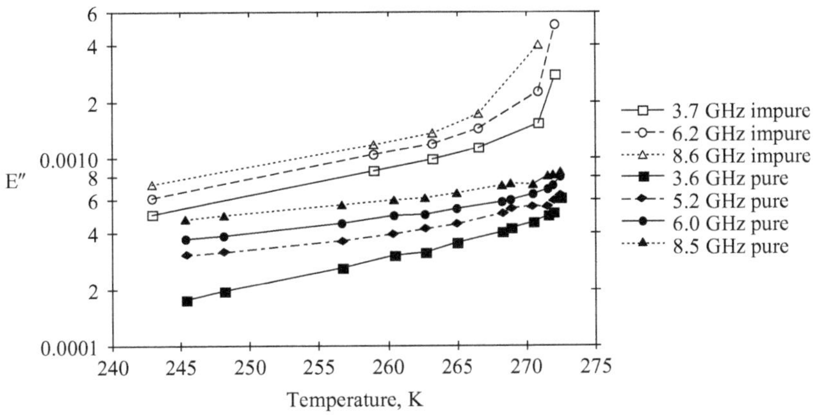

Figure 5.12 *Imaginary dielectric constants ε_i'' of pure (full symbols) and slightly saline (salinity of 13 ppm, open symbols) ice versus temperature with frequency as parameter, according to Wegmüller (1986)*

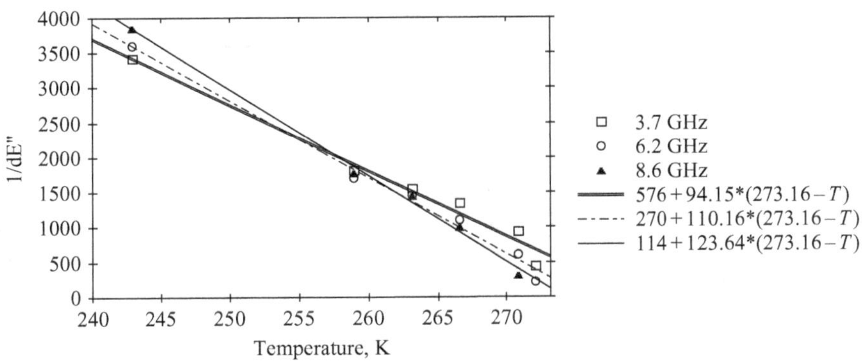

Figure 5.13 *Inverse values $1/\Delta\varepsilon_i''$ versus temperature for different frequencies. Measured data at 3.7, 6.2 and 8.6 GHz shown by symbols, and straight lines are linear fits to these data*

and ν_0 decreases. Assuming proportionality between $\Delta\varepsilon_i''$ and salinity S, we get the dielectric loss ε_{is}'' of saline ice from the parameters which have been modelled here:

$$\varepsilon_{is}'' = \varepsilon_i'' + \Delta\varepsilon_i'' S/S_0 \tag{5.38}$$

where $S_0 = 13$ ppm, ε_i'' is the dielectric loss of pure ice and $\Delta\varepsilon_i''$ is determined from Equations (5.36) and (5.37).

5.3.4 *Discussion and conclusion*

Some discussion has already been made in presenting the data and analysis. The complex dielectric constant of ice is characterised by an almost constant real part near 3.17

Table 5.6 Linear least-squares fits of $1/\Delta\varepsilon_i''$ according to (5.36), including R^2 and standard deviation σ

Frequency (GHz)	g_0	g_1	R^2	σ
3.7	576.1	94.15	0.98	167
6.2	269.6	110.16	0.99	135
8.6	113.7	123.64	0.995	104

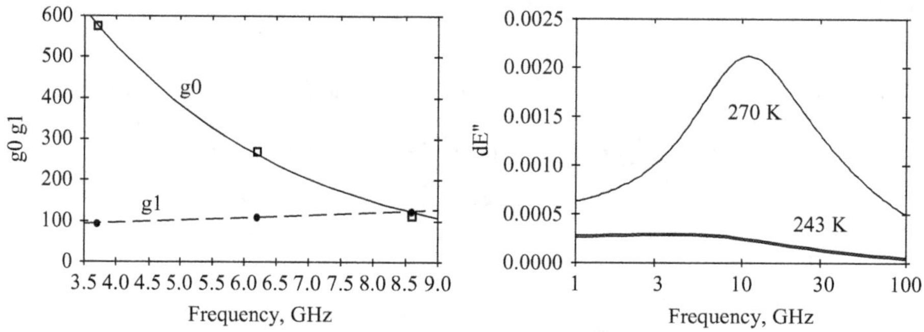

Figure 5.14 On the left: Frequency dependence of the coefficients shown in Table 5.6 (symbols) and functions of frequency fitting these coefficients (curves). On the right: Simulated spectra of $\Delta\varepsilon_i''$ for 243 K and for 270 K

with a 1 per cent anisotropy in the case of single crystals. The imaginary part ε_i'' may also be slightly anisotropic, but the effect has been neglected. With the minimum of ε_i'' near 1 GHz, ice is a very low-loss medium. At this frequency and lower, the Debye relaxation term of ice becomes noticeable. In the microwave range above 1 GHz, the theoretical model of Mishima *et al.*, (1983) was found to give the main contribution to ε_i'', as expressed by the β_M term. As this model was developed for $T < 200$ K, it cannot account for all effects. An empirically determined correction $\Delta\beta$ had to be added. This term increases with increasing temperature. Measurements over a very wide frequency range gave a consistent picture of a frequency-independent $\Delta\beta$ component. Physical reasons for this correction may be related to thermal expansion and to approximations made in the model (Mishima *et al.* 1983). A certain similarity between $\nu\Delta\beta$ and $\Delta\varepsilon_i''$ may lead to the hypothesis that very small amounts of impurities of the 'pure'-ice samples may cause a contribution to $\Delta\beta$. Although this effect cannot be excluded, there is a clear difference between the measured terms $\nu\Delta\beta$ and $\Delta\varepsilon_i''$. Since $\Delta\beta$ is frequency independent, its contribution to ε is linear with frequency. On the other hand, $\Delta\varepsilon_i''$ is clearly non-linear as shown in Figure 5.14. Therefore impurities cannot be the main source of $\Delta\beta$.

Saline ice gives rise to an increase $\Delta\varepsilon_i''$ of the dielectric loss, assumed to be proportional to salinity. This increase includes the increase of α and the increase due to the brine inclusions. As already found by Mätzler and Wegmüller (1986), the effect in the microwave range is mainly due to the brine being situated along the grain boundaries. The effect is largest near the relaxation frequency of water. For small salinity of a few parts per million, the relative contribution rapidly decreases with increasing frequency due to the increasing influence of pure-ice absorption.

5.4 Minerals and rocks *Bertrand Thomas*

We present some compiled results from measurements of dielectric properties of mineral and rock samples in the microwave and millimetre wave region. In a first part, the complex permittivity results of dry solid geological samples will be presented as well as some of their constitutive elements, the minerals. Previous work showed that the porosity/density of the samples has a dominant effect on the dielectric properties of the host material. Indeed, the volume fraction of the spaces inside solid and powdered rocks can be filled either by water and/or by air (Campbell and Ulrichs, 1969), whose dielectric constant influences significantly the global dielectric constant of the materials. In order to highlight the differences in complex permittivity from one geological species to another, only the results of solid and dry natural materials will be presented. Measurements on some powdered rocks can be found in Campbell Ulrichs (1969).

5.4.1 Dielectric properties of minerals

It is believed that the measurement of dielectric properties of structurally homogeneous and non-mixed minerals is fairly reproducible in the millimetre wave range. Minerals can be of two types: Artificial and natural. For the first type, and together with other manmade materials, an extensive survey of the dielectric properties in the millimetre and submillimetre range was covered by Lamb (1996). Additional measurements on six silicates and four metal oxides, in solid and powdered form with different densities, were performed by Nelson (1989) up to 22 GHz. Finally, recent measurements by Thomas (2004) have been performed on solid natural minerals of carbonates and silicates up to 170 GHz; other measurements have been performed on powdered carbonate and sulfate minerals (Ruffier, 2003) up to 85 GHz. Measurements concerning the dielectric properties of some natural and artificial minerals in solid state are presented in Table 5.7.

5.4.2 Dielectric properties of homogeneous rocks

An extensive study was led by Campbell and Ulrichs (1969) to determine the dielectric properties of a wide variety of terrestrial rocks in a frequency range from 450 MHz up to 35 GHz. The analysed rocks are mainly silicates, separated into two main categories: Mafic rocks (basalts, peridotite) and Felsic rocks (granites, rhyolite). Recently, additional measurements have been performed on Carbonate rocks by Thomas (2004) between 200 MHz and 170 GHz. Compiled results of both studies are presented in Table 5.8.

Table 5.7 Dielectric properties of some artificial and natural minerals reported in the literature

Pure crystals	Freq. (GHz)	ε'	ε''	References
CaF_2	110	6.7–6.9	0.015	Thomas (2004)
Calcium Fluoride	170	6.9	0.0122	Thomas (2004)
$CaCO_3$	110	8.81–9.647	0.0416–0.0517	Thomas (2004)
Calcite	170	8.74–9.233	0.045–0.0369	Thomas (2004)
$MgCO_3$	110	6.13–7.25	0.132–0.123	Thomas (2004)
Magnesite	170	7.06–7.23	0.06–0.12	Thomas (2004)
SO_4Ca-nH_2O	110	4.89	0.062	Thomas (2004)
Gypsum	170	3.7	0.2	Thomas (2004)
SiO_2	170	5.43	0.16	Thomas (2004)
Quartz monocrystal	390	4.43		Lamb (1996)
$MgAl_2O_4$	100	8.3764	0.0042	Lamb (1996)
Spinel	300	8.377	0.00963	Lamb (1996)
Al_2O_3	100	9.395–9.574	0.00446	Lamb (1996)
Alumina	350	9.393–9.57	0.0087	Lamb (1996)
Feldspar	22	5.67	0.049	Nelson *et al.* (1989)
Diamond (CVD)	200	5.669	0.0028	Lamb (1996)
Chlorite	22	6.94	0.146	Nelson *et al.* (1989)
Pyroxene	22	7.18	0.1	Nelson *et al.* (1989)
Haematite	22	10.9	0.53	Nelson *et al.* (1989)
Amphibole	22	7.03	0.027	Nelson *et al.* (1989)
Mica (muscovite)	22	8.29	0.091	Nelson *et al.* (1989)
Mica (phlogopite)	22	8.27	0.19	Nelson *et al.* (1989)

5.5 Mixing models for heterogeneous and granular media *Ari Sihvola*

5.5.1 Basic principles: The concept of effective medium

This section will provide the theoretical background for the understanding and prediction of macroscopic dielectric properties of geophysical materials. To reproduce an exact electromagnetic description for the structural complexity of materials occurring in nature is, of course, impossible. Mixing rules are an attempt at performing this task approximately. The approach in analytical mixing formulae is to idealise to the geometry and model the microstructure using simple forms, like spheres and ellipsoids. It is encouraging that, quite often, this turns out to lead to mixing rules that approximate the real world to such an extent that the mixing predictions can be used in practical remote sensing work.

Dielectric mixing rules are algebraic formulae with which the effective permittivity of the mixture can be calculated as a function of the constituent permittivities, their fractional volumes and possibly some other parameters characterising the microstructure of the mixture. The mixture can be discrete, which means that homogeneous

Table 5.8 Dielectric properties of some rock families reported in the literature

Solid rocks	Freq. (GHz)	ε'	ε''	References
Granites	0.45	5.47	0.08	Campbell and Ulrichs (1969)
(average values)	35	5.31	0.1	Campbell and Ulrichs (1969)
	170	5.31	0.197	Thomas (2004)
Basalts	0.45	7.79	0.2	Campbell and Ulrichs (1969)
(average values)	35	7.36	0.41	Campbell and Ulrichs (1969)
	170	7.67	0.45	Thomas (2004)
Carbonates	0.2	8.9	0.2	Thomas (2004)
(average values)	35	7.76	0.24	Thomas (2004)
	110	8.12	0.1	Thomas (2004)
	170	7.68	0.11	Thomas (2004)
Andesite	0.45	5.1	0.02	Campbell and Ulrichs (1969)
	35	5.0	0.07	Campbell and Ulrichs (1969)

inclusions are embedded in another homogeneous medium, or then the permittivity function can be continuous. The concept of effective or macroscopic permittivity implies that the mixture responds to electromagnetic excitation as if it were homogeneous. It may, however, be proper to recall that the dielectric constant (which term is often used synonymously with permittivity) of a material very seldom is constant with respect to temperature, frequency or any material property.

This view to the homogenisation of heterogeneous media is obviously not exact because by using electromagnetic waves of higher and higher frequencies one can always 'see' more accurately into the medium and probe its structural details. Therefore a quantity such as effective permittivity is only meaningful in the long-wavelength limit. This limit corresponds to low frequencies and, indeed, the mixing rules are very often derived using static or quasi-static arguments. The size of the inclusions in the mixture and the spatial correlation length of the permittivity function need to be small with respect to the wavelength.

The effective permittivity can be complex, $\varepsilon_{\text{eff}} = \varepsilon'_{\text{eff}} + \mathrm{i}\varepsilon''_{\text{eff}}$, where the real and imaginary parts are certain averages of the real and imaginary parts of the components' materials. To some extent, the validity of the effective permittivity can be generalised to include the first-order scattering effects of the inclusions. Because scattering entails losses, its effect contributes to the imaginary part of the effective permittivity, which can be estimated by calculating the energy that the electric dipoles radiate and which are induced within the inhomogeneities of the mixture. On the other hand, although the low-frequency character of the effective-medium description is a limitation of mixing theories, these mixing models can be extended to magnetic, anisotropic and, even, magnetoelectric materials.

When a dielectric inclusion that is exposed to an electromagnetic field is small it can be safely assumed that its momentary internal field is the same as that in the

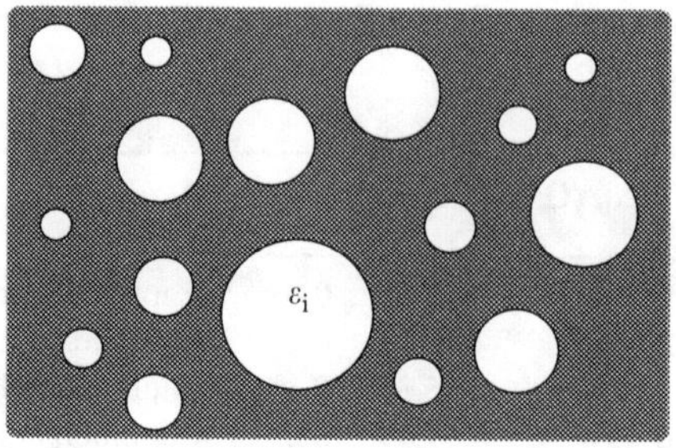

Figure 5.15 *A simple mixture: Spherical inclusions in a homogeneous background medium. The permittivity of the inclusions is ε_i and that of the environment ε_e*

problem with a static excitation. The inclusion creates a perturbation to the field which, to the lowest order, is that of an electric dipole. The polarisability of the inclusion can be enumerated by solving the Laplace equation for the field inside the scatterer, in other words, neglecting the dynamic wave processes altogether. It is not easy to give an exact upper frequency limit for the validity of the concept of effective permittivity. However, the following rule of thumb is often used: The size of an inclusion in the mixture must not exceed a tenth of the wavelength in the effective medium. In fact, this criterion is an estimate on the conservative side.

The early history of dielectric mixing rules can be traced back to the mid-1800s (Mossotti, 1850) and the turn of the century saw dielectric mixing formulae more or less in a present-day form (Maxwell, 1904). For a historical overview of homogenisation principles, see Landauer (1978) and Sihvola (1999). Note that the historical developments in the growth of understanding dielectric properties of heterogeneous materials have left traces into the terminology of dielectric mixture models. As will be seen in this section, mixing rules are known with labels like Maxwell-Garnett, Rayleigh and Bruggeman, and still earlier investigators (Mossotti, Clausius, Lorenz, and Lorentz) have their names attached to effective medium models.

5.5.2 *Polarisability of particles*

Let us start by modelling a simple mixture. The mixture to be analysed consists of the background medium where spherical inclusions are embedded according to Figure 5.15. The two components of the mixture are often called phases. The environment phase can also be termed matrix or host and the inclusion phase as guest.

The polarisability of an inclusion is a measure of its response to an incident electric field. The polarisability of a particle α is the relation between the dipole moment $\mathbf{p}$ that is induced in the inclusion by the polarisation, and the external electric field $\mathbf{E}_e$:

$$\mathbf{p} = \alpha \mathbf{E}_e \tag{5.39}$$

For a sphere the polarisability is easy to calculate. It is proportional to the internal field within the inclusion, its volume and the dielectric contrast between the inclusion and the environment. Since the electric field $\mathbf{E}_\mathrm{i}$ induced in a sphere in a uniform and static external field $\mathbf{E}_\mathrm{e}$ is also uniform, static and parallel to the external field (Jackson, 1975, Section 4.4),

$$\mathbf{E}_\mathrm{i} = \frac{3\varepsilon_\mathrm{e}}{\varepsilon_\mathrm{i} + 2\varepsilon_\mathrm{e}} \mathbf{E}_\mathrm{e} \tag{5.40}$$

the polarisability can be written immediately as

$$\alpha = V(\varepsilon_\mathrm{i} - \varepsilon_\mathrm{e}) \frac{3\varepsilon_\mathrm{e}}{\varepsilon_\mathrm{i} + 2\varepsilon_\mathrm{e}} \tag{5.41}$$

where the permittivities of the inclusion and its environment are denoted by ε_i and ε_e, respectively. The volume of the sphere is V. Note that the polarisability is a scalar. This is because the inclusion material is isotropic, and its shape is spherically symmetric.

5.5.3 *Clausius–Mossotti and Maxwell Garnett formula*

From the polarisability of a single sphere, the effective permittivity of a mixture can be calculated as a function of the number density of the spheres in the background medium with permittivity ε_e. The effective permittivity is the relation between the external field and the average electric flux density $\langle \mathbf{D} \rangle$:

$$\langle \mathbf{D} \rangle = \varepsilon_\mathrm{eff} \mathbf{E}_\mathrm{e} = \varepsilon_\mathrm{e} \mathbf{E}_e + \langle \mathbf{P} \rangle \tag{5.42}$$

where the average polarisation $\langle \mathbf{P} \rangle$ is connected to the dipole moment density in the mixture:

$$\langle \mathbf{P} \rangle = n\mathbf{p} \tag{5.43}$$

and here n is the number density of dipole moments $\mathbf{p}$ in the mixture. Note the dimensions of the quantities in SI units: $[\mathbf{D}] = [\mathbf{P}] = \mathrm{As/m^2}$, $[\mathbf{p}] = \mathrm{Asm}$, $[\mathbf{E}_\mathrm{e}] = \mathrm{V/m}$, $[\alpha] = \mathrm{Asm^2/V}$ and $[n] = \mathrm{m^{-3}}$.

In a mixture, especially when it is dense, one cannot assume the field exciting one inclusion to be the external field $\mathbf{E}_\mathrm{e}$. The surrounding polarisation increases the field effect and has to be taken into account (Kittel, 1986, p. 366–268). The field that excites one inclusion $\mathbf{E}_\mathrm{L}$ is often called the local field or Lorentzian field. It is dependent on the shape of the inclusion (Yaghjian, 1980), and for a sphere it is

$$\mathbf{E}_\mathrm{L} = \mathbf{E}_\mathrm{e} + \frac{1}{3\varepsilon_\mathrm{e}} \mathbf{P} \tag{5.44}$$

where the coefficient $1/3$ corresponds to the depolarisation factor of the sphere. Combining this equation with $\mathbf{p} = \alpha \mathbf{E}_\mathrm{L}$ leaves us with the average polarisation, and the effective permittivity can then be written (see Equation (5.42)) as

$$\varepsilon_\mathrm{eff} = \varepsilon_\mathrm{e} + \frac{n\alpha}{1 - n\alpha/3\varepsilon_\mathrm{e}} \tag{5.45}$$

The equation can often be seen in the form

$$\frac{\varepsilon_{\mathrm{eff}} - \varepsilon_{\mathrm{e}}}{\varepsilon_{\mathrm{eff}} + 2\varepsilon_{\mathrm{e}}} = \frac{n\alpha}{3\varepsilon_{\mathrm{e}}} \tag{5.46}$$

This relation carries the name Clausius–Mossotti formula, although it deserves the label Lorenz–Lorentz formula as well (Sihvola, 1991). The dilute-mixture approximation can be written by taking the limit of small n:

$$\varepsilon_{\mathrm{eff}} \approx \varepsilon_{\mathrm{e}} + n\alpha \tag{5.47}$$

In practical applications quantities like polarisabilities and scatterer densities are not always those most convenient to use. Rather, one prefers to play with the permittivities of the components of the mixture. When this is the case, it is advantageous to combine the Clausius–Mossotti formula with the polarisability expression (5.41). Then we can write

$$\frac{\varepsilon_{\mathrm{eff}} - \varepsilon_{\mathrm{e}}}{\varepsilon_{\mathrm{eff}} + 2\varepsilon_{\mathrm{e}}} = f\frac{\varepsilon_{\mathrm{i}} - \varepsilon_{\mathrm{e}}}{\varepsilon_{\mathrm{i}} + 2\varepsilon_{\mathrm{e}}} \tag{5.48}$$

where $f = nV$ is a dimensionless quantity, the volume fraction of the inclusions in the mixture. This formula is called the Rayleigh mixing formula. Note that because only the volume fraction and the permittivities appear in the mixing rule, the spheres need not be of the same size if only all of them are small compared to the wavelength.

Perhaps the most common mixing rule is the Maxwell Garnett formula which is the Rayleigh rule (5.48) written explicitly for the effective permittivity:

$$\varepsilon_{\mathrm{eff}} = \varepsilon_{\mathrm{e}} + 3f\varepsilon_{\mathrm{e}}\frac{\varepsilon_{\mathrm{i}} - \varepsilon_{\mathrm{e}}}{\varepsilon_{\mathrm{i}} + 2\varepsilon_{\mathrm{e}} - f(\varepsilon_{\mathrm{i}} - \varepsilon_{\mathrm{e}})} \tag{5.49}$$

This formula is in wide use in very diverse fields of application. The beauty of the Maxwell Garnett formula lies in its simple appearance combined with its broad applicability. It satisfies the limiting processes for vanishing inclusion phase $f \to 0$ giving $\varepsilon_{\mathrm{eff}} \to \varepsilon_{\mathrm{e}}$ and for vanishing background $f \to 1$ we have $\varepsilon_{\mathrm{eff}} \to \varepsilon_{\mathrm{i}}$.

The perturbation expansion of the Maxwell Garnett rule gives the mixing equation for dilute mixtures ($f \ll 1$):

$$\varepsilon_{\mathrm{eff}} \approx \varepsilon_{\mathrm{e}} + 3f\varepsilon_{\mathrm{e}}\frac{\varepsilon_{\mathrm{i}} - \varepsilon_{\mathrm{e}}}{\varepsilon_{\mathrm{i}} + 2\varepsilon_{\mathrm{e}}} + 3f^2\varepsilon_{\mathrm{e}}\left(\frac{\varepsilon_{\mathrm{i}} - \varepsilon_{\mathrm{e}}}{\varepsilon_{\mathrm{i}} + 2\varepsilon_{\mathrm{e}}}\right)^2 \tag{5.50}$$

Figure 5.16 shows the prediction of the Maxwell Garnett formula for different values of the dielectric contrast $\varepsilon_{\mathrm{i}}/\varepsilon_{\mathrm{e}}$. Shown is the susceptibility ratio

$$\frac{\varepsilon_{\mathrm{eff}} - \varepsilon_{\mathrm{e}}}{\varepsilon_{\mathrm{i}} - \varepsilon_{\mathrm{e}}}$$

which vanishes for $f = 0$ and is unity for $f = 1$, independently of the inclusion-to-background contrast. The figure shows clearly the fact that the effective permittivity function becomes a very non-linear function of the volume fraction for large dielectric contrasts.

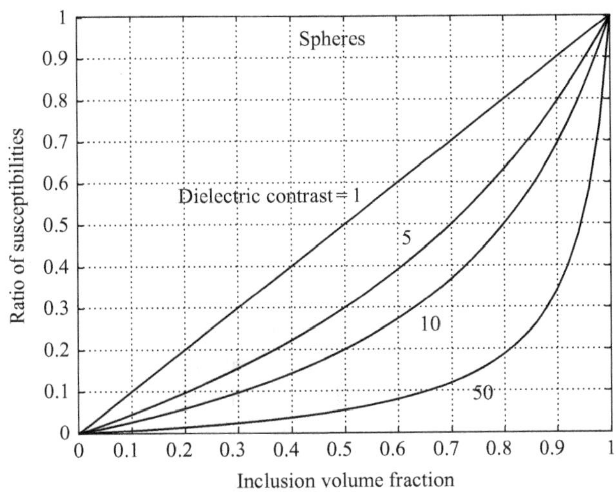

Figure 5.16 *The susceptibility ratio $(\varepsilon_{\text{eff}} - \varepsilon_e)/(\varepsilon_i - \varepsilon_e)$ for the Maxwell Garnett prediction of the effective permittivity of a mixture with spherical inclusions of permittivity ε_i in a background medium of permittivity ε_e, as a function of the volume fraction of the inclusions f and the dielectric contrast $\varepsilon_i/\varepsilon_e$*

5.5.4 Multi-phase mixtures and non-spherical inclusions

The basic treatment of a mixture as an effective medium in the above section was very idealised: A two-phase mixture with isotropic dielectric components and the inclusion geometry was spherical. There are 'materials' in the real world that follow this model quite well, for example, clouds, fog or even light rain to some extent. Fog is formed of water droplets in air that are certainly spherical, and the mixture is very dilute. The attenuation and phase delays in radio wave propagation problems in fogs and clouds can be very accurately analysed using the Maxwell Garnett principles.

But most classes of geophysical media that are encountered in remote sensing applications do not obey the assumptions of the basic Maxwell Garnett mixing. Therefore its generalisation into more complicated heterogeneities is necessary.

Wet snow, for example, is a mixture of three phases: Air, ice and liquid water. How to use the previous analysis for snow? To write down the mixing rule for mixtures where there are several components one has to choose one of the components as the host and then treat the polarisation effect of all of the guest phases separately. The previous two-phase analysis can be followed very closely. Where above the total polarisation was calculated from the individual dipole moments according to (5.43), now each guest phase contributes one such term to a sum of many. The final result is (compare (5.48))

$$\frac{\varepsilon_{\text{eff}} - \varepsilon_e}{\varepsilon_{\text{eff}} + 2\varepsilon_e} = \sum_{n=1}^{N} f_n \frac{\varepsilon_{i,n} - \varepsilon_e}{\varepsilon_{i,n} + 2\varepsilon_e} \tag{5.51}$$

where f_n is the volume fraction of the inclusions of the nth phase in the mixture, and $\varepsilon_{i,n}$ is its permittivity. And of course this can be solved for the effective permittivity:

$$\varepsilon_{\text{eff}} = \varepsilon_e + 3\varepsilon_e \frac{\sum_{n=1}^{N} f_n(\varepsilon_{i,n} - \varepsilon_e)/(\varepsilon_{i,n} + 2\varepsilon_e)}{1 - \sum_{n=1}^{N} f_n(\varepsilon_{i,n} - \varepsilon_e)/(\varepsilon_{i,n} + 2\varepsilon_e)} \tag{5.52}$$

Here again, all inclusions of all phases were assumed to be spherical.

Also, the assumption of spherical shape for the inclusions needs to be relaxed because many natural media possess inclusions of other forms. The polarisability of small particles can, of course, be calculated for any shape but in general this requires numerical effort. The only shapes for which simple analytical solutions can be found are ellipsoids. Fortunately, ellipsoids allow many practical special cases, like discs and needles, for example.

The important parameters in the geometry of an ellipsoid are its depolarisation factors. If the semi-axes of an ellipsoid in the three orthogonal directions are a_x, a_y and a_z (Figure 5.17), the depolarisation factor A_x (the factor in the a_x-direction) is

$$A_x = \frac{a_x a_y a_z}{2} \int_0^\infty \frac{ds}{(s + a_x^2)\sqrt{(s + a_x^2)(s + a_y^2)(s + a_z^2)}} \tag{5.53}$$

For the other depolarisation factor A_y (A_z), interchange a_y and a_x (a_z and a_x) in the above integral.

The three depolarisation factors for any ellipsoid satisfy

$$A_x + A_y + A_z = 1 \tag{5.54}$$

A sphere has three equal depolarisation factors of 1/3. The other two special cases are a disk (depolarisation factors $1, 0, 0$) and a needle $(0, 1/2, 1/2)$. For ellipsoids of revolution, prolate and oblate spheroids, closed-form expressions for the integral (5.53) can be found (Landau and Lifshitz, 1984).

Prolate spheroids ($a_x > a_y = a_z$) have

$$A_x = \frac{1 - e^2}{2e^3}\left(\ln\frac{1 + e}{1 - e} - 2e\right) \tag{5.55}$$

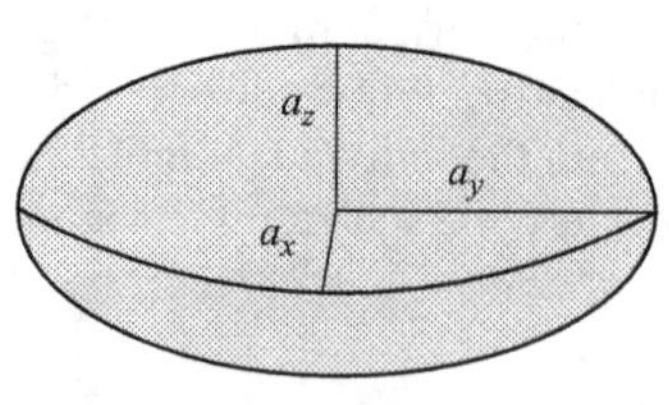

Figure 5.17 An ellipsoid, determined by its semi-axes a_x, a_y, a_z

and

$$A_y = A_z = \frac{1}{2}(1 - A_x) \tag{5.56}$$

where the eccentricity is $e = \sqrt{1 - a_y^2/a_x^2}$. For nearly spherical prolate spheroids, which have small eccentricity, the following holds:

$$A_x \simeq \frac{1}{3} - \frac{2}{15}e^2 \tag{5.57}$$

$$A_y = A_z \simeq \frac{1}{3} + \frac{1}{15}e^2 \tag{5.58}$$

For oblate spheroids ($a_x = a_y > a_z$),

$$A_z = \frac{1 + e^2}{e^3}(e - \arctan e) \tag{5.59}$$

$$A_x = A_y = \frac{1}{2}(1 - A_z) \tag{5.60}$$

where $e = \sqrt{a_x^2/a_z^2 - 1}$. For nearly-spherical oblate spheroids,

$$A_z \simeq \frac{1}{3} + \frac{2}{15}e^2 \tag{5.61}$$

$$A_x = A_y \simeq \frac{1}{3} - \frac{1}{15}e^2 \tag{5.62}$$

Figures 5.18 and 5.19 display the behaviour of the depolarisation factors of these spheroids as functions of the axial ratio of the ellipsoids.

For a general ellipsoid with three different axes, the depolarisation factors have to be calculated from the integral (5.53). Osborn and Stoner have tabulated the depolarisation factors of a general ellipsoid (Osborn, 1945; Stoner, 1945), a great achievement despite the fact that today's numerical software packages like Mathematica give these factors with very easy input efforts.

Now consider a mixture where ellipsoids of permittivity ε_i are embedded in the environment ε_e. Let all the ellipsoids be aligned. Then the effective permittivity of the mixture is anisotropic, in other words it has different permittivity components in the different principal directions. We can write the following formula for this mixture, which generalises the Maxwell Garnett mixing rule:

$$\varepsilon_{\mathrm{eff},x} = \varepsilon_\mathrm{e} + f\varepsilon_\mathrm{e}\frac{\varepsilon_\mathrm{i} - \varepsilon_\mathrm{e}}{\varepsilon_\mathrm{e} + (1 - f)A_x(\varepsilon_\mathrm{i} - \varepsilon_\mathrm{e})} \tag{5.63}$$

and for $\varepsilon_{\mathrm{eff},y}$ and $\varepsilon_{\mathrm{eff},z}$, replace A_x by A_y and A_z, respectively. This formula is sometimes named after Bohren and Battan (Mätzler, 1987; Bohren and Battan, 1982), but it was already derived in 1919 by Burger (1919).

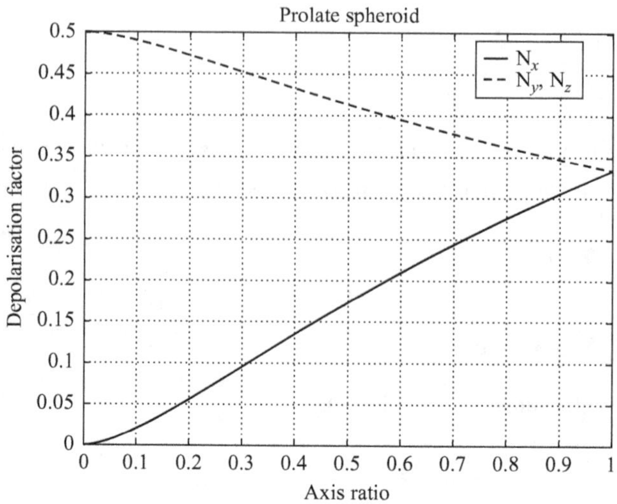

Figure 5.18 *The depolarisation factors of a prolate spheroid as a function of the axis ratio a_z/a_x. The axis of revolution is x (the longest axis of the ellipsoid compared to the two other axes). Obviously $A_x + A_y + A_z = 1$ regardless of the axis ratio*

If, on the other hand, all the ellipsoids in the mixture are randomly oriented, there is macroscopically no preferred direction. The mixture is isotropic and the effective permittivity $\varepsilon_{\mathrm{eff}}$ is a scalar:

$$\varepsilon_{\mathrm{eff}} = \varepsilon_{\mathrm{e}} + \varepsilon_{\mathrm{e}} \frac{f/3 \sum_{j=x,y,z} (\varepsilon_{\mathrm{i}} - \varepsilon_{\mathrm{e}})/(\varepsilon_{\mathrm{e}} + A_j(\varepsilon_{\mathrm{i}} - \varepsilon_{\mathrm{e}}))}{1 - f/3 \sum_{j=x,y,z} A_j(\varepsilon_{\mathrm{i}} - \varepsilon_{\mathrm{e}})/(\varepsilon_{\mathrm{e}} + A_j(\varepsilon_{\mathrm{i}} - \varepsilon_{\mathrm{e}}))} \tag{5.64}$$

Figure 5.20 shows the Maxwell Garnett predictions of the effective permittivity of isotropic two-component dielectric mixtures where the inclusion shapes are varied: They are either spheres, needles or discs. A random orientation of the inclusions guarantees that the mixture is isotropic and the permittivity is a scalar. The contrast between the inclusion and background phases is $\varepsilon_{\mathrm{i}}/\varepsilon_{\mathrm{e}} = 30$. There is a clear effect of the shape of the inclusions: Spheres give the lowest permittivity needles a larger permittivity and discs will provide the largest effect. However, if the contrast between the phases is smaller, the effect of the geometry decreases.

Finally, if the inclusions are neither aligned nor randomly oriented but rather follow an orientation distribution, the sums in (5.64) have to be replaced by terms where the dipole moment densities are weighted by the distribution function and integrated over all relevant spatial directions.

The mixture with homogeneous isotropic ellipsoidal inclusions was only one generalisation of the simplest mixture. The aligned ellipsoid case above, Equation (5.63), was an example where the geometry of the microstructure rendered

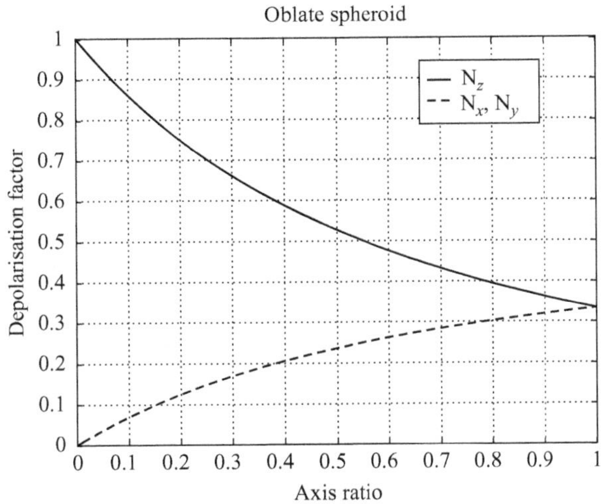

Figure 5.19 *The depolarisation factors of an oblate spheroid as a function of the axis ratio a_z/a_x. The axis of revolution is z (the shortest axis of the ellipsoid compared to the two other axes). Obviously $A_x + A_y + A_z = 1$ regardless of the axis ratio*

the macroscopic permittivity anisotropic. But the mixing principles can also be generalised to cases where one or several of the phases are anisotropic in the first place. One of the elegant ways of achieving this goal is to generalise the Maxwell Garnett formula into dyadic domain. See Sihvola (1999) for details of this process.

The previous results have been derived for discrete structures. The inclusions were assumed to be dielectrically homogeneous. However, mixtures with inhomogeneous inclusions can certainly be treated within the framework of Maxwell Garnett (and not only in the basic Maxwell Garnett case but also for more complicated approaches) mixing. The challenge with non-homogeneous scatterers is the difficulty of calculating the polarisability of such inclusions. There are, however, certain shapes for which an analytical solution can be found in the electrostatic problem.

A layered sphere is one example of such a special case. There is no restriction for the number of layers, and even the case for a radially continuous permittivity profile of the sphere has been given a solution (Sihvola and Lindell, 1989). (See also the results for mixtures with dielectrically inhomogenous ellipsoids (Sihvola and Lindell, 1990).) As an example, the following is the generalisation of the mixing rule (5.48) for the case when the inclusions are two-component spheres:

$$\frac{\varepsilon_{\text{eff}} - \varepsilon_e}{\varepsilon_{\text{eff}} + 2\varepsilon_e} = f\frac{(\varepsilon_l - \varepsilon_e)(\varepsilon_c + 2\varepsilon_l) + w(\varepsilon_c - \varepsilon_l)(\varepsilon_e + 2\varepsilon_l)}{(\varepsilon_l + 2\varepsilon_e)(\varepsilon_c + 2\varepsilon_l) + 2w(\varepsilon_c - \varepsilon_l)(\varepsilon_l - \varepsilon_e)} \quad (5.65)$$

The inclusion sphere consists of a spherical core with permittivity ε_c which is covered by a spherical shell with permittivity ε_l. The parameter $w = (b/a)^3$ is the fraction of the volume of the core from the total inclusion volume (a is the radius of the inclusion

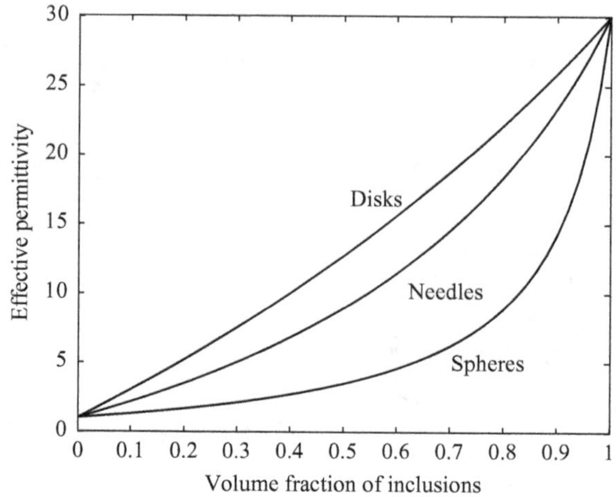

Figure 5.20 *The effective permittivity of a mixture as a function of the volume fraction of inclusions. The inclusions are randomly oriented spheres, needles or discs. The inclusion permittivity is 30 times that of the environment*

and b is the radius of the core), and f is the volume fraction of the inclusions in the mixture, as before.

If, on the other hand, the inclusions are of such a shape that they do not have a closed-form solution for the dielectric polarisability, we cannot write down a simple Maxwell Garnett formula like in the case of spheres and ellipsoids. However, if the polarisability is known, for example by numerical enumeration, the result can be used as input in the Lorenz–Lorentz formula (5.46) to calculate the effective permittivity. For example, very accurate results have been reported for the polarisability of platonic solids (tetrahedron, cube, octahedron, dodecahedron and icosahedron) (Sihvola *et al.*, 2004). These can be applied to model mixtures with more sharp-edged microstructure than the one shown in Figure 5.15.

5.5.5 *Bruggeman mixing rule and other generalised models*

There is no exact result for the effective permittivity of a mixture with random geometry. In the analysis of random media, a major difficulty comes with the problem of how to correctly take into account the interaction between the scatterers. For sparse mixtures, these effects of interaction are small and can be included by surrounding the inclusion with the average polarisation $\langle \mathbf{P} \rangle$ as was done in the above derivation of the Maxwell Garnett rule. However, when dense mixtures are treated, this approach may not be correct. In the present section, mixing rules are presented for mixtures which predict different results compared to the Maxwell Garnett rule.

One important mixing rule in the remote sensing community goes under different names: the Polder–Van Santen formula (1946), also called the Bruggeman formula

(1935) or Böttcher formula (1952), is very often encountered in material modelling studies.

The basic form of this formula for spherical scatterers is

$$(1-f)\frac{\varepsilon_e-\varepsilon_{eff}}{\varepsilon_e+2\varepsilon_{eff}}+f\frac{\varepsilon_i-\varepsilon_{eff}}{\varepsilon_i+2\varepsilon_{eff}}=0 \tag{5.66}$$

This Bruggeman formula has the special property that it treats the inclusions and the environment symmetrically. The interpretation of Equation (5.66) is that the formula balances both mixing components with respect to the unknown effective medium, using the volume fraction of each component as weight (f for the inclusions and $1-f$ for the environment). This symmetry property of (5.66) makes the radical distinction between the Maxwell Garnett rule and the Bruggeman rule. The Maxwell Garnett approach is inherently non-symmetric.

The Bruggeman formula for the case when the inclusions are randomly oriented ellipsoids is

$$\varepsilon_{eff}=\varepsilon_e+\frac{f}{3}(\varepsilon_i-\varepsilon_e)\sum_{j=x,y,z}\frac{\varepsilon_{eff}}{\varepsilon_{eff}+A_j(\varepsilon_i-\varepsilon_{eff})} \tag{5.67}$$

where now A_j are again the depolarisation factors of the inclusion ellipsoids in the three orthogonal directions.

Another well-known formula which is relevant in the theoretical studies of wave propagation in random media is the so-called coherent potential formula (Tsang *et al.*, 1985, p. 475):

$$\varepsilon_{eff}=\varepsilon_e+\frac{f}{3}(\varepsilon_i-\varepsilon_e)\sum_{j=x,y,z}\frac{(1+A_j)\varepsilon_{eff}-A_j\varepsilon_e}{\varepsilon_{eff}+A_j(\varepsilon_i-\varepsilon_e)} \tag{5.68}$$

This formula for spherical inclusions is

$$\varepsilon_{eff}=\varepsilon_e+f(\varepsilon_i-\varepsilon_e)\frac{3\varepsilon_{eff}}{3\varepsilon_{eff}+(1-f)(\varepsilon_i-\varepsilon_e)} \tag{5.69}$$

It is worth noting that for dilute mixtures ($f \ll 1$), all three mixing rules, Maxwell Garnett, Polder–Van Santen and coherent potential, predict the same results. Up to the first order in f, the formulae are the same:

$$\varepsilon_{eff}\approx\varepsilon_e+3f\varepsilon_e\frac{\varepsilon_i-\varepsilon_e}{\varepsilon_i+2\varepsilon_e} \tag{5.70}$$

A unified mixing approach (Sihvola, 1989) collects all the previous aspects of dielectric mixing rules into one family. For the case of isotropic spherical inclusions ε_i in the isotropic environment ε_e, the formula looks like

$$\frac{\varepsilon_{eff}-\varepsilon_e}{\varepsilon_{eff}+2\varepsilon_e+\nu(\varepsilon_{eff}-\varepsilon_e)}=f\frac{\varepsilon_i-\varepsilon_e}{\varepsilon_i+2\varepsilon_e+\nu(\varepsilon_{eff}-\varepsilon_e)} \tag{5.71}$$

This formula contains a dimensionless parameter ν. For different choices of ν, the previous mixing rules are recovered: $\nu=0$ gives the Maxwell Garnett rule, $\nu=2$ gives the Bruggeman formula and $\nu=3$ gives the coherent potential approximation.

The unified mixing rule (5.71) can be generalised to the case where the inclusions are ellipsoids, all randomly oriented. Then the scalar effective permittivity reads

$$\varepsilon_{\mathrm{eff}} = \varepsilon_{\mathrm{e}} + f(\varepsilon_{\mathrm{i}} - \varepsilon_{\mathrm{e}}) \frac{\sum_{k=1}^{3} \varepsilon_a/(\varepsilon_a + A_k(\varepsilon_{\mathrm{i}} - \varepsilon_{\mathrm{e}}))}{3 - f(\varepsilon_{\mathrm{i}} - \varepsilon_{\mathrm{e}}) \sum_{k=1}^{3} A_k/(\varepsilon_a + A_k(\varepsilon_{\mathrm{i}} - \varepsilon_{\mathrm{e}}))} \tag{5.72}$$

Different mixing models arise from this form by various choices of the apparent permittivity $\varepsilon_a = \varepsilon_{\mathrm{e}} + a(\varepsilon_{\mathrm{eff}} - \varepsilon_{\mathrm{e}})$:

- Maxwell Garnett: $a = 0$
- Bruggeman/Polder–Van Santen: $a = 1 - A_k, \quad k = 1, 2, 3$
- Coherent potential: $a = 1$

Of the very large set of the remaining mixing rules that are being used in the random medium theories and practical applications, the following ones deserve to be introduced.

A widely used class of mixing models is formed by the 'power-law' approximations:

$$\varepsilon_{\mathrm{eff}}^a = f\varepsilon_{\mathrm{i}}^a + (1 - f)\varepsilon_{\mathrm{e}}^a \tag{5.73}$$

For example, in the Birchak formula (1974) the parameter is $a = 1/2$, which means that the square roots of the component permittivities add up to the square root of the mixture permittivity.

Another famous formula is the Looyenga formula (1965) for which $a = 1/3$. One can also find in the literature (see, e.g. Ulaby *et al.*, 1986, p. 1080) the linear law

$$\varepsilon_{\mathrm{eff}} = f\varepsilon_{\mathrm{i}} + (1 - f)\varepsilon_{\mathrm{e}} \tag{5.74}$$

which corresponds to $a = 1$ in (5.73). This mixing rule can be given theoretical confirmation if the mixture is formed of plates or other inclusions for which no depolarisation is induced. If the depolarisation factor is $A_x = 0$, one can recover formula (5.74) from (5.63).

Other models resulting from a differential analysis are

$$\frac{\varepsilon_{\mathrm{i}} - \varepsilon_{\mathrm{eff}}}{\varepsilon_{\mathrm{i}} - \varepsilon_{\mathrm{e}}} = (1 - f)\left(\frac{\varepsilon_{\mathrm{eff}}}{\varepsilon_{\mathrm{e}}}\right)^{1/3} \tag{5.75}$$

which is sometimes called the 'Bruggeman asymmetric formula' (to distinguish it from the symmetric Bruggeman formula (5.66)), and its 'complement' (Sen *et al.*, 1981), often used in the modelling of rocks:

$$\frac{\varepsilon_{\mathrm{eff}} - \varepsilon_{\mathrm{e}}}{\varepsilon_{\mathrm{i}} - \varepsilon_{\mathrm{e}}} = f\left(\frac{\varepsilon_{\mathrm{eff}}}{\varepsilon_{\mathrm{i}}}\right)^{1/3} \tag{5.76}$$

which have the common feature of one-third powers.

There are also formulae for mixtures with spherical inclusions in a cubic array in a background matrix. These formulae can be seen as successive improvements to the classical Rayleigh result, Equation (5.48). These have been presented by Runge (1925), Meredith and Tobias (1960), McPhedran and McKenzie (1978a,b), Doyle (1978), Lam (1986), and Kristensson (2003). However, these formulae are derived for ordered mixtures, though not all necessarily for cubic-centred lattices, and from the point of view of application to random media, they suffer from the disadvantage of predicting infinite effective permittivities as the inclusions come to contact with each other.

Recently also numerical efforts have been made to calculate the effective dielectric properties of random mixtures. There is indication (Kärkkäinen *et al.*, 2001) that the macroscopic permittivity would fall between the Maxwell Garnett and Bruggeman predictions: Closer to the Maxwell Garnett if the inclusions are forced to be separate from each other within the mixture. On the other hand, if the inclusions are allowed to touch and overlap, thus forming clusters, the Bruggeman formula seems to be more valid. These mixtures have been simulated by embedding spheres into random positions within the background matrix.

5.6 Electrodynamic phenomena resulting from the heterogeneity structure *Ari Sihvola*

The mixing rules that have been discussed so far are actually rather simple algebraic formulae that combine the component permittivities and some structural parameters of the mixture. It is actually amazing how many real-life phenomena are predicted by these rules. The mixing process sometimes gives rise to qualitatively new effects such that the mixture displays properties that are absent in the component materials.

5.6.1 Frequency dependence and dispersion

Many physical mechanisms are hidden inside the material permittivity ε. A more correct description of the constitutive relation – than a plain dielectric 'constant' – between the displacement and the electric field presents the dependence as a convolution operator:

$$\mathbf{D}(t) = \varepsilon_0 \mathbf{E}(t) + \int_{-\infty}^{t} \chi(t - t')\mathbf{E}(t')\,\mathrm{d}t' \tag{5.77}$$

where the first term is the free-space response which contains the vacuum permittivity ε_0, and the material response is described by the susceptibility kernel $\chi(t)$, which, due to causality requirements, vanishes for negative arguments. In engineering applications, it is more convenient to work in the frequency-domain which means that the explicit time-dependence has been Fourier-transformed, and the permittivity is a function of frequency ω. The variation of the permittivity with respect to frequency is called dispersion.

Different physical mechanisms in the dielectric response of materials can be described by certain dispersion models. Some important ones for remote sensing applications are

- the Debye relaxation model, which is suited for fluids and other materials that contain permanent electric dipole moments. A characteristic parameter for Debye materials is the relaxation time of the moments. This time determines the relaxation frequency, and radio waves of that frequency will be attenuated in the medium.
- for solid materials, the Lorentz model describes the resonant frequencies and damping for characteristic frequencies of the molecules of the medium. Around the resonance, a very strong dispersion takes place.
- for metals, the Drude model can be used to model their free-electron type behaviour.

It is often that electromagnetic scattering analyses are interpreted with a tendency to underappreciate the richness of the dielectric response of small particles. In quasi-static modelling of mixtures, the dynamic nature of Maxwell equations is suppressed and one may think that the result is an uninteresting average of the component materials. But this is not the case as has been already pointed out above, and which is very convincingly argued in Bohren and Huffman (1983, chap 12).

The mixing process can affect strongly the dispersive characteristics of materials. For example, if metal spheres are mixed into a non-dispersive environment, the mixture becomes a Lorentz-medium with resonant behaviour. There are several others, among which another interesting example is the cloud as a mixture of small water drops in air.

5.6.2 *Transfer of range of mixing loss*

The permittivity of water obeys the Debye law (Hasted, 1973)

$$\varepsilon_{\mathrm{water}}(\omega) = \varepsilon_\infty + \frac{\varepsilon_\mathrm{s} - \varepsilon_\infty}{1 - i\omega\tau} \tag{5.78}$$

with ε_s and ε_∞ being the static and high-frequency permittivities, and τ the relaxation time. The air permittivity is constant. It turns out that clouds (water spheres in air) follow the Debye model, too, but the relaxation frequency will be shifted to a frequency far higher than that of water. The Maxwell Garnett prediction (5.49) for the effective permittivity of the mixture can be calculated, and it is

$$\varepsilon_{\mathrm{eff}}(\omega) = \varepsilon_{\infty,\mathrm{eff}} + \frac{\varepsilon_{\mathrm{s,eff}} - \varepsilon_{\infty,\mathrm{eff}}}{1 - i\omega\tau_{\mathrm{eff}}} \tag{5.79}$$

In other words, the mixture also obeys the Debye law. If the volume fraction is small ($f \ll 1$) as in the case of fog or cloud, the parameters of the mixture are

$$\varepsilon_{\infty,\mathrm{eff}} = \varepsilon_\mathrm{e} + 3f\varepsilon_\mathrm{e}\frac{\varepsilon_\infty - \varepsilon_\mathrm{e}}{\varepsilon_\infty + 2\varepsilon_\mathrm{e}} \tag{5.80}$$

$$\varepsilon_{\mathrm{s,eff}} = \varepsilon_\mathrm{e} + 3f\varepsilon_\mathrm{e}\frac{\varepsilon_\mathrm{s} - \varepsilon_\mathrm{e}}{\varepsilon_\mathrm{s} + 2\varepsilon_\mathrm{e}} \tag{5.81}$$

$$\tau_{\mathrm{eff}} = \tau\frac{\varepsilon_\infty + 2\varepsilon_\mathrm{e}}{\varepsilon_\mathrm{s} + 2\varepsilon_\mathrm{e}} \tag{5.82}$$

The result shows that the static and high-frequency permittivities of the mixture are, as expected, Maxwell-Garnett averages of the component values. But an interesting fact is that the relaxation frequency of the mixture is different from that of water. At the temperature of +20°C, the values are: $\varepsilon_\infty \approx 4.90$, $\varepsilon_s \approx 80.1$, and the relaxation time $\tau \approx 10.1$ ps. This means that the relaxation frequency (the frequency at which the imaginary part attains its maximum absolute value) is $f_r = 1/(2\pi\tau) \approx 15.8$ GHz for water, but for the mixture it is very much higher: $1/(2\pi\tau_{\text{eff}}) \approx 187$ GHz.

5.6.3 *Percolation phenomena*

The frequency dependence of the effective permittivity can be affected in a very non-linear manner if the inclusion permittivity is dispersive, as can be observed from the mixing rule (5.49). Especially in frequency ranges (like in the optical regime) where negative real parts for the permittivity can be measured, it may happen that the denominator reaches very small absolute values. It can be seen that for dilute mixures ($f \ll 1$), the condition $\varepsilon_i = -2\varepsilon_e$ gives this type of 'catastrophe'. The result is a strong absorption which explains the known fact that certain materials emit different colours in particulate form compared to the same material in bulk form.

A further effect which is predicted by dielectric mixing rules is the property called percolation. Percolation is a very non-linear phenomenon. It is an abrupt change in the behaviour of a certain parameter in the random medium as the volume fraction of the phases changes. Percolation theories have been developed in diverse fields, like ferromagnetism, forest fires and even the spread of epidemics in animal and human populations (Grimmett, 1989).

It turns out that percolation behaviour becomes pronounced if the mixture consists of components which have a strong contrast in their dielectric properties. Therefore, one can neglect percolation discussion in dry snow modelling, but if the snow is wet, percolation effects may become important. In general, moist substances are more difficult to model as mixtures, and one of the reasons is actually the strong dielectric contrast between the permittivity of water and that of ordinary dry materials.

Mixing formulas predict percolation but the threshold volume fraction is different depending on the mixing rule. It turns out that for the case of spherical high-permittivity inclusions, the percolation threshold (the volume fraction of the inclusion phase around which the sudden increase in the mixture permittivity ε_{eff} takes place) is 0.25 for the coherent potential formula, 0.333 for the Bruggeman formula and 1.0 for the Maxwell Garnett rule.

5.6.4 *Maxwell-Wagner losses and enhanced polarisation*

It is worth noting that even if the analysis in this and in the previous section have mostly been restricted to real values for the permittivity, the models apply for lossy materials. The mixing rules also take into account the losses in the case that the inclusion phase is conducting. Consider a dielectric mixture where the background material is of permittivity ε_e and the inclusion phase has conductivity σ_i, and its complex permittivity is $\varepsilon_i = \varepsilon_i' + i\sigma_i/\omega$, where ω is the angular frequency. If the volume fraction of the inclusion phase f is small, the effective conductivity $\sigma_{\text{eff}} = \omega \text{Im}\{\varepsilon_{\text{eff}}\}$ of the mixture, calculated from the Maxwell-Garnett formula, is

$$\sigma_{\text{eff}} = \frac{9\varepsilon_{\text{e}}^2 f \sigma_{\text{i}}}{(\varepsilon_{\text{i}}' + 2\varepsilon_{\text{e}})^2 + \sigma_{\text{i}}^2/\omega^2} \tag{5.83}$$

From this formula it is seen that the effective DC conductivity vanishes: $\sigma_{\text{eff}} \to 0$ as $\omega \to 0$. This is also intuitively clear: Non-contacting conducting particles in a non-conducting matrix do not make the mixture conducting.

Equation (5.83) shows that the loss factor and the imaginary part of the permittivity, that varies as ω^{-1} for the conductive inclusion material, is converted to the so called Maxwell–Wagner losses of the mixture (Wagner, 1914): losslessness in the low-frequency region, and normal conducting losses at high frequencies. The imaginary part of the effective permittivity has its maximum value at the relaxation frequency

$$\frac{\omega}{2\pi} = \frac{1}{2\pi} \frac{\sigma_{\text{i}}}{\varepsilon_{\text{i}}' + 2\varepsilon_{\text{e}}} \tag{5.84}$$

This frequency is around 200 kHz for water with conductivity 1 mS/m.

5.7 Dielectric properties of heterogeneous media *Christian Mätzler and Mike Schwank*

5.7.1 Introductory remarks and framework

This section is devoted to measured data on dielectric properties of heterogeneous and especially of granular media, and to the validation of the mixing formulas presented in the previous section. We will focus on the assumption that the medium consists of grains with dielectric constant ε_{i} embedded in a host of dielectric constant ε_{e}. Furthermore the grains have an isotropic orientation and constant axial ratios. Then the unified mixing formula of Sihvola and Kong (1988) applies (see also Sihvola, 1999, and Section 5.5 in this book). The effective dielectric constant ε can be expressed in a form suitable for direct or iterative solution

$$\varepsilon = \varepsilon_{\text{e}} + \frac{f(\varepsilon_{\text{i}} - \varepsilon_{\text{e}}) \sum_{k=1}^{3} \varepsilon_a / (\varepsilon_a + A_k (\varepsilon_{\text{i}} - \varepsilon_{\text{e}}))}{3 - f(\varepsilon_{\text{i}} - \varepsilon_{\text{e}}) \sum_{\text{k}=1}^{3} A_k / (\varepsilon_a + A_k(\varepsilon_{\text{i}} - \varepsilon_{\text{e}}))} \tag{5.85}$$

where f is the volume fraction of the grains, A_k is the depolarisation factor of the ellipsoids for the kth principal axis and ε_a is the apparent permittivity, that is the one in the immediate surroundings of the grains. The depolarisation factors are positive numbers, obeying

$$A_1 + A_2 + A_3 = 1 \tag{5.86}$$

Different mixing models are represented by different choices of ε_a varying between ε_e and ε. With

$$\varepsilon_a = \varepsilon_e + a(\varepsilon - \varepsilon_e) \tag{5.87}$$

the parameter a is defined; its value is between 0 to 1 and it is related to the parameter ν of Equation (5.71). The generalised Maxwell–Garnett formula (also called after Bohren and Battan, 1982) results from selecting $a = 0$, and the coherent potential formula from $a = 1$. The equation of Polder and van Santen (1946) is recovered from

$$a = 1 - A_k; \quad k = 1, 2, 3 \tag{5.88}$$

In Equation (5.88) a is different for each term of the sum in Equation (5.85). The Polder and Van Santen equation turned out to best describe the permittivity of dry snow (Denoth, 1982; Mätzler, 1996).

A simplification of the mixing formula is made by the assumption of spheroidal particles for which two of the three depolarisation factors are equal. This assumption reduces the number of free parameters often without limiting the potential of interpretation. By choosing the equal parameters to be $A = A_1 = A_2$, then A_3 is found from Equation (5.86). Thus, A is the only shape parameter to be determined. Small values ($A < 1/3$) represent oblate spheroids while higher ones ($0.5 \geq A > 1/3$) represent prolate spheroids. With this simplification the mixing formula reduces to a function of ε_i, ε_e, f and A. If ε_i, ε_e and f are known for the measured data, the shape parameter A can be determined. This was done for the snow data to be shown below.

A further simplification is obtained if f is very small. Then ε_a must be equal to ε_e, and the term proportional to f in the denominator of Equation (5.85) can be neglected. This leads to a linear function of f

$$\varepsilon = \varepsilon_e + \sum_{k=1}^{3} \varepsilon_k; \quad \text{where } \varepsilon_k = \frac{f(\varepsilon_i - \varepsilon_e)}{3} \cdot \frac{\varepsilon_e}{\varepsilon_e + A_k(\varepsilon_i - \varepsilon_e)} \tag{5.89}$$

It is important to note that Equation (5.89) is a bilinear transformation of ε_i. This has the consequence that if ε_i follows a Debye relaxation spectrum, each term ε_k of the sum in Equation (5.89) also follows a Debye relaxation spectrum, but with modified parameters (Mätzler, 1987):

$$\varepsilon_k = \varepsilon_{\infty k} + \frac{\varepsilon_{sk} - \varepsilon_{\infty k}}{1 - i\nu/\nu_{0k}}; \quad k = 1, 2, 3 \tag{5.90}$$

The Debye parameters are given by

$$\varepsilon_{sk} = \frac{f}{3} \frac{\varepsilon_e(\varepsilon_{si} - \varepsilon_e)}{\varepsilon_e + A_k(\varepsilon_{si} - \varepsilon_e)}; \quad \varepsilon_{\infty k} = \frac{f}{3} \frac{\varepsilon_e(\varepsilon_{\infty i} - \varepsilon_e)}{\varepsilon_e + A_k(\varepsilon_{\infty i} - \varepsilon_e)};$$

$$\nu_{0k} = \nu_{0i} \left[1 + \frac{A_k(\varepsilon_{si} - \varepsilon_{\infty i})}{\varepsilon_e + A_k \cdot (\varepsilon_{\infty i} - \varepsilon_e)} \right] \tag{5.91}$$

where ε_{si}, $\varepsilon_{\infty i}$ and ν_{0i} are the Debye parameters (static and infinite-frequency dielectric constant, and relaxation frequency) of the grains. It follows from Equation (5.91) that ν_{0k} is always larger than ν_{0i}. As we will see below, the difference can be significant.

An important behaviour of Equation (5.91) results for water and for other polar components with large values of ε_{si}. This is the only large quantity ($\gg$1) in these expressions. Depending on the value of A_k the expression for ε_{sk} may lead to large or small values. If A_k is sufficiently small, with $A_k(\varepsilon_{si} - \varepsilon_e) < 1 \leq \varepsilon_e$, then the contribution to the static dielectric constant ε_s of the mixture is significant. We call this a dominant term (Chaloupka *et al.* 1980). On the other hand if A_k is not very small, then we may have $A_k(\varepsilon_{si} - \varepsilon_e) \gg \varepsilon_e$ leading to a strongly reduced contribution to ε_s. Such a term is very small and may be negligible at frequencies below and near the relaxation frequency of the polar component. Note that there are no dominant terms for near-spherical particles with $A_k \cong 1/3$. In other words, water droplets have a strongly reduced influence on microwave radiation in comparison to water films or needles with at least one $A_k \cong 0$.

5.7.2 *Liquid-water clouds*

A simple, but important example is the description of clouds by a granular, dielectric medium. Here, the host medium is air whose dielectric constant can be assumed to be 1. The grains are small water droplets, often containing impurities. This fact could question the use of the dielectric constant of purewater for ε_i. A typical cloud drop has a diameter of about 10 μm ore more. On the other hand, cloud condensation nuclei usually are smaller than 1 μm, thus the impurity content in the cloud droplet should not exceed 0.1 per cent. But even without impurities, the cloud droplet is under enhanced pressure due to surface tension. The influence of pressure on the dielectric constant of water is negligible for situations applicable to water clouds (Kaatze, 1996). Therefore we assume that $\varepsilon_i = \varepsilon_w$, the dielectric constant of water. Equation (5.89) is applicable here, since f is of the order of 10^{-6}, and for spherical droplets, all depolarisation factors are the same ($A_k = 1/3$), leading to the dielectric constant of clouds

$$\varepsilon = 1 + 3f\frac{\varepsilon_w - 1}{\varepsilon_w + 2} \tag{5.92}$$

The real part ε' of this equation is well described by 1, and the imaginary part ε'' is given by

$$\varepsilon'' = \frac{9f\varepsilon_w''}{|\varepsilon_w + 2|^2} \tag{5.93}$$

With this result the absorption coefficient of a cloud becomes

$$\gamma_a = \frac{18\pi f\varepsilon_w''}{\lambda_0\,|\varepsilon_w + 2|^2} \tag{5.94}$$

where λ_0 is the vacuum wavelength. Instead of validating these formulae with measurements, we note that the result is identical with the respective expression for Rayleigh scattering, and with the improved Born approximation, see Equations (5.98) and (5.99) below for $A_k = 1/3$ and $\varepsilon_a = 1$. To discuss the frequency dependence of the cloud dielectric constant we choose the Debye parameters of water at $T = 0°\mathrm{C}$: $\varepsilon_{\mathrm{sw}} = 88$, $\varepsilon_{\infty\mathrm{w}} \cong 5$ and $\nu_{0\mathrm{w}} \cong 9$ GHz in Equation (5.91). For spherical droplets $(A = 1/3)$ in air $(\varepsilon_{\mathrm{e}} = 1)$, it follows that the mixture has its relaxation frequency and thus the maximum of ε'' at 120 GHz; indeed, this is much larger than 9 GHz. In addition the lack of dominant terms suppresses the dielectric losses at frequencies below about 30 GHz. This is the main reason for clouds to be transparent at these frequencies. It will be shown that water mixed with solid, hydrophilic particles, such as soil or snow, behaves quite differently because of the occurrence of dominant terms in the mixing formula.

5.7.3 Dielectric properties of snow

5.7.3.1 The real part ε' of dry snow

Results on ε' of all types of dry, alpine snow were reported by Mätzler (1996) from measurements made with a large ($V = 680\ \mathrm{cm}^3$), coaxial resonator operating near 1 GHz. The measured data are shown in Figure 5.21 together with the modelled behaviour. The dielectric constant increases in a slightly non-linear way with increasing snow density ρ or ice-volume fraction, $f = \rho/917$, where ρ is in kg/m^3. The smooth solid line in Figure 5.21 represents Equation (5.85) with (5.86) to (5.88) for oblate spheroids $(A = A_1 = A_2)$, for the dielectric constants of ice, $\varepsilon_{\mathrm{i}} = 3.185$, and of air, $\varepsilon_{\mathrm{e}} = 1$. The zoomed residuals are shown in Figure 5.21 by the noisy line whose standard deviation is 0.0064. Although the data include all types of dry snow, a single functional behaviour of the shape parameter $A(f)$ was observed (Mätzler, 1996):

$$A = \begin{cases} 0.1 + 0.5f; & 0 < f \leq 0.32 \\ 0.18 + \exp[-10(f - 0.32)]; & 0.32 < f < 0.5 \\ 1/3; & f \geq 0.5 \end{cases} \tag{5.95}$$

This is an empirical function with a physical sense. First A increases with volume fraction up to $f = 1/3$, from about 0.1 to a maximum of 0.28, close to the value of spheres. This increase is a result of the destructive metamorphism, changing the fresh, mostly oblate snow flakes (axial ratio about 1 : 7) to nearly spherical particles due to gravitational and wind forces. For $f > 1/3$ a sudden decrease is observed. It is thought that the sintering of contacting snow grains leads to an ice matrix where individual particles no longer determine the shape parameter. In fact, $f = 1/3$ is the percolation threshold (Sihvola, 1999) of the Polder and Van Santen mixing rule; above this value a more and more continuous ice matrix starts to form. The shape of this larger structure starts to dominate. The observed standard deviation of 0.0064 between the model and the measurements corresponds to the estimated measurement error which was dominated by the accuracy of the scale used to determine the snow mass. To extrapolate Equation (5.95) beyond the measured range we used the fact that for large f the medium approaches solid ice with spherical air bubbles, and thus $A = 1/3$.

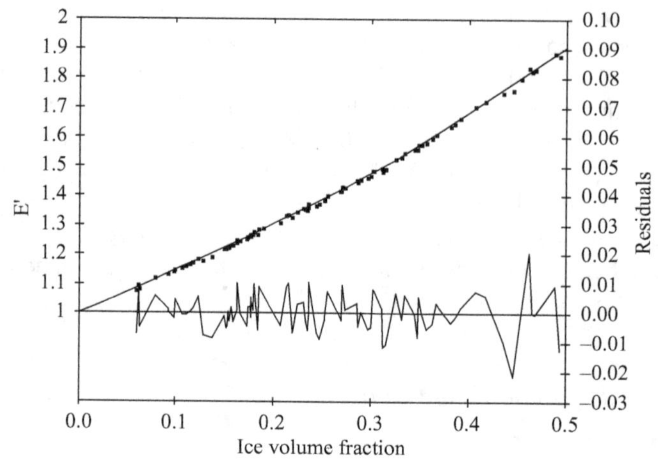

Figure 5.21 *Measured real dielectric constant E′ of dry, natural snow versus ice-volume fraction at 1 GHz according to Mätzler (1996). The solid line is the modelled behaviour according to the mixing formula of Polder and Van Santen for oblate spheroidal ice particles with a variable depolarisation factor. The noisy bottom curve shows the zoomed difference between measurements and model (residuals)*

A simpler empirical fit to directly compute $\varepsilon'(f)$ was found by Mätzler (1996) by combining the two functions of Equation (5.96):

$$\varepsilon' = \begin{cases} 1 + 1.4667f + 1.435f^3; & 0 \leq f < 0.45 \\ ((1-f) + 1.4759f)^3; & f \geq 0.45 \end{cases} \tag{5.96}$$

The standard deviation to the measurements of 0.0066 is nearly as good as for the model using Equation (5.95).

5.7.3.2 The imaginary part ε'' of dry snow

The imaginary part ε'' of the permittivity of dry snow can also be computed from the mixing formula, if the complex value of $\varepsilon_i = \varepsilon_i' + i\varepsilon_i''$ is inserted in (5.85). The resulting values of ε'' are generally smaller than ε_i''. Tiuri *et al.* (1984) found the following formula to be a good approximation to their data,

$$\varepsilon'' = (0.48 \cdot f + 0.52 \cdot f^2) \cdot \varepsilon_i'' \tag{5.97}$$

and they confirmed the formula with mixing theory, assuming $\varepsilon_i'' = 8 \cdot 10^{-4}$ at 2 GHz, corresponding to slightly impure ice. At higher frequencies scattering losses lead to higher values of ε'', sometimes exceeding ε_i'' (Surdyk and Fujita, 1995) if scattering losses are included in ε''. However, for applications in microwave radiometry it is better to reserve ε'' for absorption losses alone; this concept shall be assumed here, leading to $\varepsilon'' < \varepsilon_i''$. A simple expression is obtained for ε'' in the improved Born approximation (Mätzler, 1998)

$$\varepsilon'' = \sqrt{\varepsilon'} K^2 f \varepsilon_{\mathrm{i}}'' \tag{5.98}$$

Here K^2 is the mean-squared ratio of the electrical field inside and outside of the particles, and it is given by

$$K^2 = \frac{1}{3} \sum_{k=1}^{3} \left| \frac{\varepsilon_a}{\varepsilon_a + (\varepsilon_{\mathrm{i}} - 1)A_k} \right|^2 \tag{5.99}$$

For dry snow $K^2 \cong 0.5$ as shown in Figure 1 of Mätzler and Wiesmann (1999). Equations (5.97) and (5.98) give similar results over the typical range of ice-volume fractions. For validation the computed absorption coefficient was compared between theory and experiments (Wiesmann *et al.*, 1998). The agreement was surprisingly good.

A remark refers to potential effects due to ice-surface conductivity. Mixing theory of granular media considers its components as bulk-dielectric media without special effects at the grain surface. According to Petrenko (1993), ice surfaces show high surface conductivity, at least at low frequency. Such surfaces can be modelled by a thin coating of conducting material, and appropriate mixing formulae exist, see Section 5.5.4. One expects an increased dielectric loss of dry snow with increasing thickness of the surface layer and with increasing specific surface. The thickness of a quasi-liquid layer increases with temperature approaching 0°C. So far, this surface effect on ε has not been detected in dry snow, even for fresh snow with high specific surface and for T very close to 0°C (Mätzler *et al.*, 1997).

5.7.3.3 Wet snow

Dielectric measurements of wet snow are made with capacitance sensors, reflectometer and resonator techniques. Due to the granular nature of snow the method is limited to the lower microwave range up to a few GHz. An intercomparison experiment of such instruments gave an impression of the accuracy of the methods (Denoth *et al.*, 1984). Furthermore the experience showed that snow wetness is not at all homogeneously distributed in the snow. The most accurate relationships between the dielectric constant, snow density and volumetric liquid water content W were provided after many years of experiments by Denoth (1989). Free-wave propagation experiments to determine the dielectric constant of wet snow up to 37 GHz were made by Hallikainen *et al.* (1986), and up to 94 GHz by Mätzler *et al.* (1984). The combination of these experimental data together with dielectric mixing theory led to models of the dielectric constant of wet snow, based on the assumption that spheroidal water particles are mixed within a host medium of dry snow.

An alternative would be to assume that liquid water is distributed on the ice surface of the snow grains. This wet-skin model (Sihvola, 1999) has been avoided by snow scientists, based on arguments of thermodynamic stability (Colbeck, 1980), saying that water accumulates in veins and fillets at boundaries between contacting grains. For these shapes the spheroidal model is more suitable. However, melting snow is in a transitory and rapidly changing state. It is reasonable to assume that water forms at the ice surface where the ice phase is distorted by a quasi-liquid layer, even at $T < 0°\text{C}$. Therefore a wet-skin model should not be excluded.

5.7.3.4 Wet snow as a mixture of dry snow with prolate-ellipsoids of liquid water

The experiments of Hallikainen *et al.* (1986) and of Mätzler *et al.* (1984) showed that the imaginary dielectric permittivity has a maximum near 10 GHz. This is close to the maximum of ε''_w of water at $T = 0°\text{C}$. The maximum corresponds to the relaxation frequency. Physical mixing theory predicts, Equation (5.91), that the relaxation frequency ν_{0k} of the mixture depends on the shape of water particles, and in general is higher than the relaxation frequency $\nu_{0\text{w}}$ of water. To fit ellipsoidal water particles to the observations, it is necessary to assume strong eccentricity.

Since wet snow is considered a mixture of dry snow with isotropically oriented ellipsoidal water particles, Equation (5.85) is applicable with f replaced by the volumetric liquid water content W, ε_i replaced by the dielectric constant of liquid water ε_w, and ε_e replaced by the dielectric constant ε_d of dry snow. For small values of W, the linear approximation, Equation (5.89), can be used, leading to

$$\varepsilon = \varepsilon_\text{d} + \frac{W(\varepsilon_\text{w} - \varepsilon_\text{d})}{3} \sum_{k=1}^{3} \frac{\varepsilon_\text{d}}{\varepsilon_\text{d} + A_i(\varepsilon_\text{w} - \varepsilon_\text{d})} \tag{5.100}$$

For ε_w the three-parameter form ($\varepsilon_\text{sw}, \varepsilon_{\infty\text{w}}, \nu_{0\text{w}}$) with one Debye relaxation frequency is used. Application of the adapted Equations (5.89) gives the Debye parameters of wet snow to be inserted in Equation (5.90):

$$\varepsilon_{sk} = \frac{W}{3} \frac{\varepsilon_\text{d}(\varepsilon_\text{sw} - \varepsilon_\text{d})}{\varepsilon_\text{d} + A_k(\varepsilon_\text{sw} - \varepsilon_\text{d})}; \quad \varepsilon_{\infty k} = \frac{W}{3} \frac{\varepsilon_\text{d}(\varepsilon_{\infty\text{w}} - \varepsilon_\text{d})}{\varepsilon_\text{d} + A_k(\varepsilon_{\infty\text{w}} - \varepsilon_\text{d})};$$
$$\nu_{0k} = \nu_{0\text{w}} \left[1 + \frac{A_k(\varepsilon_\text{sw} - \varepsilon_{\infty\text{w}})}{\varepsilon_\text{d} + A_k \cdot (\varepsilon_{\infty\text{w}} - \varepsilon_\text{d})} \right] \tag{5.101}$$

The analysis showed that dominant terms in the sum of (5.100) are those with small values of A_k. For prolate particles there is one dominant term, and for oblate particles there are two such terms.

To fit the measured frequency dependence of ε' and of ε'' to the model, either prolate ellipsoids with axial ratio $\cong 1:25$, or oblate ellipsoids with axial ratio $\cong 1:150$ have to be assumed. When the ellipsoidal model is compared with dielectric measurements of wet snow in the MHz range (Denoth *et al.*, 1984), oblate ellipsoids give values that are too high. Therefore, prolate water ellipsoids with an axial ratio of about $1:25$ should to be assumed. This shape is reasonably related to the water veins. Model parameters consistent with the observations are: $A_1 = A_2 = 0.4975$, $A_3 = 0.005$.

It must be emphasised that wet snow shows high microwave absorption, so high that it can hardly be measured. In a more recent experiment to measure the absorption at 21 and 35 GHz of slightly wet snow, a larger relaxation frequency was estimated, namely about 20 GHz (Mätzler *et al.* 1997), leading to a somewhat larger value of A_3. The snow was fresh and at low density. Further measurements have to show if this difference is related to the type of snow.

With W increasing above about 0.05, the linear approximation is no longer sufficient. Measurements showing the non-linear behaviour of ε' with snow density and wetness in the MHz range were published by Denoth (1989). In terms of the volume fraction of liquid water W and density $\rho(\mathrm{g/cm^3})$ of wet snow, Denoth's formulae for ε' and for the increase $\Delta\varepsilon\prime$ due to liquid water, including its effect on density, are given by

$$\varepsilon' = 1 + 1.92\rho + 0.44\rho^2 + 18.7W + 45W^2; \quad \Delta\varepsilon' = 20.6W + 46W^2 \quad (5.102)$$

The expression for $\Delta\varepsilon'$ can be regarded as the non-linear behaviour of the static dielectric constant ε_{sk}, mainly attributable to the dominant term.

5.7.4 *Dielectric properties of vegetation*

5.7.4.1 Leaves and other green vegetation

Whereas wet snow is a medium with a small volume density of liquid water, typically $W = 0.01$ to 0.1, the opposite is true for fresh vegetation, consisting of organic material, air and water with dissolved ions and molecules, reaching $W = 0.70$–0.90. Consequently the dielectric properties of vegetation are dominated by the dielectric constant of water. As the structure of the organic matrix is highly complex, and the dielectric contrast between water and air is large, the application of conventional mixing theory is not feasible. Instead empirical models will be presented here. Future work should try to reconcile the models with physical mixing theory. This is a key problem, hampering biophysical interpretation of dielectric information.

Ulaby and El Rayes (1987) assumed linear, empirical relationships between the dielectric constants and volume fractions of the vegetation components, free water, bound water, and a residual component due to organic matter and air. Their data were based on dielectric measurements of sucrose solution to determine the spectral behaviour of bound water and of drying corn leaves made with open-ended coaxial probes at frequencies up to 20 GHz (El Rayes and Ulaby, 1987a,b). As the probes are suitable for liquids, the results on sucrose were promising. Increased errors were found for leaves due to contact problems, limiting the measurements to frequencies smaller than 10 GHz. In addition, measurements with high water content were missing. The measurements led to the dual-dispersion model of Ulaby and El Rayes (1987)

$$\varepsilon = \varepsilon_\mathrm{s} + f_\mathrm{f}\varepsilon_\mathrm{sw} + f_\mathrm{b}\varepsilon_\mathrm{b} \quad (5.103)$$

where f_f and f_b are the volume fractions of free and bound water, respectively, ε_s is a non-dispersive residual component depending on the total water content m_g expressed as the ratio of the masses of water and fresh vegetation

$$\varepsilon_\mathrm{s} = 1.7 - 0.74m_\mathrm{g} + 6.16m_\mathrm{g}^2 \quad (5.104)$$

and ε_{sw} and ε_b are the dielectric constants of saline water and bound water, respectively. For f_f and f_b the following expressions were found by Ulaby and El Rayes

$$f_f = m_g(0.55m_g - 0.76); \quad f_b = \frac{4.64m_g^2}{1 + 7.36m_g^2} \tag{5.105}$$

Later Chuah *et al.* (1999) measured the dielectric constant of tropical (rubber and oil palm) leaves with a wave-guide method over the frequency range, 4–16 GHz. They confirmed the model of Ulaby and El Rayes, however with smaller bound water content because tropical plants have lower cell-sap concentration than corn leaves. They found the following expressions for f_f and f_b:

$$f_f = m_g(0.7565m_g - 0.1333); \quad f_b = \frac{1.5306m_g - 2.5909m_g^2 + 1.4355m_g^3}{1 + 0.6m_g} \tag{5.106}$$

An experiment to measure ε of leaves at higher frequencies (21, 35 and 94 GHz), using a quasi-optical method, namely radiometric reflection and transmission measurements, was conducted by Mätzler and Sume (1989). To interpret the results, Sume *et al.* (1988) tested several physical mixing models and the linear model. The empirical, linear model was the only one with acceptable agreement with the measurements. The same technique was later applied to 33 leaves of 12 different trees, bushes and crops (Mätzler and Sume, 1989). The resulting dielectric constant, real and imaginary parts, ε' and ε'', respectively, turned out to be linear functions of the dry-matter fraction $m_d = 1 - m_g$ which is the ratio of dry mass to fresh mass. The spectral behaviour revealed the properties of liquid water. Mätzler (1994) combined the data of Ulaby and El Rayes (1987), El Rayes and Ulaby (1987b) and of Mätzler and Sume (1989) to derive a dielectric formula for vegetation over the frequency range from 1 to 100 GHz:

$$\varepsilon = 0.51 + 3.84m_d + 0.522(1 - 1.32m_d)\varepsilon_{sw} \tag{5.107}$$

where ε_{sw} is the dielectric constant of saline water with a salinity of $S = 1$ per cent, using the formula of Ulaby *et al.* (1986). The dry-matter fraction covers the interval $0.1 \leq m_d \leq 0.5$. The standard deviation of this formula to the measurements of the real and imaginary part was 1.4 and 0.8, respectively, for data points ranging from 3 to 45. Although Equation (5.107) does not include bound water explicitly, the formula can be applied at any frequency from 1 to 100 GHz to leaves in the given m_d range, covering different types of vegetation and probably other vegetation elements whose density is close to 1 g/cm^3. The fact that only two explicit variables appear makes it attractive for radiative-transfer models. The formula is implicitly a function of temperature, frequency and salinity. Only the density dependence is not included because all values considered were very close to 1 g/cm^3. It may be noted that for $m_d \to 0$, Equation (5.107) does not converge to ε_{sw}. Although the leaf mass is purewater in this case, there are still air inclusions. Physical mixing models (e.g. Polder and Van Santen, 1946) show that air bubbles in water can have a strong effect.

The temperature dependence of Equation (5.107) is significant due to the variable relaxation frequency. Since the dielectric vegetation data used to obtain this equation were from near 20°C, the predicted temperature dependence is a hypothetical one. Therefore it was tested independently over the range from $T = 11$ to $26°C$ at $\nu = 4.9$ GHz with the temperature variation of the opacity difference between a foliated and defoliated crown of a beech tree measured for two years. The model coincided well with the measurements for low wind conditions (Mätzler, 1994).

A comparison of the models of Ulaby and El Rayes (1987) and Mätzler (1994) with measurements of tropical leaves over the corresponding range of water content by Chuah *et al.* (1999) showed that the former model is slightly better for ε', whereas the latter is better for ε''. However the overall difference is quite small if Equation (5.105) is used for the free- and bound-water fractions.

5.7.4.2 Wood

Complementary information on dielectric data is needed to find the influence on ε by density variations. Density mainly varies with air content, and significant variations are found for wood. A method using coaxial probes to monitor the temporal variation of ε at about 1 GHz in living trees was developed by McDonald *et al.* (1999). Additional methods are needed on the one hand to determine the dielectric anisotropy of wood and on the other to reach the millimetre wavelength range. The anisotropy of wood is quite obvious due to the structure of parallel fibres. An example of measured emissivity at 10.4 GHz of a thick spruce board under dry and wet condition, respectively, is shown in Figure 5.22 from Künzi *et al.* (1971). The significant difference in emissivities, for instance between $e_{h\parallel}$ and $e_{h\perp}$ (with fibres being parallel and perpendicular to the plane of incidence, respectively) is due to the dielectric anisotropy. For $e_{h\parallel}$ the electric field is perpendicular to the fibres and plane of incidence, representing the effect of $\varepsilon_\perp$, and for $e_{h\perp}$ the electric field is parallel to the fibres, representing the effect of $\varepsilon_\parallel$. Applying the Fresnel formulae to the near-nadir measurements of the dry spruce board, assuming the effect of the imaginary part to be negligible, we get $\varepsilon'_\perp \cong 1.9$ and $\varepsilon'_\parallel = 2.3$, and in case of the wet sample, we find agreement for $\varepsilon'_\perp \cong 7.7$ and $\varepsilon'_\parallel = 10.8$.

Dielectric properties at room temperature of several wood types under dry condition at 10.4 GHz are presented in Table 5.9. The measurements were made with a cylindrical-cavity resonator averaging over $\varepsilon'_\perp$ and $\varepsilon''_\parallel$. The real part ε' increases with increasing wood density from spruce to oak.

5.7.5 Dielectric properties of soil

Soil is a multi-phase system with a structure as complex as the structure of vegetation. The effective dielectric constant ε of the soil is, in general, a function of (a) frequency, temperature and salinity, (b) the volumetric water content θ, (c) the volume fraction of bound and free water related to the specific soil surface area, (d) the soil bulk material and (e) the shape of the water inclusions (Dobson *et al.*, 1985a). Consequently sophisticated dielectric mixing models for soil–water mixtures are required to account for the wide diversity of the soil-structural features. Nowadays none of the existing

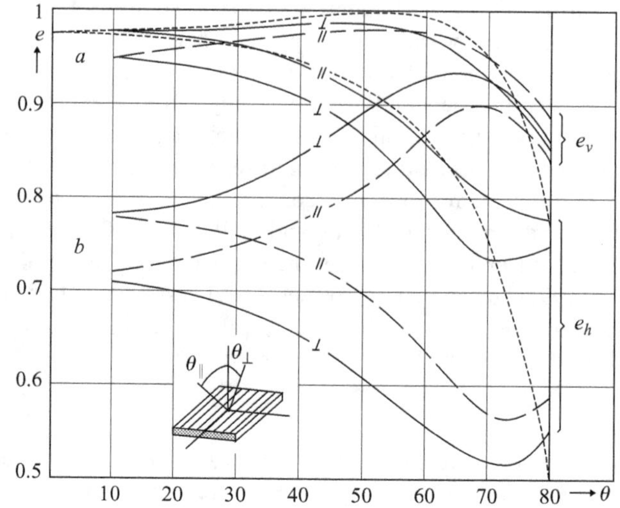

Figure 5.22 *Measured emissivities at 10.49 GHz of a thick spruce board in dry (a) and wet (b) conditions with the plane of incidence parallel (‖) and perpendicular (⊥) to the fibres, from Künzi et al. (1971). The fine-dashed lines show emissivities computed from the typical dielectric constant ($\varepsilon = 1.9 + 0.15$i) measured for dried spruce-wood samples (Table 5.9)*

Table 5.9 *Dielectric properties (averaged over $\varepsilon'_\perp$ and $\varepsilon'_\parallel$) of several dry wood types at room temperature at 10.4 GHz, from Mätzler (1970). For spruce wood, the range from oven-dried (smallest) to humid-air condition (highest values) is shown.*

Wood type	ε'	ε''
Spruce	1.6–2.1	0.07–0.4
Larch	2.3	0.3
Beech	2.4	0.38
Oak	2.5	0.3

physical models is capable of fitting experimental data exclusively based on the soil physical parameters for a wide range of soil types and moisture conditions. Consequently, various empirical and semi-empirical models are used to relate permittivity ε with soil water content θ. These soil dielectric models are the essential part of many algorithms for retrieving soil water content from microwave remote sensing and time domain reflectometry (TDR) data. At low frequencies, soil permittivity is influenced by conductivity losses. At microwave frequencies, ε mainly depends on the volume fraction of free water, and the orientational polarisation of the water molecules is predominant.

In many TDR applications the relation of Topp *et al.* (1980) between the real dielectric constant ε' and volumetric soil water content θ [m^3m^{-3}] is used:

$$\varepsilon' = 3.03 + 9.3\theta + 146.0\theta^2 - 76.7\theta^3 \tag{5.108}$$

This relation makes no attempt of any physical justification, but it proved to be performing well sufficiently below the relaxation frequency of water (<1 GHz) for coarse-textured soils with maximum specific matrix surface of 100 m^2 g^{-1} and bulk density in the range 1.35–1.5 g cm^{-1}. The main advantage of this relation is that it does not require any additional soil parameters.

In the approach proposed by Roth *et al.* (1990) the functional form of the expression relating ε to the soil water content θ is derived from an empirical multi-phase dielectric mixing model. Thereby, the aqueous, solid and gaseous soil phases are represented by their permittivities ε_{w}, ε_{s} and ε_{a}, and their volumetric contents θ, $(1 - \eta)$ and $(\eta - \theta)$:

$$\varepsilon = [\theta \cdot \varepsilon_{\mathrm{w}}^{\alpha} + (1 - \eta) \cdot \varepsilon_{\mathrm{s}}^{\alpha} + (\eta - \theta) \cdot \varepsilon_{\mathrm{a}}^{\alpha}]^{1/\alpha} \tag{5.109}$$

The porosity η is the only soil parameter which has to be determined experimentally. The parameter α was determined to be $\alpha = 0.46 \pm 0.007$ from a weighted non-linear regression applied to the measured data from eleven different field sites representing a wide range of soil types and water contents 0.08 m^3m^{-3} – 0.92 m^3m^{-3}. The uncertainty (root mean square error) of θ calculated from ε using Equation (5.109) was estimated to not exceed 0.013 m^3m^{-3} for the investigated soil types.

However, a thin water film with a paracrystalline structure might be formed between the solid and the aqueous phase due to surface interactions (Dobson *et al.*, 1985b). Because of restricted rotational freedom of the water molecules in this film its permittivity is lower than that of free water. The magnitude of this effect depends on the specific soil matrix surface and surface charge and thus on the soil texture. At the same θ this effect generally leads to a reduction of ε for fine-textured soils compared to coarse-textured ones.

The empirical model proposed by Wang and Schmugge (1980) considers textural effects in terms of an adjustable transition point θ_{t} dividing the water content range into two domains. This considers the experimental fact that for $\theta < \theta_{\mathrm{t}}$ the soil dielectric constant ε increases slowly and for $\theta > \theta_{\mathrm{t}}$ it increases more rapidly with θ. The transition point θ_{t} is higher for soils with high clay content C[kg kg^{-1}] than for soils with high sand content S[kg kg^{-1}]. Furthermore, they found that θ_{t} is strongly correlated with the wilting point θ_{wp} of the soil. In the lower θ range a linear three-component mixture is applied whereas four components are considered in the upper range. The fourth component represents the paracrystalline water in the vicinity of the solid phase, where this water phase is assumed to have the permittivity ε_{i} of ice.

ε_a, ε_w and ε_r are the permittivities of air, water and rock, respectively:

$$\varepsilon = \theta \cdot \varepsilon_x + (\eta - \theta) \cdot \varepsilon_a + (1 - \eta) \cdot \varepsilon_r \quad \text{for } \theta \leq \theta_t \tag{5.110}$$

with

$$\varepsilon_x = \varepsilon_i + (\varepsilon_w - \varepsilon_i)\frac{\theta}{\theta_t} \cdot \gamma$$

and

$$\varepsilon = \theta_t \cdot \varepsilon_x + (\theta - \theta_t) \cdot \varepsilon_w + (\eta - \theta) \cdot \varepsilon_a + (1 - \eta) \cdot \varepsilon_r \quad \text{for } \theta > \theta_t \tag{5.111}$$

with

$$\varepsilon_x = \varepsilon_i + (\varepsilon_w - \varepsilon_i) \cdot \gamma$$

From a multi-parameter regression applied to over 100 datasets of soil water content characteristics in the range $0 < \theta < 0.5$ m^3 m^{-3} measured at the frequencies 1.4 and 5 GHz they found

$$\begin{aligned} \gamma &= -0.57\theta_{wp} + 0.481 \\ \theta_t &= 0.49\theta_{wp} + 0.165 \\ \theta_{wp} &= 0.06774 - 0.00064\ S + 0.00478C \end{aligned} \tag{5.112}$$

The expression (5.112) for the parameter γ is a best fit of Equations (5.110) and (5.111) to the experimental data. At low frequency the above model needs some correction for the imaginary part ε'' to account for conductive losses (λ = wavelength in m, σ = ionic soil conductivity in ohm^{-1}m^{-1}):

$$\varepsilon'' \mapsto \varepsilon'' + 0.6\lambda\sigma \tag{5.113}$$

Another semi-empirical model based on the mixture of the four dielectric soil components solid soil, air, free water and bound water was developed by Dobson *et al.* (1985a,b). They used an empirical power-law mixing formula on the lines of equation (5.109). In addition to the soil-phase with permittivity ε_s and the free-water (fw) phase with permittivity ε_{fw}, the additional bound-water (bw) phase was considered. Since the permittivity ε_{bw} of bw is not well known, they used an approximation involving a soil-texture-dependent coefficient β allowing for eliminating ε_{bw} from the final expression:

$$\varepsilon = [1 + (1 - \eta) \cdot (\varepsilon_s^\alpha - 1) + \theta^\beta \varepsilon_{fw}^\alpha - \theta]^{1/\alpha} \tag{5.114}$$

The parameter α was determined to be $\alpha = 0.65$ by fitting the above model to dielectric measurements conducted for five different soil types at frequencies between 1.4 and 18 GHz. The soil-texture-dependent coefficients $\beta = \beta'$ and $\beta = \beta''$ were determined individually for calculating ε' and ε'', respectively:

$$\begin{aligned} \beta' &= (127.48 - 0.519S - 0.152C)/100 \\ \beta'' &= (1.33797 - 0.603S - 0.166C)/100 \end{aligned} \tag{5.115}$$

In the same article Dobson *et al.* (1985b) published a physical four-component mixing model explicitly accounting for the presence of bound water. The model is based on the mixing approach of De Loor (1968) also known as the equation of Polder and Van Santen (1946) (see Section 5.7.1). The volume fractions of the four components fw, bw, soil and air are computed from a soil physical model, and the permittivity of the fw phase is calculated assuming a Debye-type relaxation modified to account for ionic conductivity losses.

In the following we present a physical dielectric mixing model describing the idealised situation of confocal oblate ellipsoidal inclusion with semi-axes a_{grain}, b_{grain} and c_{grain} embedded in the air background. This situation might be appropriate for clay-rich soils with a substantial fraction of disc-like grains of submicron dimensions. The clay-grains represented by the ellipsoidal inclusions are assumed to be randomly oriented, and soil water is considered as shells covering the grains. To distinguish between bw and fw, two water content regimes, $\theta < \theta_{\mathrm{t}}$ and $\theta \geq \theta_{\mathrm{t}}$ (Figure 5.23) are treated separately. Below the transition point θ_{t} entire soil water θ is assumed to form bw layers of thickness s_{bw} covering the grain ellipsoids. The resulting inclusions are represented by ellipsoids covered with a layer of bw (Figure 5.23a). Above θ_{t} soil water θ is assumed to form an inner bw layer of constant thickness $s_{\mathrm{bw\,max}}$ covering the grains plus an outer fw layer of thickness s_{fw} (Figure 5.23b). The model makes use of the generalised Maxwell–Garnett formula giving ε as the result of the inclusions with number density n [m^{-3}], depolarisation factors A_i and polarisabilities α_i in the direction $i = x, y, z$:

$$\varepsilon = \varepsilon_{\mathrm{e}} + \frac{n/3 \sum_{i=x,y,z} \alpha_i}{1 - n/3\varepsilon_0 \sum_{i=x,y,z} A_i \alpha_i} \tag{5.116}$$

For ellipsoidal inclusions with semi-axes a, b and c in x-, y- and z-direction the depolarisation factor A_x is (Landau *et al.* 1984):

$$A_x = \frac{abc}{2} \int_0^{\infty} \frac{ds}{(s + a^2)\sqrt{(s + a^2)(s + b^2)(s + c^2)}} \tag{5.117}$$

For the other depolarisation factor A_y (A_z), interchange b and a (c and a) in the above integral.

An expression for the polarisabilities α_i of multi-layered dielectric ellipsoids was derived by Sihvola *et al.* (1990). The specialisation for the case of an ellipsoid (grain) covered with two layers (bw and fw) as assumed for the water content regime, $\theta \geq \theta_{\mathrm{t}}$ reads

$$\alpha_x = \varepsilon_{\mathrm{e}} V_{\mathrm{fw}} \cdot \frac{(\varepsilon_{\mathrm{fw}} - \varepsilon_{\mathrm{e}}) + [\varepsilon_{\mathrm{fw}} + A_x^{\mathrm{fw}}(\varepsilon_{\mathrm{e}} - \varepsilon_{\mathrm{fw}})] \cdot U/V}{[\varepsilon_{\mathrm{e}} + A_x^{\mathrm{fw}}(\varepsilon_{\mathrm{fw}} - \varepsilon_{\mathrm{e}})] + A_x^{\mathrm{fw}}(1 - A_x^{\mathrm{fw}})(\varepsilon_{\mathrm{fw}} - \varepsilon_{\mathrm{e}}) \cdot U/V} \tag{5.118}$$

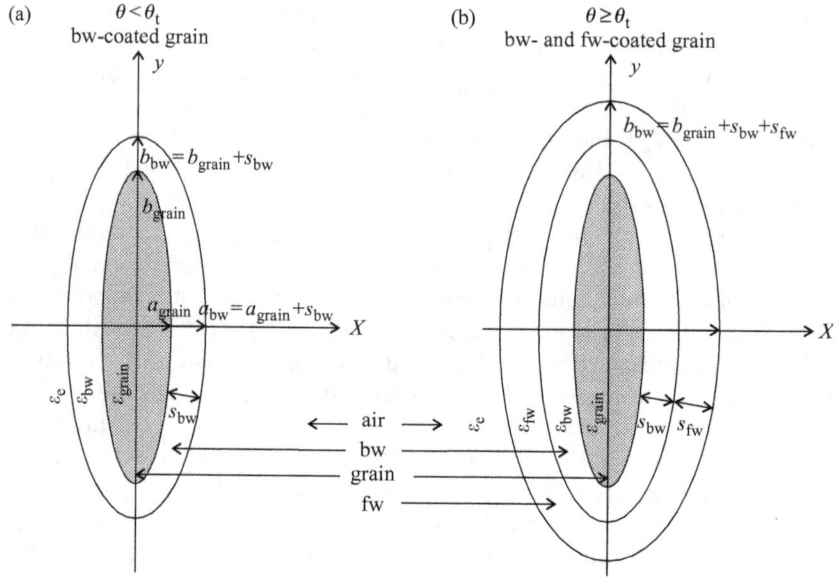

Figure 5.23 Ellipsoidal inclusions of the physical mixing model. (a) For $\theta < \theta_t$ all soil water is represented by bw layers of thickness s_{bw} covering the grains (grey). (b) For $\theta \geq \theta_t$ the grains also have an outer fw layer of thickness s_{fw}

with

$$U = (\varepsilon_{bw} - \varepsilon_{fw})\frac{V_{bw}}{V_{fw}} + [\varepsilon_{bw} + A_x^{bw}(\varepsilon_{fw} - \varepsilon_{bw})] \cdot W$$

$$V = \varepsilon_{fw} + A_x^{bw}(\varepsilon_{bw} - \varepsilon_{fw}) + A_x^{bw}(1 - A_x^{bw})(\varepsilon_{bw} - \varepsilon_{fw}) \cdot W$$

$$W = \frac{(\varepsilon_{grain} - \varepsilon_{bw})V_{grain}/V_{fw} + \varepsilon_{grain} + A_x^{grain}(\varepsilon_{bw} - \varepsilon_{grain})}{\varepsilon_{bw} + A_x^{grain}(\varepsilon_{grain} - \varepsilon_{bw}) + A_x^{grain}(1 - A_x^{grain})(\varepsilon_{grain} - \varepsilon_{bw})}$$

For calculating the polarisabilities α_y and α_z, the corresponding depolarisation factors A_y^{fw}, A_y^{bw}, A_y^{grain} and A_z^{fw}, A_z^{bw}, A_z^{grain} in the y- and z-direction have to be used. The polarisabilities α_i of an ellipsoid covered with a single bw layer as it is assumed for $\theta < \theta_t$ results from (5.116) when using $a_{fw} = a_{bw} = a_{grain} + s_{bw}$, $b_{fw} = b_{bw} = b_{grain} + s_{bw}$, $c_{fw} = c_{bw} = c_{grain} + s_{bw}$ and $\varepsilon_{fw} = \varepsilon_{bw}$.

The θ dependent thickness s_{bw} in the water content regime $\theta < \theta_t$ is calculated from

$$\theta = n(V_{bw} - V_{grain}) \tag{5.119}$$

The ellipsoid volumes $V_{grain} = (4\pi/3)a_{grain}b_{grain}c_{grain}$ and $V_{bw} = (4\pi/3)a_{bw}b_{bw}c_{bw}$ are the grain volume and the volume of the bw ellipsoid with semi-axes $a_{bw} = a_{grain} + s_{bw}$, $b_{bw} = b_{grain} + s_{bw}$, $c_{bw} = c_{grain} + s_{bw}$. Furthermore, the number density n of the grains is related to η and V_{grain} by

$$n = \frac{1-\eta}{V_{\text{grain}}} \tag{5.120}$$

The effective dielectric constant ε for the soil water regime $\theta < \theta_t$ can be calculated from (5.116) with A_i (Equation (5.117)) and α_i (Equation (5.118)) calculated for the bw-covered ellipsoidal grain representing the dielectric inclusion type. Likewise, ε for the soil water regime $\theta \geq \theta_t$ is calculated from (5.116), using α_i and A_i of the double-layer ellipsoidal inclusion and considering the bw and fw phase. The thickness $s_{\text{bw max}}$ of the inner bw layer is calculated as the maximum bw layer thickness calculated for $\theta = \theta_t$ with Equation (5.119). The thickness s_{fw} of the outer fw layer increases with θ according to

$$\theta - \theta_t = n(V_{\text{fw}} - V_{\text{bw}}) \tag{5.121}$$

$V_{\text{fw}} = (4\pi/3)a_{\text{fw}}b_{\text{fw}}c_{\text{fw}}$ is the volume of the fw ellipsoid with the semi-axes $a_{\text{fw}} = a_{\text{grain}} + s_{\text{bw max}} + s_{\text{fw}}, b_{\text{fw}} = b_{\text{grain}} + s_{\text{bw max}} + s_{\text{fw}}, c_{\text{fw}} = c_{\text{grain}} + s_{\text{bw max}} + s_{\text{fw}}$.

The soil dielectric models of Topp *et al.* (1980), Roth *et al.* (1990), Wang and Schmugge (1980), Dobson *et al.* (1985) and the physical model presented above are evaluated within the water content range $0 \leq \theta \leq 0.4$ using the model parameters given in the caption of Figure 5.24. The direct comparison between the ε' values calculated from the physical model and the ε' values resulting from the evaluation of the other models is not reasonable because the latter comprising parameters were fitted to measurements. However, the grain size parameters $a_{\text{grain}}, b_{\text{grain}}, c_{\text{grain}}$ of the physical model are chosen to represent a hypothetical pure clay soil. The parameters θ_t and η of the physical model are chosen such that the calculated ε' is smaller than the ones calculated from the other models. This reflects the experimental fact that ε' of a fine textured soil is generally smaller than ε' of coarser textured soil at the same θ. This is partly due to the higher specific surface area A_{matrix} of clay soils resulting in a higher fraction of bw. Furthermore, the increased salinity of the liquid soil phase in the presence of clay minerals leads to a reduction of ε'. Nevertheless, for all calculations the same value $\varepsilon_w = \varepsilon_{\text{fw}} = 79.7 - I6.18$ of the permittivity of water was used according to the model of Meissner *et al.* (2004) evaluated for 1.4 GHz, 20°C and zero salinity.

As can be seen the models of Topp *et al.* (1980), Roth *et al.* (1990) and Dobson *et al.* (1985) do not distinguish between two water content regimes caused by interactions between water molecules and soil-particle surface. The Wang and Schmugge (1980) model and the physical model consider such effects, leading to a slower increase of ε' for low θ regimes. The transition point of the Wang and Schmugge model is $\theta_t = 0.17\ \text{m}^3\text{m}^{-3}$; it results from Equation (5.112) for the gravimetric sand and clay content $S = 40$ per cent and $C = 20$ per cent, respectively. For the physical model we assumed $\theta_t = 0.08\ \text{m}^3\text{m}^{-3}$ leading to a steeper increase of $\varepsilon'(\theta)$ above the transition point. From θ_t, $\eta = 0.6$ and $a_{\text{grain}} = b_{\text{grain}} = 50c_{\text{grain}} = 0.5\mu\text{m}$ chosen for the physical model one can estimate the maximum bw thickness to be $s_{\text{bw max}} \approx 1$ nm

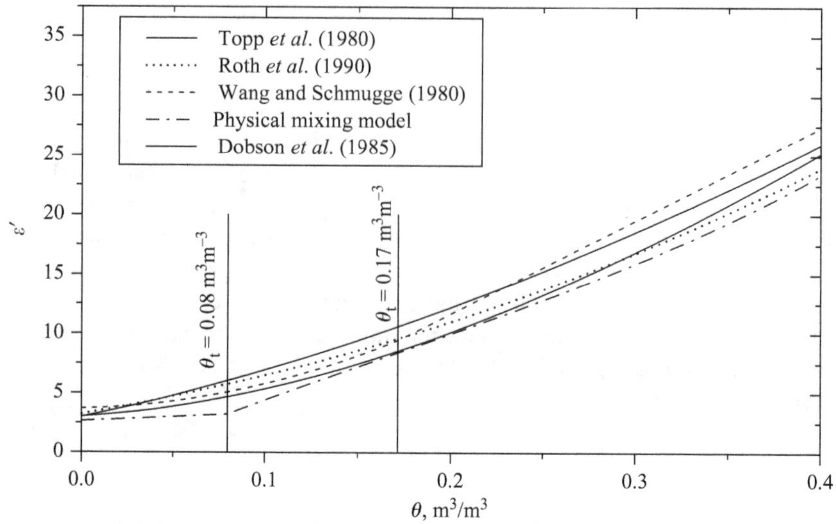

Figure 5.24 *Comparison of five soil dielectric models at 1.4 GHz versus θ for the following parameters: Topp et al. (1980): none; Roth et al. (1990): $\eta = 0.4$, $\varepsilon_a = 1$, $\varepsilon_w = 79.7 + 6.18i$, $\varepsilon_s = 5.5 + 0.2i$, Wang and Schmugge (1980): $\eta = 0.4$, $S = 40\%$, $C = 20\%$, $\varepsilon_a = 1$, $\varepsilon_w = 79.7 + 6.18i$, $\varepsilon_i = 4 + 0.1i$, $\varepsilon_r = 5.5 + 0.2i$; Dobson et al. (1985): $\eta = 0.4$, $S = 40\%$, $C = 20\%$, $\varepsilon_{fw} = 79.7 + 6.18i$, $\varepsilon_s = 4.7$; physical mixing model: $\eta = 0.6$, $\varepsilon_a = 1$, $\varepsilon_{fw} = 79.7 + 6.18i$, $\varepsilon_{bw} = 6 - I\ 0.1$, $\varepsilon_{grain} = 3 + 0.2i$, $a_{grain} = b_{grain} = 50c_{grain} = 0.5\ \mu m$, $\theta_t = 0.08\ m^3\ m^{-3}$*

(Equation 5.119) and the surface area A_{matrix} per gram of the dry clay material:

$$A_{matrix} = \frac{n A_{grain}}{\rho_{bulk}} \tag{5.122}$$

The number density n of the grains is calculated from (5.120), A_{grain} is the surface area of an ellipsoidal grain and the bulk density of the clay material is assumed to be $\rho_{bulk} = 1.5\ g\ cm^{-3}$. This leads to $A_{matrix} \approx 120\ m^2 g^{-1}$ which is a reasonable value for clay-rich soils.

References

ARCONE, S. A., GOW, A. J., and MCGREW, S.: 'Structure and dielectric properties at 4.8 and 9.5 GHz of saline ice', *J. Geophys. Res.*, 1986, **91**, pp. 14281–303

AUTY, R. P., and COLE, R. H.: 'Dielectric properties of ice and solid D_2O', *J. Chem. Phys.*, 1952, **20**, 1309–14

BARTHEL, J., BACHHUBER, K., BUCHNER, R., and HETZENHAUER, H.: 'Dielectric spectra of some common solvents: water and lower alcohols', *Chem. Phys. Lett.*, 1990, **165**, pp. 369–73

BERTOLINI, D., CASSETTARI, M., and SALVETTI, G.: 'The dielectric relaxation time of super-cooled water', *J. Chem. Phys.*, 1982, **15**, pp. 3285–90

BIRCHAK, J. R., GARDNER, L. G., HIPP, J. W., and VICTOR, J. M.: 'High dielectric constant microwave probes for sensing soil moisture', *Proc. IEEE*, 1974, **62**(1), pp. 93–8

BOHREN, C. F., and BATTAN, L. J.: 'Radar backscattering of microwaves by spongy ice spheres', *J. Atmosph. Sci.*, 1982, **39**, pp. 2623–8

BOHREN, C. F., and HUFFMAN, D. R.: 'Absorption and scattering of light by small particles' (Wiley, New York, 1983)

BÖTTCHER, C. J. F.: 'Theory of electric polarization' (Elsevier, Amsterdam, 1952)

BÖTTCHER, C. J. F., and BORDEWIJK, P.: 'Theory of electric polarization, Volume 2 Dielectrics in time-dependent fields' (Elsevier, Amsterdam, 1978, 2nd Edn (completely revised))

BÖTTCHER, C. J. F., VAN BELLE, O. C., BORDEWIJK, P., and RIP, A.: 'Theory of electric polarization, Volume 1 Dielectrics in static fields' (Elsevier, Amsterdam, 1973, 2nd Edn (completely revised))

BRACEWELL, R.: 'The fourier transform and its applications' (McGraw Hill, New York, 1965)

BRUGGEMAN, D. A. G.: 'Berechnung verschiedener physikalischer Konstanten von heterogenen Substanzen, I. Dielektrizitätskonstanten und Leitfähigkeiten der Mischkörper aus isotropen Substanzen', *Annalen der Physik*, 1935, 5. Folge, Band 24, pp. 636–64

BUCHNER, R., HEFFER, G. T., and BARTHEL, J.: 'Dielectric relaxation of aqueous NaF and KF solutions', *J. Chem. Soc. Faraday Trans.*, 1994, **90**, pp. 2475–9

BUCHNER, R., BARTHEL, J., and STAUBER, J.: 'The dielectric relaxation of water between 0°C and 35°C', *Chem. Phys. Lett.*, 1999, **306**, pp. 57–63

BURGER, H. C.: 'Das Leitvermögen verdünnter mischkristallfreier Legierungen', *Physikalische Zeitschrift*, 15 February 1919, **20**(4), pp. 73–5

CAMPBELL, M. J., and ULRICHS, J.: 'Electrical properties of rocks and their significance for lunar radar observations', *J. Geophys. Res.*, 1969, **74**(25 November 15), pp. 5867–81

CHALOUPKA, H., OSTWALD, O., and SCHIEK, B.: 'Structure independent moisture measurements', *J. Microwave Power*, 1980, **15**, pp. 221–31

CHUAH, H. T., KAM, S. W., and CHYE, Y. H.: 'Microwave dielectric properties of rubber and oil palm leaf samples: measurement and modelling', *Int. J. Remote Sens.*, 1997, **18**, pp. 2623–39

COLBECK, S. C.: 'Liquid distribution and the dielectric constant of wet snow'. Proceedings of the NASA workshop on *Microwave remote sensing of snowpack properties*, Fort Collins, CO, 20–22 May, NASA CP-2153, 1980, pp. 21–39

CORR, H., MOORE, J. C., and NICHOLLS, K.W.: 'Radar absorption due to impurities in antarctic ice', *Geophys. Res. Lett.*, 1993, **20**, 1071–4

CULKIN, F., and SMITH, N.: 'Determination of the concentration of potassium chloride solution having the same electrical conductivity, at 15°C and infinite frequency, as standard seawater of salinity 35‰', *IEEE J. Ocean. Eng.*, 1980, **5**, N° 22, p. 22

CUMMING, W. A.: 'The dielectric properties of ice and snow at 3.2 cm', *J. Appl. Phys.*, 1952, **23**(7), pp. 768–73

DE LOOR, G. D.: 'Dielectric properties of heterogeneous mixtures containing water', *J. Microwave Power*, 1968, 3–2, pp. 67–73

DEBYE, P.: 'Polar molecules' (first published by Chemical Catalog Company, New York 1929; reprinted by Dover Publications, New York, 1945; German Edition: 'Polare Molekeln', Leipzig, 1929)

DENOTH, A.: 'Effect of grain geometry on electrical properties of snow at frequencies up to 100 MHz', *J. Appl. Phys.*, 1982, **53**, pp. 7496–501

DENOTH, A.: 'Snow dielectric measurements', *Adv. Space Res.*, 1989, **9**(1) pp. (1)233-(1)243

DENOTH, A., FOGLAR, A., WEILAND, P. *et al.*: 'A comparative study of instruments for measuring the liquid-water content of snow', *J. Appl. Phys.*, 1984, **56**, pp. 2154–60

DOBSON, M. C., ULABY, F. T., HALLIKAINEN, M. T., and EL-RAYES, M. A.: 'Microwave dielectric behaviour of wet soil. Part I: Dielectric mixing models', *IEEE Trans. Geosci. Remote Sens.*, 1985a, **23**, pp. 25–34

DOBSON, M. C., ULABY, F. T., HALLIKAINEN, M. T., and El-RAYES, M. A.: 'Microwave dielectric behaviour of wet soil. Part II: Dielectric mixing models', *IEEE Trans. Geosci. Remote Sens.*, 1985b, **23**, pp. 35–44

DOYLE, W. T.: 'The Clausius-Mossotti problem for cubic array of spheres', *J. Appl. Phys.*, 1978, **49**(2), pp. 795–7

EL RAYES, M. A., and ULABY, F. T.: 'Microwave dielectric spectrum of vegetation - Part I: Experimental observations', *IEEE Trans. Geosci. Remote Sens.*, 1987a, **GE-25**, pp. 541–9

EL RAYES, M. A., and ULABY, F. T.: 'Microwave dielectric behaviour of vegetation material' NASA GSFC report, contract NAG 5-480, Jan. 1987b

ELLISON, W. J., LAMKAOUCHI, K., and MOREAU, J. M.: 'Water: A dielectric reference', *J. Mol. Liquids*, 1996a, **68**, pp. 171–279

ELLISON, W. J., BALANA, A., DELBOS, G. *et al.*: 'Study and measurement of the dielectric properties of seawater: Final report'. ESTEC/ESA, contract 11197/94/NL/CN, 1996b

ELLISON, W. J., BALANA, A., and LAMKAOUCHI, K.: 'Dielectric properties of seawater in the range 30–105 GHz: Final report'. ESTEC/ESA contract 11197/97/NL/CN, Eur. Space Agency, Paris, 1997

ELLISON, W. J., BALANA, A., DELBOS, G. *et al.*: 'New permittivity measurements of seawater', *Radio Sci.*, 1998, **33°**, pp. 639–48

ELLISON, W. J., ENGLISH, S. J., LAMKAOUCHI, K. *et al.*: 'A comparison of ocean emissivity models using the Advanced Microwave Sounding Unit, the Special Sensor Microwave Imager, the TRMM Microwave Imager, and airborne radiometer observations', *J. Geophys. Res.*, 2003, **108** (D21), pp. 4663–77

EYRING, H., and JHON, M.: 'Significant liquid structures' (J. Wiley, New York, 1969)

FRÖHLICH, H.: 'Theory of dielectrics' (Clarendon Press, Oxford, 1958)

FUJITA, S., SHIRAISHI, M., and MAE, S.: 'Measurement on the dielectric properties of acid-doped ice at 9.7 GHz', *IEEE Trans. Geosci. Remote Sens.*, 1992, **30**, 799–803

FUJITA, S., MATSUOKA, T., and MAE, S.: 'Dielectric anisotropy in ice Ih at 9.7 GHz', *Ann. Glaciol.*, 1993, **17**, 276–80

FUJITA, S., MATSUOKA, T., ISHIDA, T., MATSUOKA, K., and MAE, S.: 'A summary of the complex dielectric permittivity of ice in the megahertz range and its applications for radar sounding of polar ice sheets', in HONDOH T. (Ed.): 'Physics of ice core records' (Hokkaido University Press, 2000) pp. 185–212

GERSCHEL, A.: 'Liaisons intermoléculaires' (CNRS Editions, Paris, 1995)

GLASSTONE, S., LAIDLER, J., and EYRING, H.: 'The theory of rate processes' (MacGraw-Hill, New York, 1941)

GOUGH, S. R.: 'A low temperature dielectric cell and the permittivity of hexagonal ice to 2K', *Can. J. Chem.*, 1972, **50**, 3046–51

GRANT, E. H., and SHEPPARD, R. J.: 'Dielectric relaxation in water in the neighbourhood of 4°C', *J. Chem. Phys.*, 1974, **60**, 1792–6

GRIMMETT, G.: 'Percolation' (Springer, New York, 1989)

GUILLOU, C., ELLISON, W., EYMARD, L. *et al.*: 'Impact of new permittivity measurements on sea surface emissivity modeling in microwaves', *Radio Sci.*, 1998, **33**, pp. 649–68

GUPTA, R. R.: 'Diamagnetic susceptibility' in MADELUNG, O. (Ed.): 'Landolt-Börnstein' (New Series, Vol. 16, Springer, Berlin, 1986)

HAGGIS, G. H., HASTED, J. B., and BUCHANAN, T. J.: 'The dielectric properties of water in solutions', *J. Chem. Phys.*, 1952, **20**, pp. 1452–64

HALLIKAINEN, M., and WINEBRENNER, D. P.: 'The physical basis for sea-ice remote sensing', in CARSEY, F. D. (Ed.): 'Microwave remote sensing of sea ice' (Geophysical monograph 68, Am. Geophys. Union, Washington D.C., 1992), Ch. 3, pp. 29–46

HALLIKAINEN, M. T., ULABY, F. T., and ABDELRAZIK, M.: 'Dielectric properties of snow in the 3 to 37 GHz range', *IEEE Trans. Ant. Prop.*, 1986, **AP-34**, pp. 1329–40

HASTED, J. B.: 'Aqueous dielectrics' (Chapman & Hall, London, 1973) p. 238

HILL, N. E., VAUGHN, W. E., PRICE, A. H., and DAVIES, M.: 'Dielectric properties and molecular behaviour' (Von Nostrand, London, 1968)

HO, W., and HALL, W. F.: 'Measurement of the dielectric properties of seawater and NaCl solutions at 2.65 GHz', *J. Geophys. Res.*, 1973, **78**, pp. 6301–15

HO, W., LOVE, A. W., and VAN MELLE, M. J.: 'Measurements of the dielectric properties of seawater at 1.43 GHz'. NASA contract report CR-2458, 1974

HUFFORD, G.: 'A model for the complex permittivity of ice at frequencies below 1 THz', *Int. J. Infrared Millimeter Waves*, 1991, **12**, 677–82

JACKSON, J. D.: 'Classical electrodynamics' (New York, Wiley, 1975, 2nd edn)

JIANG, J. H., and WU, D. L.: 'Ice and water permittivities for millimeter and sub-millimeter remote sensing applications', *Atmos. Sci. Lett.*, 2004, 5, pp. 146–51

KAATZE, U.: 'Dielectric spectroscopy of aqueous solutions. Hydration phenomena and hydrogen bonded networks', *J. Molecular Liquids*, 1993, **56**, pp. 95–115

KAATZE, U.: 'Microwave dielectric properties of water', in KRASZEWSKI, A. (Ed.): 'Microwave Aquametry' (IEEE Press, Piscatawy, N. J., 1996), pp. 37–53

KÄRKKÄINEN, K., SIHVOLA, A. H., and NIKOSKINEN, K. I.: 'Analysis of a three-dimensional dielectric mixture with finite difference method', *IEEE Trans. Geosci. Remote Sens.*, May 2001, **39**(5), pp. 1013–18

KINDT, J. T., and SCHMUTTENMAER, C. A.: 'Far-infrared dielectric properties of polar liquids probed by femtosecond terahertz pulse spectroscopy', *J. Phys. Chem.*, 1996, **100**, pp. 10373–9

KITTEL, C.: 'Introduction to solid state physics', (Wiley, New York, 1986, 6th edn)

KLEIN, L. A., and SWIFT, C. T.: 'An improved model for the dielectric constant of seawater at microwave frequencies', *IEEE Trans. Antennas Propag.*, 1977, **AP25**(1) pp. 104–11

KOH, G.: 'Dielectric constant of ice at 26.5 to 40 GHz', *J. Appl. Phys.*, 1992, **71**, pp. 5119–22

KOH, G.: 'Dielectric properties of ice at millimeter wavelengths', *Geophys. Res. Lett.*, 1997, **24**(18), pp. 2311–3

KONG, J. A.: 'Electromagnetic wave theory' (John Wiley, New York, 1985)

KRASZEWSKI, A.: 'Microwave aquametry: electromagnetic wave interaction with water-containing materials' (IEEE Press, Piscataway, 1996)

KRISTENSSON, G.: 'Homogenization of spherical inclusions', *Progress Electromagnet. Res.*, 2003, **42**, pp. 1–25

KÜNZI, K., MAGUN, A., MÄTZLER, C., SCHAERER, G., and SCHANDA, E.: 'Passive microwave remote sensing at the University of Bern, Switzerland' Proceedings of the 7th international symposium on *Remote sensing environment*, University of Michigan, Ann Arbor, MI, 1971, pp. 1819–26

LAM, J.: 'Magnetic permeability of a simple cubic lattice of conducting magnetic spheres', *J. Appl. Phys.*, 1986, **60**(12), pp. 4230–5

LAMB, J.: 'Measurements of the dielectric properties of ice', *Trans. Faraday Soc.*, 1946, 42A, 238–44

LAMB, J. W.: 'Miscellaneous data on materials for millimetre and submillimetre optics', *Int. J. Infrared Millimeter Waves*, 1996, **17**(12), pp. 1997–2029

LAMKAOUCHI, K.: 'L'EAU : Étalon diélectrique'. Ph.D. thesis, Université de Bordeaux 1, N° 745, 1992

LAMKAOUCHI, K.: 'Réalisation de dispositifs large bande 3–110 GHz et études diélectriques des eaux de surface des océans en vue d'améliorer les observations des radars aéroportés et des radiomètres embarqués sur satellites'. Ph.D. thesis, Université Moulay Ismaïl, Faculté des Sciences, Meknès, Morocco, 1999

LAMKAOUCHI, K., BALANA, A., DELBOS, G., and ELLISON, W. J.: 'Permittivity measurements of lossy liquids in the range 26–110 GHz', *Measurement Sci. Technol.*, 2003, **14**, pp. 444–50

LANDAU, L. D., and LIFSHITZ, E. M.: 'Electrodynamics of continuous media', (Oxford, Pergamon Press, 1984, 2nd edn) Section 4

LANDAUER, R.: 'Electrical conductivity in inhomogeneous media'. American Institute of Physics conference proceedings. *Electrical transport and optical properties of inhomogeneous media*, (40), 1978, pp. 2–45

LEWIS, E. L.: 'Salinity: its definition and calculation', *J. Geophysical research*, 1978, **83**, p. 466

LEWIS, E. L: 'The practical salinity scale 1978 and its antecedents', *IEEE J. Ocean. Eng.*, 1980, **5**(3), p. 3

LIEBE, H. J., HUFFORD, G. A., and MANABE, T.: 'A model for the permittivity of water at frequencies below 1THz', *Int. J. Infrared Millimeter Waves*, 1991, **12**, pp. 659–75

LOOYENGA, H.: 'Dielectric constants of mixtures', *Physica*, 1965, **31**, pp. 401–6

MANABE, T., LIEBE, H. J., and HUFFORD, G. A.: 'Complex permittivity of water between 0 and 30 THz'. IEEE conference digest 12th International Conference on *IR MM Waves*, 1987, Orlando, pp. 229–30

MATSUOKA, T., FUJITA, S., and MAE, S.: 'Dielectric properties of NaCl-doped ice at 9.7 GHz'. Proceedings of the National Institute of Polar Research Symposium on Polar meteorology and glaciology, 1993, **7**, pp. 33–40

MATSUOKA, T., FUJITA, S., and MAE, S.: 'Effects of temperature on the dielectric properties of ice in the range 5-39 GHz', *J. Appl. Phys.*, 1996, **80**, pp. 5884–90

MÄTZLER, C.: 'Messung der Wärmestrahlung der Erdoberfläche im Mikrowellengebiet'. Diploma thesis, Institute of Applied Physics, University of Bern, Switzerland, April, 1970

MÄTZLER, C.: 'Applications of the interaction of microwaves with the natural snow cover', *Remote Sens. Rev.*, 1987, **2**, pp. 259–392

MÄTZLER, C.: 'Microwave (1–100 GHz) dielectric model of leaves', *IEEE Trans. Geosci. Remote Sens.*, 1994, **32**(4), pp. 947–9

MÄTZLER, C.: 'Microwave permittivity of dry snow', *IEEE Trans. Geosci. Remote Sens.*, 1996, **34**, pp. 573–81

MÄTZLER, C.: 'Improved Born approximation for scattering in a granular medium', *J. Appl. Phys.*, 1998a, **83**(11), pp. 6111–7

MÄTZLER, C.: 'Microwave properties of ice and snow', in SCHMITT, B. *et al.* (Eds): 'Solar System Ices', (Kluwer Academic Publishers, 1998b, Dordrecht) Astrophys. and Space Sci. Library, **227**, pp. 241–57

MÄTZLER, C., and SUME, A.: 'Microwave radiometry of leaves', in PAMPALONI, P. (Ed.) 'Microwave radiometry and remote sensing applications' (VSP Utrecht), 1989, pp. 133–48

MÄTZLER, C. and WEGMÜLLER, U.: 'Dielectric properties of freshwater ice at microwave frequencies', *J. Phys. D: Applied Phys.*, 1987, **20**, pp. 1623–1630; Errata, 1988, **21**, p. 1660

MÄTZLER, C., and WIESMANN, A.: 'Extension of the microwave emission model of layered snowpacks to coarse-grained snow', *Remote Sens. Environ.*, 1999, **70**(3), pp. 317–25

MÄTZLER, C., AEBISCHER, H., SCHANDA, E. *et al.*.: 'Microwave dielectric properties of surface snow', *IEEE J. Ocean. Engin.*, 1984, **OE-9**, pp. 366–71

MÄTZLER, C., STROZZI, T., WEISE, T., FLORICIOIU, D., and ROTT, H.: 'Microwave snowpack studies made in the Austrian Alps during the SIR-C/X-SAR experiment', *Int. J. Remote Sens.*, 1997, **18**, pp. 2505–30

MAXWELL-GARNETT, J. C.: 'Colours in metal glasses and metal films', *Trans. of the Royal Society*, 1904, CCIII, pp. 385–420, London

MCDONALD, K. C., ZIMMERMANN, R., WAY, J., and CHUN, W.: 'Automated instrumentation for continuous monitoring of the dielectric properties of woody vegetation: system design, implementation, and selected in situ measurements', *IEEE Trans. Geosc. Remote Sens.*, **37**(4), pp. 1880–94

MCKENZIE, D. R., MCPHERDRAN, R. C., and DERICK, G. H.: 'The conductivity of lattice of spheres. II. The body centred cubic lattices', *Proceedings of the Royal Society of London*, A., 1978, **362**, pp. 211–32

MCPHEDRAN, R. C., and MCKENZIE, D. R.: ' The conductivity of lattices of spheres. I. The simple cubic lattice', *Proc. R. Soc. London*, A, 1978a, **359**, pp. 45–63

MEISSNER, T., and WENTZ, F.: 'The complex dielectric constant of pure water and sea water from microwave satellite observations', *IEEE Trans. Geosci Remote Sens.*, 2004, **49**(9), pp. 1836–49

MEREDITH, R. E., and TOBIAS, C. W.: 'Resistance to potential flow through a cubical array of spheres', *J. Appl. Phys.*, 1960, **31**(7), pp. 1270–3

MISHIMA, O., KLUG, D. D., and WHALLEY, E.: 'The far-infrared spectrum of ice Ih in the range 8–25 cm^{-1}. Sound waves and difference bands, with application to Saturn's rings', *J. Chem. Phys.*, 1983, **78**, pp. 6399–404

MIZOGUCHI, K., HORI, Y., and TOMINAGA, Y.: 'Study on dynamical structure in water and heavy water by low frequency Raman spectroscopy', *J. Chem. Phys.*, 1992, **97**, pp. 1961–8

MOORE, J. C., and FUJITA, S.: 'Dielectric properties of ice containing acid and salt impurity at microwave and low frequencies', *J. Geophys. Res.*, 1993, **98**, pp. 9769–80

MOSSOTTI, O. F.: 'Discussione analitica sull'influenza che l'azione di un mezzo dielettrico ha sulla distribuzione dell'elettricità alla superficie di più corpi elettrici disseminati in esso', *Memorie di matematica a di fisica della Società Italiana delle scienze, residente in Modena*, 1850, **24**, Part 2, pp. 49–74

NABOKOV, G. A., and LYUBINOV, YU. A.: 'Temperature dependence of the dielectric relaxation of water along the saturation curve', *Russian J. Phys. Chem.*, 1987, **61**, pp. 106–9

NELSON, S., LINDROTH, D., and BLAKE, R.: 'Dielectric properties of selected and purified minerals at 1 and 22 GHz', *J. Microwave Power Electromagn. Energy*, 1989, **24**(4), pp. 213–20

NOYEL, G., JORAT, L. J., and DERRICHE, O.: 'Dielectric properties of normal supercooled water obtained in alcohol/water mixtures', *IEEE Trans. Elec. Insulation*, 1992, **27**, pp. 1136–43

NOYEL, G., JORAT, L. J., FERRIOL, M., COHEN-ADAD, M. T., HUCK, J., and BLANC-MIGNON, M. F.: 'Static permittivity of normal supercooled water by study of alkyhydrazines/water mixtures', *J. Mol. Liquids*, 1994, **59**, pp. 149–60

OSBORN, J. A.: 'Demagnetizing factors of the general ellipsoid', *Phys. Rev.*, 1945, **67**, 11–12, pp. 351–7

PETRENKO, V. F.: 'Electrical properties of ice' US Army Corps of Engineers Cold Regions Research and Engineering Laboratory (CRREL) special report 93–20, Hanover, New Hampshire, 1993

PETRENKO, V. F., and WHITWORTH, R. W.: 'Physics of ice' (Oxford University Press, 1999), pp. 373

POLDER, D., and VAN SANTEN, J. H.: 'The effective permeability of mixtures of solids', *Physica*, 1946, **XII**(5), pp. 257–71

PRESS, W. H., TEUKOLSKY, S. A., VETTERLING, W. T., and FLANNERY, B. P.: 'Numerical recipes in fortran: The art of scientific computing', (Cambridge University Press, 1992, 2nd edn)

REBER, B., MÄTZLER, C., and SCHANDA, E.: 'Microwave signatures of snow crusts: modelling and measurements', *Int. J. Remote Sens.*, 1987, **8**, pp. 1649–65

RØNNE, C., ÅSTRAND, P. O., WALLQVIST, A., MIKKELSEN, K. V., and KEIDING, S. R.: 'Investigation of the temperature dependence of dielectric relaxation in liquid water by THz reflection spectroscopy and molecular dynamics simulation', *J. Chem. Phys.*, 1997, **107**, pp. 5319–31

ROTH, K., SCHULIN, R., FLÜHLER, H., and ATTINGER, W.: 'Calibration of time domain reflectometry for water content measurement using a composite dielectric approach', *Water Resources Research*, 1990, **26**, pp. 2267–73

RUFFIER, G.: 'Rapport sur les mesures de poudre de carbonate'. Rapport d'activité du Programme National de Télédétection Spatiale, Observatoire de Paris, France, 2003

RUNGE, I.: 'Zur elektrischer Leitfähigkeit metallischer Aggregate', *Zeitschrift für technische Physik*, 1925, **6**(2), pp. 61–8

SEN, P. N., SCALA, C., and COHEN, M. H.: 'A self-similar model for sedimentary rocks with application to the dielectric constant of fused glass beads', *Geophysics*, 1981, **46**(5), 1981, pp. 781–95

SIHVOLA, A.: 'Self-consistency aspects of dielectric mixing theories', *IEEE Trans. Geosci. Remote Sensing.*, 1989, **27**(4), pp. 403–15

SIHVOLA, A.: 'Lorenz–Lorentz or Lorentz–Lorenz?' *IEEE Antennas and Propagation Magazine*, August 1991, **33**, p. 56

SIHVOLA, A.: 'Electromagnetic mixing formulas and applications' (IEE Publishing, London, UK, 1999)

SIHVOLA, A., and LINDELL, I. V.: 'Polarizability and effective permittivity of layered and continuously inhomogeneous dielectric spheres', *J. Electromagnet. Waves Appl.*, 1989, **3**(1), pp. 37–60

SIHVOLA, A., and LINDELL, I. V.: 'Polarizability and effective permittivity of layered and continuously inhomogeneous dielectric ellipsoids', *J. Electromagnet. Waves Appl.*, 1990, **4**(1), pp. 1–26

SIHVOLA, A., and KONG, J. A.: 'Effective permittivity of dielectric mixtures', *IEEE Trans. Geosc. Rem. Sens.*, 1988, **26**, pp. 420–9; *Errata*, 1989, **27**, pp. 101–2

SIHVOLA, A., YLÄ-OIJALA, P., JÄRVENPÄÄ, S., and AVELIN, J.: 'Polarizabilities of Platonic solids'. *IEEE Transactions on Antennas and Propagation*, Vol.52, No. 9, pp. 2226–33, September 2004

SPEEDY, R. J., and ANGELL, C. A.: 'Isothermal compressibility of supercooled water and evidence for a thermodynamic singularity at −45°C', *J. Chem. Phys.*, 1976, **63**, pp. 851–8

STOGRYN, A. P.: 'Equations for calculating the dielectric constant of saline water', *IEEE Trans Microwave Theory Tech.*, 1971, **19**, pp. 733–6

STOGRYN, A. P., BULL, H. T., RUAYI, K., and IRAVANCHY, S.: 'The microwave permittivity of sea and freshwater'. Aerojet internal report, Aerojet, Sacramento, Calif., 1996

STONER, E. C.: 'The demagnetizing factors for ellipsoids', *Philos. Mag.*, 1945, Ser. 7, **36**(263), pp. 803–21

SUME, A., MÄTZLER, C., HÜPPI, R., and SCHANDA, E.: 'Microwave radiometer and scatterometer measurements of vegetation'. FOA report C30494-3.2, ISSN 0347-3708, Swedish Defence Research Establishment, Linköping, Sweden, 1988

SURDYK, S., and FUJITA, S.: 'Microwave dielectric properties of snow: Modeling and measurements', *Geophys. Res. Lett.*, 1995, **22**, pp. 965–8

THOMAS, B.: 'Etude et réalisation d'une tête de réception hétérodyne en ondes sub-millimétriques pour l'étude des atmosphères et surfaces de planètes'. PhD report, Observatoire de Paris, Paris, France, Dec., 2004

TIURI, M. E., SIHVOLA, S., NYFORS, E. G., and HALLIKAINEN, M. T.: 'The complex dielectric constant of snow at microwave frequencies', *IEEE J. Ocean. Engin.*, 1984, **OE-9**, pp. 377–82

TOPP, G. C., DAVIS, J. L., and ANNAN, A. P.: 'Electromagnetic determination of soil water content: Measurements in coaxial transmission lines', *Water Resources Research*, 1980, **16**(3), pp. 574–82

TSANG, L., KONG, J. A., and SHIN, R. T.: 'Theory of microwave remote sensing', (Wiley, New York, 1985)

ULABY, F. T., and EL RAYES: 'Microwave dielectric spectrum of vegetation – Part II: Dual-Dispersion Model', *IEEE Trans. Geosci. Remote Sens.*, 1987, **GE-25**, pp. 550–7

ULABY, F. T., MOORE, R., and FUNG, A. K.: 'Microwave remote sensing, active and passive', Vol. II, Radar Remote Sensing and Surface Scattering and Emission Theory (1982); Vol. III, From Theory to Applications, Artech House Inc, Norwood, MA 02062, 1986

VON HIPPEL, A.: 'Dielectrics and waves' (M.I.T. Press, Boston, MA, 1954; reprinted by Artech House, Boston, MA, 1995)

WAGNER, K. W.: 'Erklärung der dielektrischen Nachwirkungsvorgänge auf Grund Maxwellscher Vorstellungen', *Archiv für Elektrotechnik*, 1914, **2**(9), pp. 371–87

WALFORD, M. E. R.: 'Field measurements of dielectric absorption in Antarctic ice and snow at very high frequencies', *J. Glaciol.*, 1968, **7**(49), pp. 89–94

WANG, J. R., and SCHMUGGE, T.: 'An empirical model for the complex dielectric permittivity of soils as a function of water content', *IEEE Trans. Geosci. Remote Sens.*, 1980, **GE-18**(4), pp. 288–95

WARREN S. T.: 'Optical constants of ice from the ultraviolet to the microwave', *Appl. Opt.*, 1984, **23**, pp. 1206–25

WEGMÜLLER, U.: 'Signaturen zur Mikrowellenfernerkundung: Bodenrauhigkeit und Permittivität von Eis' Diploma thesis, Inst. Appl. Phys., University of Bern, 3012 Bern, Switzerland, 1986

WIESMANN, A., MÄTZLER, C., and WEISE, T.: 'Radiometric and structural measurements of snow samples', *Radio Sci.*, 1998, **33**, pp. 273–89

WOSCHNAGG, K., and PRICE, P. B.: 'Temperature dependence of absorption in ice at 532 nm', *Appl. Opt.*, 2001, **40**, pp. 8–12

YAGHJIAN, A. D.: 'Electric dyadic Green's function in the source region', *Proc. IEEE*, 1980, **68**(2), pp. 248–63

ZHANG, C., LEE, K. S., ZHANG, X. C., WEI, X., and SHEN, Y. R.: 'Optical constants of ice Ih crystal at terahertz frequencies', *Appl. Phys. Lett.*, 2001, **79**(4), pp. 491–3

Appendix A

Surface emissivity data from microwave experiments at the University of Bern

Christian Mätzler and Andreas Wiesmann

Purpose

Surface-based microwave signature studies from 2 to 100 GHz were carried out over several years to provide empirical emissivity information for typical surface classes. The surface emissivity e_p (polarisation p) was determined from the effective object temperature T_{eff}, the brightness temperatures T_p of the object under investigation and of the downwelling sky radiation T_{sky} in the relevant direction, using the relationship $e_p = (T_p - T_{sky})/(T_{eff} - T_{sky})$, ranging from 0 to 1. The data presented here are based on measurements made with the PAMIR system operating at 4.9, 10.4, 21, 35 and 94 GHz (Mätzler, 1987–1994) and with a set of newer portable radiometers at 11, 21, 35, 48 and 94 GHz (Weise, 1996; Wiesmann *et al.*, 1996, 1998; Mätzler *et al.*, 1997).

Additional experiments – not reported here – were made with the RASAM instrument, a combined radiometer and scatterometer, operating over the frequency range from 2 to 12 GHz. Data collected with this instrument are reported in a separate catalogue (Wegmüller and Mätzler, 1993) which was described by Wegmüller *et al.* (1994).

Applications

To extract emissivity information, either directly from the data catalog or through a fast, empirical model, and to test microwave emissivity models.

References

MÄTZLER, C.: 'Applications of the interaction of microwaves with the natural snow cover', *Remote Sens. Rev.*, 1987, **2**, pp. 259–392

MÄTZLER, C.: 'Seasonal evolution of microwave radiation from an oat field', *Remote Sens. Environ.*, 1990, **31**, pp. 161–73

MÄTZLER, C.: 'Microwave transmissivity of a forest canopy: Experiments made with a beech', *Remote Sens. Environ.*, 1994, **48**, pp. 172–80

MÄTZLER, C.: 'Passive microwave signature catalog 1989–1992' (Vol. 1), IAP-Report, Feb. (1993), Vol. 2. May (1994), Vol. 3, July (1994)

MÄTZLER, C.: 'Passive microwave signatures of landscapes in winter', *Meteorol. Atm. Phys.*, 1994, **54**, pp. 241–60

MÄTZLER, C., STROZZI, T., WEISE, T., FLORICIOIU, D., and ROTT, H.: 'Microwave snowpack studies made in the Austrian Alps during the SIR-C/X-SAR experiment', *Int. J. Remote Sens.*, 1997, **18**, pp. 2505–30

WEGMÜLLER, U., and MÄTZLER, C.: 'Active and passive microwave signature catalogue (2–12 GHz)' (IAP-Bericht, März, 1993)

WEGMÜLLER U., MÄTZLER, C., HÜPPI R., und SCHANDA, E.: 'Active and passive microwave signature catalogue on bare soil (2–12 GHz)', *IEEE Trans. Geosci. Remote Sens.*, 1994, **32**(3), pp. 698–702

WEISE, T.: 'Radiometric data (11–94 GHz) and autocorrelation functions of dry snow samples'. Res. report no. 96-2, Institute of Applied Physics, Univ. of Bern, 1996

WIESMANN, A., STROZZI, T., and WEISE, T.: 'Passive microwave signature catalogue of snowcovers at 11, 21, 35, 48 and 94 GHz'. Res. report no. 96-8, Institute of Applied Physics, Univ. of Bern, 1996

WIESMANN, A., MÄTZLER, C., and WEISE, T.: 'Radiometric and structural measurements of snow samples', *Radio Sci.*, 1998, **33**, pp. 273–89

Instrument characteristics

PAMIR main system parameters, for more details, see Mätzler (1987); the portable radiometers have similar characteristics with an additional frequency at 48 GHz.

Variables: frequency: 5 selectable, any incidence angle selectable: 50° to 55°, full-beam width 9° to 10°, polarisation: linear h,v and intermediate values	4.9 (GHz)	10.4 (GHz)	21 (GHz)	35 (GHz)	94 (GHz)
Measurement precision, ΔT (K), at integration time of 1 s	0.1	0.1	0.1	0.1	0.2
Expected absolute error of brightness temperatures T_v, T_h (K)	1	1	1	1	2
Expected error of mean emissivity $e_p = (T_p - T_{sky})/(T_{eff} - T_{sky})$, $p = h, v$	0.005	0.005	0.005	0.005	0.01
Expected error of polarisation difference, $e_v - e_h$	0.003	0.002	0.002	0.002	0.005

Standard ground-truth parameters

WE (mm)	Water equivalent depth of snowpack (equivalent height of water column)
H_{snow} (cm)	Snow height
dw (mm)	Equivalent depth of vegetation canopy water, also known as canopy water content
h (cm)	Height of vegetation canopy
mg	Gravimetric soil moisture (water mass/dry-soil mass)
ε'	Real part of relative dielectric constant of the top 8 mm soil at about 1 GHz
ρ (g/cm^3)	Dry soil density
T_{eff} (C)	Effective temperature of surface used to compute the emissivity
T_{air}, T_{veg}, T_{soil} (C)	Air, vegetation and soil temperature, respectively

Test sites, objects and data: Object classes and test sites

Object class	Test site	Object-class description
YOUNG BARLEY	Tannacker, agricultural field in Moosseedorf, near Bern, Switzerland: Long. 7.48°E, Lat. 47.0°N, Alt. 570 m a.s.l., Ground: sandy-loam soil, dry-soil density: 0.9 to 1.2 g/cm^3, pixel size: 1 m^2	7 situations of green winter barley crop in spring 1990, height from 15 to 100 cm, $T > 3°C$.
YOUNG OAT		17 situations of green oat crop in spring 1989, height from 16 to 84 cm, $T > 7°C$ (Mätzler, 1990).
UNFROZEN GRASS		Combines the classes of unfrozen grass: SHORT, MEDIUM and HIGH.
SHORT GRASS		9 situations with grass cover, height 5–10 cm, volumetric soil moisture 13% to 60%, grass temperature above −1°C.
MEDIUM GRASS		4 situations of grass cover with height between 15 and 32 cm, $T_{veg} > 5°C$.
HIGH GRASS		6 situations of grass cover with height between 32 and 55 cm, $T_{veg} > 10°C$.
GRASS AFTER SNOW		2 situations, just after snowmelt, when short grass was still pressed to the wet ground.
FROST GRASS		3 situations of short grass with hoar frost at temperatures between −5°C and −2°C, frozen soil at −2 to −0.8°C.
FROZEN GRASS		4 situations of short, frozen grass without hoar frost at temperatures from −5 to −1.7°C, soil at −2 to 0°C (one situation with unfrozen soil).
FROZEN SOIL		2 situations of frozen, bare soil with surface temperatures of −6°C and −1°C, respectively, mostly frozen soil moisture of about 40% by volume, see low ε' at 1 GHz.
BARE SOIL		9 situations of unfrozen, bare sandy-loam soil at temperatures between 0 and 25°C, volumetric soil moisture between 12% and 45%, roughness height about 1 cm.

POWDER SNOW		5 situations of 8 to 37 cm deep powder snow of Dec. 1990 and Nov. 1993 at T from −3 to −13°C, with WE (water equivalents) up to 5 cm. The grass-covered soil was unfrozen, except for 1 case. Most of the dry snow was 1 to 7 days old.
THIN CRUST		5 situations in Nov. and Dec. 1990 of thin refrozen crusts on wet snow.
BOTTOM CRUST		2 situations of aged, refrozen snow (depths of 6 and 15 cm) on short grass and frozen ground.
WET SNOW		4 situations of wet snow on short grass and unfrozen ground.
SLF_BARE	SLF: Acronym for alpine test site of Swiss Snow and Avalanche Research Institute, Weissfluhjoch, Davos: Lat. 48.83°N, Long. 9.81°E, Alt. 2560 m a.s.l., Ground: rocky serpentine soil, pixel size: 10 m^2	9 situations of the snow-free test site in wet and dry, unfrozen condition.
SLF_BOTTOMCRUST		6 situations of a special snow situation observed in late 1984 and early 1985. An abundant snow fall in autumn 1984 metamorphosed to a 40 cm thick, hard crust and formed the bottom of the new winter snowcover. The crust was frozen to the ground, and subsequently covered by new snow.
SLF_DEEP		50 situations of deep winter snow with WE between 25 and 63 cm.
SLF_ MEDIUM		12 situations of winter snow with WE between 10 and 25 cm.
SLF_ SHALLOW		11 situations of winter snow with (WE) between 4 and 10 cm. Winter snow is dry, and it has not undergone melt metamorphism.
SLF_THICKCRUST		15 situations of wet snow covered by a 4 to 30 cm thick layer of refrozen snow.
SLF_THINCRUST		20 situations of wet snow covered by a refrozen crust whose thickness is between 1 and 3 cm.
SLF_WET		53 situations of wet snow, i.e. there was at least a wet surface layer.

Mean values of ground-truth observed in each object class

Object class	Dw (mm)	h (cm)	T_{veg} (C)	mg (%)	ρ (g/cm^3)	T_{soil} (C)	ε'_{soil}	T_{snow} (C)	H_{snow} (cm)	WE (cm)	T_{air} (C)	T_{eff} (C)
YOUNG BARLEY	1.94	47.2	11.3	30.6	1.22	11.3	22.9		0	0	11.4	11.1
YOUNG OAT	1.18	39.8	17.1	23.5	1.02	17.4	13.9		0	0	16.6	17.5
UNFROZEN GRASS	1.23	20.6	10.2	30.8	1.2	10.9	22.8		0	0	10.8	10.4
SHORT GRASS	0.71	8.9	5.9	36.1	1.21	6.6	29.4		0	0	6.1	6.2
MEDIUM GRASS	1.2	23.2	9.1	34.3	1.2	9.9	24.6		0	0	10.1	9.3
HIGH GRASS	2.23	44.1	18.1	16.8	1.18	18.7	5.4		0	0	19.9	18.3
GRASS AFTER SNOW	0.4	5	2.8	37.1	1.24	2.7	30.5		0	0	4.6	3.3
FROST GRASS	0.97	8.3	−3.7	37.8	1.23	−1.3	5.2		0.1	0.01	−2.5	−2.6
FROZEN GRASS	0.95	8.5	−3.2	44.3	1.22	−1.2	9.1		0	0	−2.7	−2.3
FROZEN SOIL	0.03	10	−4	33.3	1.15	−3.6	4.4		0.1	0.01	−5.1	−3.6
BARE SOIL	0.01	4.8	11.3	31.4	1.06	11.2	12.7		0	0	12	11.1
POWDER SNOW	0.48	7	−1.2	37.4	1.2	0	14.1	−8.4	24.1	3.65	−4.8	−1
THIN CRUST	0.4	5.8	−1.1	35	1.2	−0.2	5.2	−4.3	11	1.83	−2.4	−0.9
BOTTOM CRUST	0.75	6.5	−1.5	39.3	1.16	−0.7		−5.2	11	2.9	−1.3	−0.9
WET SNOW	0.4	8	0	35	1.2	0		0.0	14.3	3.06	2.3	0
SLF_BARE	0	0				8.4			0	0		8.4
SLF_BOTTOMCRUST	0	0						0		28.17		−3.9
SLF_DEEP	0	0						−6.6		44.94		−3.2
SLF_MEDIUM	0	0						−6		18.75		−3.2
SLF_SHALLOW	0	0						−4.1		7.32		−2
SLF_THICKCRUST	0	0						0		>10		−1.2
SLF_THINCRUST	0	0						−0.3		>10		−0.2
SLF_WET	0	0						−0.3		>10		−0.2

Standard deviations of ground-truth observed in each object class

Object class	Dw (mm)	h (cm)	T_{veg} (C)	mg (%)	ρ (g/cm^3)	T_{soil} (C)	ε'_{soil}	T_{snow} (C)	H_{snow} (cm)	WE (cm)	T_{air} (C)	T_{eff} (C)
YOUNG BARLEY	1.5	33.7	3.7	4.7	0.04	3.5	2.2		0	0	4.8	3.8
YOUNG OAT	1.43	21.6	4.9	6.3	0.02	5.1	6.3		0	0	4.7	5.2
UNFROZEN GRASS	0.84	15.7	8.7	12	0.07	8.6	11.2		0	0	8.4	8.6
SHORT GRASS	0.26	2.2	8.1	9.1	0.09	7.9	3.6		0	0	6.2	7.9
MEDIUM GRASS	0.48	6.8	2.6	3.8	0.07	2.5			0	0	3.8	2.7
HIGH GRASS	0.98	7.7	6.4	9.3	0.04	6.1			0	0	6.9	6.3
GRASS AFTER SNOW	0.01	0	1.7	3.3	0.06	1	4.9		0	0	0.6	1.1
FROST GRASS	0.12	1.2	1.5	5.9	0.05	0.6			0	0	1.3	0.9
FROZEN GRASS	0.1	1	1.5	7.1	0.05	0.9	7.9		0	0	2.1	0.8
FROZEN SOIL	0	0	4.2	5.2	0.07	3.8	0.1		0.1	0.01	3.3	3.8
BARE SOIL	0.01	4.4	4.5	4.6	0.1	4.5	5.2		0	0	5.5	4.5
POWDER SNOW	0.18	4.5	0.5	5.4	0	0.3	9.8	4.7	10.3	1.94	2.7	0.2
THIN CRUST	0	1.1	0.4	0	0	0.3		2.2	6	1.22	1.3	0.4
BOTTOM CRUST	0.49	2.1	0.7		0.16	0.6		0.3	7.1	1.56	0.4	0.4
WET SNOW	0	4.8	0	0	0	0		0.0	8.2	1.59	2.2	0
SLF_BARE	0	0				4.2			0	0		4.2
SLF_BOTTOMCRUST	0	0								4.02		0.8
SLF_DEEP	0	0						5.6		11.95		1
SLF_MEDIUM	0	0						4.8		4.94		1
SLF_SHALLOW	0	0						2.6		2.44		1.1
SLF_THICKCRUST	0	0						0.2		>10		0
SLF_THINCRUST	0	0						0.9		>10		0
SLF_WET	0	0						0.9		>10		0

Mean values of emissivities and of emissivity combinations of object classes based on PAMIR data, incidence angle 50°

Object class	4.9h	10.4h	21h	35h	94h	4.9v	10.4v	21v	35v	94v	4.9v-h	10v-h	21v-h
WATER:0–8°C	0.249	0.267	0.308	0.358	0.493	0.501	0.528	0.59	0.657	0.806	0.251	0.262	0.282
YOUNG BARLEY	0.909	0.966	0.958	0.962	0.963	0.956	0.966	0.953	0.954	0.959	0.046	0	−0.005
YOUNG OAT	0.904	0.936	0.95	0.941	0.949	0.952	0.948	0.943	0.937	0.949	0.046	0.012	−0.007
UNFROZEN GRASS	0.933	0.952	0.946	0.943	0.941	0.952	0.958	0.944	0.942	0.948	0.019	0.006	−0.001
SHORT GRASS	0.912	0.947	0.944	0.946	0.946	0.939	0.956	0.943	0.943	0.951	0.027	0.009	−0.001
MEDIUM GRASS	0.948	0.95	0.94	0.935	0.933	0.959	0.956	0.941	0.937	0.944	0.01	0.006	0.001
HIGH GRASS	0.967	0.961	0.95	0.939	0.936	0.97	0.961	0.946	0.939	0.941	0.004	0	−0.004
GRASS AFTER SNOW	0.902	0.91	0.901	0.902	0.905	0.912	0.925	0.905	0.91	0.92	0.01	0.014	0.004
FROST GRASS	0.918	0.958	0.95	0.938	0.907	0.948	0.966	0.952	0.942	0.926	0.03	0.008	0.002
FROZEN GRASS	0.919	0.962	0.959	0.955	0.937	0.95	0.967	0.956	0.952	0.948	0.031	0.006	−0.003
FROZEN SOIL	0.891	0.953	0.951	0.951	0.938	0.94	0.958	0.958	0.952	0.95	0.049	0.005	0.007
BARE SOIL	0.725	0.82	0.888	0.887	0.92	0.874	0.888	0.91	0.903	0.934	0.149	0.068	0.022
POWDER SNOW	0.895	0.921	0.928	0.909	0.777	0.94	0.954	0.952	0.943	0.841	0.046	0.033	0.024
THIN CRUST	0.877	0.861	0.816	0.787	0.631	0.923	0.941	0.91	0.853	0.658	0.046	0.08	0.094
BOTTOM CRUST	0.882	0.919	0.85	0.708	0.443	0.945	0.954	0.898	0.753	0.466	0.063	0.034	0.048
WET SNOW	0.755	0.844	0.839	0.86	0.892	0.947	0.968	0.955	0.95	0.945	0.193	0.123	0.116
SLF_BARE	0.778	0.855	0.904	0.911	0.915	0.89	0.92	0.938	0.942	0.938	0.112	0.065	0.033
SLF_BOTTOMCRUST	0.822	0.767	0.705	0.615	0.648	0.939	0.922	0.829	0.699	0.733	0.117	0.155	0.123
SLF_DEEP	0.843	0.803	0.78	0.715	0.658	0.943	0.94	0.898	0.813	0.728	0.101	0.137	0.119
SLF_MEDIUM	0.85	0.831	0.771	0.682	0.612	0.937	0.938	0.893	0.785	0.669	0.086	0.107	0.122
SLF_SHALLOW	0.865	0.847	0.833	0.768	0.672	0.934	0.938	0.921	0.848	0.688	0.069	0.092	0.088
SLF_THICKCRUST	0.909	0.952	0.878	0.677	0.472	0.983	0.99	0.931	0.709	0.486	0.073	0.038	0.053
SLF_THINCRUST	0.901	0.938	0.942	0.9	0.707	0.967	0.984	0.976	0.938	0.744	0.066	0.046	0.034
SLF_WET	0.812	0.871	0.91	0.921	0.906	0.954	0.975	0.975	0.971	0.941	0.143	0.104	0.066

The class WATER_OC-8C represents calculated omissivities of a smooth liquid-water surface. The numbers are averaged values of emissivities at 0°C and 8°C

Object class	35v-h	94v-h	10h-5h	21h-10h	35h-21h	94h-35h	10v-5v	21v-10v	35v-21v	94v-35v	94v-21v	Comb
WATER:0–8°C	0.299	0.313	0.017	0.041	0.05	0.135	0.028	0.061	0.068	0.149	0.217	0.456
YOUNG BARLEY	−0.008	−0.004	0.057	−0.008	0.004	0.001	0.01	−0.012	0.001	0.005	0.006	0.022
YOUNG OAT	−0.005	0	0.027	0.014	−0.009	0.008	−0.005	−0.005	−0.006	0.012	0.006	0.036
UNFROZEN GRASS	−0.001	0.007	0.019	−0.006	−0.003	−0.002	0.006	−0.014	−0.002	0.006	0.004	0.053
SHORT GRASS	−0.002	0.005	0.035	−0.003	0.002	−0.001	0.017	−0.013	0	0.007	0.008	0.044
MEDIUM GRASS	0.001	0.011	0.002	−0.01	−0.005	−0.003	−0.003	−0.015	−0.004	0.007	0.003	0.065
HIGH GRASS	0	0.006	−0.005	−0.011	−0.011	−0.003	−0.009	−0.015	−0.008	0.003	−0.005	0.064
GRASS AFTER SNOW	0.008	0.015	0.009	−0.009	0	0.004	0.012	−0.02	0.005	0.01	0.015	0.071
FROST GRASS	0.005	0.02	0.04	−0.008	−0.012	−0.031	0.018	−0.014	−0.009	−0.016	−0.025	0.085
FROZEN GRASS	−0.003	0.011	0.043	−0.003	−0.004	−0.018	0.018	−0.011	−0.004	−0.004	−0.008	0.045
FROZEN SOIL	0.001	0.012	0.062	−0.002	0	−0.012	0.018	0	−0.006	−0.001	−0.008	0.031
BARE SOIL	0.016	0.014	0.095	0.068	−0.002	0.033	0.014	0.022	−0.007	0.031	0.024	0.063
POWDER SNOW	0.034	0.064	0.026	0.008	−0.012	−0.161	0.013	−0.001	−0.008	−0.123	−0.112	0.126
THIN CRUST	0.066	0.027	−0.016	−0.045	−0.03	−0.156	0.018	−0.031	−0.057	−0.196	−0.252	0.503
BOTTOM CRUST	0.046	0.023	0.037	−0.07	−0.142	−0.265	0.009	−0.056	−0.145	−0.287	−0.432	0.729
WET SNOW	0.09	0.054	0.09	−0.005	0.021	0.031	0.021	−0.013	−0.004	−0.005	−0.009	0.381
SLF_BARE	0.031	0.023	0.077	0.049	0.007	0.01	0.03	0.017	0.004	0.003	0	0.065
SLF_BOTTOMCRUST	0.083	0.085	−0.055	−0.062	−0.09	0.033	−0.017	−0.093	−0.13	0.034	−0.095	1.031
SLF_DEEP	0.098	0.071	−0.04	−0.023	−0.065	−0.04	−0.004	−0.041	−0.086	−0.063	−0.158	0.734
SLF_MEDIUM	0.103	0.057	−0.019	−0.06	−0.089	−0.073	0.001	−0.045	−0.108	−0.11	−0.22	0.79
SLF_SHALLOW	0.08	0.016	−0.018	−0.014	−0.065	−0.104	0.004	−0.017	−0.073	−0.17	−0.231	0.53
SLF_THICKCRUST	0.032	0.026	0.043	−0.074	−0.202	−0.206	0.007	−0.059	−0.222	−0.223	−0.445	0.965
SLF_THINCRUST	0.038	0.036	0.037	0.004	−0.042	−0.195	0.017	−0.008	−0.037	−0.197	−0.233	0.255
SLF_WET	0.05	0.035	0.059	0.039	0.012	−0.005	0.02	0	−0.004	−0.022	−0.03	0.23

COMB: The parameter, COMB, is defined as an optimal linear combination of emissivities for snow discrimination (Mätzler, 1994): COMB $= e_v(10.4) - e_h(10.4) + e_v(21) - e_h(21) + e_v(35) - e_h(35) + 3[e_v(10.4) - e_h(35)]$

Standard deviations of emissivities and of emissivity combinations of object classes based on PAMIR data, incidence angle 50°

Object class	4.9h	10.4h	21h	35h	94h	4.9v	10.4v	21v	35v	94v	4.9v-h	10v-h	21v-h
WATER:0–8°C	0.001	−0.007	−0.018	−0.028	−0.033	0.001	−0.01	−0.026	−0.035	−0.031	0	−0.004	−0.008
YOUNG BARLEY	0.089	0.036	0.032	0.024	0.018	0.026	0.014	0.013	0.013	0.012	0.066	0.022	0.02
YOUNG OAT	0.057	0.029	0.015	0.015	0.015	0.018	0.013	0.011	0.01	0.012	0.039	0.018	0.005
UNFROZEN GRASS	0.039	0.014	0.013	0.013	0.016	0.017	0.009	0.011	0.01	0.014	0.031	0.009	0.006
SHORT GRASS	0.04	0.013	0.011	0.01	0.009	0.01	0.008	0.01	0.008	0.01	0.038	0.011	0.007
MEDIUM GRASS	0.006	0.009	0.01	0.012	0.022	0.001	0.003	0.007	0.008	0.016	0.006	0.007	0.007
HIGH GRASS	0.012	0.013	0.018	0.017	0.022	0.01	0.012	0.015	0.015	0.022	0.011	0.005	0.003
GRASSAFTER SNOW	0.013	0.013	0.004	0.004	0.006	0.007	0.007	0.002	0.002	0.005	0.006	0.006	0.006
FROST GRASS	0.007	0.006	0.004	0.004	0.008	0.008	0.005	0.004	0.001	0.006	0.009	0.002	0
FROZEN GRASS	0.022	0.008	0.004	0.006	0.009	0.012	0.006	0.004	0.004	0.008	0.015	0.004	0.002
FROZEN SOIL	0.002	0.004	0.009	0.001	0.001	0.005	0.002	0.01	0.001	0.001	0.007	0.002	0
BARE SOIL	0.07	0.052	0.036	0.039	0.031	0.033	0.03	0.022	0.026	0.021	0.039	0.023	0.015
POWDER SNOW	0.038	0.03	0.018	0.026	0.067	0.005	0.009	0.009	0.011	0.047	0.034	0.022	0.017
THIN CRUST	0.053	0.056	0.074	0.083	0.133	0.02	0.013	0.021	0.05	0.132	0.036	0.046	0.056
BOTTOM CRUST	0.01	0.018	0.08	0.097	0.015	0.004	0.002	0.042	0.088	0.002	0.006	0.016	0.038
WET SNOW	0.016	0.028	0.028	0.017	0.016	0.008	0.008	0.012	0.009	0.005	0.01	0.026	0.018
SLF_BARE	0.025	0.023	0.016	0.021	0.038	0.018	0.02	0.017	0.019	0.039	0.011	0.01	0.013
SLF_BOTTOMCRUST	0.009	0.013	0.012	0.049	0.084	0.004	0.008	0.011	0.034	0.072	0.009	0.016	0.008
SLF_DEEP	0.026	0.041	0.048	0.064	0.104	0.011	0.012	0.035	0.064	0.111	0.021	0.033	0.022
SLF_MEDIUM	0.015	0.035	0.063	0.042	0.091	0.006	0.012	0.03	0.043	0.105	0.015	0.027	0.051
SLF_SHALLOW	0.02	0.022	0.031	0.053	0.141	0.007	0.005	0.014	0.048	0.148	0.016	0.018	0.021
SLF_THICKCRUST	0.032	0.022	0.055	0.102	0.047	0.006	0.005	0.037	0.104	0.069	0.031	0.022	0.029
SLF_THINCRUST	0.054	0.031	0.023	0.043	0.136	0.023	0.01	0.008	0.039	0.14	0.044	0.025	0.02
SLF_WET	0.072	0.07	0.053	0.042	0.042	0.021	0.018	0.016	0.015	0.036	0.059	0.06	0.041

The class WATER_OC-8C represents calculated omissivities of a smooth liquid-water surface. The numbers are difference values of emissivities at 0°C and 8°C

Object class	35v-h	94v-h	10h-5h	21h-10h	35h-21h	94h-35h	10v-5v	21v-10v	35v-21v	94v-35v	94v-21v	Comb
WATER:0–8°C	−0.008	0.002	−0.007	−0.012	−0.009	−0.006	−0.011	−0.016	−0.009	0.004	−0.005	0.057
YOUNG BARLEY	0.011	0.007	0.055	0.005	0.009	0.007	0.013	0.003	0.003	0.005	0.007	0.05
YOUNG OAT	0.006	0.004	0.028	0.02	0.006	0.006	0.007	0.008	0.006	0.006	0.003	0.016
UNFROZEN GRASS	0.006	0.006	0.029	0.011	0.008	0.009	0.013	0.007	0.006	0.009	0.01	0.023
SHORT GRASS	0.005	0.004	0.029	0.01	0.006	0.009	0.007	0.007	0.004	0.008	0.009	0.018
MEDIUM GRASS	0.007	0.008	0.006	0.004	0.003	0.013	0.003	0.005	0.002	0.008	0.009	0.028
HIGH GRASS	0.004	0.003	0.012	0.015	0.007	0.01	0.009	0.01	0.007	0.011	0.008	0.028
GRASSAFTER SNOW	0.006	0.001	0	0.009	0.001	0.002	0	0.009	0	0.007	0.007	0.009
FROST GRASS	0.005	0.001	0.004	0.003	0.007	0.004	0.005	0.003	0.003	0.007	0.01	0.012
FROZEN GRASS	0.004	0.004	0.015	0.007	0.004	0.007	0.006	0.003	0.002	0.007	0.009	0.014
FROZEN SOIL	0.001	0	0.001	0.006	0.009	0	0.007	0.008	0.01	0.002	0.008	0.005
BARE SOIL	0.014	0.011	0.031	0.024	0.016	0.019	0.011	0.012	0.012	0.015	0.018	0.044
POWDER SNOW	0.015	0.022	0.023	0.024	0.02	0.037	0.005	0.005	0.018	0.02	0.042	0.06
THIN CRUST	0.034	0.008	0.016	0.025	0.024	0.054	0.012	0.014	0.037	0.084	0.12	0.249
BOTTOM CRUST	0.009	0.013	0.008	0.062	0.017	0.112	0.003	0.04	0.046	0.09	0.044	0.321
WET SNOW	0.024	0.018	0.026	0.038	0.036	0.003	0.008	0.005	0.006	0.006	0.007	0.034
SLF_BARE	0.008	0.002	0.02	0.011	0.014	0.008	0.012	0.008	0.009	0.011	0.016	0.044
SLF_BOTTOMCRUST	0.021	0.021	0.01	0.015	0.051	0.044	0.005	0.005	0.039	0.046	0.078	0.134
SLF_DEEP	0.02	0.029	0.033	0.034	0.037	0.12	0.011	0.026	0.037	0.134	0.122	0.193
SLF_MEDIUM	0.043	0.019	0.024	0.032	0.04	0.076	0.01	0.02	0.03	0.105	0.117	0.123
SLF_SHALLOW	0.015	0.028	0.021	0.029	0.039	0.141	0.007	0.013	0.041	0.131	0.144	0.162
SLF_THICKCRUST	0.011	0.028	0.042	0.045	0.059	0.109	0.005	0.036	0.07	0.113	0.078	0.331
SLF_THINCRUST	0.015	0.026	0.061	0.04	0.034	0.111	0.018	0.01	0.033	0.111	0.137	0.134
SLF_WET	0.035	0.021	0.062	0.045	0.03	0.062	0.015	0.009	0.012	0.035	0.039	0.112

Example of data taken with the portable radiometers

Date: 17 January 1996, Test-site: Weissfluhjoch (positions 1–7),
Object: dry winter snow, consisting of 3 layers
Frequency: 11, 21, 35, 94 GHz
Weather: Sunny
Air temperature (10:30): –2.1°C
Surface roughness height < 1 cm

Comments: plots of emissivity versus incidence angle (left) and versus frequency (right), average values and st dev. from seven scans at H and V polarisation, frequency in gigahertz according to key.

Snow vertical profile information:

Height (m),	Grain Shape,	Date,	Dens. (kg/m^3),	ϵ',	W (%)
0.50–0.68	/ \	15.01.96	244	1.43	0
0.25–0.50	• •	15.01.96	335	1.59	0
0.00–0.25	^ ^	15.01.96	295	1.53	0

Hsnow (radiometer position): 63 cm, Hsnow (profile): 68 cm
Temp. (°C) (Height (cm)) (14:30): –13.0° (68), –7.6° (60), –7.1° (50), –6.6° (40), –5.0° (30), –3.4° (20), –2.2° (10), –1.6° (0)

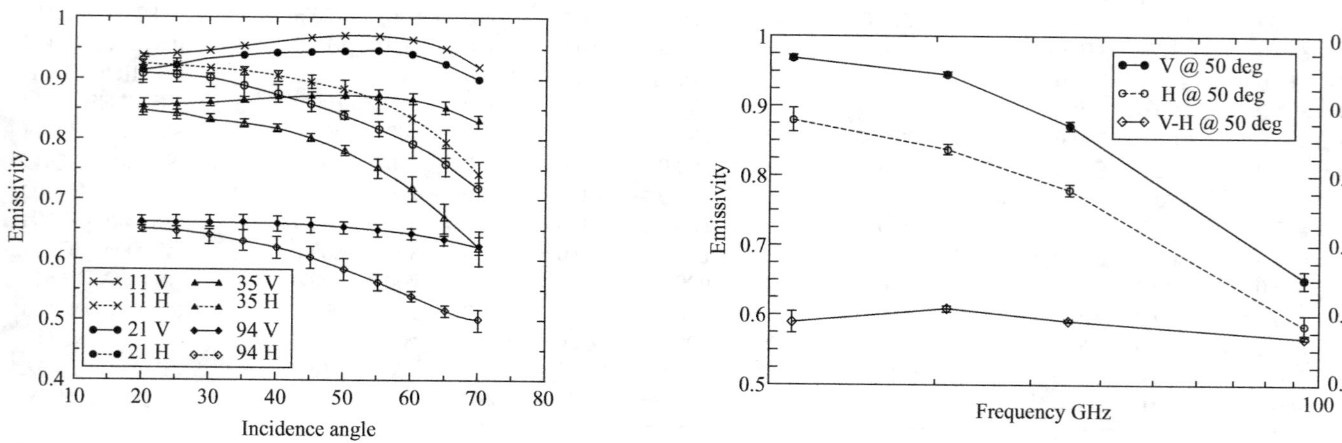

Figure 1A Example of data taken with the portable radiometers

Appendix B

Surface emissivity data from PORTOS-Avignon experiment

André Chanzy, Jean-Pierre Wigneron, Jean-Christophe Calvet, Laurent Laguerre, and K. Suresh Raju

The data displayed in the following table were collected at the INRA research centre near Avignon (43°55′N, 4°53′E, south of France) in 1993. Emissivity values were estimated from the measurements made by the PORTOS radiometer (Chanzy *et al.*, 1994). The atmospheric and sky contributions were removed from the brightness temperature measured by PORTOS by an algorithm described in Calvet *et al.* (1995). Atmospheric and sky contributions were estimated using the outputs of the Météo-France 15 levels weather forecast model PERIDOT and the cloud observations made at the experimental site along with the radiometric acquisitions. The surface effective temperature was estimated by the skin temperature measured by a thermal infrared radiometer. The expected error in microwave emissivities is ±0.015.

All the measurements were made on a silty clay loam soil (27 per cent of clay, 11 per cent of sand). For each surface type, a set of three moisture conditions (wet, medium wet and dry) were gathered in the table in order to give the range of emissivity variations due to the soil moisture. Measurements of the surface characteristics (soil moisture, surface roughness and vegetation) are given in the table for each set of microwave measurements (one line in the table) since measurements at all radiometric configurations were collected together.

The whole Avignon 93 database is available on a CD, which can be obtained upon request (contact: achanzy@avignon.inra.fr). The database includes a very large dataset in terms of incidence, moisture, surface roughness, biomass and detailed ground truth data (detailed vegetation description, soil characteristics, water balance for the wheat surface, climatic measurements).

References

CALVET J.-C., WIGNERON, J.-P., CHANZY, A., RAJU, SURESH, and LAGUERRE, L.: 'Microwave dielectric properties of a silt loam at high frequencies', *IEEE Trans. Geosci. Remote Sens.*, 1995a, **33**, pp. 634-642

CHANZY A., HABOUDANE, D., WIGNERON, J.-P., CALVET, J.-C., and GROSJEAN, O.: 'Radiométrie micro-onde sur divers types de couverts végétaux: Influence de l'humidité du sol'. Colloque International ISPRS *Mesures physiques et signatures en télédétection*, Val d'Isère (France), 1994/01/17-21, pp. 505–12, 1994

Table 1B Emissivity measurements collected over agricultural surfaces with bare soil and vegetation covers at $\theta = 40°$

Veg. type	Veg. water content (kg/m^2)	Surface observation	HRMS (mm)	Correl. length (mm)	Soil wat. content (0–5 mm) (m^3/m^3)	Emiss. 1.4 GHz H-pol	Emiss. 1.4 GHz V-pol	Emiss. 5.05 GHz H-pol	Emiss. 5.05 GHz V-pol	Emiss. 10.65 GHz H-pol	Emiss. 10.65 GHz V-pol	Emiss. 23.8 GHz H-pol	Emiss. 23.8 GHz V-pol	Emiss. 36.5 GHz H-pol	Emiss. 36.5 GHz V-pol	Emiss. 90 GHz H-pol	Emiss. 90 GHz V-pol
Bare	0	Very rough	60	72	0.247	0.83	0.91	0.80	0.84	0.82	0.84	0.84	0.87	0.89	0.91	0.89	0.90
Bare	0	Very rough	60	72	0.117	0.86	–	0.86	0.88	0.87	0.89	0.89	–	0.93	0.95	0.93	0.93
Bare	0	Very rough	60	72	0.011	0.91	0.95	0.90	0.92	0.92	0.93	0.94	0.95	0.96	0.97	0.95	0.96
Bare	0	Rough	19	66	0.311	0.75	0.88	0.79	0.86	0.82	0.85	0.83	0.85	0.83	0.84	0.89	0.90
Bare	0	Rough	19	66	0.132	0.82	0.93	0.83	0.90	0.86	0.89	0.87	0.89	0.86	0.88	0.91	0.93
Bare	0	Rough	19	66	0.014	0.90	0.95	0.93	0.95	0.90	0.94	0.93	0.94	0.94	0.96	0.95	0.96
Bare	0	Smooth	8	31	0.364	0.56	0.70	0.74	0.78	0.76	0.82	0.77	0.83	0.76	0.84	0.81	0.85
Bare	0	Smooth	8	31	0.189	0.75	–	–	0.85	–	0.85	–	0.86	–	0.89	0.85	0.91
Bare	0	Smooth	8	31	0.065	0.80	0.91	0.86	0.94	0.89	0.94	0.89	0.94	0.91	0.96	0.89	0.94
Bare	0	Very smooth	5	206	0.365	0.51	0.74	0.51	0.71	0.55	0.72	0.54	0.73	0.57	–	0.65	0.83
Bare	0	Very smooth	5	206	0.127	0.70	0.82	0.72	0.87	0.76	0.89	0.75	0.89	0.78	0.94	0.81	0.92
Bare	0	Very smooth	5	206	0.021	0.85	0.96	0.85	0.96	0.89	0.96	0.92	0.97	0.93	0.98	0.92	0.95
Wheat	1.9	green veg.	6	93	0.314	0.74	0.92	0.92	0.95	0.95	0.96	0.95	0.96	0.97	0.98	0.95	0.96
Wheat	2.6	green veg.	6	93	0.240	0.80	0.95	0.94	0.94	0.95	0.96	0.95	0.96	0.96	0.96	0.92	0.93
Wheat	2.6	green veg.	6	93	0.152	0.83	0.99	0.95	0.95	0.97	0.97	0.97	0.97	0.98	0.97	0.96	0.97
Wheat	1.2	Senescent	6	93	0.031	0.74	–	0.88	0.92	0.95	0.98	0.96	0.97	0.97	0.97	0.98	0.99
Wheat	1.1	Senescent	6	93	0.315	0.63	0.87	0.86	0.92	0.96	0.98	0.96	0.97	0.97	0.97	0.99	0.99
Wheat	1.1	Senescent	6	93	0.200	0.65	0.88	0.85	0.92	0.95	0.97	0.95	0.96	0.95	0.96	0.97	0.98
Sorgho	0.23	Sparse veg.	5	63	0.203	0.68	0.85	0.71	0.79	0.73	0.81	0.73	0.83	0.72	0.87	0.78	0.88
Sorgho	0.29	Sparse veg.	5	63	0.059	0.78	0.93	0.84	0.91	0.87	0.94	0.91	0.96	0.87	0.95	0.89	0.97
Sorgho	0.61	Sparse veg.	6	76	0.25	0.70	–	0.78	0.82	0.82	0.87	0.84	0.88	0.83	0.91	0.86	0.89
Sorgho	0.77	Sparse veg.	6	76	0.07	0.82	0.93	0.90	0.92	0.91	0.94	0.92	0.93	0.90	0.94	0.90	0.93

Appendix C

Experimental data used to construct the interpolation function for the dielectric constant of water

William Ellison

We group in this appendix all the available permittivity data for seawater. Our object in collecting it here is twofold. First, it is the data we have used in establishing the interpolation function described in 5.2.5. Second, when new permittivity data for seawater becomes available it will be a useful source of information for the construction of more precise interpolation functions.

Seawater

Table 1A is from Ellison *et al.* (1997, 2003). Tables 2A to 6A are from Ellison *et al.* (1996b). Tables 5A and 6A contain the Debye parameters that were calculated from measurements over the frequency range 3 to 20 GHz in 85 MHz steps. At each temperature and salinity the corresponding Debye function represented the data to within 1 per cent. Unfortunately, the original permittivity data was not conserved. For our function-fitting programme we recalculated permittivity values at each temperature and salinity for 1 GHz steps between 3 and 20 GHz. Tables 7A to 13A are from Stogryn *et al.* (1996).

Purewater

The data used in calculating our interpolation function is from the review article Ellison *et al.* (1996a) and is not reproduced here. Table 14A contains some permittivity data for purewater published after 1996 and Table 15A Debye parameters obtained by Rønne *et al.* (1997) from measurements over the frequency range 100–2000 GHz.

Table 1A Experimental data for the permittivity of seawater, salinity 35‰

GHz	−2°C		5°C		10°C		15°C		20°C		25°C		30°C	
	ε'	ε''	ε'	ε''	ε'	ε''	ε'	ε''	ε'	ε''	ε'	ε''	ε'	ε''
30.0	13.92	21.60	16.10	25.00	19.31	28.43	20.88	29.25	24.32	30.60	26.83	31.97	30.07	33.55
33.0	12.24	20.60	14.95	23.84	16.84	26.86	19.14	28.51	21.52	29.82	24.56	31.31	27.70	32.02
35.0	11.92	20.41	14.47	22.73	16.36	26.31	18.80	27.24	20.60	29.16	22.73	30.26	24.40	31.84
37.0	10.66	19.46	14.32	22.00	15.73	24.48	17.64	26.04	19.32	28.44	19.40	28.21	21.71	30.24
40.0	10.20	18.26	13.37	20.81	13.94	22.80	16.65	24.10	17.45	27.30	18.20	27.50	20.11	29.27
45.0	9.79	16.00	12.21	18.85	12.80	21.11	14.20	22.95	14.97	25.18	16.26	25.20	18.40	28.57
48.0	9.56	15.32	11.21	18.03	12.20	19.81	12.90	21.80	13.57	23.78	15.37	24.62	17.45	27.22
50.0	9.27	14.50	10.72	17.50	11.60	18.62	11.36	20.76	12.90	23.32	14.80	23.77	16.48	26.94
52.0	8.91	13.72	10.60	16.80	10.81	17.55	10.77	19.87	12.54	22.26	14.36	22.40	15.80	26.72
55.0	8.80	12.70	10.43	16.30	10.51	17.00	10.60	18.93	12.23	21.62	13.95	21.92	15.58	25.32
57.3	8.73	12.63	10.00	15.93	10.33	16.50	10.33	18.60	12.06	20.60	13.78	21.72	14.74	24.60
60.0	8.60	12.35	9.88	15.52	10.10	16.15	10.00	18.54	11.84	20.08	13.57	21.47	14.21	22.88
64.0	8.40	11.90	9.48	14.40	9.70	15.96	9.83	17.65	11.60	19.55	12.82	20.63	13.35	21.97
65.0	8.28	11.70	9.35	14.31	9.53	15.75	9.70	17.40	11.45	19.23	12.60	20.37	12.90	21.60
66.0	8.11	11.55	9.10	13.80	9.24	15.24	9.45	17.27	11.36	18.74	12.80	19.42	12.70	20.85
70.0	8.05	11.22	8.53	13.50	9.00	14.73	9.50	16.05	10.78	17.53	12.31	18.76	12.10	20.64
73.0	7.88	11.15	8.3	12.90	8.76	14.44	9.75	15.89	10.16	17.10	11.00	18.50	11.33	19.96
75.0	7.75	10.70	8.10	12.66	8.54	14.10	8.80	15.50	10.47	16.50	10.68	17.56	11.13	18.80
78.0	7.60	10.48	7.90	12.32	8.38	13.60	9.20	14.80	9.95	16.20	10.21	17.00	10.76	18.40
80.0	7.45	10.05	7.84	11.97	8.20	13.20	9.00	14.55	9.76	15.56	10.11	16.78	10.30	18.67
82.0	7.36	9.83	7.75	11.83	8.10	12.86	8.55	13.60	9.40	14.85	10.00	16.13	10.28	17.78
85.5	7.30	9.60	7.64	11.12	7.90	12.54	8.27	13.45	9.30	14.77	9.66	15.75	10.22	17.45
87.0	7.16	9.55	7.51	10.82	7.73	12.20	8.30	13.44	8.90	14.46	9.50	15.45	9.700	17.10
89.0	7.43	9.38	7.24	10.62	7.64	11.84	8.12	13.36	8.77	14.26	9.34	15.27	9.510	16.56
92.0	7.28	9.00	7.17	10.54	7.48	11.72	7.90	13.00	8.52	14.05	9.05	15.10	9.370	16.07
94.0	7.18	8.84	7.13	10.35	7.35	11.22	7.60	12.94	8.12	13.23	8.90	14.75	9.120	15.77
95.0	6.82	8.60	6.83	10.15	7.40	11.11	7.50	12.44	8.00	13.68	8.80	14.70	9.00	15.70
97.0	7.15	8.55	7.07	9.90	7.52	10.92	7.65	12.31	7.95	13.40	8.40	14.44	8.91	14.93
100.0	7.05	8.26	7.05	9.60	7.25	10.54	7.30	11.88	7.86	12.85	8.34	13.80	8.62	15.12
103.0	6.93	8.11	7.06	9.48	7.31	10.36	7.41	11.56	7.73	13.00	8.25	13.48	8.48	14.61
105.0	6.73	8.01	7.00	9.38	7.20	10.16	7.56	11.18	7.67	12.60	8.12	13.70	8.74	14.88

Dr. Rønne kindly supplied the original experimental permittivity data at $T = -2$, 5.6, 15.3, 19.3, 30, 41.8, 50.9, 57.9, 70.2, and 93.5°C. These are given in Tables 16A, 17A and 18A.

Table 2A Permittivity data, Atlantic–Gironde estuary

Salinity = 23.2‰							Salinity = 28‰						
T (°C)	6.8 GHz		10.65 GHz		18.7 GHz		T(°C)	6.8 GHz		10.65 GHz		18.7 GHz	
	ε'	ε''	ε'	ε''	ε'	ε''		ε'	ε''	ε'	ε''	ε'	ε''
−1.990	51.19	40.33	34.46	39.43	18.03	30.17	−2.01	50.93	40.80	35.10	39.18	19.35	29.53
−0.960	52.53	40.14	35.85	39.77	19.00	30.78	−1.01	52.05	40.53	36.30	39.42	20.07	30.14
−0.010	53.54	39.72	37.00	39.96	19.62	31.51	−0.02	53.12	40.20	37.49	39.61	20.80	30.78
1.010	54.76	39.50	38.36	40.19	20.64	32.04	0.98	54.10	39.86	38.64	39.74	21.55	31.36
2.020	55.80	39.09	39.61	40.30	21.48	32.69	1.99	55.05	39.49	39.81	39.83	22.33	31.93
3.030	56.77	38.63	40.84	40.36	22.25	33.27	2.97	55.87	39.13	40.94	39.84	23.08	32.45
4.010	57.66	38.19	42.01	40.36	23.14	33.82	3.97	56.68	38.72	42.04	39.82	23.86	32.97
5.000	58.49	37.75	43.16	40.32	24.04	34.32	4.99	57.46	38.30	43.13	39.76	24.70	33.45
6.000	59.30	37.29	44.32	40.22	24.96	34.82	5.97	58.15	37.87	44.14	39.65	25.50	33.91
6.990	60.00	36.85	45.38	40.09	25.88	35.19	6.97	58.81	37.42	45.15	39.50	26.30	34.34
7.980	60.68	36.38	46.45	39.90	26.85	35.55	7.98	59.45	36.97	46.14	39.31	27.16	34.73
9.000	61.32	35.92	47.48	39.68	27.77	35.89	8.98	59.99	36.52	47.07	39.08	27.97	35.09
10.01	61.87	35.47	48.44	39.41	28.76	36.22	9.98	60.48	36.09	47.94	38.82	28.78	35.42
11.01	62.38	35.01	49.36	39.11	29.70	36.42	10.97	60.93	35.67	48.77	38.54	29.62	35.70
11.99	62.80	34.58	50.19	38.77	30.69	36.64	11.97	61.34	35.25	49.58	38.21	30.44	35.97
12.98	63.16	34.17	50.95	38.39	31.55	36.73	12.98	61.71	34.86	50.33	37.86	31.29	36.16
13.99	63.46	33.74	51.66	37.96	32.49	36.87	13.97	62.02	34.48	51.02	37.48	32.10	36.36
15.00	63.70	33.36	52.29	37.51	33.38	36.93	14.99	62.36	34.14	51.72	37.09	32.97	36.53
15.98	63.85	33.03	52.80	37.03	34.23	36.90	15.98	62.63	33.80	52.35	36.65	33.78	36.64
17.00	64.03	32.66	53.36	36.51	35.15	36.91	16.98	62.83	33.46	52.90	36.22	34.54	36.73
17.98	64.16	32.34	53.84	36.04	35.95	36.86	17.97	63.01	33.14	53.42	35.79	35.30	36.80
19.01	64.30	32.00	54.33	35.62	36.84	36.91	18.98	63.17	32.82	53.93	35.37	36.05	36.84
19.99	64.36	31.72	54.69	35.22	37.68	36.79	19.98	63.26	32.49	54.37	34.93	36.82	36.79
21.02	64.46	31.37	55.14	34.81	38.51	36.71	20.97	63.37	32.21	54.79	34.56	37.54	36.73
22.00	64.52	31.05	55.55	34.39	39.25	36.61	21.99	63.46	31.94	55.21	34.17	38.32	36.64
23.01	64.56	30.72	55.94	33.94	40.05	36.45	22.96	63.51	31.70	55.55	33.80	39.00	36.54
24.00	64.58	30.41	56.29	33.52	40.72	36.32	23.97	63.55	31.45	55.89	33.43	39.70	36.38
25.01	64.58	30.11	56.62	33.08	41.46	36.15	25.05	63.60	31.21	56.25	33.03	40.43	36.24
26.00	64.54	29.81	56.91	32.64	42.04	35.97	25.94	63.63	31.01	56.52	32.72	41.04	36.10
27.00	64.49	29.54	57.18	32.22	42.69	35.76	26.96	63.68	30.83	56.83	32.40	41.73	35.91
28.00	64.42	29.27	57.40	31.80	43.26	35.56	27.98	63.67	30.63	57.10	32.05	42.38	35.72
28.97	64.35	29.02	57.60	31.41	43.82	35.33	29.01	63.61	30.42	57.30	31.70	42.93	35.54
30.01	64.23	28.75	57.78	31.00	44.34	35.05	30.02	63.59	30.24	57.52	31.39	43.49	35.36

Table 3A Permittivity data, East Mediterranean

Salinity = 38.024‰							Salinity = 38.893‰						
T (°C)	6.8 GHz		10.65 GHz		18.7 GHz		T (°C)	6.8 GHz		10.65 GHz		18.7 GHz	
	ε'	ε''	ε'	ε''	ε'	ε''		ε'	ε''	ε'	ε''	ε'	ε''
−2.00	49.68	41.47	34.02	39.3t	19.23	29.51	−2.01	50.79	41.02	36.77	39.26	20.05	29.71
−1.02	50.75	41.30	35.18	39.65	19.96	30.15	−1.03	51.78	40.78	37.89	39.50	20.74	30.33
0.00	51.78	41.07	36.33	39.89	20.74	30.74	−0.04	52.70	40.49	38.98	39.68	21.41	30.94
1.00	52.72	40.83	37.44	40.07	21.50	31.33	0.99	53.56	40.15	40.06	39.80	22.11	31.51
1.98	53.61	40.56	38.54	40.19	22.25	31.84	1.97	54.37	39.85	41.10	39.89	22.83	32.03
2.99	54.43	40.27	39.58	40.28	23.04	32.38	2.97	55.14	39.52	42.12	39.95	23.56	32.54
4.01	55.22	39.96	40.64	40.32	23.82	32.85	3.96	55.84	39.19	43.10	39.97	24.31	33.03
5.00	55.93	39.65	41.62	40.31	24.61	33.33	4.98	56.50	38.84	44.07	39.95	25.07	33.48
6.02	56.61	39.33	42.60	40.26	25.44	33.80	5.990	57.12	38.49	45.01	39.88	25.86	33.93
7.00	57.22	39.02	43.54	40.18	26.27	34.15	6.990	57.67	38.15	45.89	39.81	26.66	34.34
8.01	57.78	38.68	44.43	40.06	27.05	34.58	8.000	58.20	37.82	46.75	39.69	27.43	34.67
9.00	58.29	38.38	45.28	39.91	27.86	34.88	9.010	58.65	37.45	47.55	39.54	28.22	35.03
10.00	58.77	38.07	46.10	39.74	28.69	35.21	9.990	59.05	37.12	48.31	39.37	28.96	35.32
11.01	59.21	37.76	46.91	39.52	29.55	35.48	10.96	59.43	36.82	49.06	39.20	29.72	35.60
11.99	59.58	37.49	47.63	39.29	30.35	35.71	11.99	59.74	36.50	49.78	38.97	30.50	35.83
13.02	59.96	37.19	48.36	39.02	31.17	35.91	12.98	60.04	36.20	50.44	38.75	31.24	36.05
14.00	60.28	36.95	49.02	38.72	31.95	36.10	13.96	60.30	35.92	51.07	38.52	31.99	36.24
15.01	60.58	36.72	49.69	38.39	32.81	36.23	14.99	60.53	35.63	51.68	38.27	32.75	36.37
16.02	60.84	36.49	50.29	38.04	33.63	36.39	15.99	60.75	35.38	52.26	38.02	33.51	36.50
17.00	61.07	36.29	50.84	37.72	34.41	36.48	16.98	60.94	35.15	52.81	37.75	34.24	36.60
18.00	61.27	36.08	51.37	37.39	35.17	36.54	17.98	61.12	34.95	53.33	37.47	35.00	36.62
19.01	61.44	35.87	51.87	37.08	35.88	36.59	18.97	61.29	34.77	53.80	37.20	35.75	36.68
20.01	61.60	35.67	52.33	36.78	36.64	36.59	19.98	61.42	34.60	54.21	36.91	36.47	36.70
21.00	61.73	35.49	52.77	36.48	37.36	36.60	20.98	61.52	34.48	54.50	36.64	37.16	36.72
22.02	61.82	35.31	53.18	36.18	38.04	36.54	21.98	61.58	34.34	54.71	36.27	37.84	36.71
23.00	61.91	35.15	53.54	35.88	38.75	36.50	22.96	61.64	34.21	55.01	35.92	38.48	36.65
23.99	61.96	34.99	53.88	35.59	39.36	36.43	23.98	61.68	34.09	55.33	35.58	39.12	36.59
24.98	61.98	34.85	54.19	35.30	39.98	36.34	24.98	61.71	33.98	55.62	35.22	39.79	36.46
26.00	61.99	34.70	54.45	34.99	40.55	36.24	25.97	61.71	33.88	55.88	34.90	40.34	36.35
27.01	61.96	34.57	54.70	34.69	41.12	36.12	26.98	61.71	33.79	56.10	34.57	40.92	36.22
28.01	61.89	34.41	54.90	34.39	41.64	36.01	27.98	61.68	33.73	56.29	34.24	41.50	36.06
28.98	61.77	34.26	55.03	34.09	42.09	35.88	28.98	61.65	33.67	56.45	33.93	42.02	35.91
30.01	61.84	34.24	55.29	33.91	42.68	35.68	30.00	61.58	33.62	56.56	33.60	42.53	35.73

Table 4A Permittivity data, North Polar sample

Salinity 30.255‰

T (°C)	6.8 GHz		10.65 GHz		18.7 GHz	
	ε'	ε''	ε'	ε''	ε'	ε''
−2.04	51.06	40.99	34.84	39.61	18.83	29.79
−1.02	52.20	40.72	36.07	39.89	19.53	30.47
−0.02	53.24	40.42	37.27	40.08	20.32	31.10
0.98	54.20	40.10	38.40	40.25	21.07	31.66
1.98	55.10	39.75	39.54	40.37	21.82	32.24
2.97	55.95	39.35	40.63	40.41	22.57	32.79
3.96	56.72	38.96	41.69	40.40	23.38	33.27
4.98	57.45	38.53	42.74	40.37	24.17	33.78
6.00	58.14	38.12	43.77	40.28	24.97	34.27
6.98	58.75	37.70	44.75	40.13	25.78	34.68
8.01	59.32	37.25	45.70	39.95	26.57	35.05
8.99	59.84	36.84	46.57	39.72	27.37	35.42
9.98	60.31	36.43	47.43	39.50	28.17	35.74
10.98	60.73	36.02	48.25	39.26	28.97	36.05
11.99	61.11	35.60	49.03	38.98	29.80	36.31
12.98	61.46	35.23	49.76	38.69	30.62	36.51
13.99	61.75	34.84	50.47	38.37	31.46	36.75
14.99	62.00	34.49	51.10	38.04	32.19	36.91
15.98	62.25	34.16	51.71	37.71	33.03	36.97
16.99	62.46	33.85	52.28	37.37	33.79	37.05
17.99	62.64	33.59	52.81	37.01	34.57	37.12
19.00	62.81	33.32	53.34	36.62	35.34	37.08
19.98	62.92	33.06	53.76	36.18	36.07	37.09
20.99	63.03	32.82	54.18	35.76	36.81	37.06
22.00	63.12	32.60	54.58	35.34	37.49	37.03
22.99	63.18	32.38	54.94	34.94	38.17	36.97
23.99	63.24	32.20	55.28	34.56	38.86	36.86
24.98	63.28	32.01	55.59	34.18	39.51	36.80
25.95	63.31	31.84	55.90	33.83	40.12	36.66
26.98	63.32	31.68	56.16	33.46	40.77	36.51
27.99	63.31	31.54	56.40	33.10	41.36	36.37
28.97	63.29	31.41	56.61	32.76	41.94	36.21
29.98	63.26	31.31	56.80	32.42	42.50	35.96

Table 5A Debye parameters for the Atlantic–Gironde estuary and N. Polar samples

Salinity 23.2‰				Salinity 28‰				Salinity 30.255‰			
T (°C)	ε_s	ε_∞	τ (ps)	T (°C)	ε_s	ε_∞	τ (ps)	T (°C)	ε_s	ε_∞	τ (ps)
−1.990	79.052	5.146	18.330	−2.01	79.06	6.657	18.5988	−2.04	79.191	6.606	18.767
−0.960	79.088	5.307	17.690	−1.01	78.646	6.593	17.8734	−1.02	78.840	6.577	18.041
−0.010	78.426	4.985	16.910	−0.02	78.261	6.524	17.1770	−0.02	78.491	6.555	17.367
1.010	78.595	5.265	16.400	0.98	77.861	6.446	16.5324	0.98	78.176	6.559	16.7608
2.020	78.260	5.174	15.760	1.99	77.480	6.381	15.9094	1.98	77.806	6.536	16.1607
3.030	77.848	5.015	15.130	2.97	77.162	6.339	15.3817	2.97	77.426	6.487	15.582
4.010	77.487	4.892	14.560	3.97	76.843	6.293	14.8429	3.96	77.028	6.428	15.0355
5.000	77.263	4.894	14.070	4.99	76.467	6.227	14.3101	4.98	76.636	6.375	14.5074
6.000	77.029	4.875	13.570	5.97	76.075	6.142	13.8178	6.00	76.256	6.326	14.0028
6.990	76.796	4.890	13.140	6.97	75.686	6.017	13.3226	6.98	75.885	6.275	13.5355
7.980	76.545	4.891	12.700	7.98	75.334	5.942	12.861	8.01	75.484	6.234	13.0864
9.000	76.313	4.913	12.280	8.98	74.943	5.840	12.4237	8.99	75.112	6.164	12.667
10.010	76.080	4.962	11.900	9.98	74.566	5.730	12.0102	9.98	74.744	6.110	12.2633
11.010	75.849	5.021	11.540	10.97	74.232	5.645	11.6245	10.98	74.380	6.045	11.8724
11.990	75.621	5.111	11.210	11.97	73.887	5.593	11.2610	11.99	74.023	5.999	11.5051
12.980	75.364	5.204	10.900	12.98	73.561	5.558	10.9221	12.98	73.686	5.998	11.1687
13.990	75.090	5.317	10.600	13.97	73.234	5.499	10.5954	13.99	73.296	5.89	10.8141
15.000	74.856	5.525	10.350	14.99	73.066	5.670	10.3238	14.99	72.973	5.917	10.5217
15.980	74.634	5.820	10.130	15.98	72.824	5.739	10.0528	15.98	72.685	5.933	10.2336
17.000	74.399	6.083	9.899	16.98	72.493	5.697	9.77636	16.99	72.40	5.995	9.96605
17.980	74.160	6.333	9.690	17.97	72.226	5.767	9.53111	17.99	72.152	6.108	9.7233
19.010	73.974	6.622	9.478	18.98	71.951	5.827	9.28657	19.00	71.893	6.231	9.48288
19.990	73.820	7.033	9.319	19.98	71.578	5.587	8.99658	19.98	71.521	6.157	9.23691
21.020	73.535	7.202	9.100	20.97	71.353	5.968	8.8386	20.99	71.188	6.166	9.00095
22.000	73.302	7.467	8.911	21.99	71.159	6.230	8.65058	22.00	70.862	6.159	8.77210
23.010	73.033	7.695	8.714	22.96	70.944	6.451	8.47739	22.99	70.581	6.227	8.56697
24.000	72.750	7.898	8.528	23.97	70.696	6.639	8.29527	23.99	70.312	6.322	8.37687
25.010	72.459	8.121	8.350	25.05	70.500	6.962	8.12591	24.98	70.051	6.442	8.19461
26.000	72.144	8.303	8.175	25.94	70.339	7.221	7.98959	25.95	69.804	6.548	7.85295
27.000	71.839	8.511	8.008	26.96	70.239	7.683	7.86933	26.98	69.555	6.699	7.69440
28.000	71.508	8.685	7.846	27.98	70.045	7.983	7.72571	27.99	69.300	6.843	8.01918
28.970	71.208	8.890	7.700	29.01	69.72	8.057	7.55567	28.97	69.062	7.016	7.55004
30.010	70.853	9.080	7.545	30.02	69.513	8.336	7.42354	29.98	68.810	7.183	7.40674

Table 6A Debye parameters for 'standard' seawater and for the East Mediterranean sample

Salinity 35‰				Salinity 38.024‰				Salinity 38.893‰			
T (°C)	ε_s	ε_∞	τ (ps)	T (°C)	ε_s	ε_∞	τ (ps)	T (°C)	ε_s	ε_∞	τ (ps)
−2.04	79.548	8.264	18.58	−2.00	77.106	7.116	18.66	−2.01	77.591	7.680	18.32
−1.00	79.762	8.639	17.96	−1.02	76.865	7.146	17.98	−1.03	77.218	7.626	17.62
0.03	78.914	8.271	17.14	0.00	76.617	7.181	17.33	−0.04	76.762	7.518	16.95
0.97	78.038	7.837	16.40	1.00	76.354	7.204	16.72	0.99	76.260	7.379	16.28
2.00	77.710	7.814	15.81	1.98	76.084	7.226	16.15	1.97	75.904	7.331	15.71
2.98	77.432	7.830	15.27	2.99	75.807	7.245	15.60	2.97	75.522	7.259	15.15
4.01	77.132	7.827	14.74	4.01	75.515	7.264	15.07	3.96	75.164	7.205	14.64
5.03	76.806	7.788	14.22	5.00	75.235	7.283	14.58	4.98	74.810	7.158	14.14
6.00	76.404	7.674	13.73	6.02	74.968	7.328	14.12	5.99	74.463	7.112	13.66
7.00	76.099	7.644	13.28	7.00	74.707	7.370	13.68	6.99	74.121	7.068	13.22
7.99	75.868	7.671	12.88	8.01	74.406	7.390	13.25	8.00	73.799	7.046	12.76
9.02	75.555	7.633	12.46	9.00	74.134	7.424	12.86	9.01	73.403	6.950	12.37
10.02	75.207	7.550	12.06	10.00	73.874	7.488	12.49	9.99	73.057	6.890	11.99
11.03	74.878	7.492	11.68	11.01	73.627	7.572	12.13	10.96	72.733	6.822	11.63
12.02	74.561	7.433	11.34	11.99	73.38	7.651	11.80	11.99	72.305	6.695	11.25
13.02	74.229	7.368	11.00	13.02	73.112	7.74	11.47	12.98	71.967	6.661	10.92
14.02	73.993	7.385	10.69	14.00	72.865	7.857	11.19	13.96	71.632	6.597	10.60
15.01	73.790	7.464	10.42	15.01	72.684	8.006	10.91	14.99	71.295	6.556	10.29
16.01	73.566	7.522	10.16	16.02	72.479	8.166	10.65	15.99	70.980	6.523	10.01
17.00	73.365	7.606	9.908	17.00	72.243	8.294	10.400	16.98	70.725	6.582	9.751
18.03	73.126	7.656	9.655	18.00	72.020	8.466	10.160	17.98	70.479	6.652	9.502
19.01	72.921	7.761	9.431	19.01	71.797	8.656	9.931	18.97	70.252	6.738	9.272
20.02	72.731	7.912	9.216	20.01	71.56	8.839	9.714	19.98	69.959	6.731	9.028
21.02	72.529	8.084	9.026	21.00	71.334	9.036	9.506	20.98	69.641	6.616	8.793
22.00	72.366	8.332	8.857	22.02	71.090	9.252	9.307	21.98	69.357	6.622	8.581
22.99	72.171	8.541	8.690	23.00	70.875	9.485	9.124	22.96	69.058	6.636	8.375
23.96	72.021	8.812	8.541	23.99	70.623	9.693	8.945	23.98	68.791	6.710	8.181
25.02	71.845	9.129	8.394	24.98	70.357	9.866	8.766	24.98	68.565	6.904	8.011
26.00	71.716	9.531	8.280	26.00	70.047	10.045	8.588	25.97	68.299	6.998	7.841
27.01	71.654	10.069	8.188	27.01	69.730	10.202	8.413	26.98	68.029	7.116	7.674
28.02	71.502	10.546	8.102	28.01	69.329	10.204	8.217	27.98	67.764	7.240	7.155
29.03	71.331	10.876	7.990	28.98	68.833	9.978	8.001	28.98	67.510	7.405	7.369
29.98	70.909	10.900	7.855	30.01	68.851	10.711	7.953	30.00	67.223	7.520	7.219

Table 7A Data from Stogryn (1996), temperature T = 0°C

GHz		Salinity									
		0	2.09	3.92	7.17	11.2	15.46	20.14	22.47	31.68	35.96
7	ε'	55.66	55.44	55.27	54.86	54.53	53.99	53.67	53.33	52.34	51.96
	ε''	39.97	39.78	39.81	40.05	40.21	40.36	40.70	40.77	41.13	41.47
8	ε'	50.93	50.82	50.64	50.38	50.16	49.72	49.47	49.25	48.40	48.12
	ε''	40.41	40.53	4053	40.70	40.78	40.86	41.07	41.09	41.30	41.48
9	ε'	46.13	45.92	45.88	45.60	45.45	45.13	44.02	44.80	44.18	44.00
	ε''	40.80	40.88	40.89	40.95	40.99	41.03	41.18	41.14	41.26	41.37
10	ε'	41.72	41.72	41.64	41.43	41.42	41.15	41.09	40.90	40.42	40.31
	ε''	40.38	40.41	40.47	40.47	40.60	40.54	40.64	40.56	40.60	40.67
11	ε'	37.93	37.98	37.92	37.80	37.75	37.64	37.58	37.47	37.14	37.09
	ε''	39.69	39.69	39.68	39.70	39.77	39.74	39.81	39.74	39.75	39.79
12	ε'	34.69	34.72	34.70	34.59	34.60	34.50	34.52	34.39	34.14	34.13
	ε''	38.64	38.76	38.75	38.76	38.85	38.82	38.86	38.81	38.80	38.82
13	ε'	31.86	31.88	31.91	31.83	31.83	31.85	31.87	31.80	31.64	31.65
	ε''	37.70	37.77	37.70	37.74	37.72	37.78	37.83	37.77	37.69	37.77
14	ε'	29.15	29.60	29.65	29.57	29.61	29.60	29.85	29.56	29.47	29.56
	ε''	36.42	36.42	36.48	36.44	36.46	36.52	36.52	36.52	36.46	36.58

Table 8A Data from Stogryn (1996), temperature T = 4°C

GHz		Salinity									
		0	2.09	3.92	7.17	11.2	15.46	20.14	22.47	31.68	35.96
7	ε'	60.55	60.17	59.93	59.47	59.04	58.16	57.64	57.34	56.04	55.65
	ε''	37.33	37.33	37.60	37.76	38.29	38.48	38.83	39.04	39.69	40.07
8	ε'	55.83	55.93	55.67	55.32	54.94	54.18	53.81	53.48	52.43	52.11
	ε''	38.76	38.68	38.87	39.03	39.27	39.46	39.67	39.82	40.29	40.52
9	ε'	51.21	51.34	51.12	50.84	50.52	49.94	49.62	49.36	48.53	48.33
	ε''	39.60	39.70	39.84	39.82	40.08	40.16	40.27	40.37	40.64	40.78
10	ε'	47.99	47.19	47.01	46.84	46.54	46.06	45.81	45.62	44.93	44.78
	ε''	39.96	39.99	40.08	40.09	40.24	40.25	40.29	40.36	40.49	40.59
11	ε'	43.24	43.35	43.21	43.07	42.91	42.49	42.30	42.19	41.66	41.58
	ε''	39.77	39.87	39.90	39.91	40.00	39.96	39.98	40.03	40.11	40.17
12	ε'	39.75	39.98	39.84	39.81	39.65	39.31	39.18	39.07	38.64	38.65
	ε''	39.24	39.43	39.45	39.48	39.53	39.45	39.48	39.50	39.53	39.56
13	ε'	36.77	36.94	36.83	36.78	36.72	36.46	36.42	36.36	35.92	35.99
	ε''	38.71	38.78	38.80	38.81	38.87	38.82	38.80	38.87	38.86	38.85
14	ε'	34.08	34.33	34.28	34.22	34.06	33.94	33.82	33.89	33.60	33.60
	ε''	37.95	37.79	37.86	37.81	37.90	37.83	37.79	37.89	37.81	37.76

Table 9A Data from Stogryn (1996), temperature $T = 10°C$

GHz		Salinity										
		0	2.09	3.92	7.17	11.2	15.46	20.14	22.47	31.68	35.96	38.00
7	ε'	65.33	64.68	64.24	63.57	62.84	61.98	61.30	60.84	59.38	58.58	58.34
	ε''	33.29	33.45	33.65	34.15	34.74	35.22	35.95	36.28	37.42	38.06	38.13
8	ε'	61.39	61.32	60.88	60.30	59.74	59.09	58.54	58.06	56.86	56.06	55.91
	ε''	35.38	35.49	35.63	35.92	36.31	36.69	37.19	37.37	38.14	38.66	38.66
9	ε'	57.27	57.34	56.95	56.41	55.89	55.33	54.83	54.41	53.34	52.68	52.52
	ε''	36.80	37.15	37.24	37.43	37.70	37.97	38.35	38.47	39.00	39.39	39.32
10	ε'	53.61	53.44	53.11	52.71	52.24	51.76	51.36	50.93	50.04	49.48	49.31
	ε''	37.84	38.14	38.21	38.31	38.52	38.70	38.93	39.00	39.34	39.56	39.55
11	ε'	49.91	49.80	49.55	49.23	48.75	48.33	48.02	47.71	46.94	46.48	46.33
	ε''	38.55	38.66	38.66	38.75	38.87	38.96	39.16	39.17	39.42	39.56	39.55
12	ε'	46.53	46.50	46.34	46.06	45.66	45.30	45.03	44.80	44.16	43.76	43.68
	ε''	38.78	38.70	38.72	38.78	38.84	38.88	39.01	39.04	39.22	39.33	39.29
13	ε'	43.43	43.30	43.15	42.90	42.60	42.30	42.09	41.85	41.32	41.02	40.89
	ε''	38.82	38.83	38.78	38.80	38.85	38.80	38.91	38.90	39.04	39.09	39.12
14	ε'	40.47	40.55	40.50	40.27	39.98	39.76	39.54	39.39	39.02	38.70	38.62
	ε''	38.47	38.29	38.24	38.24	38.23	38.21	38.24	38.29	38.29	38.35	38.33

Table 10A Data from Stogryn (1996), temperature $T = 15°C$

GHz		Salinity										
		0	2.09	3.92	7.17	11.02	15.46	22.47	28.49	35.00	35.33	38.00
7	ε'	67.88	67.40	66.80	65.92	65.70	64.29	63.92	63.12	60.60	60.55	60.16
	ε''	29.71	30.33	30.41	31.11	32.09	32.88	33.53	34.73	36.26	36.43	36.75
8	ε'	64.70	63.89	64.19	63.42	62.38	61.84	60.61	59.81	58.67	58.66	58.33
	ε''	32.22	32.58	32.60	33.13	33.84	34.50	35.22	36.01	36.88	36.96	37.25
9	ε'	61.26	60.51	60.75	60.06	59.15	58.68	57.42	56.66	55.69	55.62	55.33
	ε''	34.14	34.44	34.46	34.82	35.28	35.93	36.54	37.02	37.71	37.81	37.96
10	ε'	57.73	57.27	57.37	56.80	55.99	55.39	54.37	53.67	52.85	52.79	52.51
	ε''	35.58	35.91	35.75	36.04	36.40	36.86	37.51	37.78	38.19	38.21	38.43
11	ε'	54.48	54.15	54.12	53.46	52.92	52.40	51.43	50.84	50.06	50.04	49.77
	ε''	36.56	36.99	36.71	36.92	37.21	37.55	38.11	38.27	38.44	38.53	38.62
12	ε'	51.41	51.15	50.90	50.32	49.94	49.40	48.63	48.18	47.43	47.33	47.19
	ε''	37.36	37.67	37.34	37.42	37.70	37.89	38.35	38.50	38.44	38.52	38.55
13	ε'	48.38	48.29	48.04	47.45	47.03	46.65	45.95	45.69	44.77	44.74	44.57
	ε''	37.82	37.97	37.80	37.88	37.88	38.29	38.22	38.47	38.60	38.61	38.59
14	ε'	45.34		45.26	44.75		44.08			42.53	42.49	42.38
	ε''	37.76		37.64	37.74		37.89			38.09	38.17	38.15

Table 11A *Data from Stogryn (1996), temperature $T = 20°C$*

GHz		Salinity									
		0	2.09	3.92	7.17	11.20	15.46	20.14	22.47	35.00	38.00
7	ε'	69.04	68.47	68.03	67.18	66.15	65.36	64.51	64.04	61.68	61.15
	ε''	26.49	27.08	27.56	28.65	29.78	30.62	31.67	32.24	35.06	35.62
8	ε'	66.71	66.22	65.79	65.11	64.23	63.52	62.73	62.23	60.04	59.64
	ε''	29.08	29.56	29.82	30.70	31.56	32.25	33.08	33.41	35.53	35.92
9	ε'	63.74	63.26	62.89	62.20	61.39	60.72	60.01	59.55	57.51	57.16
	ε''	31.14	31.49	31.67	32.38	33.03	33.59	34.25	34.48	36.26	36.56
10	ε'	60.79	60.41	60.07	59.43	58.68	58.01	57.40	57.07	54.96	54.77
	ε''	32.82	33.10	33.22	33.81	34.35	34.67	35.18	35.40	36.71	36.98
11	ε'	57.80	57.41	57.17	56.55	55.80	55.24	54.65	54.31	52.57	52.34
	ε''	34.13	34.45	34.53	34.96	35.37	35.62	36.05	36.19	37.15	37.36
12	ε'	54.96	54.54	54.45	53.79	53.04	52.56	52.02	51.68	50.09	49.90
	ε''	35.30	35.50	35.63	35.93	36.19	36.38	36.70	36.75	37.44	37.53
13	ε'	52.12	51.72	51.60	51.06	50.46	50.00	49.53	49.28	47.88	47.71
	ε''	36.07	36.13	36.20	36.38	36.63	36.76	37.02	37.07	37.65	37.83
14	ε'	49.58	49.13	49.12	48.61	48.10	47.65	47.17	47.02	45.66	45.54
	ε''	36.30	36.41	36.44	36.64	36.77	36.90	37.06	37.17	37.52	37.56

Table 12A *Data from Stogryn (1996), temperature $T = 25°C$*

GHz		Salinity										
		0	2.09	7.17	11.2	15.46	20.14	22.47	30.19	31.68	35.33	37.98
7	ε'	69.81	69.19	68.06	67.13	66.35	65.55	64.87	63.82	63.13	62.32	62.22
	ε''	23.61	24.28	25.86	27.27	28.44	29.62	30.44	32.20	33.30	34.00	34.30
8	ε'	67.56	67.22	66.30	64.98	64.46	63.62	62.89	62.30		60.84	60.57
	ε''	25.87	26.39	27.83	28.75	29.93	30.79	31.38	32.97		34.21	34.67
9	ε'	65.11	64.67	63.87	62.81	62.19	61.40	60.84	60.16	59.21	58.67	58.37
	ε''	28.19	28.53	29.55	30.53	31.25	31.96	32.52	33.70	34.30	34.74	35.12
10	ε'	62.73	62.33	61.61	60.40	59.99	59.20	58.66	58.06		56.59	56.27
	ε''	29.89	30.30	31.05	31.83	32.35	32.99	33.32	34.53		35.14	35.47
11	ε'	60.31	59.98	59.35	58.26	57.71	56.97	56.45	56.01	55.27	54.64	54.30
	ε''	31.49	31.63	32.60	32.79	33.42	33.92	34.10	35.21	35.26	35.52	35.80
12	ε'	58.06	57.56	57.08	55.84	55.32	54.75	54.29	53.81		52.66	52.24
	ε''	32.73	33.00	33.65	33.89	34.18	34.67	34.78	35.72		35.88	36.16
13	ε'	55.78	55.04	54.67	53.71	53.10	52.56	52.20	51.50	51.07	50.57	50.09
	ε''	34.05	33.74	34.59	34.55	34.98	35.34	35.45	36.26	36.02	36.41	36.52

Table 13A Data from Stogryn (1996), temperature $T = 30°C$

GHz		Salinity									
		0	2.09	7.17	11.2	15.46	20.14	22.47	31.68	35.96	37.98
7	ε'	69.63	68.93	67.86	67.06	66.46	65.09	64.84	62.79		61.97
	ε''	20.91	21.66	23.47	24.99	26.46	28.40	28.96	31.84		33.59
8	ε'	68.01	67.75	66.70	65.82	65.28	64.08		61.99	61.15	61.11
	ε''	23.02	23.92	25.29	26.34	27.40	29.02		31.63	32.78	32.97
9	ε'	66.03	65.67	64.71	63.90	63.39	62.22	62.02	60.26	59.42	59.44
	ε''	25.25	25.84	27.01	27.91	28.78	30.06	30.49	32.43	33.19	33.38
10	ε'	63.93	63.59	62.70	61.98	61.45	60.34	60.15	58.43	57.66	57.65
	ε''	27.03	27.55	28.52	29.27	29.99	31.04	31.37	32.85	33.57	33.74
11	ε'	61.85	61.46	60.61	59.91	59.47	58.39	58.25	56.54	55.87	55.82
	ε''	28.68	29.04	29.80	30.51	31.10	31.99	32.23	33.50	34.01	34.21
12	ε'	59.72	59.37	58.50	57.86	57.36	56.40	56.31	54.67	54.10	54.01
	ε''	30.10	30.35	30.99	31.63	32.07	32.81	33.01	34.09	34.45	34.65
13	ε'	57.68	57.17	56.34	55.82	55.46	54.47	54.40	52.81	52.29	52.22
	ε''	31.36	31.56	32.10	32.63	33.00	33.58	33.75	34.60	34.80	35.08
14	ε'	55.49	54.84	54.03	53.46	52.97	52.22	52.10	50.51	50.22	50.09
	ε''	32.42	32.48	32.92	33.34	33.63	34.13	34.23	35.02	35.08	35.27

Table 14A Permittivity data from Lamkaouchi (2003) for purewater

T (°C)	GHz	ε'	% error ε'	ε''	% error ε''
0	90	7.10	3	9.40	3
20	35	19.92	3	29.10	3
20	90	8.50	3	13.80	3
25	35	22.80	3	30.50	3
25	75	10.02	3	17.10	3
25	79	10.28	3	17.28	3
25	89	8.75	3	15.76	3
25	90	8.80	3	15.30	3
30	35	25.45	3	32.20	3
30	90	9.10	3	16.10	3

Table 15A Debye parameters from Rønne et al. (1997) for purewater over the frequency range 100–1000 GHz

T (°C)	ε_s	ε_1	τ_1	τ_2	ε_∞
−2.0	88.72	4.99	15.70	0.22	3.8
5.7	85.64	5.40	13.10	0.26	3.7
15.4	81.91	5.25	9.67	0.18	3.4
19.2	80.50	5.20	8.50	0.17	3.3
30.1	76.57	5.40	7.00	0.200	3.4
41.9	72.54	4.76	5.50	0.077	2.1
51.0	69.57	4.76	4.48	0.072	1.6

Table 16A Experimental purewater permittivity data (mean and s.d.) for T = −2, 5.6, 15.3 and 19.3°C, kindly supplied by Dr. C. Rønne

GHz	T = −2.0°C				T = 5.6°C				T = 15.3°C				T = 19.3°C			
	ε'	sd ε'	ε''	sd ε''	ε'	sd ε'	ε''	sd ε''	ε'	sd ε'	ε''	sd ε''	ε'	sd ε'	ε''	sd ε''
97.6	4.99	0.49	8.81	0.10	8.53	1.59	9.02	1.04	8.05	0.80	12.89	1.14	9.91	0.59	14.30	0.93
122.0	5.71	0.30	6.93	0.46	8.78	2.85	8.64	2.49	7.80	1.20	9.62	2.86	10.98	2.35	10.05	0.80
146.4	5.69	0.38	6.20	0.20	5.30	0.71	7.91	0.79	5.43	0.44	9.72	0.92	5.60	0.74	10.41	0.47
170.8	4.77	0.28	4.84	0.09	5.74	0.36	6.48	0.26	5.85	0.07	7.52	0.34	6.09	0.36	8.59	0.24
195.2	5.07	0.08	4.04	0.02	5.35	0.07	4.84	0.23	5.79	0.12	5.88	0.33	6.09	0.17	6.71	0.19
219.6	5.14	0.06	3.92	0.18	5.39	0.04	5.05	0.34	5.51	0.12	6.17	0.07	5.61	0.10	6.70	0.25
244.0	4.51	0.13	3.76	0.08	5.58	0.20	4.55	0.17	5.76	0.19	5.67	0.23	6.02	0.15	6.24	0.11
268.4	4.51	0.11	3.62	0.12	5.38	0.20	4.24	0.18	5.71	0.12	5.18	0.24	5.66	0.12	5.57	0.11
292.8	4.99	0.11	3.35	0.01	5.12	0.08	4.19	0.14	5.38	0.13	5.00	0.16	5.37	0.14	5.37	0.09
317.2	4.78	0.03	3.64	0.07	4.95	0.09	3.61	0.12	5.20	0.17	4.58	0.11	5.20	0.07	4.90	0.03
341.6	4.37	0.03	3.21	0.06	5.22	0.13	3.48	0.09	5.31	0.09	4.37	0.09	5.35	0.05	4.62	0.09
366.0	4.74	0.10	2.81	0.10	5.14	0.11	3.53	0.06	5.31	0.06	4.35	0.12	5.21	0.08	4.62	0.06
390.4	4.95	0.07	2.78	0.06	5.05	0.11	3.34	0.07	5.14	0.19	4.02	0.16	5.16	0.07	4.20	0.11
414.8	4.75	0.02	2.67	0.02	4.99	0.10	2.99	0.03	4.97	0.06	3.68	0.09	5.05	0.05	3.92	0.06
439.1	4.71	0.05	2.63	0.04	4.81	0.12	3.02	0.12	4.84	0.04	3.63	0.12	4.88	0.05	3.90	0.11
463.5	4.71	0.05	2.43	0.10	4.78	0.10	3.01	0.12	4.71	0.09	3.64	0.10	4.78	0.04	3.91	0.14
487.9	4.71	0.05	2.38	0.03	4.71	0.07	2.71	0.10	4.61	0.16	3.33	0.13	4.80	0.03	3.61	0.04
512.3	4.64	0.04	2.23	0.04	4.72	0.03	2.78	0.07	4.61	0.13	3.31	0.06	4.76	0.04	3.58	0.02
536.7	4.77	0.06	2.24	0.10	4.63	0.12	2.59	0.13	4.67	0.11	3.08	0.02	4.70	0.03	3.43	0.11
561.1	4.68	0.07	2.27	0.09	4.69	0.08	2.57	0.19	4.67	0.11	2.92	0.09	4.74	0.03	3.22	0.09
585.5	4.58	0.05	2.07	0.06	4.55	0.08	2.44	0.10	4.64	0.22	2.90	0.06	4.68	0.03	3.13	0.06
609.9	4.51	0.08	2.07	0.11	4.61	0.03	2.36	0.14	4.69	0.07	2.82	0.06	4.64	0.01	3.04	0.07
634.3	4.58	0.06	1.99	0.03	4.55	0.03	2.32	0.21	4.62	0.16	2.74	0.05	4.58	0.05	3.01	0.06
658.7	4.56	0.08	1.98	0.09	4.45	0.12	2.31	0.12	4.64	0.14	2.77	0.07	4.55	0.06	2.98	0.11
683.1	4.47	0.04	1.88	0.1	4.44	0.14	2.25	0.09	4.58	0.20	2.81	0.09	4.50	0.03	2.98	0.04
707.5	4.46	0.03	1.85	0.1	4.52	0.03	2.16	0.12	4.61	0.1	2.77	0.07	4.52	0.05	2.87	0.08
731.9	4.41	0.06	1.8	0.08	4.41	0.02	2.18	0.17	4.57	0.15	2.7	0.06	4.52	0.06	2.82	0.06
756.3	4.34	0.03	1.71	0.1	4.45	0.06	2.09	0.1	4.46	0.18	2.58	0.09	4.44	0.10	2.76	0.09
780.7	4.33	0.01	1.67	0.11	4.43	0.07	2.04	0.19	4.45	0.12	2.56	0.06	4.41	0.04	2.66	0.08
805.1	4.35	0.07	1.68	0.06	4.36	0.08	1.98	0.2	4.41	0.15	2.42	0.04	4.30	0.04	2.57	0.12

Table 16A (Continued)

GHz	$T = -2.0°C$				$T = 5.6°C$				$T = 15.3°C$				$T = 19.3°C$			
	ε'	sd ε'	ε''	sd ε''	ε'	sd ε'	ε''	sd ε''	ε'	sd ε'	ε''	sd ε''	ε'	sd ε'	ε''	sd ε''
829.5	4.33	0.08	1.63	0.06	4.32	0.07	2.02	0.15	4.35	0.1	2.43	0.07	4.37	0.07	2.58	0.11
853.9	4.28	0.06	1.58	0.09	4.27	0.04	1.95	0.15	4.3	0.13	2.4	0.08	4.35	0.05	2.55	0.10
878.3	4.31	0.08	1.52	0.12	4.31	0.07	1.97	0.13	4.34	0.08	2.4	0.06	4.36	0.02	2.54	0.10
902.7	4.26	0.05	1.52	0.12	4.33	0.02	2.01	0.14	4.38	0.09	2.37	0.05	4.42	0.08	2.53	0.22
927.1	4.26	0.06	1.52	0.18	4.27	0.1	1.88	0.22	4.36	0.04	2.28	0.09	4.36	0.03	2.45	0.13
951.5	4.25	0.06	1.5	0.15	4.18	0.07	1.88	0.1	4.22	0.04	2.27	0.04	4.28	0.05	2.45	0.11
975.9	4.18	0.03	1.46	0.19	4.13	0.06	1.77	0.23	4.29	0.07	2.11	0.14	4.26	0.09	2.36	0.07
1000.3	4.24	0.05	1.47	0.16	4.28	0.09	1.84	0.19	4.33	0.14	2.12	0.14	4.27	0.07	2.34	0.13
1024.7	4.26	0.06	1.43	0.10	4.1	0.17	1.72	0.16	4.19	0.1	2.19	0.13	4.20	0.07	2.35	0.12
1049.1	4.18	0.06	1.38	0.13	4.07	0.04	1.68	0.12	4.18	0.03	2.08	0.14	4.18	0.06	2.34	0.18
1073.5	4.2	0.07	1.34	0.17	4.20	0.17	1.77	0.20	4.15	0.16	2.07	0.12	4.23	0.07	2.29	0.11
1097.9	4.26	0.10	1.26	0.12	4.20	0.15	1.63	0.24	4.20	0.08	2.01	0.12	4.23	0.06	2.20	0.17
1122.3	4.20	0.04	1.31	0.16	4.22	0.05	1.68	0.19	4.13	0.10	2.09	0.15	4.20	0.06	2.27	0.13
1146.7	4.23	0.14	1.33	0.11	4.16	0.14	1.66	0.15	4.15	0.12	2.00	0.07	4.09	0.03	2.21	0.18
1171.1	4.28	0.09	1.37	0.17	4.06	0.16	1.59	0.21	4.17	0.02	1.94	0.08	4.18	0.07	2.12	0.09
1195.5	4.19	0.09	1.26	0.15	4.17	0.18	1.50	0.17	4.04	0.10	1.88	0.13	4.17	0.04	2.01	0.17
1219.9	4.16	0.02	1.24	0.10	4.05	0.06	1.64	0.25	4.14	0.14	1.96	0.19	4.05	0.03	2.18	0.15
1244.3	4.21	0.10	1.18	0.14	4.00	0.16	1.43	0.18	4.02	0.13	1.80	0.16	4.00	0.03	2.11	0.23
1268.6	4.12	0.06	1.27	0.14	4.19	0.05	1.33	0.24	4.14	0.13	1.81	0.13	4.08	0.08	2.09	0.16
1293.0	4.20	0.05	1.18	0.12	4.10	0.05	1.47	0.31	4.15	0.23	1.82	0.25	4.17	0.13	2.02	0.29
1317.4	4.10	0.11	1.17	0.18	4.07	0.11	1.36	0.10	4.07	0.17	1.81	0.14	4.16	0.06	2.12	0.09
1341.8	4.04	0.09	1.17	0.22	4.09	0.09	1.47	0.15	4.09	0.19	1.73	0.15	4.08	0.08	2.08	0.23
1366.2	4.03	0.04	1.12	0.17	4.08	0.11	1.43	0.14	4.16	0.07	1.76	0.2	4.16	0.11	1.92	0.18
1390.6	4.02	0.11	1.16	0.23	4.14	0.15	1.50	0.15	4.04	0.15	1.77	0.28	4.02	0.04	1.99	0.19
1415.0	4.03	0.04	1.02	0.14	4.05	0.12	1.26	0.29	3.97	0.03	1.66	0.25	3.99	0.11	1.91	0.22
1439.4	4.05	0.07	1.11	0.18	4.03	0.07	1.43	0.34	3.96	0.13	1.64	0.17	4.00	0.1	2.02	0.21
1463.8	4.10	0.11	1.05	0.29	4.13	0.35	1.30	0.29	3.87	0.09	1.76	0.31	3.93	0.05	1.99	0.26
1488.2	4.07	0.12	1.05	0.17	4.04	0.14	1.18	0.15	3.96	0.13	1.66	0.22	4.08	0.15	2.08	0.20
1512.6	4.07	0.13	1.00	0.20	4.11	0.15	1.27	0.08	3.9	0.17	1.69	0.15	4.06	0.08	1.98	0.23
1537.0	4.03	0.11	1.01	0.33	4.09	0.16	1.34	0.22	3.94	0.10	1.58	0.43	3.98	0.22	1.86	0.15
1561.4	4.02	0.14	1.02	0.20	3.94	0.30	1.17	0.05	3.94	0.11	1.71	0.13	4.04	0.15	1.95	0.17
1585.8	4.03	0.09	1.02	0.17	4.03	0.34	1.34	0.20	3.91	0.19	1.80	0.06	4.02	0.08	1.70	0.29
1610.2	3.99	0.04	1.01	0.22	3.96	0.14	1.16	0.31	3.83	0.06	1.71	0.23	4.08	0.13	1.93	0.23
1634.6	3.93	0.26	0.95	0.20	3.91	0.24	1.01	0.17	3.88	0.19	1.68	0.24	3.91	0.19	1.98	0.18
1659.0	4.07	0.14	0.89	0.18	3.94	0.20	1.01	0.51	3.92	0.10	1.71	0.30	4.09	0.31	1.90	0.14
1683.4	3.98	0.22	0.82	0.25	3.96	0.31	1.21	0.34	3.82	0.17	1.55	0.28	4.22	0.29	1.76	0.34
1707.8	3.86	0.03	0.98	0.35	3.59	0.14	1.5	0.32	3.87	0.28	1.76	0.48	4.15	0.23	1.90	0.22
1732.2	3.86	0.07	0.76	0.19	4.01	0.17	1.13	0.50	3.8	0.15	1.44	0.50	4.02	0.10	1.68	0.44
1756.6	3.81	0.20	0.89	0.21	4.12	0.31	1.17	0.26	3.96	0.54	1.53	0.30	3.93	0.22	1.91	0.32
1781.0	4.04	0.16	1.02	0.23	4.07	0.31	1.35	0.34	3.88	0.39	1.59	0.17	4.25	0.38	1.90	0.53
1805.4	3.91	0.13	0.88	0.19	4.02	0.48	1.60	0.33	3.93	0.19	1.79	0.17	3.80	0.14	1.87	0.17
1829.8	4.01	0.13	0.66	0.16	3.9	0.27	1.20	0.53	3.89	0.18	1.91	0.22	3.99	0.18	1.67	0.24
1854.2	3.98	0.28	0.81	0.31	4.09	0.49	0.96	0.55	3.97	0.50	1.72	0.16	4.15	0.15	1.71	0.32
1878.6	3.81	0.25	0.74	0.27	3.94	0.45	1.14	0.40	3.66	0.23	1.70	0.34	3.95	0.68	1.73	0.19
1903.0	4.03	0.14	0.60	0.43	3.98	0.43	0.92	0.27	3.62	0.5	1.62	0.18	3.98	0.06	1.65	0.53
1927.4	4.18	0.24	0.87	0.48	3.83	0.46	1.41	0.35	4.02	0.28	1.81	0.37	3.84	0.17	1.62	0.40
1951.8	4.06	0.31	0.59	0.39	3.23	0.55	1.23	0.61	3.90	0.28	1.55	0.20	4.27	0.66	2.12	0.30
1976.2	4.03	0.26	0.49	0.33	3.47	0.62	1.12	0.72	4.03	0.7	1.64	0.38	3.87	0.4	1.58	0.74
2000.6	4.01	0.54	0.79	0.32	3.64	0.56	0.82	0.54	4.23	0.21	1.69	0.59	3.94	0.19	1.96	0.43
2025.0	3.95	0.23	0.57	0.16	4.22	0.58	2.12	1.28	3.66	0.34	1.80	0.62	4.15	0.33	1.59	0.73
2049.4	4.11	0.35	0.42	0.23	3.90	0.49	1.36	0.83	3.70	0.30	1.62	0.19	3.73	0.46	1.35	0.45

Table 17A Experimental purewater permittivity data (mean and s.d.) for T = 30, 41.8 and 50.9°C, kindly supplied by Dr. C. Rønne

GHz	T = 30°C				T = 41.8°C				T = 50.9°C			
	ε	sd ε'	ε''	sd ε''	ε'	sd ε'	ε''	sd ε''	ε'	sd ε'	ε''	sd ε''
97.6	11.32	1.29	12.79	1.86	9.09	0.97	17.00	3.65	8.22	1.07	21.00	1.70
122.0	13.71	6.57	10.46	6.91	7.64	0.88	13.99	3.61	9.59	1.33	17.36	0.94
146.4	5.26	2.53	14.13	1.53	8.11	0.90	13.92	3.49	10.06	1.47	13.91	0.40
170.8	7.21	0.65	9.78	0.55	5.30	0.43	10.65	1.11	6.53	0.17	12.71	0.56
195.2	6.31	0.31	8.10	0.36	6.25	0.19	8.86	0.99	6.61	0.35	10.43	0.10
219.6	5.98	0.33	7.73	0.37	6.12	0.25	8.81	0.98	6.82	0.17	10.48	0.20
244.0	6.14	0.17	7.06	0.23	4.98	0.37	8.31	0.57	5.38	0.15	9.57	0.30
268.4	6.00	0.35	6.62	0.47	4.89	0.25	7.65	0.68	5.05	0.18	9.01	0.30
292.8	5.56	0.24	6.25	0.14	5.69	0.07	6.83	0.48	6.08	0.1	8.24	0.15
317.2	5.46	0.32	5.47	0.22	5.35	0.10	7.11	0.69	5.48	0.14	8.32	0.01
341.6	5.52	0.12	5.57	0.14	4.77	0.12	6.32	0.47	4.85	0.10	7.35	0.13
366.0	5.28	0.04	5.45	0.26	5.18	0.06	6.02	0.41	5.41	0.06	6.98	0.08
390.4	5.31	0.19	5.00	0.16	5.27	0.17	5.79	0.43	5.40	0.10	6.61	0.10
414.8	5.14	0.16	4.69	0.25	5.12	0.11	5.48	0.36	5.18	0.09	6.32	0.09
439.1	4.99	0.12	4.48	0.23	5.09	0.10	5.28	0.26	5.09	0.08	6.13	0.07
463.5	4.79	0.10	4.52	0.10	4.89	0.05	5.09	0.31	5.01	0.08	5.84	0.04
487.9	4.89	0.08	4.20	0.13	4.89	0.06	4.98	0.19	4.96	0.05	5.66	0.04
512.3	4.95	0.25	4.28	0.09	5.02	0.09	4.64	0.26	5.01	0.05	5.39	0.04
536.7	4.81	0.11	4.01	0.12	5.05	0.06	4.65	0.23	5.06	0.08	5.27	0.07
561.1	4.85	0.12	3.83	0.08	4.92	0.09	4.53	0.24	4.95	0.06	5.15	0.01
585.5	4.77	0.10	3.69	0.13	4.70	0.10	4.32	0.17	4.78	0.07	4.86	0.03
609.9	4.81	0.16	3.54	0.05	4.80	0.05	4.20	0.18	4.80	0.05	4.83	0.08
634.3	4.59	0.05	3.64	0.31	4.74	0.10	4.20	0.20	4.84	0.07	4.74	0.04
658.7	4.64	0.12	3.38	0.07	4.66	0.05	3.96	0.14	4.67	0.09	4.58	0.08
683.1	4.49	0.16	3.39	0.19	4.59	0.05	3.83	0.20	4.61	0.03	4.36	0.07
707.5	4.49	0.10	3.27	0.10	4.57	0.05	3.79	0.14	4.56	0.04	4.25	0.10
731.9	4.47	0.15	3.27	0.19	4.62	0.07	3.68	0.09	4.52	0.01	4.16	0.08
756.3	4.48	0.04	3.14	0.13	4.5	0.05	3.56	0.07	4.46	0.06	4.10	0.02
780.7	4.49	0.07	3.11	0.10	4.46	0.09	3.51	0.13	4.47	0.08	3.98	0.06
805.1	4.41	0.19	3.02	0.24	4.42	0.07	3.43	0.11	4.37	0.04	3.98	0.03
829.5	4.36	0.09	3.06	0.13	4.35	0.06	3.39	0.10	4.39	0.05	3.85	0.04
853.9	4.34	0.08	2.96	0.36	4.36	0.07	3.35	0.13	4.36	0.03	3.77	0.06
878.3	4.29	0.16	2.86	0.25	4.34	0.04	3.29	0.15	4.36	0.05	3.65	0.04
902.7	4.43	0.09	2.92	0.08	4.32	0.06	3.16	0.04	4.29	0.02	3.63	0.07
927.1	4.33	0.15	2.86	0.05	4.38	0.07	3.18	0.05	4.29	0.02	3.59	0.04
951.5	4.25	0.09	2.81	0.15	4.31	0.03	3.15	0.03	4.23	0.06	3.58	0.04
975.9	4.39	0.11	2.71	0.27	4.29	0.09	3.16	0.13	4.24	0.03	3.6	0.04
1000.3	4.34	0.16	2.60	0.11	4.24	0.11	3.10	0.11	4.27	0.03	3.45	0.09
1024.7	4.35	0.25	2.64	0.18	4.26	0.10	3.01	0.26	4.25	0.06	3.41	0.07
1049.1	4.18	0.09	2.63	0.14	4.17	0.09	3.10	0.07	4.20	0.03	3.41	0.06

Table 17A (Continued)

GHz	$T = 30°C$				$T = 41.8°C$				$T = 50.9°C$			
	ε	sd ε'	ε''	sd ε''	ε'	sd ε'	ε''	sd ε''	ε'	sd ε'	ε''	sd ε''
1073.5	4.22	0.12	2.52	0.16	4.2	0.06	2.98	0.05	4.22	0.04	3.30	0.05
1097.9	4.21	0.09	2.35	0.22	4.31	0.05	2.90	0.06	4.18	0.03	3.21	0.04
1122.3	4.15	0.07	2.58	0.27	4.17	0.12	2.91	0.16	4.22	0.06	3.34	0.06
1146.7	4.25	0.15	2.46	0.12	4.24	0.13	2.86	0.03	4.20	0.03	3.22	0.04
1171.1	4.38	0.18	2.42	0.15	4.24	0.02	2.89	0.12	4.24	0.03	3.26	0.08
1195.5	4.28	0.32	2.34	0.41	4.14	0.04	2.77	0.17	4.19	0.04	3.12	0.05
1219.9	4.19	0.29	2.29	0.21	4.12	0.04	2.75	0.09	4.13	0.03	3.18	0.07
1244.3	4.01	0.27	2.48	0.28	4.13	0.16	2.69	0.17	4.10	0.02	3.10	0.05
1268.6	4.24	0.36	2.43	0.38	4.15	0.07	2.74	0.08	4.14	0.05	3.12	0.04
1293.0	4.22	0.30	2.38	0.34	4.06	0.10	2.80	0.06	4.10	0.05	3.08	0.03
1317.4	4.04	0.21	2.29	0.22	4.11	0.14	2.68	0.06	4.12	0.04	3.05	0.06
1341.8	4.17	0.24	2.29	0.55	4.03	0.05	2.75	0.12	4.02	0.07	2.99	0.11
1366.2	4.05	0.28	2.29	0.35	4.09	0.13	2.60	0.19	3.94	0.04	3.00	0.07
1390.6	4.17	0.18	2.25	0.38	4.14	0.10	2.59	0.08	3.98	0.07	3.04	0.05
1415.0	4.00	0.21	2.07	0.55	3.93	0.06	2.59	0.11	3.85	0.04	3.07	0.10
1439.4	3.93	0.28	2.13	0.22	4.01	0.10	2.55	0.20	3.96	0.09	2.97	0.11
1463.8	3.88	0.40	2.18	0.44	3.96	0.16	2.60	0.05	3.91	0.06	2.89	0.08
1488.2	4.23	0.24	2.29	0.08	3.92	0.10	2.43	0.05	4.01	0.09	2.95	0.08
1512.6	4.31	0.28	2.28	0.42	4.04	0.08	2.68	0.26	3.98	0.03	2.91	0.08
1537.0	3.82	0.13	2.16	0.11	3.91	0.14	2.50	0.12	3.90	0.10	2.85	0.14
1561.4	4.13	0.39	2.05	0.48	4.07	0.10	2.56	0.16	3.82	0.10	2.85	0.05
1585.8	4.07	0.34	2.06	0.14	3.85	0.09	2.35	0.11	3.80	0.18	2.98	0.05
1610.2	3.99	0.21	2.13	0.54	3.81	0.26	2.46	0.10	3.85	0.06	2.92	0.11
1634.6	4.06	0.06	2.16	0.38	3.96	0.21	2.49	0.19	3.92	0.13	2.98	0.03
1659.0	4.27	0.32	2.04	0.36	3.88	0.28	2.75	0.09	3.95	0.09	2.88	0.08
1683.4	3.77	0.44	1.99	0.75	3.76	0.17	2.50	0.30	3.63	0.08	2.77	0.16
1707.8	3.78	0.51	2.44	0.37	3.92	0.17	2.70	0.32	3.62	0.15	3.11	0.09
1732.2	4.03	0.26	2.13	0.28	3.73	0.20	2.47	0.06	3.69	0.04	2.77	0.12
1756.6	3.79	0.24	2.17	0.28	3.71	0.19	2.40	0.30	3.77	0.13	3.11	0.15
1781.0	3.64	0.38	1.85	0.14	3.71	0.27	2.48	0.37	3.82	0.11	2.90	0.14
1805.4	4.35	0.75	2.30	0.55	3.92	0.31	2.52	0.19	3.93	0.18	2.68	0.36
1829.8	4.13	0.32	1.62	0.19	3.78	0.60	2.26	0.32	3.64	0.06	2.75	0.22
1854.2	3.85	0.62	2.14	0.61	3.84	0.32	2.30	0.34	3.70	0.15	2.63	0.29
1878.6	3.53	0.30	2.05	0.60	3.64	0.36	2.46	0.24	3.77	0.19	2.85	0.14
1903.0	4.35	0.69	1.81	0.60	3.85	0.47	2.22	0.23	3.65	0.07	3.03	0.17
1927.4	4.01	1.06	1.68	0.69	3.55	0.25	2.58	0.23	3.74	0.19	2.72	0.14
1951.8	4.23	1.30	1.42	0.49	3.60	0.42	2.37	0.18	3.86	0.21	2.95	0.28
1976.2	4.29	0.95	2.76	1.36	4.00	0.45	2.61	0.90	3.81	0.49	2.81	0.24
2000.6	4.72	0.50	1.38	0.81	3.64	0.36	2.09	0.31	3.51	0.10	2.82	0.33
2025.0	3.74	0.40	1.54	0.54	3.75	0.49	2.22	0.39	3.38	0.13	2.89	0.14
2049.4	4.63	1.14	1.56	0.89	3.46	0.32	2.54	0.56	3.50	0.54	2.87	0.12

Table 18A Experimental purewater permittivity data (mean and s.d.) for T = 57.9, 70.2 and 93.5°C, kindly supplied by Dr. C. Rønne

GHz	$T = 57.9°C$				$T = 70.2°C$				$T = 93.5°C$			
	ε'	sd ε'	ε''	sd ε''	ε'	sd ε'	ε''	sd ε''	ε'	sd ε'	ε''	sd ε''
97.6	9.07	1.73	25.36	1.83	12.16	5.25	29.57	2.70	17.23	2.38	27.12	2.98
122.0	9.17	1.5	20.68	0.54	12.58	2.47	25.11	5.20	14.23	1.82	24.94	2.13
146.4	9.72	2.25	19.29	1.62	8.18	6.35	18.59	3.52	11.01	2.89	21.34	1.86
170.8	6.66	0.77	13.92	1.12	6.48	1.71	15.54	0.91	8.43	0.83	17.94	0.09
195.2	6.52	0.34	11.6	0.37	7.46	1.72	13.94	0.90	8.66	0.60	16.00	0.49
219.6	7.24	0.70	11.51	0.57	6.51	2.22	13.19	1.33	8.16	1.26	14.94	0.30
244.0	5.15	0.49	10.66	0.34	5.76	0.57	11.38	0.52	6.54	0.85	14.03	0.42
268.4	5.37	0.67	9.81	0.15	4.47	1.60	11.22	1.09	5.96	0.43	12.90	0.36
292.8	6.07	0.31	8.79	0.18	5.48	2.22	9.41	0.73	6.48	0.34	11.86	0.42
317.2	5.37	0.32	9.06	0.28	4.86	1.79	9.39	0.54	5.91	0.23	12.01	0.37
341.6	4.60	0.24	8.08	0.10	4.66	1.49	8.72	0.64	5.14	0.38	10.97	0.29
366.0	5.35	0.16	7.38	0.16	5.09	1.89	8.14	0.37	5.61	0.23	9.76	0.28
390.4	5.48	0.23	7.11	0.07	5.45	1.46	7.50	0.71	5.86	0.34	9.37	0.26
414.8	5.19	0.25	6.95	0.15	5.39	1.23	7.21	0.58	5.43	0.23	8.97	0.40
439.1	5.21	0.27	6.70	0.09	5.17	0.67	7.32	0.58	5.39	0.19	8.73	0.29
463.5	4.96	0.12	6.23	0.06	5.05	0.45	6.66	0.50	5.18	0.24	8.09	0.34
487.9	5.02	0.28	6.08	0.06	5.04	0.28	6.29	0.33	5.11	0.24	7.83	0.33
512.3	5.02	0.12	5.74	0.11	5.14	0.48	5.96	0.28	5.05	0.15	7.30	0.33
536.7	5.12	0.19	5.63	0.07	5.36	0.48	5.71	0.32	5.14	0.16	7.13	0.39
561.1	4.96	0.22	5.55	0.08	5.31	0.28	5.78	0.31	5.11	0.13	6.97	0.31
585.5	4.77	0.21	5.18	0.12	5.10	0.27	5.39	0.42	4.85	0.21	6.60	0.42
609.9	4.86	0.12	5.21	0.12	5.15	0.41	5.36	0.28	4.77	0.14	6.37	0.34
634.3	4.84	0.08	5.03	0.12	5.29	0.36	5.43	0.19	4.78	0.17	6.15	0.37
658.7	4.62	0.19	4.89	0.05	4.89	0.16	5.29	0.34	4.66	0.16	6.01	0.43
683.1	4.58	0.15	4.64	0.04	4.72	0.16	5.01	0.23	4.57	0.12	5.75	0.43
707.5	4.59	0.17	4.56	0.13	4.95	0.28	4.90	0.15	4.54	0.11	5.56	0.44
731.9	4.54	0.13	4.45	0.10	4.81	0.29	4.90	0.17	4.50	0.09	5.45	0.45
756.3	4.51	0.19	4.36	0.07	4.71	0.15	4.76	0.31	4.37	0.10	5.29	0.42
780.7	4.53	0.23	4.19	0.10	4.70	0.17	4.54	0.11	4.28	0.11	5.11	0.50
805.1	4.43	0.12	4.18	0.06	4.76	0.22	4.50	0.21	4.34	0.10	5.00	0.52
829.5	4.40	0.11	4.13	0.08	4.66	0.27	4.45	0.26	4.25	0.08	4.88	0.41
853.9	4.38	0.18	3.90	0.06	4.50	0.11	4.33	0.08	4.26	0.13	4.79	0.48
878.3	4.36	0.11	3.89	0.04	4.66	0.29	4.12	0.14	4.22	0.04	4.62	0.44
902.7	4.36	0.17	3.82	0.08	4.62	0.25	4.04	0.10	4.22	0.07	4.57	0.46
927.1	4.32	0.08	3.76	0.08	4.56	0.15	4.10	0.02	4.23	0.05	4.48	0.49
951.5	4.33	0.17	3.78	0.08	4.47	0.21	4.13	0.12	4.14	0.16	4.36	0.54
975.9	4.26	0.16	3.76	0.03	4.56	0.22	4.03	0.11	4.16	0.10	4.24	0.54
1000.3	4.26	0.14	3.67	0.06	4.46	0.18	3.95	0.04	4.20	0.09	4.16	0.49
1024.7	4.22	0.11	3.64	0.06	4.53	0.19	3.97	0.07	4.18	0.06	4.07	0.53
1049.1	4.27	0.15	3.62	0.06	4.44	0.19	3.83	0.18	4.12	0.04	3.99	0.53

Table 18A (Continued)

GHz	$T = 57.9°C$				$T = 70.2°C$				$T = 93.5°C$			
	ε'	sd ε'	ε''	sd ε''	ε'	sd ε'	ε''	sd ε''	ε'	sd ε'	ε''	sd ε''
1073.5	4.21	0.16	3.46	0.09	4.48	0.17	3.76	0.11	4.11	0.07	3.92	0.55
1097.9	4.25	0.11	3.35	0.09	4.43	0.19	3.47	0.10	4.04	0.08	3.65	0.53
1122.3	4.22	0.10	3.51	0.12	4.61	0.24	3.74	0.19	4.08	0.07	3.79	0.55
1146.7	4.25	0.14	3.34	0.06	4.48	0.15	3.46	0.07	4.09	0.15	3.62	0.57
1171.1	4.25	0.13	3.36	0.03	4.75	0.18	3.71	0.18	4.26	0.08	3.61	0.60
1195.5	4.21	0.08	3.30	0.10	4.45	0.22	3.55	0.20	4.09	0.13	3.57	0.56
1219.9	4.14	0.09	3.39	0.12	4.50	0.25	3.70	0.19	4.09	0.06	3.57	0.56
1244.3	4.16	0.19	3.32	0.11	4.39	0.23	3.56	0.11	4.01	0.11	3.45	0.48
1268.6	4.20	0.17	3.32	0.04	4.26	0.19	3.45	0.16	4.02	0.09	3.33	0.63
1293.0	4.18	0.09	3.18	0.10	4.35	0.27	3.37	0.04	4.04	0.16	3.29	0.62
1317.4	4.04	0.07	3.21	0.08	4.27	0.18	3.45	0.18	3.96	0.07	3.31	0.62
1341.8	4.00	0.21	3.27	0.05	4.34	0.20	3.33	0.18	3.96	0.02	3.27	0.57
1366.2	4.00	0.11	3.17	0.06	4.46	0.20	3.58	0.20	3.95	0.16	3.31	0.65
1390.6	4.02	0.05	3.11	0.09	4.21	0.32	3.39	0.04	3.97	0.10	3.21	0.75
1415.0	3.85	0.05	3.13	0.08	4.05	0.30	3.36	0.20	3.82	0.11	3.24	0.66
1439.4	3.93	0.08	3.16	0.09	4.27	0.19	3.30	0.06	3.96	0.13	3.07	0.76
1463.8	3.92	0.07	3.09	0.06	4.19	0.34	3.16	0.26	3.96	0.12	3.01	0.66
1488.2	3.95	0.12	2.93	0.13	4.32	0.18	3.31	0.40	3.98	0.15	3.04	0.55
1512.6	4.00	0.07	2.94	0.19	4.18	0.32	3.19	0.25	4.01	0.13	3.00	0.59
1537.0	4.01	0.18	3.07	0.25	4.33	0.23	3.31	0.14	3.95	0.17	2.91	0.58
1561.4	4.05	0.08	3.09	0.09	4.38	0.15	3.12	0.29	4.02	0.07	2.98	0.67
1585.8	3.90	0.13	3.09	0.12	4.30	0.36	3.20	0.23	3.96	0.16	2.86	0.76
1610.2	3.93	0.08	3.05	0.14	4.04	0.28	3.23	0.24	3.81	0.12	2.94	0.63
1634.6	3.86	0.15	2.98	0.15	4.29	0.08	3.10	0.50	3.93	0.06	2.94	0.54
1659.0	4.00	0.19	2.83	0.16	3.97	0.31	3.41	0.28	3.96	0.08	2.93	0.71
1683.4	3.73	0.18	3.05	0.20	3.79	0.31	3.30	0.24	3.64	0.13	3.03	0.75
1707.8	3.81	0.17	3.05	0.23	4.48	0.40	3.78	0.42	3.99	0.07	2.97	0.55
1732.2	3.89	0.27	3.08	0.21	3.86	0.19	3.21	0.37	3.62	0.34	3.04	0.71
1756.6	3.91	0.48	2.96	0.28	4.21	0.39	3.55	0.52	3.99	0.03	2.72	0.59
1781.0	4.19	0.20	2.97	0.37	4.12	0.38	3.02	0.20	4.01	0.23	2.67	0.64
1805.4	4.21	0.33	3.00	0.32	4.32	0.31	3.09	0.40	4.08	0.13	2.67	0.64
1829.8	3.83	0.37	2.73	0.09	4.24	0.31	2.93	0.42	3.97	0.03	2.45	0.57
1854.2	3.74	0.46	2.71	0.38	3.87	0.80	2.86	0.19	3.61	0.09	2.42	0.67
1878.6	3.77	0.19	2.80	0.22	3.97	0.50	3.12	0.92	3.83	0.70	2.56	0.83
1903.0	3.91	0.28	2.96	0.13	4.24	1.01	2.90	0.50	3.84	0.35	2.47	0.91
1927.4	3.71	0.28	2.72	0.41	4.31	0.53	2.44	0.53	3.92	0.33	2.28	0.68
1951.8	4.19	0.53	2.78	0.34	4.05	0.47	2.51	0.55	3.61	0.28	2.58	0.89
1976.2	3.87	0.72	2.88	0.17	4.28	1.15	2.99	0.58	3.52	0.50	2.56	0.55
2000.6	3.81	0.29	2.69	0.41	3.82	0.77	2.59	0.37	3.99	0.29	2.32	0.83
2025.0	3.88	0.25	2.40	0.31	3.74	0.34	3.00	0.88	3.58	0.31	2.43	0.86
2049.4	3.42	0.30	2.77	0.72	3.86	0.75	2.85	0.54	3.80	0.01	2.63	1.12

Appendix D

Useful mixing formulae

Ari Sihvola

This Appendix repeats some of the important mixing rules and other formulae. The geometry of the mixture is the following: in an environment with permittivity ε_e, inclusions (permittivity ε_i) occupy a volume fraction f. The homogenised effective permittivity is denoted by ε_{eff}.

Maxwell Garnett rule

Isotropic materials, spherical inclusions

$$\varepsilon_{eff} = \varepsilon_e + 3f\varepsilon_e \frac{\varepsilon_i - \varepsilon_e}{\varepsilon_i + 2\varepsilon_e - f(\varepsilon_i - \varepsilon_e)} = \varepsilon_e + \frac{n\alpha}{1 - (n\alpha/3\varepsilon_e)} \tag{D.1}$$

Also in the form

$$\frac{\varepsilon_{eff} - \varepsilon_e}{\varepsilon_{eff} + 2\varepsilon_e} = f\frac{\varepsilon_i - \varepsilon_e}{\varepsilon_i + 2\varepsilon_e} = \frac{n\alpha}{3\varepsilon_e} \tag{D.2}$$

Polarisability of a dielectric sphere, referred to the environment

$$\alpha = V(\varepsilon_i - \varepsilon_e)\frac{3\varepsilon_e}{\varepsilon_i + 2\varepsilon_e} \tag{D.3}$$

d-dimensional spheres

$$\frac{\varepsilon_{eff} - \varepsilon_e}{\varepsilon_{eff} + (d-1)\varepsilon_e} = f\frac{\varepsilon_i - \varepsilon_e}{\varepsilon_i + (d-1)\varepsilon_e} \tag{D.4}$$

Multi-phase mixture, isotropic spheres

$$\varepsilon_{eff} = \varepsilon_e + 3\varepsilon_e \frac{\sum_{i=1}^{K} f_i(\varepsilon_i - \varepsilon_e)/(\varepsilon_i + 2\varepsilon_e)}{1 - \sum_{i=1}^{K} f_i(\varepsilon_i - \varepsilon_e)/(\varepsilon_i + 2\varepsilon_e)} \tag{D.5}$$

Aligned ellipsoids

$$\varepsilon_{\text{eff},x} = \varepsilon_{\text{e}} + f\varepsilon_{\text{e}}\frac{\varepsilon_{\text{i}} - \varepsilon_{\text{e}}}{\varepsilon_{\text{e}} + (1-f)A_x(\varepsilon_{\text{i}} - \varepsilon_{\text{e}})} \tag{D.6}$$

Randomly oriented ellipsoids

$$\varepsilon_{\text{eff}} = \varepsilon_{\text{e}} + \varepsilon_{\text{e}}\frac{(f/3)\sum_{j=x,y,z}(\varepsilon_{\text{i}} - \varepsilon_{\text{e}})/(\varepsilon_{\text{e}} + A_j(\varepsilon_{\text{i}} - \varepsilon_{\text{e}}))}{1 - (f/3)\sum_{j=x,y,z}A_j(\varepsilon_{\text{i}} - \varepsilon_{\text{e}})/(\varepsilon_{\text{e}} + A_j(\varepsilon_{\text{i}} - \varepsilon_{\text{e}}))} \tag{D.7}$$

Depolarisation factors of ellipsoids ($A_x + A_y + A_z = 1$)

$$A_x = \frac{a_x a_y a_z}{2}\int_0^\infty \frac{ds}{(s + a_x^2)\sqrt{(s + a_x^2)(s + a_y^2)(s + a_z^2)}} \tag{D.8}$$

Prolate spheroids: ($a_x > a_y = a_z$, $e = \sqrt{1 - a_y^2/a_x^2}$)

$$A_x = \frac{1 - e^2}{2e^3}\left(\ln\frac{1+e}{1-e} - 2e\right) \tag{D.9}$$

Oblate spheroids: $\left(a_z < a_x = a_y,\ e = \sqrt{a_x^2/a_z^2 - 1}\right)$

$$A_z = \frac{1 + e^2}{e^3}(e - \tan^{-1} e) \tag{D.10}$$

Bruggeman (Polder–Van Santen)

Spheres

$$(1-f)\frac{\varepsilon_{\text{e}} - \varepsilon_{\text{eff}}}{\varepsilon_{\text{e}} + 2\varepsilon_{\text{eff}}} + f\frac{\varepsilon_{\text{i}} - \varepsilon_{\text{eff}}}{\varepsilon_{\text{i}} + 2\varepsilon_{\text{eff}}} = 0 \tag{D.11}$$

Randomly oriented ellipsoids

$$\varepsilon_{\text{eff}} = \varepsilon_{\text{e}} + \frac{f}{3}(\varepsilon_{\text{i}} - \varepsilon_{\text{e}})\sum_{j=x,y,z}\frac{\varepsilon_{\text{eff}}}{\varepsilon_{\text{eff}} + A_j(\varepsilon_{\text{i}} - \varepsilon_{\text{eff}})} \tag{D.12}$$

Coherent potential

Spheres

$$\varepsilon_{\text{eff}} = \varepsilon_{\text{e}} + f(\varepsilon_{\text{i}} - \varepsilon_{\text{e}})\frac{3\varepsilon_{\text{eff}}}{3\varepsilon_{\text{eff}} + (1-f)(\varepsilon_{\text{i}} - \varepsilon_{\text{e}})} \tag{D.13}$$

Randomly oriented ellipsoids

$$\varepsilon_{\text{eff}} = \varepsilon_{\text{e}} + \frac{f}{3}(\varepsilon_{\text{i}} - \varepsilon_{\text{e}}) \sum_{j=x,y,z} \frac{(1 + A_j)\varepsilon_{\text{eff}} - A_j\varepsilon_{\text{e}}}{\varepsilon_{\text{eff}} + A_j(\varepsilon_{\text{i}} - \varepsilon_{\text{e}})} \tag{D.14}$$

Unified mixing formula, spheres

$$\frac{\varepsilon_{\text{eff}} - \varepsilon_{\text{e}}}{\varepsilon_{\text{eff}} + 2\varepsilon_{\text{e}} + \nu(\varepsilon_{\text{eff}} - \varepsilon_{\text{e}})} = f\frac{\varepsilon_{\text{i}} - \varepsilon_{\text{e}}}{\varepsilon_{\text{i}} + 2\varepsilon_{\text{e}} + \nu(\varepsilon_{\text{eff}} - \varepsilon_{\text{e}})} \tag{D.15}$$

Maxwell Garnett ($\nu = 0$), Bruggeman ($\nu = 2$), Coherent potential ($\nu = 3$).

Index

www.ingramcontent.com/pod-product-compliance
Lightning Source LLC
LaVergne TN
LVHW011856200926
848484LV00002B/461
* 9 7 8 0 8 6 3 4 1 5 7 3 9 *